T0189540

Communications in Computer and Information Science 1422

Xingming Sun · Xiaorui Zhang ·
Zhihua Xia · Elisa Bertino (Eds.)

Advances in Artificial Intelligence and Security

7th International Conference, ICAIS 2021
Dublin, Ireland, July 19–23, 2021
Proceedings, Part I

Editors
Xingming Sun
Nanjing University of Information Science and Technology
Nanjing, China

Xiaorui Zhang
Nanjing University of Information Science and Technology
Nanjing, China

Zhihua Xia
Jinan University
Guangzhou, China

Elisa Bertino
Purdue University
West Lafayette, IN, USA

ISSN 1865-0929 ISSN 1865-0937 (electronic)
Communications in Computer and Information Science
ISBN 978-3-030-78614-4 ISBN 978-3-030-78615-1 (eBook)
https://doi.org/10.1007/978-3-030-78615-1

This Springer imprint is published by the registered company Springer Nature Switzerland AG
The registered company address is: Gewerbestrasse 11, 6330 Cham, Switzerland

Preface

The 7th International Conference on Artificial Intelligence and Security (ICAIS 2021), formerly called the International Conference on Cloud Computing and Security (ICCCS), was held during July 19–23, 2021, in Dublin, Ireland. Over the past six years, ICAIS has become a leading conference for researchers and engineers to share their latest results of research, development, and applications in the fields of artificial intelligence and information security.

We used the Microsoft Conference Management Toolkits (CMT) system to manage the submission and review processes of ICAIS 2021. We received 1013 submissions from authors in 20 countries and regions, including the USA, Canada, the UK, Italy, Ireland, Japan, Russia, France, Australia, South Korea, South Africa, Iraq, Kazakhstan, Indonesia, Vietnam, Ghana, China, Taiwan, and Macao, etc. The submissions covered the areas of artificial intelligence, big data, cloud computing and security, information hiding, IoT security, multimedia forensics, encryption and cybersecurity, and so on. We thank our Technical Program Committee (TPC) members and external reviewers for their efforts in reviewing the papers and providing valuable comments to the authors. From the total of 1013 submissions, and based on at least three reviews per submission, the Program Chairs decided to accept 183 papers to be published in three Communications in Computer and Information Science (CCIS) volumes and 122 papers to be published in two Lecture Notes in Computer Science (LNCS) volumes, yielding an acceptance rate of 30%. This volume of the conference proceedings contains all the regular, poster, and workshop papers.

The conference program was enriched by a series of keynote presentations, and the keynote speakers included Michael Scott, MIRACL Labs, Ireland, and Sakir Sezer, Queen's University of Belfast, UK. We enjoyed their wonderful speeches.

There were 49 workshops organized as part of ICAIS 2021 which covered all the hot topics in artificial intelligence and security. We would like to take this moment to express our sincere appreciation for the contribution of all the workshop chairs and their participants. We would like to extend our sincere thanks to all authors who submitted papers to ICAIS 2021 and to all TPC members. It was a truly great experience to work with such talented and hard-working researchers. We also appreciate the external reviewers for assisting the TPC members in their particular areas of expertise. Moreover, we want to thank our sponsors: Association for Computing Machinery; Nanjing University of Information Science and Technology; Dublin City University; New York University; Michigan State University; University of Central Arkansas; Université Bretagne Sud; National Nature Science Foundation of China; Tech Science Press; Nanjing Normal University; Northeastern State University; Engineering

Research Center of Digital Forensics, Ministry of Education, China; and ACM SIGWEB China.

April 2021

Xingming Sun
Xiaorui Zhang
Zhihua Xia
Elisa Bertino

Organization

General Chairs

Martin Collier	Dublin City University, Ireland
Xingming Sun	Nanjing University of Information Science and Technology, China
Yun Q. Shi	New Jersey Institute of Technology, USA
Mauro Barni	University of Siena, Italy
Elisa Bertino	Purdue University, USA

Technical Program Chairs

Noel Murphy	Dublin City University, Ireland
Aniello Castiglione	University of Salerno, Italy
Yunbiao Guo	China Information Technology Security Evaluation Center, China
Suzanne K. McIntosh	New York University, USA
Xiaorui Zhang	Engineering Research Center of Digital Forensics, Ministry of Education, China
Q. M. Jonathan Wu	University of Windsor, Canada

Publication Chairs

Zhihua Xia	Nanjing University of Information Science and Technology, China
Zhaoqing Pan	Nanjing University of Information Science and Technology, China

Workshop Chair

Baowei Wang	Nanjing University of Information Science and Technology, China

Organization Chairs

Xiaojun Wang	Dublin City University, Ireland
Genlin Ji	Nanjing Normal University, China
Zhangjie Fu	Nanjing University of Information Science and Technology, China

Technical Program Committee Members

Saeed Arif	University of Algeria, Algeria
Anthony Ayodele	University of Maryland, USA
Zhifeng Bao	Royal Melbourne Institute of Technology, Australia
Zhiping Cai	National University of Defense Technology, China
Ning Cao	Qingdao Binhai University, China
Paolina Centonze	Iona College, USA
Chin-chen Chang	Feng Chia University, Taiwan, China
Han-Chieh Chao	Taiwan Dong Hwa University, Taiwan, China
Bing Chen	Nanjing University of Aeronautics and Astronautics, China
Hanhua Chen	Huazhong University of Science and Technology, China
Xiaofeng Chen	Xidian University, China
Jieren Cheng	Hainan University, China
Lianhua Chi	IBM Research Center, Australia
Kim-Kwang Raymond Choo	University of Texas at San Antonio, USA
Ilyong Chung	Chosun University, South Korea
Robert H. Deng	Singapore Management University, Singapore
Jintai Ding	University of Cincinnati, USA
Xinwen Fu	University of Central Florida, USA
Zhangjie Fu	Nanjing University of Information Science and Technology, China
Moncef Gabbouj	Tampere University of Technology, Finland
Ruili Geng	Spectral MD, USA
Song Guo	Hong Kong Polytechnic University, Hong Kong
Mohammad Mehedi Hassan	King Saud University, Saudi Arabia
Russell Higgs	University College Dublin, Ireland
Dinh Thai Hoang	University of Technology Sydney, Australia
Wien Hong	Nanfang College of Sun Yat-sen University, China
Chih-Hsien Hsia	National Ilan University, Taiwan, China
Robert Hsu	Chung Hua University, Taiwan, China
Xinyi Huang	Fujian Normal University, China
Yongfeng Huang	Tsinghua University, China
Zhiqiu Huang	Nanjing University of Aeronautics and Astronautics, China
Patrick C. K. Hung	University of Ontario Institute of Technology, Canada
Farookh Hussain	University of Technology Sydney, Australia
Genlin Ji	Nanjing Normal University, China
Hai Jin	Huazhong University of Science and Technology, China
Sam Tak Wu Kwong	City University of Hong Kong, China
Chin-Feng Lai	Taiwan Cheng Kung University, Taiwan, China
Loukas Lazos	University of Arizona, USA

Sungyoung Lee	Kyung Hee University, South Korea
Chengcheng Li	University of Cincinnati, USA
Feifei Li	Utah State University, USA
Jin Li	Guangzhou University, China
Jing Li	Rutgers University, USA
Kuan-Ching Li	Providence University, Taiwan, China
Peng Li	University of Aizu, Japan
Yangming Li	University of Washington, USA
Luming Liang	Uber Technology, USA
Haixiang Lin	Leiden University, Netherlands
Xiaodong Lin	University of Ontario Institute of Technology, Canada
Zhenyi Lin	Verizon Wireless, USA
Alex Liu	Michigan State University, USA
Guangchi Liu	Stratifyd Inc., USA
Guohua Liu	Donghua University, China
Joseph Liu	Monash University, Australia
Quansheng Liu	University of South Brittany, France
Xiaodong Liu	Edinburgh Napier University, UK
Yuling Liu	Hunan University, China
Zhe Liu	University of Waterloo, Canada
Daniel Xiapu Luo	Hong Kong Polytechnic University, Hong Kong
Xiangyang Luo	Zhengzhou Science and Technology Institute, China
Tom Masino	TradeWeb LLC, USA
Suzanne K. McIntosh	New York University, USA
Nasir Memon	New York University, USA
Sangman Moh	Chosun University, South Korea
Yi Mu	University of Wollongong, Australia
Elie Naufal	Applied Deep Learning LLC, USA
Jiangqun Ni	Sun Yat-sen University, China
Rafal Niemiec	University of Information Technology and Management, Poland
Zemin Ning	Wellcome Trust Sanger Institute, UK
Shaozhang Niu	Beijing University of Posts and Telecommunications, China
Srikant Ojha	Sharda University, India
Jeff Z. Pan	University of Aberdeen, UK
Wei Pang	University of Aberdeen, UK
Chen Qian	University of California, Santa Cruz, USA
Zhenxing Qian	Fudan University, China
Chuan Qin	University of Shanghai for Science and Technology, China
Jiaohua Qin	Central South University of Forestry and Technology, China
Yanzhen Qu	Colorado Technical University, USA
Zhiguo Qu	Nanjing University of Information Science and Technology, China

Yongjun Ren	Nanjing University of Information Science and Technology, China
Arun Kumar Sangaiah	VIT University, India
Di Shang	Long Island University, USA
Victor S. Sheng	University of Central Arkansas, USA
Zheng-guo Sheng	University of Sussex, UK
Robert Simon Sherratt	University of Reading, UK
Yun Q. Shi	New Jersey Institute of Technology, USA
Frank Y. Shih	New Jersey Institute of Technology, USA
Biao Song	King Saud University, Saudi Arabia
Guang Sun	Hunan University of Finance and Economics, China
Jianguo Sun	Harbin University of Engineering, China
Krzysztof Szczypiorski	Warsaw University of Technology, Poland
Tsuyoshi Takagi	Kyushu University, Japan
Shanyu Tang	University of West London, UK
Jing Tian	National University of Singapore, Singapore
Yoshito Tobe	Aoyang University, Japan
Cezhong Tong	Washington University in St. Louis, USA
Pengjun Wan	Illinois Institute of Technology, USA
Cai-Zhuang Wang	Ames Laboratory, USA
Ding Wang	Peking University, China
Guiling Wang	New Jersey Institute of Technology, USA
Honggang Wang	University of Massachusetts Dartmouth, USA
Jian Wang	Nanjing University of Aeronautics and Astronautics, China
Jie Wang	University of Massachusetts Lowell, USA
Jin Wang	Changsha University of Science and Technology, China
Liangmin Wang	Jiangsu University, China
Ruili Wang	Massey University, New Zealand
Xiaojun Wang	Dublin City University, Ireland
Xiaokang Wang	St. Francis Xavier University, Canada
Zhaoxia Wang	A-Star, Singapore
Sheng Wen	Swinburne University of Technology, Australia
Jian Weng	Jinan University, China
Edward Wong	New York University, USA
Eric Wong	University of Texas at Dallas, USA
Shaoen Wu	Ball State University, USA
Shuangkui Xia	Beijing Institute of Electronics Technology and Application, China
Lingyun Xiang	Changsha University of Science and Technology, China
Yang Xiang	Deakin University, Australia
Yang Xiao	University of Alabama, USA
Haoran Xie	The Education University of Hong Kong, China
Naixue Xiong	Northeastern State University, USA

Wei Qi Yan	Auckland University of Technology, New Zealand
Aimin Yang	Guangdong University of Foreign Studies, China
Ching-Nung Yang	Taiwan Dong Hwa University, Taiwan, China
Chunfang Yang	Zhengzhou Science and Technology Institute, China
Fan Yang	University of Maryland, USA
Guomin Yang	University of Wollongong, Australia
Qing Yang	University of North Texas, USA
Yimin Yang	Lakehead University, Canada
Ming Yin	Purdue University, USA
Shaodi You	Australian National University, Australia
Kun-Ming Yu	Chung Hua University, Taiwan, China
Weiming Zhang	University of Science and Technology of China, China
Xinpeng Zhang	Fudan University, China
Yan Zhang	Simula Research Laboratory, Norway
Yanchun Zhang	Victoria University, Australia
Yao Zhao	Beijing Jiaotong University, China

Organization Committee Members

Xianyi Chen	Nanjing University of Information Science and Technology, China
Zilong Jin	Nanjing University of Information Science and Technology, China
Yiwei Li	Columbia University, USA
Yuling Liu	Hunan University, China
Zhiguo Qu	Nanjing University of Information Science and Technology, China
Huiyu Sun	New York University, USA
Le Sun	Nanjing University of Information Science and Technology, China
Jian Su	Nanjing University of Information Science and Technology, China
Qing Tian	Nanjing University of Information Science and Technology, China
Yuan Tian	King Saud University, Saudi Arabia
Qi Wang	Nanjing University of Information Science and Technology, China
Lingyun Xiang	Changsha University of Science and Technology, China
Zhihua Xia	Nanjing University of Information Science and Technology, China
Lizhi Xiong	Nanjing University of Information Science and Technology, China

Contents – Part I

Artificial Intelligence

Contents – Part II

Big Data

Contents – Part III

Information Hiding

IoT Security

Artificial Intelligence

A Load Balancing Mechanism for a Multi-controller Software Defined WiFi Network

Sohaib Manzoor[1](✉), Hira Manzoor[2], Mahak Manzoor[3], and Anzar Mahmood[1]

[1] Department of Electrical, Mirpur University of Science and Technology (MUST), Mirpur 10250, (AJK), Pakistan
sohaib.ee@must.edu.pk

[2] School of Electronic Information and Communications, Huazhong University of Science and Technology, Wuhan 430074, China
hira@hust.edu.cn

[3] Department of CS and IT, Mirpur University of Science and Technology (MUST), Mirpur 10250, (AJK), Pakistan
mahak.csit@must.edu.pk

Abstract. Software defined WiFi networks (SD-WiFi) lead to a promising evolution path, by supporting changes in network traffic, providing centralized network provisioning and allowing flexible resource allocation. Load balancing still remains a challenging issue due to the increased number of WiFi users. In this paper, we propose an efficient algorithmic approach to solve the load imbalance issue in SD-WiFi. Traffic generated by users constituting the load, arrive at WiFi access points (APs). Support vector machine (SVM) segregates the traffic into high priority class (HP) and low priority class (LP) by considering flow deadlines and flow types. Controllers are arranged in two-tiers: global and local controllers. The local controllers (LC) update their load information to the global controller (GC) periodically. Binary search tree (BST) is employed in the GC to find the least loaded LC for flow processing. OMNeT++ simulator is used to perform extensive simulations for performance evaluation of the proposed SVM-BST scheme. The proposed SVM-BST scheme outperforms load information strategy and EASM schemes in throughput by 50–200% and in response time by 20–30% respectively.

Keywords: WiFi · SDN · Load balancing · Multi-controllers

1 Introduction

Software defined network (SDN) is a new emerging network paradigm that provides a decoupled data and control plane to solve various issues in a traditional wireless network [1]. Efficiency aware greedy algorithm has been deployed in SDN architecture to perform switch migration in order to balance load among multiple

X. Sun et al. (Eds.): ICAIS 2021, CCIS 1422, pp. 3–14, 2021.
https://doi.org/10.1007/978-3-030-78615-1_1

controllers [2]. Load balancing is achieved in high density software defined WiFi networks by considering AP associations [3]. Flow classification has helped in achieving load balancing in an SDN enhanced two-tier WiFi network [4]. Traffic transfer between WiFi and WiMAX support load balancing in a WiFi Network [5]. The handover is dependent upon the admission control in the access point (AP) and base stations (BS), bandwidth reserved in AP and BS and finally the class aware load balancing. Fuzzy logic and particle swarm optimization aid in achieving load balancing in a multi controller WiFi network based on SDN platform [6].

A wildcard method together with SDN policies is used for achieving load balancing in a server [7]. The wild card method take advantage of SDN OpenFlow rules to set load balancing priorities. The networking devices such as forwarding and access devices are simplified in a wireless network incorporating SDN capabilities to behave in accordance to a centralized controller [8]. A round robin technique is used to install specific rules over OpenFlow to perform corresponding actions in order to address the load imbalance issue in WiFi networks [9]. The role of wireless load from a healthcare perspective in a software defined environment is presented in [10]. The load among multiple controllers is balanced through a load information strategy [11], which uses average message arrival rate as the load balancing parameter. The controller instructs the switches to perform migration. WiFi optimization and traffic aware load balancing is performed to prioritize the healthcare traffic using probability distribution [12,13].

In this paper, we propose a novel approach to resolve the load imbalance issue in the SD-WiFi. The architecture consists of multi-controllers where the GC is responsible to take network wide decisions as a centralized controller and to balance the load among all the LCs. The network traffic generated by wireless users arrive at the APs where it is classified into two classes as HP and LP depending upon the flow types and deadlines. The classification is achieved by support vector machine (SVM), which is a machine learning technique. HP traffic class is prioritized over LP class as it is more sensitive to delay and so is forwarded to the switch earlier. The switch on arrival of a new flow, interrogates the GC regarding the flow policies. The flow requests are then forwarded by the GC to LCs. The forwarding decision is made on the current load condition of the LCs. Load monitoring of all LCs is done in terms of hard disk, memory and CPU utilization periodically. BST allows the GC to forward the flow requests to the least loaded LC. Once the most appropriate LC is chosen by the GC, the flow policies are installed back in the OpenFlow switch to ensure quick request processing at an improved response time.

The rest of the paper is organized as follows. In Sect. 2 we present the literature review on load balancing algorithms in SD-WiFi. In Sect. 3 we provide the details of the SVM-BST algorithm to achieve load balancing in SD-WiFi. In Sect. 4 we report the results obtained from simulation experiments which reflect the performance evaluation. Finally we conclude our contribution in Sect. 5.

2 Related Work

Flow stealer aims at balancing load among multiple controllers [14]. The workload of overloaded controllers is balanced by idle controllers by sharing tasks and resources. The method fails to address the overloading issue of idle controllers while sharing tasks of overloaded controllers. Genetic algorithms are used to balance the load among multiple controllers [15]. The switches are classified according to the flow entries and the switch with maximum flows are assigned the least loaded controller. The method incorporates complexity as mutation rate is needed to be initialized each time the algorithm triggers. Finite state machine (FSM) load balancing method make use of specific policies to forward packets in the networks [16]. The exact degree of load balancing is ignored. Collaborative platform make use of the distributed SDN architecture to balance the load among the multiple controllers [17]. A data collection algorithm is used to collect the load information of the controllers. The shifting of control permission of the switches among multiple controllers help in load balancing. The load metric is only considered in the source controller. A semi-matched load balancing scheme is used to balance the load among the APs [18]. Channel busy time ratio (CBTR) is used to compute the load on each AP. The modeling of AP is done through a weighted bipartite graph. The model increases the computation complexity due to several computations.

A two-tier approach is used to balance the load among multiple controllers [19]. The load balancing is performed at both the AP end and the controller end by making use of the OpenFlow protocol. The controller after receiving the capacity, load and association table reports from the APs, make suitable calculations for AP association and de-associations. The AP de-associations are made using Last-In-First-Out (LIFO), which increases the time duration for the wireless users that made the first de-association request. Load among APs is balanced using handovers through a centralized approach [20]. Three modules (1) decision conversion module, (2) metric monitoring module and (3) load control module are designed to achieve the load balancing. If a certain AP is overloaded then the SDN controller shifts the load from that AP to the neighboring AP. The load of the neighboring AP was not considered. SDN is also used to balance the load in the Long-term evaluation (LTE) networks [21]. The switches intimate the centralized controller, each time a new Packet_In message is received. The OpenFlow switches help in determining the load of the gateway by using static information about traffic per volume. Scalability of control plane is not taken into the account as only one SDN controller is used to process all the tasks.

3 Method

3.1 Overview

In this section we discuss the overall architecture of SD-WiFi. The WiFi network consists of the wireless stations such as smart phones, tablets etc. and APs. The data plane consists of OpenFlow switches that forward the user requests to controllers. The control plane consists of multiple controllers such as a GC and LCs. The application plane helps to support the SVM-BST load balancing algorithm, which solves the load imbalance issue in SD-WiFi, as shown in Fig. 1. User traffic is generated from wireless stations in form of flow requests that constitute the network load. These wireless stations are part of the access networks such as smart homes and wireless local area networks (WLAN). The flow requests and the consequent network load increases with the increase in wireless devices. On reaching the AP the traffic is classified in to HP and LP class and is prioritized based on flow types and flow deadlines using SVM. The flows are forwarded to controllers via OpenFlow switches. LCs update their load information to GC periodically and BST allows the GC to select the least loaded LC for flow processing. The goals of SVM-BST are two fold (1) to classify and prioritize traffic to minimize response time for the traffic whose deadlines are not met before they reach the destination and (2) to balance load among controllers to enhance throughput and to achieve optimal delay performance.

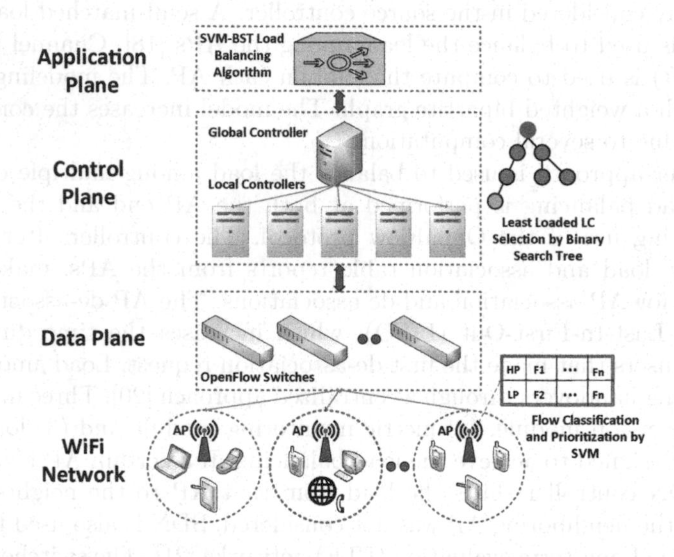

Fig. 1. The proposed SD-WiFi architecture employing load balancing mechanism in multiple controllers.

Traffic Classification and Prioritization. The load balancing is achieved at the data plane by segregating the flows into two classes which are high priority class (HP) and low priority class (LP). The HP is delay sensitive and need to be processed at top priority. In order to minimize the latency factor, it is very important to determine the class of incoming flows. The HP and LP are determined through a machine learning technique called SVM, which is an efficient binary classifier. SVM makes use of flow size and delay constraint as the primary features for flow classification.

SVM is a supervised machine learning approach that makes use of the training data in order to classify the incoming flows. The segregation is achieved by searching an optimal hyperplane. The flows in the SVM are treated as support vectors, where each point can be laid on a hyperplane. If we have n training observations $(\vec{a}_1, ..., \vec{a}_n) \in \mathbb{R}^z$, where $\mathbb{R}^z$ is the real valued $\mathcal{Z}$ dimensional space, and n class labels $z_1, ..., z_n \in \{-1, 1\}$, SVM, in subject to the constraints tends to build an optimal hyperplane boundary as shown in Eq. 1.

$$z_i \left(\vec{b} \cdot \vec{a} + b_0\right) \geq M(1 - \epsilon_i), \quad \forall i = 1, ..., n. \tag{1}$$

Where z_i represents the specific number of flow such as HP i^{th} or LP i^{th} flow, the delay constraint and flows size values are represented by $\vec{a}$, the hyperplane is defined by $(\vec{b} \cdot \vec{x} + b_0)$, the maximum possible margin between the two flows is represented by M. The non-separable cases are handled by the slack variables ϵ_i. SVM aims to reduce errors by introducing a control variable called *budget*, represented by C, which makes a trade off between the separation margin and training errors.

$$\epsilon_i \geq 0, \quad \sum_{i=1}^{n} \epsilon_i \leq C. \tag{2}$$

The relation between C and ϵ_i is represented in Eq. 2. The limits to which the ϵ_i can be modified to violate the margin is controlled by C. The more the C, the more the margin values. $C = 0$ implies $\epsilon_i = 0$, where $\epsilon_i = 0$ means that the ith observation (a_i) is on the correct side of margin, $\epsilon_i > 1$ mean that a_i is on the wrong side of the hyperplane and $\epsilon_i > 0$ means a_i is on the wrong side of the margin.

SVM reduces the over fitting, utilizes the memory efficiently, learns highly complex classification quickly and is robust to number of variables. The clear margin of separation achieved is almost 99.9% efficient. The flows larger in size and with higher deadlines are treated as LP flows and the flows with smaller size and lower deadline are classified as HP flows. After the flow classification, the APs forward the flows to GC with the help of OpenFlow.

Load Balancing Among Controllers. The control plane comprises of the GC and LCs. The GC is responsible to manage the load among the LCs. The packets that arrive to the GC from the OpenFlow switches are assigned to the LCs through BST. Two traffic queues, such as HP and LP are maintained by the GC. The HP is given higher priority in terms of processing than the LP. The overloaded LCs are balanced by shifting the flows to the underloaded LCs.

The GC maintains the load i of each LC through 5 metrics such as hard disk utilization $hd_{(i)}$, memory utilization $mem_{(i)}$, and the CPU utilization $cpu_{(i)}$, number of flows in the buffer$fb_{(i)}$ and the number of flows processing $fp_{(i)}$. Load values LC_i are updated to the GC by the LCs periodically. The load calculation of each LC is given in Eq. 3.

$$LC_{(i)} = (Sy_m, Us_m) \tag{3}$$

Sy_m denotes the system metric and Us_m denotes the user metrics, these metrics are formulated as,

$$Sy_m = hd_{(i)} + mem_{(i)} + cpu_{(i)} \tag{4}$$

$$U_m = fp_{(i)} + fb_{(i)} \tag{5}$$

The load of each LC defines its states of being overloaded or underloaded and helps the GC to balance the load among each LC. The LCs with load values equal to or above 0.5 are treated as overloaded and those with load values below 0.5 are treated as underloaded. The load status of each controller, for three different time periods T_1, T_2 and T_3 is depicted in Table 1. For each time period the number of packets injected are varied such as in T_1 the packets injected are 2000 per second, in T_2 packets injected are 4000 per second and in T_3 the packets injected are 6000 per second. In the proposed SVM-BST scheme, even if the LCs are overloaded initially, they still have the capability to process the packets.

Table 1. Load of local controllers managed by global controller.

Time period	Controller ID	Actual capacity used (%)	Current load	Status
T1	LC1	40	0.4	Underloaded
	LC2	70	0.7	Overloaded
	LC3	20	0.2	Underloaded
	LC4	10	0.1	UnderLoaded
T2	LC1	50	0.5	Overloaded
	LC2	40	0.4	UnderLoaded
	LC3	30	0.3	UnderLoaded
	LC4	40	0.4	UnderLoaded
T3	LC1	30	0.3	UnderLoaded
	LC2	60	0.6	Overloaded
	LC3	40	0.4	UnderLoaded
	LC4	20	0.2	UnderLoaded

The GC, based on the load states of the local controllers, creates a BST. The incoming packets are analyzed in the GC and assigned to the most suitable LC in terms of load conditions. The BST is created from the probability load values of each of the LC and these load values correspond to a specific load state. The BST is constructed from the probability values ranging from [0–1]. The BST starts its operation from the root node i.e., the parent node followed by the right and left child nodes.

The load values from the LCs, for instance are considered as $\{0.4, 0.6, 0.7, 0.2, 0.5,\ 0.3,\ 0.1\}$ and the BST is constructed as shown in Fig. 2. In-order traversal is followed by the BST, in which the least loaded LC is assigned the packets first followed by increasing values of the load as shown by the dotted lines in Fig. 2. The LC with load value of 0.1 is assigned the packet first, next the LC with load value of 0.2 is assigned the packet for processing and so on. BST aims in balancing the load among the controllers by assigning the packet to the least loaded controllers. The load status of the LCS is updated periodically.

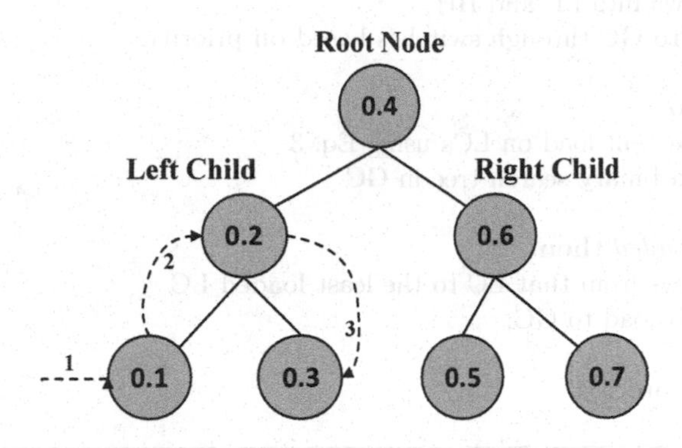

Fig. 2. Binary search tree in global controller.

Algorithm 1 explains the process of our proposed SVM-BST load balancing mechanism. The flow classification at data plane and binary search tree in global controller helps to improve the degree of load balancing in SD-WiFi.

4 Performance Evaluation

In this section, we present the simulation topology and parameters to evaluate the performance of the proposed SVM-BST scheme. The simulation setup is presented first followed by the comparative results.

4.1 Simulation Setup

We used OMNeT++ to implement our simulation model of SD-WiFi using OpenFlow. The simulation model consists of SDN controller, OpenFlow enabled switches, WiFi access points and wireless stations. OMNeT++ provides the flexibility of creation of virtual network nodes such as access points, mobile hosts etc. and supports the graphical user interface. The load balancing design is implemented in C++, where as the SDN controller used is NOX. Emulation of the wireless links of different capacities is performed through a control tool tc. The QoS parameters are monitored through the iPerf tool. The simulation topology is depicted in Fig. 3.

Algorithm 1. The SVM-BST Load Balancing Algorithm

1: Receive all flows from users
2: **for** *allflows* **do**
3: Initialize flow size, delay constraint.
4: Direct flows into SVM classifier
5: Classify flows into LP and HP
6: Send flows to GC through switches based on priority
7: **end for**
8: **for** *allLCs* **do**
9: Compute current load on LCs using Eq. 3
10: Construct a binary search tree in GC
11: **end for**
12: **if** *LCisoverloaded* **then**
13: Migrate flows from that LC to the least loaded LC
14: Update LCs load to GC;
15: **else**
16: Repeat the process;
17: **end if**

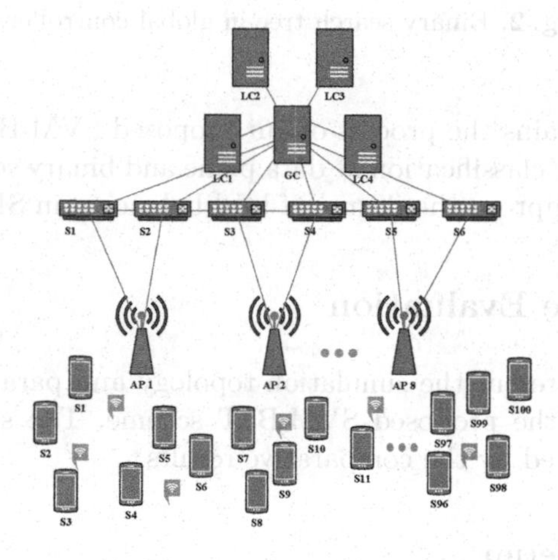

Fig. 3. SVM-BST simulation topology.

The topology consists of one global controller, 4 local controllers, 5 OpenFlow enabled switches, 16 access points and 100 wireless stations. The simulation topology shows a data center with a two-dimensional area of 1000 * 1000 m. The experiments are performed on a Dell PC, a core i7, 2.4 GHz CPU and 16 GB of RAM with windows 10 operating system. The key simulation parameters are depicted in Table 2.

Table 2. Simulation parameters

Parameters	Value
Number of controllers	Global controller 1
	Local controllers 4
OpenFlow enabled switches	6
WiFi access points	8
Wireless stations	100
Number of CPUs	4
Memory	16 GB
Hard disk	500 GB
Connection speed	300 Mbps
Flow size	100 Bytes
Flow request interval	0.001 s
Flow table size	600 Entries
Flow timeout	0.3 s
Transmission range	1000*1000 m
Simulation time	30 min

4.2 Comparative Results

In this section, we compare the performance evaluation of our proposed SVM-BST load balancing mechanism to EASM [2] and load information strategy [11] schemes.

Throughput. Throughput, which is one of the most important metrics to be discussed for any network evaluation, is used for comparative analysis. Throughput is the measure of total number of packets received in ratio to the total number of packets transmitted. The throughput comparison of the proposed SVM-BST to load information strategy and EASM schemes is depicted in Fig. 4a.

EASM is a switch migration scheme, comprising of complex mathematical computations which include routing overhead, state synchronization and data interaction overhead. A trigger factor is used to balance the load and probability distribution is used to select the switch for migration. Multiple mathematical computations degrade the throughput performance. Load information strategy uses the average message arrival rate to balance the load among multiple distributed controllers. The switches are arranged in descending order based on the message arrival rate and migrated accordingly. To choose the switch with minimum number of flows overburdens the controller due to unnecessary message exchange between the switches and the controller, hence decreasing the throughput.

Throughput decreases due to improper load balancing method and complex mathematical computations. On increasing the number of controllers the throughput changes as more network resources share the workload. SVM-BST outperforms EASM scheme by 200% and load information scheme by 50% in throughput performance.

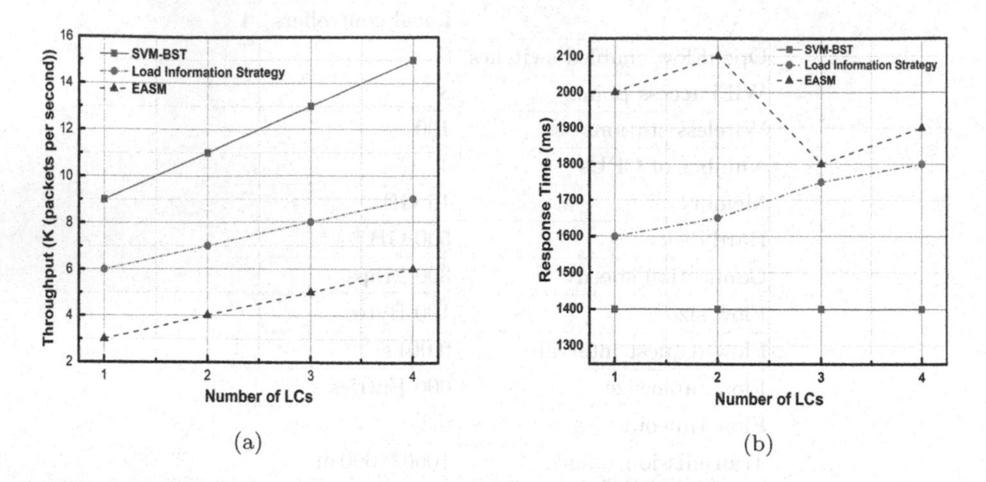

(a) (b)

Fig. 4. (a) Throughput performance. (b) Response time performance.

Response Time. A decreased response time intimates better performance of the proposed SVM-BST scheme as depicted in Fig. 4b. Response time depends on the performance of the algorithms involved in the network. Complex mathematical computations and unnecessary message exchange between switches and the controllers increase the response time in EASM and load information strategy, respectively. SVM-BST exhibits less response time, as the flow prioritization in the data plane and the binary search tree in the GC, enhance the flow processing capabilities, which degrades the computational latency. With the increase of LCs, the packets to be processed also increase such as from 1000 packets per second to 6000 packets per second. It is clearly observed from the Fig. 4b, that even with the increase of packet flows the flow handling capabilities of the LCs increase in SVM-BST and a 20–30% reduction of response time is achieved as compared to load information strategy and EASM schemes respectively.

5 Conclusion

In this paper, a novel load balancing mechanism is proposed for multiple controllers in order to balance the load efficiently in SD-WiFi. The proposed multi-controller architecture aims to balance the load at data plane and control plane. The flow requests generated from wireless user stations arrive at APs where they are classified into two priority classes by using SVM. One is the HP and the other is the LP. HP is delay sensitive traffic and processed first. The LCs update their load to GC in terms of hard disk, memory and CPU utilization periodically. The

BST classifies the LCs in terms of load, and helps the GCs to direct the flows to the least loaded LC. The simulation results prove the efficiency of the proposed SVM-BST scheme in terms of throughput and response time when compared to EASM and load information strategy schemes. The proposed scheme can be applied to cloud based software define networks employing real time applications with heavy loading.

Acknowledgment. We thank our guide Dr. Xiaojun Hei, Associate Professor at school of Electronic Information and Communications, Huazhong University of Science and Technology, Wuhan, China, for providing us the advanced lab facilities for computational experiments.

References

1. Manzoor, S., Chen, Z., Gao, Y., Hei, X., Cheng, W.: Towards QoS-aware load balancing for high density software defined Wi-Fi networks. IEEE Access **8**, 117623–117638 (2020)
2. Hu, T., Guo, Z., Lan, J., Zhang, J., Zhao, W.: EASM: efficiency-aware switch migration for balancing controller loads in software-defined networking (2017). arXiv preprint arXiv:1711.08659
3. Chen, Z., Manzoor, S., Gao, Y., Hei, X.: Achieving load balancing in high-density software defined WiFi networks. In: International Conference on Frontiers of Information Technology (FIT), December 2017
4. Manzoor, S., Hei, X., Cheng, W.: Towards dynamic 2-tier load balancing for software defined WiFi networks. In: International Conference on Communication and Information Systems (ICCIS), November 2017
5. Sarma, A., Chakraborty, S., Nandi, S.: Deciding handover points based on context-aware load balancing in a WiFi-WiMAX heterogeneous network environment. IEEE Trans. Veh. Technol. **65**(1), 348–357 (2016)
6. Manzoor, S., Hei, X., Cheng, W.: A Multi-controller load balancing strategy for software defined WiFi networks. In: Sun, X., Pan, Z., Bertino, E. (eds.) ICCCS 2018. LNCS, vol. 11066, pp. 622–633. Springer, Cham (2018). https://doi.org/10.1007/978-3-030-00015-8_54
7. Lin, T.L., Kuo, C.H., Chang, H.Y., Chang, W.K., Lin, Y.Y.: A parameterized wildcard method based on SDN for server load balancing. In: 2016 International Conference on Networking and Network Applications (NaNA), pp. 383–386. IEEE (2016)
8. Yang, M., Li, Y., Jin, D., Zeng, L., Wu, X., Vasilakos, A.V.: Software-defined and virtualized future mobile and wireless networks: a survey. Mob. Netw. Appl. **20**(1), 4–18 (2015)
9. Al-Najjar, A., Layeghy, S., Portmann, M.: Pushing SDN to the end-host, network load balancing using openflow. In: 2016 IEEE International Conference on Pervasive Computing and Communication Workshops (PerCom Workshops), pp. 1–6. IEEE (2016)
10. Manzoor, S., Zhang, C., Hei, X., Cheng, W.: Understanding traffic load in software defined WiFi networks for healthcare. In: IEEE International Conference On Consumer Electronics - Taiwan (ICCE-TW), pp. 1–2. IEEE (2019)

11. Yu, J., Wang, Y., Pei, K., Zhang, S., Li, J.: A load balancing mechanism for multiple SDN controllers based on load informing strategy. In: 2016 18th Asia-Pacific Network Operations and Management Symposium (APNOMS), pp. 1–4. IEEE (2016)
12. Manzoor, S., Yin, Y., Gao, Y., Hei, X., Cheng, W.: A systematic study of IEEE 802.11 DCF network optimization from theory to testbed. IEEE Access **8**, 154114–154132 (2020)
13. Manzoor, S., Karmon, P., Hei, X., Cheng, W.: Traffic aware load balancing in software defined WiFi networks for healthcare. In: Information Communication Technologies Conference (ICTC), Nanjing, China, IEEE, May 2020
14. Song, P., Liu, Y., Liu, T., Qian, D.: Flow stealer lightweight load balancing by stealing flows in distributed SDN controllers. Sci. China Inf. Sci. **60**(3), 1–16 (2017)
15. Kang, S.B., Kwon, G.I.: Load balancing of software-defined network controller using genetic algorithm. Contemp. Eng. Sci. **9**(18), 881–888 (2016)
16. Zhou, Y., et al.: A load balancing strategy of SDN controller based on distributed decision. In: 2014 IEEE 13th International Conference on Trust, Security and Privacy in Computing and Communications (TrustCom), pp. 851–856. IEEE (2014)
17. Zhong, H., Sheng, J., Cui, J., Xu, Y.: SCPLBS: a smart cooperative platform for load balancing and security on SDN distributed controllers. In: 2017 3rd IEEE International Conference on Cybernetics (CYBCONF), pp. 1–6. IEEE (2017)
18. Lei, T., Wen, X., Lu, Z., Li, Y.: A semi-matching based load balancing scheme for dense IEEE 802.11 WLANs. IEEE Access **5**, 15332–15339 (2017)
19. Lin, Y.-D., Wang, C.C., Lu, Y.-J., Lai, Y.-C., Yang, H.-C.: Two-tier dynamic load balancing in SDN-enabled Wi-Fi networks. Wirel. Netw. **24**(8), 2811–2823 (2017). https://doi.org/10.1007/s11276-017-1504-3
20. Kiran, N., Changchuan, Y., Akram, Z.: AP load balance based handover in software defined WiFi systems. In: 2016 IEEE International Conference on Network Infrastructure and Digital Content (IC-NIDC), pp. 6–11. IEEE (2016)
21. Adalian, N., Ajaeiya, G., Dawy, Z., Elhajj, I.H., Kayssi, A., Chehab, A.: Load balancing in LTE core networks using SDN. In: IEEE International Multidisciplinary Conference on Engineering Technology (IMCET), pp. 213–217. IEEE (2016)

HGC: Hybrid Gradient Compression in Distributed Deep Learning

Kaifan Hu(✉), Chengkun Wu, and En Zhu

School of Computer, National University of Defense Technology, Changsha 410073, Hunan, China
{kaifan_hu,chengkun_wu,enzhu}@nudt.edu.cn

Abstract. Distributed stochastic gradient descent (SGD) algorithms are widely deployed in training large-scale deep learning models, while the communication overhead among workers becomes the new system bottleneck. Recently, two major categories of gradient compression techniques were proposed, including gradient quantization and sparsification. At best, the gradient quantization technique can obtain a compression ratio of 32, with little impact on model convergence accuracy. The gradient sparse technique can achieve a much higher compression ratio with some loss of model accuracy. To obtain a higher communication compression ratio with the minimum model accuracy loss, we proposed a mixed compression strategy named Hybrid Gradient Compression (HGC), which combines the merits of both quantization and sparsification. We validated the efficiency of our Hybrid Gradient Compression method by testing some complex models with millions of parameters (e.g., ResNet, VGG, LSTM, etc.) on the datasets including CIFAR-10, CIFAR-100 and Penn TreeBank on a GPU cluster. According to our tests, HGC can achieve a much higher gradient compression ratio at the cost of a small accuracy loss.

Keywords: Gradient compression · Machine learning · Distributed computing · Distributed optimization

1 Introduction

Deep learning is widely used in various fields. Challenging deep learning tasks on increasingly large-scale datasets require complex Deep Neural Networks (DNNs) models. To train such a model, it generally requires a highly computationally efficient method with the ability to harness high-performance computing power. The distributed stochastic gradient descent algorithms excel as a group of competitive candidates. A data parallelism strategy of SGD is to make model copies in computing workers and train in parallel on different subsets of data. It can be formulated as below, where w_t is the parameter of the model in iteration t, η_t is the learning rate or the step size. And $g^p(x)$ is gradient on a data batch $x \in \mathcal{X}_p$, in which the $\mathcal{X}_p$ is a subset of dataset $\mathcal{X}$ computed by the local computation worker p.

X. Sun et al. (Eds.): ICAIS 2021, CCIS 1422, pp. 15–27, 2021.
https://doi.org/10.1007/978-3-030-78615-1_2

$$w_{t+1} = w_t - \eta_t \frac{1}{P} \sum_{p=1}^{P} g^p(x)$$

Increasing the number of workers can reduce computation time dramatically in the beginning. However, as the scale of the distributed system grows up, especially in parameter server framework [10], which need all workers to communicate with parameter server node for gradient synchronization, the extensive gradient and parameter synchronization prolong the communication time and even amortize the efficiency of the computation process. As the number of workers increasing, a communication bottleneck has appeared and significantly affects the efficiency of communication in distributed training. A more generic and important research topic is accelerating the distributed training of dense models by utilizing compression techniques. Two typical types of compression techniques include sparsification and quantization. For instance, Aji and Heafield [1] proposed to sparsify the gradient vector by dropping out small values in order to reduce the time cost during gradient transmission, which is usually known as a famous sparsification algorithm - Top-K. For the same purpose, quantizing the elements of gradient vector into low-precision values with a smaller bit to storage has also been extensively studied.

To note, quantization techniques could obtain a slight loss on convergence accuracy with a compression ratio of 32 at most. In contrast, sparsification techniques could achieve a higher compression ratio but make the convergence accuracy significantly reduce. Our research aims to maintain the model accuracy and achieve a higher compression ratio for communication-efficient at the same time.

Based on that, we proposed a hybrid method named Hybrid Gradient Compression (HGC), which incorporates the advantages of both gradient quantization and sparsification, aiming at getting a higher compression ratio and lower accuracy loss simultaneously. Our contributions can be summarized as (1) we perform a theoretical analysis on the gradient compression and communication efficiency; (2) we proposed an efficient gradient compression method that can reduce communication overhead at a small accuracy loss.

2 Related Work

Generally, gradient compression techniques can be divided into three categories (Quantization, Sparsification, and Low-Rank Decomposition).

2.1 Quantization

Quantization is a method that quantizes the elements in the gradient vector to low-precision values with little space to storage (less than 32 bit). Quantization can reduce the communication bandwidth between the workers and alleviate communication bottlenecks. There are many works proposed recently. 1-bit SGD [12] is to quantize the gradients into 1 bit to transfer, and the experiment in which it used the 1-bit SGD to train a speech model achieved a higher speedup.

After 1-bit SGD, 8-bit quantization [6] has been proposed. It maps each $Float32$ gradient to 8 bit: 1 bit for sign, 3 bit for the exponent, and the other 4 bit for the mantissa. Quantized SGD (QSGD) [2] proposed another method which used stochastic rounding to estimate the gradients and could quantize the gradients into 4 or 8 bit. Similar to QSGD, there is another approach named TernGrad, which uses 3-level values (-1, 0 and 1) to represent a gradient, could quantize the gradients into 2 bit. QSGD and TernGrad also give a theoretical analysis to demonstrate the convergence of their quantization algorithm. Moreover, both of them also do lots of experiments to demonstrate that their algorithm works a lot. SignSGD [4] transmits the sign of gradient elements by quantizing the negative components to -1 and the others to 1.

2.2 Sparsification

Sparsifing the whole gradients to a part of them is another way to solve the communication bottleneck. Threshold-v [16] selects the elements by giving a threshold v that whose absolute values are larger than a fixed defined threshold value, which is difficult to choose in practice, and uses zeros to represent the other elements. Unlike the threshold-v, Top-K [1] selects the k largest values of gradient vector in absolute value, and there is also a random version named Random-K [15]. To further realize a lower loss on the accuracy, DGC [11] apply a local update inspired by the momentum SGD, and a warm-up step for the selection of hyperparameter k. After finding that the variance of the gradient affects the convergence rate, Wangni et al. [18] proposed an unbiased sparse coding to maximize sparsity and control the variance to ensure the convergence rate. Concurrently, Adacomp [5] has been proposed to automatically modify the compression rate depending on the local gradient activity.

2.3 Low-Rank Decomposition

DNNs are over-parameterized and exhibit low-rank structure [3]. Based on this observation, low-rank methods represent the gradient as a matrix M and factorize it into two low-rank matrices P and R that are smaller than M. We then achieve a gradient compression by just transmitting the matrix P and R rather than matrix M. Typically, the factorization is approximate. ATOMO [17] minimizes the variance of the quantized stochastic gradient. Let $\tilde{g}$ be an unbiased estimator of stochastic gradient g that has atomic decomposition to get a low-rank approximation of g.

Due to the high computational complexity of the matrix decomposition (SVD have a computational complexity of $\mathcal{O}(n^3)$), the low-rank decomposition approach is rarely used in a large-scale situation. Therefore, researchers pay more attention to the other two methods. In our work, we also focus on the other two methods.

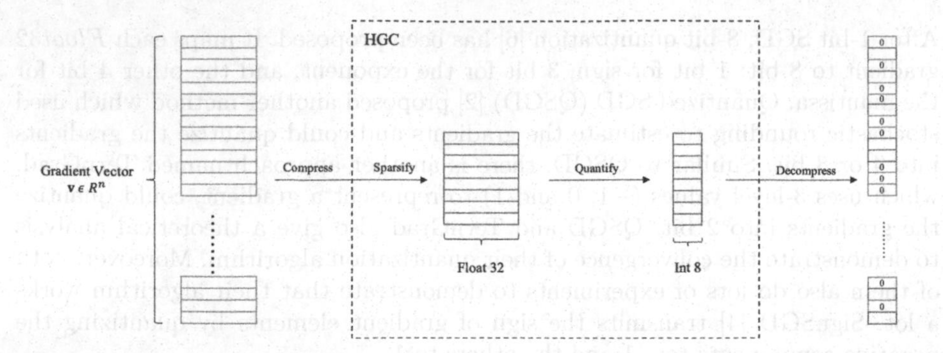

Fig. 1. The framework of the Hybrid Gradient Compression Algorithm, which has two phases for gradient vector compression. For a gradient vector $\mathbf{v} \in \mathbb{R}^n$, HGC firstly sparsify it, and then quantify the element to a low precision representation, where we just use *8bit* to present the gradient instead of the *Float32*.

3 Hybrid Gradient Compression

An overview of the Hybrid Gradient Compression method is shown in Fig. 1. As we can see, the HGC algorithm can be divided into two main steps, sparsification and quantization. The key idea behind HGC is that we employ a small sparsity to avoid gradient staleness and maintain more information for less accuracy loss and further improve the efficiency by the quantization step to achieve a higher total compression ratio.

3.1 Sparsification Step

First, we reduce the communication bandwidth requirement by transferring only the important elements in the gradient vector (sparse update). Like the Top-K approach, We use the gradient magnitude as a simple heuristics for importance: only gradients whose absolute value larger than a threshold are transmitted. To maintain the information, we accumulate the rest of the gradients locally in a residual vector. And then, in each step, these residual gradients will be accumulated to the gradient vector locally computed. Thus, we send the large gradients immediately and eventually send all of the gradients over time. As shown in Algorithm 1, step 3 and 9–12 is the overview of the sparsification step.

To get rid of gradient top-k selection with a high computational complexity $\mathcal{O}(n + klogn)$, where n is the number of the gradient elements, and the k is the number of gradient element we want to get after sparsification, we compute the mean of the gradient vector $|G|$ as a threshold value (a complexity of $\mathcal{O}(n)$) and we change the threshold value by multiplying $3/4$ or $5/4$ if the number of gradient elements whose absolute value larger than the threshold is less or larger than k. After several iterations, it could self-adapt into an appropriate value and get nearly k gradient elements. Thus the computation complexity of the sparsification step in HGC is $\mathcal{O}(n)$ rather than $\mathcal{O}(n + klogn)$. Hierarchically calculating the threshold significantly reduces selection time because the layer-wise gradient vector has a smaller size than the full gradient vector. In practice, total extra computation time is negligible compared to network communication time, which is usually from hundreds of milliseconds to several seconds depending on the network bandwidth.

Algorithm 1. Hybrid Gradient Compression on worker k

Require: dataset $\mathcal{X}$
Require: minibatch size b per worker
Require: the number of worker N
Require: distributed optimization function SGD
Require: init layer-wise parameters $w = \{w[0], \ldots, w[L]\}$, where L is the number of layer in the model

```
1: G^k ← 0
2: for t = 0, 1, ... do
3:   G_t^k ← G_{t-1}^k
4:   for i = 1, ..., b do
5:     Sample data x from the dataset X
6:     G_t^k ← G_t^k + (1/b)∇f(x; w_t)
7:   end for
8:   for j = 0, 1, ..., L do
9:     Initialize threshold and self-adjust it: thr ← mean of |G_t^k[j]|
10:    Mask ← |G_t^k| > thr
11:    G̃_t^k[j] ← G_t^k[j] ⊙ Mask
12:    G_t^k[j] ← G_t^k ⊙ ¬Mask
13:    G̃_t^k[j] ← Q(G̃_t^k[j])
14:  end for
15:  All − reduce G_t^k : G_t ← Σ_{k=1}^N Q^{-1}(G̃_t^k)
16:  w_{t+1} ← SGD(w_t, G_t)
17: end for
```

3.2 Quantization Step

After a sparse pick on the gradients, we now consider a general, parametrizable lossy-compression scheme for stochastic gradient vectors. More specifically, we make some optimizations base on the QSGD [2]. The quantization function is denoted with Q(v). We defined uniformly distributed levels between 0 and 1, to which each value is quantized in a way which preserves the value in expectation, and introduces a smaller variance compared to QSGD.

For any $\mathbf{v} \in \mathbb{R}^n$ with $\mathbf{v} \neq \mathbf{0}$, $Q(\mathbf{v})$ is defined as

$$Q(v_i) = \|\mathbf{v}\|_\infty \cdot sgn(v_i) \cdot \varphi_i(\mathbf{v}, \mathbf{s})$$

where $\varphi_i(\mathbf{v}, s), i = 1, 2 \ldots$ are independent random variables defined as follow. Let $0 \leq l < s$ be an integer such that $|v_i| / \|\mathbf{v}\|_\infty \in [l/s, (l+1)/s]$. That is, $[l/s, (l+1)/s]$ is the quantization interval corresponding to $|v_i| / \|\mathbf{v}\|_\infty$. Then have

$$\varphi_i(\mathbf{v}, s) = \begin{cases} l/s & \text{with the probability } 1 - p(\frac{|v_i|}{\|\mathbf{v}\|_\infty}, s) \\ (l+1)/s & \text{otherwise} \end{cases}$$

Here, $p(a, s)$ is computed by $as - l$ for any $a \in [0, 1]$. if $\mathbf{v} = \mathbf{0}$, then we define $Q(\mathbf{v}, s) = \mathbf{0}$.

Firstly, we should compute $|v_i - \max_{1<i<d} |v_i||, i = 1, \ldots, d$, where the v_i is the ith element of gradient vector $\mathbf{v}$ after the sparsification step. And then we adopt the result into quantization step to make a further compression.

There is the main difference between our quantization method and QSGD that we use the infinite norm instead of the $l2$-norm. The reason can be interpreted as follow.

Consider a vector $\mathbf{v} \in \mathbb{R}^n$ and $v_i, v_2, \ldots, v_n$ are the elements in this vector. Without loss of generality, give a sequence as follows,

$$v_1 \approx v_2 \approx \cdots \approx v_k < v_{k+1} < \cdots < v_n$$

The $l2$-norm of that vector is about $\sqrt{k}\,\|\mathbf{v}\|_\infty = \sqrt{k}|v_t|, t = 1, \ldots, k$. And we could get $|v_i| / \|\mathbf{v}\|_\infty < 1/\sqrt{k} < 1$. That means we will only used $[0, \frac{1}{\sqrt{k}}]$ to quantify such a vector $\mathbf{v}$ and make a lot of coding space wasted. To avoid that situation, we used the $\|\mathbf{v}\|_\infty$ instead of the $\|\mathbf{v}\|_2$ to make full use of the encoding space and get a better variance bound at the same time.

Lemma 1. For any vector $\mathbf{v} \in \mathbb{R}^n$, we could have three properties as follow, (i) $\mathbb{E}[Q(\mathbf{v})] = \mathbf{v}$ (unbiasedness), (ii) $\mathbb{E}[\|\mathbf{Q_s}(\mathbf{v}) - \mathbf{v}\|_2^2] \leq min(n/s^2, \sqrt{n}/s)\,\|\mathbf{v}\|_\infty^2$ (variance bound), and (iii) $\mathbb{E}[\|\mathbf{Q}[\mathbf{v}]\|_0] \leq s(s + \sqrt{n})$ (sparsity).

Proof. The first and third claim obviously holds. We thus turn our attention to the second claim of the lemma. We first note the following bound:

$$\begin{aligned} \mathbb{E}[\varphi_i(\mathbf{v}, s)^2] &= \mathbb{E}[\varphi_i(\mathbf{v}, s)]^2 + \mathbb{E}[(\varphi_i(\mathbf{v}, s) - \mathbb{E}[\varphi_i(\mathbf{v}, s)])^2)] \\ &= \frac{v_i^2}{\|\mathbf{v}\|_\infty} + \frac{1}{s^2} p(\frac{|v_i|}{\|\mathbf{v}\|_\infty}, s)(1 - p(\frac{|v_i|}{\|\mathbf{v}\|_\infty}, s)) \\ &\leq \frac{v_i^2}{\|\mathbf{v}\|_\infty} + \frac{1}{s^2} p(\frac{|v_i|}{\|\mathbf{v}\|_\infty}, s) \end{aligned}$$

Using this bound, we have

$$\begin{aligned}
\mathbb{E}[\|\mathbf{Q}(\mathbf{v},\mathbf{s})\|^2] &= \sum_{i=1}^{n} \mathbb{E}[\|\mathbf{v}\|_\infty^2 \varphi_i(\frac{|v_i|}{\|\mathbf{v}\|_\infty}, s)^2] \\
&\leq \|\mathbf{v}\|_\infty^2 \sum_{i=1}^{n} [\frac{|v_i|^2}{\|\mathbf{v}\|_\infty^2} + \frac{1}{s^2} p(\frac{|v_i|}{\|\mathbf{v}\|_\infty}, s)] \\
&= (1 + \frac{1}{s^2} \sum_{i=1} np(\frac{|v_i|}{\|\mathbf{v}\|_\infty}, s))\|\mathbf{v}\|_\infty^2 \\
&\overset{(1)}{\leq} (1 + min(\frac{n}{s^2}, \frac{\|\mathbf{v}\|_1}{s\|\mathbf{v}\|_\infty}))\|\mathbf{v}\|_\infty^2 \\
&\leq (1 + min(\frac{n}{s^2}, \frac{\sqrt{n}}{s}))\|\mathbf{v}\|_\infty^2
\end{aligned}$$

where (1) follows from the fact that $p(a, s) \leq 1$ and $p(a, s) \leq as$. This immediately implies the (ii) in Lemma 1.

Compared to the QSGD, our quantization approach have a smaller variance bound. It means that our algorithm loses less information when quantizing the gradient vector.

In our algorithm, we fix the number $s = 127$, and then we only need 8 bit to encode the gradient in the gradient vector $\mathbf{v}$ (Fig. 2).

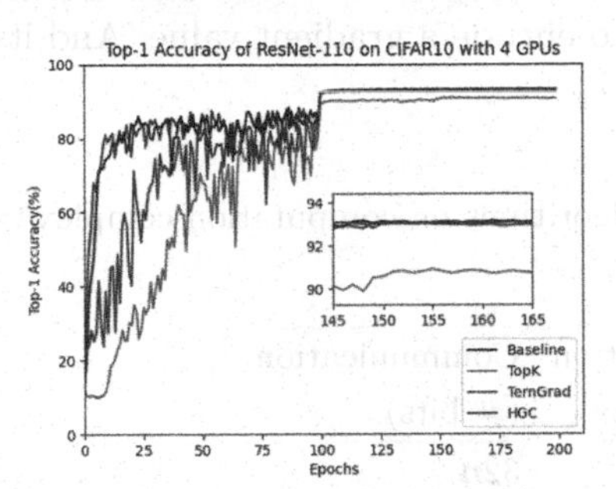

(a) Top-1 accuracy of ResNet-110 with 4 GPUs

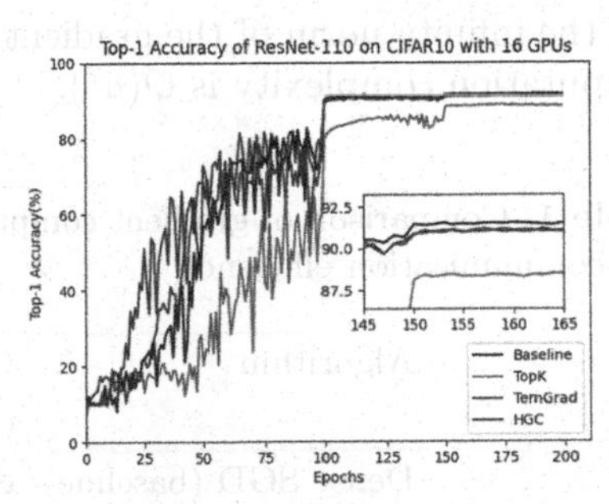

(b) Top-1 accuracy of ResNet-110 with 16 GPUs

Fig. 2. Learning curves of ResNet-110 in image classification task with worker size 4 and 16.

3.3 Communication Efficiency

Note that our algorithm includes the sparsification and quantization, we could quickly get the computation complexity and communication cost in each gradients' transmission.

The HGC process consists of two main parts for a gradient vector $g \in \mathbb{R}^n$:

- the communication cost of transmitting gradient vector in sparsification step use about $32\gamma n$ bits to encode the gradient, where γ is the sparsity for sparsification step. We compute the mean of gradient vector (a complexity of $\mathcal{O}(n)$) and then set it as the initial threshold for computing an approximate threshold to pick the gradient elements. Thus its total computation complexity is $\mathcal{O}(n)$.
- the communication cost of transmitting a gradient vector in the quantization step could use $8n + 2 \times 32$ bits. In this step, we should compute the infinite norm (a complexity of $\mathcal{O}(n)$) and then apply quantization for each gradient selected from step 1. The total computation complexity is $\mathcal{O}(n^2)$.

From the above analysis, we could get the HGC could use $8k + 64$ bits to encode the gradient vector where the k is the number of gradient after sparse selection with a complexity of $\mathcal{O}(n^2)$.

As we can see from Table 1, The baseline (Dense SGD) does not process local gradients and has a computation complexity of $\mathcal{O}(1)$. Thus the baseline needs $32n$ bits to encode as a gradient value needs $Float32$ to storage. The Top-K algorithm needs to find the k gradient with a larger absolute value and has a complexity of $\mathcal{O}(n + klogn)$ in the implementation of PyTorch. And then, the Top-K algorithm uses $Float32$ to encode these k gradients. Thus the total communication cost is $32k$ bits. TernGrad has a total communication cost of $2n + 32$ because it uses 3-level value (-1, 0, and 1), which only needs 2 bits, and the infinity norm of the gradient vector to encode a gradient value. And its computation complexity is $\mathcal{O}(n^2)$.

Table 1. Comparison of gradient compression algorithms on computation complexity and communication efficiency

Algorithm	Computation complexity	Communication (# bits)
Dense SGD (baseline)	$\mathcal{O}(1)$	$32n$
TernGrad	$\mathcal{O}(n^2)$	$2n + 32$
Top-K	$\mathcal{O}(n + klogn)$	$32k$
HGC (ours)	$\mathcal{O}(n^2)$	$8k + 64$

4 Experiment

4.1 Experiment Setup

Experiment Environment. We completed all the experiments in this paper on GPU-V100, a GPU cluster. Each computing node in the GPU-V100 cluster contains 2 Intel(R) Xeon(R) Gold 6132 14-core multi-core CPUs and 4 NVIDIA

Table 2. Training ResNet-110 with a worker size range from 4 to 16. The baseline above means no compression method used, and we fixed the sparsification ratio as 99% for Top-K because the model is difficult to converge when using a higher compression ratio. We fixed the sparsification ratio as 98% to achieve a 200×'s compression. Batch size in total means the number of data that all worker process per iteration.

# GPUs	Batch size	Compression method	Top-1 accuracy	Training time per iter. (s)
4	512	Baseline	93.22%	0.316
		Top K	90.90% (−2.32%)	0.364
		TernGrad	93.07% (−0.15%)	0.412
		HGC	93.19% (−0.03%)	0.425
8	1024	Baseline	92.42%	0.446
		Top K	88.75% (−3.66%)	0.404
		TernGrad	92.36% (−0.06%)	0.433
		HGC	92.34% (−0.08%)	0.434
16	2048	Baseline	91.66%	0.568
		Top K	86.88% (−4.78%)	0.451
		TernGrad	91.18% (−0.48%)	0.483
		HGC	91.28% (−0.38%)	0.448

Table 3. Comparison of gradient compression ratio on CIFAR-100 Dataset for different models with 8 workers.

Model	# Params	Training method	Top-1 Acc.	Top-5 Acc.	Gradient size	Compression ratio
VGG16	14.73 M	Baseline	66.96%	89.81%	56.18 MB	
		TernGrad	62.03% (−4.93%)	87.38% (−2.43%)	3.51 MB	16×
		HGC	62.09% (−4.87%)	87.55% (−2.26%)	0.42 MB	133×
AlexNet	2.47 M	Baseline	55.32%	82.37%	9.43 MB	
		HGC	51.09% (−4.23%)	80.08% (−2.29%)	0.07 MB	135×

Tesla V100 GPUs, with 256GB of RAM (shared by 2 CPUs). The equipped GPU model is NVIDIA Tesla V100 SXM2, the peak double-precision floating-point calculation capability is up to 7.8 TFLOPS, the single-precision floating-point capability is 15.7 TFLOPS, the Tensor performance is 125 TFLOPS, and the card is equipped with 16GB HBM2 device memory. The GPU-V100 cluster operating system is CentOS Linux release 7.6.1810, and the GPU driver is version 418.87.00. Furthermore, all nodes in the cluster are connected with a high-bandwidth 100-Gbps InfiniBand network.

Implementation. We implement our algorithm using the PyTorch Framework by using GRACE [20], which is a gradient compression framework for distributed deep learning. Dependent software libraries include CUDA-10.0, cuDNN-7.6.4, NCCL-2.4.6, PyTorch-1.4.0, OpenMPI-4.0.1, and Horovod-0.18.2 [13], which are kept the same for all evaluated algorithms.

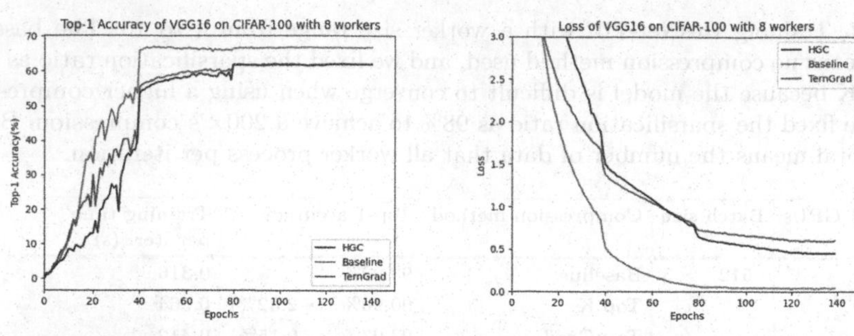

(a) Top-1 accuracy of VGG16 with 8 workers (b) Training loss of VGG16 with 8 workers

Fig. 3. Learning curves of VGG16 in image classification task with 8 workers. (Color figure online)

Image Classification. We train some network models with millions of parameters (e.g. VGG [14], ResNet [7], AlexNet [9], etc.) on CIFAR-10 dataset or CIFAR-100 dataset. We train the models with momentum SGD following the training schedule in [8].

Natural Language Processing. The Penn Treebank corpus (PTB) dataset consists of 923,000 training, 73,000 validation and 82,000 test words. We adopt the 2-layer LSTM model with 1500 hidden units per layer using the vanilla SGD with gradient clipping.

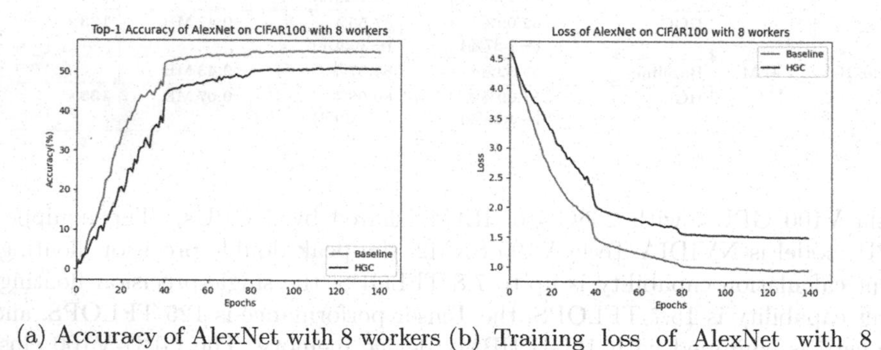

(a) Accuracy of AlexNet with 8 workers (b) Training loss of AlexNet with 8 workers

Fig. 4. Learning curves of AlexNet in image classification task with 8 workers.

4.2 Convergence Accuracy and Scalability

Convergence accuracy is one of the essential indicators for evaluating algorithm performance. We compare HGC's performance with dense SGD (without compression), one sparsification techniques Top-K [1], and one quantization technique TernGrad [19]. We use the accuracy of dense SGD as a baseline and compare the other three algorithms.

Firstly, we use all four algorithms to train ResNet110 with worker size 4, 8, and 16. We set the gradient sparsity of 99% (only 1% is non-zero) for the Top-K because the model is difficult to converge if we use a higher sparsity. Compared with the Top-K, to maintain more information from gradient, a sparsity of 98% is chosen for the sparsification step in the HGC algorithm. Table 2 presents the details for accuracy performance with different worker size. Figure 3(a) and 3(b) are the Top-1 accuracy of ResNet-110 on CIFAR-10 with four workers. The learning curve of Top-K (red) is worse than the baseline due to gradient staleness. With a lower sparsity to maintain more information and further compression by quantization, the learning curve of HGC (blue) converges slightly faster than Top-K, and the accuracy is much closer to the baseline (black). We could found that when worker size increases, Top-K has poor scalability cause of an obvious accuracy loss, but HGC is much closer to the baseline. HGC also has a higher compression ratio compared to TernGrad (green). The accuracy of ResNet-110 is well maintained by the Hybrid Gradient Compression method.

When scaling to a different model and dataset, the accuracy is still comparable to the baseline. Table 3 shows the results of AlexNet and VGG16 training on the CIFAR-100 dataset with eight workers. Compared with the quantization method such as TernGrad, which achieves a compression ratio no more than 16×, HGC gives a better performance on compression (100+) than TernGrad with no much accuracy loss. More details about the model accuracy could be found in Fig. 3 and 4. Due to the gradient staleness, the learning curve of HGC (black) converges slightly slower than baseline (red). However, HGC could achieve a slight accuracy loss finally.

For language modeling, Table 4 shows the perplexity and compression ratio of the language model trained with 8 workers. As shown in Table 4, Hybrid Gradient Compression compresses the gradient by 82× with a slight loss in perplexity.

In conclusion, the HGC algorithm has good scalability with comparison to the Top-K and TernGrad. It could get a close accuracy to the baseline while training with a different model, dataset, and worker size.

Table 4. Training results of language model with 8 workers.

Task	Train LSTM-PTB Model		
Train method	Perplexity	Gradient Size	Compression Ratio
baseline	150.47	194.68 MB	1×
HGC	162.34	2.37 MB	82×

4.3 Compression Ratio and Execution Time

In this section, we would have a discussion on the communication cost, which we used compression ratio to represent cause of no access to record the exact time cost during communication between workers in the GRACE, a distributed deep learning framework. We also measure the average time cost per iteration as another indicator for evaluating algorithm performance.

As shown in Table 2, we could find that our algorithm has achieved a maximum compression ratio of 200×, which means our algorithm has fewer communication costs than others. Cause of a high computation complexity of $\mathcal{O}(n^2)$, The average training time per iteration of HGC is the largest among three compression algorithms when worker size less than 8. More specifically, the worker size (no more than 8) in our experiments is such small that it is difficult to reproduce the communication bottleneck in distributed training of a large cluster. In other words, the local computation cost rather than the communication overhead dominates the model training time. Cause of an additional computation complexity of $\mathcal{O}(n^2)$, in our experiments, our algorithm has a higher training time cost than others while the worker size less than 8. When worker size scaling to 16, HGC can get better performance than others. Nevertheless, We believe our algorithm could be more efficient when applying in a larger cluster training. Besides, Our algorithm also has a good compression ratio while training other models. i.e., 133× and 135× for VGG16 and AlexNet, and 82× for LSTM-PTB.

5 Conclusion

In this paper, we proposed a hybrid compression algorithm (HGC) to reduce the communication between workers, achieving a higher compression ratio with a slight accuracy loss during the model training. We also give a theoretical analysis on the quantization and communication efficiency of HGC. A number of validation experiments demonstrate the effectiveness of our algorithm. In the future, we would focus on how to decrease the computation complexity and further improve the applicability of our algorithm.

References

1. Aji, K.A.F., Heafield: sparse communication for distributed gradient descent. In: Proceedings of the 2017 Conference on Empirical Methods in Natural Language Processing, pp. 440–445. Copenhagen, Denmark (2017). https://doi.org/10.18653/v1/d17-1045
2. Alistarh, D., Grubic, D., Li, J., Tomioka, R., Vojnovic, M.: QSGD: communication-efficient SGD via gradient quantization and encoding. Adv. Neural Inf. Process. Syst. **30**, 1709–1720 (2017)
3. Arora, S., Ge, R., Neyshabur, B., Zhang, Y.: Stronger generalization bounds for deep nets via a compression approach (2018)
4. Bernstein, J., Wang, Y.X., Azizzadenesheli, K., Anandkumar, A.: signSGD compressed optimisation for non-convex problems. In: Dy, J., Krause, A. (eds.) Proceedings of the 35th International Conference on Machine Learning, pp. 560–569. Stockholmsmässan, Stockholm, Sweden (2018)

5. Chen, C., Choi, J., Brand, D., Agrawal, A., Zhang, W., Gopalakrishnan, K.: Adacomp: adaptive residual gradient compression for data-parallel distributed training. In: 32nd AAAI Conference on Artificial Intelligence, AAAI 2018, pp. 2827–2835, January 2018
6. Dettmers, T.: 8-Bit approximations for parallelism in deep learning. In: International Conference on Learning Representations (2016)
7. He, K., Zhang, X., Ren, S., Sun, J.: Deep residual learning for image recognition. In: IEEE Conference on Computer Vision and Pattern Recognition, pp. 770–778 (2016). https://doi.org/10.1109/cvpr.2016.90
8. Idelbayev, Y.: Proper ResNet implementation for CIFAR10/CIFAR100 in PyTorch. https://github.com/akamaster/pytorch_resnet_cifar10
9. Krizhevsky, A., Sutskever, I., Hinton, G.E.: Imagenet classification with deep convolutional neural networks. In: Pereira, F., Burges, C.J.C., Bottou, L., Weinberger, K.Q. (eds) Advances in Neural Information Processing Systems, vol. 25, pp. 1097–1105 (2012)
10. Li, M., Andersen, D.G., Smola, A.J., Yu, K.: Communication efficient distributed machine learning with the parameter server. Adv. Neural Inf. Process. Syst. **27**, 19–27 (2014)
11. Lin, Y., Han, S., Mao, H., Wang, Y., Dally, W.J.: Deep gradient compression: reducing the communication bandwidth for distributed training. In: International Conference on Learning Representations, May 2018
12. Seide, F., Fu, H., Droppo, J., Li, G., Yu, D.: 1-bit stochastic gradient descent and application to data-parallel distributed training of speech DNNs. In: Interspeech 2014 (2014)
13. Sergeev, A., Balso, M.D.: Horovod: fast and easy distributed deep learning in TensorFlow (2018). arXiv preprint: arXiv:1802.05799
14. Simonyan, K., Zisserman, A.: Very deep convolutional networks for large-scale image recognition. In: International Conference on Learning Representations, May 2015
15. Stich, S.U., Cordonnier, J.B., Jaggi, M.: Sparsified SGD with memory. Adv. Neural Inf. Process. Syst. **31**, 4447–4458 (2018)
16. Strom, N.: Scalable distributed DNN training using commodity GPU cloud computing. In: INTERSPEECH, pp. 1488–1492 (2015)
17. Wang, H., Sievert, S., Liu, S., Charles, Z., Papailiopoulos, D., Wright, S.: ATOMO: communication-efficient learning via atomic sparsification. Adv. Neural Inf. Process. Syst. **31**, 9850–9861 (2018)
18. Wangni, J., Wang, J., Liu, J., Zhang, T.: Gradient sparsification for communication-efficient distributed optimization. Adv. Neural Inf. Process. Syst. **31**, 1299–1309 (2018)
19. Wen, W., et al.: TernGrad: ternary gradients to reduce communication in distributed deep learning. Adv. Neural Inf. Process. Syst. **30**, 1509–1519 (2017)
20. Xu, H., et al.: Compressed Communication for Distributed Deep Learning: Survey and Quantitative Evaluation. Technical report, KAUST, April 2020

Application of Lidar INS Integrated Positioning in Automatic Driving

Zhiguo Li, Ying Wang(✉), and Zhenti Xiong

Hunan Automotive Engineering Vocational College, Zhuzhou 412000, China

Abstract. Due to factors such as ground blockage, electromagnetic interference and malicious attacks, the satellite-inertial navigation system cannot provide stable and reliable high-precision positioning information. Therefore, it is necessary to provide an autonomous and reliable navigation system for the self driving vehicle as a supplement. In this study, vehicle-mounted radar is used to obtain point cloud information around the self-driving car, and the above point cloud information is filtered to find a road that can be driven, and a least squares calculation is performed on the area to obtain a fitted path. Finally, the real-time travel trajectory of autonomous vehicles is planned. Based on the above method, an unmanned experimental vehicle was used for engineering verification. The results show that the radar-inertial navigation mode can achieve stable and reliable L4 automatic driving.

Keywords: Autopilot · Positioning · Radar-inertial navigation · Trajectory planning

1 Introduction

Autopilot technology originated from military and special applications. From the 1980s to the early 1990s, DARPA has carried out independent land vehicle project research through the strategic computing plan [1].

Due to the advantages of strong adaptability to the physical environment, small risk, low cost, non-contact, no casualties, long endurance, multi-function, autonomous control and many other aspects, unmanned vehicles not only have irreplaceable advantages in the field of military operations, but also have been widely used in social security, emergency rescue and other fields [2].

In the work of public security, fire protection, disaster relief and counter-terrorism, through the technology of man-machine cooperation, independent operation and self-organized team, the unmanned vehicle can be on duty all day, all day, all day, and reduce the casualties in the dangerous scene of fire protection, disaster relief and counter-terrorism. At the same time, it can effectively make up for the shortage of personnel and achieve rapid response [3].

High precision satellite positioning and machine vision are common methods for trajectory planning of unmanned vehicle, especially the trajectory planning technology based on high precision satellite positioning and digital map, which is widely used in

X. Sun et al. (Eds.): ICAIS 2021, CCIS 1422, pp. 28–38, 2021.
https://doi.org/10.1007/978-3-030-78615-1_3

the field of automatic driving. However, in the real scene, there are some problems, such as signal occlusion, electromagnetic interference, light shortage and so on. The availability of satellite positioning and visual data is poor, which makes the unmanned vehicle unable to plan and control the trajectory in the conventional way. Literature [4] provided a kind of lidar navigation technology applied to intelligent agricultural machinery. This technology was based on the single line lidar to scan crops on both sides to form road feature points. The least square method was used to fit the point cloud data to form the driving path. Literature [5] Studied and optimized the lidar road edge detection algorithm, proposed a navigation method based on the passable area, and used vector field histogram + (VFH +) to plan the local path of vehicles, effectively solved the navigation failure problem caused by a single path planning method in harsh environment.

Aiming at the navigation problems faced by the unmanned vehicle in the complex scene, this paper has validated a kind of navigation technology based on lidar, which can realize obstacle recognition, trajectory planning and vehicle control in the complex scene, and improve the environmental adaptability and reliability of the unmanned vehicle.

2 Unmanned Vehicle Sensing System

2.1 Principle of Unmanned Vehicle System

The driverless system can be divided into four parts: environment perception, intelligent planning and decision-making, adaptive control and vehicle chassis wire control. The sensing system obtains the state information of vehicle surroundings and itself through lidar, vision, ins and other sensors, and forms driving situation map after data processing and fusion, which is sent to the planning and decision-making module [6].

The planning and decision-making system generates planning trajectories and decision-making commands according to the driving situation map, mission planning and its own state. The intelligent control system realizes the real-time and accurate tracking of the track through the horizontal control and longitudinal control, and forms the steering, driving and braking commands, which are sent to the bottom executive parts of the vehicle. The executive part of the vehicle adapts according to the vehicle chassis, running state, road condition, load, weather and other factors, and then executes the command sent by the control module. The schematic structure of the driverless system is shown in Fig. 1 [7].

2.2 Principle of Unmanned Vehicle Sensing System

As shown in Fig. 1, the left side is the automatic driving system sensor. Through point cloud data, lidar can realize obstacle recognition and surrounding environment mapping; Millimeter wave radar and ultrasonic radar are mainly used for obstacle detection; Visual sensing is mainly used for obstacle recognition, color and semantic information recognition of surrounding environment, such as lane line, traffic light, speed limit plate, etc.; INS/GPS/BDS is mainly used for vehicle attitude perception and positioning. Figure 2 shows a sensor configuration scheme for unmanned sightseeing/shuttle bus on campus.

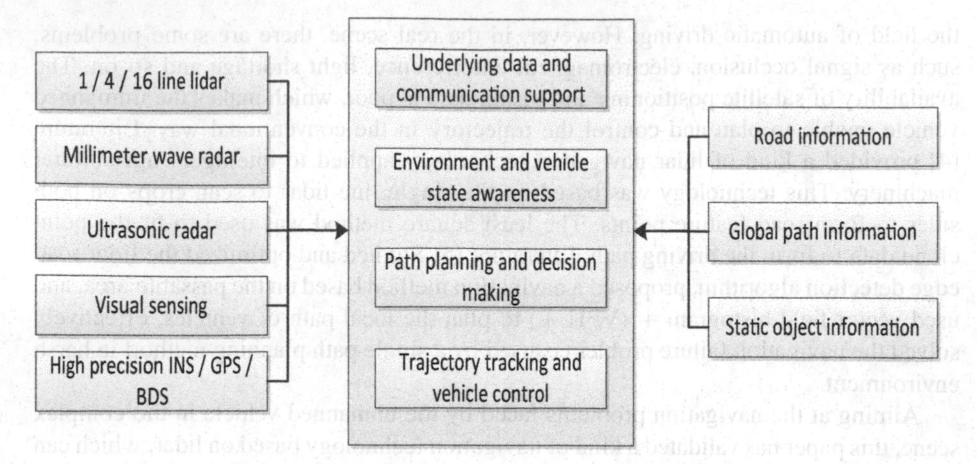

Fig. 1. Principle Structure Diagram of Autopilot System.

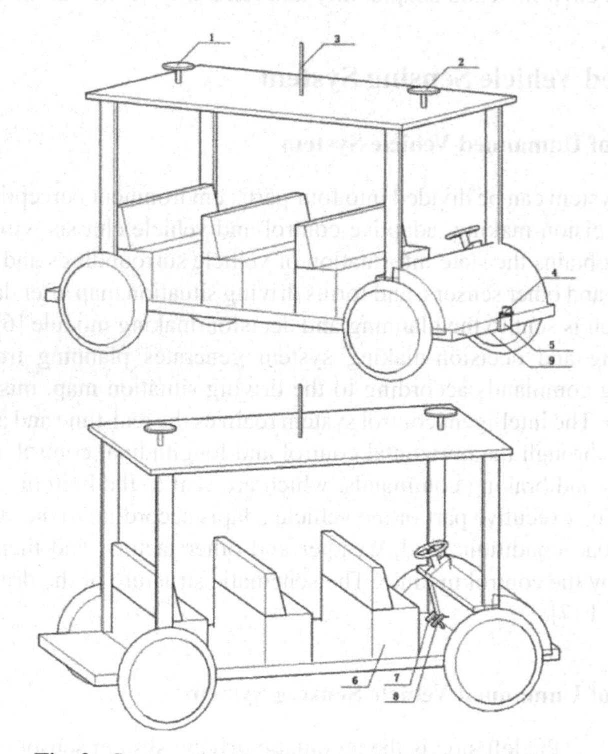

Fig. 2. Sensor scheme of unmanned sightseeing vehicle.

Different sensors have different implementation principles and data characteristics, so they are suitable for different environments. Therefore, multi-sensor data fusion is an effective way to improve the performance of sensing system. Autonomous tracking, obstacle recognition and obstacle avoidance are the basic functions of unmanned vehicle.

Among them, autonomous tracking mainly depends on GPS/BDS positioning. Obstacle recognition and avoidance mainly rely on millimeter wave radar, lidar and machine vision. In the L4 level urban automatic driving scene, there are often scenes in which satellite positioning is not available, such as tunnels, dense trees, urban canyons, and so on. At the same time, the vision sensor is easy to be affected by the light, and its reliability is difficult to meet the requirements of unmanned vehicle operation [8]. This paper will study the navigation technology of unmanned vehicle based on 16 line lidar. Using this technology, unmanned vehicle can realize trajectory planning and control in complex environment where high-precision positioning and vision are not available.

3 Technology of Lidar Navigation

Lidar can locate without receiving electromagnetic signals from outside, and is not affected by light. Therefore, it can work stably in the environment where the satellite signal is blocked and the light condition is poor. In this study, according to the LIDAR point cloud data, road features are obtained to form trajectory points, and then path fitting and trajectory planning are carried out.

3.1 Road Feature Extraction

There are clear road edge information such as green belt and buildings on both sides of the road. Therefore, in this study, firstly, the road feature information in LIDAR point cloud data is extracted, and then the polar coordinate system data of lidar is transformed into vehicle body coordinate system to obtain the position information of vehicle body.

In this study, the origin of the vehicle body coordinate system is located at 20 cm above the ground. The y-axis direction is taken as the front direction, and the x-axis direction is on the right-hand side. The definition of z-axis conforms to the right-hand coordinate system. The lidar is installed in front of the roof. The y-axis direction is the front direction of the car, and there is an inclination angle with the horizontal plane. The coordinate of LIDAR point cloud data is polar coordinate system, which needs to be converted to Cartesian coordinate system. The conversion relationship is shown in Formula 1.

$$\begin{cases} x_L = r\cos(\omega)\sin(\alpha) \\ y_L = r\cos(\omega)\cos(\alpha) \\ z_L = r\sin(\omega) \end{cases} \tag{1}$$

γ is the measured distance, ω is the vertical angle of the point cloud, α is the horizontal rotation angle of the point cloud, x_L, $\mathrm{y_L}$, $\mathrm{z_L}$ is the projection from polar coordinate system to Cartesian coordinate system.

The conversion relationship between radar coordinate system and vehicle body coordinate system is shown in formula 2.

$$\begin{bmatrix} x_V \\ y_V \\ z_V \end{bmatrix} = \begin{bmatrix} 1 & 0 & 0 \\ 0 & \cos(\alpha) & \sin(\alpha) \\ 0 & -\sin(\alpha) & \cos(\alpha) \end{bmatrix} \begin{bmatrix} x_L \\ y_L \\ z_L \end{bmatrix} + \begin{bmatrix} \Delta X \\ \Delta Y \\ \Delta H \end{bmatrix} \tag{2}$$

α is the angle between the lidar Y axis and the horizontal plane, ΔX, ΔY, ΔH are the relative position relationship between the lidar coordinate system origin and the vehicle body coordinate system origin, and x_V, y_V, z_V are the projection of LIDAR point cloud data in the vehicle body coordinate system [9].

Through formula (1) and formula (2), the LIDAR point cloud data coordinates under the car body coordinate system can be obtained, and the point cloud data can be filtered out to obtain stable and reliable road feature points.

3.2 Path Fitting and Trajectory Planning

According to the road feature points obtained in Sect. 2.1, trajectory planning based on lidar navigation is carried out. Two cases of obvious road features on both sides and obvious road features on one side are selected to design trajectory planning algorithm. For the first case, the trajectory points are calculated by formula (3).

$$\begin{cases} x_{ij} = \dfrac{x_{Ri} + x_{Lj}}{2} \\ y_{ij} = \dfrac{y_{Ri} + y_{Lj}}{2} \end{cases} \quad (3)$$

(x_{Li}, y_{Li}), (x_{Ri}, y_{Ri}) are the coordinates of the left and right feature points respectively, and (x_{ij}, y_{ij}) are the coordinates of the middle road track points obtained from the left and right feature points. The least square method is used to fit the multi-level function of the trajectory, and the trajectory curve can be obtained to realize the lidar navigation [10].

For the second case, the least square method is used to directly fit the road feature points, and then the trajectory function is horizontally translated along the x-axis direction of the car body coordinate system to obtain the travel trajectory curve.

Figure 3 shows two typical scene trajectory fitting diagrams.

In fact, the road curve in reality is relatively gentle, so this study carries out quadratic curve fitting for the trajectory. The main calculation rules are shown in formulas (4) to (7).

Suppose that the quadratic curve expression of vehicle trajectory is shown in formula (4):

$$f(x) = a_0 + a_1 x + a_2 x^2 \quad (4)$$

According to formula (4) and the above trajectory point data, the trajectory fitting variance is calculated, and the optimal trajectory function is obtained. The calculation process is shown in formulae (5), (6) and (7).

$$\delta(a_0, a_1, a_2) = \sum_{i=1}^{m} (f(x_i) - y_i)^2 = \sum_{i=1}^{m} (a_0 + a_1 x_i + a_2 x_i^2 - y_i)^2 \quad (5)$$

$$\begin{cases} \dfrac{\partial \delta}{\partial a_0} = 2\sum\limits_{i=1}^{m} (a_0 + a_1 x_i + a_2 x_i^2 - y_i) = 0 \\ \dfrac{\partial \delta}{\partial a_1} = 2\sum\limits_{i=1}^{m} (a_0 + a_1 x_i + a_2 x_i^2 - y_i) x_i = 0 \\ \dfrac{\partial \delta}{\partial a_2} = 2\sum\limits_{i=1}^{m} (a_0 + a_1 x_i + a_2 x_i^2 - y_i) x_i^2 = 0 \end{cases} \quad (6)$$

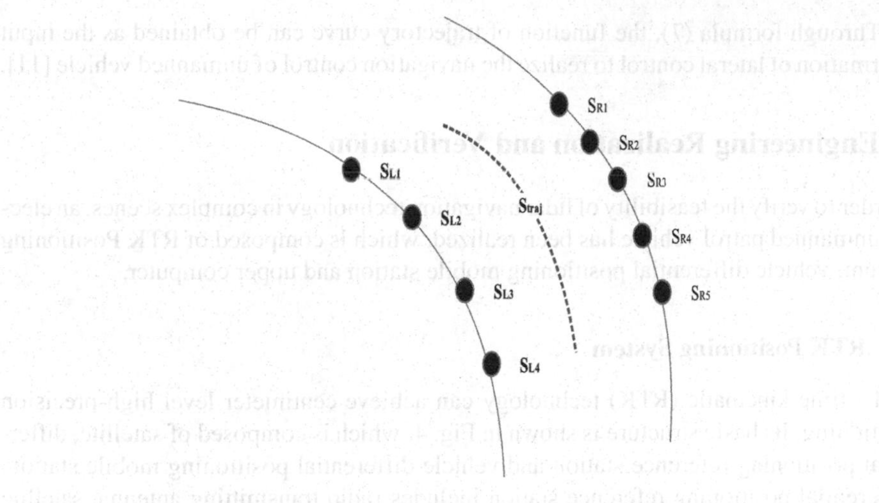

(a) Planning of road edge track on both sides

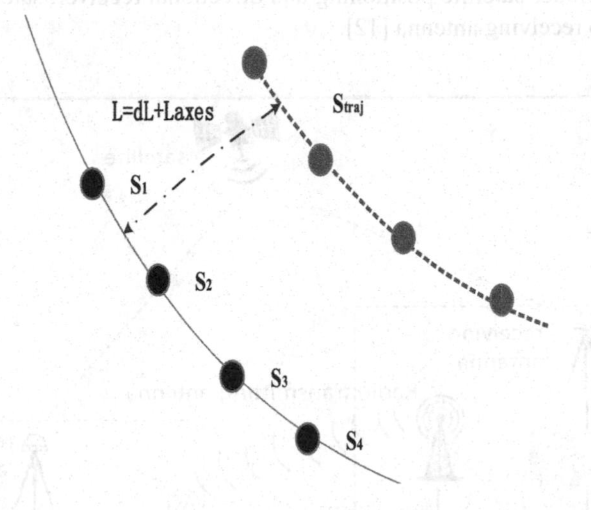

(b) Planning of road edge track on one sides

Fig. 3. Typical scene trajectory fitting diagram.

$$\begin{bmatrix} m & \sum_{i=1}^{m} x_i & \sum_{i=1}^{m} x_i^2 \\ \sum_{i=1}^{m} x_i & \sum_{i=1}^{m} x_i^2 & \sum_{i=1}^{m} x_i^3 \\ \sum_{i=1}^{m} x_i^2 & \sum_{i=1}^{m} x_i^3 & \sum_{i=1}^{m} x_i^4 \end{bmatrix} \begin{bmatrix} a_0 \\ a_1 \\ a_2 \end{bmatrix} = \begin{bmatrix} \sum_{i=1}^{m} y_i \\ \sum_{i=1}^{m} x_i y_i \\ \sum_{i=1}^{m} x_i^2 y_i \end{bmatrix} \tag{7}$$

Through formula (7), the function of trajectory curve can be obtained as the input information of lateral control to realize the navigation control of unmanned vehicle [11].

4 Engineering Realization and Verification

In order to verify the feasibility of lidar navigation technology in complex scenes, an electric unmanned patrol vehicle has been realized, which is composed of RTK Positioning System, vehicle differential positioning mobile station and upper computer.

4.1 RTK Positioning System

Real - time kinematic (RTK) technology can achieve centimeter level high-precision positioning. Its basic structure is shown in Fig. 4, which is composed of satellite, differential positioning reference station and vehicle differential positioning mobile station. Differential positioning reference station includes radio transmitting antenna, satellite receiver, radio station and satellite receiving antenna. Vehicle differential positioning mobile station includes satellite positioning and directional receiver, satellite receiving antenna and radio receiving antenna [12].

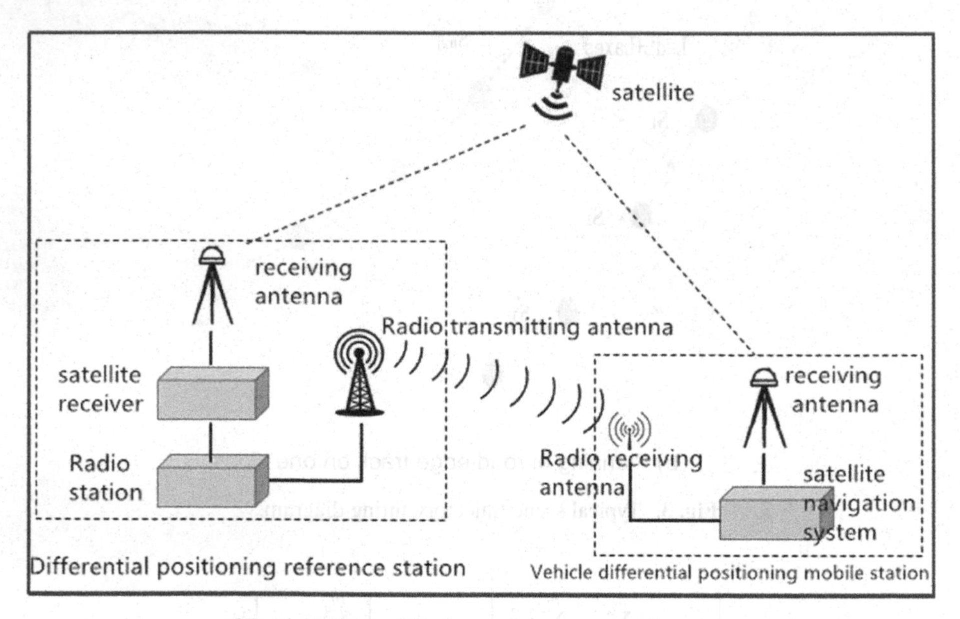

Fig. 4. RTK positioning system structure diagram.

The working principle of the positioning system is that a satellite receiver is installed on the reference station to continuously observe the satellite, and the carrier phase correction is transmitted to the vehicle differential positioning mobile station through the radio station and radio transmitting antenna in real time [13]. When the vehicle differential positioning mobile station receives the satellite signal, it receives the differential

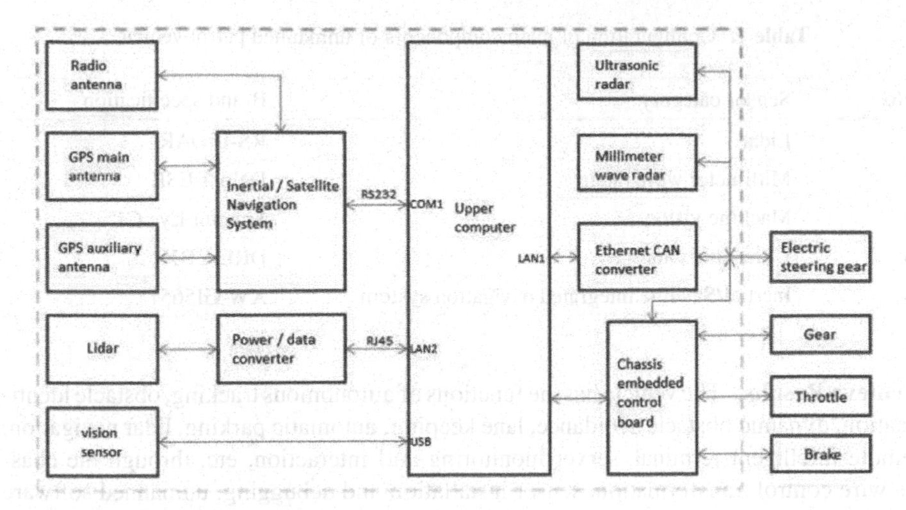

Fig. 5. Multi sensor fusion automatic vehicle driving system structure diagram.

positioning information of the reference station through the radio. According to the principle of relative positioning, it processes data in real time and gives three-dimensional coordinates of vehicle differential positioning mobile station with centimeter accuracy, and sends these data to the upper computer of vehicle automatic driving positioning and control system through serial communication interface.

4.2 Autopilot System

Unmanned Vehicle System Structure. According to the structure diagram of the driverless system in Fig. 1, the vehicle control part controls the unmanned vehicle through the chassis embedded control panel, and the system structure is shown in Fig. 5.

As shown in Fig. 5, the automatic driving system includes environment and vehicle state perception module, vehicle chassis control module, bottom data and communication support module. The environment and vehicle state perception module consists of upper computer, inertial/satellite integrated navigation system, radio antenna, GPS main antenna, vision sensor, laser radar, millimeter wave radar and ultrasonic radar. The underlying data and communication support module is realized by Ethernet CAN converter. Millimeter wave radar, ultrasonic radar, chassis embedded control board and vehicle electric steering gear communicate with upper computer through it.

The vehicle chassis control module is realized by the chassis embedded control board. The chassis embedded control board is connected with the gear, accelerator and brake system of the vehicle respectively. Under the control of the upper computer, the vehicle gear, accelerator and brake are controlled. The electric steering gear controls the steering of the vehicle under the control of the upper computer [14].

Configuration of Unmanned Patrol Vehicle. According to the system structure shown in Fig. 5, the configuration of sensors and navigation of unmanned vehicle is shown in Table 1.

Table 1. Configuration of main components of unmanned patrol vehicle.

No	Sensor category	Brand specification
1	Lidar	RS-LiDAR
2	Millimeter wave radar	Delphi-ESR
3	Machine vision	Smarter Eye C1
4	Ultrasonic radar	DRE-CBX
5	Inertial/Satellite integrated navigation system	XW-GI5651

Achieve Results. The vehicle has the functions of autonomous tracking, obstacle identification, dynamic obstacle avoidance, lane keeping, automatic parking, lidar navigation, remote intelligent terminal, server monitoring and interaction, etc. through the chassis wire control transformation, sensor installation and debugging, unmanned software development and testing, the vehicle has realized L4 level automatic driving in closed scenes.

Using this unmanned vehicle, this research has carried out the trajectory planning and control technology verification based on lidar. The test shows that when the satellite positioning and vision are invalid, relying on the lidar unmanned vehicle can achieve stable and reliable path planning and navigation control. The test vehicle, test scene and effect picture are shown in Figs. 6 and 7.

Fig. 6. Test vehicle and test site.

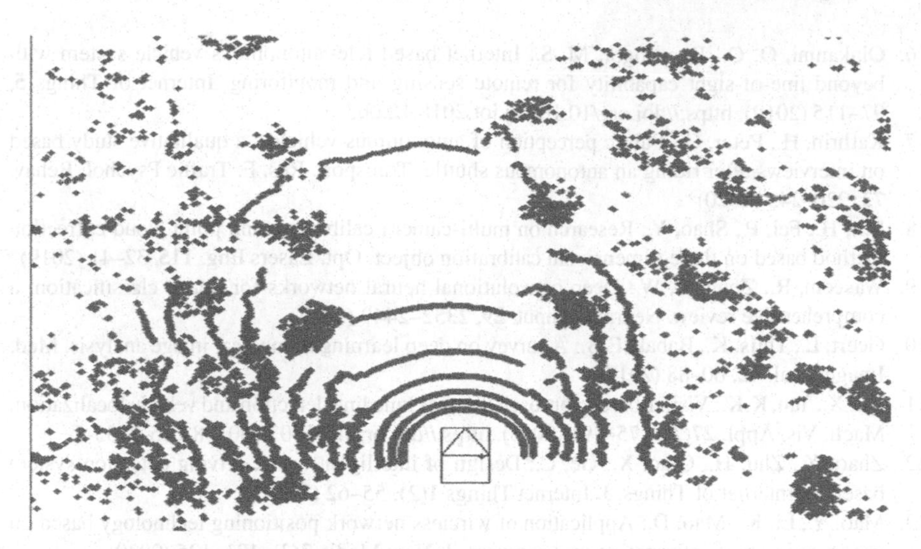

Fig. 7. Real time operation effect of test vehicle.

5 Conclusion

In this study, a laser radar aided unmanned vehicle (UAV) fusion positioning technology for complex scenes has been proposed. In the scene where satellite navigation can not locate normally, the least square algorithm is used to fit the filtered radar point cloud data to obtain the driverless vehicle road area, and then the trajectory planning is carried out to realize the motion control of the unmanned vehicle. In this study, the technology has been implemented and verified in the electric unmanned patrol vehicle, and the RTK Positioning System and automatic driving control system have been constructed, and the unmanned driving software has been developed. The test results show that when the satellite RTK Positioning is not available, the stable and reliable unmanned vehicle navigation can be achieved by relying on the lidar.

Funding Statement. This research is supported by the Natural Science Foundation of Hunan Province in 2020 (2020JJ6093).

References

1. Innovation at DARPA.Defense Advanced Research Projects Agency, http://www.darpa.mil/attachments/DARPAInnovation2016.pdf. Accessed 5 Dec 2017
2. Michael, I.: Autopilot: automatic data center management. ACM SIGOPS Oper. Syst. Rev. **41**(2), 60–67 (2007)
3. Albright, B.: Going on autopilot. Desktop Eng. **24**(9), 18–22 (2019)
4. Zhang, S.: Navigation of An Agricultural Autonomous Vehicle Based on Laser Radar. M.S. dissertation, Nanjing Agricultural University, Nanjing (2014)
5. Zheng, X., Stphane, G., Tu, X.: Obstacle avoidance model for UAVs with joint target based on multi-strategies and follow-up vector field. Procedia Comput. Sci. **170**, 257–264 (2020)

6. Olakanmi, O. O., Benyeogor, M. S.: Internet based tele-autonomous vehicle system with beyond line-of-sight capability for remote sensing and monitoring. Internet of Things **5**, 97–115 (2019). https://doi.org/10.1016/j.iot.2018.12.003
7. Kathrin, H., Peter, G.: Public perception of autonomous vehicles: a qualitative study based on interviews after riding an autonomous shuttle. Transport. Res. F: Traffic Psychol. Behav. **72**, 226–243 (2020)
8. Lin, H., Fei, P., Shao, Y.: Research on multi-camera calibration and point cloud correction method based on three-dimensional calibration object. Opt. Lasers Eng. **115**, 32–41 (2019)
9. Waseem, R., Zenghui, W.: Deep convolutional neural networks for image classification: a comprehensive review. Neural Comput. **29**, 2352–2449 (2017)
10. Geert, L., Thijs, K., Babak, E.B.: A survey on deep learning in medical image analysis. Med. Image Anal. **42**, 60–88 (2017)
11. Du, X., Tan, K.K.: Vision-based approach towards lane line detection and vehicle localization. Mach. Vis. Appl. **27**(2), 175–191 (2015). https://doi.org/10.1007/s00138-015-0735-5
12. Zhao, X., Zhu, H., Qian, X., Ge, C.: Design of intelligent drunk driving detection system based on Internet of Things. J. Internet Things **1**(2), 55–62 (2019)
13. Mao, Y., Li, K., Mao, D.: Application of wireless network positioning technology based on gps in geographic information measurement. J. New Media **2**(3), 131–135 (2020)
14. Peng, C., Yu, Z.: Modeling analysis for positioning error of mobile lidar based on multibody system kinematics. Intell. Autom. Soft Comput. **25**(4), 827–835 (2019)

A Review on Named Entity Recognition in Chinese Medical Text

Lu Zhou[1], Weiguang Qu[1,2(✉)], Tingxin Wei[2,3], Junsheng Zhou[1], Yanhui Gu[1], and Bin Li[2]

[1] School of Computer and Electronic Information School of Artificial Intelligence, Nanjing Normal University, Nanjing 210023, China
[2] School of Chinese Language and Literature, Nanjing Normal University, Nanjing 210097, China
[3] International College for Chinese Studies, Nanjing Normal University, Nanjing 210097, China

Abstract. In this paper, a survey is done to introduce the named entity recognition task in Chinese medical text and its practical significance. First, the existing datasets for the named entity recognition task of Chinese medical text are presented, then the survey is given on the algorithms for this task, mainly from the perspectives on matching and sequence labeling. Finally, the future development of named entity recognition in Chinese medical text is discussed.

Keywords: Named entity recognition · Chinese medical text

1 Introduction

As a fundamental step in Text Mining System (TMS), Named Entity Recognition (NER) plays a considerable role in text pre-processing which locates the entity spans in text and classifies them into predefined categories automatically. Identification of named entity in text is the prerequisite for many other text mining technologies, such as Information Extraction and Text Classification. Therefore NER has always been a research hotspot in the field of natural language processing and artificial intelligence.

Chinese medical NER task is to extract entity spans from medical texts and classify them into predefined categories such as diseases and drugs in Chinese. For example, in the sentence "延髓和脊髓受损，则可引起各种类型的瘫痪。" (Damage to the medulla oblongata and spinal cord can lead to various types of paralysis.), "延髓和脊髓受损" (damage to the medulla oblongata and spinal cord) and "瘫痪" (paralysis) are two symptom entities, meanwhile, "延髓和脊髓受损" (damage to the medulla oblongata and spinal cord) overlaps two body parts entities "延髓" (medulla oblongata) and "脊髓" (spinal cord). With the rapid development in medical science and increase of medical literature, a huge amount of new medical terminologies have been straight emerged. Since medical terms are usually in the form of abbreviations and have many synonyms, manual identification is time-consuming and laborious, which makes the automatic NER particularly important.

X. Sun et al. (Eds.): ICAIS 2021, CCIS 1422, pp. 39–51, 2021.
https://doi.org/10.1007/978-3-030-78615-1_4

The NER task in Chinese medical text makes it accessible to mine medical knowledge from large scale of medical texts which helps discover and prevent infectious and sudden diseases, and explore undiscovered medical knowledge for medical researchers. Besides, it also can effectively promote the standardization of entity naming and the spread of medical knowledge, provide the technical assistance for downstream tasks such as the relationship extraction and the medical knowledge graph construction, and work as a part of the clinic electronic medical record system, the clinical decision support system, and so on.

The paper is structured in four sections. Datasets supplied for NER task in Chinese medical text are presented in Sect. 2. Section 3 discusses the approaches for NER in Chinese medical text based on matching and sequence labeling. Finally, Sect. 4 summarizes the development and some future directions that may promote researches in this field.

2 Datasets

The resources for Chinese medical NER at present can be categorized as Electronic Medical Record (EMR) and Biomedical Literatures. In the medical field, there is very few fine-labeled resources available for NER task in Chinese as in English. It is difficult to carry out researches without unified named entity annotation standards and a large scale of medical datasets which usually were acquired from medical institutes like hospitals and authoritative networks and annotated by experts. Fortunately, China Conference on Knowledge Graph and Semantic Computing (CCKS) has released the annotated Chinese electronic medical record corpus since 2017, and China Health Information Processing Conference (CHIP) released the task and resource for NER in Chinese medical text which provided annotated Chinese medical texts in 2020.

2.1 CCKS Dataset

CCKS released the task of medical entities recognition and attribute extraction in Chinese EMR annually from 2017 to 2020. Meanwhile an annotated Chinese EMR corpus were released and kept being revised for 4 years which had 6292 entities of 6 categories in its vocabulary. In this corpus, a document is taken as a unit. Table 1 shows the number of different categories of entities in the datasets released annually from 2017 to 2019. For brevity, INE indicates "Inspection and examination", SNS to "Symptoms and signs", DND to "Diseases and diagnosis", TRE to "Treatment", EXA to "Examination", INS to "Inspection", OPE to "Operation" and ANA to "Anatomical regions".

2.2 CHIP Dataset

CHIP is an annual seminar that focuses on medical treatment, health, and biological information processing. The clinical terminology standardization task released by CHIP-2019 aims to deal with issues of terminology standardization, which is to carry out semantic normalization for the surgical entities extracted from Chinese EMR, in other words, offering the normalized term corresponding to the given original operative word.

Table 1. Categories of entities in datasets released by CCKS for the task of NER in Chinese EMR from 2017 to 2019

2017	Text	INE	SNS	DND	TRE	Body	—	Total
	400	12689	10142	1255	1513	13740	—	39359
2018	Text	INE	SNS	DND	TRE	Body	—	Total
	600	844	807	315	464	1081	—	3511
2019	Text	EXA	INS	DND	OPE	Drug	ANA	Total
	1379	313	511	2798	905	719	1933	7179

CHIP-2019 released the Yidu-N4K dataset, containing original words and corresponding normalized words. For example, all the initial words can be normalized to the same word, as shown in Table 2.

Table 2. Examples of standardization in the Yidu-N4K dataset

Original words	Standard words
肝血管瘤切除术 (Resection for liver hemangioma)	肝病损切除术 (Hepatectomy)
肝右叶占位切除术 (Resection for focal right hepatic lobe lesions)	
肝癌切除术 (Hepatic carcinectomy)	
肝肿瘤切除术 (Hepatoma resection)	
右肝癌切除术 (Right hepatic carcinectomy)	
肝病损切除术 (Hepatectomy)	

In 2020, CHIP released the NER task in Chinese medical text with a Chinese medical text dataset of 1.64 million characters and 26,903 sentences. The corpus is from clinical pediatrics. There are 15,000 samples in the training set and 5,000 in the validation set. Each sample records the medical text and the positions, including the start and the end, and categories of the entities in the text. Specifically, the information of each entity in a sample is separated from another one by "|||", and further, the positions and the category are 4-space apart. As the example, "脑水肿时给脱水剂。" (When brain edema, give dehydrant.) shown in Fig. 1 below, the entity "脑" (brain) of the body category denoted by "0 0 bod", which represents the position index from 0 to 0 in this sentence is a "bod" entity, is nested within the symptom entity "脑水肿" (brain edema) tagged as "sym", and the entity "脱水剂" (dehydrant) is a drug entity tagged as "dru". Table 3 shows the number of entities of nine categories in the CHIP-2020 dataset.

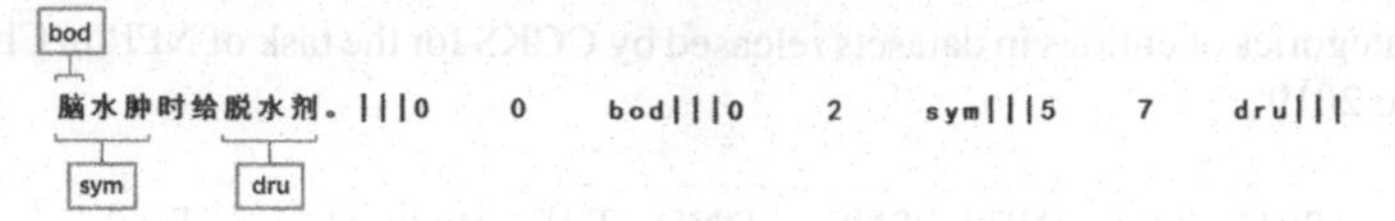

Fig. 1. An example of the CHIP-2020 dataset

Table 3. Statistics of entities in the CHIP-2020 dataset

Category	Tag	Number	Proportion
Disease	Dis	8018	23.75%
Symptom	Sym	10512	31.13%
Medical procedural	Pro	3911	11.58%
Medical equipment	Equ	560	1.66%
Drug	Dru	2046	6.06%
Medical examination item	Ite	1488	4.41%
Body part	Bod	6356	18.82%
Department	Dep	103	0.31%
Microbe	Mic	771	2.28%
Total		33765	100%

2.3 Summary

With the development of medical information technologies and conferences like CCKS and CHIP, there has boomed many standardized Chinese medical text datasets. However, there are still many challenges in this field.

Firstly, there is the lack of unified standard annotation and benchmark datasets. Although CCKS and CHIP have released annotated datasets, the set of tags and annotation format and rules are different and changed every year. Besides these datasets still have some defects that limit the scale of the corpus for Chinese medical text NER task. For example, the corpus released by CCKS-2019 did not take nested entities and modifiers into consideration, like ignoring the word "无" (no) in the beginning of the span "无发热" (no fever) to modify "发热" (fever).

Secondly, Traditional Chinese Medicine (TCM) which is a special kind of Chinese medical text represents the experience and wisdom of doctors in ancient times. However, the long history can be a two-edged sword, as the difference in language characteristics between the past and present, the lack of unified writing standardization make it harder to do NER task in TCM text.

Last but not the least, more knowledge resources in the medical field need to be excavated urgently, and more kinds of Chinese medical text like medical papers are needed to be annotated since the available datasets for this task are only from clinic medical texts and electronic medical records.

3 Research Approaches

There are two main research approaches for the NER task in Chinese medical text. One is matching-based and the other is sequence-labeling-based.

3.1 Approaches Based on Matching

Considering that medical texts are highly professional with medical terminology and sentence structures, there are two main approaches based on matching: dictionary-based approaches and rule-based approaches.

The dictionary-based approaches do string matching work referring to the external medical named entity dictionary. However, it cannot deal with the medical entities outside the dictionaries which are springing up with the rapid development of the medical science. It would take many experts much time and energy to update the external authoritative medical entity dictionary, let alone it is almost impossible to cover all the medical entities. The quality of dictionaries is difficult to guarantee, which restricts the effect of recognition. Therefore, the dictionary-based approach is usually combined with other approaches.

The rule-based approaches induce the sentence structure and entity presentations in medical texts to design rules for extracting and classifying entities. Obviously, manually designed rules cannot include all the linguistic patterns, so that it relies on continuously adding rules by experts. What's worse, rules sometimes don't work that well which is led by the difficulty of drafting a unified set of rules. Zhao et al. [1] used jieba, the word segmentation tool for initial segmentation, and then recognized entities by the rule description language (RDL) and entropy expansion to overcome the defect that jieba is not targeted to the medical field. It was helpful to identify new words, reduce irrelevant word appearances, and extract and describe the labels.

3.2 Approaches Based on Sequence Labeling

Nowadays, most works regarding NER in Chinese medical text considers it as a sequence labeling issue, while the pure dictionary-based and rule-based approaches are integrated in the post-processing. Take the "BMESO" (Begin, Middle, End, Single, Other) label set as an example, Table 4 shows a character-level labeled sample "脑水肿时给脱水剂。" (When brain edema, give dehydrant.), which is the same as that in Fig. 1.

Table 4. An example of labeling as "BMESO" in character-level

Character	脑	水	肿	时	给	脱	水	剂	。
Label	S-bod B-sym	M-sym	E-sym	O	O	B-dru	M-dru	E-dru	O

As there are no apparent boundaries of Chinese words as that in English, following are three frequently used levels of granularity for sequence labeling in the process of obtaining sentence representation vectors.

The Word-Level Labeling Approach. Mostly entities in Chinese medical text are words in linguistics, so the first reaction is to get word expressions and then labels with Chinese word segmentation toolkits. Up to now, PKUSEG (Luo et al. [2]) is the only segmentation toolkit targeting at multi-domain, including medicine.

The Character-Level Labeling Approach. Due to the mistakes of word segmentation, error propagation in detecting entity boundaries and predicting categories are inevitable. Therefore many approaches carry out Chinese Medical NER task at the character level, which is also accompanied by the defect of ignoring the information of word boundaries. Sun et al. [2] designed a feature extraction approach where the segment label of a character was combined with the word feature it located to get the segmentation information of both the text and manually labeling. For example, the noun "脱水剂" (Dehydrant) is Tagged as "n" for part-of-speech (POS) and its character POS sequence was "n, n, n". To maintain the segmentation, the sequence was replaced by "B-n, M-n, E-n" with the "BMESO" Label Set.

The Dictionary-Combined Character-Level Labeling Approach. Due to the issues of exclusive word-level or character-level labeling approaches, many researchers explore in combining the word vector referring to incorporated dictionaries and the character-level input sequence vector. This approach avoids word segmentation error as well as makes explicit use of information out of words and their order. Wan et al. [3] took the word-segmented sentence as the original input and then constructed the joint vector of word and character. While according to each Chinese character, Wang et al. [4] set 5 schemes based on dictionaries and the content to create the feature vector. The input in Li et al. [5] was the concentration of trained character-level vectors of words with formal characteristics and word-level vectors with semantic features obtained from a massive background corpus.

To be specific, the sequence labeling is to pre-process an original text sentence as a sequence and give each element in it a predefined label. The model input is the sequence and the output is the prediction of entities and the category of each. The formalization of the NER task can be defined as follows. Let S be a text sentence.

$$S = [c_1, c_2, \ldots, c_n] \tag{1}$$

Let c_i be a character, i = 1, 2, $\cdots$, n and let n be the length of S.

$$X = [x_1, x_2, \cdots, x_m] \tag{2}$$

Equation (2) is the pre-processed sequence of S, where x_i is a token in sequence X, $i = 1, 2, \cdots, m$, m is the length of X. Note that m is not always equal to n due to the granularity and m = n when character-level labeling approach is applied. The approach based on sequence labeling is to train a model that can give each token in the sequence X the most appropriate label,y_i, within Y, the defined set of labels, such as "B-bod" in the character-level labeling way, which denotes the beginning token of a "body parts" entity. The prediction is calculated by Eq. (3).

$$\hat{y} = \operatorname{argmax}(p(y|x)) \tag{3}$$

Machine Learning-Based Approaches. Machine learning approaches are mainly used in determining the entity boundary and predicting the category of the entity, which usually requests labeled training data. The effect of machine learning models, to a large extent, depends on the quality of manual feature extraction schemes. When it comes to the NER task in Chinese medical text, Hidden Markov Model (HMM), Conditional Random Field Model (CRF), Maximum Entropy (ME) and mixed models are the common approaches. Figure 2 shows the general process of recognition.

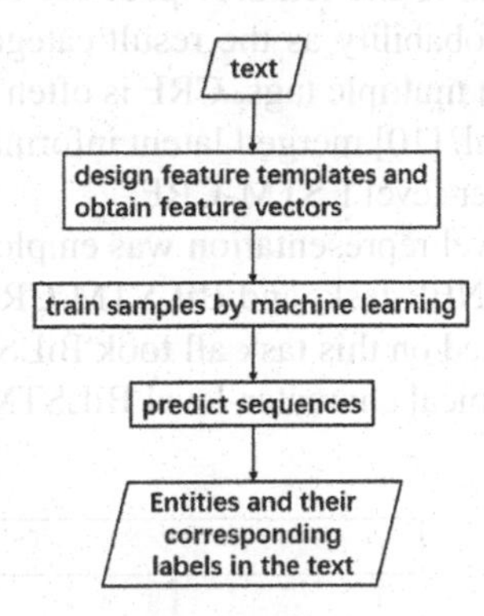

Fig. 2. The general process of recognition in machine learning approaches

HMM. This model utilizes the content information and the Viterbi algorithm to generate the most feasible labeled sequence. It is confirmed that HMM is rapid to train and recognize but hard to reach the presupposition that variables are mutually independent. According to the characteristics of TCM clinic records, Wei et al. [6] improved the sequence labeling to make HMM suitable for classifying serialized data and identifying symptom entities in TCM.

ME. ME picks up the model with maximum information entropy from the ones that meet constraints. This kind of model settles the parameter smoothing problem, while its training process costs much with slow convergence rate and label-bias problem.

CRF. If the observation sequence and marker sequence are given, all probabilities can be calculated by scanning forward once and backward once to obtain the forward vector and backward vector respectively. It can be combined with manually designed features that make full use of the content and solve the unrealized independent assumption in HMM and the label-bias problem in ME to further improve the performance. Yang. [7] Firstly built a lexicon of disease names from PharmGKB before recognizing disease entities by combining the dictionary-based approach and crf, and then adjusted the result with full and abbreviated name pairs in the content. Zhai et al. [8] extracted features, such as keywords and lexicons, to do recognition by CRF, and then optimized outputs with rules to improve the performance of mere CRF.

Mixed Model. It allows existing models complementary to each other. Qu. [9] formulated standards to build an annotated corpus. In it, there were multiple ensemble classifiers based on ME, CRF, SVM and fusion classification algorithms.

Neural Network-Based Approaches. To overcome the defects of machine learning and feature selection and reduce the cost both of the labor and training, researchers exploit neural networks to push NER task in Chinese medical text forward. neural network models employ deep network structures to obtain high dimensional feature representation vectors automatically. Long-Short Term Memory (LSTM) is of common use in this field, especially its two important variants: Bi-directional LSTM (BiLSTM) and Gated Recurrent Unit (GRU). LSTM can remove or add information to the cell state in long distance and map hidden states to the feature space for classification, then select the dimension with the highest probability as the result category. For LSTM is unable to learn the relationships between multiple tags, CRF is often combined. To make full use of word information, Wang et al. [10] merged latent information of words, the output of lattice-LSTM, into the character-level LSTM-CRF.

BiLSTM with character-level representation was employed in all 7 papers included in CCKS-2017 Chinese EMR NER task, and BiLSTM-CRF in 6 of them. Moreover, 3 papers that CCKS-2019 included on this task all took BiLSTM-CRF as the recognition module. Figure 3 shows the typical character-level BiLSTM-CRF model structure.

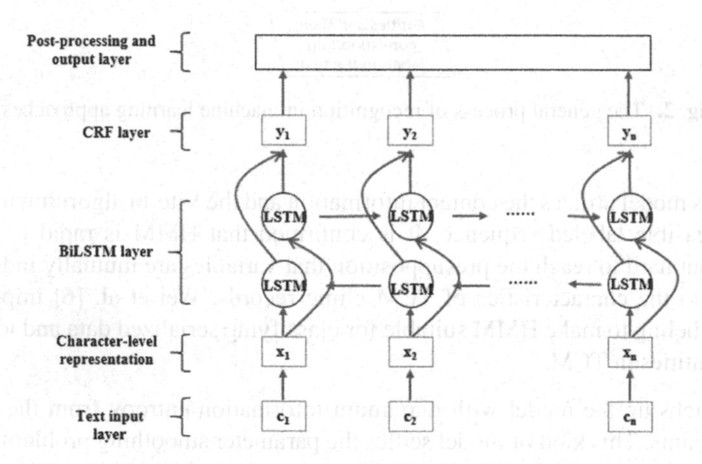

Fig. 3. The typical character-level BiLSTM-CRF model structure

On the dataset released by CCKS-2017, Qiu et al. [11] proposed a residual convolutional neural network-CRF (RN-CNN-CRF) model which had been proved to have higher training speed and the F1 value. Wu et al. [12] intentionally not split the entity span labeled in the training data in the process of word segmentation, to some extent, avoiding the segmentation error, and got the best performance. Ouyang et al. [13] designed many features, such as segment, POS and medical terms. Xia and Wang [14] used Active Learning to get improved by manually labeling. To utilize unlabeled data and catch deep text features, Tang et al. [15] proposed a framework absorbed the language model and attention, which input sentence vectors into Bi-directional GRU (BiGRU) and pre-trained language model separately, sent the two outputs to another BiGRU and multi-attention module, and finally predicted the label sequence with CRF. Zhang et al. [16] designed four features for CRF: BOC, POS, CT (the character type) and POICS (relative position of the character), and made the comparison of the performance between CRF and BiLSTM-CRF, which proved improvement in BiLSTM and economy of the manually feature selection. Compare with Geng. [17], who compared CRF with BiLSTM too, the F1 score decreased by 2.3%. Considering the limited dataset and specific Chinese medical filed, the more and the better rules are made, the better performance the model will get. However, in the long run, approaches based on deep learning neural networks can outbreak these limitations and adapt to the continuously development of the medical science. Table 5 shows the evaluation results with various approaches on the CCKS-2017 dataset.

Table 5. Evaluation results with various approaches on the CCKS-2017 dataset

	Models	F1 (%)
Hu et al. [18]	Vote by recognitions based on rules, CRF and BiLSTM separately	91.08
Wu et al. [12]	BiLSTM-CRF	93.41
Ouyang et al. [13]	BiLSTM-CRF	88.85
Xia et al. [14]	BiLSTM-CRF + Active learning	89.88
Li et al. [19]	BiLSTM-CRF	87.95
Geng. [17]	CRF	92.73
Zhang et al. [16]	CRF + 4 features	89.49
	BiLSTM-CRF	90.43
Tang et al. [15]	Language model + Attention	91.34

BiLSTM and CRF models were still the hottest in CCKS-2018 to settle the sequence labeling issue. External dictionaries were helpful to build the lexicon due to the entity naming format. The popular approach was still integrating multiple models and constructing rules to enrich representations. CRF, BiLSTM-CRF, BiLSTM-CNN-CRF were the main models. Based on BiLSTM-CRF, Sheng et al. [20] introduced CNN to further extract semantic features from the given text to border entities. It first determined the domain of the entity and then built the model on the corresponding corpus to get the fine-grained category. Specifically, it defined 5 categories of entities in the medical field, including hospitals, diseases, departments, drugs and symptoms. Experiments showed that application of CNN can extract useful information for NER.

Recently, BERT (Devlin et al. [21]) has attracted the attention of researchers. It can obtain more and deeper information than traditional embedding generation in a direct way and has been widely used in CCKS-2019 NER task in Chinese EMR. Qiao et al. [22] used "BMESO", the character-level labeling set, designed and fused three models: BERT + TimeDistributed_Dense (BT/FBT), BERT + BiLSTM + TimeDistributed_Dense (BBT/ FBB-T) and BERT + BiLSTM + CRF (BBC/ FBBC), and built constraint rules through algorithms like Apriori. Zhao et al. [23] took the voting result by three models as the prediction and post-process it according to rules. Models can be divided as follows:

(1) Ones introduced more fine-grained labels from multiple aspects and dimensions, output NER results from two layers which were both BiLSTM-CRF but with different parameters and then obtained labels by filtering aliasing conditions; (2) Ones employed Lattice LSTM (Zhang et al. [24]) and improved CharCNN + BiLSTM + CRF (Ma et al. [25]); (3) Ones introduced BERT to enhance semantic character representations and then extracted labels by BiLSTM-CRF. In the post-processing, they built a dictionary consisting of 30,000 medical terms and manual rules for contents of entities. Compare with base BiLSTM-CRF, Li et al. [26] found that BERT did not make better performance because it was pre-trained on the corpus in the general domain. So they trained a stroke ELMo (Matthew et al. [27]) model to get character representations and BiLSTM-CRF as upper layers. Besides, they applied Transfer Learning that treated CCKS-2018 dataset as the source domain and CCKS-2019 dataset as the target domain of three structures, and finally voted on the prediction. Chen et al. [28] proposed a KNN-BERT-BiLSTM-CRF model based on Transfer Learning on NER task in liver cancer QA text to deal with the lack of the corpus.

All the three papers in CCKS-2019 Task 1 took character-level labeling, BERT or ELMo for deeper knowledge instead of word2vec, and BiLSTM-CRF. Table 6 shows evaluation results with various approaches on the CCKS-2019 dataset.

Table 6. Evaluation results with various approaches on the CCKS-2019 dataset

	Models	F1(%)
Qiao et al. [22]	BBC + BBT + FBBC + rules	85.62
Zhao et al. [23]	Lattice	92.6
	BERT base	92.9
	Mixed model + rules	93.7
Li et al. [26]	BERT	81.59
	Lattice LSTM	83.38
	BiLSTM-CRF	82.81
	BiLSTM-CRF + ELMo	84.55
	FS-TL(Ensemble)	86.03

3.3 Summary

This section discusses the research approaches for the NER task in Chinese medical text mainly from the perspectives of approaches based on matching and on sequence labeling. Most of research approaches do not handle nested entities because available datasets are not annotated that way.

4 Conclusion

The evaluation conferences and datasets mentioned above have strongly promoted the development of NER in Chinese medical text, while the recognition accuracy and application efficiency are urged to be increased. The present developmental tendency is the multi-feature extraction and multi-model integration. To be specific, for text representation, deep learning neural networks are used such as CNN, LSTM and attention, to get deeper features, and add rules and manual feature templates according to the certain field at the same time; for label sequence prediction, CRF is the major approach for multi-label tasks.

Here are some prospects we expect. Firstly, establish a large scale of available unified annotated Chinese medical text. It will act a fundamental role to help further study and intuitive comparison among approaches. Secondly, build a standard Chinese medical dictionary. It will be crucial for approaches that integrated lexicons. Thirdly, take use of semi-supervised learning, unsupervised learning and deep learning on the NER task with less or even no annotated corpus to address the issue of the lack of corpus. For example, Transfer Learning improves the performance by prior knowledge of pre-trained models and external annotated corpus. Li et al. [29] applied Machine Reading Comprehension (MRC) framework for NER and designed an extra loss function, span_loss, to address the issue of nested entities. The core idea was to treat the NER task as a start-end positon pair extraction problem. BERT + MRC has been applied on experimented on the general domain for location, person names and organizations. It has not been proved whether new approaches can achieve good performance on NER in Chinese medical text. Last but not the least, apply mixing multi-features and models in Chinese NER, as multi-feature extraction and multi-model integration are the developmental tendency and proved to improve the performance.

Acknowledgements. The authors would like to thank all anonymous reviewers for their constructive suggestions which have resulted in improvement on the presentations. This research is supported by the National Science Foundation of China (grant 61772278, author: Qu, W.; grant number: 61472191, author: Zhou, J. http://www.nsfc.gov.cn/), the National Social Science Foundation of China (grant number:18BYY127, author: Li B. http://www.cssn.cn) and Jiangsu Higher Institutions' Excellent Innovative Team for Philosophy and Social Science (grant number: 2017STD006, author: Qu, W. http://jyt.jiangsu.gov.cn).

Conflicts of Interest:. The authors declare that they have no conflicts of interest to report regarding the present study.

References

1. Zhao, J., Zhang, Y., Cai, K.: Medical entity recognition and label extraction based on natural language processing. Comput. Technol. Develop. **29**(9), 18–23 (2019). (in Chinese)
2. Luo, R.X., Xu, J.J., Zhang, Y., Ren, X.C., Sun, X.: PKUSEG: a toolkit for multi-domain Chinese word segmentation. arXiv preprint: arXiv:1906.11455 (2019)

3. Sun, A., Yu, Y.X., Luo, Y.G., Wang, Q.: Research on feature extraction scheme of Chinese-character granularity in sequence labeling model: a case study about clinical named entity recognition of CCKS2017: Task2. Libr. Inf. Serv. **62**(11), 103–111 (2018). (in Chinese)
4. Wan, L., Luo, Y.R., Li, Z., Qi, X.R.: The recognition of naming entity of Bi-LSTM Chinese electronic medical records based on the joint training of Chinese characters and words. China Digital Med. **14**(2), 54–56 (2019). (in Chinese)
5. Wang, Q., Xia, Y.H., Zhou, Y.M., Ruan, T., Gao, D.Q., He, P.: Incorporating dictionaries into deep neural networks for the Chinese clinical named entity recognition. J. Biomed. Inf. **92**, 103–133 (2019)
6. Li, L.S., Guo, K.Y.: Biomedical named entity recognition with CNN-BLSTM-CRF. J. Chin. Inf. Process. **32**(1), 116–122 (2018). (in Chinese)
7. Wei, Z.Q., Shu, H.P., Wang, Y.Q.: A study on the named entity recognition method based on sequence labeling on symptom names in traditional Chinese medical. Shandong Industrial Technology **8**, 237–238 (2015). (in Chinese)
8. Yang, Y.: Research on disease name recognition and disease normalization in biomedical literature. M.S. thesis, Dalian University of Technology, China (2015). (in Chinese)
9. Zhai, J.Y., Chen, C.Y., Zhang, Y., Chen, Y., Liu, Y.W.: A study on the named entity recognition of Chinese electronic medical record based on combination of CRF and rules. Journal of Baotou Medical College **33**(11), 124–125, 130 (2017). (in Chinese)
10. Qv, C.Y.: Research on named entity recognition for Chinese electronic medical records. M.S. thesis, Harbin Institute of Technology, China (2015). (in Chinese)
11. Wang, B.R., Lin, X., Lin, X.D., Zhu, W.L., Ma, X.H.: Chinese name language entity recognition (NER) using Lattice LSTM in medical language. Chin. J. Heal. Inf. Manage. **16**(1), 84–88 (2019). (in Chinese)
12. Qiu, J.H., Wang, Q., Zhou, Y.M., Ruan, T., Gao, J.: Fast and accurate recognition of Chinese clinical named entities with residual dilated convolutions. In: 2018 IEEE International Conference on Bioinformatics and Biomedicine, pp. 935–942. IEEE, Madrid (2018)
13. Wu, J.H., Hu, X., Zhao, R.S., Ren, F.L., Hu, M.H.: Clinical named entity recognition via bi-directional LSTM-CRF model. In: Proceedings of the Evaluation Task at the China Conference on Knowledge Graph and Semantic Computing (CCKS-Tasks 2017), pp. 37–42. CEUR Workshop Proceedings, Chengdu (2017)
14. Ouyang, E., Li, Y.X., Jin, L., Li, Z.F., Zhang, X.Y.: Exploring n-gram character presentation in bidirectional RNN-CRF for Chinese clinical named entity recognition. In: Proceedings of the Evaluation Task at the China Conference on Knowledge Graph and Semantic Computing (CCKS-Tasks 2017), pp. 37–42. CEUR Workshop Proceedings, Chengdu (2017)
15. Xia, Y.H., Wang, Q.: Clinical named entity recognition - ECUST in the CCKS-2017 shared task 2. In: Proceedings of the Evaluation Task at the China Conference on Knowledge Graph and Semantic Computing (CCKS-Tasks 2017), pp. 43–48. CEUR Workshop Proceedings, Chengdu (2017)
16. Tang, G.Q., Gao, D.Q., Ruan, T., Ye, Q., Wang, Q.: Clinical electronic medical record named entity recognition incorporating language model and attention mechanism. Comput. Sci. **47**(3), 211–216 (2020). (in Chinese)
17. Zhang, Y., Wang, X.W., Hou, Z., Li, J.: Clinical named entity recognition from Chinese electronic health records via machine learning methods. JMIR Med. Inf. **6**(4), e50 (2018)
18. Geng, D.W.: Clinical name entity recognition using conditional random field with augmented features. In: Proceedings of the Evaluation Task at the China Conference on Knowledge Graph and Semantic Computing (CCKS-Tasks 2017), pp. 61–68. CEUR Workshop Proceedings, Chengdu (2017)

19. Hu, J.L., Shi, X., Liu, Z.J., Wang, X.L., Chen, Q.C., Tang, B.Z.: HITSZ_CNER: a hybrid system for entity recognition from Chinese clinical text. In: Proceedings of the Evaluation Task at the China Conference on Knowledge Graph and Semantic Computing (CCKS-Tasks 2017), pp. 25–30. CEUR Workshop Proceedings, Chengdu (2017)
20. Li, Z.Z., Zhang, Q., Liu, Y., Feng, D.W., Huang, Z.: Recurrent neural networks with specialized word embedding for Chinese clinical named entity recognition. In: Proceedings of the Evaluation Task at the China Conference on Knowledge Graph and Semantic Computing (CCKS-Tasks 2017), pp. 55–60. CEUR Workshop Proceedings, Chengdu (2017)
21. Sheng, J., Xiang, Z.P., Qin, B., Liu, M., Wang, L.F.: Fine-grained named entity recognition for multi-scenario. J. Chin. Inf. Process. **33**(6), 80–87 (2019). (in Chinese)
22. Devlin, J., Chang, M.W., Lee, K., Toutanova, K.: BERT: pre-training of deep bidirectional transformers for language understanding. In: Proceedings of the 2019 Conference of the North American Chapter of the Association for Computational Linguistics: Human Language Technologies, vol. 1, pp. 4171–4186. Association for Computational Linguistics, Minneapolis (2019)
23. Qiao, R., Yang, X.R., Huang, W.K.: Medical named entity recognition based on BERT and model incorporation. In: Proceedings of the Evaluation Tasks at the China Conference on Knowledge Graph and Semantic Computing (CCKS 2019). Chinese Information Processing Society of China, Hangzhou (2019). (in Chinese)
24. Liu, M., Zhou, X., Cao, Z., Wu, J.: Team MSIIP at CCKS 2019 Task 1. In: Proceedings of the Evaluation Tasks at the China Conference on Knowledge Graph and Semantic Computing (CCKS 2019). Chinese Information Processing Society of China, Hangzhou (2019)
25. Zhang, Y., Yang, J.: Chinese NER using Lattice LSTM. In: Proceedings of the 56th Annual Meeting of the Association for Computational Linguistics, vol. 1, pp. 1554–1564. Association for Computational Linguistics, Melbourne (2018)
26. Ma, X.Z., Hovy, E.H.: End-to-end sequence labeling via Bi-directional LSTM-CNNs-CRF. In: Proceedings of the 54th Annual Meeting of the Association for Computational Linguistics, vol. 1, pp. 1064–1074. Association for Computational Linguistics, Berlin (2016)
27. Li, N., Luo, L., Ding, Z.Y., Song, Y.W., Yang, Z.H., Lin, H.F.: DUTIR at the CCKS-2019 Task1: improving Chinese clinical named entity recognition using stroke ELMo and transfer learning. In: Proceedings of the Evaluation Tasks at the China Conference on Knowledge Graph and Semantic Computing (CCKS 2019). Chinese Information Processing Society of China, Hangzhou (2019)
28. Peters, M., et al.: Deep contextualized word representations. In: Proceedings of the 2018 Conference of the North American Chapter of the Association for Computational Linguistics: Human Language Technologies, vol. 1, pp. 2227–2237. Association for Computational Linguistics, New Orleans (2018)
29. Chen, M.S., Xia, C.X.: Identifying entities of online questions from cancer patients based on transfer learning. Data Anal. Knowl. Disc. **3**(12), 61–69 (2019)
30. Li, X.Y., Feng, J.R., Meng, Y.X., Han, Q.H., Wu, F., Li, J.W.: A unified MRC framework for named entity recognition. In: Proceedings of the 58th Annual Meeting of the Association for Computational Linguistics, pp. 5849–5859. Association for Computational Linguistics, Florence (2020)

Robust Liver Vessel Extraction Using DV-Net with D-BCE Loss Function

Jun Su[1], Zhe Liu[1(✉)], Yuqing Song[1], Wenqiang Wang[1], Kai Han[1], Yangyang Tang[1], and Xuesheng Liu[2]

[1] School of Computer Science and Communications Engineering, Jiangsu University, Zhenjiang 212013, China
1000004088@ujs.edu.cn

[2] Department of Anesthesiology, the First Affiliated Hospital of Anhui Medical University, Key Laboratory of Anesthesiology and Perioperative Medicine of Anhui Higher Education Institutes, Anhui Medical University, Hefei, China

Abstract. Recently, liver vessel segmentation has aroused widespread interests in medical image analysis. Accurately extracting blood vessels from the liver is a difficult task due to the complex vessel structures and image noise. To make the network better adapt to this complex feature, a deeper network is needed to fit this nonlinear transformation. In this work, we introduce the dense block structure into the V-net to construct a new Dense V-Net (DV-Net) and use data augmentation to segment the liver vessels from abdominal CT volumes with few training samples. Besides, we propose the D-BCE loss function to cope with the problem of dynamic changes in pixel ratio caused by 3D segmentation random patches, which can also control the trade-off between false positives and negatives. In this way, the proposed DV-Net structure can acquire a more powerful discrimination capability between vessel areas and non-vessel areas. Our method is tested on the public datasets from 3Dircadb. The average dice and sensitivity on the 3Dircadb dataset were 74.76% and 75.27% respectively. Experiments are also conducted in another public dataset from Medical Segmentation Decalthon and also obtain much higher accuracies. Moreover, our approach is automatic, accurate and robust for liver vessel extraction and can enjoy better convergence properties, making it more efficient and reliable in practice.

Keywords: Liver vessel segmentation · Deep learning · D-BCE loss · DV-Net

1 Introduction

The segmentation of liver vessels is of great significance to liver surgery planning. Accurate liver vessel images can help doctors diagnose and treat liver diseases [1]. However, due to the low contrast, high noise and complex structure of liver vessels, automatic liver vessel extraction is still a challenge. Nowadays, accurate liver vessel extraction mainly relies on manual extraction by experts, which is time-consuming and also a tedious and repeated task for experts. Therefore, it is great necessary and urgent to extract hepatic vessels accurately and automatically from the liver in clinical setting.

X. Sun et al. (Eds.): ICAIS 2021, CCIS 1422, pp. 52–61, 2021.
https://doi.org/10.1007/978-3-030-78615-1_5

Recently, lots of studies on liver vessel segmentation have been proposed. For example, Zeng et al. [2] utilized the 3D region growing and hybrid active contour model to automatically segment liver vessel. Zeng et al. [3] applied the extreme learning machine to segment liver vessels. But these methods need to balance different parameters and carefully design the vessel features. Meanwhile, with the significant improvement of computer hard ware, deep neural networks have been widely used in medical image segmentation [4]. Many scholars have tried to leverage 3D convolution for liver vessel segmentation. Hang et al. [5] proposed a 3D U-net solution with improved loss function to extract liver vessels with few training samples. But this method is based on a complex loss function. Kitrungrotsakul et al. [6] uses multi deep convolution network to segment the hepatic vessels from computed tomography (CT) image. Yu et al. [7] proposes 3D Residual U-Net for liver vessel segmentation, the method first reduces the original image and then performs training, which will largely destroy the original information. Obviously, in order to train without losing information in the limitations of GPU memory, 3D segmentation can only apply random patch training (a random area is selected in the original image for each iteration). Since the selected area each time is random, we cannot guarantee that the foreground pixels of each selected area are much smaller than the background pixels, and there may be other situations (e.g., the background and foreground pixels are similar in proportion). This is the problem that people ignore in 3D segmentation at this stage. Therefore, the loss function (e.g., Dice) that people usually utilize for category imbalance may not be a fantastic strategy. We have to consider the problem of class imbalance and also consider curve smoothing.

In this paper, we make the following two contributions:

- We proposed a 3D Dense architecture based on V-Net to improve the accuracy of liver vessels segmentation. Besides, in order to get further performance improvement and improve the speed of the convergence process, we integrate the multi-resolution feature maps obtained from the segmentation layer to generate the final output.
- In view of the dynamic change of foreground and background pixel ratio caused by random patch training, we added binary cross entropy loss as a regular term on the basis of Dice loss, which can leverage the flexibility of dice loss of class imbalance and at same time use cross-entropy for curve smoothing. We compared with some classic networks (e.g., 3D U-Net [8], V-Net [9], 3D DenseNet [10]) and different loss functions, our method shows better results.

2 Methods

The method we proposed is mainly divided into three steps. a) Considering the limitation of GPU memory and avoiding the interference of other abdominal tissues, the area is first reduced according to the liver mask, and then the data is standardized and normalized, which is the preprocessing part of the data. b) The original 3D data is divided into a series of $96 \times 96 \times 96$ area blocks, and the segmentation result of each patch is obtained through the designed DV-Net model. c) Combine the segmentation results of each patch into a complete liver area, and then remove the noise through connected domain detection to obtain the final segmentation result.

2.1 Datasets and Preprocessing

In the experiments, the datasets we used are the public datasets 3Dircadb (https://www.ircad.fr/research/3d-ircadb01/). The datasets contain 20 abdominal CT images with different resolutions, intensity distributions, and vascular structures. The pixel spacing varied from 0.56 to 0.87 mm, slice thickness varied from 1 to 4 mm and slice number varied from 74 to 260. We selected 14 cases from 3Dircadb datasets with annotated data as the training data, and another 6 cases as the testing data. To further verify the robustness of our model, we also tested our algorithm on a part of another public dataset from Medical Segmentation Decathlon [17]. Rotation and mirror operations are used in the training phase for data augmentation to make more effective use of data and prevent overfitting.

Preprocessing is to make it easier for neural networks to learn effective features and ensure the effectiveness of training. It mainly including three steps. (1) In order to avoid the influence of the tissues around the liver and reduce the amount of calculation, CT images and annotated images are cropped to the liver area based on the liver mask which is obtained by liver segmentation method [11], and adjusted to the size of 288 × 288 × 96. (2) CT value is limited to [0, 400] to better focus on liver area. (3) All images are standardized by removing the mean value and dividing the standard deviation, and finally normalized to make all the data map to the range of [0, 1]. It can be denoted as follows:

$$x_i = \frac{x_i - \frac{1}{n}\sum_{i=1}^{n} x_i}{\sqrt{\frac{1}{n}\sum_{i=1}^{n}\left(x_i - \frac{1}{n}\sum_{i=1}^{n} x_i\right)}} \quad (1)$$

$$x_i = \frac{x_i - min(x_1, \cdots, x_i, \cdots, x_n)}{max(x_1, \cdots, x_i, \cdots, x_n) - min(x_1, \cdots, x_i, \cdots, x_n)} \quad (2)$$

Where x represents the pixel value and n is the total number of pixels.

2.2 Model Architecture

V-Net [12] is a popular 3D fully convolutional neural network. It is a symmetric structure, including encoding and decoding parts, the encoding part is used to learn features from the input data, and the decoding part is used to reconstruct the features to obtain the final segmentation result. Skip connection fuses low-layer and high-layer to retain more semantic information. 5 × 5 × 5 convolutional kernels are used in the V-Net to extract spatial-dimension features and channel-dimension features. In addition, the residual structure solves the problem of network degradation.

Our architecture is shown in Fig. 1 In the network, we leverage two 3 × 3 × 3 convolution kernels instead of 5 × 5 × 5 convolution kernels, which provides better segmentation results with less parameters and larger receptive field size. Moreover, this operation can increase the depth of the network, and the ability to express non-linear characteristics will also be enhanced. [13] also proved that when the total number of neurons is equal, increasing the depth of the network produce more linear regions. However, the deepening of the network will lead to the loss of gradient, which makes the network unable to get effective training. To solve this problem and enhance the transfer of features, we introduce densely connected structures, which means the input of each layer comes from the output of all previous layers in the dense block.

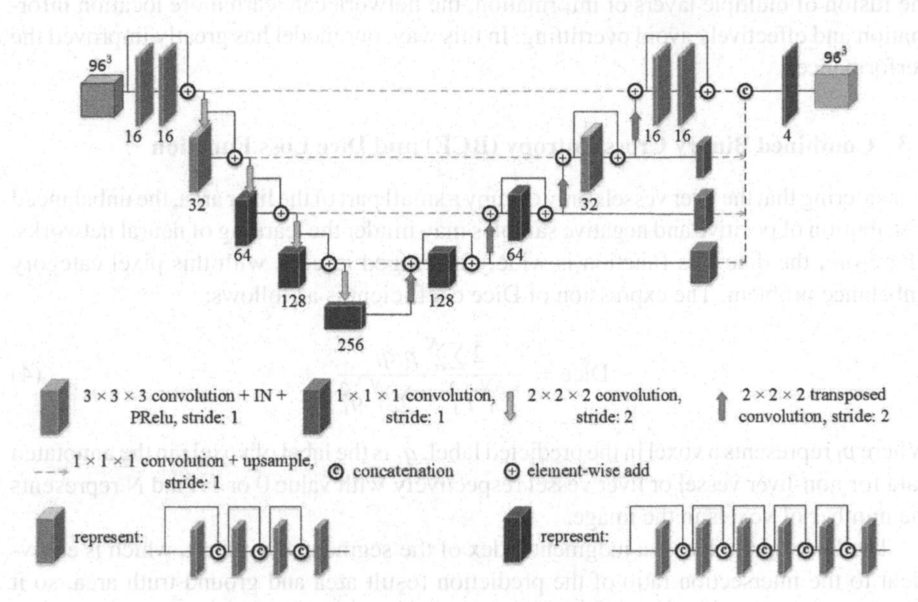

Fig. 1. Architecture of DV-Net

After convolution, a batch normalization (BN) layer is often followed to achieve the effect of regularization to stabilize the distribution of input data of the network, and accelerate the model learning speed in many practices. But due to the huge amount of parameters and memory limitations of 3D segmentation, we have to set a smaller batch size. While BN is the mean and variance calculated in the (mini-)batch to normalize the features. In other words, a small batch size can lead to inaccurate estimates of batch statistics and increase the model error. To address this problem, many scholars have proposed some solutions, such as Layer Normalization (LN) [14] and Instance Normalization (IN) [15], which avoid normalizing along the batch dimension. The experiments in [15, 16] show that IN works much better than BN with a small batch size. We also confirm this view in our experiments. Besides, we select the Parametric Rectified Linear Unit (PReLU) as the activation function.

In the decoding stage, in order to obtain more information, the feature maps of each layer are mapped to obtain an output, and these outputs are channel-stacked. In the last layer of the network, $1 \times 1 \times 1$ convolution is used for channel fusion to obtain the final output. The feature fusion formula is as follows:

$$\mathrm{H} = \sum_{i=1}^{4} \eta_i x_i \tag{3}$$

Where H is the feature map after fusion, x_i is the i-th feature. η_1 represents the weight of the i-th feature, and $i \in \{1, 2, 3, 4\}$.

As we all know, the deeper the information, the higher the representativeness. Therefore, when fusing channels, the deeper the feature map, the greater the weight. According to this principle, we set $\eta_1 = 0.2$, $\eta_1 = 0.4$, $\eta_1 = 0.6$, $\eta_1 = 0.8$, respectively. Through the fusion of multiple layers of information, the network can learn more location information and effectively avoid overfitting. In this way, our model has greatly improved the performance.

2.3 Combined Binary Cross Entropy (BCE) and Dice Loss Function

Considering that the liver vessels only occupy a small part of the liver area, the unbalanced distribution of positive and negative samples may hinder the learning of neural networks. Therefore, the dice loss function is widely leveraged to cope with this pixel category imbalance problem. The expression of Dice coefficient is as follows:

$$\text{Dice} = \frac{2\sum_i^N p_i q_i}{\sum_i^N p_i^2 + \sum_i^N q_i^2} \tag{4}$$

Where p_i represents a voxel in the predicted label. q_i is the label of voxel i in the annotated data for non-liver vessel or liver vessel respectively with value 0 or 1. And N represents the number of voxels in the image.

The Dice coefficient is a judgment index of the segmentation effect, which is equivalent to the intersection ratio of the prediction result area and ground truth area, so it calculates loss by taking all pixels of a category as a whole. Dice loss directly leverages the segmentation effect evaluation index as loss function to supervise the network, and also ignores a large number of background pixels when calculating the intersection and ratio, which solves the problem of imbalance between positive and negative samples, so the convergence speed is very fast. However, the patch selected by the random patch training method is not necessarily a situation where the ratio of foreground pixels to background pixels is large, and there are also cases where the ratio of foreground pixels to background pixels is similar. In this case, the learning of background pixels can play a positive role in feature extraction.

To alleviate the extreme problem of the dice loss function, we introduce the BCE loss function as a regular term. The BCE loss function is defined as:

$$L_{BCE} = mean\{l_0, \cdots\cdots l_{N-1}\} \tag{5}$$

$$l_n = -(p_i log(q_i) + (1 - p_i) log(1 - q_i)) \tag{6}$$

It can be drawn from the formula that the loss function of cross entropy separately evaluates the class prediction of each pixel vector, and then averages all pixels, so we can think that the pixels in the image are learned equally. In addition, the term $(1 - p_i)\log(1 - q_i)$ penalizes false positives as it is zero when the prediction is correct.

The combination of the Dice loss function and the BCE loss function makes the network not completely ignore the association with background pixels in the process of focusing on foreground pixel learning. Therefore, the training of the network will not be dominated by background pixels, it is also easy to learn the characteristics of smaller objects, and can solve the problem of dynamic changes in the ratio of front and background pixels caused by random patches. Experimental results show that the combined loss function can better improve the segmentation accuracy. The final loss function is defined as follows:

$$\mathrm{L} = 1 - Dice + \lambda L_{BCE} \tag{7}$$

As the coefficient of the regular term, λ is used to adjust the balance of the sample category ratio, which may have a certain correlation with the dataset. The proposed algorithm can greatly improve the segmentation accuracy by selecting a proper λ experimentally.

2.4 Post-processing

Due to the limitation in the memory of GPU, we have to intercept the input data into a series of patches during training and prediction. After a complete predicted process, these patches are spliced back into the original shape. Afterwards, the morphological operation is applied to remove some noises caused by classification (regions less than 100 mm^3 are removed).

3 Experiments and Results

3.1 Setting and Platform

Our proposed DV-Net apply Dropout [18] with a rate of 0.5 before each up-sampling block and down-sampling block. Kaiming initialization [19] is used for weight initialization to make the network converge faster. To train the model, we leverage small randomly cropped patches with a size of 96 × 96 × 96. In this way, we can obtain enough training data and reduce the need for memory. The batch size is set to 2. The Adam optimizer [20] with a learning rate of 0.001 is employed to perform the gradient descent algorithm.

All experiments of our proposed method were performed on a workstation with Intel Core i9 9900X @ 3.5 GHz, 128 GB RTX, NVIDIA 2080 Ti GPU (11 GB memory), Ubuntu 18.04.1, Python 3.7 and Pytorch 1.6.

3.2 Evaluation Metric

In order to compare the performance of different methods, three different metrices have been used in our experiments for evaluation, there are Dice similarity coefficient (DSC) $\left(DSC = \frac{2\times TP}{2\times TP+FP+FN}\right)$, Accuracy (ACC) $\left(ACC = \frac{TP+TN}{TP+FN+TN+FP}\right)$ and Sensitivity (SEN) $\left(SEN = \frac{TP}{TP+FN}\right)$, TP, TN, FP and FN represent the number of true positive, true negatives, false positives and false negatives, respectively.

Table 1. Evaluation of the proposed network with different loss functions

Method	DSC (%)	SEN (%)	ACC (%)
Focal	68.59	64.31	98.78
Dice	71.43	68.92	98.88
BCE	72.86	68.01	98.95
D-BCE	**74.76**	**75.27**	**98.95**

3.3 Comparisons of Loss Function

The loss function in our model was D-BCE, and parameter λ in (7) was integrated into the proposed similarity metric. Figure 2 shows the average dice, sensitivity and accuracy of the proposed network without post-processing on 20 3Dircadb datasets with λ ranging from 0.1 to 1. A change in λ could affect the attention of background pixels. Considering the segmentation accuracy and sensitivity, λ was set to 0.3 in the experiment. The average dice value and sensitivity of our method were 74.76% and 75.27% respectively, compared to 71.43% and 68.92% for the network using the dice loss function ($\lambda = 0$).

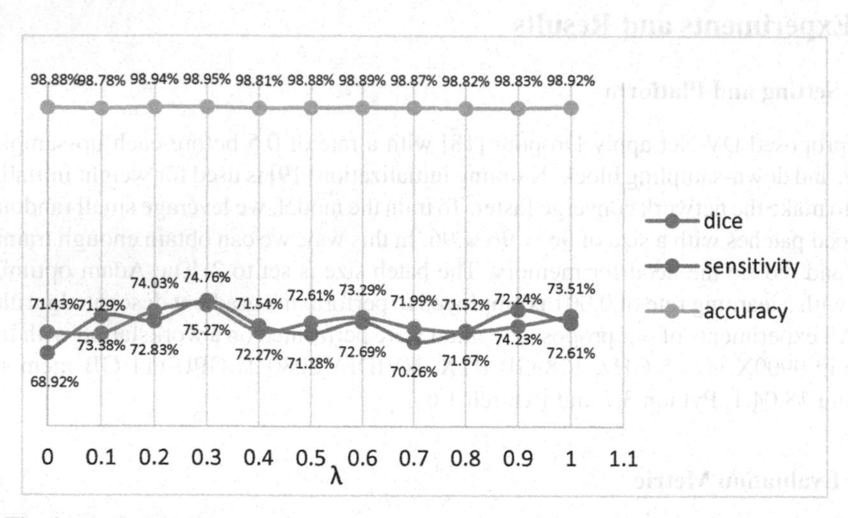

Fig. 2. Evaluation of the proposed network on 3Dircadb datasets with different λ value

Table 1 shows the accuracy comparison of different loss functions with the proposed DV-Net. It can be drawn from the table that the Focal loss function has the lowest score for each evaluation index, which further shows that the loss function we usually use to solve the extreme imbalance of the sample is not suitable for the random patch training method. D-BCE loss function can obtain better segmentation and sensitivity (a measure of the classifier's ability to recognize positive examples). Figure 3 shows the performances of the DV-Net on slices with different loss functions. The yellow mark indicates the misclassified voxels. It is obvious that the liver vessels extracted by the

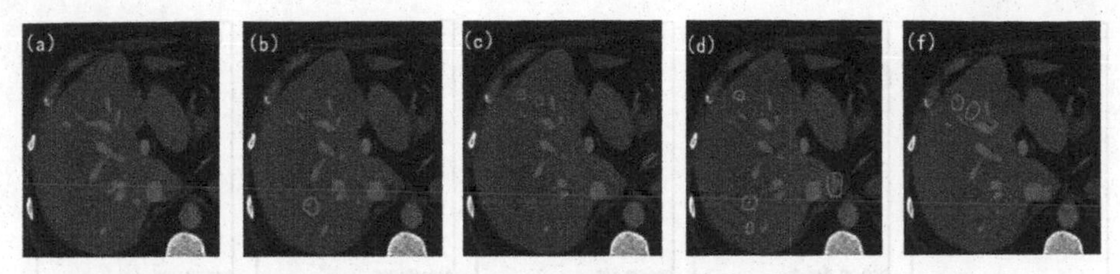

Fig. 3. Examples of performances of the DV-Net on slices with different loss functions. (a) Ground truth; (b) D-BCE; (c) BCE; (d) Dice; (e) Focal.

D-BCE loss function have the least misclassified voxels and the contours are closer to the real contours.

3.4 Results on 3Dircadb Dataset

Results are summarized in Table 2 In this experiment, we select 14 cases for training, and 6 cases for testing. To improve the segmentation performance and effectively prevent overfitting during training, we have used data augmentation to increase the amount of training, which includes rotation and mirror transformation. We compare our results with popular 3D models 3D U-Net, V-Net and 3D DenseNet to verify that our model is effective, and it turns out our results are consistently better than baselines. Figure 4 shows some 3D visualization results. We can observe that the results of the other three methods have many misclassified voxels and the vascular structure is messy. The blood vessels extracted by our method are closer to the real vessels.

Table 2. Comparative evaluation with some classic networks

Methods	DSC	SEN	ACC
3D U-Net	70.43%	70.89%	98.77%
3D Dense-Net	65.21%	62.72%	96.23%
V-Net	68.54%	68.99%	98.70%
Ours	**74.76%**	**75.27%**	**98.92%**

3.5 Results on the MSD Dataset

To verify the robustness of our model, we further apply our approach on a public dataset – hepatic vessel segmentation in Medical Segmentation Decathlon 17. Considering the training time of the model, we only select 42 cases as the training data and 18 cases as the testing data. The average dice value, sensitivity and accuracy of the proposed method on the datasets were 71.46%, 73.26% and 99.73% respectively. Figure 5 shows some examples of performances on MSD datasets with the software ITK-SNAP [21]. We can notice that for small and scattered blood vessels, our model is equipped with stronger segmentation capabilities and contain less noise.

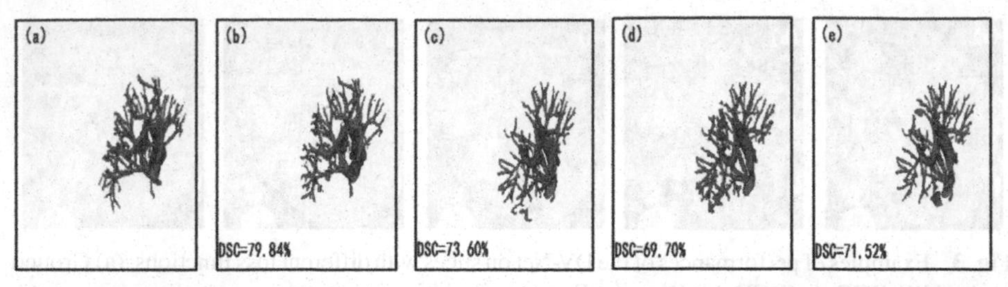

Fig. 4. 3D visualization examples of different methods on 3Dircadb datasets. (a) Ground truth; (b) Performance of our algorithm; (c) (d) (e) Results of 3D U-Net, 3D Dense-Net and V-Net.

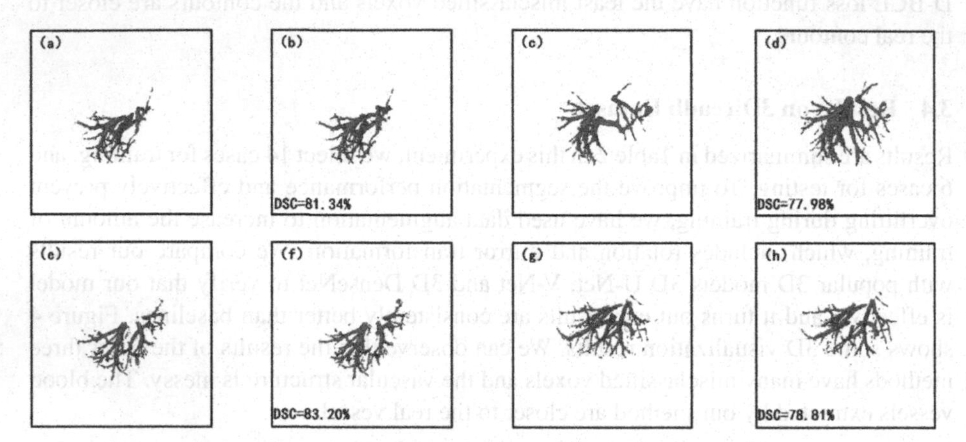

Fig. 5. Examples of performances on MSD datasets. (a) (c) (e) (g) Ground truth; (b) (d) (f) (h) Results of segmentation.

4 Conclusion

In this paper, we have considered that the random patch training method will bring about the problem of dynamic changes in the ratio of foreground and background pixels. The dice loss function we usually use (for unbalanced sample categories) may not be suitable for this training method. So we develop the D-BCE loss function with the BCE loss function as a regular term. In addition, we have made some improvements on the basis of V-Net, making it greater for the network to learn complex vascular features. Compared with the classic networks (3D U-Net, 3D Densenet and V-Net) and different loss functions, our method achieves better segmentation results. Our method has been tested on 20 3Dircadb datasets and 60 MSD datasets, and the experimental results show that our proposed method is efficient for liver vessel segmentation.

Acknowledgements. This work was supported by the National Natural Science Foundation of China (61976106, 61572239), Zhenjiang Key Deprogram "Fire Early Warning Technology Based on Multimodal Data Analysis" (SH2020011) Jiangsu Emergency Management Science and Technology Project "Research on Very Early Warning of Fire Based on Multi-modal Data Analysis and Multi-Intelligent Body Technology" (YJGL-TG-2020-8).

References

1. Selle, D., Preim, B., Schenk, A., et al.: Analysis of vasculature for liver surgical planning. IEEE Trans. Med. Imaging **21**(11), 1344–1357 (2002). https://doi.org/10.1109/tmi.2002.801166
2. Zeng, Y., Liao, S., Tang, P., et al.: Automatic liver vessel segmentation using 3D region growing and hybrid active contour model. Comput. Biol. Med. **97**, 63–73 (2018). https://doi.org/10.1016/j.compbiomed.2018.04.014
3. Zeng, Y.Z., et al.: Liver vessel segmentation based on extreme learning machine. Physica Medica **32**(5), 709–716 (2016). https://doi.org/10.1016/j.ejmp.2016.04.003
4. Litjens, G., Kooi, T., Bejnordi, B.E., et al.: A survey on deep learning in medical image analysis. Med. Image Anal. **42**, 60–88 (2017). https://doi.org/10.1016/j.media.2017.07.005
5. Huang, Q., Sun, J., Ding, H., Wang, X., Wang, G.: Robust liver vessel extraction using 3d u-net with variant dice loss function. Comput. Biol. Med. **101**, 153–162 (2018)
6. Kitrungrotsakul, T., Han, X.H., Chen, Y.W.: Robust hepatic vessel segmentation using multi deep convolution network. In: SPIE Medical Imaging, p. 1013711 (2017)
7. Wei, Y., et al.: Liver vessels segmentation based on 3d residual U-NET. In: 2019 IEEE International Conference on Image Processing (ICIP) IEEE (2019)
8. Çiçek, Ö., Abdulkadir, A., Lienkamp, S.S., Brox, T., Ronneberger, O.: 3d u-net: learning dense volumetric segmentation from sparse annotation, pp. 424–432 (2016). https://doi.org/10.1007/978-3-319-46723-8_49
9. Milletari, F., Navab, N., Ahmadi, S.A.: V-Net, Fully convolutional neural networks for volumetric medical image segmentation. In: Fourth International Conference on 3D Vision (3DV), pp. 565–571 (2016)
10. Huang, G., Liu, Z., Kilian, Q.W.: Densely connected convolutional networks. In: IEEE Conference on Computer Vision & Pattern Recognition (2017). https://doi.org/10.1109/cvpr.2017.243
11. Li, X., Chen, H., Qi, X., et al.: H-DenseUNet: hybrid densely connected UNet for liver and tumor segmentation from CT volumes. IEEE Trans. Med. Imaging **37**(12), 2663–2674 (2018)
12. Milletari, F., Navab, N., Ahmadi, S.-A.: V-Net: fully convolutional neural networks for volumetric medical image segmentation. In: 2016 Fourth International Conference on 3D Vision, pp. 565–571 (2016)
13. Montúfar, G., Pascanu, R., Cho, K., et al.: On the number of linear regions of deep neural networks. Adv. Neural Inf. Process. Syst. **24**, 415–423 (2014)
14. Ba, J.L., Kiros, J.R., Hinton, G.E.: Layer normalization. arXiv:1607.06450 (2016)
15. Ulyanov, D., Vedaldi, A., Lempitsky, V.: Instance normalization: the missing ingredient for fast stylization. arXiv:1607.08022 (2016)
16. Ulyanov, D., Vedaldi, A., Lempitsky, V.: Improved texture networks: maximizing quality and diversity in feed-forward stylization and texture synthesis. In: Proceedings of the IEEE Conference on Computer Vision and Pattern Recognition, pp. 6924–6932 (2017)
17. Simpson, A.L., et al.: A large annotated medical image dataset for the development and evaluation of segmentation algorithms, vol. 2, p. 8. arXiv:1902.09063 (2019)
18. Srivastava, N., Hinton, G., Krizhevsky, A., Sutskever, I., Salakhutdinov, R.: Dropout: a simple way to prevent neural networks from overfitting. J. Mach. Learn. Res. **15**(1), 1929–1958 (2014)
19. He, K., Zhang, X., Ren, S., et al.: Delving Deep into Rectifiers: Surpassing Human-Level Performance on ImageNet Classification (2015)
20. Kingma, D.P., Ba, J.: Adam: A method for stochastic optimization. arXiv:1412.6980 (2014)
21. Yushkevich, P.A.: User-guided 3D active contour segmentation of anatomical structures: Significantly improved efficiency and reliability. Neuroimage **31**(3), 1116–1128 (2006). https://doi.org/10.1016/j.neuroimage.2006.01.015

Cloud Computing-Based Graph Convolutional Network Power Consumption Prediction Method

Yong Ma[1(✉)], Honglei Sheng[2,3], Shang Wu[1], Shuai Gong[1], and Hang Cheng[1]

[1] Information and Communication Branch of State Grid, Anhui Electric Electricity Co., Ltd., Hefei 230009, People's Republic of China

[2] NARI Group Corporation/State Grid Electric Electricity Research Institute, Nanjing 211000, People's Republic of China

[3] Nari Information Communication Technology Co., Ltd., Nanjing 210003, People's Republic of China

Abstract. With the continuous increase of electricity consumption data in smart grids, the data storage and data analysis capabilities of traditional single-node data mining algorithms can no longer meet the requirements of electricity consumption forecasting. This paper designs a electricity consumption forecasting method based on cloud computing and graph convolutional network. The method first proposes a GCN-based electricity consumption forecasting model, then builds a Hadoop platform, uses MapReduce to parallelize and iteratively trains the GCN model on the platform, and then uses the trained model to predict electricity consumption. In addition, the prediction accuracy of this method is verified through experiments.

Keywords: Cloud computing · Hadoop · GCN · Data prediction

1 Introduction

Electricity consumption forecasting has always been an important work in electricity decision-making in smart grids [1]. Correct data prediction can help the electricity grid to carry out reasonable resource scheduling, reduce electricity loss in electricity transmission, and improve the operating efficiency of the electricity system [2]. However, with the development of smart grids, electricity consumption data continues to increase, and the data storage and data analysis capabilities of traditional single-node data mining algorithms can no longer meet the forecasting requirements [3]. New methods need to be studied to improve the forecasting speed and level of electricity consumption forecasting.

Based on this status, this paper designs a electricity consumption forecasting method based on cloud computing and graph convolutional network. Cloud computing has become the most popular business computing model due to its low cost, scalable scale, strong availability, and high resource utilization [4]. In the electricity industry, cloud computing has been fully researched and applied in electricity system monitoring and dispatching, grid intelligent early warning analysis, and grid electricity loss [5, 6]. This

X. Sun et al. (Eds.): ICAIS 2021, CCIS 1422, pp. 62–69, 2021.
https://doi.org/10.1007/978-3-030-78615-1_6

paper uses the distributed storage and parallel computing capabilities of cloud computing technology to solve the problem of transforming massive electricity consumption data. Compared with the traditional single-node data prediction method, the use of cloud computing makes this method more suitable for the processing of power grid big data, and under the same amount of data, the running time of this method is shorter. Graph Convolutional Neural Networks (GCN for short) are developed from Convolutional Neural Networks and were proposed by Thomas Kpif in the paper "Semi-supervised classification with graph convolutional networks" in 2017. The GCN is very good at capturing the interdependence between instances [7]. Compared with BP (Back Propagation) neural network [8], SVM (Support Vector Machines) [9] and ARIMA (Autoregressive Integrated Moving Average mode) [10], it is more suitable for modeling. consumption objects with complex dependencies and dynamic changes. Because the residents who conduct electricity consumption have geographical distribution, the graph convolutional network can better analyze the dependence of electricity consumption between users, so as to obtain higher prediction accuracy.

This paper uses the daily electricity consumption of a residential area in a city of Jiangsu Province as the experimental data to verify the accuracy of the method. The experimental results show that the accuracy of the method can reach more than 90%. Compared with the single node graph convolution network method, it is proved that this method has smaller average power consumption percentage error.

2 Preparation

2.1 Cloud Computing Architecture

Cloud computing is developed from distributed computing and parallel computing. It decomposes the data processing programs on the "cloud" into many small programs through the network, and then a large number of servers process these small programs, and after the processing results are obtained, they are returned to the user. By distributing its calculations and data in a large number of distributed computers, the computing electricity and storage of cloud computing have gained strong scalability [11]. Hadoop is an open source distributed computing framework [12], which is a technical realization of cloud computing and big data. Running programs on clusters through Hadoop can improve data processing speed and efficiency [12]. This paper uses the MapReduce computing model in the Hadoop framework. MapReduce is composed of Map and Reduce functions. There is one Jobtracker and multiple Tasktrackes during the execution of a job. MapReduce accepts job requests through job submission nodes, and then Mapper and Recucer are sent to the idle tasktracker process by the jobtracker process, and the tasktracker process receives the task and executes the task until it is completed [13].

The workflow of the MapReduce model is shown in Fig. 1. The Input phase in the figure is a job consisting of four letters A, B, C, and D. The Splitting phase divides the job into three pieces. Then in the Mapping phase, call the MAP function for processing. The figure shows the number of occurrences of each letter in the data block. Afterwards, intermediate results are generated in the Shuffling stage, and the framework will sort these intermediate results according to the key. The figure shows the sorting in alphabetical order. In the Reducing stage, the intermediate results are divided into four Reduce nodes

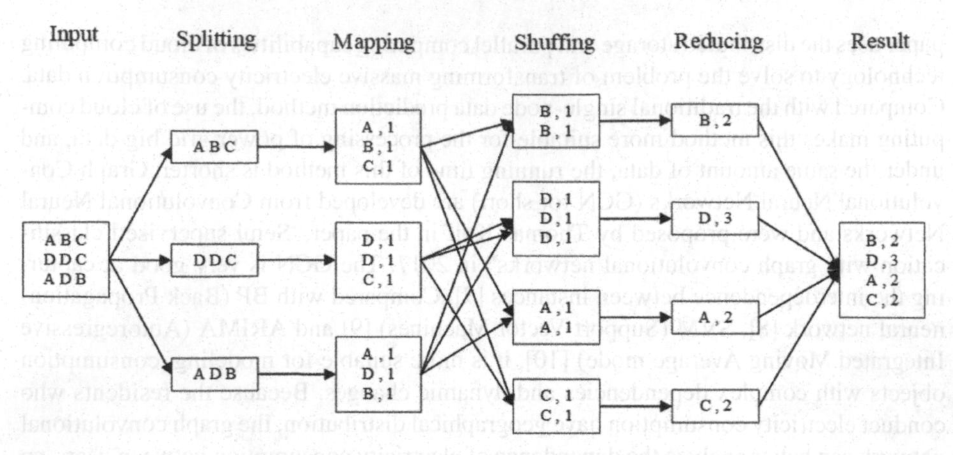

Fig. 1. The workflow of the MapReduce model.

by value, and the Reduce function is used to perform aggregation calculations and output the final results.

2.2 GCN

In recent years, graph neural networks have received very high attention due to their electricityful processing capabilities in many fields. A graph is different from traditional data signals, videos, etc. As a data form, the number of neighbors between its vertices can be different, and its edges do not need to be defined and explained. Due to the various points of the graph, it can be used to represent the relationship network in many fields. Usually, the nodes in the graph are defined as individual individuals, and the edges represent the relationship between individuals.

Define graph $G = (V, E)$, where V is the set of vertices of the graph, E is the set of edges of the graph, and set $T = \{(e_{ij})\}_{i=1,j=1}^{|V|,|V|}$ is the weighted adjacency matrix of graph G, where e_{ij} represents the connection strength between points v_i and v_j.

The graph convolution calculation first needs to calculate the Laplacian matrix L of the graph, and its formula is as follows:

$$L = G'^{-\frac{1}{2}}(G' - T)G'^{-\frac{1}{2}} \tag{1}$$

Where G' is the diagonal matrix of G.

From the formula, we can see that the Laplace matrix L is a real symmetric matrix, so there is a diagonal matrix $\Lambda = diag(\lambda_1, \lambda_2, \ldots, \lambda_{|V|})$

Then we can deduce $L = R^T \Lambda R$.

As shown in the following formula, the graph convolution operation is:

$$R^T(g \times Gh) = R^T g \otimes R^T h \tag{2}$$

Among them, $g \times Gh = R\left(R^T g \otimes R^T x\right)$, x is the input, and g is the convolution kernel.

further:

$$g \times Gx \approx \left(G'^{-\frac{1}{2}} G' G'^{-\frac{1}{2}} + G'^{-\frac{1}{2}} A G'^{-\frac{1}{2}} \right) x = \left(\tilde{G}'^{-\frac{1}{2}} \tilde{A} \tilde{G}'^{-\frac{1}{2}} \right) x \tag{3}$$

Using the above formula for the graph convolutional network, the output obtained is as follows:

$$F^l = \xi(\tilde{G}'^{-\frac{1}{2}} \tilde{T} \tilde{G}'^{-\frac{1}{2}} F^{l-1} W^l) \tag{4}$$

Where F^l is the output of the l-th neural network, and W^l is the weight parameter of the l-th neural network.

3 Model Building

3.1 GCN Model

To predict the electricity consumption of electricity grid users, specifically, it is to predict the electricity consumption of users in the future through the electricity collected by N meter collectors of residential buildings. Therefore, the electricity meter collector network can be defined as $G = (V, E)$, where $V = \{v_1, v_2, \ldots, v_N\}$ is the node set, where N meter collectors are represented here, and $|V| = N$; E is the edge set, which is used to indicate the strength of the connection relationship between the nodes in the network. Here, it indicates the distance relationship between the electric meter and the electric meter.

The problem of predicting residential electricity consumption can be defined as inputting time series data $X = (X_1, X_2, \ldots, X_{t-1})$, where $X_t = \left\{\left(k_i^t\right)\right\}_{i=1}^N$ represents the time series data collected by all meter collectors at time t, and predicting the output at time t through function F:

$$X_t = f(X_1, X_2, \ldots, X_{t-1}) \tag{5}$$

Input the data X into a graph convolutional neural network contains 2 convolutional layers, and the forward propagation formula is as follow:

$$F^l = \xi(\tilde{G}'^{-\frac{1}{2}} W^l \tilde{T} \tilde{G}'^{-\frac{1}{2}} F^{l-1}) \tag{6}$$

The loss function formula for model training is as follows:

$$Loss = \|True_t - \mathrm{Pre}_t\| + \partial \mathrm{Reg} \tag{7}$$

$True_t$ represents the true value at time t, Pre_t represents the predicted value of the model at time t, ∂ is a parameter, and Reg is a regular term.

3.2 GCN Model Based on Cloud Computing

Under the MapReduce framework, the graph convolutional network is trained by batch update and data parallelism. The training process is divided into two parts: Map and Reduce. The Map part is responsible for receiving input and generating intermediate key-value pairs of network weight value w changes; the Reduce part is responsible for summarizing the local changes to obtain the global changes and output them for batch update.

The Map process algorithm is shown in Algorithm 1.

Algorithm 1 Map algorithm

Input: <*key*, *value*>

Output: $<key = w,\ \text{value} = \Delta w>$

Step 1: Initialize the network;

Step 2: Call the map() function to receive the <*key*, *value*> key-value pair;

Step 3: Through forward propagation and back propagation, the weight value and offset of each layer of GCN are obtained locally;

Step 4: Output $<key = W,\ value = \Delta W>$ key-value pairs.

The Reduce process algorithm is shown in Algorithm 2, where *Gg* represents the global gradient change.

Algorithm 2 Reduce algorithm

Input: $<key = W,\ value = \Delta W>$

Output: $< key = W, value = Gg >$

Step 1: Accept the output <*key*, *value*> of the Map process as input;

Step 2: Traverse the input local gradient change and calculate the global gradient change *Gg*;

Step 3: Output $<key = W, value = Gg>$;

Step 4: Use the output to update the GCN in batches.

Repeat the MapReduce task several times until the GCN error is within the threshold or the maximum number of iterations is reached. At this time, the training of GCN model is completed, and the trained model can be used to predict the power consumption data.

On Hadoop, the basic process of MapReduce work is:

1. Split divides the training data into data blocks of the same size as HDFS blocks and user operating programs.
2. The user job program is submitted to the Master node, and the Master node finds suitable Map and Reduce nodes for it and transmits the data to the found Map and Reduce nodes.
3. The Master node starts the Map node execution program, and the Map node reads the local data block for calculation.
4. Each Map node processes and organizes the data blocks it reads, and places the intermediate results locally. At the same time, the Map node sends a work completion

notification to the Master node and informs the Master node of the storage location of the intermediate results.

5. After all the Map work is completed, the Master node starts the Reduce node. The Reduce node obtains the storage location of the intermediate results through the Master node, and then remotely reads the intermediate results.
6. The Reduce node processes the intermediate results, and then outputs the results to a file.

4 Experiment Analysis

The development platform selected for this experiment is Eclipse. In order to verify the computational efficiency of the graph convolutional network algorithm based on cloud computing under the Hadoop platform, this article compares the experimental verification with the single-server graph convolutional network. Use the same experimental data to compare the performance of the models. The experimental data uses the daily electricity consumption data collected from October 2019 to March 2020 in a community in a city in Jiangsu Province, which is divided into 24 sequences by time.

Use 70% of the collected electricity consumption data as the training set and the remaining 30% as the test set to test the accuracy of the graph convolutional network algorithm based on cloud computing. As shown in the figure: its accuracy can reach more than 90% (Fig. 2).

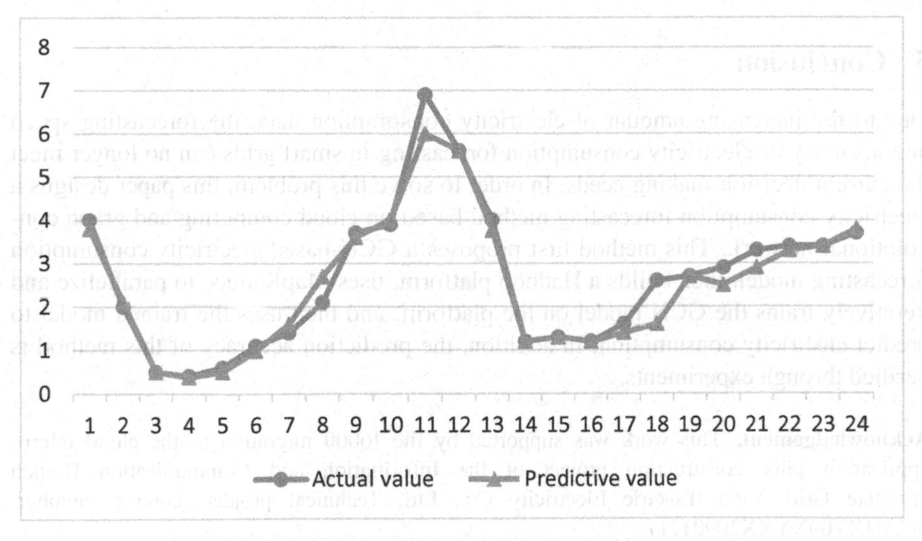

Fig. 2. Accuracy of model forecasting electricity consumption.

We use the average electricity consumption percentage error MAPE to measure the comparison of prediction experiments between a single-server graph convolutional network and a cloud-based graph convolutional network algorithm.

$$MAPE = \sum_{t=1}^{n} \left| \frac{obs_t - pre_t}{obs_t} \right| \times \frac{100}{n} \tag{8}$$

Among them, obs_t is the true value at time t, and pre_t is the predicted value at time t. The following table shows the comparison chart of MAPE (Table 1)

Table 1. Comparison chart of the average electricity consumption error for a week predicted by the model

Data	Traditional graph convolutional network	Graph convolutional network algorithm based on cloud computing
12.11	3.55	3.21
12.12	4.5	4.02
12.13	3.66	2.99
12.14	5.43	5.33
12.15	8.76	7.34
12.16	14.56	13.55
12.17	17.9	16.93

5 Conclusion

Due to the increasing amount of electricity consumption data, the forecasting speed and accuracy of electricity consumption forecasting in smart grids can no longer meet the current decision-making needs. In order to solve this problem, this paper designs a electricity consumption forecasting method based on cloud computing and graph convolutional network. This method first proposes a GCN-based electricity consumption forecasting model, then builds a Hadoop platform, uses MapReduce to parallelize and iteratively trains the GCN model on the platform, and then uses the trained model to predict electricity consumption. In addition, the prediction accuracy of this method is verified through experiments.

Acknowledgement. This work was supported by the I6000 migration to the cloud micro-application pilot construction project of the Information and Communication Branch of State Grid Anhui Electric Electricity Co., Ltd. Technical project (contract number: SGAHXT00XYXX2000121).

References

1. Wang, Y., et al.: Electricity consumption forecast of provincial electricity companies based on VEC Model. In: 2019 IEEE 3rd Conference on Energy Internet and Energy System Integration (EI2), pp. 1514–1518. Changsha, China (2019)

2. Liu, H., et al.: The application of a modified bp artificial neural network in the prediction of electricity consumption. Autom. Technol. Appl. **26**(12), 3–5 (2007)
3. Allahvirdizadeh, Y., Moghaddam, M.P., Shayanfar, H.A.: A survey on cloud computing in energy management of the smart grids. vol. 29, no. 10 (2019)
4. Yu, J., Yang, Z., Zhu, S., Xu, B., Li, S., Zhang, M.: A bibliometric analysis of cloud computing technology research. In: 2018 IEEE 3rd Advanced Information Technology, Electronic and Automation Control Conference (IAEAC), pp. 2353–2358. Chongqing (2018)
5. Kaur, A., Singh, V.P., Singh Gill, S.: The future of cloud computing: opportunities, challenges and research trends. In: 2018 2nd International Conference on I-SMAC (IoT in Social, Mobile, Analytics and Cloud) (I-SMAC)I-SMAC (IoT in Social, Mobile, Analytics and Cloud) (I-SMAC), pp. 213–219. Palladam, India (2018)
6. Sun, D., Zeng, J., Yang, B., Wang, N., Bo, Q., Yang, J.: Study of decision-making system for electricity grid consistency planning. In: 2019 IEEE 4th International Conference on Cloud Computing and Big Data Analysis (ICCCBDA), pp. 160–164. Chengdu, China (2019)
7. Ye, H., Cao, B., Chen, J., Liu, J., Wen, Y., Chen, J.: A web services classification method based on GCN. In: 2019 IEEE International Conference on Parallel & Distributed Processing with Applications, Big Data & Cloud Computing, Sustainable Computing & Communications, Social Computing & Networking (ISPA/BDCloud/SocialCom/SustainCom), pp. 1107–1114. Xiamen, China (2019)
8. Wang, B., et al.: Back propagation (BP) neural network prediction and chaotic characteristics analysis of free falling liquid film fluctuation on corrugated plate wall. Ann. Nuclear Energy **148**, 107711 (2020)
9. Duan, M.: Short-time prediction of traffic flow based on PSO optimized SVM. In: 2018 International Conference on Intelligent Transportation, Big Data & Smart City (ICITBS), pp. 41–45. Xiamen (2018)
10. Wang, Z., Lou, Y.: Hydrological time series forecast model based on wavelet de-noising and ARIMA-LSTM. In: 2019 IEEE 3rd Information Technology, Networking, Electronic and Automation Control Conference (ITNEC), pp. 1697–1701. Chengdu, China (2019)
11. Gill, S.S., Buyya, R.: Failure management for reliable cloud computing: a taxonomy, model, and future directions. Comput. Sci. Eng. **22**(3), 52–63 (2020)
12. Sontakke, V., Dayanand, R.B.: Optimization of hadoop mapreduce model in cloud computing environment. In: 2019 International Conference on Smart Systems and Inventive Technology (ICSSIT), pp. 510–515. Tirunelveli, India (2019)
13. Kulkarni, O., Jena, S., Sankar, V.R.: MapReduce framework based big data clustering using fractional integrated sparse fuzzy C means algorithm. In: IET Image Processing (2020)

Intelligent Intrusion Detection of Measurement Automation System Based on Deep Learning

Tao Liu[1], Shaocheng Wu[1], Ziyv Guo[2(✉)], Jie Zhao[1], Wenlong Sun[1], and Xiaohong Cao[1]

[1] Shenzhen Power Supply Bureau Co. Ltd., Shenzhen 518001, Guangdong, China
[2] Beijing University of Posts and Telecommunications, Beijing 100876, China
zyguo@bupt.edu.cn

Abstract. In the smart grid with measurement automation system, the intrusion detection system judges the intrusion event by analyzing the transmission data in the grid. Aiming at the characteristics of traditional intrusion detection, such as loss of features, low detection efficiency and poor adaptability, an intrusion detection method based on stacked denoising convolutional autoencoders is proposed, which combine convolutional neural network and denoising autoencoder to strengthen feature recognition ability, using Dropout and regularization methods to prevent overfitting, and using Adam algorithm to obtain optimal parameters. Finally, the NSL-KDD data set is used to verify the proposed method. Experimental results show that the overall recognition rate of this method is 97.25%, which is 11.59%, 9.63% and 4.07% higher than the existing NN, SVM, and ICNN accuracy rates respectively.

Keywords: Measurement automation system · Intrusion detection · Convolutional neural network · Autoencoder

1 Introduction

In recent years, with the continuous expansion of the grid scale and the integration of different types of energy, the centralized control grid is no longer suitable for the needs of today's society. In order to solve the challenges faced by the existing power grid, a new concept of smart grid has emerged in the power industry [1]. Smart grid is considered as the infrastructure of modern power grid. Through automatic control, high-power converter, modern communication means, sensor metering technology, modern energy management and other technologies, based on the demand, the reliability of energy network is optimized, so as to improve the efficiency and reliability [2].

As one of the cores of smart grid, measurement automation system undertakes the important functions of data acquisition and analysis [3]. With the deepening of the application of measurement automation systems, measurement automation terminals have adopted a large number of wireless communication methods to connect to the smart grid [4]. The wireless communication channel carries important business information such as remote control, telemetry, and remote signaling. Wireless attacks can cause information

X. Sun et al. (Eds.): ICAIS 2021, CCIS 1422, pp. 70–83, 2021.
https://doi.org/10.1007/978-3-030-78615-1_7

tampering or leakage, which can lead to the loss of accuracy and credibility of grid control, cause cascading grid failures or equipment damage, and cause economic losses, and endanger personal and social safety [5]. The complex network environment and endless attack methods also make the metering automation system face many challenges, and related security defense issues have gradually become current research hotpots.

Intrusion Detection System (IDS) is a security management system used to detect network intrusions. It can monitor the network transmission data in real time and take intrusive actions without affecting the internal network [6]. It can take measures such as monitoring, analysis, and early warning to improve the network's ability to respond to external threats. IDS anomaly detection is a behavior-based detection technology that detects unknown attacks by checking whether the actual network behavior deviates from its normal behavior. Many researches on machine learning have developed intrusion detection technologies with machine intelligence, and achieved good results, such as support vector machines, artificial neural networks and genetic algorithms [7–9]. However, because the measurement automation system needs to consider the complexity of wireless communication combined with power service detection, the use of traditional intrusion detection models often has the following problems: (1) The collected network traffic data is high-dimensional, and manual selection of features is not effective enough and has little basis. Important features may be lost and redundant features may be retained. (2) Poor self-adaptation ability. As the network operating environment and structure changes, it is necessary to continuously update the model to detect new and unknown attacks. (3) The model fitting ability is poor. The traditional machine learning model has a simple structure and limited feature extraction and learning capabilities. When faced with a large-scale data set, it cannot form an effective non-linear mapping of the data distribution.

Therefore, in response to the problems of traditional methods, this paper designs and implements a learning method that can automatically extract intrusion features and analysis, and proposes an intrusion detection model based on stacked denoising convolutional autoencoding. The traffic data is processed as two-dimensional gray image as input. Combine the denoising autoencoder with the convolutional neural network, use the convolutional characteristics of the Convolutional Neural Network (CNN), fully learn the network features, solve the model by reconstructing the error, and use the adaptive algorithm Adam to optimize the network, so that it can learn the intrusion characteristics as much as possible and get a better recognition effect.

2 Related Work

Many machine learning algorithms have been applied to intrusion detection, using different machine learning algorithms to reduce the false alarm rate and detect abnormal network behaviors, currently. Shon et al. [10] proposed an intrusion detection classifier that combines genetic algorithm and support vector machine (SVM), which has a wide range of adaptability to real environment data sets; Zhao [11] proposed a Least Squares Support Vector Machine (LSSVM) model for network intrusion detection; Hussain et al. [12] proposed a two-stage hybrid classification method. In the first stage, SVM is used for anomaly detection, and in the second stage, artificial neural network is used for misuse detection. Traditional machine learning methods are very effective in intrusion detection and have a certain accuracy rate, but they cannot adjust the parameters of preprocessing and feature extraction independently. They need the manual participation of experts to complete the learning and classification goals, and the model performance depends on the quality of parameter tuning and selection features.

In order to solve the above problems, researchers have introduced deep learning technology. In recent years, deep learning has been widely used in speech recognition, image recognition and natural language processing. Since deep learning can automatically extract features from raw data and can effectively process large, complex and multi-dimensional network data, many scholars have also begun to combine deep learning methods with the field of intrusion detection, and have achieved good results. Erfani et al. [13] proposed a hybrid model that combines a deep belief network (DBN) with a single-class vector machine, and first uses restricted Boltzmann machine (RBM) to eliminate the negative effects of noise and abnormal data on the network and then use a single-class vector machine to achieve the classification task. Staudemeyer [14] pointed out for the first time that an LSTM recurrent neural network can be used for intrusion detection. LSTM can learn to look back in time and find some associations from a time perspective. Javaid et al. [15] combined the coding layer of the Sparse Autoencoder (used for feature extraction) and the soft-max function (used for class probability estimation), and designed a "self-learning" classification mechanism for NSL-KDD classification. Khan et al. [16] used CNN-based residual network (ResNet) and GoogleNet model for malware detection. Although the method based on deep learning has improved sample recognition ability and performance, it is prone to over fitting problems in the network training process, and there are many parameters, the training time is long, and the detection accuracy and efficiency need to be further improved.

3 Intrusion Detection Model of Measurement Automation System

The model framework of this paper is an intrusion detection model based on stacked denoising convolutional autoencoding neural network. The overall framework of the model is shown in Fig. 1.

3.1 The Overall Architecture of the Intrusion Detection Model

It can be seen from Fig. 1 that the model mainly includes the following steps for the identification of intrusion detection in the measurement automation system:

Data Acquisition. The measurement automation system environment is built to obtain real-time network traffic data by monitoring and recording network traffic, including source address, target address, connection attribute and other related information.

Data Preprocessing. Process the data into a constructed and processable format. First, the character attributes are mapped to numerical attributes through one-hot encoding, and then the data is normalized to the [0,1] interval to eliminate the influence of the large dimension of different features in the network connection on the training of the intrusion detection model, and finally map the data to a two-dimensional grayscale image.

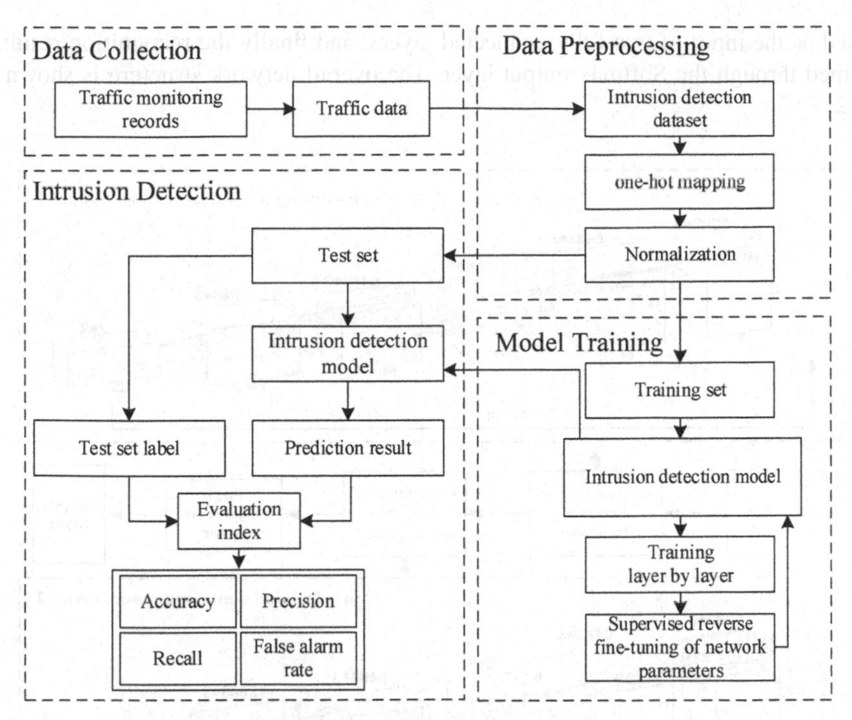

Fig. 1. Intrusion detection model of measurement automation system

Model Building and Training. The stacked denoising convolutional autoencoder model was built to extract and analyze features. It pre-trains and adjusts the parameters of the standard data set to achieve the optimal extraction of standard data features. The input layer of the model is a two-dimensional gray image processing format, and the hidden layer is composed of the encoding and decoding of the convolution layer, the pooling layer and the full connection layer, the activation function of the convolutional layer adopts ReLU to learn feature information independently, the fully connected layer introduces the Dropout method to prevent over-fitting, and the output layer uses the soft-max classifier to output classification decisions.

3.2 Stacked Denoising Convolutional Autoencoding Network

The convolutional denoising autoencoder combines the convolution and pooling operations of the convolutional neural network on the basis of the autoencoder, so as to realize feature extraction and better solve the redundancy and distortion of various data information in the measurement automation system, effectively improve the detection rate. The stacked denoising convolutional autoencoder network designed in this paper first constructs an improved convolutional denoising autoencoder which let the data go through two convolution operations and then perform a pooling operation to strengthen the feature learning ability of the network. Then stack two convolutional denoising autoencoders, the first of which retains the complete structure, and the second retains only the encoding part. The output of the second convolutional denoising autoencoder is used as the input of two fully connected layers, and finally the recognition result is obtained through the Softmax output layer. The overall network structure is shown in Fig. 2.

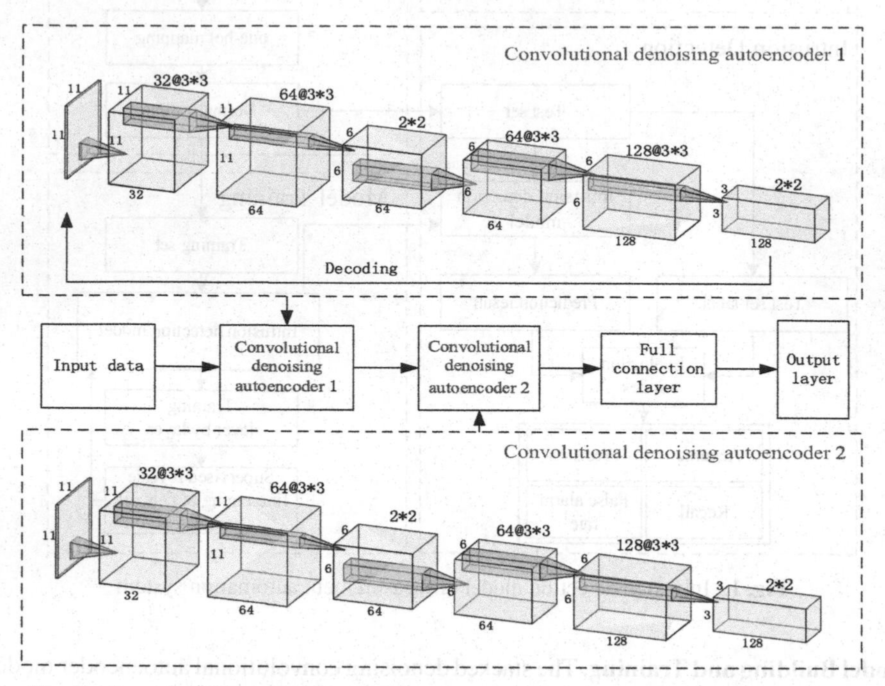

Fig. 2. Overall architecture of stacked denoising convolutional autoencoding network

Convolutional Autoencoding Network. The detailed encoding and decoding process is derived as follows:

Encoding Process:

The output of the convolutional layer can be expressed as:

$$h_1 = f\left(x \otimes W'_{11} + b'_{11}\right) \tag{1}$$

Where x represents the input feature vector, $\otimes$ is a convolution operation, W'_{11} represents the first layer weight, and b'_{11} represents the first layer bias, f is a nonlinear activation function, such as Sigmoid, Tanh, and ReLU. Compared with other activation functions, ReLU can make the network converge faster and reduce the training time. Therefore, this paper adopts the ReLU activation function, which is:

$$f(x)_{ReLU} = \begin{cases} 0 \ (x \leq 0) \\ x \ (x > 0) \end{cases} \tag{2}$$

The output of the pooling layer can be expressed as:

$$h_2 = pool(h_1) = f\left(down(x) + b'_{11}\right) \tag{3}$$

Where *pool* represents the pooling operation. This paper uses maximum pooling to reduce the information redundancy caused by the convolution operation and $down(\bullet)$ represents downsampling.

Decoding Process:

$$h'_2 = f\left(h_2 \otimes W'_{22} + b'_{22}\right) \tag{4}$$

$$h'_1 = upsample\left(h'_2\right) = f\left(upsample(x) + b'_{22}\right) \tag{5}$$

$$x' = f\left(h'_1 \otimes W'_{21} + b'_{21}\right) \tag{6}$$

Where x' is the reconstructed x, W'_{22} and b'_{22} are the weight and offset of the first layer of convolution during decoding, h'_2 is decoded convolution output, *upsample* is upsampling, h'_1 is decoded pooled output, f is decoding activation function.

Training of Convolutional Autoencoding Network. The training process of the stacked denoising convolutional autoencoding network is as follow.

Forward Propagation:

① Randomly select batch input from the standard data set, and the data parameter dimension is (batch_size, h, w, c).

② Enter the convolutional denoising autoencoding neural network shown in Fig. 2, and the convolution operation and pooling operation are as shown in Eqs. (7) and (8) respectively.

$$h_j^l = f\left(\sum_{i\in M_J} h_j^{l-1} \otimes W_{ij}^l + b_j^l\right) \tag{7}$$

$$Z_j^l = \beta\left(W_j^l down\left(Z_j^{l-1}\right) + b_j^l\right) \tag{8}$$

③ Using the conv2d_transpose function and upsample function in Tensorflow to perform deconvolution pooling decoding, and input to the fully connected layer to output the result. The fully connected layer (FC) is calculated as follows:

$$y_j^l = f\left(\sum\nolimits_{i\in M_J} y_j^{l-1} \otimes W_{ij}^l + b_j^l\right) \tag{9}$$

④ Solve the reconstruction error and use Softmax for data classification.

Back Propagation:

① Calculate the overall loss function $J(\omega, b)$ according to the classification results of the training set samples.

② Back propagate the weights and biases of the training network until convergence. In the model training process, in order to speed up the convergence time and improve the convergence accuracy, this paper uses Adam [17] to update the network model parameters. This method solves the problem of slow convergence and easy local optimization, and saves computer resources.

The loss function of this model is:

$$J(\omega, b) = J\left(\omega, b; x^i, y^i\right) + \frac{\lambda}{2}\sum\nolimits_{l=1}^{n_l-1}\sum\nolimits_{i=1}^{s_i}\sum\nolimits_{j=1}^{s_j+1}(\omega_{ji}^l)^2 \tag{10}$$

$$J\left(\omega, b; x^i, y^i\right) = \frac{1}{N}\sum\nolimits_{i=1}^{N} y\ln a + (1-y)\ln(1-a) \tag{11}$$

Where $J\left(\omega, b; x^i, y^i\right)$ is the cross-entropy loss function, which can reduce the difficulty of different classification problems in unbalanced data. $\frac{\lambda}{2}\sum\nolimits_{l=1}^{n_l-1}\sum\nolimits_{i=1}^{s_i}\sum\nolimits_{j=1}^{s_j+1}(\omega_{ji}^l)^2$ is a regularization term, that is, weight attenuation. Its purpose is to reduce the weight range and prevent training from overfitting. $a = \sigma(h)$, $h = \omega * x + b$. The activation function Sigmoid and its derivatives are shown in Eqs. (12) and (13):

$$\sigma(h) = \frac{1}{1+e^{-h}} \tag{12}$$

$$\sigma'(h) = \frac{e^{-h}}{(1+e^{-h})^2} = \sigma(h)\left(1-\sigma'(h)\right) \tag{13}$$

The cross-entropy derivation of weights and biases is as follows:

$$\frac{\partial J\left(\omega, b; x^i, y^i\right)}{\partial \omega_j} = \frac{1}{N}\sum\nolimits_{i=1}^{N}\frac{\sigma'_{\omega_j}(h)x_j}{\sigma(h)(1-\sigma(h))}(\sigma(h)-y) \tag{14}$$

$$\frac{\partial J\left(\omega, b; x^i, y^i\right)}{\partial b_j} = \frac{1}{N}\sum\nolimits_{i=1}^{N}\frac{\sigma'_{b_j}(h)}{\sigma(h)(1-\sigma(h))}(\sigma(h)-y) \tag{15}$$

By substituting into Eq. (13), it can be obtained that:

$$\frac{\partial J\left(\omega, b; x^i, y^i\right)}{\partial \omega_j} = \frac{1}{N}\sum\nolimits_{i=1}^{N} x_j(\sigma(h)-y) \tag{16}$$

$$\frac{\partial J\left(\omega, b; x^i, y^i\right)}{\partial b_j} = \frac{1}{N}\sum\nolimits_{i=1}^{N}(\sigma(h)-y) \tag{17}$$

In this paper, Adam optimization algorithm is used to update the weight and bias. The algorithm is shown in Algorithm 1.

Algorithm 1. Adam algorithm

Require: α: Step size
Require: β_1, $\beta_2 \in [0,1)$: Exponential decay rates for the moment estimates
Require: $f(\theta)$ Stochastic objective function with parameters θ
Require: θ_0: Initial parameter vector
$m_0 \leftarrow 0$ (Initialize 1st moment vector)
$v_0 \leftarrow 0$ (Initialize 2nd moment vector)
$t \leftarrow 0$ (Initialize time step)
while θ_t not converged do
$t \leftarrow t + 1$
$g_t \leftarrow \nabla_\theta f_t(\theta_{t-1})$ (Get gradients w.r.t. stochastic objective at time step t)
$m_t \leftarrow \beta_1 \cdot m_{t-1} + (1 - \beta_1) \cdot g_t$ (Update biased first moment estimate)
$v_t \leftarrow \beta_2 v_{t-1} + (1 - \beta_2) \cdot g_t^2$ (Update biased second raw moment estimate)
$\widehat{m}_t \leftarrow m_t/(1 - \beta_1^t)$ (Compute bias-corrected first moment estimate)
$\widehat{v}_t \leftarrow v_t/(1 - \beta_2^t)$ (Compute bias-corrected second raw moment estimate)
$\theta_t \leftarrow \theta_{t-1} - \alpha \cdot \widehat{m}_t/(\sqrt{\widehat{v}_t} + \varepsilon)$ (Update parameters)
end while
return θ_t (Resulting parameters)

In general, $\alpha = 0.001$, $\beta_1 = 0.9$, $\beta_2 = 0.999$, $\varepsilon = 10^{-8}$.

4 Experimental Data and Evaluation Index

4.1 Experimental Data Set

In this paper, the NSL-KDD [18] data set is selected as the experimental benchmark data. It is an optimized version of KDDCup99, which solves the problem of redundant data in the KDDCup99 data set. Its original training set KDDTrain contains 125,973 pieces of data, and the original test set KDDTest contains 22,544 pieces of data. This article uses 25,192 pieces of data of KDDTrain +20% as experimental data. Each row of data in the data set has 41 characteristic attributes and 1 label attribute, which mainly include 4 types of attacks: Dos (Denial of Service Attack), Probe (Port Vulnerability Scanning Attack), R2L (Remote Illegal Access Attack), U2R (unauthorized access attack).

4.2 Data Preprocessing

The NSL-KDD data set contains 41 feature attributes, including symbolic features (tcp, udp, icmp, …) and numerical features. The data needs to be standardized before it can be applied to the detection algorithm.

Character Data Mapping Numeric Data

"0, udp, ftp_data, SF, 491, 0, 0, 0, 0, 0, 0, 0, 0, 0, 0, 0, 0, 0, 0, 0, 0, 0, 2, 2, 0, 0, 0, 0, 1, 0, 0, 150, 25, 0.17, 0.03, 0.17, 0, 0, 0, 0.05, 0, Normal" is a piece of data in the data set. The analysis shows that the 2, 3 and 4 dimensional values are character types and need to be converted to numeric types, for example, the second dimension has three types (tcp, udp, icmp), and the third dimension has ('auth', 'bgp', 'courier', etc.) 70 types, the 4th dimension has 11 types ('OTH', 'REJ', 'RSTO', etc.), which are processed according to one-hot encoding, and finally 41-dimension is converted into 122-dimension attribute, one-hot encoding is shown in the Fig. 3 shown.

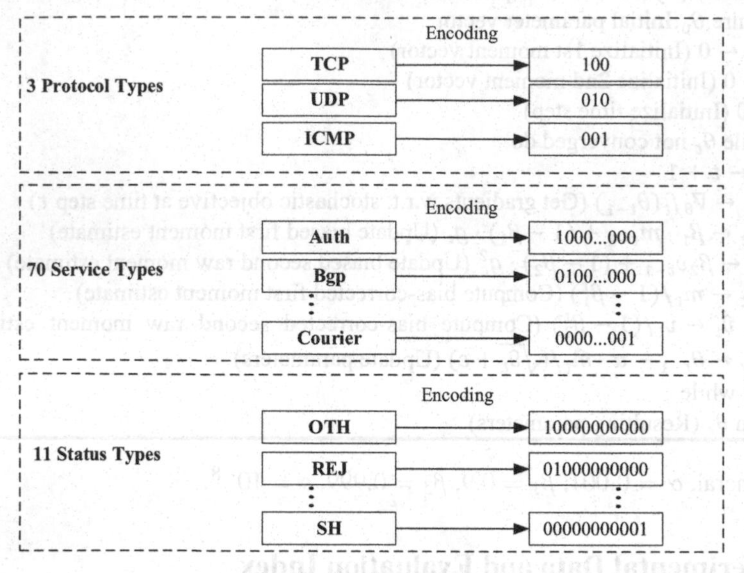

Fig. 3. One-hot encoding

Numerical Normalization. In feature vectors, different features generally have different dimensions and magnitudes. This situation will cause large differences in the size of the different feature values of each sample, which will affect the performance of the model. Features with a large order of magnitude will greatly affect the results of the model classification. Therefore, it is necessary to adopt Min-Max standardized processing to eliminate the influence of different orders of magnitude on the experimental results.

$$X_{normal} = \frac{x - x_{min}}{x_{max} - x_{min}} \tag{18}$$

Where x represents the original value of the sample feature, x_{min}, x_{max} represents the minimum and maximum value of the data, and X_{normal} represents the new characteristic value after normalization of each data.

4.3 Evaluation Index

This paper uses a confusion matrix to measure the experimental results, as shown in Table 1.

The evaluation indexes are as follows:

Accuracy (ACC): The ratio of the number of correctly classified samples to the total number of samples.

$$ACC = \frac{TP + TN}{TP + TN + FP + FN} \tag{19}$$

Table 1. Intrusion detection confusion matrix

Actual	Variable value	
	Normal	Attack
Normal	TN	FP
Attack	FN	TP

Precision (P): The data correctly judged as intrusion/normal accounts for the total number of data predicted as intrusion/normal.

$$P = \frac{TP}{TP + FP} \tag{20}$$

Detection rate/Recall rate (Recall, R): The data correctly judged as intrusion/normal accounted for the total number of intrusion/normal data.

$$R = \frac{TP}{TP + FN} \tag{21}$$

False Alarm Rate (FAR): The number of normal data predicted incorrectly accounts for the total number of normal data.

$$FAR = \frac{FP}{FP + TN} \tag{22}$$

F1-Score: This indicator is the harmonic average of Precision and Recall.

$$F1 - Score = \frac{2 \times P \times R}{P + R} = \frac{2 \times TP}{2 \times TP + FP + FN} \tag{23}$$

5 Experiment and Result Analysis

In order to verify the advantages of the intrusion detection method proposed in this paper under the measurement automation system, the intrusion detection model of this paper is simulated, and evaluation indicators are designed to test the performance.

5.1 Experimental Environment and Parameter Selection

This experiment uses Tensorflow to carry out the experiment simulation and chooses the Python programming language. The computer hardware configuration is Inter(R) Core(TM)i7-6700CPU@2.60 GHz processor, 16 GB memory, and the operating system is 64-bit Windows10. The main parameter variables in the model include convolutional autoencoding network structure parameters, learning rate, connection probability, and training times. The specific values of the parameters are shown in Table 2.

Table 2. Experimental variable parameters

Variable name	Variable value
Input layer	Input = (11,11,1)
Convolutional layer 1	Conv1 = (3,3,32), ReLU
Convolutional layer 2	Conv2 = (3,3,64), ReLU
Pool layer 1	Pool1 = Max pooling
Convolutional layer 3	Conv3 = (3,3,64), ReLU
Convolutional layer 4	Conv4 = (3,3,128), ReLU
Pool layer 2	Pool2 = Max pooling
Upsampling layer	Upsample = (6,6)
Convolutional layer 5	Conv5 = (3,3,32), ReLU
Convolutional layer 6	Conv6 = (3,3,64), ReLU
Pool layer 3	Pool3 = Max pooling
Convolutional layer 7	Conv7 = (3,3,64), ReLU
Convolutional layer 8	Conv8 = (3,3,128), ReLU
Pool layer 4	Pool2 = Max pooling
Full connection layer	Number of nodes 512, ReLU
Output layer	Number of nodes 5, Softmax
Dropout	0.6
Learning rate	0.001

5.2 Result Analysis

In order to evaluate the performance of the model in this paper, the model in this paper is compared with the classic network model NN [19], SVM [19], and the improved convolutional neural network model in literature [20]. The results are shown in Table 3 respectively. Figure 4 is a comparison chart of the accuracy of each type of attack in different models.

Table 3. Results of comparison with other models

Model	Accuracy	Recall	FAR	F1-score
This paper	97.25	96.77	0.97	96.90
NN [14]	85.66	84.18	2.31	84.52
SVM [14]	87.62	86.33	1.96	86.75
ICNN [15]	93.18	92.89	1.16	92.91

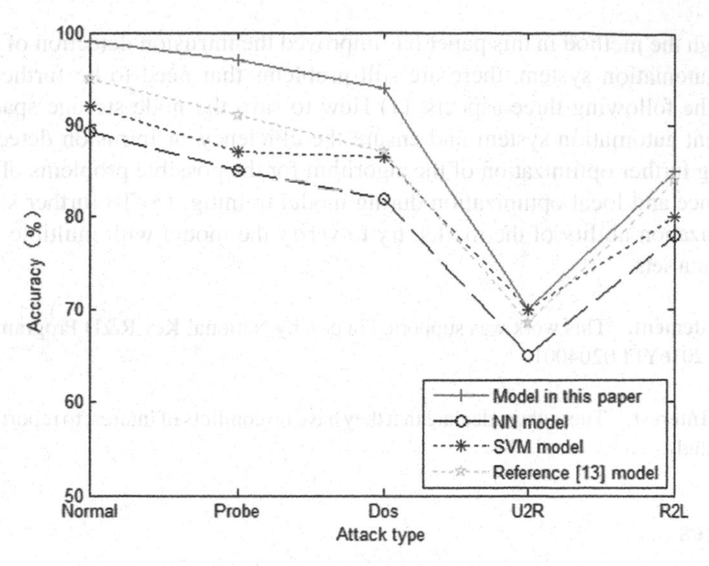

Fig. 4. Comparison of accuracy of various attacks

It can be seen from Table 3 that the accuracy of the model proposed in this paper is 11.59% higher than the NN model, 9.63% higher than the SVM model, 4.07% higher than the improved convolutional neural network model in [13], and the detection rate is respectively improved by 12.59%, 10.44% and 3.88%, it is also better than other models in false alarm rate and F1-Score performance. It can be seen from Fig. 4 that the model proposed in this paper is significantly better than the NN model, SVM model and the model in the literature [13] in the recognition of Normal, Probe, and Dos attacks, and there has also been a slight improvement in the identification of U2R and R2L attacks. In summary, it can be concluded that the model proposed in this paper combining the characteristics of convolutional neural network and autoencoding network can be well applied to the intrusion detection system of the measurement automation system, and has good classification and detection performance.

6 Conclusion

At present, in view of the problems of low recognition efficiency, severe feature loss, and poor adaptive ability in the intrusion detection technology of measurement automation systems, this paper proposes an intrusion detection model based on stacked convolutional denoising autoencoding network. Compared with traditional models, this model can learn the internal characteristics of data more fully. The model in this paper uses Dropout and regularization to avoid the occurrence of over-fitting, and uses Adam to optimize the reconstruction error, speed up the convergence speed, and avoid local optimization. Compared with other models, the accuracy and detection rate of the model proposed in this paper are significantly improved, reaching 97.25% and 96.77% respectively, and the false alarm rate is also slightly reduced.

Although the method in this paper has improved the intrusion detection of the measurement automation system, there are still problems that need to be further solved, mainly in the following three aspects: (1) How to save the node storage space in the measurement automation system and ensure the efficiency of intrusion detection; (2) Considering further optimization of the algorithm for the possible problems of gradient disappearance and local optimization during model training; (3) To further strengthen the generalization ability of the model, try to verify the model with multiple intrusion detection data sets.

Funding Statement. This work was supported in part by National Key R&D Program of China under Grant 2016YFF0204001.

Conflicts of Interest. The authors declare that they have no conflicts of interest to report regarding the present study.

References

1. Farhangi, H.: The path of the smart grid. IEEE Power Energy Mag. **8**(1), 18–28 (2010)
2. Gungor, V.C., Sahin, D., Kocak, T., et al.: Smart grid technologies: communication technologies and standards. IEEE Trans. Indust. Inf. **7**(4), 529–539 (2011)
3. Wang, Y., Yang, J.F., Shen, Y.H.: Design and application of metering automation system in large power supply enterprise. Electr. Measure. Instrum. **48**(011), 63–66 (2011)
4. Wu, S.C., Zhou, S.L.: Design and development of smart diagnosis system for station faults in metrology automation. GUANGDONG Electric Power **24**(8), 59–62 (2011)
5. Tang, Y.Q., Li, W., Zhang, L.J.: Endogenous secure communication based on characteristics of wireless channel. Radio Commun. Technol. **46**(274)(02), 31–39 (2020)
6. Allen, J.H., Christie, A., Fithen, W., Willke, B.: State of the Practice of Intrusion Detection Technologies. Carnegie Mellon Software Engineering Institute (2000)
7. Zhang, J.S., Liang, S.J., Zuo, J.L.: Research on Internet of Things intrusion detectionby optimizing SVM using Grey Wolf Optimization algorithm. Inf. Technol. Netw. Secur. **39**(10), 44–48 (2020)
8. Liu, K.K., Xie, F., Guo, X.X.: Cloud computing environment intrusion detection technology based on bp neural network. Comput. Dig. Eng. **12**, 141–145 (2014)
9. Liu, F.F.: Network intrusion detection based on ant colony optimization algorithm selecting parameters of neural network. Mod. Electr. Tech. **21**, 80–83 (2017)
10. Shon, T., Moon, J.: A hybrid machine learning approach to network anomaly detection. Inf. Sci. **177**(18), 3799–3821 (2007)
11. Zhao, F.: Detection method of LSSVM network intrusion based on hybrid kernel function. Mod. Elect. Tech. **38**(21), 96–99 (2015)
12. Hussain, J., Lalmuanawma, S., Chhakchhuak, L.: A two-stage hybrid classification technique for network intrusion detection system. Int. J. Comput. Intell. Syst. **9**(5), 863–875 (2016)
13. Erfani, S.M., Rajasegarar, S., Karunasekera, S., Leckie, C.: High-dimensional and large-scale anomaly detection using a linear one. Pattern Recogn. **58**(C), 121–134 (2016)
14. Staudemeyer, R.C.: Applying long short-term memory re-current neural networks to intrusion detection. South Afr. Comput. J. **56**(1), 136–154 (2015)
15. Javaid, A., Niyaz, Q., Sun, W., et al.: A deep learning approach for network intrusion detection system. In: Proceedings of the 9th EAI International Conference on Bio-inspired Information and Communications Technologies, pp. 21–26. Springer, Cham (2016)

16. Khan, R.U., Zhang, X.S., Kumar, R.: Analysis of ResNet and GoogleNet models for malware detection. J. Comput. Virol. Hack. Tech. **15**, 29–37 (2018)
17. Kingma, D., Ba, J.: Adam: A Method for Stochastic Optimization. Computerence (2014)
18. Tavallaee, M., Bagheri, E., Lu, W., et al.: A detailed analysis of the KDD CUP 99 data set. In: Proceedings of the 2009 IEEE Symposium on Computational Intelligence for Security and Defense Applications, pp. 53–58. Ottawa, ON, Canada (2009)
19. Gao, N., Gao, L., Gao, Q., et al.: An intrusion detection model based on deep belief networks. In: 2014 Second International Conference on Proceedings of Advanced Cloud and Big Data (CBD), pp. 247–252, [S. l.]. IEEE (2014)
20. Hang, M.X., Chen, W., Zhang, R.J.: Abnormal flow detection based on improved one-dimensional convolutional neural network. J. Comput. Appl.

LSTM-Exploit: Intelligent Penetration Based on LSTM Tool

Ximin Wang, Luyi Huang, Junlan Zhu, Wenbo He, Zhaopeng Qin, and Ming Yuan(✉)

Jiangsu Police Institute, Nanjing 210031, China
yuanming@jspi.edu.cn

Abstract. LSTM-Exploit based on the cyclic neural network "LSTM" will analyze existing exploit rules, explore the internal relationships between payloads applicable to different systems to design and create intelligent penetration tools.

Keywords: Intelligent penetration · LSTM · Long short-term memory network · Nmap · Metasploit

1 Background Overview

According to "Cyber Attack Trends: Report For The First Half Of 2020", phishing and malware attacks increased sharply in the first half of 2020, from less than 5,000 per week in February to more than 200,000 per week in late April. In addition, in May and June, as countries began to lift anti-epidemic measures, attackers intensified their attacks related to COVID-19. Compared with March and April, at the end of June, global cyber attacks of all types increased by 34%.

Frequent network attacks have exposed more security issues: on the one hand, the attackers' targets are shifting from traditional system vulnerabilities to various mobile infection vectors; on the other hand, the trend of traditional distributed denial of service attack has slowed down, and the trend of advanced persistent threat attack rose. Attackers often obtain possible attack points of the target through a series of organized and planned information gathering activities, and then carry out large-scale, multi-method cyber attacks, which can easily break through the target defense and directly lead to the leakage of sensitive information even system failure.

In the face of increasingly complex network environments, penetration testing has attracted much attention as an important part of product security testing.

Current mainstream penetration testing requires a lot of manual work. Manually mining system or application vulnerabilities is helpful to discover new vulnerabilities, but if the target opens multiple ports, you need to perform multiple tests on each port. The port-based testing tools are like Nmap and Metasploit, in the environment where many ports are open, after scanning with Nmap, you need to manually test in Metasploit. In the terminal of Metasploit, use the command "search" to search for the target exploit according to the service name scanned by Nmap. After the option of exploit is set, set the option of test target system through the command "set target", the command

X. Sun et al. (Eds.): ICAIS 2021, CCIS 1422, pp. 84–93, 2021.
https://doi.org/10.1007/978-3-030-78615-1_8

"show payloads" shows payloads for different protocols and systems, and the command "set" set the option of payload. Then set the other corresponding options according to the selected object. Then you need to traverse these data to test. However, due to the limitation of the target system and port service version, a lot of invalid test procedures are often generated. In view of this situation, popular artificial intelligence algorithms can be introduced for prediction to reduce unnecessary testing process.

The existing smart tools include "GyoiThon" and "Deep Exploit".

"GyoiThon" is used to implement attacks on vulnerable websites. Machine learning is embodied in the use of Naive Bayes Algorithm to predict web fingerprints and the information of servers. After obtaining the information about the website, data is directly transmitted to Metasploit for execution. It lacks the study of automatic attack process in Metasploit. The learning and prediction of data is only reflected in the information collection process. When the Metasploit command is executed, the payloads used are still judged according to ordinary conditional statements, which cannot achieve the goal of predicting correct payload from a large collection of exploits and targets. The efficiency of penetration work is still low, but can only gather information more widely.

"Deep Exploit" is a tool released at the 2018 DEFCON Hacking Conference. LSTM-Exploit learned its execution process and optimized its following problems: Deep Exploit cannot be tested against port 445 of Windows because of static file configuration problems. LSTM-Exploit optimizes such static file. Deep Exploit uses the algorithm "A3C" and a multi-threaded model to improve the prediction speed. However, it does not save the trained model, but saves the successfully tested data, so that the model needs to be retrained every time when an IP is retested. And since the results of such algorithm are often local optimal solutions, the tunnel protocol and operating system specified in the options of target payload cannot be completely affected by the entire prediction. Therefore, LSTM-Exploit uses the cyclic neural network LSTM to consider the global prediction results and saves the trained model to reduce the time loss caused by training.

2 Intelligent Penetration

2.1 Introduction to Intelligent Penetration Tool

Penetration testing is essentially a confrontation between developers and security testers. In order to allow security testers to conduct penetration testing more efficiently, the development of mainstream penetration tools mostly points to vulnerability scanners or malicious code detection tools, which lacks precise location of the target environment and automatically executing malicious code after vulnerabilities are discovered. LSTM-Exploit is based on the long and short-term memory neural network "LSTM", which realizes the fully automatic process from vulnerability discovery to malicious code execution.

In view of the common use of Nmap and Metasploit in penetration testing, LSTM-Exploit choose to integrate the two through the neural network "LSTM" to complete the penetration process from the acquisition of the information of port and services to payload execution.

LSTM-Exploit uses the built-in exploit module in Metasploit as a set of vulnerabilities. For different Metasploit versions, start the service "msfrpcd" and open the port

"55553" of the service as the application programming interface for LSTM-Exploit and Metasploit to connect and communicate.

The tool saves the connected port and address, attackable services and systems, and other information in the static file for direct reference.

The early information collection uses Nmap to facilitate the detection of host information and open ports and other information. By saving the Nmap scan results as special file, it is more convenient to read the scan information in more detail. After reading the special file information, detecting the ports for websites, and performing fingerprint verification on the website, the website template or middleware information used by the website is obtained, and all the above information is integrated and interacted with Metasploit to select the exploit.

The training data required by the "LSTM" model for prediction can be obtained in the learning mode by attacking the two target machines of "Metasploitable2" and "Metasploitable3". After obtaining the successfully attacked data, save it as the result file. Then use the training data to train the "LSTM" model, save the trained model to provide a trained "LSTM" model for the real penetration environment in the testing mode.

When the tool is in the testing mode, there is no need to interact with Metasploit to get the payload. You only need to use the collected data to predict through the existing model to get the payload. After completing a penetration test, the successful data will be updated to the result file again.

2.2 LSTM Neural Network and Penetration Testing

The LSTM neural network has *Gated Recurrent Units* that can better save and delete different data. For data with greater relevance, the impact of previous data on subsequent data can be fully considered, and this feature can be used to achieve memory of different payloads in Metasploit.

Forget Gate:

$$C_{t-1} = sigmoid(h_{t-1}, x_t) \; * \; C_{t-1} \tag{1}$$

In the forget gate, the output state of the upper layer, such as "reverse_http", will be processed with the input vector of this layer. Some features, such as the HTTP protocol, are deleted, and the method of establishing a communication tunnel like reversion is retained. The principle is as follows:

Combine the input vector x_t and the input state of the previous layer h_{t-1} to obtain the threshold vector f_t after the activation function $sigmoid(x_t, \; h_{t-1})$, and perform the dot product operation with the shared state C_{t-1}, so that the part of the previous layer state and the input of this layer that need to be discarded can be deleted.

Incoming Gate:

$$C_t = sigmoid(h_{t-1}, x_t) \; * \; tanh(h_{t-1}, x_t) + C_{t-1} \tag{2}$$

In the incoming layer, the input of this layer will be processed with the input state of the previous layer, and the content that needs to be saved in the previous layer is retained.

For example, the payload "reverse_http" can retain the HTTP protocol and the reverse rebound method. The principle is as follows:

Combine the input vector x_t and the input state of the previous layer h_{t-1} to obtain a new vector i_t after the activation function. The input vector x_t and the input state of the upper layer h_{t-1} are combined to obtain a new vector l_t after the activation function $tanh(h_{t-1}, x_t)$, which represents the part of the upper layer state and the input of this layer that need to be passed in. Do the dot product operation with l_t and i_t, and add the result to the shared state C_{t-1}, which means that the information that needs to be passed in is memorized by the shared cell state. At this time, C_{t-1} has been completed and becomes C_t, waiting for the output operation of the output gate.

Output Gate:

$$h_t = sigmoid(h_{t-1}, x_t) * tanh(C_{t-1}) \quad (3)$$

At the output gate, confirm the information that needs to be output for this prediction. For example, output the new payload that has been retained and deleted: "reverse_tcp", which means that the previous HTTP protocol is forgotten, but the reverse rebound is remembered. The principle is as follows:

Combine the input vector x_t and the input state of the previous layer h_{t-1} to obtain a new vector O_t after the activation function $sigmoid(x_t, h_{t-1})$, and perform the dot product operation on O_t and the updated C_t after the activation function $tanh(C_t)$ to confirm which information can be output, update h_{t-1} to h_t, pass h_t and C_t to repeat the operations above for the next layer.

In general, the *Gated Recurrent Units of* LSTM will memorize the necessary information, and is more inclined to this type of payload in the next prediction, and the success rate will be higher. Because the test is a single target IP, if the test subject can use a certain protocol and a certain way to establish a single connection, other types of vulnerability tests may also be applicable to this connection establishment method.

2.3 LSTM Long Short-Term Memory Network Implementation

After updating the training set file, the neural network can be trained. By dividing the different columns of the training set file, the training set data is obtained. The LSTM module chooses "relu" as the activation function, and uses the algorithm of Adaptive moment estimation back propagation to adjust the two values to reduce the loss value. LSTM-Exploit uses the LSTM model encapsulated in the library whose name is "tensorflow.keras" to implement long and short-term memory neural network construction.

Use such package to build a single-layer unidirectional neural network:

```
from tensorflow import keras


def create_network(lstm_shape):
    model = keras.models.Sequential(
        [
            keras.layers.LSTM(activation='relu',
                              input_shape=(lstm_shape,)),
            keras.layers.Dense(1)
        ]
    )

    model.compile(optimizer='adam',
                  loss='binary_crossentropy',metrics=['accuracy'])
    return model
```

3 Tool Design and Implementation

3.1 Overall Architecture

The tool architecture is shown in Fig. 1. The tool consists of three functions: gather information, explore modules, and execute testing.

In Learning mode, the first step is to use Nmap to scan the target IP and import the result into the result file. Get the open port for websites and use the Spider to crawl the web page content, and the Content Manage System and server information enabled by the website can be obtained by regular matching. And integrate with the content of result file.

The second step is to send a command "search exploit" to Metasploit to get the exploit, target and payload, and filter the matching data based on necessary parameters and the name of payload and exploit to reduce the number of data.

The third step is to send test instructions to Metasploit and return the test results. The successful test results are sorted locally into a training set and trained to obtain the "LSTM" model which is provided for further prediction in the test mode.

In Test mode. The first two steps are the same as the Learning mode. In the third step, the "LSTM" module trained in the Learning mode is used to directly obtain the payload without communicating with Metasploit to obtain all the payloads. The fourth step is to directly use the exploit and payload which are obtained by prediction for testing, save the successful result to the training set, retrain and save the model. Then save the successfully tested data to the training set file for easy viewing. In this mode, existing trained module is necessary.

Learning mode is used for early model training, and Test mode is suitable for real penetration testing environments, and can update training data after each training to improve the prediction accuracy of the next model.

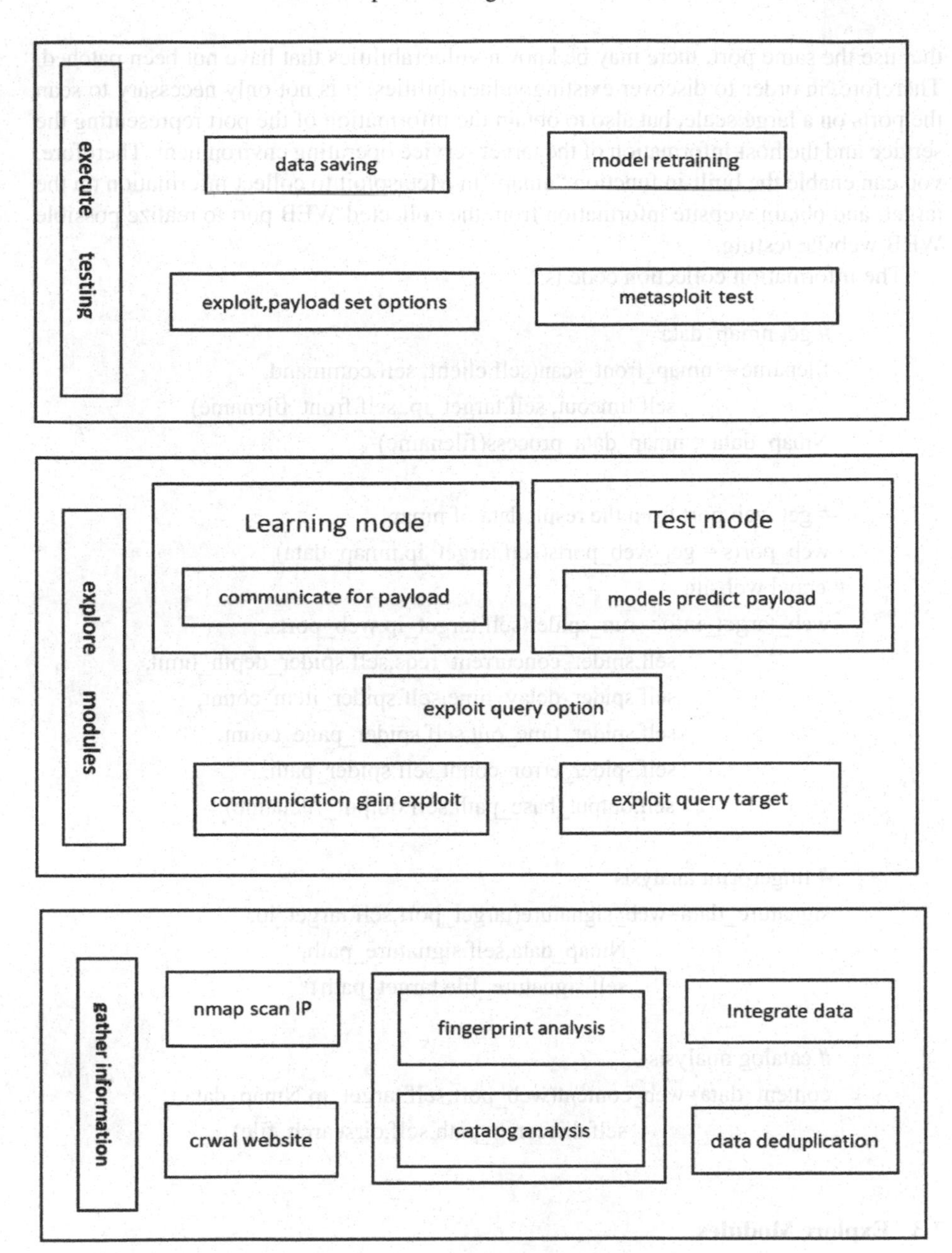

Fig. 1. Architecture diagram of tool

3.2 Gather Information

Products need to open different ports in different environments to achieve various mutually coordinated or independent functions. The open ports can be Web server and middleware ports, database ports and tunnel protocol ports. For different versions of services that use the same port, there may be known vulnerabilities that have not been patched. Therefore, in order to discover existing vulnerabilities, it is not only necessary to scan the ports on a large scale, but also to obtain the information of the port representing the service and the host information of the target service operating environment. Therefore, you can enable the built-in function "nmap" in Metasploit to collect information on the target, and obtain website information from the collected WEB port to realize possible WEB website testing.

The information collection code is:

```
# get nmap_data
filename = nmap_front_scan(self.client, self.command,
                self.timeout, self.target_ip, self.front_filename)
Nmap_data = nmap_data_process(filename)

# get web port from the result data of nmap
web_ports = get_web_ports(self.target_ip,nmap_data)
# crawl website
web_target_info = run_spider(self.target_ip,web_ports,
                self.spider_concurrent_reqs,self.spider_depth_limit,
                self.spider_delay_time,self.spider_item_count,
                self.spider_time_out,self.spider_page_count,
                self.spider_error_count,self.spider_path,
                self.output_base_path,self.output_filename)

# fingerprint analysis
signature_data=web_signature(target_port,self.target_ip,
                    Nmap_data,self.signature_path,
                    self.signature_file,target_path)

# catalog analysis
content_data=web_content(web_port,self.target_ip,Nmap_data,
                    self.dirsearch_path,self.dirsearch_file)
```

3.3 Explore Modules

According to the collected information, with product and the type of operation system as parameters, communicate with Metasploit to execute the command "search product" and the command "show targets" to get the set of exploits that can be executed by each service, and the target set corresponding to each exploit in the exploit set. And according to the the type of operation system delete the options of exploit that do not meet the type of operating system, and choose different methods to obtain the payload in the Learning mode and Test mode.

The Learning mode communicates with Metasploit. According to the options of different exploits which has been set, the code as follows:

```
# Get all the exploits and payloads in metasploit
        exploit_list = get_exploit_list(self.client)
        payload_list = get_payload_list(self.client)

        # Get the exploit based on the information collection results
        nmap_to_exploit=nmap_get_exploit(self.client,self.nmap_data,
                        exploit_list,self.allow_os_types,self.allow_services)

        # Get target and payload according to the exploit
        exploits_to_payloads=exploit_to_payloads(self.client,
                        nmap_to_exploit,exploit_list)

        # Combine the above information to get complete matching data
        execute_nmap_data,new_path_index=nmap_exploit_to_payloads(
                        self.nmap_data,exploits_to_payloads,payload_list,
                        Self.path_index)
```

In Test mode, use the existing model to predict the payload. The code is as follows:

```
# gather information and explorc modules
exploit_list = get_exploit_list(self.client)
payload_list = get_payload_list(self.client)
Nmap_to_exploit=Nmap_get_exploit(client,self.Nmap_data,
                exploit_list,self.allow_os_types,
                self.allow_services)
exploit_to_targets=get_exploit_to_targets(self.client,
                Nmap_to_exploit,exploit_list)
Nmap_exploit_targets,tmp_path_index=get_Nmap_exploit_targets(
                Nmap_to_exploit,exploit_to_targets,
                self.tunnel,self.path_index)
# predict payload
lstm_result = lstm_predict(Nmap_exploit_targets,
                self.model_path,self.model_name)
```

3.4 Execute Testing

According to the detected information and Metasploit communication to obtain the options that the test module needs to set, write the configuration which is in the local static file into the corresponding options, and test the corresponding port service by communicating with Metasploit. If the test is successful, save the successful data to the training set file which can be updated after each test. Then train the neural network by importing the training set file and update the LSTM model to improve the prediction accuracy after each test.

Code shows as below:

```
# Get the options that need to be set for each exploit and payload
exploit_options=get_options_from_etp(self.client,
                    self.execute_nmap_data,self.exploit_list,
                    self.allow_options)

# Set the obtained option to the dictionary
tmp_execute_nmap_data,tmp_path_index=
                  set_option_to_data(self.client,
                  self.execute_nmap_data,exploit_options,
                  self.path_index,self.allow_payloads,
                  self.payload_list)

# Perform the test and save the successful result to a csv file
success_data=start_attack(client,target_ip,
                  tmp_execute_nmap_data,self.path_index,
                  self.exploit_list,self.payload_list,self.local_ip)
success_data=success_to_train(success_data,exploit_list,
                  payload_list)
train_csv=save_to_csv(success_data,self.train_csv_path,
                  self.train_csv_filename,payload_list)

# Train module
model_file=train_new_model(train_csv,self.epochs,
                  self.model_path,self.model_name)
```

4 Conclusion

There is a one-click penetration script "auto-pwn.rb" in the enterprise version of Metasploit. LSTM-Exploit and this scripts have the same function, that is: they can automatically penetrate, but traditional scripts use conditional statement, but LSTM-Exploit uses the LSTM model for testing, which can significantly improve efficiency, and with the increase of testing, the accuracy of prediction will also be improved. In the face of different Metasploit version, you only need to add static port information to achieve information loading, without the need to modify the code to be compatible with Metasploit's vulnerability set. In order to communicate with the Metasploit framework, the library "pymetasploit3" is used to directly access all the data built in Metasploit, which simplifies the code of communication between client and Metasploit.

However, LSTM-Exploit also has some shortcomings. This tool can only test static ports set in advance, and can only use vulnerability payloads whose parameters need to be set for testing. In the Learning mode, due to excessive access to Metasploit's open service and the difference of the execution speed of the exploit module, it is easy to cause concurrent errors and cause the service to crash. There are still some inconveniences in the process of practice. In the future, the tool still needs continuous improvement.

References

1. "Cyber Attack Trends 2020 Mid Year Report 2020" written by global network security solution provider CHKP. http://www.springer.com/lncs. Accessed 15 Sept 2020
2. GyoiThon v0.0.3-beta releases: growing penetration test tool using Machine Learning. https://secuirtyonline.info/gyoithon/
3. Deep Exploit fully automated penetration test tool. https://github.com/13o-bbr-bbq/machine_learning_security/tree/master/DeepExploit/doc/DEFCON_AIVillage_20180802.pdf
4. A python based tool that automates several different types of attacks including Metasploit's Autopwn and SQL Injection. http://securestate.com/pages/free-tools.aspx

Pig Target Detection from Image Based on Improved YOLO V3

Dan Yin, Wengsheng Tang(✉), Peng Chen, and Bo Yang

College of Information Science and Engineering, Hunan Normal University, Changsha 410000, China
tangws@hunnu.edu.cn

Abstract. Smart farming has always been one of the current research hotspots. Reflected by the behaviors and moves, the physiological conditions of pigs can be detected. The inability to detect the behaviors of pigs at scale has become the urgent issues. This makes the recognition of pigs an extremely significant problem. Building on the prior work on picture-based recognition of target detection, this paper put forward an improved YOLO V3 to detect pigs from image. To overcome the lack of pig's pictures training data, transfer learning is used. To improve the accuracy of algorithm, attention mechanism is introduced into YOLO V3. Results show the algorithm we exploited can efficiently complete the task for pig real-time detection. Compared with the classical YOLO V3, the improved YOLO V3 has better metrics on precision, recall, F1 score and average precision. The improved model achieves result: 94.12% AP. The result is encouraging enough to make people collect more labeled pig's picture data to improve the generalization capability of the algorithm.

Keywords: Intelligent farming · Target detection · YOLO v3

1 Introduction

Smart pig farming is inseparable part of the application of computer vision. However, the application of computer vision technology in smart pig farming is still in exploratory stage. All we need to do is obtaining and analyzing the behavior of pigs by using of computer vision to get the current physiological state of the pig. Therefore, the fundamental task is to detect pigs from images. A fast and accurate algorithm is required.

Traditional target detection algorithms, including HOG [1] and DPM [2], etc. The algorithm can be divided into three stages [3] when carrying out target detection tasks: region selection, feature extraction and feature classification. On the one hand, in the stage of region selection, the sliding window algorithm is adopted in the traditional algorithms. However, generating a large number of redundant boxes, this algorithm has a poor performance. In the stage of feature extraction, feature extractor is manually designed with low robustness and low quality. On the other hand, the support vector machine is used in traditional algorithms for feature classification. Unfortunately, the

X. Sun et al. (Eds.): ICAIS 2021, CCIS 1422, pp. 94–104, 2021.
https://doi.org/10.1007/978-3-030-78615-1_9

algorithm has a large amount of computation in the classification of multidimensional data, reducing the accuracy of the algorithm.

In recent years, the development of deep neural network has greatly promoted the study of target detection algorithms. Many target detection algorithms based on deep neural network are proposed. Compared with the traditional algorithms, the new algorithms have advantage over accuracy and rapidity. The performance of the neural network directly affects the accuracy of the algorithm. At present, the target detection algorithms combined with deep neural network are mainly divided into two categories [3]: (1) The First-order target detection algorithm, (2) The Second-order target detection algorithm. The Second-order target detection algorithm is divided into two steps. The first step focuses on finding the location of the target, and the second step predicts the category of target. The representative algorithm includes Fast RCNN [4], Faster RCNN [5], Mask RCNN [6], etc.; The first-order target detection algorithm mainly combines two steps of the second-order target detection algorithm, finding out the target location and recognizing the target category at the same time. The representative algorithm includes YOLO [7], YOLO V2 [8], SSD [9], YOLO V3 [10], etc.

The paper is organized as follows. Section 2 briefly compares the advantage of the second-order target detection algorithm and the first-order target detection algorithm, and introduces the application prospect. Section 3 introduces the structure of YOLO V3 and summarizes improvement methods of YOLO V3 network. An improved YOLO V3 algorithm is proposed. Section 4 compares SSD, Faster RCNN, YOLO V3, and improved YOLO V3 on the pig picture dataset, analyzing the performance of the improved YOLO V3. We conclude in Sect. 5.

2 Related Work

As a kind of information carrier, rich information is contained in images. How to acquire the information from image is the research hotspot of computer vision. On the other hand, as one of the fields of computer vision, many fields such as image classification, image segmentation and target tracking all rely on target detection. Therefore, target detection occupies an important position in computer version. At present, mainstream target detection algorithms can be divided into two categories: first-order target detection algorithms and second-order target detection algorithms. The first-order target detection algorithm has better real-time performance, while the second-order target detection algorithm has higher accuracy.

2.1 Second-Order Target Detection Algorithms

At present, combined with deep neural network, target detection has been widely applied in the fields of driver-assistance system [12], robotics, intelligent monitoring of industrial Internet of Things [13]. In the target detection algorithm, Faster RCNN is widely used in the second-order target detection algorithm because of its high target recognition accuracy. Li H Y [30] proposes an improved Faster RCNN network to detect pedestrians around the vehicle. On the INRIA data set, the improved network detection accuracy reaches 88.8%. Xue D X [31] raises an improved artificial neural network recognition model IHCDM based on Faster RCNN. Experimental results show that the improved network recognition accuracy reaches 99.5%, which is 0.5% higher than that before the improvement. Bai J [32] uses the Faster RCNN network to identify and detect students' classroom behavior, reflecting the student's listening state during class. The experimental detection accuracy of this network reached 76.32%.

However, Mask RCNN is more used to segment the images. It is more applied to the detection of targets in medical images. In order to automatically extract the prostate cancer focus area from the multi-parameter magnetic resonance (mp-MRI) prostate area, Huang Y F [33] proposes a new deep convolutional neural network model SE-Mask-RCNN based on Mask RCNN. Comparing with Mask RCNN, the improved network gets a better performance. To obtain a better coronary artery segmentation method, Shao K [34] raises a new network by merging geometric features into the Mask RCNN network. The average accuracy of the improved method reaches 83%.

2.2 First-Order Target Detection Algorithms

In the first-order target detection algorithm, YOLO V3 has high accuracy with maintaining good real-time. Therefore, YOLO V3 is increasingly applied to the military and civilian areas [22–26]. The YOLO V3 network application research can be divided into two categories. One is improving the YOLO V3 network for the detection of difficult targets, including small targets and blocked targets; The other is combining YOLO V3 network with other networks. Such as using YOLO V3 network to quickly locate the target in the image to make it easier for the subsequent research.

In the first of research, in order to solve the problem of low recognition rate of occlusion targets and small targets in the detection of vehicle targets, Wen H B [14] improves the data pre-processing stage based on YOLO V3. By marking the difficult sample in advance and establishing the difficult sample dataset, the average accuracy of the algorithm is improved by 17%. To improve the YOLO V3 algorithm on small target recognition accuracy, Hu Y Y [15] puts forward an improved algorithm of YOLO V3, optimizing the YOLO V3 network from the network structure. A new output layer is added to the classic network. And the receptive field of new output feature map is smaller. The feature information of medium and small targets in the image can better be found. Considering many similar patterns and lacking of feature information for identification, He D J and Liu S Z [16, 17] propose an improved YOLO V3 algorithm for the detection of individual cows. The accuracy of improved network reached 95.91%.

In the second type of research, in order to find out whether there is a missing nut on the high-speed railway contact network in time, Chen Q [18] combines YOLO V3 network with Senet network. Experimental data shows that the recognition accuracy of algorithm reaches 88.24%. To solve the problem of low accuracy in the classification task of fine-grained images carried out by traditional recognition algorithm, Li S Y [19] combines migration learning with YOLO V3. The experiment proves that the accuracy of the improved network model is 68.45%, which is 1.05% higher than that before segmentation. In order to mine the multi-scale feature information of clothing images, Wang Z W [20] proposes the concept of three-scale image and the FSF formula. Firstly, the YOLO V3 algorithm was used to quickly extract the three-level scale image from the garment image, reducing the influence of background and other interference information.

Secondly, the convolutional neural network is used to extract the features of the three-level size image, obtaining three-level feature information by using the FSF formula to fuse the feature information of the image on different scales. To deal with the occlusion problem in the current multi-target tracking process, Li X C [21] proposes a new multi-target pedestrian tracking algorithm on the basis of the YOLO V3 algorithm.

3 Methodology

The first-order target detection algorithm obtains to the target location information and target category at the same time. Faster detection speed would enable it to be widely used in tasks with higher requirement in real-time. By using many current excellent detection frameworks, including residual network and feature fusion, YOLO V3 network gets better accuracy. It conducts a lot of research [10] on how to improve the detection accuracy of the network with ensuring real-time performance. Great real-time performance and well accuracy would enable us to choose the YOLO v3 algorithm to experiment on pig images dataset.

3.1 Network Structure

YOLO V3 [10] network adopts the network structure of Darknet53, containing 53 convolutional layers. Darknet53 borrows from the idea of residual neural network [11], composed of 5 residual structures. To avoid the gradient disappearance caused by excessively deep network, each residual structure is composed of multiple residual units. The residual unit is used to increase the network depth. On the other hand, the convolution kernel with the step size of 2 is used for convolution instead of pooling operation in to reduce the feature map's size. YOLO V3 network carried out a total of 5 times of subsampled on the input image, outputing the feature map containing 3 scales after the last 3 times of subsampled. Then, through upsampling and channel splicing-together operation, the features of deep and shallow layers were fused with each other. The feature map of 3 sizes was finally output for target detection. Due to the small size of the feature map in the deep layer and the large receptive field, it is more conducive to detect large-size objects. While the feature map in the shallow layer, on the contrary, is better at detecting small-size objects. Finally, since Softmax function can suppress each other in output prediction of multiple categories. Logistic function classifiers are independent of each other. YOLO V3 network uses Logistic function instead of Softmax function to process prediction scores of categories, realizing the task of multi-category prediction.

3.2 Attention Mechanism

At present, the attention mechanism can be divided into two categories: the channel attention mechanism in the SE network and the Convolutional Block Attention Module (CBAM) attention mechanism. In the channel attention mechanism, the feature map which extracted from the backbone network as the input. The global average pooling is used. And the global feature compression matrix of the feature map is obtained. Then, the feature weights of the feature map are acquired through the two full connection layers, and the weighted feature map is output. By using the attention model, the characteristic response value of each channel is adaptively calibrated. The weight of each characteristic channel is automatically acquired. According to the weight of the channel, useful features will be enhanced, unimportant features will be suppressed. The accuracy of the network detection of the target area is improved. The CBAM Channel Attention Module consists of two sub-modules: Channel Attention Module (CAM) and Spartial Attention Module (SAM). The difference between the CAM structure and the SE network attention mechanism is that a parallel maximum pooling operation is added to the input feature layer. SAM structure takes the feature map output by CAM module as the input. First, a global maximum pooling and a global average pooling are made on the feature map, two results are splitted on the channel. Then through a convolution operation, the dimension is reduced to one channel. Spatial attention feature weight is generated by Sigmoid. Multiplying the feature weight and the input feature of the module obtains the weighted feature map in finally.

In YOLO V3 algorithm, the extracted convolution feature will not be weighted. Each region of the entire feature map is treated equally. However, in real life scenes, there is often complex and rich contextual information around the object to be detected in the diagram. Weighting the features of the target area can make it better positioned on the features to be detected. The network performance is improved. By using the attention mechanism of SENet network, Liu Dan proposed the Gaussian-Yolo V3 [27] network. Xu C J [28] also proposed to add attention mechanism in YOLO V3 network, with adding each attention mechanism into the YOLO V3 network for comparative experiments. The experimental results show that best-performing combination sequence is the method in which the channel attention module is directly connected to the spatial attention module in the Convolutional Block Attention Module (CBAM).

3.3 Transfer Learning

To get better training results with a small amount of data, transfer learning is used on all networks. Definition of transfer learning [11]: given source domain Ds and corresponding task Ts, given target domain Dt and corresponding task Tt. Transfer learning is when Ds $\neq$ Dt or Ts $\neq$ Tt, use the knowledge in Ds and Ts to help learn the prediction function ft () on Dt. The advantages of transfer learning: firstly, the use of transfer learning can save a lot of training time and a lot of computing resources. Secondly, the network can also achieve a relatively good accuracy rate with small data sets. At the same time, it can solve the problem of network underfitting caused by the small amount of data. This paper uses the pre-trained network on the VOC 2007 data set for transfer learning.

3.4 Improved YOLO V3

Due to the small amount of training pictures, transfer learning is used to improve the performance of the network. The network weight files obtained from the training of the network on the VOC 2007 data set is used to initialize the network parameters. In the first few rounds of training, the parameters of the backbone feature extraction network are frozen, only the feature pyramid network layer is trained. Then the entire network is unfrozen and trained. On the other hand, we add SE network attention mechanism to the classic YOLO v3 model, getting a better network. A new and improved YOLO v3 network was finally formed. The architecture of pig detection is shown in Fig. 1.

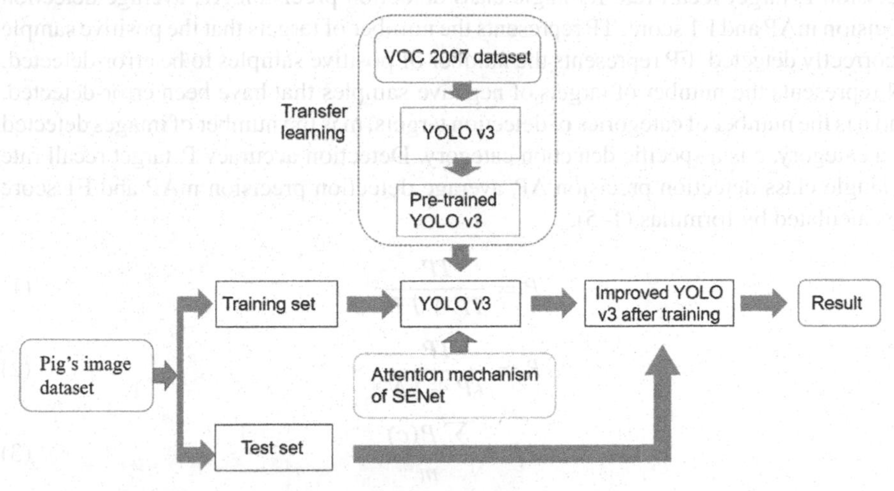

Fig. 1. The architecture of pig detection

4 Experiment

In the field of target detection, the most important thing is the accuracy and the speed of detection target. By analyzing the advantages of the improved network model, the improved network model is compared with some classical networks. The effectiveness of the improved network model is proved.

4.1 Experiment Setup

The experimental computer hardware configuration: dual-core Intel (R) Core (TM) i7-9700 CPU @ 3.00 GHz, memory size is 16 GB, graphics card is RTX 2060. The software system configuration: CUDA 10.0, CUDNN 7.4, Python3.6 programming language.

Building pig picture dataset by the Internet. Two first-order target detection algorithms are selected: SSD and YOLO V3. A second-order target detection algorithm is selected: Faster RCNN. And an improved algorithm is selected: YOLO V3 network with added attention mechanisms. SSD, Faster RCNN, YOLO v3 and the improved YOLO V3 were constructed respectively. The performance of all networks is compared on the pig picture dataset. Small dataset, transfer learning is used. The network is initialized with weight files which obtained by pre-training on the VOC 2007 data set. Perform comparison is experimented on SSD, Faster RCNN, and YOLO v3 networks to analyze the performance of each network. YOLO v3 and the improved YOLO v3 are compared to analyze the feasibility and effectiveness of the improvement of the YOLO v3 network on the pig picture dataset.

4.2 Performance Indicators

In the field of target detection, evaluation criteria are often used, including detection precision P, target recall rate R, single class detection precision AP, average detection precision mAP and F1 score. TP represents the number of targets that the positive sample is correctly detected. FP represents the number of positive samples to be error-detected. FN represents the number of targets of negative samples that have been error-detected. And n is the number of categories of detection targets, m is the number of images detected by a category, c is a specific detection category. Detection accuracy P, target recall rate R, single class detection precision AP, average detection precision mAP and F1 score are calculated by formulas (1–5).

$$P = \frac{TP}{TP + FP} \tag{1}$$

$$R = \frac{TP}{TP + FN} \tag{2}$$

$$AP = \frac{\sum P(c)}{m} \tag{3}$$

$$mAP = \frac{\sum AP(c)}{n} \tag{4}$$

$$F1 = \frac{2 \times P \times R}{P + R} \tag{5}$$

4.3 Experimental Results

The threshold score of four networks is 0.5. The experimental performances of each network are shown in Table 1 In the first-order target detection algorithm, the performance indicators of YOLO V3 network are superior to SSD network. The average precision of YOLO V3 network is 4.09% higher than SSD network. Compared with the first-order detection algorithm, the second-order target detection algorithm Faster RCNN has higher recall rate. Its average precision is only 0.87% less than that of YOLO V3 network. The P-R curves of each network are shown in Fig. 2, Fig. 3, and Fig. 4.

Because the first-order target detection algorithm has higher real-time performance, and the average precision of the YOLO v3 network is slightly higher than Faster RCNN. Therefore, the YOLO v3 network was selected as the main target detection network to identify pigs.

After adding attention mechanism, the network's performance has been improved. The improved network precision is 0.9% higher than before, and the recall is 1.75% higher than before. The average precision of the improved network has been improved by 1.16%. The Table 2 shows the performances. The Precision curves of networks are shown in Fig. 5, Fig. 6. The Recall curves of each network are shown in Fig. 7, and Fig. 8.

Adding the attention mechanism, the network will suppress the information of non-target in the graph, reduce the weight of the non-target image in the feature map. The

Table 1. The experimental performance of each network.

	SSD	Faster RCNN	YOLO v3
F1 score	0.85	0.79	0.89
AP	88.87%	92.09%	92.96%

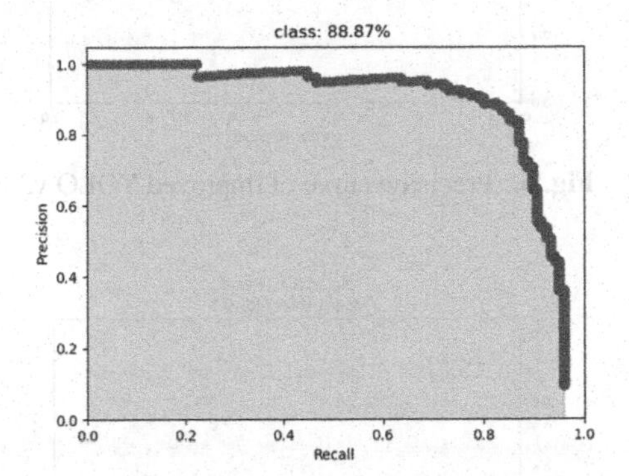

Fig. 2. P-R curve of SSD

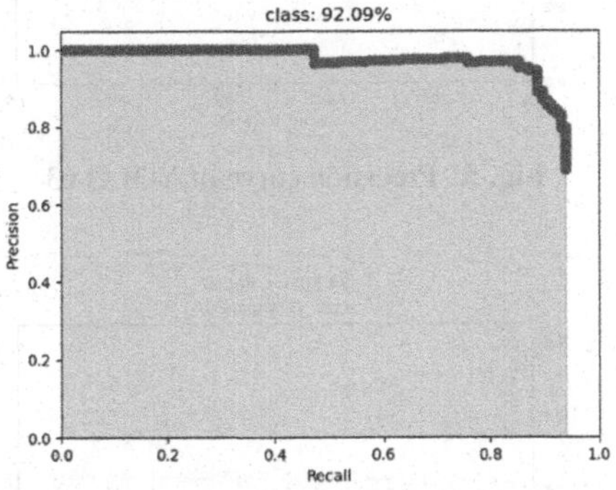

Fig. 3. P-R curve of Faster RCNN

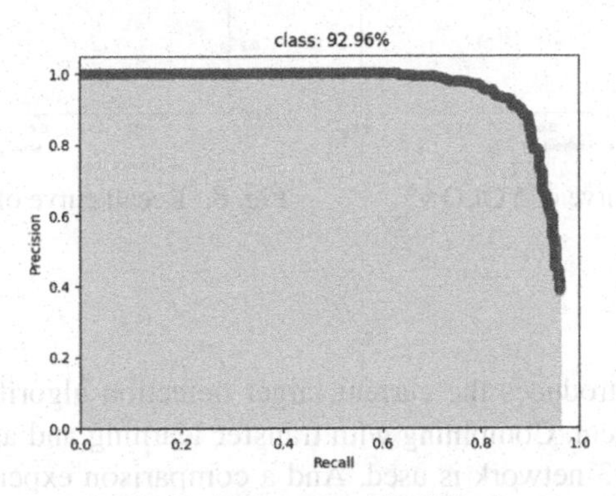

Fig. 4. P-R curve of YOLO v3

Table 2. The experimental performances of YOLO v3 and improved YOLO v3.

	Precision	Recall	F1 score	AP
YOLO v3	93.11%	84.50%	0.89	92.96%
Improved YOLO V3	94.01%	86.25%	0.90	94.12%

weight of the target image when extracting the target graphic features is improved. Detecting the target object and extracting the category information of the target object is easier. The overall network performance has improved.

93.11% = Precision
score_threhold=0.5
Precision
Score_Threhold

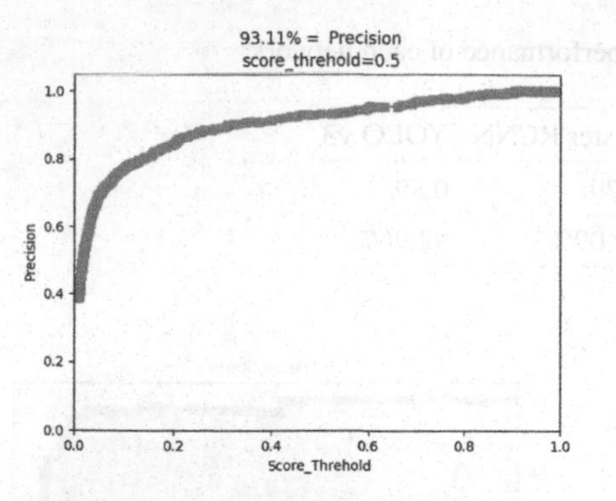

Fig. 5. Precision curve of YOLO v3

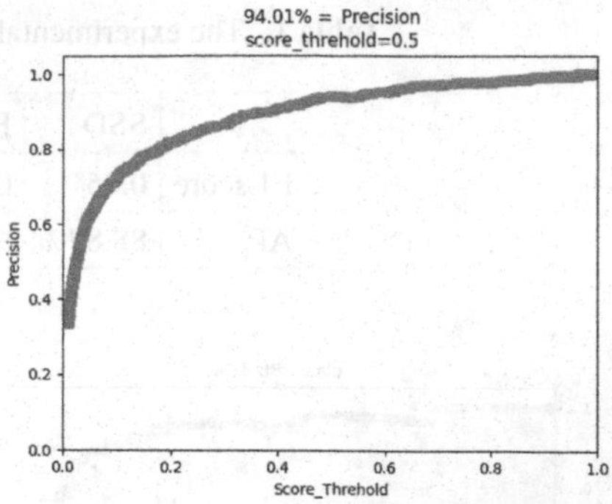

94.01% = Precision
score_threhold=0.5
Precision
Score_Threhold

Fig. 6. Precision curve of improved YOLO v3

84.50% = Recall
score_threhold=0.5
Recall
Score_Threhold

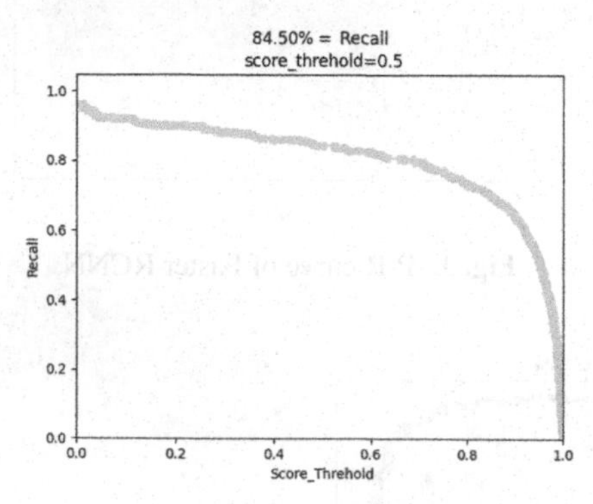

Fig. 7. Recall curve of YOLO v3

86.25% = Recall
score_threhold=0.5
Recall
Score_Threhold

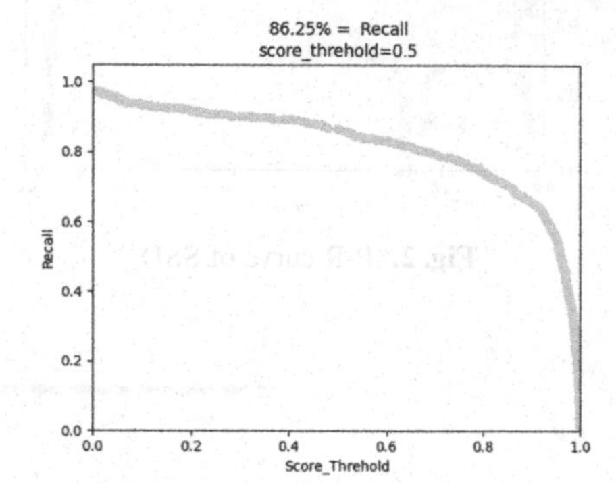

Fig. 8. Recall curve of improved YOLO v3

5 Conclusions

This article briefly introduces the current target detection algorithms, and select the YOLO V3 to experiment. Combining with transfer learning and attention mechanism, an improved YOLO v3 network is used. And a comparison experiment is carried out to analyze the performance of the YOLO v3 network and the improved YOLO v3 network. The result shows that the improved model has better accuracy. The feasibility and effectiveness of the improvement of YOLO v3 network on the pig picture dataset is proved. However, there are still some problems in current.

(1) After adding the attention mechanism, the network performance is not improved very much.

This problem may be caused by the small dataset. Although the transfer learning is used to improve network performance. Due to the small dataset, both networks may well learn the target feature information. The less performance improvement is resulted. We hope to collect more pig images and build a larger dataset in follow-up experiments.

(2) How to continue to improve the YOLO V3 network.

Firstly, new clustering function and loss function will be proposed to get more appropriate prediction box size and numbers, improving the performance of the network. Secondly, whether the new network module with better performance can be proposed to reduce the loss of data in network transmission; Thirdly, how to combine the advantages of the existing network module with YOLO V3 network.

(3) How to better apply computer vision technology in smart pig farming.

At present, most applications of computer vision technology in smart pig farming are still in the exploratory stage [11]. This experiment only uses the target detection algorithm to detect pig images. Behavior analysis, weight estimation, and disease prediction of pigs all require the application of computer vision technology. Therefore, exploring the application of computer vision technology in these aspects should be emphasized in following research.

Acknowledgements. This work is supported by some people and company. I would like to thank my teachers, Mr. Tang and Mr. Yang, for their guidance and help in my thesis writing. Finally, thanks for the help of the company—"Hunan Baodong agriculture and animal husbandry company".

Funding Statement. This work was supported in part by the Hunan Province's Strategic and Emerging Industrial Projects under Grant 2018GK4035, in part by the Hunan Province's Changsha Zhuzhou Xiangtan National Independent Innovation Demonstration Zone projects under Grant 2017XK2058.

Conflicts of Interest. The authors declare that they have no conflicts of interest to report regarding the present study.

References

1. Hu, Z.P., Zhou, S.: Research on curvature - HOG target detection algorithm. J. Sig. Process, 36–41 (2013)
2. Yu, X.G.: Improved DPM and its application in pedestrian detection. National University of Defense Technology (2013)
3. J. H, G. Z.: A review of target detection algorithms for deep convolutional neural networks. Comput. Eng. Appl. **56**(17), 12–23 (2020)
4. Girshick R.: Fast R-CNN. Computer Science (2015)
5. Ren, S., He, K., Girshick, R., et al.: Faster R-CNN: towards real-time object detection with region proposal networks. IEEE Trans. Pattern Anal. Mach. Intell. **39**(6), 1137–1149 (2017)
6. He, K., Gkioxari, G., Dollar, P., Girshick, R.: Mask R-CNN. IEEE Trans. Pattern Anal. Mach. Intell. **42**(2), 386–397 (2020). https://doi.org/10.1109/TPAMI.2018.2844175
7. Redmon, J., Divvala, S., Girshick, R., et al.: You only look once: unified, real-time object detection. In: Computer Vision & Pattern Recognition, pp. 799–788. IEEE (2016)
8. Redmon, J., Farhadi, A.: YOLO9000: better, faster, stronger. In: IEEE Conference on Computer Vision & Pattern Recognition, pp. 6517–6525. IEEE (2017)
9. Liu, W., et al.: SSD: single shot multibox detector. In: Leibe, B., Matas, J., Sebe, N., Welling, M. (eds.) ECCV 2016. LNCS, vol. 9905, pp. 21–37. Springer, Cham (2016). https://doi.org/10.1007/978-3-319-46448-0_2
10. Redmon, J., Farhadi, A.: YOLOv3: an incremental improvement. arXiv e-prints (2018)

11. Qing, Y.: Computer-generated image recognition method based on CNN. J. Southwest China Normal Univ. (Nat. Sci. Edn.) **44**(05), 109–114 (2019)
12. Wang, H.L., Qi, X.L., Wu, G.S.: Research progress of target detection technology based on deep convolutional neural network. Comput. Sci. **45**(09), 11–19 (2018)
13. Lu, H.T., Zhang, Q.C.: A review of the application of deep convolutional neural networks in computer vision. Data Acquisition Process. **31**(1), 1–17 (2016)
14. Wen, H.B., Zhang, G.H.: Target detection method of driving video based on YOLO. Automob. Sci. Technol. **01**, 73–76 (2019)
15. Hu, Y.Y., Wu, X.J., Zheng, G.D.: Object detection of UAV for anti-UAV based on improved YOLO v3. In: The 38th China Control Conference, pp. 885–889 (2019)
16. He, D.J., Liu, J.M., Xiong, H.T., et al.: An individual identification method for milking cows based on improved YOLO V3 model. J. Agric. Mach. **51**(04), 250–560 (2020)
17. Liu, S.Z., Li, C.R., Liu, T.J., et al.: Cow target detection based on YOLO V3 model. J. Tarim Univ. **31**(02), 85–90 (2019)
18. Chen, Q., Liu, L., Han, R., et al: Image identification method on high speed railway contact network based on YOLO v3 and SENet. In: The 38th China Control Conference (2019)
19. Li, S.Y., Liu, Y.H., Zhang, R.F.: A method for dog identification based on transfer learning and model fusion. Intell. Comput. Appl. **9**(06), 101–106 (2019)
20. Wang, Z.W., Pu, Y.Y., Wang, X.: Multi-scale garment image precision retrieval based on multi-feature fusion. Chin. J. Comput. **4**(43), 740–752 (2020)
21. Li, X.C., Liu, X.M., Chen, X.N.: Multi-target tracking algorithm based on YOLO detection. Comput. Eng. Sci. **42**(4), 665–672 (2020)
22. Li, J., Gu, J., Huang, Z., et al.: Application research of improved YOLO V3 algorithm in PCB electronic component detection. Appl. Sci. **9**(18), 3750 (2019)
23. Wei, H.B., Zhang, R.J., Du, G.M., et al.: Vegetable recognition algorithm based on improved YOLO V3. J. Zhengzhou Univ. (Eng. Sci. Edn.) **41**(2), 7–12+31 (2020)
24. Jiao, T.C., Li, Q., Lin, M.S., et al.: Combination of anti-residual block and YOLOv3 target detection method. Transducer Microsyst. Technol. **38**(9), 144–146+156 (2019)
25. Shi, H.: Research on vehicle Detection Algorithm based on deep convolutional neural network. Hunan Normal University (2019)
26. Ju, M.R., Luo, H.B., Wang, Z.B., et al.: Improved YOLO V3 algorithm and its application in small target detection. J. Opt. **39**(7), 245–525 (2019)
27. Liu, D., Wu, Y.J., Luo, N.C., et al.: Gaussian-yolov3 target detection with embedded attention and feature interwoven modules. Comput. Appl. **40**(8), 2225–2230 (2020)
28. Xu, C.J., Wang, X.F., Yang, Y.D.: Attention-yolo: YOLO detection algorithm with Attention mechanism. Comput. Eng. Appl. **55**(6), 13–23+125 (2019)
29. Yan, J.T.: Research progress of computer vision technology in intelligent pig raising. Chin. J. Anim. Sci. **55**(12), 38–42 (2019)
30. Li, H.Y.: Pedestrian recognition method based on Faster R-CNN, pp. 23–26. China Society of Automotive Engineers (2020)
31. Xue, D.X.: Research on image recognition model of white moth based on Faster R-CNN. J. Environ. Entomol. **42**(06), 1502–1509 (2020)
32. Bai, J.: Student classroom behavior detection based on Faster R-CNN and transfer learning based on multi-channel feature fusion. J. Guangxi Normal Univ. (Nat. Sci. Edn.) **38**(05), 1–11 (2020)
33. Huang, Y.F.: SE-Mask-RCNN: multi-parameter MRI prostate cancer segmentation method. J. Zhejiang Univ. (Eng. Sci.) (2021)
34. Shao, K.: A coronary artery segmentation method based on Mask RCNN fusion geometric features. J. Taiyuan Univ. Technol. (2020)

ICPM: An Intelligent Compound Prediction Model Based on GA and GRNN

Fang Chen and Cong Zhang(✉)

School of Mathematics and Computer Science,
Wuhan Polytechnic University, Wuhan 430000, China

Abstract. In order to reduce the prediction error of the heavy metal content of farmland soil by General Regression Neural Network (GRNN), an Intelligent Compound Prediction Model (ICPM) was proposed. As the result of Genetic Algorithm optimization is good or bad, it mainly depends on whether it can guarantee the diversity of the population in the optimization process. Based on this, an Improved Genetic Algorithm (IGA) is proposed. IGA introduces the probability adjustment of the sine function transformation and the better gene replacement criterion into the Genetic Algorithm (GA). In the process of IGA's optimization, the crossover probability and mutation probability continue to increase, ensuring the continuity of the diversity of the population. ICPM is a combined forecasting model of GRNN and IGA. It optimizes the smoothing factor of GRNN through IGA. The process of repeated optimization of IGA is also the process of repeated learning of existing knowledge by GRNN. ICPM not only ensures the continuous optimization of the population, but also the diversity of the population. Combined with the simulation prediction of the content of heavy metals Cr, Cu and Pb in farmland soil in Dongxihu District of Wuhan City, it proved that ICPM has better prediction performance and better generalization performance than GRNN and other models.

Keywords: Genetic Algorithm · Generalized regression neural network · Parameter optimization · Heavy metal content prediction · Small sample prediction

1 Introduction

We are in the age of information technology. It is obvious that data is very important for us to understand the world. Data has become an irreplaceable core component of modern information. Predictive analysis of data can help us better solve practical problems. However, when encountering actual problems, we often encounter the inability to make accurate judgments on actual problems due to the inability to obtain a large amount of data. Artificial Intelligence provides us with new possibilities to solve practical problems in a small sample situation. We can use Artificial Intelligence to learn from known small sample data, and then use the learned wisdom to predict unknown data.

The traditional methods used for data prediction are Logistic Regression (LR) [1], Support Vector Machines (SVM) [2], etc. However, due to the complexity of actual

X. Sun et al. (Eds.): ICAIS 2021, CCIS 1422, pp. 105–118, 2021.
https://doi.org/10.1007/978-3-030-78615-1_10

problems and the uncertainty of characteristic variables, the effects of these methods in the actual process are not ideal. Neural Network is a mathematical model or calculation model that imitates the structure and function of biological Neural Network, and is used to estimate or approximate functions. Neural Network has excellent non-linear mapping ability and is an excellent helper of Artificial Intelligence for data prediction. Common Neural Networks used for data prediction include Back Propagation Neural Networks (BPNN) [3], Radial Basis Function Neural Network (RBFNN) [4], Wavelet Neural Network (WNN) [5] and General Regression Neural Network (GRNN) [6], etc. They have good prediction effects in practical application problems, such as wind turbine blade strain prediction [7], adjustment of bus arrival time [8], prediction of network traffic in smart substations [9] and intelligent fracturing warning model [10]. Among common Neural Networks, GRNN is a feed-forward Neural Network that does not require reverse iterative training. Therefore, compared with other common neural networks, GRNN is simpler and has a better prediction effect on small sample data.

When training GRNN, the quality of the model is closely related to the smoothing factor of the model. The selection method of smoothing factor that we often use is trial and error method, but due to the large selection range of smoothing factor, trial and error method will have more work, and its optimization result may not reach the ideal we expect value. In order to solve this problem, many Swarm Intelligence algorithms are used to optimize GRNN. Through Swarm Intelligence, GRNN can repeatedly train the model to obtain the values we expect. Common Swarm Intelligence algorithms include: Genetic Algorithm (GA) [11], Particle Swarm Optimization (PSO) [12], Artificial Fish Swarm Algorithm (AFSA) [13], etc. This paper uses GA to optimize GRNN. Genetic Algorithm takes genetic codes as individuals. Its optimization process will be towards a single extreme.

With the continuous evolution of the population, GA will develop towards a single optimization direction, and the population diversity will gradually disappear. The most important thing for the GA optimization results is to ensure the continuity of the population optimization and the diversity of the population. Many scholars have proposed improvements to GA based on this idea [14–16]. We have improved GA according to the optimization its characteristics: Through better gene replacement criteria to ensure the continuity of the population optimization; Adaptive cross-probability genetic algorithm adjustment and non-linear increase in the probability of mutation ensure the diversity of the population. We combined the Improved Genetic Algorithm (IGA) and GRNN into an Intelligent Composite Prediction Model (ICPM), and used ICPM to predict the content of soil heavy metals Cr, Cu and Pb in the Dongxihu District of Wuhan.

The main contributions of this paper are as follows: 1) Propose a model based on soil heavy metal content prediction; 2) Propose an improved method of GA; 3) Propose an Intelligent Compound Prediction Model (ICPM) combining GA and GRNN; 4) Use optimization process of GA to obtain the optimal hyperparameters of GRNN; 5) Prediction experiments of soil heavy metal content were carried out, and there is an acceptable match between the predicted value and the measured value.

The rest of this paper is organized as follows: Sect. 2 introduces the related work of current data prediction; Sect. 3 introduces the related theories of GA and GRNN; Sect. 4 improves GA and constructs ICPM; Sect. 5 uses ICPM to predict the content of three heavy metals in Dongxihu District, Wuhan; Sect. 6 summarizes the work content of this paper. Subsequent paragraphs, however, are indented.

2 Related Work

Machine Learning provides a powerful means for data prediction. Many scholars apply Machine Learning to data prediction in various industries.

Support Vector Machines (SVM), Random Forests (RF), and Bayesian optimization are all common methods used in data prediction. There are some examples: In 2019, Tang, Jinjun applied SVM to predict traffic flow [17]; In 2020, Xue, Liang applied RF for predicting shale gas production [18]; In 2016, Maurya, Abhinav applied Bayesian optimization to predict rare internal failures in manufacturing processes [19].

Neural network is a simulation that reflects the human brain biological system and has good mapping capabilities. During data training, Neural Network can learn from existing knowledge and use the learned ability to predict new knowledge. There are some examples: In 2016, Sun, Wei, M. Ye, and Y. Xu applied BPNN to predict carbon dioxide emissions[20]; in 2017, Shen, Limin, Y. Wen, and X. Li applied RBFNN to cooling Load forecasting [21]; In 2016, Agbossou applied WNN to predict wind speed [22]; In 2016, Yang, Gao applied GRNN to predict ultra-short-term wind speed [23]. Although Neural Network has a better ability to map data, the performance of the neural network is closely related to its own hyperparameter selection. Therefore, when applying Neural Networks for data prediction, we need to pay attention to the selection of hyperparameters.

Previous research results show that the basic principle of improved GA is to ensure the diversity of the population in the process of GA optimization. This paper proposes an Improved Genetic Algorithm (IGA). The improvements are as follows. After the population is updated, the individual compares before and after the gene update according to the better gene replacement criteria, and only retains the better genes. The crossover probability is changed with the fitness value. The crossover probability increases within a certain range, that is, as the number of iterations increases, the more genes that perform the crossover operation. Make the mutation probability increase nonlinearly within a certain range with the increase of the number of iterations. That is, as the number of iterations increases, execute the more genes involved in mutation operations, the greater the possibility of sustained population diversity. ICPM combines GRNN and GA to coordinate intelligent prediction. Within a certain range, ICPM has the ability to predict unknown variables with the same data characteristics.

3 System Model and Definitions

3.1 Genetic Algorithm

Genetic Algorithm (GA) [24, 25], is a highly parallel, random and adaptive global optimization algorithm based on "survival of the fittest". It expresses the solution of the problem as a "chromosome" survival process of the fittest. The solution process is through the iterative evolution of the "chromosome" group, including operations such as duplication, crossover, and mutation, until certain performance indicators and convergence conditions are met, so as to obtain the optimal or satisfactory solution to the problem.

The main algorithm steps of GA are as follows:

1) The coding expresses the problem to be solved as a chromosome or individual in the genetic space;
2) Calculate the individual fitness value and evaluate the code;
3) Selection operation to select operators with higher fitness in the population. Commonly used selection operators include: fitness ratio method, random traversal sampling method, and local selection method;
4) Crossover operation, the operation of replacing and reorganizing the partial structure of two parent individuals according to the crossover probability to generate a new individual. The crossover probability is generally a fixed value set artificially;
5) Mutation operator, which changes the genetic code of individuals in the population according to the mutation probability, which is generally a fixed value set artificially;
6) Repeat steps 3 to 5, when the conditions for stopping iteration or the maximum number of iterations are met, the iteration is stopped, and then the optimal individual of the population is output.

3.2 Generalized Regression Neural Network (GRNN)

Generalized Regression Neural Network (GRNN) is a kind of radial basis function neural network. It has strong nonlinear mapping ability, flexible network structure, high fault tolerance and robustness, and is suitable for solving nonlinear problems. Even if the number of training samples is limited, GRNN has excellent performance in terms of approximation ability and fast learning speed, so it is very suitable for small samples and belongs to the study of prediction. GRNN is improved on the basis of the radial basis function neural network, and the generalized regression neural network is established through the radial basis function neuron and linear neuron. GRNN consists of four layers, which are input layer, mode layer, summing layer and output layer. The network structure diagram is shown in the Fig. 1 below [26, 27].

(1) Input Layer
The input layer parameters of this article include: longitude, latitude, height, crop type.
(2) Pattern Layer
The number of neurons in the pattern layer is the number of neurons in the input layer, and the transfer parameters of the pattern layer are:

$$P_i = \exp\left(-\frac{(X - X_i)^T (X - X_i)}{2\sigma^2}\right)(i = 1, 2, \cdots, n) \tag{1}$$

In the formula, X, X_i are test sample vectors and training sample vectors; n is the number of input layer variables; P_i is the output value of the model layer; σ is the smoothing

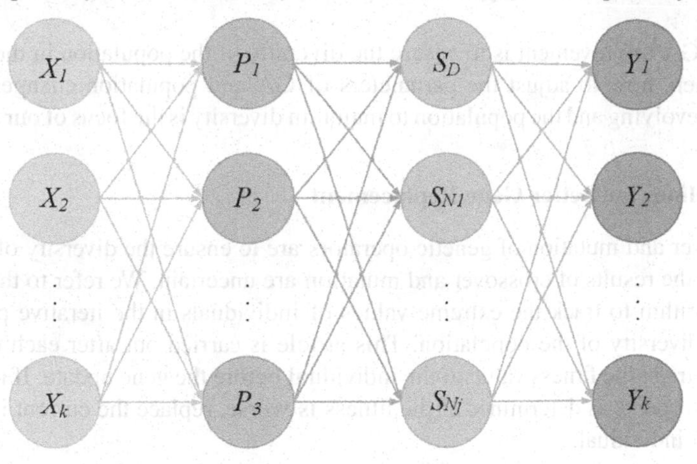

Fig. 1. Generalized regression neural network structure diagram

factor, which is a key parameter that affects the accurate performance of the GRNN model.

(3) Summation Layer
In this article, the summation layer contains two types of neurons. The first type of neuron is used to sum the outputs of the neurons in all the pattern layers, which can be expressed as:

$$S_D = \sum_{i=1}^{n} P_n \tag{2}$$

The second type of neuron is used to perform a weighted summation of the neurons in all model layers, and the formula is expressed as:

$$S_{Nj} = \sum_{i=1}^{n} Y_{ij} P_i (j = 1, 2, \cdots, k) \tag{3}$$

In the formula, Y_{ij} is the expected expected value; k is the dimension of the output vector in the learning sample.

(4) Output Layer
The output layer divides the weighted sum S_{Nj} of each neuron by the output S_D of each summing neuron, thereby outputting the predicted value Y_j.

$$Y_j = \frac{S_{Nj}}{S_D} (j = 1, 2, \cdots, k) \tag{4}$$

4 Improved Genetic Algorithm (IGA)

The key to GA improvement is to ensure the diversity of the population in the iterative process. Then, how to adjust the parameters of GA and population changes to keep individuals evolving and the population to maintain diversity is the focus of our attention.

4.1 Guidelines for Better Gene Replacement

The crossover and mutation of genetic operators are to ensure the diversity of the population, but the results of crossover and mutation are uncertain. We refer to the particle swarm algorithm to track the extreme values of individuals in the iterative process to ensure the diversity of the population. This article is carried out after each crossover mutation to track the fitness value of the individual before the gene update. If the fitness is better, the update is determined If the fitness is worse, replace the current individual with the last individual.

4.2 Crossover Probability

In the process of crossover operation to update the population, the crossover probability is often a fixed value artificially set. In the optimization process, due to the premise of the better gene replacement criterion, the optimization feature of the entire population is continuous. In the iterative process of the population, the individual keeps track of the position of the last individual, and makes better replacement adjustments, so that the entire population develops in a better direction. In this process, the diversity of the population may disappear to a certain extent, and with the increase of the number of iterations, the disappearance of diversity is more obvious. Therefore, it is necessary to update the genome to varying degrees according to the development characteristics of the population. To solve this problem, the fitness value of the population can be combined to make it non-linearly increase within a certain range. The design formula is as follows:

$$p_c = 0.7 \times \sin\left(0.5 \times \pi \times \frac{f_{min} - f_{max}}{f_{max}}\right) + 0.9 \tag{5}$$

Where: f_{min} is the minimum fitness value after each iteration; f_{max} is the maximum fitness value after each iteration.

During the whole iteration process, the gap between f_{min} and f_{max} keeps decreasing, so the size of $(f_{min} - f_{max})/f_{max}$ is (−1, 0), and as the number of iterations increases, the nonlinearity increases. According to the calculation formula of p_c, p_c will increase nonlinearly between (0.2, 0.9).

4.3 Mutation Probability

In the process of performing mutation operations to update population individuals, the mutation probability is often a fixed value artificially set. Under the principle of better gene replacement, the new gene produced is likely to be replaced by the old better gene. Although the continuity of the optimization of the population can be guaranteed, it may also lead to the disappearance of the diversity of the population. If the set mutation probability is too large, it will increase the complexity of the algorithm and require more optimization time. If the set mutation probability is too small, the continuity of population diversity cannot be guaranteed. In order to solve this problem, we propose a formula strategy in which the mutation probability increases nonlinearly as the number of iterations increases. The formula for the probability of mutation is as follows:

$$p_m = 0.7 \times \sin\left(0.5 \times \pi \times \frac{t-T}{T}\right) + 0.9 \quad (6)$$

5 The Establishment of Intelligent Compound Predict Mode (ICPM)

Intelligent Compound Prediction Model (ICPM) is based on GRNN. The parameter smoothing factor of GRNN is the optimization target of IGA. GRNN repeatedly learned known data through the process of IGA optimization of super parameters. Mean Square Error (MSE) between the predicted value and the true value is the fitness value of IGA. After many times of training, the model can get the desired parameters, and the trained model is saved as ICPM. ICPM can be used to predict unknown data.

The fitness formula of IGA is as follows:

$$fitness = \frac{1}{n}\sum\nolimits_{i=1}^{n}\left(y_i^{'} - y_i\right)^2 \quad (7)$$

Where y_i is the expected actual value, $y_i^{'}$ is the predictive value.

Framework of ICPM is described as follows:

1: The Smoothing factor is regarded as individuals in the IGA population, which are the optimization goals of IGA;
2: for each individual do
3: Perform an GRNN calculation;
4: Calculate fitness value;
5: Criteria for better gene replacement
6: if meet the optimization conditions then
7: End;
8: else:
9: Repeat steps 2-5;
10: end if

Based on the above method, ICPM was established. The specific algorithm steps of the model are as follows:

1) Collect raw data, perform data preprocessing on the raw data, and divide the processed data into training set and test set. The training set is used for hyperparameter

training of the model, and the test set is used to verify the prediction performance of the model.

2) Taking the hyperparameters of the model as the optimization goal of IGA, initialize the population individuals.
3) Perform selection operations.
4) Update the cross probability and perform the cross operation.
5) Update the mutation probability and perform the mutation operation.
6) Perform GRNN prediction and calculate the fitness value of IGA
7) Update the population according to more gene replacement guidelines
8) Repeat steps 3–7
9) If the expected error value is reached, or the maximum number of iterations is reached, the update will stop.
10) Save the last trained model as ICPM to predict the unknown data in the test set.

The specific algorithm flow chart is shown as Fig. 2.

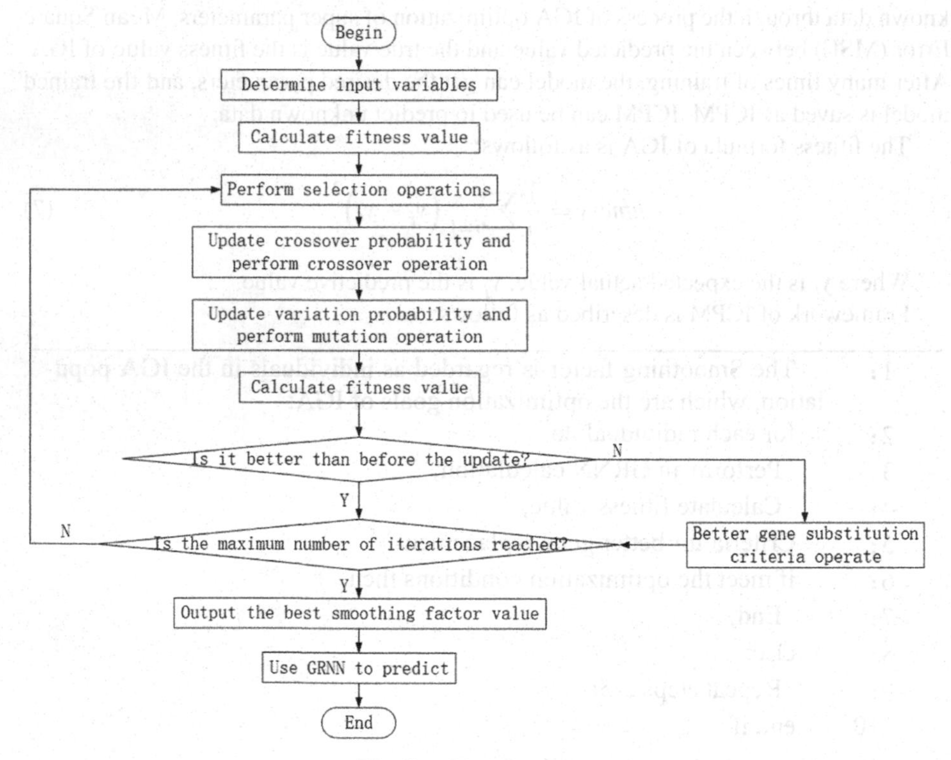

Fig. 2. ICPM flow chart

6 ICPM Realization Process

6.1 Experimental Data

In order to verify the feasibility and superiority of ICPM, soil heavy metals Cr, Cu and Pb in Dongxihu District of Wuhan were selected for content prediction. The longitude, latitude, sampling depth, and crop type code of the sample are used as the input variables of the model, and the heavy metal content is used as the output variable of the model. There are 82 sets of data for each metal, of which 62 sets are used as training data and 20 sets are used as test data. According to the different types of crops in the data set, the test data is uniformly selected from the data set with heavy metal content from low to high, and the unselected data is used as the training data. First train the model through the training data, and then use the trained model as the ICPM to predict the heavy metal content of the test set. Before the experiment, first normalize the experimental data, and select the maximum and minimum values as the normalization method.

The parameter settings of the six prediction models are shown in Table 1. *iter* is the number of iterations of the model; C is the penalty factor of SVM; σ is the kernel parameter of SVM; l is the learning rate of BPNN; s is the smoothing factor of GRNN; M is the population number.

Table 1. Parameter settings

Model	Parameter
SVM	$C = 25, \sigma = 10$
BP	$l = 0.01$, $iter = 100$
GRNN	$s = 0.5$, $iter = 100$
PSO-GRNN	$M = 20$, $iter = 100$
GA-GRNN	$M = 20$, $iter = 100$
ICPM	$M = 20$, $iter = 100$

6.2 Parameter Changes

The left graph of Fig. 3 is a graph of crossover probability. Observation shows that the changes in the crossover probability of the three metals are different. On the whole, there is a stepwise nonlinear increase between (0.2, 0.9). Among them, the crossover probability of Cr increases between (0.5, 0.9), the crossover probability of Cu increases between (0.3, 0.5), and the crossover probability of Pb increases between (0.5, 0.8). It can be seen from Formula 5 that the size of the crossover probability is related to the change of the individual fitness value of IGA. Initially, the smaller the crossover probability, the greater the difference between the maximum and minimum individual fitness; after multiple iterations, the greater the crossover probability, the smaller the difference between the maximum and minimum individual fitness. The increase in the uncertainty of the crossover probability can ensure that the diversity of the population does not disappear with the increase in the number of iterations.

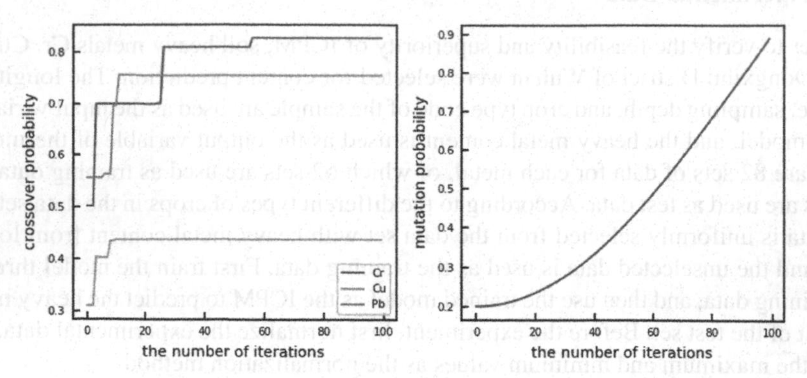

Fig. 3. Change in crossover probability and mutation probability

The right graph of Fig. 3 is a graph of mutation probability. Observation shows that the mutation probability of the three metals is the same, within the range of (0.2, 0.9). It increases with the number of iterations. Under the principle of better gene replacement, the ever-increasing probability of mutation can expand the optimization range of the population.

6.3 Prediction for Soil Heavy Metal Content

The errors between the predicted and actual values of the three soil heavy metal contents by the six models are shown in Fig. 4. From the analysis in Fig. 4, we can see that there are certain errors in the predictions of the six models for the three metals. Prediction of SVM for three metals is more error than others. This may be because the SVM model lacks the process of learning and training. The prediction effect of BPNN on the three metals is average. The reason may be that the effect of BPNN training and learning is not good, or the generalization performance of the model obtained by training is poor. The prediction error results of GRNN, PSO-GRNN and GA-GRNN are relatively similar. This may be because the initial parameter settings of GRNN are similar to the parameters obtained by PSO-GRNN and GA-GRNN training, or have similar influence on the model. On the whole, the prediction error of ICPM is the smallest, which shows that ICPM has better prediction performance and generalization performance. This also indirectly proves that the parameter optimization process of IGA helps GRNN obtain excellent hyperparameters, and the generalization performance of the trained model is good.

MAE, RMSE, MAPE and RMSLE are the error values of four different calculation methods. Mean Absolute Error (MAE) reflects the average deviation between the predicted value and the true value. If the data value is larger, the calculated MAE value will be larger. When predicting the same set of data, the smaller MAE value, the better the prediction effect. Root Mean Square Error (RMSE) is used to measure the deviation between the true value and the predicted value. Mean Absolute Percentage Error

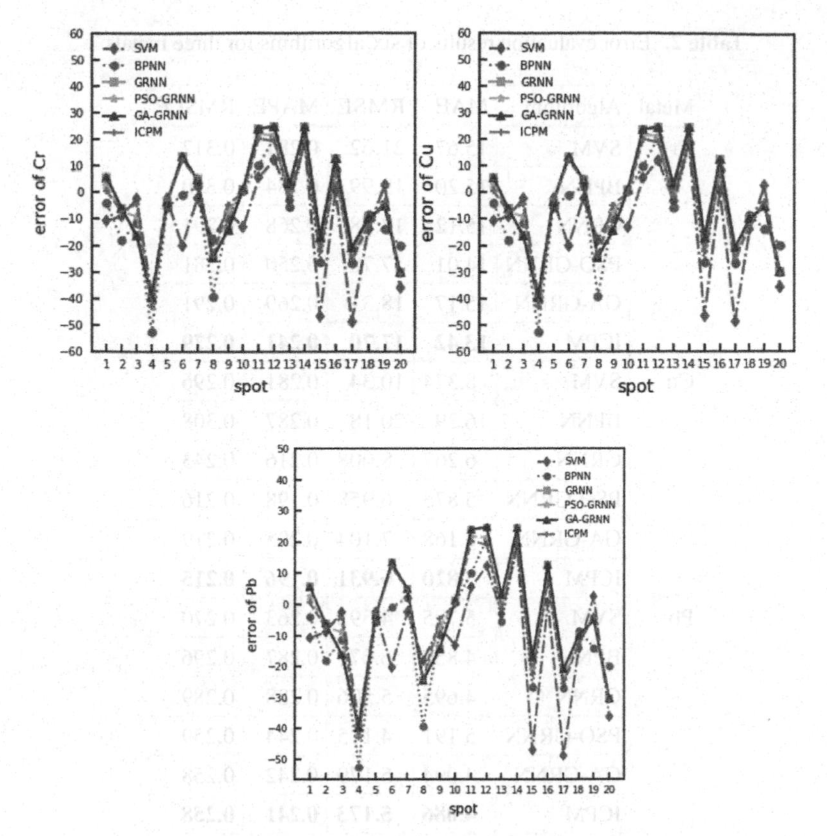

Fig. 4. The prediction error value of the test set of three metals

(MAPE) reflects the average prediction error between the predicted value and the true value, and can evaluate the quality of the prediction result. In the calculation process, MAPE uses the actual value as the denominator to make the calculated value more convincing. Root Mean Square Logarithmic Error (RMSLE) is a method of measuring error rate. The smaller RMSLE value, the higher the prediction accuracy of the model. In the comparative experiment in this article, we calculated the MAE, RMSE, MAPE and RMSLE of each model. In the experiment, for the same set of data, the smaller the error value, the higher the prediction effect.

The calculation formula for the four groups of error values is as follows:

$$MAE = \frac{1}{n}\sum\nolimits_{i=1}^{n}\left|y_i^{\prime} - y_i\right| \tag{8}$$

$$RMSE = \sqrt{\frac{1}{n}\sum\nolimits_{i=1}^{n}\left(y_i^{\prime} - y_i\right)^2} \tag{9}$$

$$MAPE = \frac{1}{n}\sum_{i=1}^{n}\frac{|y_i - y_i^{\prime}|}{y_i} \tag{10}$$

Table 2. Error evaluation results of six algorithms for three metals

Metal	Algorithm	MAE	RMSE	MAPE	RMSLE
Cr	SVM	15.67	21.62	0.281	0.312
	BPNN	15.20	19.99	0.284	0.310
	GRNN	15.12	18.28	0.268	0.291
	PSO-GRNN	14.01	17.78	0.250	0.281
	GA-GRNN	15.17	18.31	0.269	0.291
	ICPM	**13.42**	**17.70**	**0.241**	**0.279**
Cu	SVM	8.374	10.34	0.281	0.296
	BPNN	16.29	20.18	0.287	0.308
	GRNN	6.267	8.008	0.216	0.243
	PSO-GRNN	5.873	6.958	0.198	0.216
	GA-GRNN	6.168	7.104	0.205	0.219
	ICPM	**5.820**	**6.931**	**0.196**	**0.215**
Pb	SVM	5.795	4.592	0.263	0.270
	BPNN	4.858	6.071	0.287	0.296
	GRNN	4.695	5.506	0.288	0.289
	PSO-GRNN	5.191	4.115	0.244	0.259
	GA-GRNN	4.094	5.179	0.242	0.258
	ICPM	**4.086**	**5.173**	**0.241**	**0.258**

$$RMSLE = \sqrt{\frac{1}{n}\sum\nolimits_{i=1}^{n}\left(\log(y_i' + 1) - \log(y_i + 2)\right)^2} \tag{11}$$

The error evaluation results of the six algorithms for the three metals are shown in Table 2. It can be seen from the results of 4 kinds of errors in Table 2 that ICPM has 4 sets of smaller error values compared with other 5 models, indicating that the ICPM prediction model has certain advantages and the model has stable generalization performance.

7 Conclusion and Future Work

This paper proposes an intelligent composite prediction model (ICPM) for soil heavy metal content. ICPM is based on GRNN, through GA's parameter optimization of GRNN, GRNN adjusts the model parameters repeatedly. In this paper, three metals in Dongxihu District of Wuhan City are used as experimental objects, and the prediction performance and generalization performance of the model are judged by predicting the content of heavy metals and comparing errors. Comparing ICPM with SVM, BPNN, GRNN, PSO-GRNN, GA-GRNN, the experimental results show that the MAPE of Cr is reduced by

4%, 4.3%, 2.7%, 0.9%, and 2.8%, respectively; the MAPE of Cu is reduced respectively 8.5%, 9.1%, 2.0%, 0.2%, 0.9%; MAPE of Pb decreased by 2.2%, 4.6%, 4.7%, 0.3%, 0.1%, respectively. In the future work, for the model proposed in this paper, we can consider continuing to improve GA to improve the optimization speed and accuracy of the model. For example, GA can be improved by combining other intelligent algorithms. Or, by optimizing the structure of the ICPM, the prediction time or prediction error can be reduced. For example, the effective training times of the monitoring model can not only ensure the optimization of the required parameters, but also avoid unnecessary optimization time. Or, you can try to build a model by combining other swarm intelligence algorithms and machine learning algorithms. In short, improving the generalization performance and prediction accuracy of prediction models is the focus of future work.

Funding Statement. This work was supported in part by the Major Technical Innovation Projects of Hubei Province under Grant 2018ABA099, in part by the National Science Fund for Youth of Hubei Province of China under Grant 2018CFB408, in part by the Natural Science Foundation of Hubei Province of China under Grant 2015CFA061, in part by the National Nature Science Foundation of China under Grant 61272278, and in part by Research on Key Technologies of Intelligent Decision-making for Food Big Data under Grant 2018A01038.

Conflicts of Interest. They have no conflicts of interest to report regarding the present study.

References

1. Sohn, S.Y., Kim, D.H., Yoon, J.H.: Technology credit scoring model with fuzzy logistic regression. Appl. Soft Comput. **43**, 150–158 (2016)
2. Zeng, N., et al.: A new switching-delayed-PSO-based optimized SVM algorithm for diagnosis of Alzheimer's disease. Neurocomput. **320**, 195–202 (2018)
3. Zhao, H., et al.: The pollutant concentration prediction model of NNP-BPNN based on the INI algorithm, AW method and neighbor-PCA. J. Ambient Intell. Humanized Comput. **10**(8), 059–3065 (2019)
4. Kanirajan, A.P., Suresh Kumar, V.: Power quality disturbance detection and classification using wavelet and RBFNN. Appl. Soft Comput. **35**, 470–481 (2015)
5. Shen, X., et al.: Prediction of entering percentage into expressway service areas based on wavelet neural networks and genetic algorithms. IEEE Access **7**, 54562–54574 (2019)
6. Zhou, P., et al.: Modeling error PDF optimization based wavelet neural network modeling of dynamic system and its application in blast furnace ironmaking. Neurocomputing **285**, 167–175 (2018)
7. Liu, X., et al.: PSO-BP neural network-based strain prediction of wind turbine blades. Materials **12**(12), 1889 (2019)
8. Wang, L., et al.: Bus arrival time prediction using RBF neural networks adjusted by online data. Procedia Soc. Behav. Sci. **138**(14), 67–75 (2014)
9. Xia, X., Liu, X., Lou, J.: A network traffic prediction model of smart substation based on IGSA-WNN. ETRI J. **42**(3), 366–375 (2020)
10. Liang, H., et al.: An sand plug of fracturing intelligent early warning model embedded in remote monitoring system. IEEE Access **7**, 47944–47954 (2019)
11. Zheng, Z.Y., et al.: Artificial neural network – genetic algorithm to optimize wheat germ fermentation condition: application to the production of two anti-tumor benzoquinones. Food Chem. **227**, 264 (2017)

12. Junior, F.E.F., Yen, G.G.: Particle swarm optimization of deep neural networks architectures for image classification. Swarm Evol. Comput. **49**, 62–74 (2019)
13. Sathya, D.J., Geetha, K.: Hybrid ANN optimized artificial fish swarm algorithm based classifier for classification of suspicious lesions in breast DCE-MRI. Pol. J. Med. Phys. Eng. **23**(4), 81–88 (2017)
14. Deng, Y., Liu, Y., Zhou, D.: An improved genetic algorithm with initial population strategy for symmetric TSP. Math. Probl. Eng. **2015**, 1–6 (2015)
15. Li, D.-J., Li, Y.-Y., Li, J.-X., Fu, Y.: Gesture recognition based on BP neural network improved by chaotic genetic algorithm. Int. J. Autom. Comput. **15**(3), 267–276 (2018). https://doi.org/10.1007/s11633-017-1107-6
16. Sun, N., Lu, Y.: A self-adaptive genetic algorithm with improved mutation mode based on measurement of population diversity. Neural Comput. Appl. **31**, 1–9 (2018)
17. Tang, J., et al.: Traffic flow prediction based on combination of support vector machine and data denoising schemes. Phys. A Stat. Mech. Appl. **534**(15), 120642 (2019)
18. Xue, L., et al.: A data-driven shale gas production forecasting method based on the multi-objective random forest regression. J. Petrol. Scie. Eng. **196**, 107801 (2020)
19. Maurya, A.: Bayesian optimization for predicting rare internal failures in manufacturing processes. In: 2016 IEEE International Conference on Big Data (Big Data). IEEE (2016)
20. Sun, W., Ye, M., Xu, Y.: Study of carbon dioxide emissions prediction in Hebei province, China using a BPNN based on GA. J. Renew. Sustain. Energy. **8**(4), 043101 (2016)
21. Shen, L., Wen, Y., Li, X.: Improving prediction accuracy of cooling load using EMD. PSR and RBFNN. J. Phy. Conf. **887**, 012016 (2017)
22. Agbossou, K., et al.: Time series prediction using artificial wavelet neural network and multi-resolution analysis: application to wind speed data. Renew. Energy **92**, 202–211 (2016)
23. Yang, G., et al.: Research on ultra short-term wind speed prediction based on GRNN all information neural network. Measur. Control Technol. **35**(4), 149–152 (2016)
24. Szustakowski, J.D., Weng, Z.: Protein structure alignment using a genetic algorithm. Proteins Struct. Funct. Bioinform. **38**(4), 428–440 (2015)
25. Elkelesh, A., et al.: Decoder-tailored polar code design using the genetic algorithm. IEEE Trans. Commun. **67**, 4521–4534 (2019)
26. Li, G., et al.: Application of general regression neural network to model a novel integrated fluidized bed gasifier. Int. J. Hydrogen Energy **43**(11), 5512–5521 (2018). https://doi.org/10.1016/j.ijhydene.2018.01.130
27. Ghritlahre, H.K., Prasad, R.K.: Exergetic performance prediction of solar air heater using MLP, GRNN and RBF models of artificial neural network technique. J. Environ. Manage. **223**, 566–575 (2018)

A New Method of Halftoning and Inverse Halftoning Based on GAN Network

Jianjin Gu and Li Li(✉)

School of Computer Science, Hangzhou Dianzi University, Hangzhou 310018, China
Lili2008@hdu.edu.cn

Abstract. Halftoning is a method of quantifying a continuous tone image into a binary image. Halftone methods can be categorized into dithering, error diffusion and iterative method. We present a new method different from traditional halftoning algorithms for learning the mapping between continuous images and halftone images using conditional generative adversarial networks (conditional GANs). We regard halftoning and inverse halftoning as the process of image-to-image translation, and use the classic pix2pixHD network. In this work, we use multi-scale generator and discriminator architectures to perform both halftoning and its structural reconstruction. The experimental results show that this method can better fit some classical halftoning algorithms to realize halftoning and inverse halftoning. Compared with the existing methods, our method can better realize reconstruction of halftone images.

Keywords: Halftoning · Inverse Halftoning · GAN

1 Introduction

With the rapid development of technology, the printing industry is gradually moving towards digitalization. Compared with traditional printing, digital printing is much more manageable, flexible, and eco-friendly. It has effectively improved printing efficiency and printing quality.

Halftone technology refers to the technology of quantifying a continuous tone image into a binary image or a color image with only a few colors, so that the visual effect of the quantified image at a certain distance is similar to the original image. Digital halftoning [1] takes advantage of the low-pass characteristic of the human eye. When observing at a certain distance, the human eye regards the spatially close part of the image as a whole. The local average gray value of halftone image observed by the human eye is approximate to the local average gray value of the original image, thus forming a continuous tonal effect on the whole. Inverse halftone technology is a special digital image recovery technology to convert a halftone image into a continuous tone image. It is the key link of halftone image conversion screening, digital file management and high-precision image recognition.

The purpose of this paper is to realize the mapping between continuous tone image and halftone image. In 2020, Guo et al. used Cyclegan network to learn the mapping

X. Sun et al. (Eds.): ICAIS 2021, CCIS 1422, pp. 119–131, 2021.
https://doi.org/10.1007/978-3-030-78615-1_11

between continuous tone images and halftone images, and the image reconstruction results of the method were consistent with those of the existing methods. Inspired by pix2pixHD [2], in this paper we propose a method of halftoning and inverse halftoning based on GAN network to learn the process of halftone. The experimental results are similar to those of the traditional halftoning algorithm, and this method can well fit the output of the classic halftoning algorithm. In addition, compared with the traditional inverse halftone method our method not only can achieve the reconstruction of halftone images, but also has better effect and higher accuracy compared with the method of Guo et al. [3].

2 Related Work

2.1 Ordered Dithering

In the early development of digital printing, ordered dithering algorithm [4] is the most widely used. It shakes the matrix pieced together to form a large matrix, so it can cover the original continuous tone image. It then checks whether the latter pixel is greater than the former pixel, and gives an ink on the corresponding position of halftone image.

2.2 Error Diffusion

In 1975, Floyd and Steinberg proposed the error diffusion algorithm. This algorithm and ordered dithering algorithm are essentially different. Ordered dithering algorithm generates halftone images through "dot process", so that the current pixel processing will not affect other pixels in the image. By contrast, the error diffusion algorithm is a "neighborhood process". It generates halftone images by quantizing pixels sequentially and diffusing the quantization errors to the adjacent pixels via a heuristically defined weight matrix.

Figure 1 shows the grayscale image of Lena, the halftone image of Lena after error diffusion algorithm. In order to better demonstrate the effect, we enlarge the face region.

2.3 Iterative Method

The iterative method is a search method that finds the best binary output by minimizing the error between the original image and the halftone image. The halftone image obtained by this algorithm is of good quality, but at the same time, it consumes a lot of time and computation. The most representative iterative method is DBS (Direct Binary Search) algorithm. DBS uses a visual model to reduce visual error between halftone images and original continuous images by constantly inverting pixels and exchanging them with the adjacent 8 neighborhoods.

2.4 Halftoning Image Quality Assessment

Quality evaluation of halftone image is different from that of continuous tone image. The pixel value of continuous image ranges from 0 to 255, while the pixel value of halftone image is only 0, 1. The local average gray value of a halftone image is approximately equal to a continuous tone image. As a result, various grays of the output image are generated as a whole, forming a continuous tonal effect visually. If we still calculate the error between the individual pixel values, the result will be a big error. In 2012, Chen et al. proposed an image quality evaluation method based on halftone characteristics [5], which divide the original image and halftone image into several small areas of the same size, regarding all pixels in each small area as a whole, and the gray value was the arithmetic average value of pixel values in the small area. Then we calculate the PSNR and SSIM of the original image and the evaluated image.

Fig. 1. Gray scale image, halftone image, enlarged face region.

2.5 H-GAN

Generative Adversarial Networks (GANs) is a deep learning model, and is one of the most promising methods of unsupervised learning used in recent years. It can not only be used to generate various images and natural language data, but also inspire and promote the development of various semi-supervised and unsupervised learning tasks [6–10]. Generative Adversarial Networks (GANs) [11, 12] have achieved remarkable results in image generation among others [13–17]. Recent methods adopt the same idea for conditional image generation applications [18–22]. Guo et al. proposed a new method to realize halftoning and inverse halftoning by using Cyclegan. Because of the use of Cyclegan, this method can learn both halftoning and inverse halftoning simultaneously. The result shows that it yield better image quality than do traditional inverse halftoning algorithms.

3 Our Method

In contrast to traditional halftoning algorithm and inverse halftoning algorithm, we propose a new method based on GAN. We regard halftoning and inverse halftoning processes as a kind of "image translation" [23, 24]. We use the classic pix2pixHD network, an improved version of the pix2pix network. Because the pix2pixHD network is unidirectional mapping, separate training is needed to realize halftoning and inverse halftoning. We can realize both halftoning and inverse halftoning by learning-based methods.

3.1 Pix2pix

The pix2pix is proposed for image-to-image translation. It is a conditional GAN framework. The purpose of conditional GAN is to model the distribution of real images given the input labels: $\underset{G}{min}\,\underset{D}{max}\,\mathcal{L}_{GAN}(G, D)$, where $\mathcal{L}_{GAN}(G, D)$ is expressed as, $s\mathbb{E}_{(s,x)}[\log D(s, x)] + \mathbb{E}_{s}[\log(1 - D(s, G(s)))]$, x represents the continuous grayscale image, x represents its corresponding halftone image.

It consists of a generator G and a discriminator D. Unlike the original Conditional GAN, the input of the pix2pix generator has no noise information, only conditional information. The generator is structured as Unet. Some GAN use encoder-decoder model as generator, but in comparison, Unet adds skip connections on the basis of encoder-decoder, so the generated pictures have better effects. It utilizes PatchGAN as the discriminator. Instead of determining whether the picture is real or not, PatchGAN determines whether each N × N patch is true, and averages the decisions (Fig. 2).

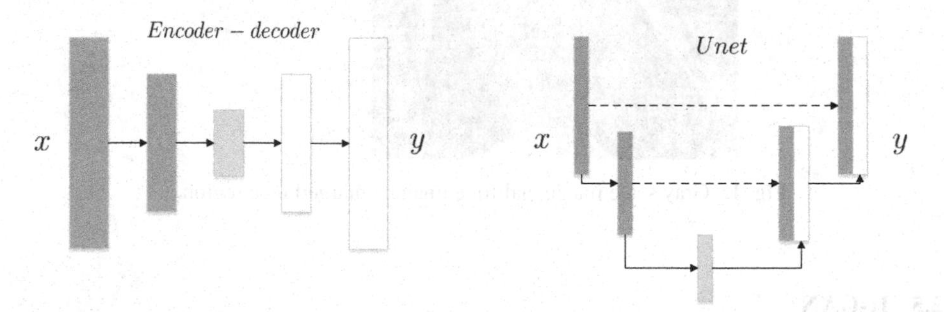

Fig. 2. Two choices for the architecture of the generator.

3.2 Our Network

We use the pix2pixHD network proposed by Wang et al. in 2018 [2]. It is composed of a coarse-to-fine generator and a multi-scale discriminator architecture.

We use the generator network of pix2pixHD. As shown in Fig. 3, the generator is composed of two sub-networks: G1 and G2. G1 is the global generator network and G2 is the local enhancer network. Images are first subsampled by generator G2, and then generator G1 is used to generate low-resolution images. The results obtained are

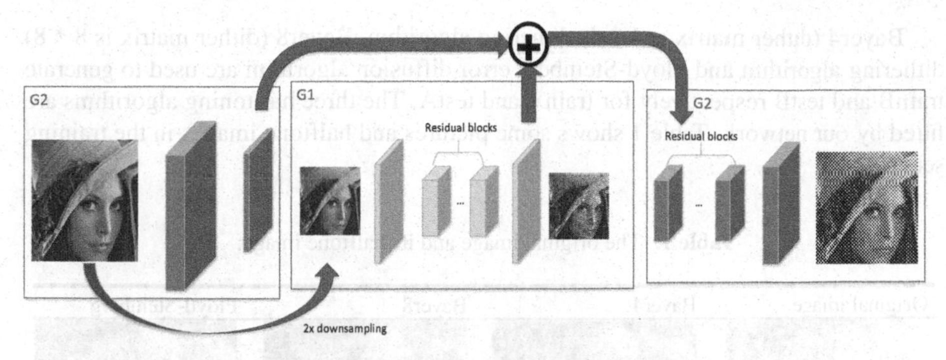

Fig. 3. Network architecture of our generator.

added to the images just sampled, which, as the input of the subsequent network of G2, generate high-resolution images.

We use two discriminators with the same network structure. Although two discriminators have the same structure, they handle different input size. The purpose of designing multi-scale discriminator is to use the discriminator with small input size to promote the synthesis of the whole image and the discriminator with large size to promote the synthesis of image details. Because of the multi-scale discriminator, the minimax game becomes

$$\min_{G} \max_{D_1, D_2} \sum_{K=1,2} L_{GAN}(G, D_k) \tag{1}$$

In order to contribute to the stability of the training process, it adds the feature matching loss based on discriminator characteristics:

$$L_{FM}(G, D_k) = \mathbb{E}_{(s,x)} \sum_{i=1}^{T} \frac{1}{N_i} \left[\left\| D_k^{(i)}(s, x) - D_k^{(i)}(s, G(s)) \right\|_1 \right] \tag{2}$$

S represents the continuous grayscale image, represents its corresponding halftone image, represents the number of layers of the network and represents the number of elements in each layer. The loss function is shown below:

$$\min_{G} \left(\max_{D, D2} \sum_{k=1,2} L_{GAN}(G, D_k) + \lambda \sum_{k=1,2} L_{FM}(G, D_k) \right) \tag{3}$$

4 Experiments

In the experiment, the learning rate is 0.0002, the number of epochs is 100, we use Adam optimizer, and the number of ResnetBlocks is 9. Training set: Select 10,000 pictures from COCO dataset and grayscale them to 256 * 256, which is called trainA. Test set: Select 2500 images from COCO dataset, grayscale them and cut them to 256 * 256, which is called testA.

Bayer4 (dither matrix is 4 * 4) dithering algorithm, Bayer8 (dither matrix is 8 * 8) dithering algorithm and Floyd-Steinberg error diffusion algorithm are used to generate trainB and testB respectively for trainA and testA. The three halftoning algorithms are fitted by our network, Table 1 shows some pictures and halftone images in the training set.

Table 1. The original image and its halftone image.

Original image	Bayer4	Bayer8	Floyd-Steinberg

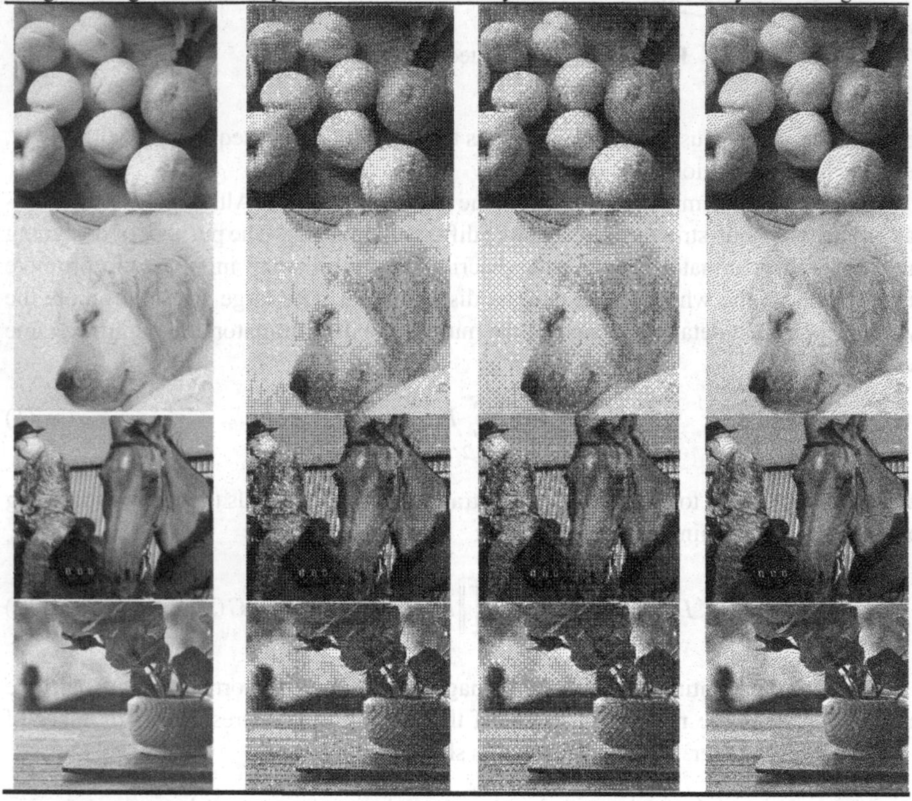

5 Result

Tables 2, 3 and 4 show some original images in the test set and halftone images generated by the halftoning algorithm and the halftone images generated by the corresponding network. The result shows that there is almost no difference between the two halftone images.

Table 2. Halftone images generated by Bayer4 and our method.

Original image	Bayer4	Our method

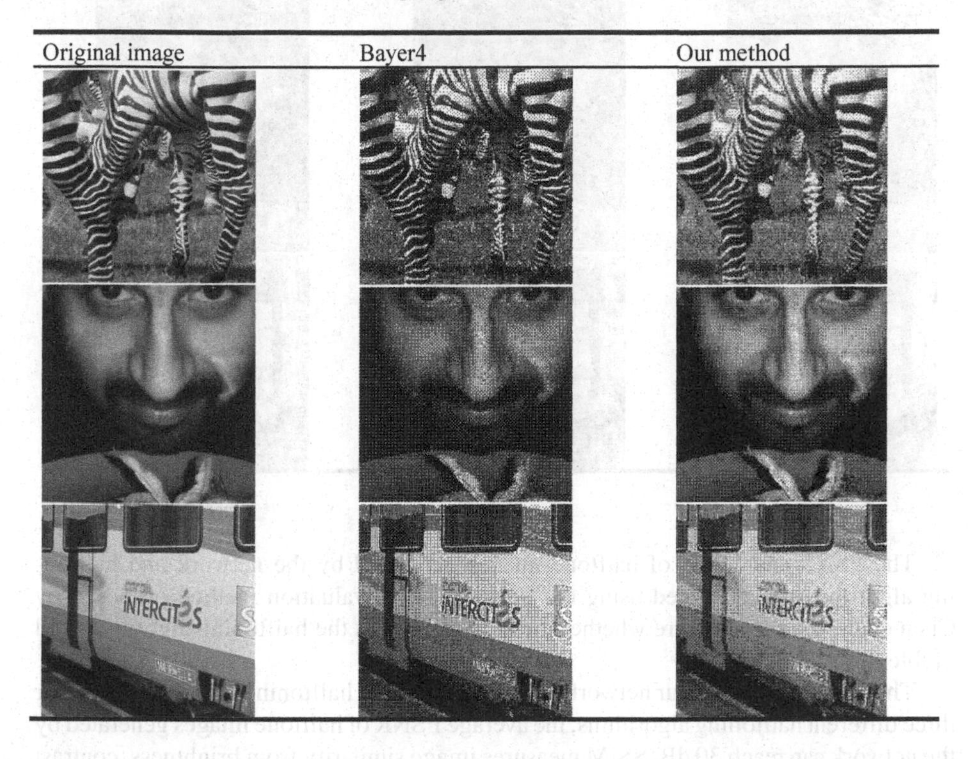

Table 3. Halftone images generated by Bayer8 and our method.

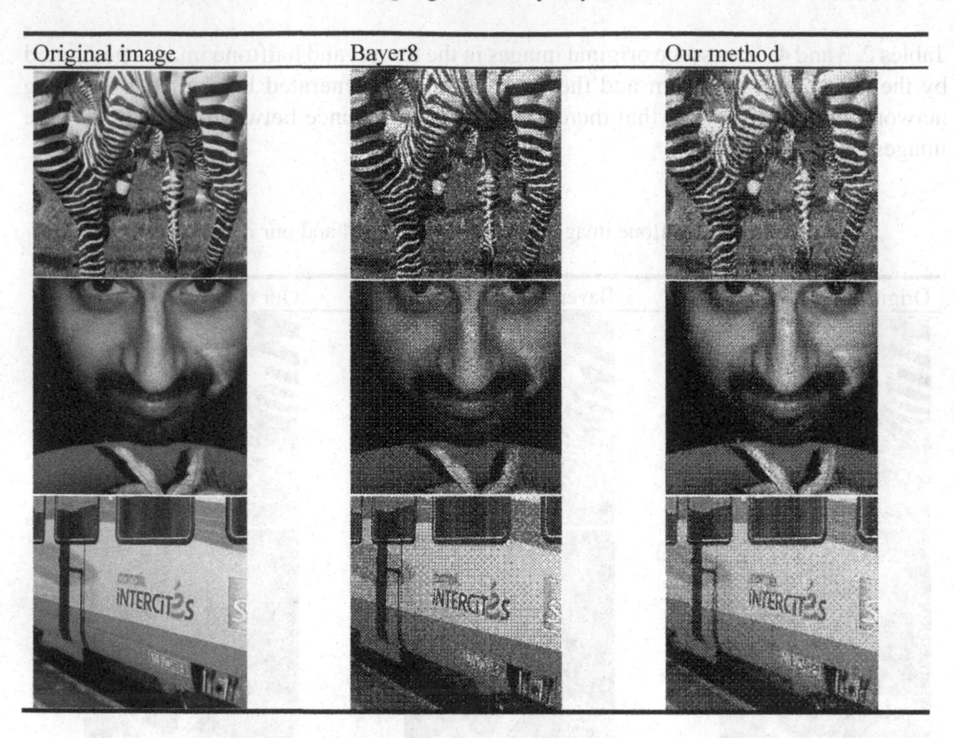

The PSNR and SSIM of halftone images generated by the network and halftoning algorithm are calculated using the image quality evaluation method proposed by Chen et al., so as to measure whether the network can fit the halftoning algorithm well (Table 5).

The results show that our network can fit the output of halftoning algorithm well. For three different halftoning algorithms, the average PSNR of halftone images generated by the network can reach 30 dB. SSIM measures image similarity from brightness, contrast and structure. The value range of SSIM is [0, 1]. The larger value represents smaller image distortion. The average SSIM can reach 0.973.

In addition, we also carry out inverse halftoning on the halftone images obtained by three halftoning algorithms, and compare the difference between the continuous images generated by the network and the original continuous gray images. Table 6 shows PSNR and SSIM of the inverse halftone images and the original continuous grayscale images respectively.

Table 4. Halftone images generated by Floyd-Steinberg and our method.

Original image	Floyd-Steinberg	Our method

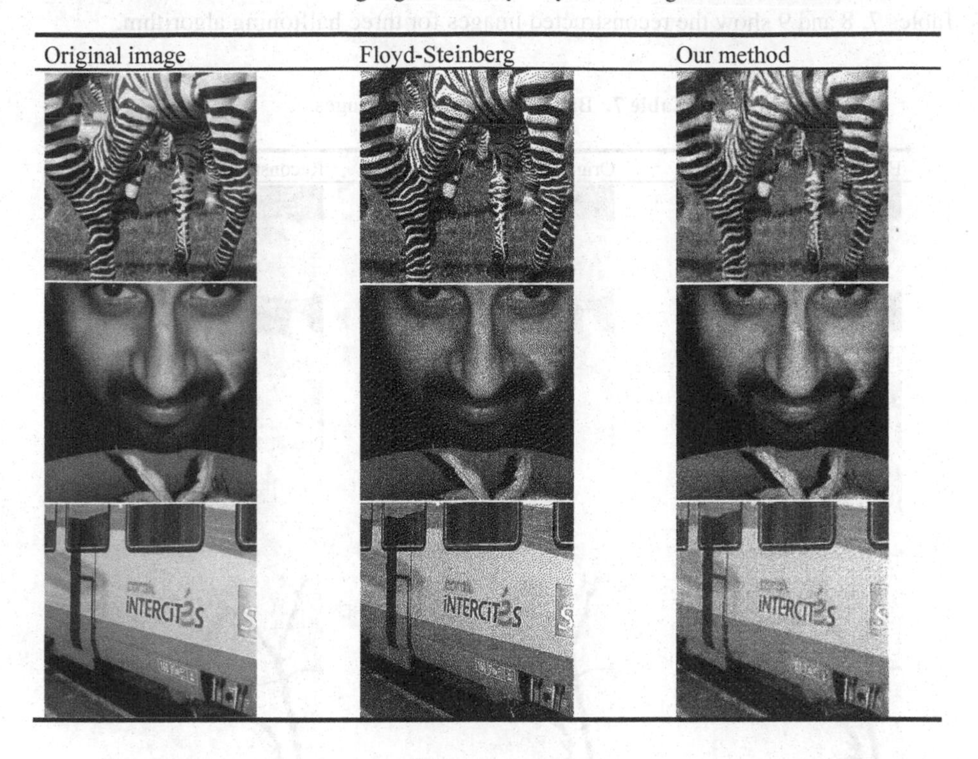

Table 5. Metrics of halftone images generated by our method and halftoning algorithm.

Metrics	Bayer4	Bayer8	Floyd-Steinberg
PSNR	29.96	30.08	30.51
SSIM	0.97	0.97	0.98

Table 6. Halftone reconstruction.

Metrics	Bayer4	Bayer8	Floyd-Steinberg
PSNR	28.24	29.54	31.43
SSIM	0.89	0.90	0.91

The experimental results show that our network achieves inverse halftoning well. Tables 7, 8 and 9 show the reconstructed images for three halftoning algorithm.

Table 7. Bayer4 reconstructed images.

Bayer4 halftone image	Original continuous image	Reconstructed image

Compared with the method of Guo et al. [3], our method has a better effect in the reconstruction of halftone images and generates higher quality continuous grayscale images. The comparison results are shown in Table 10.

Table 8. Bayer8 reconstructed images.

Bayer8 halftone image	Original continuous image	Reconstructed image

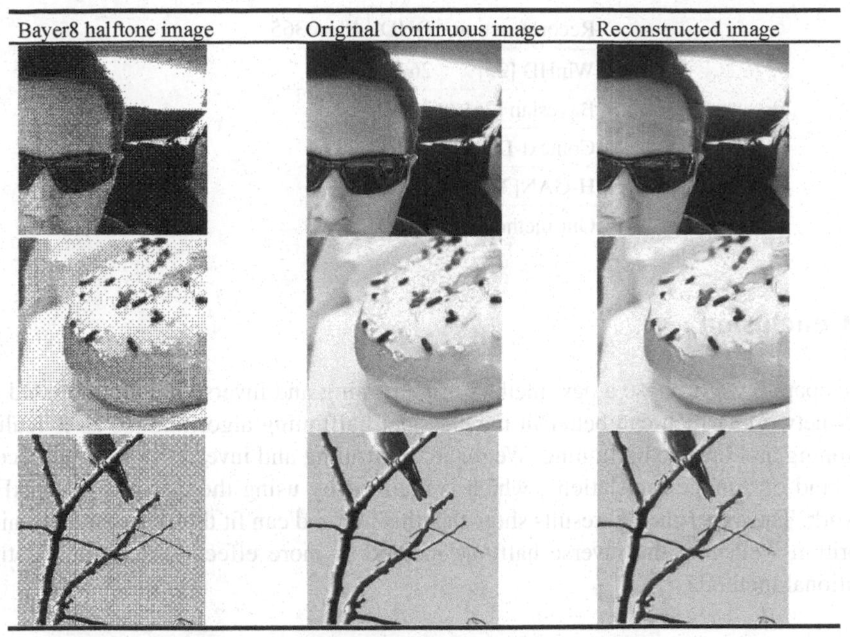

Table 9. Floyd-Steinberg reconstructed images.

Floyd-Steinberg halftone image	Original continuous image	Reconstructed image

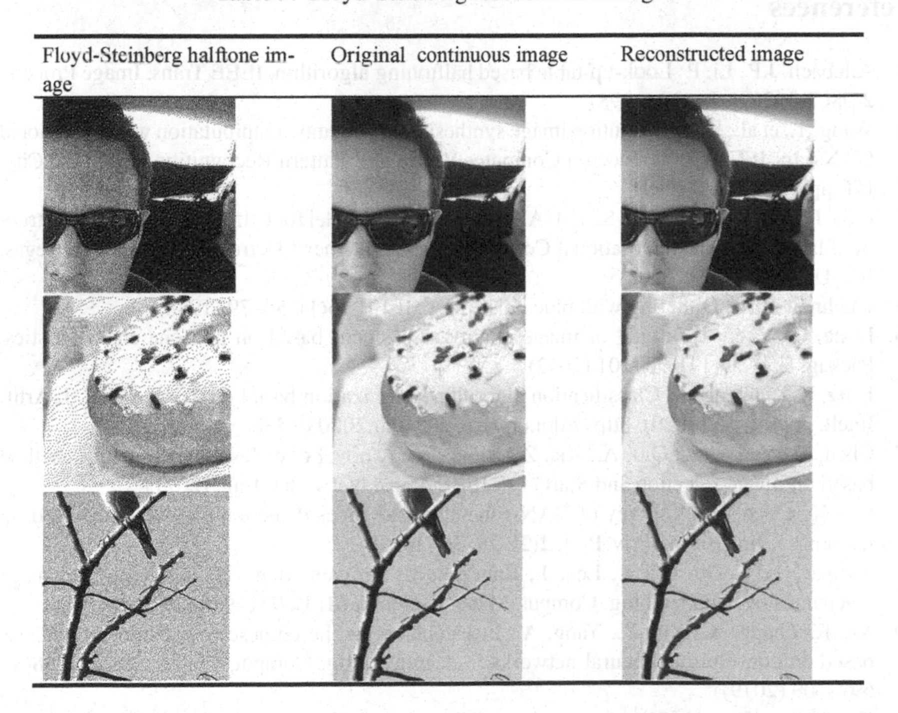

Table 10. Reconstructed halftone image comparison.

Reconstruction	DHD	Place365
WinHD [25]	26.81	27.88
Bayesian [26]	26.12	27.36
Context-Driven	26.55	28.66
H-GAN[3]	26.95	29.06
Our method	**28.33**	**30.93**

6 Conclusion

In this paper, we propose a new method of halftoning and inverse halftoning based on GAN network, which can better fit the classical halftoning algorithm, so as to realize halftoning and inverse halftoning. We regard halftoning and inverse halftoning process as a kind of "image translation", which is realized by using the classical pix2pixHD network. The experimental results show that this method can fit the classical halftoning algorithm well, and the inverse halftone method is more effective than the existing traditional methods.

References

1. Allebach, J.P., Li, P.: Look-up-table based halftoning algorithm. IEEE Trans. Image Process. **2**(98CB36269), 34–38 (1998)
2. Wang, T., et al.: High-resolution image synthesis and semantic manipulation with conditional GANs. In: IEEE Conference on Computer Vision and Pattern Recognition, Salt Lake City, UT, pp. 8798–8807 (2018)
3. Guo, J., Sankarasrinivasan, S.: H-GAN: deep learning model for halftoning and its reconstruction. In: 2020 IEEE International Conference on Consumer Electronics (ICCE), Las Vegas, NV, USA, pp. 1–2 (2020)
4. Ulichney, R.A.: Dithering with blue noise. Proc. IEEE **76**(1), 56–79 (1988)
5. Li-na, C., Zhen, L.: Study of image quality assessment based on halftone characteristics. Packag. Eng. **33**(17), 98–101 (2012)
6. Fang, K., Ouyang, J.: Classification algorithm optimization based on Triple-GAN. J. Artif. Intell. **2**(1), 1–15 (2020). https://doi.org/10.32604/jai.2020.09738
7. Chen, X., Zhang, Z., Qiu, A., Xia, Z., Xiong, N.: A novel coverless steganography method based on image selection and StarGAN. IEEE Trans. Netw. Sci. Eng. (2021)
8. Liu, X., Chen, X.: A survey of GAN-generated fake faces detection method based on deep learning. J. Inf. Hiding Priv. Prot. **2**(2), 29–36 (2020)
9. Nguyen, L.H., Do, D.T.T., Lee, J., Rabczuk, T., Nguyen-Xuan, H.: Forecasting damage mechanics by deep learning. Comput. Mater. Continua **61**(3), 951–978 (2019)
10. Xu, F., Zhang, X., Xin, Z., Yang, A.: Investigation on the chinese text sentiment analysis based on convolutional neural networks in deep learning. Comput. Mater. Continua **58**(3), 697–709 (2019)
11. Goodfellow, I., et al.: Generative adversarial nets. In: NIPS 2014 (2014)

12. Zhao, J., Mathieu, M., LeCun, Y.: Energy-based generative adversarial network. In: ICLR 2017 (2017)
13. Denton, E.L., Chintala, S., Fergus, R., et al.: Deep generative image models using a laplacian pyramid of adversarial networks. In: NIPS 2015 (2015)
14. Radford, A., Metz, L., Chintala, S.: Unsupervised representation learning with deep convolutional generative adversarial networks. In: ICLR 2016 (2016)
15. Zhu, J.-Y., Krähenbühl, P., Shechtman, E., Efros, A.A.: Generative visual manipulation on the natural image manifold. In: Leibe, B., Matas, J., Sebe, N., Welling, M. (eds.) ECCV 2016. LNCS, vol. 9909, pp. 597–613. Springer, Cham (2016). https://doi.org/10.1007/978-3-319-46454-1_36
16. Salimans, T., Goodfellow, I., Zaremba, W., Cheung, V., Radford, A., Chen, X.: Improved techniques for training GANs. In: NIPS 2016 (2016)
17. Mathieu, M.F., Zhao, J., Ramesh, A., Sprechmann, P., LeCun, Y.: Disentangling factors of variation in deep representation using adversarial training. In: NIPS 2016 (2016)
18. Reed, S., Akata, Z., Yan, X., Logeswaran, L., Schiele, B., Lee, H.: Generative adversarial text to image synthesis. In: ICML 2016 (2016)
19. Pathak, D., Krahenbuhl, P., Donahue, J., Darrell, T., Efros, A.A.: Context encoders: feature learning by inpainting. In: CVPR, Las Vegas, NV, pp. 2536–2544 (2016)
20. Mathieu, M., Couprie, C., LeCun, Y.: Deep multiscale video prediction beyond mean square error. In: ICLR 2016 (2016)
21. Vondrick, C., Pirsiavash, H., Torralba, A.: Generating videos with scene dynamics. In: NIPS 2016 (2016)
22. Wu, J., Zhang, C., Xue, T., Freeman, B., Tenenbaum, J.: Learning a probabilistic latent space of object shapes via 3d generative-adversarial modeling. In: NIPS 2016 (2016)
23. Isola, P., Zhu, J.-Y., Zhou, T., Efros, A.A.: Image-to-image translation with conditional adversarial networks. In: CVPR, Honolulu, HI, pp. 5967–5976 (2017)
24. He, K., Zhang, X., Ren, S., Sun, J.: Deep residual learning for image recognition. In: IEEE Conference on Computer Vision and Pattern Recognition (CVPR), pp. 770–778 (2016)
25. Neelamani, R., et al.: WInHD: wavelet-based inverse halftoning via deconvolution. IEEE Trans. Image Process. (2002)
26. Liu, Y.F., et al.: Inverse halftoning based on the Bayesian theorem. IEEE Trans. Image Process. **20**(4), 1077–1084 (2011)

Sow Estrus Diagnosis from Sound Samples Based on Improved Deep Learning

Peng Chen, Wengsheng Tang(✉), Dan Yin, and Bo Yang

College of Information Science and Engineering, Hunan Normal University, Changsha 410000, China

tangws@hunnu.edu.cn

Abstract. The inability to diagnose sow estrus at scale has become the burning issues in the livestock farming. An intelligent screening tool would be a game changer. Changing the dependence on the traditional method of artificial detection, this paper proposes, develops and uses an Artificial Intelligence-based screening solution for sow estrus detection. Sow estrus call is one of the significant calls which can reflect the physiological conditions. This makes the recognition of sow estrus call an exceedingly challenging problem. This problem is addressed by investigating the distinctness of Mel spectrum in the sound system generated in estrus when compared to other pig sounds. To overcome the lack of sow estrus sounds training data, transfer learning is exploited. To reduce the risk of misdiagnosis we raise a two-pronged AI architecture, one is collecting and transmitting sound to the detector, the other is identifying calls in sow estrus diagnosis system. Results show the AI architecture can distinguish among sow estrus calls and other non-sow-estrus-calls. The performance is good enough to encourage a large-scale collection of labeled sow estrus call data to improve the generalization capability of the AI architecture.

Keywords: Speech recognition · Deep learning · Transfer learning

1 Introduction

With the enhancement of computer processing capabilities, the wide application of speech recognition technology has gradually changed the way we live and the way we work. The application of speech recognition technology in livestock farming can greatly improve the efficiency. The different sounds made by pigs can reflect different physiological conditions [1]. Having a good number of auditory training, experienced breeders can judge the physiological conditions by the sounds. For example, the cough of pigs may indicate that they are suffering from respiratory diseases [2, 3], and the deep and long calls of sows may imply that they are in estrus [4].

There are three ways to identify the estrus of sows by manually [5]: (1) External observation: this method is judged by the external changes of the vaginal orifice. When the vulva is red, swollen and glossy with mucus flowing out, sows will reach the peak period of estrus; (2) Back pressing method: the sow's waist is pressed by man's hands.

X. Sun et al. (Eds.): ICAIS 2021, CCIS 1422, pp. 132–143, 2021.
https://doi.org/10.1007/978-3-030-78615-1_12

If the sow stands still without any resistance, it means that the sow's estrus may come. (3) External hormone therapy: This method is a new method used in some developed countries to identify estrus of sow. It is to use synthetic male sex pheromone to spray directly on the nose of the tested sow. Having some specific symptoms such as standing still, the sows are considered to be in estrus. Simple and convenient, this method is especially suitable for pig farms.

However, these methods also have their disadvantages. What the most remarkable disadvantage is that, with the increase of the number of pigs, artificial screening not only consumes huge man-power, but also has the problem of recognition accuracy [14]. Therefore, it is significant to use artificial intelligence technology instead of manual methods to identify the estrus of sow.

Few having studied the sound of sow estrus, there are researchers studying other behaviors of pigs (such as pig cough) with speech recognition technology. Based on pig speech recognition, scholars have carried out relevant studies on pig disease detection [12, 29], sound type identification [10–12], and sound monitoring in different states [6–9].

The steps of speech recognition include signal preprocessing, feature extraction and acoustic modeling [20]. At present, the commonly used feature vectors for pig voice recognition include power spectral density (PSD) [21], Mel-Frequency Ceptral Coefficients (MFCC) [22], Linear Predictive Cepstral Coding (LPCC) [23], etc. These feature vectors have good discrimination for some human voice signals, but for some sounds such as cough, scream and metal sound in piggery, the discrimination haven't showed a good performance [24]. There are two methods for pig sound recognition [25]: based on machine learning and based on deep learning. For pig sound recognition based on machine learning, many researchers have explored SVM model and HMM model, the classical models for speech recognition and classification. For pig speech recognition based on deep learning, researchers have tried on AlexNet network, deep belief network and bi-long short term memory network respectively, all of these network structure have received ideal results [13].

2 Related Work

Humans can express various emotional changes through voice and intonation. Most animals also express their state through voice or communicate with other members of the population. Detection of animal sounds helps people understand their complex behaviors and states. Therefore, more and more scholars begin to analyze and detect animal behaviors and states from the perspective of sound [26].

Currently, there are two main modeling methods for recognizing animal sounds, one is based on machine learning, and the other is based on deep learning. Figure 1 shows the processing steps of two different methods. The steps of the machine learning method: take the sound signal as input after processing, and then extract features manually, learn weights through some classic machine learning algorithms (such as SVM), and finally make predictions. The difference between deep learning and machine learning lies in the steps of extracting features. Deep learning uses neural networks to extract features, while machine learning extracts features manually [27, 28].

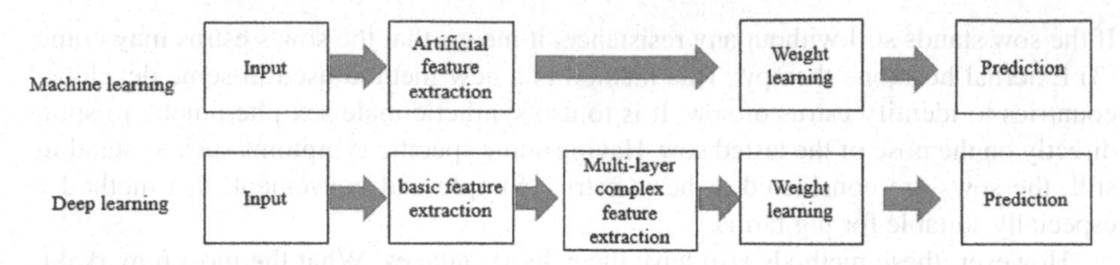

Fig. 1. Comparison of traditional machine learning and deep learning processes

2.1 Machine Learning Model

Related research abroad is relatively early. Exadaktylos [29] et al. proposed a real-time algorithm for identifying the coughing sound of sick pigs based on signal frequency domain features, using fuzzy mean clustering method to train the feature reference quantity of each sound, and using Euclidean distance to get the classification threshold is used to classify and recognize pig cough sounds and other sounds, with a total recognition rate of 85%. Yongwha Chung [30] et al. used the secondary classification structure composed of support vector data domain description (SVDD) and sparse representation classifier (SRC) in the automatic detection of swine wasting diseases as an early abnormality detector and respiratory disease classifier. The experimental results show that the method is better in terms of economy and practicability.

Domestic research is relatively late (as early as 2015). Although HMM is a relatively mature mathematical model, it is minimal to study the model itself to try to improve the recognition rate. In order to improve the efficiency of the quantization codebook, Liu Zhenyu et al. [15] performed vector quantization VQ processing on the front end of the HMM, and then adopted the codebook design. It solves a series of problems including recognition problem, character sequence matching state sequence determination problem, model training problem, etc., analyzes and recognizes pig cough sound, and provides important data basis for distinguishing pig respiratory diseases. Many domestic researchers have explored the SVM model. Zhang Qiming [10] and others have used the training classification function of libsvm algorithm to train the model, and the recognition rate of pig cough is 93.8%. Han Leilei [8] used decision tree-support vector machine to preliminarily classify the voice signals collected from piggery, which received 94% performance in recognition rate. Zhang Sunan [12] compared SVM model, HMM model and AdaBoost model, and used these models to automatically recognize pig voice in different states. Optimized the results based on three model, they have concluded that the recognition accuracy of multi model was higher than that of single one. Yan Li [16] proposed a clustering algorithm using skewness coefficient to reduce the dimension of features, which improves the recognition accuracy of SVM. In addition, Xu Yanni [6] used fuzzy c-means clustering algorithm for speech recognition, which get the accuracy of 83% for cough sound and scream of pigs by optimizing the objective function and getting the membership degree of each sample point to all class centers.

2.2 Deep Learning Model

Deep learning is a branch of machine learning research [30], which is an algorithm for modeling the complex relationship between data through multi-layer structure. Deep structure is relative to shallow structure [31]. At present, most machine learning methods, such as regression and classification, are generally shallow structure algorithms. In the case of limited labeled samples and a large number of unlabeled samples, their generalization ability is restricted to a certain extent [32]. And the shallow model also has an important feature, that is, it needs to rely on human experience to extract the features of the sample, the quality of feature selection has become a key part which can influent the whole system performance. It requires not only a lot of manpower to explore better features but also a lot of time to adjust, which is time-consuming and laborious. The deep learning network is just like the learning mechanism of human brain. In the face of a large amount of perceptual information, it forms more abstract high-level features through the combination of low-level features, and learns the distributed features of data. Therefore, it can realize the hierarchical representation of input information to represent the attributes or categories of information like human brain [33]. Compared with the machine learning of shallow structure, the extracted features are more effective. It is automatically completed by the network, and does not need manual participation.

Compared with the widespread application of machine learning in pig sound recognition, researchers have less research on deep learning-based live pig voice recognition modeling. The literature [17] proposed a pig cough recognition model based on 5-layer deep belief network DBN, and introduced the deep belief network DBN into the field of pig speech recognition for the first time, and achieved better results. In literature [18], a bidirectional long-term and short-term memory network connection time series classification model (BLSTM-CTC) was proposed to learn the change rule of continuous pig voice through BLSTM network, and an end-to-end continuous pig voice recognition system was realized by CTC, with the highest recognition rate of 94%. In literature [27], convolutional neural network was used to recognize pig cough. The AlexNet network was fine-tuned, and the sound signal spectrum was input into the neural network. The output result indicated a good performance of 97.5% in accuracy.

3 Methodology

3.1 Convolutional Neural Network and Transfer Learning

This paper used a convolutional neural network and transfer learning to identify the sow's estrus. In the first architecture, we used convolutional neural network we proposed to recognize pig sounds. In the second architecture, we used transfer learning to solve the problem of insufficient data set, and we recognized sow estrus in the pig sounds.

3.2 System Structure

In order to identify the estrus sounds of sows in real pig farms, in this section, we design a dual recognition system architecture to accomplish the following two goals: (1) Detect pig sounds from the different and complex environments. (2) Diagnose sow estrus from different types of pig sounds.

Figure 2 shows the entire system architecture. The sound from the pig farm will be collected and sent to the server. First, it will pass a pig sound detector to determine whether it is a pig sound. If the pig sound is not detected, it will return "not detected". If a pig sound is detected, the next estrus diagnosis is activated. The sound will be sent to a binary classifier based on deep transfer learning and the diagnosis result will be reported.

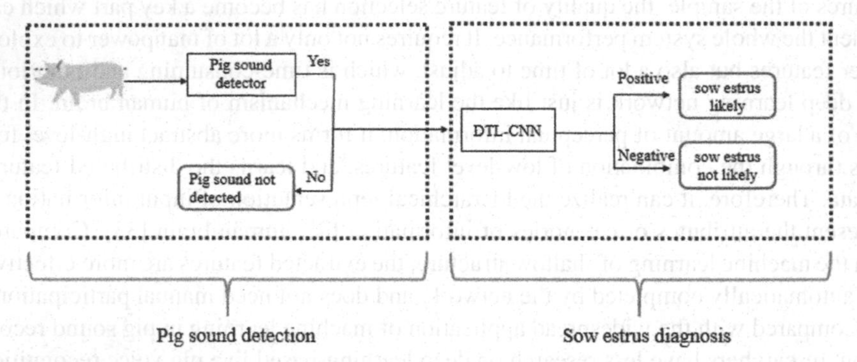

Fig. 2. System structure

The detailed information of the detection and diagnosis classifier is shown below.

(1) Pig sound detection: The collected sound samples will first calculate its Mel spectrogram, and then this image will be adjusted and transformed into a unified image dimension, to obtain a $320 \times 240 \times 1$ dimensional image. Then input the obtained image to our convolutional neural network classifier to determine whether the recorded input sound is a pig sound.

An overview of the CNN structure we used is shown in Fig. 3. The Mel spectrogram pass a layer consisting of two blocks, each block consisting of two convolutional layers, followed by a 2×2 max pooling layer and a dropout value of 0.15 to prevent the model from overfitting. The convolution layer of the first block uses 16 filters and a 5×5 convolution kernel, and the convolution layer of the second block uses 32 filters and a 5×5 convolution kernel. The complex features learned from these 4 convolutional layers are flattened, and then passed to the fully connected layer of 256 neural institutes, followed by a dropout of 0.3 to prevent overfitting. Finally, the output layer is composed of 2 neurons and a softmax activation function to distinguish whether the given input is a pig sound. In this model, Relu is used as the activation function of all convolutional layers, and Adam is used as the optimizer due to its better efficiency and flexibility.

(2) Diagnosis of sow estrus: When a pig sound is detected, it will be sent to the diagnosis of estrus to further diagnose whether the sound is estrus. Due to the limited data set, in order to obtain the most reliable results, we introduce the transfer learning model into our estrus diagnostic device to train our data set.

When the data set is small, it is basically impossible to train a deep CNN network from 0. But for the parameters of the trained model, after simple adjustment and training,

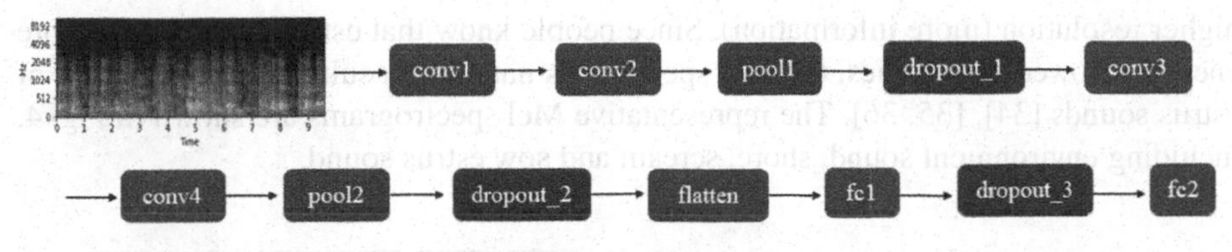

Fig. 3. Pig sound detector

it can be well transferred to other different data sets, and at the same time, it does not require a large amount of computing power. The reason is that although the data sets are different, but the underlying features are mostly common. In this experiment, we used the VGG16 model as a pre-training model for migration learning. The model uses the parameters that have been trained on ImageNet, removes the top layer, and outputs an $8 \times 8 \times 2048$ tensor, while adding a new output Layer, first use the GlobalAveragePooling2D function to convert the $8 \times 8 \times 2048$ output into a 1×2048 tensor. Followed by a fully connected layer of 1024 nodes, a 2-node output layer and the softmax activation function are used to classify between sow estrus and not sow estrus.

4 Experiment and Result

4.1 Datasets

(1) Data sets for training the pig sound detector: In order to recognize the pig sound in the place where various background noises may exist, we design a pig sound detector to our artificial intelligence engine. This detector is used as a filter to distinguish between pig sounds and 50 common types of environmental noise before diagnosis. In order to train and test this detector, we used the ESC50 data set [31] and some pig sounds from the Freesound Website [32]. The ESC50 data set is a publicly available data set that provides a large collection of environmental sounds. Freesound is a free audio data sharing platform that provides approximately 40,000 independent visits per day, and 3.5 million registered users visit more than 160,000 times to upload sounds. All sounds in Freesound are manually reviewed by a group of users (Freesound users) to check whether the description is correct and whether the sound is illegal. We used 400 samples of pig sounds and 1600 samples of non-pig sounds to train and test our pig sound detection system.

(2) Data sets for training the estrus diagnostic: In order to train our estrus diagnosis system, we collected the samples of the sow's estrus sound. In addition, we collected the pig's snoring sound, and screaming sound. Among them, some sound samples are from Freesound, and some sound samples are collected from real pig farms.

All audio data samples are saved as wav format audio with a duration of 5 s (for audio samples greater than 5 s will be equidistantly cut into 5 s audio samples), the sampling rate is 44.1 kHz, all audio will be converted into the Mel spectrum image for further processing. The reason why the Mel spectrum is used is that compared with the same s frequency band in the normal spectrum, the Mel scale in the Mel spectrum has unequal spacing in the frequency band and is at a lower frequency, which provides a higher resolution (more information). Since people know that estrus sounds have more energy at lower frequencies, the Mel spectrum is naturally a suitable representative of estrus sounds [34], [35, 36]. The representative Mel spectrograms are shown in Fig. 4, including environment sound, snore, scream and sow estrus sound.

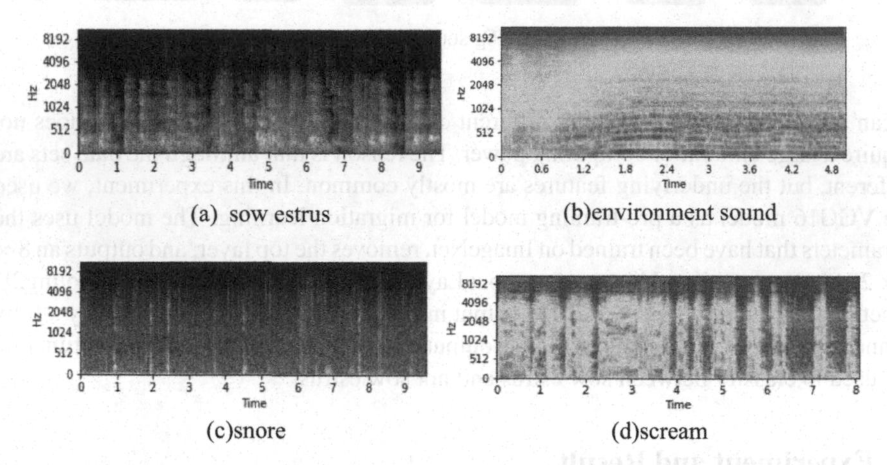

Fig. 4. Four different spectrograms

4.2 Result Analysis

Evaluation indicators: Accuracy, precision and F1 scores commonly used in pig sound detection are considered as evaluation criteria. TP indicates the number of pig sound correctly predicted, FP indicates that the number of non-pig sound was incorrectly predicted as pig sound, FN indicates the number of pig sound that were incorrectly predicted to be non-pig sound, and TN indicates the number of non-pig sound that were predicted non-pig sound. Precision, accuracy and F1 scores are calculated by formulas (1), (2) and (3).

$$accuracy = \frac{TP + TN}{TP + TN + FP + FN} \tag{1}$$

$$precision = \frac{TP}{TP + FP} \tag{2}$$

$$F1 = 2 * precision * \frac{recall}{precision + recall} \tag{3}$$

(1) Pig sound detector

For the training and validation data sets, the average loss based on the convolutional neural network is shown in Fig. 6. It can be seen that after 35 rounds of training, the curve tends to converge. The loss function of the training curve and the test curve indicate that the model is not over-fitting.

To evaluate the model, we used F1 score, precision and accuracy on the validation set to evaluate the performance of the model. These indicators are widely used in machine learning models. We used 5-fold cross validation method in the training process, which is very suitable for machine learning models.

The normalized confusion matrix and performance metrics for detection algorithm are shown in Fig. 5 and Table 1. The experimental results show that our pig sound detection algorithm can distinguish between pig sounds and non-pig sounds, with an overall accuracy rate of 96.62%.

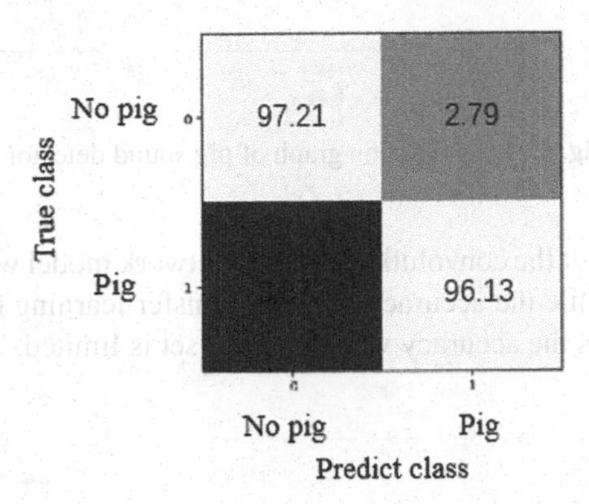

Fig. 5. Normalized mean confusion matrix (in percentage) using 5-fold cross validation

Table 1. Performance metrics for Pig sound detector

Accuracy	F1 score	Precision
96.62%	96.65%	97.18%

(2) Sow estrus sound diagnosis

We used F1 score, precision and accuracy on the validation set to evaluate the performance of the model. The performance metrics for sow estrus diagnosis algorithm are shown in Table 2. In the case of limited data, the recognition rate of the classifier based on deep transfer learning is 95.33%. This shows that our algorithm can meet the basic requirements and has high accuracy.

Figure 7 shows the average loss of the DTL-CNN classifier (sow estrus diagnosis) for the training and validation data sets. The two curves start to stabilize at about 25 epochs, which indicates that this is a reasonable learning time and there is no overfitting. Table 2 reports the performance indicators of the DTL-CNN classifier. The normalized confusion matrix and performance metrics for detection algorithm are shown in Fig. 8.

We also compared the convolutional neural network using transfer learning and without using transfer learning. The evaluation indicators are shown in Table 3. Experiments

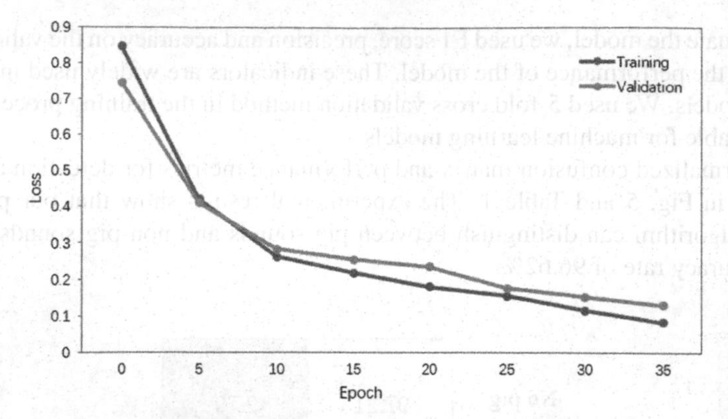

Fig. 6. Loss function graph of pig sound detector

show that the accuracy of the convolutional neural network model without transfer learning is only 77.03%, while the accuracy of using transfer learning has reached 95.33%, which greatly improves the accuracy when the data set is limited.

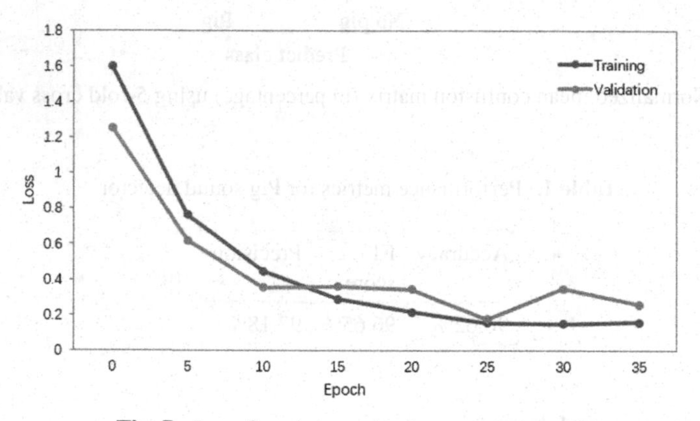

Fig. 7. Loss function graph of sow estrus sound

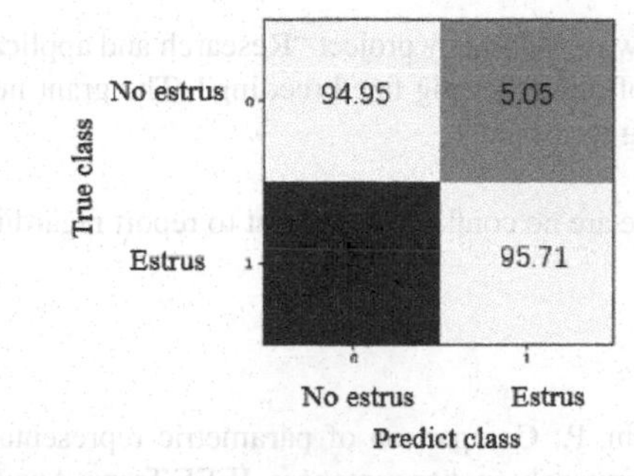

Fig. 8. Normalized mean confusion matrix (in percentage) using 5-fold cross validation

Table 2. Performance Metrics for sow estrus diagnosis

Accuracy	F1 score	Precision
95.33%	95.32%	94.95%

Table 3. Comparison of transfer learning and no transfer learning

	Transfer Learning	Without Transfer Learning
Accuracy	95.33%	77.03%

5 Conclusion

Lack of dataset, cost and long turnaround time of manual observation are key factors behind the low accuracy of recognition rate of sow estrus. Inspired by the actual requirement, this paper proposes a detection method for sow estrus recognition. This method uses transfer learning, achieving a high recognition rate under a limited data set. The result shows that the algorithm we exploited can efficiently complete the task for sow estrus diagnosis. Using two-pronged AI architecture, the algorithm we proposed improved the accuracy of recognition. In the first architecture, we used a fine-tuned 4-layer convolutional neural network, filtering the pig sound from 50 kinds of different environment noise. The accuracy indicates the algorithm is correct and fast. In the second architecture, in order to address the problem of lacking dataset, we used transfer learning in the algorithm. The result shows that using transfer learning is much better than not using it. In the future, it will be the research trend of overlapping multiple pig sounds. We expect it to be more mature and prosperous.

Acknowledgements. This work is supported by some people and company. I would like to thank my teachers, Mr. Tang and Mr. Yang, for their guidance and help in my thesis writing. Finally, thanks for the help of the company—"Hunan Baodong agriculture and animal husbandry company".

Funding Statement. This work is fund by project "Research and application of new sensor and key technology of Internet of things for pig fine breeding". The grant number is 2018GK4035, and the project leader is Tang wensheng.

Conflicts of Interest. There are no conflicts of interest to report regarding the present study.

References

1. Davis, S.B., Mermelstein, P.: Comparison of parametric representations for monosyllabic word recognition in continuously spoken sentences. IEEE Trans. Acoust. Speech Sig. Process. **28**(4), 357–366 (1980)
2. Han, W., Chan, C.F., Choy, C.S., et al.: An efficient MFCC extraction method in speech recognition. In: 2006 IEEE International Symposium on Circuits and Systems, Island of Kos, pp. 4–16 (2006)
3. Yan, Z.J., Huo, Q., Xu, J.: A Scalable Approach to Using DNN-Derived Features in GMM-HMM Based Acoustic Modeling for LVCSR. American Mathematical Society (2013)
4. Hinton, G., Deng, L., Yu, D., et al.: Deep neural networks for acoustic modeling in speech recognition: the shared views of four research groups. IEEE Sig. Process. Mag. **29**, 82–97 (2012)
5. https://wenku.baidu.com/view/846c2173a417866fb84a8e42.html
6. Xu, Y.N., Shen, M., Yan, L., Liu, L.S., Chen, C.R., Xu, P.Q.: Research on cough recognition algorithm of waiting Meishan sows. J. Nanjing Agric. Univ. **20**(3), 681–687 (2016)
7. Xu, Y.N.: Research and implementation of sow cough monitoring system based on speech recognition technology. Nanjing Agricultural University (2016)
8. Han, L.L., Tian, J.Y., Zhang, S.N., Li, J.L.: Abnormal sound recognition of pigs based on decision tree support vector machine and fuzzy inference. Anim. Husbandry Vet. Med. **12**(20), 38–44 (2019)
9. Han, L.L.: Pig abnormal sound recognition and monitoring system based on support vector machine. Taiyuan University of Technology (2019)
10. Zhang, Q.M., Yuan, R.L., Fan, F., Wang, F.: Research on pig voice recognition based on SVM algorithm. Comput. Knowl. Technol. **5**(13), 162–164 (2017)
11. Ishi, C.T., Hirose, K., Minematsu, N.: Mora F0 representation for accent type identification in continuous speech and considerations on its relation with perceived pitch values. Speech Commun. **41**(2–3), 441–453 (2003)
12. Zhang, S.: Pig behavior detection and analysis based on video tracking and multi-model voice recognition, M.S. Taiyuan University of Technology (2016)
13. Tanaka, Y., Nakano, I., Obayashi, T.: Environmental sound recognition after unilateral subcortical lesions. Cortex **38**(1), 69–76 (2002)
14. Liu, Z.Y., He, X.Y., Sang, J., Li, Y.T.: Research on pig cough sound recognition based on hidden Markov model. Inf. Technol. Branch Chin. Anim. Husbandry Vet. Soc. **3**(23), 107–112 (2015)
15. Li, Y., Shao, Q., Wu, W.X., Xie, Q.J., Sun, X., Wei, C.B.: Sound feature extraction and classification of lactating sows based on skewness clustering. Trans. Chin. Soc. Agric. Mach. **7**(12), 300–306 (2016)
16. Oanes, P.: The design and operation of the mechanical speech recognizer at university college London. J. Br. Inst. Radio Eng. **13**(54), 211–229 (1959)
17. Li, X., Zhao, J., Gao, Y., Liu, W.H., Lei, M.G., Tan, H.Q.: Recognition of continuous cough in pigs based on continuous speech recognition technology. Trans. Chin. Soc. Agric. Eng. **6**(14), 174–180 (2019)

18. Mitchell, T.: Machine Learning. McGraw-Hill (2003)
19. Zhou, H.F.: Research on embedded speech recognition system based on HMM, M.S. Guangdong University of Technology (2011)
20. Deng, L., Li, X.: Machine learning paradigms for speech recognition: an overview. IEEE Trans. Audio Speech Lang. Process. **21**(5), 1060–1089 (2013)
21. Gales, M.J.F., Ragni, A., Zhang, A., et al.: Structured discriminative models for speech recognition. In: Symposium on Machine Learning in Speech and Language Processing (2012)
22. Zhu, Z.Y., Zhang, B.: Speech recognition method based on fuzzy support vector machine. Comput. Eng., 180–182 (2014)
23. Bhahramani, Z.: an introduction to hidden Markov models and Bayesian networks. Int. J. Pattern Recognit. Artif. Intell. **15**(1), 9–42 (2001)
24. Cho, K.Y.: Improved learning algorithms for restricted Boltzmann machines. Aalto University, Espoo (2011)
25. Chen, J., Deng, L.: A primal-dual method for training recurrent neural networks constrained. Echo-State Property **2013**(420), 629201 (2014)
26. Hayashi, T., Watanabe, S., Toda, T., et al.: BLSTM-HMM hybrid system combined with sound activity detection network for polyphonic sound event detection. In: Speech and Signal Processing (2017)
27. Exadaktylos, V., Silva, M., Aerts, J.M., et al.: Real-time recognition of sick pig cough sounds. Comput. Electron. Agric. **63**(2), 207–214 (2008)
28. Chung, Y., Seunggeun, O., Lee, J., et al.: Automatic detection and recognition of Pig wasting diseases using sound data in audio surveillance systems. Sensors **13**(10), 12929–12942 (2013)
29. Lin, Y.K., Su, M.C., Hsieh, Y.Z.: The application and improvement of deep neural networks in environmental sound recognition. Appl. Sci. **10**(17), 15–29 (2020)
30. https://freesound.org
31. Imran, A., Posokhova, I., Qureshi, H.N., et al.: AI4COVID-19: AI enabled preliminary diagnosis for COVID-19 from cough samples via an app. arXiv e-prints (2020)
32. Chen, M., Pan, J., Zhao, Q., et al.: Multi-task learning in deep neural networks for Mandarin-English code-mixing speech recognition. IEICE Trans. Inf. Syst. **99**(10), 2554–2557 (2016)
33. Xuan, L.I., Jian, Z., Yun, G., et al.: Recognition of Pig Cough Sound Based on Deep Belief Nets, Transactions of the Chinese Society for Agricultural Machinery (2018)
34. Guarino, M., Jans, P., Costa, A., et al.: Field test of algorithm for automatic cough detection in pig houses. Comput. Electron. Agric. **62**(1), 22–28 (2008)

Research on YOLO Model and Its Application in Fault Status Recognition of Freight Trains

Xueqi Li, Qing Liu(✉), Tongcai Liu, and Jingbo Wang

Wuhan University of Technology, Wuhan 430070, People's Republic of China

Abstract. Compared with other object detection and recognition methods, YOLO is an end-to-end algorithm that integrates target region prediction and target category prediction into a single neural network model. It can realize rapid target detection and recognition with high accuracy, which is more suitable for field application. The network models of YOLOV3 and YOLOV4 are studied, aiming at the application environment of fault state detection in railway transportation and freight. In view of the shortcomings of YOLOV3 algorithm, the Dense block-BC is used to replace Darknet-53 to build the YOLOV3 backbone network. CIoU is used to replace the MSE to calculate the boundary box loss, and Focal loss is used to redesign the loss function to improve the YOLOV3 model. The improved YOLOV3 and YOLOV4 are respectively used to detect the fault state in railway locomotive transportation, and the better detection results are obtained.16,660 pictures are collected directly from the railway transportation site as train set, and other 3,000 pictures are used as test set. The omission factor can be reduced to 4.164%, and the error rate can be decreased to 4.218%.The detection speed of each picture is 0.4s, which can meet the requirements of non-stopping fault state detection in railway transportation.

Keywords: Fault status recognition of freight trains · YOLOV3 · YOLOV4 · Object detection

1 Introduction

The current object detection algorithms can be divided into two categories. One is the two-stage algorithm, which has lower error rate and omission factor. But its speed is slower than one-stage algorithm, which cannot meet the real-time detection scene. The other is the one-stage detection algorithm, which only needs a single test to directly obtain the detection results, so it has a faster detection speed.

The pioneering work of one-stage target detection algorithm, YOLOv1 [1], is proposed by Joseph Redmon, Santosh Divvala et al., in 2015. Although the detection accuracy of YOLOv1 is not high, it greatly improves the detection speed when compared with other two-stage algorithms. Then Joseph Redmon proposes YOLOv2 [2] in 2017, and YOLOv3 [3] in 2018. The improvement of YOLO version mainly lies in the enhancement of its performance, including detection accuracy and recall rate. Apart from having obvious advantages in detection speed, YOLOv3 also greatly improves the detection

X. Sun et al. (Eds.): ICAIS 2021, CCIS 1422, pp. 144–156, 2021.
https://doi.org/10.1007/978-3-030-78615-1_13

accuracy. Because of its high precision and excellent speed, it has been widely applied in various fields of social production and life, such as architecture [4], medical treatment [5], ship [6], traffic [7], pedestrian detection [8] and so on. At the beginning of 2020, Alexey Bochkovskiy proposes YOLOv4 [9] algorithm. In YOLOv4, a large number of the latest object detection tricks are summarized and the best combination is found through a large number of experiments.YOLOv4 is released earlier this year and has yet to be seen in practical applications. In view of the deficiencies of YOLOV3, several improvements are proposed. Meanwhile, the differences between YOLOV4 network structure and YOLOV3 are analyzed.

In the railway freight transportation, covered goods wagon and the tank car are the most basic two kinds of train types. In the process of railway freight transport, when the boxcars' door or window open, foreign body suspension, tank car bonnet open, and violative hitching, etc., it is a fault status of locomotive operation, which will bring potential safety hazard to railway transport. The traditional way of handling is to rely on the station staff to visually observe the train to find out whether there is a fault state. Obviously, such a method has many problems such as heavy workload and serious missing inspection. Therefore, the paper focuses on the visual detection technology to realize the identification of train fault state, promote the development of visual identification in railway transportation fault state, improve the safety of railway freight and raise the level and accuracy rate of fault state detection. Train fault status detection is to collect images and start the detection when the train enters the station, and to complete the detection of all train carriages before the train leaves the station. The train images collected are of large size, with three pictures on both sides and top of each carriage. To achieve rapid detection and meet the requirements of engineering application, YOLO detection algorithm is an effective approach. In this paper, the improved Yolov3 and Yolov4 detection algorithms are used to detect and identify train fault status respectively, and their detection effect and performance are analyzed.

2 The YOLOv3 Model and Its Improvements

The feature extraction network of Yolov3 is darknet-53 without pooling layer and full connected layer. After the extracted feature map is processed by convolution, upsampling and splicing, three feature maps of different scales are output, as shown in the red box in the figure. Multi-scale is adopted to detect targets of different sizes, and the finer the grid cell is, the finer the object can be detected.

The excellent performance of Yolov3 enables it to meet many conditions of target detection and identification. But it still has the place which can improve, the following detailed analysis yolov3 deficiency and the improvement method.

2.1 A Subsection Sample

In YOLOv3, Darknet-53 is used as the backbone network, and the core part is residual module. The feature of residual module is that it can avoid the gradient disappearing while deepening the network, thus enhancing the network performance. However, while improving the performance of the network, the residual module has a low utilization rate for shallow features. Therefore, to improve network performance, it is necessary to improve the utilization rate of network features and enhance the feature extraction capability of each layer of network, so as to significantly reduce the required training parameters and reduce the complexity of the network.

In order to improve the performance of the backbone network, to reduce training parameters, to avoid the low efficiency of feature extraction in the original backbone network, residual modules are not used to construct the network. Instead, dense block -BC is put forward to build the backbone network, which is more efficient in the feature extraction.

Dense Block -BC is based on the Dense block. The core idea lies in the feature reuse, which ensures the maximum amount of information obtained by each network layer. The configuration of the Dense block [10] is shown in Fig. 1.

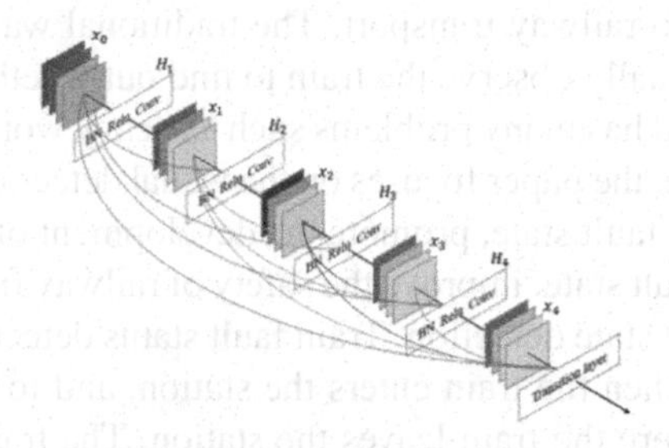

Fig. 1. Dense Block structure.

It is assumed that the network has L layers, each containing a nonlinear operation $H_l(\cdot)$, that is, a combination of convolution operation, BN operation, Relu function, etc. If a picture x_0 is entered into the Dense block network for forward propagation, x_l represents the output of layer L. The connection mode of the network can be expressed by formula (1).

$$x_l = H_l([x_0, x_1, \ldots, x_{l-1}]) \quad (1)$$

Where $[x_0, x_1, \ldots, x_{l-1}]$ represents merging the feature graph output from layer $0, \ldots, l-1$. The feature maps are merged on the channel dimension in the dense block, instead of addition like in the residual module. Its advantage is to avoid the information interference in the process of addition and is more conducive to the downward transmission of shallow features.

Ordinary convolutional layers are changed into Bottleneck layers in Dense Block-BC, and a 1×1 convolutional layer is added before 3×3 convolution for dimension reduction, which is intended to reduce the number of input feature maps and improve computing efficiency. The feature is further compressed in the transition layer between two dense block modules, which further reduces the training parameters of dense block.

The advantages of Dense block-BC are as follows: (1) Compared with other traditional network structures, there is no need to repeat learning redundant features, so the training parameters are less. (2) To improve the transmission of information and gradient in the network, each layer can directly use the gradient information at the bottom level and the original input information, that is, the middle layer does not simply extract features. (3) It avoids the problem of gradient disappearance and gradient explosion, makes the network easier to train, and alleviates the overfitting phenomenon caused by small training set.

The new backbone network based on Dense Block-BC is shown in Table 1. A total of 4 Dense blocks are used in new backbone. The number of network layers within each Dense block is 6, 12, 24 and 12 in turn. Each Dense block is connected by a transition layer and each transition layer contains a 1×1 convolution and a 3×3 convolution with step size of 2. Set the growth rate k to 12, that is, each 3×3 convolution in the Dense block will output 12 feature maps. The coefficient of compression θ is set to 1. In other words, the feature map is not compressed at the transition layer, because the number of feature maps is small and no further compression is required.

Table 1. Improved backbone network structure.

Type	Filters	Size	Output
Convolutional	32	7x7/2	208x208
Convolutional	64	3x3/2	104x104
[Convolutional	48	1x1]	
Convolutional	12	3x3]	
x 6			
Denseblock			104x104
Transition Layer	128	1x1 , 3x3/2	52x52
[Convolutional	48	1x1]	
Convolutional	12	3x3]	
x 12			
Denseblock			52x52
Transition Layer	256	1x1 , 3x3/2	26x26
[Convolutional	48	1x1]	
Convolutional	12	3x3]	
x 24			
Denseblock			26x26
Transition Layer	512	1x1 , 3x3/2	13x13
[Convolutional	48	1x1]	
Convolutional	12	3x3]	
x 12			
Denseblock			13x13

2.2 Use CIoU Instead of MSE to Calculate the Bounding Box Loss

MSE is taken as the bounding box loss function in YOLOv3 to conduct regression for four parameters that describe the target bounding box's center point coordinates (x, y) and the length and width (w, h) of the bounding box. However, MSE is often very sensitive to the size of the bounding box. Therefore, it is hoped that the loss function can be designed with a mathematical index that can not only measure the coincidence relationship between the predicted and the real bounding box, but is not sensitive to the scale size.

IOU can describe the measurement index of coincidence relation between bounding boxes better, which is a ratio, so it is less affected by the scale size of bounding boxes. However, using IOU as loss function calculation is still inadequate. When they do not overlap, the IOU cannot measure the distance relationship between them. In this case, the prediction box closer to the real target box should be better than the one far away from the real box.

In order to solve the deficiency of IOU, GIoU [11] is proposed. However, when the target box completely covers the prediction box, the values of IOU and GIoU are the same. At this time, GIoU degrades into IOU and its relative position relationship cannot be distinguished.

To solve the above problems, CIoU is introduced in the paper. Three important geometric factors are taken into account in CIoU Loss, including overlapping area, distance of center point and aspect ratio, so its convergence accuracy is higher. The penalty term in CIoU [12] is R_{CIoU}, and CIoU Loss function is L_{CIoU}.Its definition is shown in formula (2)–(5).

$$R_{CIoU} = \frac{\rho^2(b, b^{gt})}{c^2} + \alpha v \tag{2}$$

$$v = \frac{4}{\pi^2}(\arctan\frac{\omega^{gt}}{h^{gt}} - \arctan\frac{\omega}{h})^2 \tag{3}$$

$$\alpha = \frac{v}{(1 - \mathrm{IoU}) + v} \tag{4}$$

$$L_{CIoU} = 1 - \mathrm{IoU} + \frac{d^2}{c^2} + \alpha v \tag{5}$$

α is used as trade-off parameters. v is used to measure the aspect ratio of consistency. $\rho(\cdot)$ is the Euclidean distance. b and b^{gt} represent the center point of prediction box B and target box B^{gt}, respectively. d is the Euclidean distance between the two centers. c represents the diagonal distance of the smallest convex shapes mentioned in GIoU.

CIoU Loss can directly optimize the distance between the two detection boxes, increase the Loss of the detection box scale and the Loss of the length and width. Its convergence speed is faster and accuracy is higher.

2.3 Redesign the Loss Function with Focal Loss

In YOLOv3, confidence loss is more easily affected by sample imbalance than object category loss. Confidence is used to distinguish whether there are objects in the bounding box. It is a process of dichotomizing foreground and background. As most of the area in the picture belongs to the background, the sample imbalance is likely to occur in this process. To solve this problem, the original confidence loss is changed to the one based on Focal loss.

Focal Loss [13] is an improved loss function of traditional cross entropy loss, aiming at solving extreme imbalance between the foreground and background of one-stage object detection algorithm during training.

Focal loss function adds a scaling factor with adjustable focusing parameter $\lambda > 0$, which is defined as follows in formula (6).

$$FL = \begin{cases} -\alpha(1-\hat{y})^{\lambda}\log\hat{y}, & y = 1 \\ -(1-\alpha)\hat{y}^{\lambda}\log(1-\hat{y}), & y = 0 \end{cases} \quad (6)$$

As can be seen from Formula (6), with the join of λ, the loss of simple samples is gradually reduced, and the network focuses more on the classification of difficult and misclassified samples. In this way, Focal loss ensures that the contribution of difficult samples in loss is much more important than that of simple samples, which reduces the impact of simple samples. During model training, difficult and easy samples are not fixed. With the enhancement of network resolution, the original difficulty sample may become simple sample. At this time, because of the change in the output predicted value, the scaling factor can also dynamically adapt to the change of the network, ensuring the stability of the network training process.

2.4 Comparative Analysis of Experiments

The experimental standard data set used in this paper is PASCAL VOC 2007, which contains 20 categories. The detection results of improved YOLOV3 with dense block BC, Ciou and Focal loss are compared with the original YOLOV3. The comparison is shown in Fig. 2 below. The dense block BC structure is introduced into the original

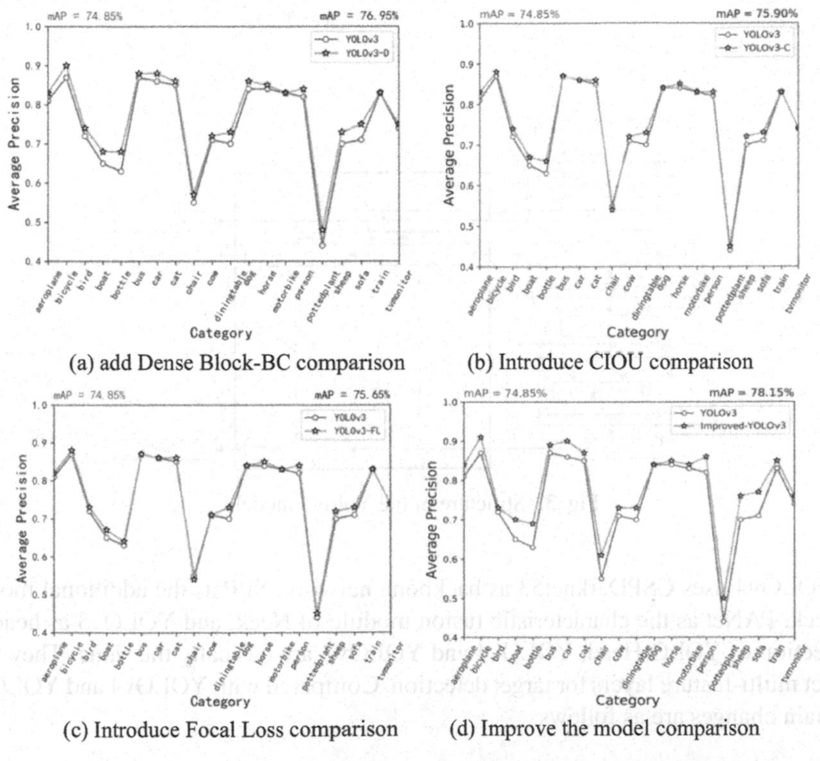

(a) add Dense Block-BC comparison (b) Introduce CIOU comparison

(c) Introduce Focal Loss comparison (d) Improve the model comparison

Fig. 2. Comparison of improved YOLOv3 and original model.

framework in YOLOv3-D. CIoU is introduced in YOLOv3-C. Focal Loss framework is introduced into the original framework in YOLOv3-FL.Improved-YOLOv3 refers to the improved model with all the above improvements introduced into the original model.

The mAP of each algorithm is shown in Table 2.

Table 2. Detection results on standard data set PASCAL VOC 2007.

algo	YOLOV3	YOLOv3-D	YOLOv3-C	YOLOv3-FL	Improved YOLOV3
mAP	74.85%	76.95%	75.90%	75.65%	78.15%

Table 2shows that the mAP improved by 2.1% after the addition of the Dense block-BC. With the addition of CIOU, mAP is improved by 1.05%. With the addition of Focal loss, mAP is improved by 0.8%. After three improvements are combined, the mAP of Improved-YOLOv3 is increased by 3.3% to 78.15%.The improved algorithm of YOLOv3 in the paper greatly improves the detection accuracy.

3 YOLOv4

In 2020, Alexey Bochkovskiy proposes YOLOv4. The structure is shown in Fig. 3 below.

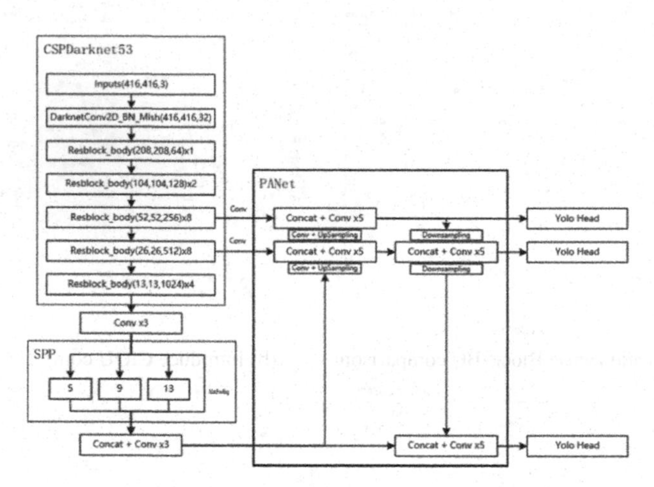

Fig. 3. Structure of the Yolov4 model.

YOLOv4 uses CSPDarknet53 as backbone network, SPP as the additional module of Neck, PANet as the characteristic fusion module of Neck, and YOLOv3 as head. In the section of YOLO Head, YOLOv4 and YOLOv3 are basically the same. They both extract multi-feature layers for target detection. Compared with YOLOv4 and YOLOv3, the main changes are as follows:

(1) Improvement on the input during training: including Mosaic data enhancement, cmBN normalization and SAT self-confrontation training;

(2) Improvement of the main network: including the introduction of CSPDark-net53, Mish activation function, Dropblock operation;

The stack of the original residual blocks is split into left and right parts in the CSP module. The main part continues to stack the original residual blocks. The other part, like a residual edge, is directly connected to the end after a small amount of processing. It improves the information flow of dense block and transition layer, optimizes the gradient back propagation path, and improves the learning ability of the network. At the same time, it improves the processing speed and reduces the memory.

(3) Introduction of Neck module: including SPP module and FPN + PAN structure;

The SPP structure is located after CSPDarknet53. After three convolutions of the last feature layer of CSPDarknet53, four different scales of maximum pooling are performed. The sizes of maximum pooling cores are 13×13, 9×9, 5×5 and 1×1, respectively. After being processed by SPP, the receptive field of the network can be greatly increased and the most significant contextual characteristics can be separated.

PANet [15] has a very important feature: repeated extraction of features. In FPN, the top-down feature fusion method transfers the high-level semantic features and enhances the whole pyramid, but only enhances the semantic information and does not transfer the positioning information. Therefore, the bottom-up feature pyramid is also needed to be implemented. Such operation is a supplement to FPN to pass on the low-level positioning features. In YOLOV4, PANet structure is mainly used on three effective feature layers.

(4) The loss function is improved as CIOU loss. This idea coincides with the improvement of YOLOV3 in this paper. And the NMS of prediction box screening is improved as DIOU nms.

The above improvements have improved the transmission efficiency of information flow in the network, reduced the loss of feature information in each layer of the network, increased the network's receptive field and feature learning ability, and significantly improved its recognition speed and accuracy when compared with YOLOV3 in the case of consistent input parameters.

4 Application of Train Fault Status Recognition

4.1 Image Characteristics of Railway Train Fault Status

There are many types of train fault status in railway transportation, mainly including door or window open of boxcars, foreign body suspension, vehicle body damage, cargo leakage, commodity autoignition, tank car bonnet open, violative hitching and so on. After field investigation, there will be three types of fault states with high frequency, which are foreign body suspension of the boxcar, bonnet open of the tank car and violative hitching. These types are as the main targets of the train fault status detection and recognition algorithm, namely, the research target of the paper. Figures 4 respectively includes example pictures of foreign body suspension of the boxcar, violative hitching and bonnet open of the tank car, in which foreign body suspension include weeds, ropes, sticks, cloth strips and plastic garbage bags. In order to ensure the integrity and comprehensiveness of visual inspection, when the locomotive enters the station, the on-site cameras will trigger to shoot both sides and the top of the car body. There are three pictures for each car, and each picture only contains one car. Due to the various fault states of the train body, the algorithm in the paper needs to conduct a comprehensive detection of the possible fault types on each carriage.

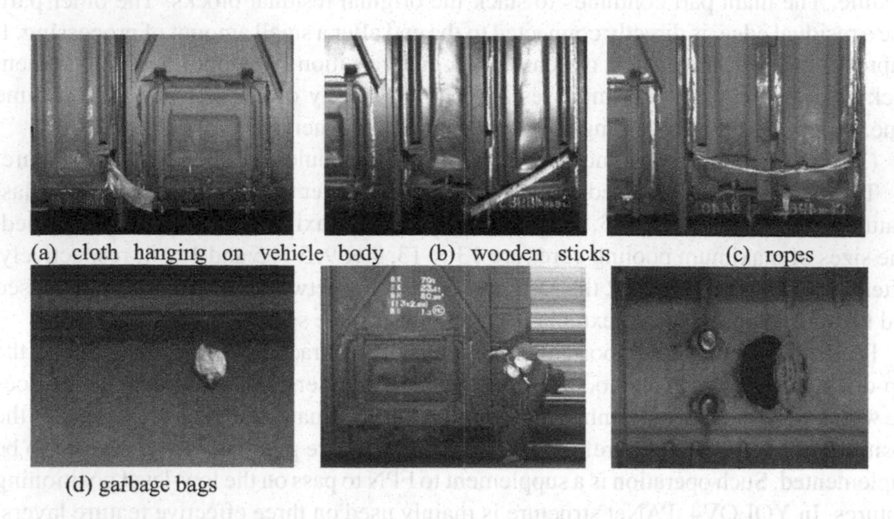

(a) cloth hanging on vehicle body (b) wooden sticks (c) ropes (d) garbage bags

Fig. 4. Foreign body suspension, Violative hitching, Bonnet open of tank car

Due to the relatively complex environment on the railway site, some adverse environmental factors may affect the shooting effect of the camera, and thus adversely affect the detection effect of the algorithm. Adverse factors include inclement weather, such as rainy days and foggy days, which may cause unclear picture and cause a certain degree of interference. It also includes that there is no stable light source on the scene, which cannot ensure that every picture taken can get sufficient light. If the picture taken is under insufficient light, it will greatly weaken the visibility of fault status on the train and increase the difficulty of detection. At the same time, the electronic equipment will inevitably produce a certain degree of noise, which will also interfere with the detection. In addition, due to the high safety requirements of the railway system, the appearance of the fault condition is relatively less, making it difficult to collect samples and the number of samples is very small. A small number of fault state images cannot be directly processed into training set to directly train the improved YOLOV3 and YOLOV4. Otherwise it is easy to enter the over-fitting state and cannot achieve ideal engineering applications. Aiming at the situation of few samples in fault status, transfer learning is adopted in the paper to carry out pre-training of YOLO model.

4.2 Pre-training of YOLO Model with Transfer Learning

Transfer learning can greatly reduce the dependence of the network on the number of target samples, so that it can fully train the network in the case of small samples. In the paper, the Imagenet data set is used as the source data domain to conduct pre-training on YOLO network, so that the network can gradually acquire the ability to analyze data and extract features. The convolutional layer weight of the backbone is saved. Then the pre-trained YOLO network is transferred to the train fault status data set in the target domain for secondary training, so that the YOLO network can learn the characteristics of the new target more quickly and adapt to the train fault status detection task more quickly. The following describes the training methods and steps for improved YOLOv3.

In the pre training stage, the Dense Block-BC backbone of the improved YOLOv3 network is extracted independently, and an average pooling layer and a full connection layer are introduced. The Imagenet data which has 1000 classes is used for training, and the convolution layer weight of the backbone network is saved. After the pre-training, the network is transferred to the train fault status data set for the formal training.Before the formal training, the parameters of the pre-training backbone network should be loaded in the complete improved YOLOv3 network, and the remaining output layer parameters should be initialized. The configuration of network training parameters for train fault status data set in target domain is shown in Table 3.

Table 3. Parameter setting of transfer learning.

Training parameter	Setting value
Activation function	LeakyRelu
Learning rate	0.001
Batch_size	64
Epoch	50
Optimization algorithm	Adam $\beta_1 = 0.9, \beta_2 = 0.9, \epsilon = 10^{-8}$

As shown in Table 3, the LeakyRelu activation function is selected to ensure the effectiveness of gradient descent in the negative axis. The Adam algorithm is used to optimize the gradient descent algorithm, which can accelerate the convergence of the model. The super parameters required by Adam algorithm are given in Table 3. Batch_size is set to 64 to ensure the accuracy of gradient descent direction while reducing the burden of computer memory.

4.3 Comparative Analysis of Identification Results

In the paper, the improved YOLOV3 and YOLOV4 are tested for locomotive fault status recognition. The training set consists of 16,660 samples, including 1,215 violative hitching samples, 4,514 foreign body suspension samples, 173 bonnet open samples, and 10,758 negative samples, which are trained 250,000 times. The test set consists of 3000 untrained samples, including 500 violative hitching samples, 500 foreign body suspension samples, 30 bonnet open samples, and 1970 failure-free samples.

The comparison of improved YOLOV3 and YOLOV4 recognition results is shown in Table 4. Among them,

$$omission\ rate = \frac{\text{Number of missing fault samples}}{\text{Number of total fault samples}} \times 100\%;$$

$$error\ rate = \frac{\text{Number of error detection samples}}{\text{Number of total normal samples}} \times 100\%;$$

$$\text{recall rate} = 1 - omission\ rate;\ \text{accuracy} = 1 - error\ rate;$$

Table 4. Omission and error rate of improved YOLOV3 and YOLOV4.

	Loss	Omission rate	Error rate	Recall rate	Accuracy
Improved-YOLOV3	0.1212	8.229%	6.645%	91.771%	93.355%
YOLOV4	0.6996	4.164%	4.218%	95.836%	95.782%

As can be seen from Table 4, the recall rate of improved YOLOV3 is 91.771%, and the accuracy is 93.355%. The recall rate and accuracy rate are both >90% in the case of fewer samples of train fault status, which can meet the requirements of engineering application. The data set, training times and other conditions remain unchanged. After replacing the model with YOLOV4, the recall rate and accuracy rate rise to 95.836% and 95.782%, respectively, with an increase of 4.065% and 2.427% compared with the improved YOLOV3.In the engineering system, GPU acceleration is carried out in YOLOV4. NVIDIA GeForce GTX 1060 with 6G Video Memory is used in the system, and the detection sample size is 1248 * 416. The average detection time is 0.4 s per image, which shows that YOLOV4 further improved the performance of YOLO model in fault status identification in railway freight.

5 Conclusion

In the paper, the deficiencies of the YOLOv3 model are analyzed and improved. The Dense block-BC module is introduced in the feature extraction section. The CIoU is introduced to calculate the boundary box loss, and the Focal loss is used to redesign the loss function. The improved model is validated against standard data sets. The improved Yolov3 model is applied to train fault status detection and identification, and a complete set of railway train fault identification system is designed to implement successfully on the engineering application. Then in 2020, YOLOv4 is applied to train fault status identification. The author uses the keyword "YOLOv3 + YOLOV4 + railway locomotives" to retrieve in international and domestic well-known database SCI, EI, SPRINGER, IEEE, CNKI, Wan fang and VIP. At present, there is hardly relevant literature on visual detecting train faults at home and abroad. After retrieving the above database, there is a literature about windows open status detection for boxcar in railway transportation based on traditional computer vision [16], which is the work of the authors' team. There is also a literature on the detection of high speed railway Contact network [17]. In this paper, the identification scheme of U-shaped ring nut of high-speed railway catenary is studied. Yolov3 is used for positioning and SENet is used for classification. The recognition accuracy is 88.24%, but the two models are not improved. The work of the paper makes a meaningful exploration on the research of fault state detection algorithm in non-stop railway transportation and realizes the engineering application of locomotive fault state detection in railway transportation. The improved YOLOv3 proposed in this paper can accurately identify and display a variety of train fault types. With the launch of YOLOV4, it provides more effective model identification for locomotive fault state detection.

Based on the framework research and engineering application practice of Yolov4, the following improvement directions are proposed for subsequent optimization and improvement.

(1) CSPDarknet53 is used in Yolov4. CSP is used to enhance the learning ability of Darknet53. However, in fact, in a trained deep neural network, it usually contains rich or even redundant feature maps to ensure a comprehensive understanding of the input data, which contains many similar feature maps. This kind of feature map does not need convolution operation, but can be generated with lower cost operation. The Ghost module proposed in GhostNet [14] can achieve this purpose and achieve model acceleration and lightweight. This module can be applied to the feature extraction part of Yolov4 to enhance its timeliness and accuracy.

(2) In order to solve the sample imbalance that is easy to occur in practical problems, the improvement of yolov3 in the paper can be used in Yolov4. That is, the confidence loss is changed to Focal loss to solve the extreme imbalance between the foreground and background of the one-stage target detection algorithm during training.

(3) Although optimized operations such as sparse connectivity and weight sharing can effectively alleviate the contradiction between model complexity and expression capacity, its information "memory" ability is not strong. For example, the model of cyclic neural network is difficult to learn the long distance dependence in the input sequence. Therefore, the Attention mechanism can be used to further improve the ability of neural network in information processing. Channel wise attention and spatial attention can be introduced respectively. Their advantages are: (a) increasing the corresponding features' weight of the categories to be predicted; (b) assigning more weight to the key parts, so that the attention of the model can be more focused on these parts.

References

1. Redmon, J., Divvala, S., Girshick, R., et al.: You only look once: unified, real-time object detection. In: Proceedings of the CVPR, pp. 779–788. IEEE (2016)
2. Redmon, J., Farhadi,A.: YOLO9000: better, faster, stronger. In: Proceedings of the CVPR, pp. 6517–6525. IEEE (2017)
3. Redmon, J., Farhadi, A.: Yolov3: an incremental improvement. arXiv preprint arXiv:1804.02767 (2018)
4. Ma, H., Liu, Y., Ren, Y., Jingxian, Y.: Detection of collapsed buildings in post-earthquake remote sensing images based on the improved YOLOv3. Remote Sens. **12**(1), 44 (2019)

5. Liu, J., Wang, X.W.: Early recognition of tomato gray leaf spot disease based on MobileNetv2-YOLOv3 model. Plant Meth. **16**(1), 83 (2020)
6. Wu, Z., Li, L., Gao, Y.: Rotary Convolution Integration YOLOv3 Model for remote sensing Image ship detection. Comput. Eng. Appl. **55**(22), 146–151 (2019)
7. Li, X., Zhang, J., Xie, Z., Wang, J.: Rapid detection algorithm for traffic signs based on three-scale nested residual structure. J. Comput. Res. Develop. **57**(5), 1022–1036 (2020)
8. Liu, L., Zheng, Y., Fu, D.: Blocking pedestrian detection algorithm based on improved YOLOv3 network structure. Pattern Recogn. Artif. Intell. **33**(6), 568–495 (2020)
9. Bochkovskiy, A., Wang, C.Y., Liao, H.Y.M.: YOLOv4: optimal speed and accuracy of object detection (2020)
10. Huang, G., Liu, Z., Maaten, L.V.D., et al.: Densely connected convolutional networks. In: Proceedings of the CVPR. IEEE Computer Society (2017)
11. Rezatofighi, H., Tsoi, N., Gwak, J.Y.: Generalized Intersection over Union: A Metric and A Loss for Bounding Box Regression (2019)
12. Zheng, Z., Wang, P., Liu, W., Li, J., Ye, R., Ren, D.: Distance-IoU loss: faster and better learning for bounding box regression. Proc. AAAI Conf. Artif. Intell. **34**(07), 12993–13000 (2020). https://doi.org/10.1609/aaai.v34i07.6999
13. Lin, T.-Y., Goyal, P., Girshick, R.: Focal loss for dense object detection. IEEE Trans. Pattern Anal. Mach. Intell. **99**, 2999–3007 (2017)
14. Han, K., Wang, Y., Tian, Q., et al.: GhostNet: more features from cheap operations. In: 2020 IEEE Conference on Computer Vision and Pattern Recognition (CVPR). IEEE (2020)
15. Wang, K., Liew, J.H., Zou, Y., et al.: PANet: few-shot image semantic segmentation with prototype alignment (2019)
16. Yan, W., Liu, Q., Guo, J.: Windows open status detection for boxcar in railway transportation based on computer vision. In: International Conference on Industrial Informatics-computing Technology. IEEE Computer Society (2017)
17. Chen, Q., Liu, L., Han, R., Qian, J., Qi, D.: Image identification method on high speed railway contact network based on YOLO v3 and SENet. In: 2019 Chinese Control Conference (CCC), Guangzhou, China, pp. 8772–8777 (2019)
18. Chen, M., Wang, X., He, M., Jin, L., Javeed, K., Wang, X.: A network traffic classification model based on metric learning. Comput. Mater. Continua **64**(2), 941–959 (2020)

Feedforward Deep Neural Network-Based Model for the Forecast of Severe Convective Wind

Yu Jing, Nan Wang(✉), Qiang Zhao, Pingyun Li, and Qiyuan Hu

Shaanxi Meteorological Observatory, Xi'an, China

Abstract. In this study, we compared and analyzed the environmental characteristics in Shaanxi between 2016–2018 (before and after the occurrence of convective wind) based on both severe and non-severe wind samples extracted from the fifth generation of European Centre for Medium-Range Weather Forecasts (ECMWF) Reanalysis (ERA5) data and automatic weather station data from the China Meteorological Administration. Through the establishment of deep feedforward neural networks, these environmental characteristics were then used to define a model for the forecast of severe convective wind events. For ~75% (or > 75%) severe convective wind samples, the differences between 1 h after the occurrence of severe convective wind and when it occurred, 1 h after it occurred and 1 h before it occurred in MLCAPE, MUCAPE, SBCAPE, MLCIN, MUCIN, SBCIN, MLEL, MUEL, SBEL, SBLCL and temperature were negative, while those in MLLI, MULI, SBLI, sea level pressure (SLP), and precipitation were positive. This suggests that, for ~75% (or > 75%) severe convective wind samples, atmospheric stability increased or SLP increased, or temperature decreased; moreover, precipitation was found to occur after severe convective wind. Finally, a certain degree of differentiation was noted between the parameters associated with the severe and non-severe convective wind samples. We designed two schemes (each containing three kinds of experiments) and trained a feedforward deep neural network to predict severe convective wind events. The network experiment including the higher number of elements was found to provide the highest threat score (Ts).

Keywords: Severe convective wind · Environmental characteristics · Feedforward deep neural network

1 Introduction

Severe convective wind caused by strong convective weather is a fundamental element in operational weather forecasting [1]. Previous research has focused on the analysis of storm types producing severe convective winds [2, 3], including squall lines, bow echoes, and supercells. In particular, some researchers have analyzed the evolution of severe-convective-wind-producing systems [4–7] (e.g., large-scale bow echo convective systems leading to extensive wind damage [4]). Furthermore, others have analyzed the connection between atmospheric conditions [8–15] and severe convective winds. For

X. Sun et al. (Eds.): ICAIS 2021, CCIS 1422, pp. 157–167, 2021.
https://doi.org/10.1007/978-3-030-78615-1_14

example, convective wind forecasting in weak-, moderate-, and strong-shear environments have been reviewed [8], differences between the variables of three MCS categories (classified based on the production of severe surface winds) have been identified and discussed [9], climatic features and environmental parameters associated with severe convective winds have been analyzed [11], and environments that support damaging wind leading to convection have been examined [13]. Severe convective winds are often triggered by small- and meso-scale convective systems, which are highly localized. The severe convective winds forecasting is still challenging [16–18]. The downdraft convective available potential energy (DCAPE), convective inhibition (CIN), vertical wind shear between 0–3 km, PWAT, and other parameters can provide some indication on the occurrence of severe convective winds [19, 20]; therefore, it is very important to study the environmental conditions linked to severe convective winds. In recent years, deep learning technology has been gradually applied to the meteorology field (e.g., precipitation nowcasting [21], short-time heavy rainfall recognition [22], weather phenomenon recognition [23], and radar quantitative precipitation estimation [24]).

In this study, we analyzed the environmental parameters associated with severe and non-severe convective wind (SCW and NSCW, respectively) events in order to determine the environmental conditions before and after the two types of wind events and the corresponding differences. Furthermore, these environmental parameters were used to train a feedforward deep neural network model.

2 Data and Methods

The Severe Weather Prediction Center of the National Meteorological Center (China) defines any wind gust with speed ≥ 17.2 m s^{-1} as a severe wind event. SCW events can be identified based on cloud-to-ground lightning [25]. Based on this method, we identified a total of 406 SCW events that occurred between 2016–2018. Since severe convective winds are often triggered by small- and meso-scale convective systems [17], we selected only NSCW events that occurred at sites $\geq$ 200 km away from SCW events (at the time of their measurement), for a total of 13,692 NSCW events. High-altitude data were obtained from ECMWF Reanalysis (ERA5) hourly reanalysis data collected between April and September from 2016 to 2018 and having a resolution of 0.1° × 0.1°. The ERA5 reanalysis data were interpolated to the SCW and NSCW sites respectively, and only those referring to heights above the ground were retained. The surface elements of the automatic weather station were combined with the interpolated ERA5 elements; then, the merged data were used to calculate the 0–1 km, 0–3 km, and 0–6 km shears, as well as the MLCAPE, MLCIN, MLEL, MLLCL, MLLI, MUCAPE, MUCIN, MUEL, MULCL, MULI, SBCAPE, SBCIN, SBEL, SBLCL, SBLI, PWV, DCAPE (in these abbreviations, "ML", "MU", and "SB" stand for "mixed-layer parcel", "the most unstable layer parcel", and "surface-based parcel", respectively, which were used to calculate the convective available potential energy (CAPE), the CIN, the equilibrium level (EL), the lifting condensation level (LCL), the lifted index (LI), the precipitable water vapor (PWV), and the DCAPE). In all cases, the calculation time started 1 h before the occurrence of the SCW events and ended 1 h after them. For each parameter, we calculated the differences between the values registered: during the occurrence of the

SCW events and 1 h before them, 1 h after the SCW events and during their occurrence, and 1 h after the SCW events and 1 h before their occurrence. The characteristics of the environmental parameters of the SCW and NSCW events were compared and, finally, were used to train a feedforward deep neural network model for their prediction.

3 Environmental Analyses

As described in Sect. 2, the ERA5 hourly reanalysis and automatic weather station data were combined to calculate the environmental parameters associated with the SCW and NSCW events.

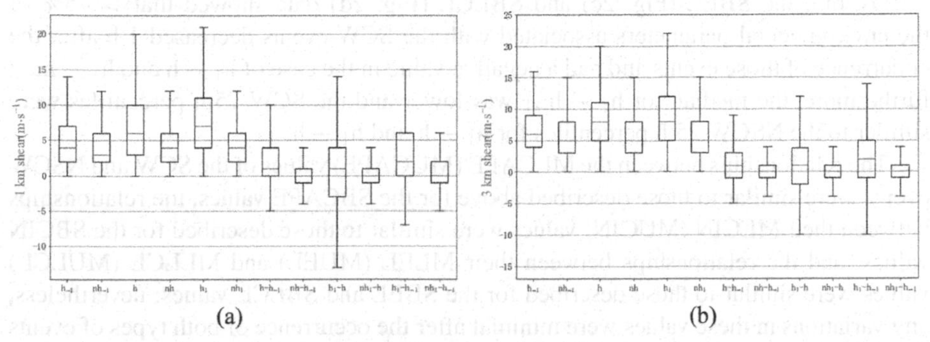

Fig. 1. Box plot showing the 1-km (a), 3-km (b) shears of the SCW and NSCW events. The character "n" identifies the environmental parameters associated with the NSCW events, Meanwhile, "h_{-1}" indicates the values of the environmental parameters 1 h before the SCW events, "h" their values during the SCW events, "h_1" their values 1 h after the SCW events, "$h - h_{-1}$" the difference between the environmental parameter values registered during the SCW events and 1 h before them, "h_1_h" the difference between the environmental parameter values registered 1 h after the SCW events and during their occurrence, and "h_1_h_{-1}" the difference between the environmental parameter values registered 1 h after the occurrence of SCW events and 1 h before them (the same below).

Figure 1 shows how the median, 25th percentile, 75th percentile, and upper whisker 1 km-shear SCW values (between h_{-1} and h_1 in Fig. 1a) were greater than the correspondent NSCW values; moreover, the 25th, 50th, and 75th percentile 1-km shear SCW values increased over time. Additionally, the positive value ratios of the 1 km-shear SCW differences $h - h_{-1}$, $h_1 - h$, and $h_1 - h_{-1}$ were greater than the correspondent NSCW values, and the median of the 1-km shear SCW differences was similar to the 75th percentile of the 1-km shear NSCW differences $h - h_{-1}$, $h_1 - h$, and $h_1 - h_{-1}$. The relationship between the 3-km shear SCW and NSCW differences (Fig. 1b) was similar to that between the correspondent 1-km shear differences; however, the positive value ratios of the 3-km shear SCW differences $h - h_{-1}$ and $h_1 - h_{-1}$ were lower than those of the correspondent 1-km shear SCW differences. Finally, the 6-km shear SCW and NSCW differences were very similar to each other (not shown in the figures).

Figure 2a shows how the median of the SBCAPE values of the SCW events (SCW SBCAPE) for h_1 was significantly lower than that of the NSCW events (NSCW

SBCAPE); moreover, the 75th percentile of the SCW SBCAPE was much lower than that of the NSCW SBCAPE. Notably, the negative value ratio of the SCW SBCAPE gradually increased over time (from $h - h_{-1}$ to $h_1 - h$ and $h_1 - h_{-1}$). The median of the SCW SBCAPE for $h - h_{-1}$, $h_1 - h$, and $h_1 - h_{-1}$ gradually decreased, while the median of the NSCW SBCAPE for $h - h_{-1}$ and $h_1 - h$ showed no obvious changes; furthermore, the 75th percentile of the SCW SBCAPE was similar to the 25th percentile of the NSCW SBCAPE for $h_1 - h_{-1}$.

The median of the SCW CIN (Fig. 2b) was relatively high for h_{-1}, but relatively low for h_1. Importantly, > 50% of the SCW CIN values for h_1 were < -200 J kg^{-1}, while 75% of the SCW CIN values for $h_1 - h_{-1}$ were negative and the median value of the SCW CIN was lower than the lower whisker of the NSCW CIN for $h_1 - h_{-1}$.

Overall, the SBEL (Fig. 2c) and SBLCL (Fig. 2d) data showed that > 75% of the environmental parameters associated with the SCW events decreased 1 h after the occurrence of those events and had a negative value in the case of $h_1 - h$ and $h_1 - h_{-1}$; furthermore, the median for $h_1 - h_{-1}$ was lower and the SCW 75th percentiles were similar to the NSCW 25th percentiles for $h_1 - h$ and $h_1 - h_{-1}$.

The relationships between the MLCAPE (MUCAPE) values of the SCW and NSCW events were similar to those described above for the SBCAPE values, the relationships between their MLCIN (MUCIN) values were similar to those described for the SBCIN values, and the relationships between their MLEL (MUEL) and MLLCL (MULCL) values were similar to those described for the SBEL and SBLCL values; nevertheless, any variations in these values were minimal after the occurrence of both types of events and the differences between them were quite small (not shown in the figures).

Interestingly, > 75% of the SCW SBLI values increased 1 h after the occurrence of the correspondent events (Fig. 2e). Also, ≥ 75% of the SCW SBLI values were positive for $h - h_{-1}$, $h_1 - h$, and $h_1 - h_{-1}$ and their proportion gradually increased over time. In particular, the SCW SBLI 25th percentile value of the was similar to the NSCW SBLI upper whisker value for $h_1 - h_{-1}$. Moreover, ~75% of the SCW MLLI values were positive for $h - h_{-1}$ and $h_1 - h$, and the SCW MLLI 25th percentile value was close to the upper whisker NSCW MLLI value for $h - h_{-1}$ and $h_1 - h$ (Fig. 2f). The variations in MULI were similar to those in SBLI; however, the change in MULI 1 h after the occurrence of the SCW events was smaller and the differences between the SCW and NSCW MULI values were smaller (not shown in the figures).

It was noted that ~75% of the SCW DCAPE values were negative for $h_1 - h_{-1}$, and that the median of the SCW DCAPE values was similar to the NSCW DCAPE 25th percentile value for $h_1 - h_{-1}$ (Fig. 2g). Additionally, > 50% of the SCW PWV values were positive for $h_1 - h_{-1}$, and the difference between the SCW and NSCW PWV values was small (not shown in the figures).

Of the SCW sea level pressure (SLP) and precipitation values, > 75% were positive for $h_1 - h$ and $h_1 - h_{-1}$, while > 75% of the SCW temperature values were negative for $h_1 - h$ and $h_1 - h_{-1}$. This means that > 75% of the SCW SLP values should have increased, or that the temperature values should have decreased, or that precipitation should have occurred after the SCW events (Fig. 3). The proportion of positive SCW SLP values gradually increased from $h - h_{-1}$ to $h_1 - h$ and $h_1 - h_{-1}$: ~50% of the NSCW SLP values were positive for $h_1 - h$, while < 50% of them were positive for

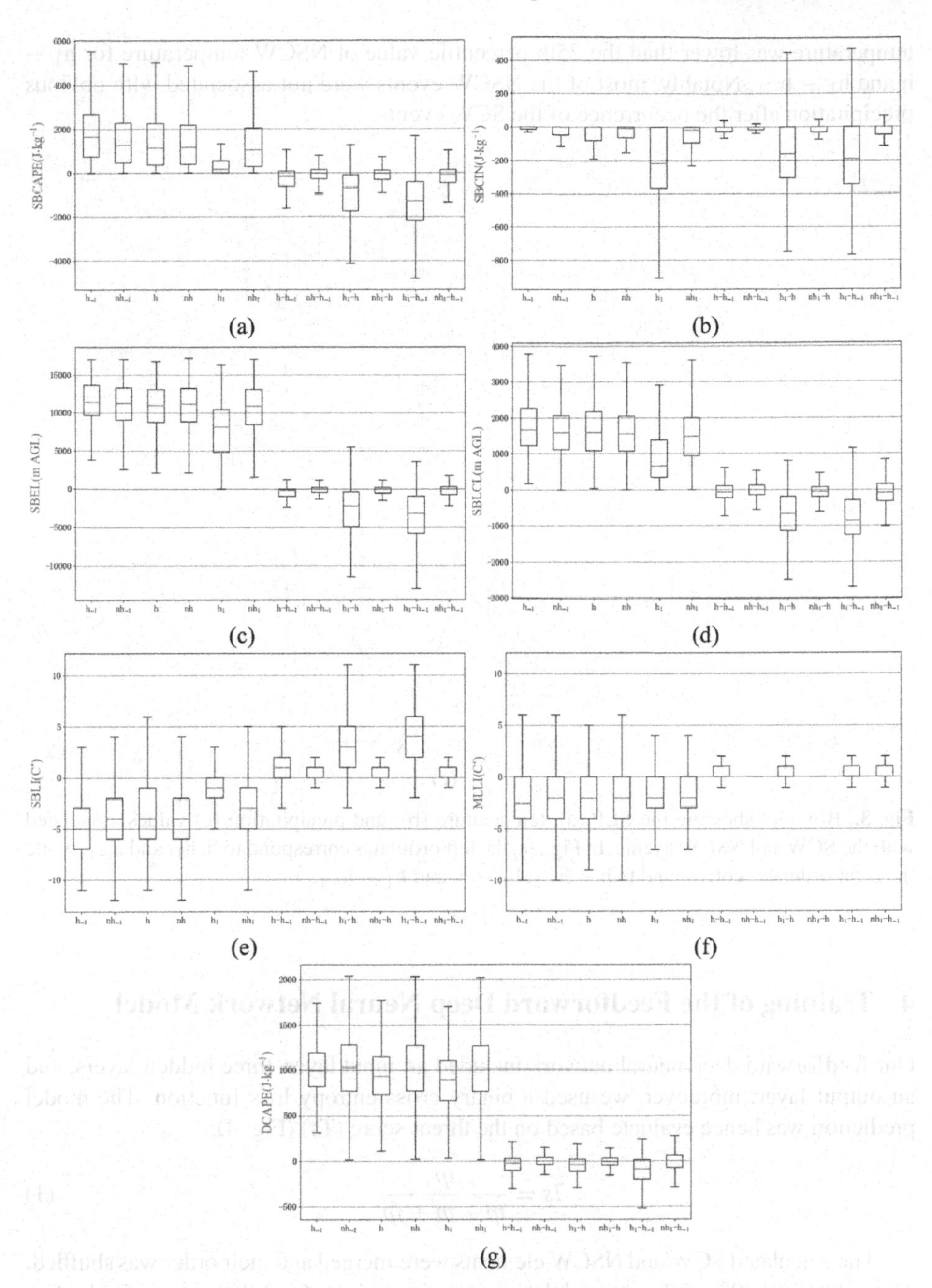

Fig. 2. Box plot showing the SBCAPE (a), SBCIN (b), SBEL (c), SBLCL (d), SBLI (e), MLLI (f), and DCAPE (g) values associated with the SCW and NSCW events.

$h_1 - h_{-1}$. Meanwhile, the proportion of negative SCW temperature values gradually increased from $h - h_{-1}$ to $h_1 - h$ and $h_1 - h_{-1}$: the 75th percentile value of SCW temperature was lower than the 25th percentile value of NSCW temperature for $h_1 - h$ and $h_1 - h_{-1}$. Notably, most of the NSCW events were not associated with obvious precipitation after the occurrence of the SCW events.

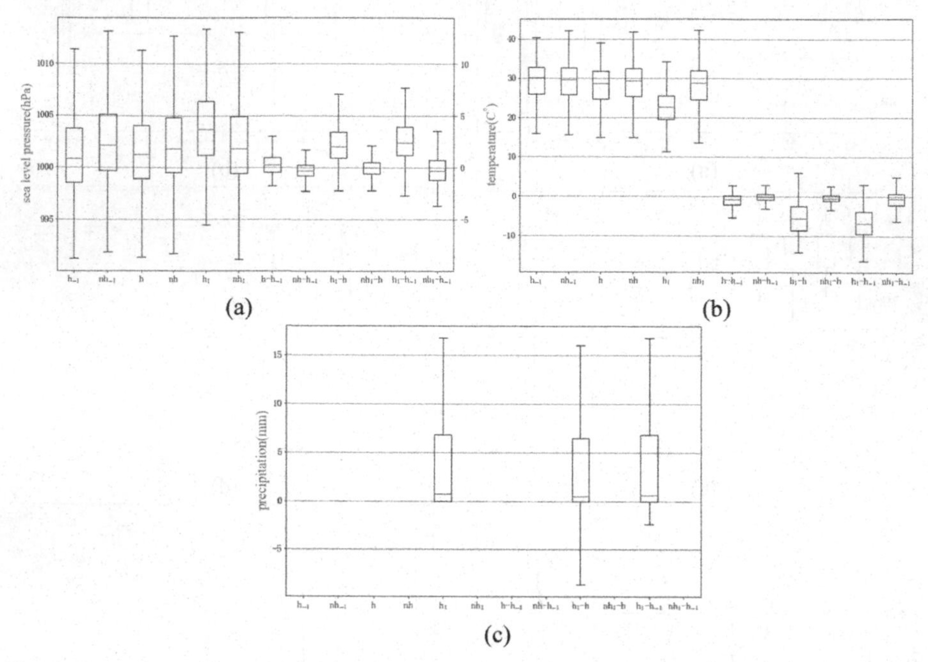

Fig. 3. Box plot showing the SLP (a), temperature (b), and precipitation (c) values associated with the SCW and NSCW events. In Fig. 3a, the left ordinates correspond to h, h_1, and h_{-1}, while the right ordinates correspond to $h - h_{-1}$, $h_1 - h$, and $h_1 - h_{-1}$.

4 Training of the Feedforward Deep Neural Network Model

Our feedforward deep neural network included an input layer, three hidden layers, and an output layer; moreover, we used a binary cross-entropy loss function. The model prediction was hence evaluate based on the threat score (Ts) (Fig. 4):

$$Ts = \frac{tp}{tp + fn + fp} \tag{1}$$

The calculated SCW and NSCW elements were merged and their order was shuffled. Then, 80% and 20% of the merged data were set as train and test data, respectively; furthermore, 20% of the train data were used as validation data. Through the feedforward deep neural network, we conducted a severe convective wind prediction experiment, in which the SCW and NSCW events represented positive and negative examples, respectively. Two schemes were then designed. In Scheme 1, the number of neurons in the hidden layer decreased layer by layer, while in Scheme 2 each layer of the hidden layer

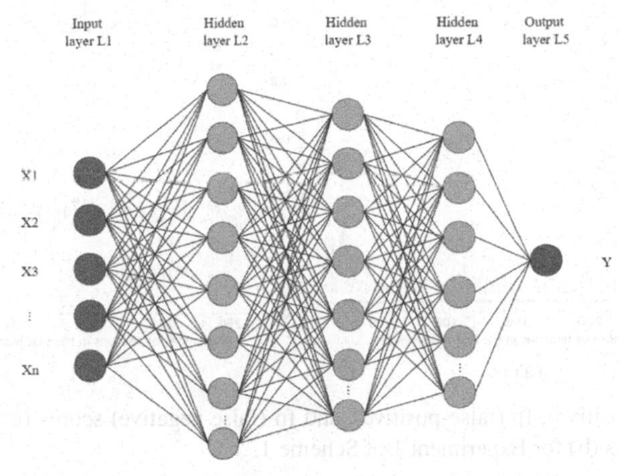

Fig. 4. Representation of the feedforward deep neural network framework used in this study.

included the same number of neurons. Notably, each of these schemes included three experiments. In Experiment 1, the values of 23 environmental parameters for h_{-1}, h, h_1, $h - h_{-1}$, $h_1 - h$, and $h_1 - h_{-1}$ (for a total of 138 elements) were entered for the model training. In Experiment 2, the environmental parameters with the highest degree of distinction between the SCW and NSCW events (i.e., the SBCAPE, SBCIN, SBEL, SBLCL, SBLI, SLP, temperature, and precipitation values for h_{-1}, h, h_1, $h - h_{-1}$, $h_1 - h$, and $h_1 - h_{-1}$, for a total of 48 elements) were entered for the model training. In Experiment 3, the environmental parameters with the highest degree of distinction between the SCW and NSCW events (i.e., SLP (for $h_1 - h$), temperature (for h_1, $h_1 - h$, and $h_1 - h_{-1}$), precipitation (for h_1, $h_1 - h$, and $h_1 - h_{-1}$), MLLI (for $h - h_{-1}$ and $h_1 - h$), MULI (for $h_1 - h_{-1}$), SBCAP (for $h_1 - h_{-1}$), SBCIN (for h_1, $h_1 - h$, and $h_1 - h_{-1}$), SBEL (for $h_1 - h$ and $h_1 - h_{-1}$), SBLCL (for $h_1 - h$), and SBLI (for $h_1 - h$ and $h_1 - h_{-1}$), for a total of 19 elements) were entered for the model training. Notably, in all experiments, the number of neurons in the first hidden layer decreased from 138 $\times$ 2.

Figures 5, 6, 7, 8, 9 and 10 show how, when the threshold setting for a positive class of output layer was 0.5, in both schemes Experiment 1 resulted in fewer false-positive scores and higher Ts, while Experiment 3 resulted in a higher number of false-positive scores and lower Ts. Notably, the highest Ts was reached in Experiment 1 of Scheme 1 (when the number of neurons in the first hidden layer was 179), while the second highest Ts was reached in Experiment 2 of Scheme 2 (when the number of neurons in the first hidden layer was 163).

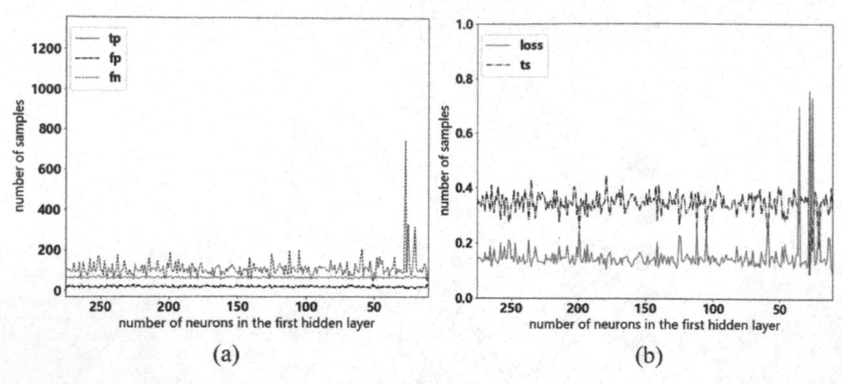

(a) (b)

Fig. 5. Tp (true-positive), fp (false-positive), and fn (false-negative) scores (a); validation data loss and test data Ts (b) for Experiment 1 of Scheme 1.

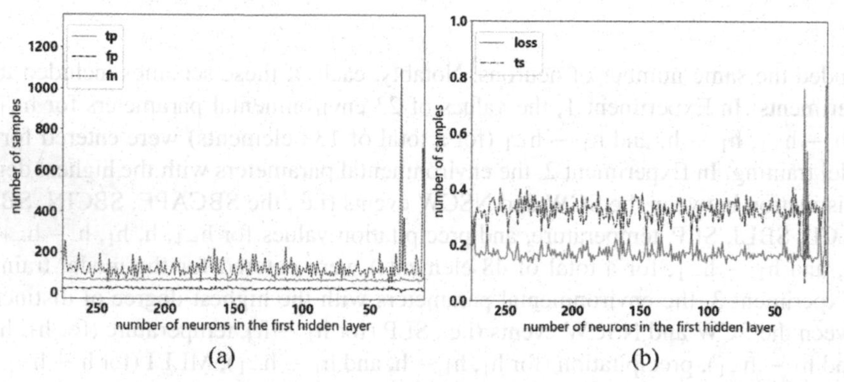

(a) (b)

Fig. 6. Tp, fp, and fn scores (a); validation data loss and test data Ts (b) for Experiment 2 of Scheme 1.

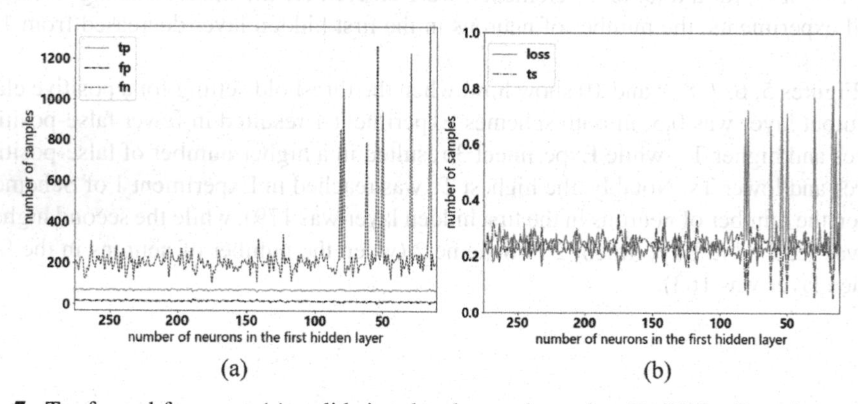

(a) (b)

Fig. 7. Tp, fp, and fn scores (a); validation data loss and test data Ts (b) for Experiment 3 of Scheme 1.

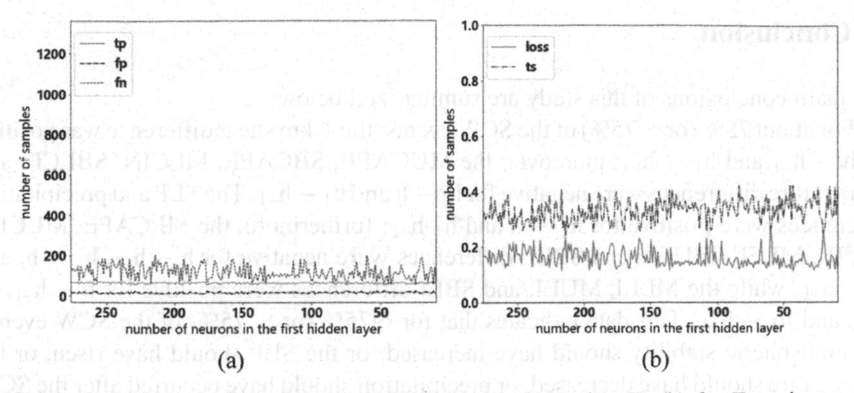

Fig. 8. Tp, fp, and fn scores (a); validation data loss and test data Ts (b) for Experiment 1 of Scheme 2.

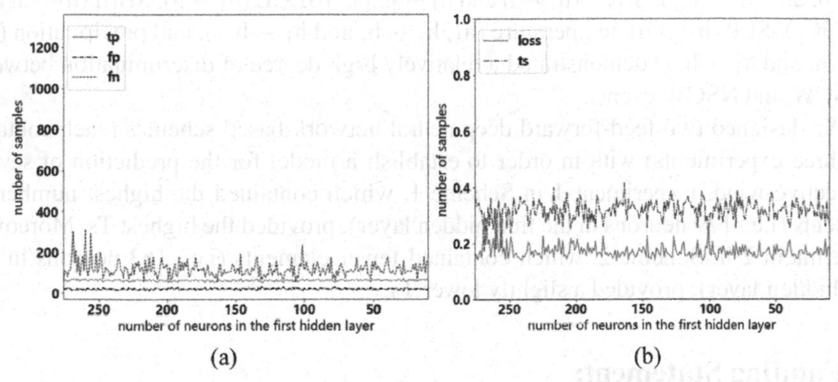

Fig. 9. Tp, fp, and fn scores (a); validation data loss and test data Ts (b) for Experiment 2 of Scheme 2

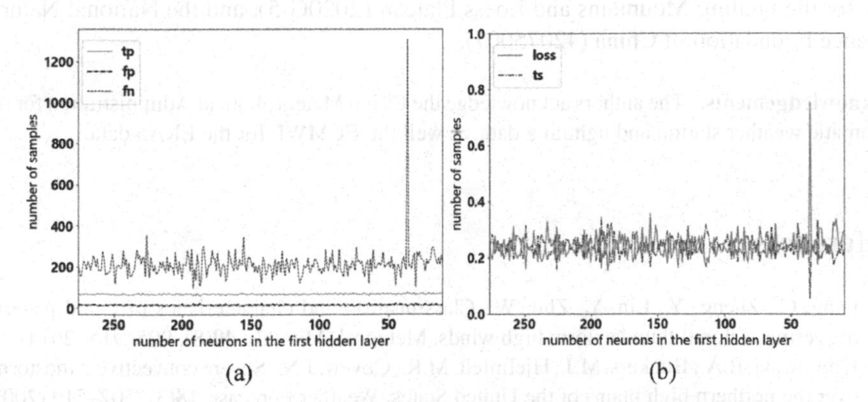

Fig. 10. Tp, fp, and fn scores (a); validation data loss and test data Ts (b) for Experiment 3 in Scheme 2.

5 Conclusion

The main conclusions of this study are summarized below.

For about 75% (or > 75%) of the SCW events, the 1-km shear difference was positive for $h - h_{-1}$ and $h_1 - h_{-1}$; moreover, the MUCAPE, SBCAPE, MLCIN, SBLCL, and temperature differences were negative for $h_1 - h$ and $h_1 - h_{-1}$. The SLP and precipitation differences were positive for $h_1 - h$ and h_1-h_{-1}; furthermore, the MLCAPE, MUCIN, SBCIN, MLEL, MUEL, and SBEL differences were negative for $h - h_{-1}$, $h_1 - h$, and $h_1 - h_{-1}$, while the MLLI, MULI, and SBLI differences were positive for $h - h_{-1}$, $h_1 - h$, and $h_1 - h_{-1}$. This data indicates that for ~75% (or > 75%) of the SCW events, the atmospheric stability should have increased, or the SLP should have risen, or the temperature should have decreased, or precipitation should have occurred after the SCW events.

The MLLI ($h - h_{-1}$ and $h_1 - h$), MULI ($h_1 - h_{-1}$), SBCAP ($h_1 - h_{-1}$), SBCIN (h_1, $h_1 - h$, and $h_1 - h_{-1}$), SBEL ($h_1 - h$ and $h_1 - h_{-1}$), SBLCL ($h_1 - h$), SBLI ($h_1 - h$ and $h_1 - h_{-1}$), SLP ($h_1 - h$), temperature (h_1, $h_1 - h$, and $h_1 - h_{-1}$), and precipitation (h_1, $h_1 - h$, and $h_1 - h_{-1}$) demonstrated a relatively high degree of discrimination between the SCW and NSCW events.

We designed two feed-forward deep neural network-based schemes (each containing three experiments) with in order to establish a model for the prediction of severe convective wind. Experiment 1 in Scheme 1, which contained the highest number of elements (i.e., 179 neurons in the first hidden layer), provided the highest Ts. Moreover, Experiment 2 in Scheme 2, which contained fewer elements (i.e., 163 neurons in the first hidden layer), provided a slightly lower Ts.

6 Funding Statement:

This work was supported by the Key Laboratory of Eco-Environment and Meteorology for the Qinling Mountains and Loess Plateau (2020G-5), and the National Natural Science Foundation of China (42075007).

Acknowledgements. The authors acknowledge the China Meteorological Administration for the automatic weather station and lightning data, as well the ECMWF for the ERA5 data.

References

1. Fang, C., Zheng, Y., Lin, Y., Zhu, W.: Classification and characteristics of cloud patterns triggering regional thunderstorm high winds. Meteorol. Monthly **40**(8), 905–915 (2014)
2. Klimowski, B.A., Bunkers, M.J., Hjelmfelt, M.R., Covert, J.N.: Severe convective windstorms over the northern high plains of the United States. Weather Forecast. **18**(3), 502–519 (2003)
3. Schoen, J.M., Ashley, W.S.: A Climatology of fatal convective wind events by storm type. Weather Forecast. **26**(1), 109–121 (2011)
4. Przybylinski, R.W.: The bow echo: observations, numerical simulations, and severe weather detection methods. Weather Forecast. **10**(2), 203–218 (1995)

5. Kanak, J., Benko, M., Simon, A., Sokol, A.: Case study of the 9 May 2003 windstorm in southwestern Slovakia. Atmos. Res. **83**(2), 162–175 (2007)
6. Lennard, C.: Simulating an extreme wind event in a topographically complex region. Bound.-Layer Meteorol. **153**(2), 237–250 (2014). https://doi.org/10.1007/s10546-014-9939-x
7. Wang, X., Zhou, X., Yu, X.: Comparative study of environmental characteristics of a windstorm and their impacts on storm structures. Acta Meteor. Sin. **71**(5), 839–852 (2013)
8. Johns, R.H., Doswell, C.A., III.: Severe local storms forecasting. Weather Forecast. **7**(4), 588–612 (1992)
9. Cohen, A.E., Coniglio, M.C., Corfidi, S.F., Corfidi, S.J.: Discrimination of mesoscale convective system environments using sounding observations. Weather Forecast. **22**(5), 1045–1062 (2007)
10. Meng, Z., Yan, D., Zhang, Y.: General features of squall lines in East China. Mon. Weather Rev. **141**(5), 1629–1647 (2013)
11. Fei, H., Wang, X., Zhou, X., Yu, X.: Climatic characteristics and environmental parameters of severe thunderstroms gales in China. Meteorol. Monthly **42**(12), 1513–1521 (2016)
12. Chen, G.T.J., Chou, H.C.: General characteristics of squall lines observed in TAMEX. Mon. Weather Rev. **121**(3), 726–733 (1993)
13. Kuchera, E.L., Parker, M.D.: Severe convective wind environments. Weather Forecast. **21**(4), 595–612 (2006)
14. Bentley, M.L., Mote, T.L., Byrd, S.F.: A synoptic climatology of derecho producing mesoscale convective systems in the North-Central Plains. Int. J. Climatol. **20**(11), 1329–1349 (2000)
15. Ashley, W.S., Mote, T.L., Bentley, M.L.: On the episodic nature of derecho-producing convective systems in the United States. Int. J. Climatol. **25**(14), 1915–1932 (2010)
16. Tang, W., Zhou, Q., Liu, X., Zhu, W., Mao, X.: Analysis on verification of national severe convective weather categorical forecasts. Meteorol. Monthly **43**(1), 67–76 (2017)
17. Yang, X., Sun, J., Lu, R., Zhang, X.: Environmental Characteristics of severe convective wind over south China. Meteorol. Monthly **43**(7), 769–780 (2017)
18. Pantillon, F., Lerch, S., Knippertz, P., Corsmeier, U.: Forecasting wind gusts in winter storms using a calibrated convection-permitting ensemble. Q. J. R. Meteorol. Soc. **144**(715), 1864–1881 (2018)
19. Ma, S., Wang, X., Yu, X.: Environmental parameter characteristics of severe wind with extreme thunderstorm. J. Appl. Meteorol. Sci. **30**(3), 292–301 (2019)
20. Fang, C., Wang, X., Sheng, J., Cao, Y.: Temporal and spatial distribution of North China thunder-gust winds and the statistical analysis of physical characteristics. Plateau Meteorol. **36**(5), 1368–1385 (2017)
21. Franch, G., Nerini, D., Pendesini, M., Coviello, L., Jurman, G., Furlanello, C.: Precipitation nowcasting with orographic enhanced stacked generalization: improving deep learning predictions on extreme events. Atmosphere **11**(3), 267 (2020)
22. Lu, Z., Ren, Y., Sun, X., Jia, H.: Recognition of short-time heavy rainfall based on deep learning. J. Tianjin Univ. (Sci. Technol.) **51**(2), 111–119 (2018)
23. Wang, C., Liu, P., Jia, K., Jia, X.: Identification of weather phenomena based on lightweight convolutional neural networks. Comput. Mater. Continua **64**(3), 2043–2055 (2020)
24. Wang, S., Yu, X., Liu, L., Huang, J., Wong, T., Jiang, C.: An approach for radar quantitative precipitation estimation based on spatiotemporal network. Comput. Mater. Continua **65**(1), 459–479 (2020)
25. Yang, X., Sun, J., Zheng, Y.: A 5-yr climatology of severe convective wind events over China. Weather Forecast. **32**(4), 1289–1299 (2017)

Asynchronous Compatible D-ML System for Edge Devices

Guangfu Ge[1](✉), Feng Zhu[1,2], and Yuchen Huang[2]

[1] Nanjing Research Institute of Electronic Engineering, Nanjing, China
[2] Nanjing University of Aeronautics and Astronautics, Nanjing, China

Abstract. The parameter server system (Parameter Server, PS) is the most widely used distributed machine learning system. Its core module is the communication module, which mainly includes three parts: communication content, communication frequency, and communication pace. The communication scheme of the existing PS system will have problems such as waste of computing power resources and a large proportion of communication overhead. This paper hopes to design and implement a distributed set that saves computing power and has a smaller communication overhead based on the communication module of the original PS system. Machine learning system, this article mainly has the following contributions: First, through the analysis and experiment of Bulk Synchronous Parallel (BSP) communication pace, based on the superiority of BSP algorithm level, it overcomes the waste of computing power of BSP at the system level, and designs and implements a new set of Communication pace, the new communication pace under the same training task can save more computing power and achieve higher model accuracy. Secondly, this article combines the quantization compression technology with the parameter server system, and uses the quantization technology to compress and decompress the communication content in the parameter server system, so that the volume of the communication content during the operation of the parameter server system is significantly reduced, thereby reducing the communication overhead and improving Speedup ratio of the system. Finally, based on the analysis and experiment of the change trend of the model accuracy rate, this paper designs and implements a dynamic communication frequency adjustment scheme based on the change trend of the model accuracy rate, which reduces the overall communication frequency during the operation of the parameter server system and reduces the parameter server. T Experiments show that the system designed in this paper is superior to the existing mainstream solutions in terms of communication pace, communication content, and communication frequency. When training the same model, this system compared with the existing parameter server system in model convergence time, model There is a noticeable improvement in the highest accuracy.

Keywords: Distributed · Machine learning · Communication frequency · Communication content · Communication pace · Quantization Compression · BSP

X. Sun et al. (Eds.): ICAIS 2021, CCIS 1422, pp. 168–179, 2021.
https://doi.org/10.1007/978-3-030-78615-1_15

1 Introduction

1.1 A Subsection Sample

In recent years, artificial intelligence has made unprecedented progress, and the key to promote its obvious progress is two aspects: large-scale data (big data), large scale computing power (strong computing power). When we have enough data and strong computing power, we can train a machine learning model (large model) with enough scale and high complexity, which can be closer to the decision boundary, so its accuracy and efficiency are much higher than that of traditional artificial intelligence technology. While big data and big model have positive effects, they inevitably lead to many new challenges.

First of all, for enterprises, a lot of datas are stored on users' mobile phones or electronic devices. To train a good model with these datas, the most direct way is to upload the user's data to the enterprise's server for model training. However, if the user's private data, including but not limited to pictures, recordings, videos, etc., there are often big legal issues and privacy issues, most of the time, enterprises do not have the right or ability to upload user's data to the centralized server.

Secondly, The more data you have, the more computing power you need. In industry, the magnitude of data is often TB or even greater. Such a large data scale is a great challenge to the computing power. On the one hand, the memory of a single machine is often limited, so it is difficult to aggregate such large-scale data into one machine for training. On the other hand, the computing power of a single machine is also limited, Although there are GPU and even TPU as special machine learning chips, the development and growth of these chips are difficult to meet the requirements of training the required models quickly and accurately in the face of huge data scale. In order to solve this kind of problem, distributed machine learning emerges as the times require.

Distributed machine learning applies the idea of distributed computing to machine learning, but compared with the traditional distributed computing, it has obvious characteristics.

First of all, although machine learning relies on data, its purpose is to learn rules or models from the data, rather than accurately store or index the original data. Machine learning has strong robustness for small differences in data details. Therefore, unlike other distributed computing tasks, which requires the execution of thc computing process on a single machine and a cluster is completely consistent, but disturbed machine learning only needs that the learned laws and models are statistically consistent. In other words, even if the distributed implementation of a machine learning task differs from its stand-alone implementation at the execution level, as long as the rules and models of the final output are not significantly different from those of the stand-alone version, then the distributed implementation is a good implementation.

Secondly, the ultimate goal of machine learning is to use the learned model to complete classification, regression and other tasks. If the distributed implementation of machine learning blindly pursues acceleration and loses the accuracy of the model, the result is not worth the loss. But on the other hand, because machine learning is concerned with the expected accuracy on the test set rather than the empirical accuracy on the training set, Therefore, if we introduce some small noise into the distributed implementation, it will probably increase the generalization ability of the learning process and bring better test results.

One of the mainstream implementations of distributed machine learning is the parameter server (PS) [1]. The core principle of the parameter server system is to distribute data to the computer cluster (multiple computing nodes, workers). Each computing node conducts sub training locally, and then uploads the results of the sub training to the parameter server then update the overall model, and then get the latest model on the parameter server to the local, and then carry out the next round of training based on the latest overall model, and so on until the model achieves the expected effect.

2 Related Work

Distributed optimization and inference is becoming popular for solving large scale machine learning problems. Using a cluster of machines overcomes the problem that no single machine can solve these problems sufficiently rapidly, due to the growth of data in both the number of observations and parameters. Implementing an efficient distributed algorithm, however, is not easy. Both intensive computational workloads and the volume of data communication demands careful system design.

It is worth noting that our system targets situations that go beyond the typical cluster-compute scenario where a modest number of homogeneous, exclusively- used, and highly reliable is exclusively available to the researcher. That is, we target cloud-computing situations where machines are possibly unreliable, jobs may get preempted, data may be lost, and where network latency and temporary workloads lead to a much more diverse performance profifile. For instance, it is understood that synchronous operations may be significantly degraded due to occasional slowdowns, reboots, migrations, etc. of individual servers involved. In other words, we target real cloud computing scenarios applicable to Google, Baidu, Amazon, Microsoft, etc. rather than low utilization-rate, exclusive use, high performance supercomputer clusters. This requires a more robust approach to computation.

There are several general distributed machine learning systems. Mahout is based on Hadoop and MLI, based on spark, and adopts iterative MapReduce framework. Although spark is much better than Hadoop MapReduce because of its state preserving and optimized execution strategy, both methods use synchronous iterative communication mode. This makes them vulnerable to the uneven performance distribution of iterative machine learning algorithms, that is, machines may be slow at any given time. To overcome this limitation, distributed graphlab [2] uses graph abstraction to schedule communication asynchronously. However, it lacks the flexibility of the framework based on mapreduce and relies on coarse-grained snapshots for recovery. In addition, global variable synchronization is not the first type of primitive. Of course, in addition to these general frameworks, many application specific systems have been developed, such as [3–6].

3 System Design-Communication Content Compression Based on Quantization

Parameter or parameter updating is the main content of communication between work node and parameter server node in PS system. Most of these communication contents are floating-point data, and floating-point number will occupy a large amount of memory, that is to say, it will generate large communication development, The most direct way to solve this problem is to compress the content of communication. For the compression of floating-point data, the most effective method is to use quantization technology [7, 8].

Quantization is a good compression and coding method. The purpose of quantization is to reduce the storage volume of the model and accelerate the operation. The core of quantization is to convert floating-point storage (operation) to integer storage (operation), That is to say, a weight needs to be represented by float32, and quantization needs make the data converted to int8. In this way, the memory occupied by the model can be greatly reduced, thus reducing the communication bandwidth and communication overhead.

Quantization can be divided into two categories according to whether the mapping function is linear or not: linear quantization and nonlinear quantization. The quantization technique used in this paper is linear quantization. The linear quantization process can be expressed by formula (1).

$$r = \mathrm{Round}(S(q - Z)) \tag{1}$$

Among them, Q represents the original float32 type value, Z represents the offset of float32 value, which is also called zero point in many places, S refers to the scaling factor of float32, which is also called scale in many places. Round() represents the mathematical function of rounding approximate rounding. In addition to rounding, it is also possible to use up or down rounding, R represents the quantized result (an integer value). According to whether Z is 0, linear quantization can be divided into symmetric quantization and asymmetric quantization.

As shown in Fig. 1, it shows the linear quantization diagram. The so-called symmetric quantization is to use a mapping formula to map the input data to the range of [128, 127]. In the graph, -max (| XF |) represents the minimum value of input data, and max (| XF |) represents the maximum value of input data. The core of symmetric quantization is the processing of zeros. The mapping formula needs to ensure that the zeros in the original input data still correspond to the zeros in the [128, 127] interval after passing through the mapping formula. In a word, symmetric quantization maps the input data in the range of [128, 127] through mapping relations. For mapping relations, we need to solve the parameters, namely Z and s. In symmetric quantization, R is represented by a signed integer value (int8), where z = 0, and q = 0, there is exactly r = 0. In symmetric quantization, we can take z = 0, and the value of S can be calculated by formula (2).

$$S = \frac{2^{n-1} - 1}{\max\ (|x|)} \tag{2}$$

Therefore, the specific steps of quantification are as follows:

(1) The corresponding Min_Value and Max_value is counted in the input data (usually the weight or activation value);

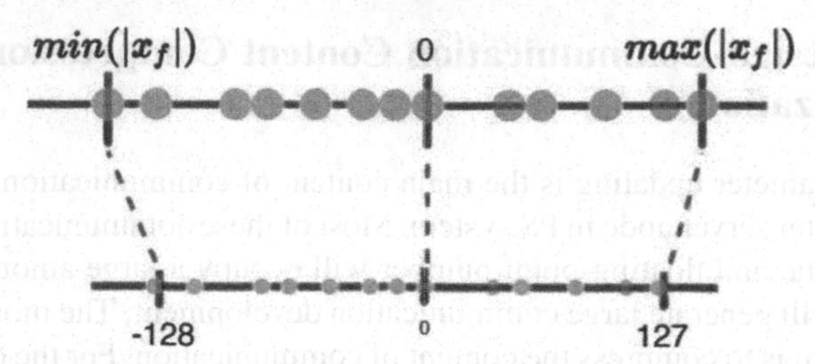

Fig. 1. Schematic diagram of linear quantization.

(2) Choose the appropriate quantization type, namely symmetric quantization (int8) or asymmetric quantization (uint8);
(3) According to the type of quantification, Min_Value and Max_Value get to calculate the quantified parameters Z/zero point and S/scale;
(4) According to the calibration data, the model is quantized from fp32 to int8;

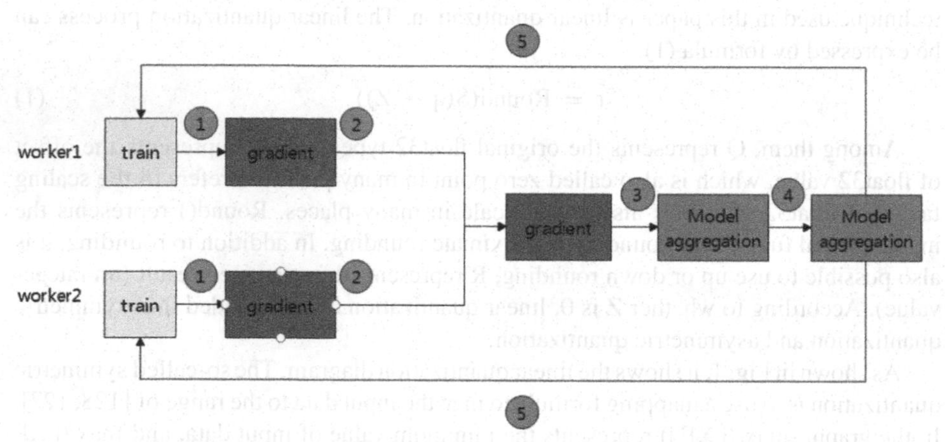

Fig. 2. Schematic diagram of core operation steps of parameter server system before quantification.

As shown in Fig. 2, the core steps of the parameter server system are as follows:

(1) The work node trains locally and generates a new gradient (model) after completing a round of iteration (training);
(2) The work node sends (push) the gradient to the parameter server node, that is, the parameter server node receives the gradient transmitted by the work node;
(3) The parameter server node aggregates the gradient received from the work node with the global model maintained locally to obtain a new global model (gradient);
(4) The work node obtains the latest global model (pull) from the parameter server;

The work node starts the next iteration based on the new global model obtained from the parameter server node;

From the above core process analysis, we can know that the most core communication steps in the parameter server system are push and pull. The main time consumption of communication is also these two parts, and gradient (model) is the communication content of these two parts.

Therefore, in order to reduce communication time and communication cost, we need to compress the communication content of these two parts, namely ladder Degree (model). The core step is to quantify the communication content before the work node push, and then reverse quantify the received content on the parameter server. The parameter server node quantifies the new global model before it is ready to send it to the work node. The work node receives the model and then reverse quantization, and then carries out the next round of iteration and training.

Specifically, as shown in Fig. 3, it is followed that the core process of the server system operation after introducing the quantified parameters:

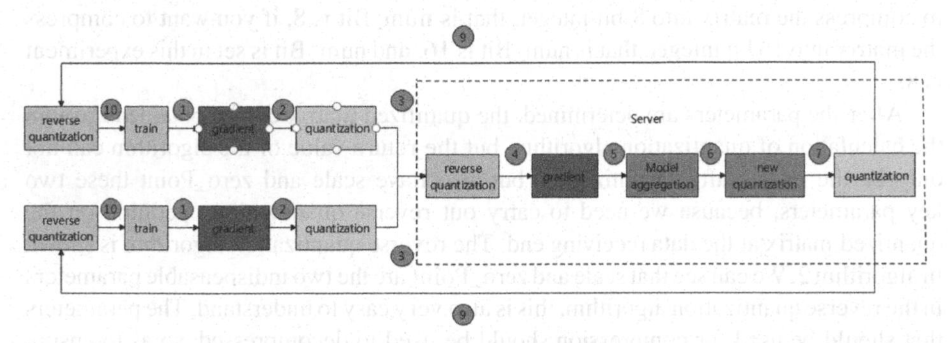

Fig. 3. Schematic diagram of core operation steps of parameter server system after introducing quantization.

4 System Experimental Implementation

The software system of this experiment is based on the ray [9, 10] framework. Ray is a new high-performance distributed execution framework developed by UC Berkeley riselab. It uses a different architecture from the traditional distributed computing system and abstracts distributed computing, and has better computational performance than spark. Ray is a distributed execution engine based on python. The same code can run on a single machine to achieve efficient multiprocessing, and can be used for a large number of calculations on the cluster. We can use ray to implement the parameter server system.

In the experiment, we use convolution neural network to train the model. In addition, the key factor to determine the efficiency of neural network is the super parameter. This paper mainly involves the following core parameters:

- num-Workers: the number of work nodes. This paper is limited by the number of hardware processor cores. Take the value between 1 and 5 for the test;

- sleep-Gap: Superparameters used to simulate the difference of computing power of different nodes. The unit is s. in this paper, 0.1 is taken as the starting point and 0.1 is added for each test;
- min-Lock: the maximum communication times difference mainly used to control the difference of communication times between working nodes with different computing power in the SSP communication pace, which is set as 5 in this paper;
- iterations: the maximum number of iterations is set as 100 in this paper; the above are the specific parameter settings of this experiment.

The core of communication content compression is mainly the realization of quantization and reverse quantization. For example, algorithm 1 is the core algorithm to realize quantization in this system. The algorithm needs two parameters, one is the matrix X to be quantized, which is the communication content to be compressed in the experiment, and the other parameter is the number of bits expected to be quantized, that is, if you want to compress the matrix into 8-bit integer, that is num_Bit is 8, if you want to compress the matrix into 16 bit integer, that is num_Bit is 16, and num_Bit is set in this experiment is 8.

After the parameters are determined, the quantized matrix can be obtained through the calculation of quantization algorithm, but the return value of the algorithm can not only be the matrix after quantization, but also have scale and zero_Point these two key parameters, because we need to carry out reverse quantization operation on the quantized matrix at the data receiving end. The reverse quantization algorithm is shown in algorithm 2. We can see that scale and zero_Point are the two indispensable parameters in the reverse quantization algorithm, this is also very easy to understand, The parameters that should be used for compression should be used to decompressed, so as to ensure the authenticity of the data.

Algorithm 1 quantization algorithm

Input: Matrix to be quantized, X;Expected quantized bits, NUM_BIT;
Output: Quantized matrix vector, q_Tensor
1: qmin=0
2: qmax $=2^{NUM\_BIT}$-1
3: min_val, max_val = X.min(), X.max()
4: scale $= \frac{max\_val - min\_val}{qmax - qmin}$
5: initial-zero-point = qmin - $\frac{min\_val}{scale}$
6: $zero\_point = 0$
7: if $initial\_zero\_point <$ qmin then
8: $zero\_point$ = qmin
9: else if $initial\_zero\_point >$ qmax then
10: zero_point = qmax
11: else
12: zero_point = initial_zero_point
13: end if
14: zero_point= int(zero_point)
15: Q_X = zero_point $+\frac{x}{scale}$
16: Q_X.clamp(qmin, qmax).round()
17: Q_X = Q_X.round().byte()
18: Q_T ensor=QTensor(tensor=q_x, scale=scale, zero_point=zero_point)

Algorithm 2 reverse quantization algorithm

Input: Quantized matrix vector, Q_Tensor;
Output: Matrix to be quantized, X;Expected quantized bits, NUM_BIT;
1: scale=Q_T ensor.scale
2: x=Q_T ensor.tensor.float()
3: zero_point=Q_T ensor.zero_point
4: temp=Q_T ensor.tensor.float() - Q_T ensor.zero_point
5: X=Q_T ensor.scale × temp

---

5 Experimental Results of Communication Content Compression

Taking BSP communication pace as an example, the computational power difference of the working nodes in the control system is 0.1, and the total number of iterations of the system is 200. In order to make the experimental results more convincing, we use 25 work nodes to conduct comparative experiments in each dimension. The experimental results are as follows:

As shown in Fig. 4, it shows the comparison of communication time before and after the introduction of quantization under different number of work nodes. The horizontal axis represents the number of work nodes, and the vertical axis represents the communication time required for training the same model under the corresponding number of work nodes. It can be seen that when the number of work nodes changes from 2 to 5, the communication time before and after quantization is increased. This is because if the number of iterations is fixed, the number of communication times will increase, so

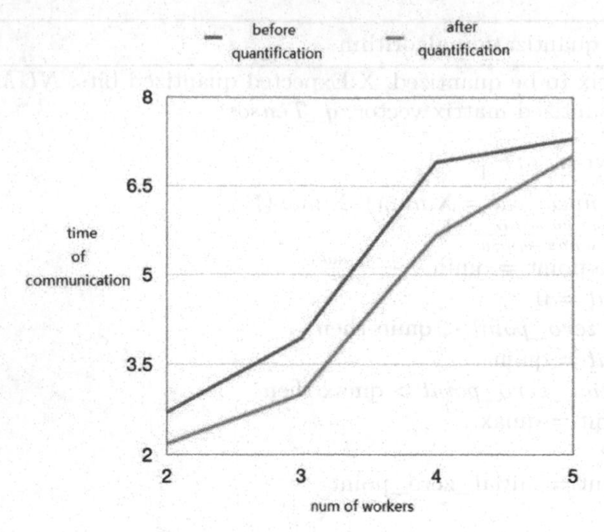

Fig. 4. Comparison of communication time before and after quantization.

the total communication time will also increase. however, after introducing quantization, the time required for communication is lower than that before quantization, which shows that the introduction of quantization reduces the communication overhead, which is because the introduction of quantization compresses the communication content and increases the communication speed.

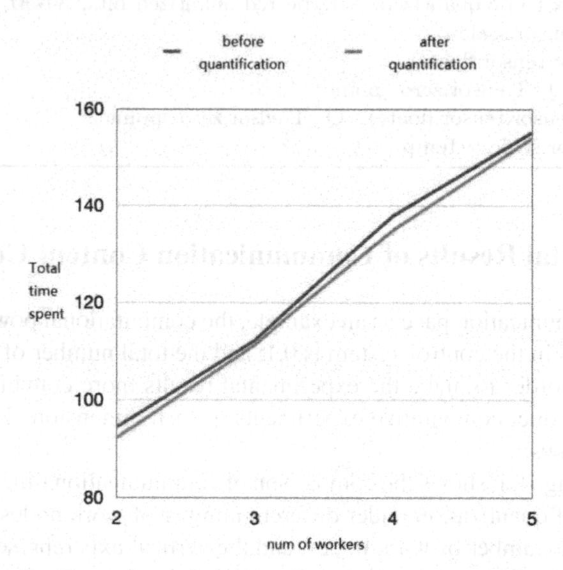

Fig. 5. Comparison of total time before and after quantification.

Although the introduction of quantization can reduce the communication time, it will increase the additional quantization and reverse quantization costs, that is, the additional time cost of compression and decompression. In order to measure the disparity of communication time reduction and compression time cost, as shown in Fig. 5, we compare the total time consumption of the system before and after the introduction of quantization under different number of work nodes. It can be seen that although the introduction of quantization increases the time cost of quantization, the total time required for training the same model after introducing quantization is lower than that before introducing quantization, that is to say, the benefits introduced by quantization technology are greater than the losses, that is to say, it is feasible to use quantization compressed communication content.

Another reason for introducing quantization is to reduce the acceleration ratio of the system. The so-called acceleration ratio refers to the proportion of communication time in the total time consumption during the system operation. As shown in Fig. 6, it shows the proportion of communication time in the total time before and after quantization. The horizontal axis represents the number of work nodes, and the vertical axis represents the proportion of communication time in the total time. It can be seen that the proportion of communication overhead after quantization is lower than that before quantization. This is because we reduce the communication time consumption of the system by introducing quantization while other time-consuming of the system remains unchanged. Therefore, the introduction of quantization technology is conducive to reducing the acceleration ratio of the system.

Finally, although the introduction of quantization can reduce the communication time and reduce the communication cost, we also need to pay attention to whether it will have a negative impact on the accuracy of the model. As shown in Fig. 7, the comparison chart of accuracy before and after quantization is shown when the number of work nodes is 4. The horizontal axis represents the number of iterations, and the vertical axis represents

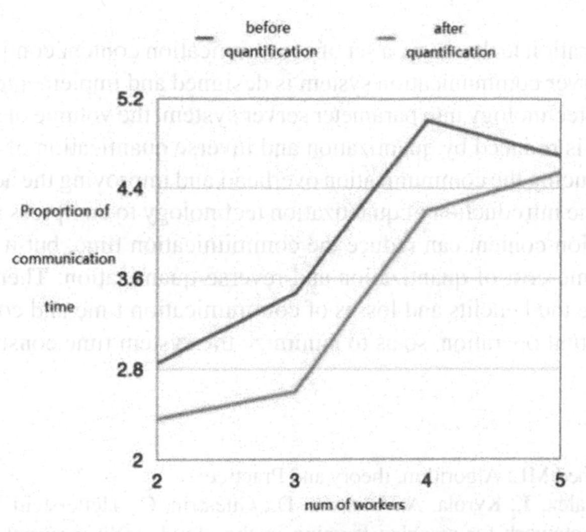

Fig. 6. Communication time in total time before and after quantization.

the accuracy rate. From the figure, we can see that the accuracy curve of the model before and after the introduction of quantization is approximately coincident, that is, the introduction of quantification will not have a significant negative impact on the accuracy of the model.

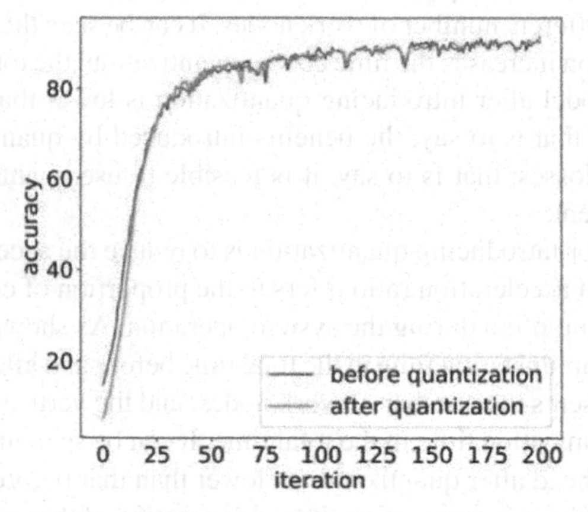

Fig. 7. Comparison of model accuracy before and after quantification.

In conclusion, the quantization compression of communication content can reduce the communication overhead and the proportion of communication time consumption without negative impact on the accuracy of the model, so as to improve the system efficiency, which is an effective measure to improve the communication efficiency of the system.

6 Conclusion

Based on quantization technology, a set of communication content compression scheme for parameter server communication system is designed and implemented. By introducing quantization technology into parameter server system, the volume of system communication content is reduced by quantization and inverse quantization of communication content, thus reducing the communication overhead and improving the acceleration ratio of the system; The introduction of quantization technology to compress and decompress the communication content can reduce the communication time, but it also introduces the additional time cost of quantization and reverse quantization. Therefore, it is necessary to balance the benefits and losses of communication time and compression time reasonably in actual operation, so as to minimize the system time consumption;

References

1. Liu, Y.: Disturbed ML: Algorithm, theory and Practice
2. Low, Y., Gonzalez, J., Kyrola, A., Bickson, D., Guestrin, C., Hellerstein, J.M.: Distributed graphlab: a framework for machine learning in the cloud. arXiv preprint arXiv:1204.6078 (2012)

3. Ahmed, A., Shervashidze, N., Narayanamurthy, S., Josifovski, V., Smola, A.J.: Distributed large-scale natural graph factorization. In: Proceedings of the 22nd international conference on World Wide Web, pp. 37–48 (2013)
4. Dean, J., et al.: Large scale distributed deep networks. In Advances in neural information processing systems, pp. 1223–1231 (2012)
5. Chen, W.-Y., Song, Y., Bai, H., Lin, C.-J., Chang, E.Y.: Parallel spectral clustering in distributed systems. IEEE Trans. Patt. Anal. Mach. Intell. **33**(3), 568–586 (2010)
6. Agarwal, A., Wainwright, M.J., Duchi, J.C.: Distributed dual averaging in networks. In: Advances in Neural Information Processing Systems, pp. 550–558 (2010)
7. Han, S., Mao, H., Dally, W.J.: Deep compression: compressing deep neural networks with pruning, trained quantization and Huffman coding. arXiv preprint arXiv:1510.00149 (2015)
8. Gong, Y., Liu, L., Yang, M., Bourdev, L.: Compressing deep convolutional networks using vector quantization. arXiv preprint arXiv:1412.6115 (2014)
9. Berkeley University of California. Ray. [EB/OL]. https://ray.io/

Ball K-Medoids: Faster and Exacter

Qiao Peng, Shibin Zhang(✉), Jinquan Zhang, Yuanyuan Huang, Boyi Yao, and Haozhe Tang

School of Cybersecurity, Chengdu University of Information Technology, Chengdu 610200, China
cuitzsb@cuit.edu.cn

Abstract. Cluster analysis can be viewed as a result of the natural evolution of the vast amount of data from daily life, and can discover invisible feature information to contribute to the analysis. K-means algorithm is one of the wide data clustering methods in a variety of real-world applications thanks to its simpleness. However, the k-means is sensitive to noise and outlier data points because a small number of such data can substantially influence the mean value of the cluster. In light of this, the k-medoids algorithm selects a point as a new center that minimizes the sum of the dissimilarities in the cluster, to diminish such sensitivity to outliers. Nevertheless, the line of the k-medoids algorithm is limited by its amounts of computation and not to handle with data efficiently. To this end, we present a novel k-medoids algorithm motivated by the theory of ball cluster, relationship between clusters and partitioning cluster for assigning samples into their nearest medoids efficiently, called ball k-medoids, which drop the distance calculation of sample-medoid significantly. Moreover, a threshold is inferenced by the rollback method for reducing computation of medoid-medoid distance and accelerating clustering. Experiments finally demonstrate that the performance of ball k-medoids achieves more efficient in comparison with other k-medoids algorithms, and it performs exacter accuracy compared with k-means.

Keywords: Cluster analysis · Ball cluster · Ball K-Medoids

1 Introduction

Cluster analysis is a significant technique in data mining and machine learning [1, 2], and it is usually that given a set of samples is partitioned into different subsets so that samples within a subset, called a cluster, are similar to each other. What the related machine learning develops rapidly [3, 4] promotes cluster analysis to achieve widespread success in a variety of domains to discover intrinsically unknown knowledge, such as recommender systems [5], pattern recognition [6], image annotation [7], and medical analysis [8]. In light of this situation, many excellent studies thrive the area, and can be roughly divided into multiple categories, namely partitioning method, such as k-means [9] and k-medoids [10], hierarchical method, such as BIRCH [11] and Chameleon [12], density-based method, such as DBSCAN [11] and OPTICS [14], and grid-based method, for instance, STING [15]. Especially, to some extent, the partitioning method usually outperforms other approaches because of its scalability and easiness.

X. Sun et al. (Eds.): ICAIS 2021, CCIS 1422, pp. 180–192, 2021.
https://doi.org/10.1007/978-3-030-78615-1_16

The primary version of k-means is proposed by Stuart Lloyd in 1982 [9], which is referred to as the Lloyd algorithm generally. Llyod mainly consists of two steps in each iteration. To be specific, one step is assignment so as to partitioning each sample into the nearest center according to a certain distance metric, e.g. Euclidean distance, and another step updates the center of each cluster by mean principle. The aforementioned two steps repeat until the algorithm converges. Due to the performance of simplicity and flexibility compared with other algorithms, the k-means is one of the widely common partitional clustering algorithms [1]. However, the Lloyd algorithm is sensitive to outliers because of the updating step according to mean theory, which results in adapting noisy ground data ineffectively. To this end, the medoid theory is proposed by Kaufman and Rousseeuw [16], and they designed PAM (Partitioning Around Medoids) [10] to mitigate the influence of outliers effectively by clustering noisy data around medoids, instead of mean centers, and simultaneously developed CLARA (Clustering LARge Applications) to deal with a larger dataset. Besides, recent advances based on k-medoids are proposed by researchers, such as CLARANS based on PAM and CLARA in [17, 18]. In addition, the Voronoi iteration algorithm gets rid of the theory of PAM and fuses the Lloyd algorithm in [17]. Our approach is based on the improvement of the latter.

The theory of ball clustering is proposed to focus on reducing the computation of sample-center, and improves the traditional k-means algorithm efficiently in [20]. We obtain a faster and exacter ball k-medoids model motivated by incorporating ball cluster theory into the variant of k-means—k-medoids. The specific contributions of our work are listed as follows:

- Designing a novel k-medoids model by introducing ball clustering, neighbor relationship and partitioning cluster, called ball k-medoids.
- Our algorithm can decrease the computation of sample-medoid and medoid-medoid effectively in each iteration, and experiments demonstrate that our approach is not only more efficient than other k-medoids, but also exacter than k-means.
- An in-depth analysis for the influence of size and dimension of the dataset to our algorithm.

The remaining parts of this paper are organized as follows: Sect. 2 states k-medoids methods of other researchers. Section 3 illustrates the difference between k-means and k-medoids and the related theory of ball cluster in detail. We evaluate the performance of ball k-medoids in Sect. 4. Finally, Sect. 5 summarizes this paper and discusses future works.

2 Related Work

The line of *k*-medoids algorithm is a variant class of *k*-means, it differs in choosing a sample, which minimizes the sum of dissimilarities between itself and other samples within the cluster, as a new center for each cluster, often called medoid, rather than a mean center of samples within the cluster.

What PAM (Partitioning Around Medoids) was developed by Kaufman and Rousseeuw [10] is one of the earliest *k*-medoids methods. We assume that the dataset represented by D is divided into medoids and non-medoids which are represented by M and P respectively, and a medoid is denoted as m belongs to M $(m \in M)$, and a non-medoid is denoted as p within P $(p \in P)$. We denote the operation as E_{mp} which swaps a pair of medoid and non-medoid, the sum of reduced dissimilarities is then represented by DE_{mp} with respect to the swap. PAM algorithm attempts to find the $\min\{DE_{mp}\}$ of the entire dataset iteratively. If the $\min\{DE_{mp}\} < 0$, it indicates that the non-medoid is fit to be a medoid, and the swap should be performed, so that medoids are optimized progressively until $\min\{DE_{mp}\} \geq 0$ to obtain optimal medoids. PAM, however, is a class of global optimization algorithm, which results in overloaded time complexity due to all pairs of medoid and non-medoid being traversed iteratively. Perhaps it works satisfactorily for small datasets, but is not to deal with large datasets efficiently. Furthermore, Han et al. [21] proved the drawback of PAM for large datasets by reason of its exceeding time complexity. To this end, Kaufman and Rousseeuw [10] also proposed CLARA (Clustering LARge Applications) which focuses on dropping the size of the dataset by applying random sampling on the origin dataset. CLARA draws multiple small samples from the dataset randomly for better performances and it performs PAM on these small samples so as to obtain the best clustering as output. Because sampling is sufficiently random, we think that it traverses the entire dataset approximately and avoids high time complexity. Complementary to PAM, CLARA performs efficiently and can handle large datasets, but Lucasius, Dane and Kateman [22] reported that the performance of CLARA suffers substantially with increasing k, and they indicated that it is not efficient to handle larger k. Due to the motivation of PAM and CLARA, Ng and Han abstracted the above algorithms into a graph structure and designed CLARANS (Clustering Large Application based on RANdomized Search) algorithm [17, 18], which obviously improves the quality of clustering over CLARA and outperforms PAM efficiently. Park and Jun [19] fused primary the Lloyd algorithm, instead of traditional PAM theory, to propose a novel k-medoids, which alternates the update step, the center of the cluster is a sample that minimizes the sum of dissimilarities, instead of a mean of samples within the cluster, and it is efficient and effective, called as Voronoi iteration algorithm.

Inspired by ball cluster and Voronoi, we propose a novel k-medoids algorithm, called ball k-medoids, and experiments indicate that it outperforms other algorithms.

3 Ball K-Medoids Theory

In this section, we first compare the Lloyd with Voronoi iteration algorithm in brief, then illustrate the related theory of ball cluster, and present our algorithm in detail. Suppose that a given dataset of n samples which are a d-dimension vector represented by D^d $(x_1, \ldots x_n \in D^d)$. k denotes the number of clusters, and the set of medoids then consist of k medoids which is the same as samples represented by C^d $(c_1, \ldots c_k \in C^d)$.

3.1 Differences Between the Lloyd and the Voronoi

Llyod algorithm initializes centers of clusters randomly and repeats the assignment step and the update step until convergence. The assignment step assigns samples into their nearest cluster according to distance metric (usually Euclidean distance) between medoids and non-medoids. New centers of clusters are recomputed by the mean of al samples within the cluster in the update step. Clusters stabilize progressively, and the internal dissimilarities of each cluster and the similarities among clusters drop iteratively until clusters are no longer changing. Thus, current centers are optimal.

In contrast to the Lloyd algorithm, the Voronoi iteration algorithm only alternates the update step and chooses the sample which minimizes the sum of dissimilarities as a new center rather than the mean of samples within the cluster. It results in the center exchanged by another non-center so that the Voronoi iteration algorithm is less sensitive to the outliers in comparison with the Lloyd algorithm.

Voronoi iteration algorithm is robust over the Lloyd for small datasets, but it is not adaptive to handle large datasets. This is not surprising if we analyze its time complexity. It searches the most centrally located sample iteratively and results in $O(n^2)$ in each iteration, but the Lloyd is of $O(n)$ due to mean theory. Many advances need to work with respect to the Voronoi iteration algorithm.

3.2 Ball K-Medoids Theory

Ball k-medoids achieves the sublinear level progressively of time complexity by introducing the concepts of ball cluster, neighbor ball and partitioning cluster, which drops calculations between iterations. The related concepts and references are as follows, and the related distance calculations is based on the Euclidean distance.

Ball Clustering. We describe a cluster as a ball structure by defining its ball center and radius. the ball center is the most central sample of the cluster, which minimizes the sum of dissimilarities between itself and other non-centers in the current iteration. The radius is the maximum distance between the ball center and another non-center. They are characterized by Eq. (1) and Eq. (2) mathematically as follows:

$$c = \operatorname{argmin}_{i \in C} \sum_{i'=1}^{N} ||x_i - x_{i'}||^2 \tag{1}$$

$$r = max(||x_i - c||^2) \tag{2}$$

Where C denotes the cluster, and x_i is a sample of cluster C, and N is the number of samples within the cluster C, then c and r represent ball center and radius respectively.

Neighbor Relationships. We expect that the amount of computation decreases progressively. The computation mainly consists of updating the ball center, reassigning samples and calculating medoid-medoid distance. We avoid redundant calculations between sample and far medoid by an ideal approach, which can drop computations of each sample in the assignment step and then result in decreasing the total computations in each iteration, and it achieves the goal that is to improve the algorithm's efficiency gradually.

Then, we introduce the concept of the neighbor ball according to be motivated by the above theory, so that avoid redundant computations in assigning samples. The relationship of clusters is defined by neighbor relationships according to measuring medoid-medoid distances. We think that if the relationship of two clusters satisfies Eq. (3), the ball cluster is a neighbor ball of another ball cluster.

$$||c_i - c_j||^2 < 2r_i \tag{3}$$

Where c_i and c_j denote the ball medoid of C_i and C_j, respectively. If c_i and c_j satisfy Eq. (3), C_j is a neighbor cluster of C_i.

We need to compute medoid-medoid distance so as to measure neighbor relationships among medoids after introducing neighbor ball. By now, we expect to discover a way that can decrease medoid-medoid distance effectively for dropping the algorithm's calculation gradually. The neighbor ball set of every cluster is updated in each iteration, and we wish to eliminate redundant computations of non-neighbor balls with respect to the cluster by studying calculational information from the preview iteration, so that drop the computations successfully and improve efficiency greatly in the end.

We denote the movement of medoids as $\delta_i^t = \left\|c_i^t - c_i^{t-1}\right\|^2$ between iterations, where t represents the tth iteration, and the distance of two medoids is represented by $dist\left(c_i^t - c_j^t\right)$ at t^{th} iteration. We can obtain a threshold by applying the rollback method as follows:

1. If two ball medoids satisfy Eq. (4), C_j is not a neighbor ball of C_i at the tth iteration according to Eq. (3).

$$dist\left(c_i^t - c_j^t\right) \geq 2r_i \tag{4}$$

2. Eq. (5) must be satisfied between iterations, because the minimum of $dist\left(c_i^t - c_j^t\right)$ is equal to the value of $dist\left(c_i^{t-1} - c_j^{t-1}\right)$ subtracted by the sum of δ_i^t and δ_j^t at the tth iteration.

$$dist\left(c_i^t - c_j^t\right) \geq dist\left(c_i^{t-1}, c_j^{t-1}\right) - \delta_i^t - \delta_j^t \tag{5}$$

3. The difference between Eq. (4) and Eq. (5) is Eq. (6) as follows:

$$dist\left(c_i^{t-1}, c_i^{t-1}\right) \geq 2r_i^t + \delta_i^t + \delta_j^t \tag{6}$$

If C_i and C_j satisfy Eq. (6), then C_j is not a neighbor ball of C_i, and the involved calculations ought to be avoided, and each term of the equation is computed before Eq. (6) and no extra calculations afterwards. We can avoid unnecessary calculation of medoid-medoid in advance Eq. (6) and achieve the goal of reducing medoid-medoid distance aforementioned. Algorithm 1 demonstrates the procedure in detail.

```
Algorithm 1 Dropping calculation of each pair medoids
progressively
Input: the set of medoids
Output: distance matrix of medoids
1: for i ← 1 to K do
2:   for j ← 1 to K do
3:     if satisfying Eq. (6) then
4:       dist(c_i^t, c_j^t) is the difference between dist(c_i^{t-1}, c_i^{t-1})
and δ_i^t + δ_j^t
5:     else
6:       calculating dist(c_i^t, c_j^t)
7:     end if
8:   end for
9: end for
```

Partitioning Cluster. A sample of a cluster is only computed distances with medoid within the cluster and medoids of neighbor clusters belonging to the cluster in the assignment step after combining the neighbor ball's theory, not all medoids, and the sample is then divided to which cluster by measuring these distances at the current iteration. Furthermore, we wish to decrease sample-medoid calculations. A ball cluster is partitioned into several areas according to the distance between its medoid and its neighbor balls. Distances of the cluster's medoid c_i correspond to neighbor ball cluster's medoids $c_j\left(c_j \in N_{C_i}\right)$ are calculated and sorted by distance metric, and the static area is defined by a ball region, where c_i is the ball center, and the radius is represented by $min_{c_j \in N_{C_i}}\{||c_i - c_j||^2/2\}$, and remaining region refer to the dynamic area, which is further partitioned into different annular regions precisely by orders of $\left\{||c_i - c_j||^2/2\right\}$. Samples of the cluster are divided by Eq. (7) as follows:

$$||x - c_i||^2 \leq min_{c_j \in N_{C_i}}\left\{||c_i - c_j||^2/2\right\} \tag{7}$$

Where x is a sample within the cluster C_i, and N_{C_i} is the neighbor ball set of C_i. By now, samples within the cluster must be divided into two states by Eq. (7) as follows:

1. The sample satisfies Eq. (7) and is divided into the static area.
2. In contrast, these samples in the dynamic area are not to meet Eq. (7).

Samples of the static area stay in the origin cluster in the assignment step, but a sample of the dynamic area is computed distance with medoids belonging to the i-closest annular regions. For example, x is a sample of the ith annual region, then x needs to calculate distances to medoids of the i-closest annual regions, and sample-medoid distances related to annual regions are compared with the origin medoid, and x is divided into other neighbor clusters or stayed in the origin cluster by a sample-medoid distance metric. The computation is further decreased by the partitioning cluster in the assignment step.

Algorithm 2 reassigning samples by cluster partitioning

Input: the cluster C_i, neighbor cluster $\{N_{C_i}\}$;
Output: the result of the assignment with respect to samples of C_i
1: Sorting $\{N_{C_i}\}$ by ascending distance
2: **for** traversing sample x_i within C_i **do**
3: **if** meeting Eq. (7) **then**
4: continue
5: **else**
6: Distinguishing the annual region of x_i, calculating sample-medoid distances and assigning x_i
7: **end if**
8: **end for**

3.3 Ball K-Medoids Algorithm

The process of ball k-medoids algorithm is shown in Algorithm 3 as follows:

Algorithm 3 Ball k-medoids algorithm

Input: dataset D^d, initial centroids C^d, clusters K
Output: the optimal medoids $c_1, \ldots, c_k$
1: Assigning samples by Euclidean distance metric
2: **do**
3: Updating medoid c_i^t of clusters via Eq. (1) and (2), and calculating $\delta(c_i^t)$
4: Updating distance matrix of medoids via Alg.1
5: **for** $i \leftarrow 1$ **to** K **do**
6: Clearing $\{N_{C_i}^t\}$
7: **for** $j \leftarrow 1$ **to** K **do**
8: **if** meeting Eq. (3) **then**
9: Append c_j to $\{N_{C_i}\}$
10: **end if**
11: **end for**
12: Reassigning samples of C_i via Alg.2
13: **end for**
14: t increased by 1
15: **while** clusters stabilize

4 Experiments

In this section, we firstly list various datasets related to experiments in brief, and present comparative results about the line of k-medoids, and demonstrate the efficiency and accuracy of our algorithm, and analyze the influence of dimension on the performances of our algorithm finally.

4.1 Experiment Setup

These involved datasets are mainly divided into two categories, one is clustering datasets which is utilized for comparative experiments of the line of k-medoids as shown in Table 1, and another one is to evaluated classification accuracy with respect to the latest k-means and our algorithm (see Table 2).

Table 1. Datasets of clustering

Dataset	Instances	Attributes
Stoneflake	79	8
Sobar	72	19
User-Knowledge	403	5
Four-class	862	3
Absenteeism	740	21

Table 2. Datasets of classification

Dataset	Instances	Attributes	categories
Diagnosis	120	7	2
Iris	150	4	3
Four-class	862	3	2
CMC	1473	9	3
Phoneme	5404	5	2

We select other K-medoids algorithms in comparison with our method for analyzing performances, and include early PAM (Partitioning Around Medoids) [10], CLARA (Clustering LARge Applications) [10], CLARANS (Clustering Large Application based on RANdomized Search) [17, 18], Voronoi iteration algorithm [19] and Gray-Medoids [23]. We test PAM, Voronoi iteration algorithm and our algorithm on the same dataset with the same initial medoids which are sampled uniformly in experiments, and thereby obtain the same result for fairness. However, CLARA and CLARANS apply random theory, and then generate the optimal medoids. For the best performance, the batch and the size of sampling are set to 5 and $\sqrt{n}+2k$ respectively in CLARA. Besides, *the localnum* and the *maxneighbor* are equal to 5 and $k*(n-k)$ respectively in CLARANS according to analysis in [18]. We use Intel(R) Core(TM) i5–9400 CPU with 2.90 GHz and 8 GB DRAM in experiments.

4.2 Clustering and Classification Results

Our algorithm reduces the distance computation of sample-medoid by fusing ball clustering, neighbor ball and partitioning ball, and drops the distance computation of medoid-medoids between iterations by inferencing a threshold value. Table 3 and Table 4 demonstrate that our algorithm provides consistent speedup over other k-medoids algorithms and drops calculation remarkably on some widely clustering datasets in most cases. Some classification tasks are performed in Table 5, which shows that our algorithm achieves more precisely in comparison with ball k-means which applies the same theory of ball cluster, due to the insensitivity of the medoid.

Table 3. Comparison of runtime in datasets (ms)

Dataset	k	PAM	CLARA	CLARANS	Voronoi	Gray-Medoids	Ours
Stoneflakes	5	2.91	1.79	4.84	0.05	**0.02**	0.04
	10	7.90	16.87	7.89	**0.02**	0.67	0.03
	20	22.02	59.43	19.31	**0.01**	3.07	0.02
Sobar	5	3.13	2.09	5.15	0.06	0.07	**0.05**
	10	6.71	17.66	8.41	**0.02**	2.29	0.03
	20	15.14	47.85	26.19	**0.01**	3.03	0.02
User-Knowledge	5	33.12	3.40	29.98	0.33	3.27	**0.13**
	10	142.46	31.96	57.98	0.21	**6.84**	**0.12**
	20	521.42	339.00	193.70	0.21	**10.71**	**0.12**
Four-class	5	3323.00	5.20	178.61	3.33	**0.05**	3.09
	10	1615.18	35.66	463.19	2.57	28.24	**1.98**
	20	8882.72	472.16	1365.37	1.01	52.69	**0.79**
Absenteeism	5	518.04	11.24	331.24	9.74	86.22	**7.69**
	10	1736.56	93.67	993.79	3.63	163.79	**3.32**
	20	7888.45	1029.14	2615.36	1.96	320.57	**1.14**

4.3 The Influence of Dimension

The dimension of the sample is vital to the clustering algorithm, especially for the complicated applying scenes, which is a challenge for the clustering task. We should

Table 4. Comparison of distance calculations in datasets

Dataset	k	PAM	CLARA	CLARANS	Voronoi	Gray-Medoids	Ours
Stoneflakes	5	6.85E + 05	3.75E + 05	5.48E + 05	1.90E + 03	7.40E + 02	**5.91E + 02**
	10	2.86E + 06	5.63E + 06	1.43E + 06	1.90E + 03	4.83E + 04	**6.16E + 02**
	20	9.75E + 06	2.46E + 07	4.18E + 06	1.90E + 03	7.08E + 04	**6.34E + 02**
Sobar	5	4.49E + 05	2.48E + 05	5.61E + 05	4.32E + 03	**6.70E + 02**	7.38E + 02
	10	1.54E + 06	3.78E + 06	1.16E + 06	2.88E + 03	3.91E + 04	**7.81E + 02**
	20	4.33E + 06	1.32E + 07	5.41E + 06	1.44E + 03	5.51E + 04	**5.21E + 02**
User-Knowledge	5	9.60E + 06	8.13E + 05	8.00E + 06	3.10E + 03	3.21E + 05	**7.99E + 02**
	10	6.77E + 07	1.36E + 07	2.46E + 07	3.10E + 03	6.18E + 05	**1.76E + 03**
	20	3.40E + 08	2.07E + 08	1.13E + 08	9.29E + 03	1.14E + 06	**3.42E + 03**
Four-class	5	1.29E + 08	1.71E + 06	7.35E + 07	8.62E + 03	8.57E + 03	**8.01E + 03**
	10	1.02E + 09	1.96E + 07	2.90E + 08	1.38E + 04	7.27E + 06	**1.07E + 04**
	20	3.36E + 09	3.72E + 08	1.13E + 09	8.62E + 03	1.42E + 07	**7.06E + 03**
Absenteeism	5	8.10E + 07	1.53E + 06	5.40E + 07	8.14E + 04	2.70E + 06	**6.67E + 03**
	10	4.76E + 08	2.32E + 07	2.66E + 08	4.88E + 04	5.34E + 06	**5.86E + 03**
	20	3.29E + 09	3.96E + 08	1.04E + 09	4.88E + 04	1.04E + 07	**7.42E + 03**

not be limited to process the low dimension information, but evaluate our method on a variety of high dimensions, real-world datasets, Fig. 1(above) demonstrates the influence of dimension to the neighbor ball ratio, which approaches 1 with increasing dimension. However, the number of clusters mitigates this case in part extent, especially if k = 100, then the neighbor ball ratio is not in excess of 60%. It is obvious that the large k achieves more efficient on the same dimension as shown in Fig. 1(left), although results in a mass of distance calculations in the algorithm as shown in Fig. 1(right). It proves that our algorithm is adaptive to perform the large k problem.

Table 5. Comparison of accuracy (%)

Dataset	Ball K-Means	Ours
Diagnosis	55.00	**60.32**
Iris	88.66	**92.67**
Four-class	63.67	**66.59**
CMC	71.90	**77.19**
Phoneme	66.78	**76.02**

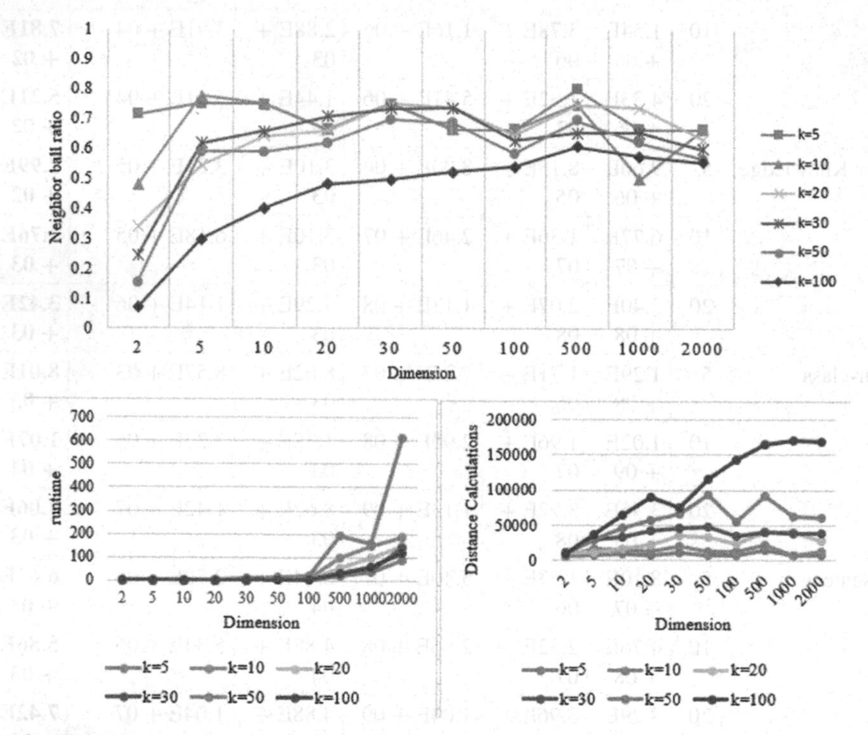

Fig. 1. The influence of dimension sampled from RNA-Seq randomly. (above) the neighbor ball ratio. (left) runtime. (right) distance calculations.

5 Conclusion and Future Work

In this paper, we design a novel ball k-medoids algorithm, which is faster and exacter. It is not limited to swap medoid and non-medoids pairwise in comparison with early PAM, CLARA and CLARANS algorithm, but fuses the idea of Voronoi iteration algorithm and the ball clustering theory, which approaches to processing a variety of clustering problems efficiently. Experiments demonstrate that the performance of ball k-medoids outperforms other k-medoids algorithms.

For the moment, ball k-medoids has the same drawback as other k-medoids algorithms, which works inefficiently in updating medoids step, and it results in limiting the development of k-medoids. We should put efforts in improving the efficiency of updating medoids further so as to generate a new k-medoids clustering method.

Acknowledgement. The authors really appreciate the handling associate editor and all innominate reviews for their valuable comments. This work is supported by the National Natural Science Foundation of China (No. 62076042, No. 61572086), the Key Research and Development Project of Sichuan Province (No. 2020YFG0307, No. 2018TJPT0012), the Key Research and Development Project of Chengdu (No. 2019-YF05–02028-GX), the Innovation Team of Quantum Security Communication of Sichuan Province (No. 17TD0009), Sichuan Science and Technology Program under Grants 2018RZ0072, 20ZDYF0660, the Foundation of Chengdu University of Information Technology under Grant J201707, the National Key R&D Program of China under Grant (No. 2017YFB0802300), the Key Research and Development Project of Chengdu (No. 2019-YF05–02028-GX).

References

1. Aggarwal, C., Reddy, C.: Data Clustering: Algorithms and Applications. Chapman & Hall/CRC Press, Boca Raton, FL, USA (2013)
2. Han, J., Kamber, M., Pei, J.: Data Mining: Concepts and Techniques. Morgan Kaufmann Publishers, San Francisco, CA, USA (2012)
3. Xiao, Z., Xu, X., Xing, H., Chen, J.: RTFN: Robust Temporal Feature Network. arXiv preprint arXiv:2008.07707 (2020)
4. Huang, Y., Wang, D., Sun, Y., Hang, B.: A fast intra coding algorithm for HEVC by jointly utilizing naive Bayesian and SVM. Multimedia Tools Appl. **79**(45–46), 33957–33971 (2020). https://doi.org/10.1007/s11042-020-08882-x
5. Shepitsen, A., Gemmell, J., Mobasher, B., Burke, R.: Personalized recommendation in social tagging systems using hierarchical clustering. In: Proceedings of the ACM Conference on Recommender Systems (RECSYS), New York, MY, USA, pp. 259–266 (2008)
6. Chien, Y.: Pattern classification and scene analysis. IEEE Trans. Autom. Control **19**(4), 462–463 (1974)
7. Tao, D., Li, X., Gao, X.: Large sparse cone non-negative matrix factorization for image annotation. In: ACM Transactions on Intelligent Systems and Technology (TIST), vol. 8, no. 3, p. 37 (2017)
8. Medan, G., Shamul, N., Joskowicz, L.: Sparse 3D radon space rigid registration of CT scans: method and validation study. IEEE Trans. Med. Imaging **36**(2), 497–506 (2017)
9. Lloyd, S.: Least squares quantization in PCM. IEEE Trans. Inf. Theory **28**(2), 129–137 (1982)
10. Kaufman, L., Rousseeuw, P.J.: Finding Groups in Data: an Introduction to Cluster Analysis. John Wiley & Sons, Hoboken (1990)
11. Zhang, T., Ramakrishnan, R., Livny, M.: BIRCH: a new data clustering algorithm and its applications. Data Min. Knowl. Disc. **1**(2), 141–182 (1997)
12. Karypis, G., Han, E.-H., Kumar, V.: Chameleon: hierarchical clustering using dynamic modeling. Computer. **32**(8), 68–75 (1999)
13. Ester, M., Kriegel, H.-P., Sander, J., Xu, X.: A density-based algorithm for discovering clusters in large spatial databases with noise. In: Proceedings ACM Conference on Knowledge Discovery and Data Mining, Oregon, Portland, pp. 226–231 (1996)

14. Ankerst, M., Breunig, M., Kriegel, H.-P., Sander, J.: OPTICS: ordering points to identify the clustering structure. In: Proceedings of the ACM SIGMOD international conference on Management of data, New York, NY, USA, pp. 49–60 (1999)
15. Wang, W., Yang, J., Muntz, R.R.: STING: a statistical information grid approach to spatial data mining. In: Proceedings of the International Conference on Very Large Data Bases, San Francisco, CA, USA, pp 186–195 (1997)
16. Kaufman, L., Rousseeuwm, P.J.: Clustering by means of medoids. In: Statistical Data Analysis Based on the L1 Norm and Related Methods, pp. 405–416 (1987)
17. Ng, R.T., Han, J.: Efficient and effective clustering methods for spatial data mining. In: Proceedings of the International Conference on Very Large Data Bases, San Francisco, CA, USA, pp. 144–155 (1994)
18. Ng, R.T., Han, J.: CLARANS: a method for clustering objects for spatial data mining. In: IEEE Transactions on Knowledge and Data Engineering, vol. 14, no. 5, pp. 1003–1016 (2002)
19. Park, H.-S., Jun, C.-H.: A simple and fast algorithm for k-medoids clustering. Expert Syst. Appl. **36**(2), 3336–3341 (2009)
20. Xia, S., et al.: A fast adaptive k-means with no bounds. In: IEEE Transactions on Pattern Analysis and Machine Intelligence, p. 1 (2020)
21. Han, J., Kamber, M., Tung, A.K.H.: Spatial clustering methods in data mining: a survey. In: Geographic Data Mining and Knowledge Discovery (2001)
22. Lucasius, C.B., Dane, A.D., Kateman, G.: On k-medoid clustering of large data sets with the aid of a genetic algorithm: background, feasibility and comparison. Anal. Chim. Acta **282**(3), 647–669 (1993)
23. Gao, S., Zhou, X., Li, S.: Improved K-medoids clustering based on Gray association rule. In: Patnaik, S., Jain, V. (eds.) Recent Developments in Intelligent Computing, Communication and Devices. AISC, vol. 752, pp. 349–356. Springer, Singapore (2019). https://doi.org/10.1007/978-981-10-8944-2_41

GVNP: Global Vectors for Node Representation

Dongao Zhang[1], Ziyang Chen[1], Peijia Zheng[1,2(✉)], and Hongmei Liu[1]

[1] School of Computer Science and Engineering Guangdong Key Laboratory of Information Security Technology Sun Yat-Sen University, Guangzhou 510006, China
{zhangdao,chenzy33}@mail2.sysu.edu.cn, {zhpj,isslhm}@mail.sysu.edu.cn

[2] State Key Laboratory of Information Security, Institute of Information Engineering, Chinese Academy of Sciences, Beijing 100093, China

Abstract. Learning low-dimensional embeddings of nodes in networks is an effective way to solve the network analytic problem, from traffic network to recommender systems. However, most existing approaches are inherently transductive, their framework is built on a single fixed graph. Inspired by node2vec, we optimize the random walk strategy and propose GVNP, an unsupervised method that can learn continuous feature representations for nodes and leverage node feature information to efficiently generate node embeddings for previously unseen data in networks. Experimental results demonstrate that GVNP performs well on the transductive and inductive task.

Keywords: Data mining · Graph learning · Deep networks

1 Introduction

Research on graphs has attracted a lot of attention, such as [4,9,14,18,21]. Node classification aims to assign a class label to each node in a graph based on the rules learned from the labeled nodes. Intuitively, "similar" nodes have the same labels. It is one of the most common applications discussed in graph embedding literature. In general, each node is embedded as a low-dimensional vector. Node classification is conducted by applying a classifier on the set of labeled node embedding for training. Nodes that are "close" in the graph are embedded to have similar vector representation [3]. For example, in a social network, users who have the same role will be embedded very similarly. Low-dimensional vector embedding of nodes is often used as an input in various prediction and graph analysis tasks [2,8,10], which has proved to be very effective. Previous work has shown that embedding works well. DeepWalk develops an algorithm that learns representations of a graph's vertices by modeling a stream of short random walks, then use SkipGram and hierarchical softmax to process vertex sequence and get embeddings of vertices. Node2vec optimizes DeepWalk's walk strategy, it performs DFS and BFS walk strategies on the graph to obtain vertex sequences,

X. Sun et al. (Eds.): ICAIS 2021, CCIS 1422, pp. 193–202, 2021.
https://doi.org/10.1007/978-3-030-78615-1_17

through changing the parameters to bias the walk towards different network exploration strategies. GCN presents a scalable approach for semi-supervised learning on graph-structured data that is based on an efficient variant of convolutional neural networks that operate directly on graphs. It aggregates the neighbors and features of a vertex.

However, DeepWalk and node2vec only use the graph information and ignores the vertex feature. They use SkipGram to process vertex sequence, that it cannot fully utilize vertex sequence. GCN makes full use of graph and vertex information to state-of-art performance, but it needs an adjacent matrix to get normalized graph Laplacian as input, which means GCN is inherently transductive, same with the former.

To overcome the above limitations, we propose GVNP, a semi-supervised algorithm for general embedding and inductive node embedding based on the former. We optimize random walk strategy based on transition graph, to generate (sample) network neighborhoods for nodes with similar structures and features. Inspired by prior work [12], we use a custom graph-based objective function to process node sequence. Our key contribution is in proposing an efficient algorithm for generating embedding for new nodes, which doesn't require a learning process, our algorithm performs strong baselines on this inductive node-classification benchmarks.

1.1 Problem Formulation

We use $G = (V, E, X)$ to denote an undirected network with $V \in \mathbb{R}^{|V|}$ as the node set of n nodes and $E \in \mathbb{R}^{|V|\times|V|}$ as the edge set, $X \in \mathbb{R}^{|V|\times N}$ denotes feature matrix, each row X_i is a N-dimensional vector for v_i. Given a specific node v_i, the set of neighbor nodes is defined as $\mathcal{N}_i = \{v_j | v_j \in V, (v_i, v_j) \in E\}$. In addition, we denote transition graph of G as $T \in \mathbb{R}^{|V|\times|V|}$, $t_{ij} \in T$ denotes the probability that node V_i transfers to node V_j. In a later section, we will show how to compute the transition graph in this paper.

Given a network $G = (V, E, X)$, the problem of network embedding aims to learn a mapping function $f_1 : V \mapsto R^d$ that projects each node to a d-dimensional space ($d \ll |V|$) to capture the structural properties of the network.

Previous studies focus on a single fixed graph, it is shown that the learned node representations can benefit various graph mining tasks, one major challenge is that they can't handle new nodes. For the unseen nodes and the edges that are connected to them, we also have a mapping function $f_2 : V(unseen) \mapsto R^d$.

2 GVNP: Non-transductive Graph Embedding

In this section, we present GVNP, an algorithmic framework which can learn continuous feature representations for nodes and leverage node feature information to efficiently generate node embeddings for previously unseen data in networks. First, it calculated an special transition graph based on the network, which is directed graph such as (Fig. 1). Second, the random walk generator takes a transition graph T and samples a random vertex v_i as the root of the random walk

W_{v_i}, then use GloVe to process W_v and get the output: embeddings of nodes. It also have very efficient ways to generate embeddings for new nodes.

Algorithm 1. Learning representation

Input: $G(V, E, X)$
Parameter: Embedding size d, Walks per node r, Walk length l, return p, In-out q, GloVe iter n
Output: Node representation $\Phi \in \mathbb{R}^{|V| \times d}$

```
1: Initialize walks to Empty
2: Build transition graph T*
3: for iter=1 to r do
4:     for all nodes u ∈ V do
5:         walk=GVNPWalk(G*, u, l)
6:         Append walk to Walks
7:     end for
8: end for
9: Initialize co-occurrence matrix C
10: for iters=1 to n do
11:     for each C_ij > 0 do
12:         loss=J(Φ, C_ij)
13:         Φ = Φ − δ * ∂loss/∂Φ
14:     end for
15: end for
```

Algorithm 2. GVNPWalk(T^*, u, l)

```
1: Initialize walk to u
2: for walk_i=1 to l do
3:     node=u
4:     node_neighbors=Getneighbors(G*,u)
5:     v=AlasSample(node_neighbors,G*)
6:     append v to walk
7: end for
```

2.1 Transition Graph

Formally, given a $G = (V, E, X)$, we construct the directed transition diagram $T \in \mathbb{R}^{|V| \times |V|}$ such as (Fig. 1), $t_{ij} \in T$ denotes the normalized probability that node V_i transfers to node V_j when we simulate a random walk of fixed length l. Let n_i denote the ith node in the walk, starting with $n_0 = u$, and t_{vx} denotes:

$$P(n_i = x | n_{i-1} = v) = t_{vx} \tag{1}$$

To fully retain the structural information and features of the similarity, nodes are directly linked with the anchor node (first-order proximity), share the same

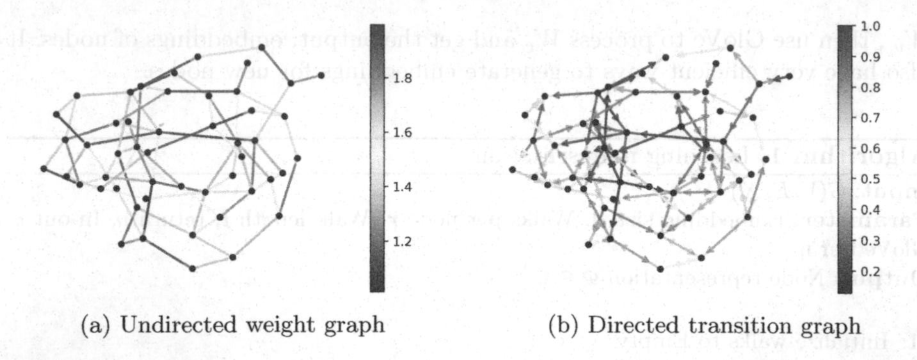

(a) Undirected weight graph

(b) Directed transition graph

Fig. 1. The undirected weight graph (a) and directed transition graph(b), the weights of different edges correspond to different colors

neighbor node with the anchor node (second-order proximity) or have similar feature should be assigned a bigger t value.

Before we get to t, we update the weights of each directed edge for G. If G is an undirected graph, we convert it to a directed graph. We use W_{vx} denotes the weight of directed edge $<v, x>$ leading from v, and W_{newvx} denotes updated weight:

$$W_{newvx} = W_{vx} \cdot (\beta \cdot X_v X_w^T + 1) \cdot \sum_{t \in \mathcal{N}_v} \frac{\alpha_{tx}}{l} \tag{2}$$

In this equation, $X_v X_w^T$ denotes the product of node v's feature and node w's feature, β is a hyper-parameter which be used to control the weight of feature factors. In this way, we guide the walk by the similarity of features. Inspired by node2vec, we also use α_{tx} to guide the walk by the similarity of structure. Consider a random walk that just traversed edge (t, v) and now resides at node v, the walk now needs to decide on the next step so it evaluates the α_{tx} on edges (v, x) leading from v, d_{tx} denotes the shortest path from t to x where:

$$\alpha_{pq}(t, x) = \begin{cases} \frac{1}{p} & \text{if } d_{tx} = 0 \\ 1 & \text{if } d_{tx} = 1 \\ \frac{1}{q} & \text{if } d_{tx} = 2 \end{cases} \tag{3}$$

There are two hyper-parameter, the return parameter$p$, and the In-out parameter$q$, look into [5] for details, to control the walking strategy. Setting p and q to a high value ensures that we are more likely to sample the node which shares the same node and less likely to sample an already visited node in the following two steps. In this paper, we're taking the average value. $\mathcal{N}_v$ denotes set of node v's neighborhoods, l denotes the number of nodes in the set, Fig. 1 shows the process.

After we update all weight of edges, we normalize the weight, then get all $t_{ij} \in T, < i, j > \in E$, where:

$$t_{ij} = \frac{W_{newij}}{\sum_{x \in \mathcal{N}_i} W_{newix}} \tag{4}$$

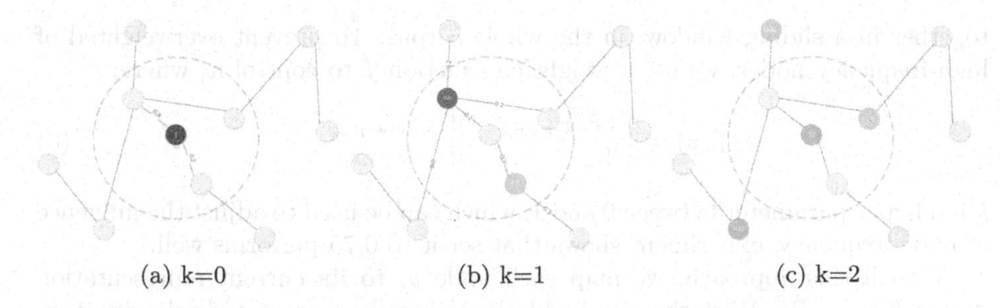

Fig. 2. The transition of vertex, (a) denotes original state of new node, (b) and (c) denote the transition state at different k values

At this point, the transition graph is calculated, and we will implement a random walk on it.

2.2 Transductive Task

To use the language model approach, then it needs to construct our corpus. In this paper, we also create our corpus, sampling the sequence of nodes on the transition graph. The algorithm consists of two main components: random walk generator and an update procedure.

Random Walk Generator. We take a transition graph and samples uniformly a random node v_i as the root of random walk $\mathcal{W}_{v_i}$. Then we will use the alias sample method to sample from the neighbors of the last node visited by transition probability t until reaching the length we set.

As is shown in Algorithm 1, in each of the γ loops we will start random walks at each node and set the fixed length of each walk $\mathcal{W}$.

Update Procedure: GolVe. We use GolVe to efficiently leverages statistical information by training only on the nonzero elements in a node-node co-occurrence matrix, rather than on the entire sparse matrix or individual context windows in a large corpus.

We seek to optimize the following objective function, which minimizes the loss function J:

$$J = \sum_{i,j}^{N} f\left(C_{i,j}\right)\left(v_i^T v_j + b_i + b_j - \log\left(C_{i,j}\right)\right)^2 \tag{5}$$

in this equation, N is the size of the node set, v_i and v_j denote the vector of node i and j respectively, which is the embedding we aim to learn. b_i and b_j are additional bias. $C \in \mathbb{R}^{|V|\times|V|}$ denotes the co-occurrence matrix, which is constructed by corpus $\mathcal{W}$. And $C_{i,j} \in C$ denotes times that nodes i and j appear together in a sliding window, in the whole corpus. To prevent overweighted of high-frequency nodes, we use a weighting function f to control it, where:

$$f(x) = \begin{cases} (x/x_{max})^k, & \text{if } x < x_{max} \\ 1, & \text{if } x >= x_{max} \end{cases} \tag{6}$$

k is a hyper-parameter between 0 and 1, which can be used to adjust the influence of node frequency, experiment shows that set it to 0.75 performs well.

Trough this approach, we map each node v_j to its current representation vector $\Phi(v_j) \in \mathbb{R}^d$. With these embeddings we can get a great node classification performance.

2.3 New Embedding Aggregator

The most previous work focus on a fixed graph, some work transforms the whole adjacency matrix of the graph in the spatial or spectral domain, and then use it as input to train the model, which results in their lack of ability to deal with new nodes. In this section, we propose a new approach to handle new nodes, an efficient embedding generator that can map a new node to its appropriate representation vector.

We assume that new node v_i is connected to the graph G, which means there exist at least one edge (i, j) and $v_j \in G$. New node v_i also have feature $X_i \in \mathbb{R}^N$. For new node, first we use Eq. 2 and Eq. 4 to update transition probability between node v_i and v_s's neighborhoods, then we get the new transition matrix T, each row t_i denotes the probability that node i moves to another node.

To generate embedding for new nodes, we let nodes aggregate information from their k-order neighbors. There is a parameter k, which determines k-order neighbors are been aggregated. In this paper, we set k to 2. Firstly, we let the new node propagate twice among neighbors, like Fig. 2, then take the average probability that it gets to each vertex. Secondly, we weighted the embedding of these vertices and get the embedding for the new node.

$$\mathbf{V}_i \leftarrow \text{MEAN}\left(\{p \cdot \mathbf{V}_i^{k=0}\} \cup \{p \cdot \mathbf{V}_i^{k=1}\}\right) \tag{7}$$

2.4 Complexity Analysis

We analyze complexity in three parts, first part, generating a transition graph. We calculate the transfer probability for every two connected nodes, because real graphs are always sparse graphs, our computational complexity in this part is close to $O(|E|)$, in the worst case, $O(V^2)$, but that is impossible to reach it in a social network.

The second part, generating embedding. As can be seen from Eq. 5 and the explicit form of the weighting function $f(X)$, the computational the complexity of the model depends on the number of nonzero elements in the matrix C. The complexity of the model is much better than the worst-case $O(V^2)$, and in fact, it does somewhat better than the on-line window-based methods which scale like $O(|C|)$.

The third part, embedding aggregator. For each new node, we only need to perform a constant number of calculations, we just need to transfer between k-order neighbors.

3 Experiments

In this section, we present an experimental analysis of our method. We test it in some experiments of document classification in a citation network.

Dataset. We utilize the three citation network datasets: Citeseer, Cora, and Pubmed. The datasets consist of Please check and confirm if the inserted citation of sparse bag-of-words feature vectors for each document and a list of citation links between documents. Links can be treated as edges (Table 1).

Table 1. Dataset statistics

Dataset	Nodes	Edges	Classes	Features
Citeseer	3327	4372	6	3703
Cora	2708	5429	7	1433
Pubmed	19717	44338	3	500

Experimental Set-Up. For the transductive task in this paper:

- Cora: β =0.02, p =1, q =2, $windowsize$ = 40, $xmax$ = 15, d = 70, $walkiter_r$ = 15, $walk_l$ = 80, $GloVe_iter_n$ = 15
- Citeseer: β = 0.35, p = 1.2, q = 2, $windowsize$ = 35, $xmax$ = 15, d = 300, $walkiter_r$ = 15, $walk\_l$ = 80, $GloVe\_iter\_n$ = 15
- Pubmed: β = 1, p = 1.2, q = 1.5, $windowsize$ = 35, $xmax$ = 15, d = 200, $walkiter\_r$ = 15, $walk\_l$ = 80, $GloVe\_iter\_n$ = 15

For the inductive task, we train the 80% data set to get the embeddings, the rest is used as a new node to make predictions.

Baseline. We compare against the same baseline methods as in DeepWalk [13] and node2vec [5]. The setting of node2vec's random walk parameters is consistent with this paper. SkipGram's parameter of DeepWalk and node2vec is set to the same. Setting embedding d to 100, window size to 5.

3.1 Result

The node feature representations are input to a one-vs-rest logistic regression classifier with L2 regularization. For each dataset, we randomly split train and test data and run it a hundred times, then you take the average of micro-F1 and macro-F1. The experimental results are shown in Tabels 2, 3 and 4. **Our** denotes the transductive result of this paper, **Our*** denotes the inductive result of this paper.

Table 2. Cora dataset

	% Labeled nodes	10%	20%	30%	40%	50%	60%	70%	80%	90%
Micro-F1(%)	DeepWalk	61.93	70.45	72.40	75.41	76.88	77.03	77.39	78.12	78.83
	node2vec	64.18	71.48	74.58	76.67	78.21	78.91	79.28	80.24	80.47
	our	**72.13**	**77.79**	**80.74**	**82.58**	**84.01**	**85.34**	**86.22**	**86.92**	**87.59**
	our*	**71.01**	**73.93**	**76.36**	**80.15**	**82.68**	**83.16**	**85.37**	**84.32**	**86.81**
Macro-F1(%)	DeepWalk	59.27	68.66	71.07	74.23	75.17	75.80	76.01	76.96	77.31
	node2vec	61.51	69.86	73.36	75.65	77.27	78.05	78.50	79.53	79.61
	our	**69.44**	**75.98**	**79.16**	**81.25**	**82.75**	**84.16**	**85.08**	**85.86**	**86.44**
	our*	**68.41**	**71.52**	**75.46**	**79.33**	**81.13**	**82.02**	**83.91**	**83.01**	**85.13**

Table 3. Citeseer dataset

	% Labeled nodes	10%	20%	30%	40%	50%	60%	70%	80%	90%
Micro-F1(%)	DeepWalk	41.35	43.76	46.03	48.99	52.53	54.16	55.46	56.65	57.02
	node2vec	47.93	50.63	53.59	55.99	57.87	59.31	60.34	60.59	61.21
	our	**52.25**	**58.57**	**62.30**	**64.61**	**66.60**	**67.19**	**69.29**	**71.96**	**72.06**
	our*	**49.81**	**55.46**	**60.17**	**61.33**	**64.82**	**66.72**	**67.83**	**68.03**	**68.41**
Macro-F1(%)	DeepWalk	29.89	32.31	33.99	36.70	40.76	41.57	43.08	43.13	42.40
	node2vec	34.40	37.21	39.58	42.93	45.49	47.55	49.28	50.05	50.95
	our	**50.34**	**54.38**	**58.32**	**60.68**	**62.67**	**63.27**	**64.18**	**64.73**	**64.61**
	our*	**46.27**	**50.53**	**55.91**	**56.63**	**60.22**	**62.18**	**63.27**	**63.15**	**63.43**

Table 4. Pubmed dataset

	% Labeled nodes	10%	20%	30%	40%	50%	60%	70%	80%	90%
Micro-F1(%)	DeepWalk	66.73	68.15	68.62	70.35	71.71	72.43	72.86	73.56	73.89
	node2vec	70.85	71.45	71.99	72.58	73.04	73.33	74.13	74.87	75.54
	our	**81.80**	**82.77**	**83.31**	**83.43**	**84.19**	**84.83**	**85.88**	**86.93**	**86.86**
	our*	**80.55**	**81.36**	**82.07**	**82.73**	**83.22**	**83.52**	**83.73**	**84.13**	**84.61**
Macro-F1(%)	DeepWalk	56.31	60.42	61.30	63.83	64.47	65.40	66.33	67.13	67.75
	node2vec	65.32	66.18	66.90	67.91	68.62	69.03	49.28	69.49	70.78
	our	**78.93**	**80.08**	**80.68**	**80.82**	**81.08**	**81.24**	**81.30**	**81.32**	**79.27**
	our*	**77.21**	**79.27**	**79.85**	**80.03**	**80.38**	**80.75**	**81.01**	**80.47**	**80.89**

4 Related Work

The recent advances in network embedding can roughly fall into three categories: matrix factorization based methods such as RF-Semi-NMF-PCA [1], HOPE [11], SE [20]; random walk based methods, such as DeepWalk [13], node2vec [5], Planetoid [19]; and neural network based methods, such as SAE [15], GCN [7], GAT [16], HAN [17]. Commonly, these methods focus on an upstream task that aims to find the right embedding space. The learned embeddings can be used as features to boost the results of downstream learning tasks. Graphsage [6] aims to solve inductive learning.

5 Conclusions

We propose GVNP, an efficient unsupervised approach for learning latent social representations of nodes. By using the biased random walk based on the transition graph and the global information from it as input, our method learns a representation that retains structural and node features. What is important is that GVNP is scalable. We can add new nodes to the transition graph and get embeddings efficiently.

Acknowledgement. This work was supported in part by the Natural Science Foundation of Guangdong (2019A1515010746, 2019A1515011549), in part by the SYSU Youth Teacher Development Program (19lgpy218), in part by the National Science Foundation of China (61972430).

References

1. Allab, K., Labiod, L., Nadif, M.: A semi-NMF-PCA unified framework for data clustering. IEEE Trans. Knowl. Data Eng. **29**(1), 2–16 (2016)
2. Bojchevski, A., Günnemann, S.: Adversarial attacks on node embeddings via graph poisoning. In: International Conference on Machine Learning, pp. 695–704 (2019)
3. Cai, H., Zheng, V.W., Chang, K.C.C.: A comprehensive survey of graph embedding: problems, techniques, and applications. IEEE Trans. Knowl. Data Eng. **30**(9), 1616–1637 (2018)
4. Feng, A., Gao, Z., Song, X., Ke, K., Xu, T., Zhang, X.: Modeling multi-targets sentiment classification via graph convolutional networks and auxiliary relation. Materials Continua Comput. (2020)
5. Grover, A., Leskovec, J.: node2vec: scalable feature learning for networks. In: Proceedings of the 22nd ACM SIGKDD International Conference on Knowledge Discovery and Data Mining, pp. 855–864. ACM (2016)
6. Hamilton, W., Ying, Z., Leskovec, J.: Inductive representation learning on large graphs. In: Advances in Neural Information Processing Systems, pp. 1024–1034 (2017)
7. Kipf, T.N., Welling, M.: Semi-supervised classification with graph convolutional networks. arXiv preprint arXiv:1609.02907 (2016)
8. Liu, N., Tan, Q., Li, Y., Yang, H., Zhou, J., Hu, X.: Is a single vector enough? Exploring node polysemy for network embedding. arXiv preprint arXiv:1905.10668 (2019)
9. Lu, W., et al.: Graph-based Chinese word sense disambiguation with multi-knowledge integration. Comput. Mater. Continua **61**(1), 197–212 (2019)
10. Nikolentzos, G., Meladianos, P., Vazirgiannis, M.: Matching node embeddings for graph similarity. In: Thirty-First AAAI Conference on Artificial Intelligence (2017)
11. Ou, M., Cui, P., Pei, J., Zhang, Z., Zhu, W.: Asymmetric transitivity preserving graph embedding. In: Proceedings of the 22nd ACM SIGKDD International Conference on Knowledge Discovery and Data Mining, pp. 1105–1114. ACM (2016)
12. Pennington, J., Socher, R., Manning, C.: GloVe: global vectors for word representation. In: Proceedings of the 2014 Conference on Empirical Methods in Natural Language Processing (EMNLP), pp. 1532–1543 (2014)
13. Perozzi, B., Al-Rfou, R., Skiena, S.: DeepWalk: online learning of social representations. In: Proceedings of the 20th ACM SIGKDD International Conference on Knowledge Discovery and Data Mining, pp. 701–710. ACM (2014)
14. Polat, N.C., Yaylali, G., Tanay, B.: A method for decision making problems by using graph representation of soft set relations (2019)

15. Tian, F., Gao, B., Cui, Q., Chen, E., Liu, T.Y.: Learning deep representations for graph clustering. In: Twenty-Eighth AAAI Conference on Artificial Intelligence (2014)
16. Veličković, P., Cucurull, G., Casanova, A., Romero, A., Lio, P., Bengio, Y.: Graph attention networks. arXiv preprint arXiv:1710.10903 (2017)
17. Wang, X., et al.: Heterogeneous graph attention network. In: The World Wide Web Conference, pp. 2022–2032. ACM (2019)
18. Wang, X., Zheng, Q., Zheng, K., Sui, Y., Zhang, J.: Semi-GSGCN: social robot detection research with graph neural network. CMC-Comput. Mater. Continua **65**(1), 617–638 (2020)
19. Yang, Z., Cohen, W.W., Salakhutdinov, R.: Revisiting semi-supervised learning with graph embeddings. arXiv preprint arXiv:1603.08861 (2016)
20. Yao, L., et al.: Incorporating knowledge graph embeddings into topic modeling. In: Thirty-First AAAI Conference on Artificial Intelligence (2017)
21. Yu, X., Tian, Z., Qiu, J., Su, S., Yan, X.: An intrusion detection algorithm based on feature graph. Comput. Mater. Continua **1**(61), 255–274 (2019)

Investigation on Loan Approval Based on Convolutional Neural Network

Mingli Wu(✉), Chunlai Du(✉), Yafei Huang, Xianwei Cui, and Jianyong Duan

School of Information Science and Technology, North China University of Technology, Beijing 100144, China
{wuml,duchunlai}@ncut.edu.cn

Abstract. With the economic development, loan business has rapidly developed in China. The risk that customers can't repay their loans on time has increased. Therefore it is an important problem for financial organizations to approve the customers' loan application or not. Typical machine learning methods for classification can be employed to mine customers' financial information and give valuable judgments. However, these learning methods rely on shallow features, and the relationships between these features are not well studied. We investigate the function of Convolutional Neural Network (CNN) in this work, as it is successful in field of image recognition, speech recognition and natural language processing. We investigate four different CNN models. Experiments show that the fourth model with stochastic gradient descent algorithm and momentum achieves the best performance. Its accuracy and recall are 0.95 and 0.26 respectively.

Keywords: Machine learning · Classification · Convolutional neural network · Loan approval

1 Introduction

With the rapid development of network, new applications are continuing to emerge, which provide great convenience for people. However, they bring a series of security risks [1–3]. For example, with the economic development, loan business has increased rapidly in China. At the same time, the risks that customers can't repay their loans on time are obvious. It is an important problem to predict whether customers can repay their loan on time. In traditional credit evaluation methods, customers' educational background, occupation, personal income, and other factors are exploited. However, the weights of these factors are difficult to set. Compared with traditional methods, machine learning methods can adjust parameters on training data sets. Machine learning methods combine multiple features with well-adjusted parameters to give more accurate decisions than traditional ones. In recent years, researchers have used machine learning classification methods to solve loan approval problems. They found that multivariate regression and random forest methods performed better than credit scoring methods. Naïve Bayes, Decision Tree, KNN and neural networks have also been used to predict loan repayments. These methods rely on well-designed manual features, and the skill of designers will influence experimental results heavily.

X. Sun et al. (Eds.): ICAIS 2021, CCIS 1422, pp. 203–216, 2021.
https://doi.org/10.1007/978-3-030-78615-1_18

Deep learning methods do not require prior knowledge of features and manual extraction of features. It does not result in inappropriate extraction features due to experience. At present, deep learning technology has attracted extensive attention in the fields of image recognition, speech recognition, text mining, etc. Researchers have found that deep learning technology outperforms traditional neural networks in these areas. However, deep learning has not been well studied for loan approval. In this work, we study deep neural networks and apply them to loan approval. We discuss the network structure, parameter optimization, and compare it to typical classification methods on the standard dataset.

We study convolutional neural networks for credit risk prediction in our experiments. Experiments are conducted on the data set from Chinese Rong360 financial Platform. We judge loan applications based on the customers' basic attributes, bank transaction details and credit card bills, etc. After comparing decision tree, KNN, Naïve Bayes, Random Forest, DNN, and CNN, we found that CNN is best model when using different weight initialization, stochastic gradient descent and momentum. The accuracy and recall of CNN are approximately 0.95 and 0.26, respectively.

The rest of the paper is organized as follows: In Sect. 2, we briefly discuss the related work. Section 3 describes the CNN framework. The network structure and parameters of the model are described in detail here. Section 4 gives a description of the experiments and evaluations. Finally, the conclusions are given in Sect. 5.

2 Related Work

Researchers studied loan application and credit risk with machine learning methods in the past [4–6]. Shih et al. [7] employed data mining methods for classification based on customer lending behavior. They developed a target market model for commercial banks, using logistic regression, decision tree, neural network and support vector machine methods. They reported that the neural network model is the best. Serrano-Cinca and Gutiérrez-Nieto [8] reported that the profit scoring system using multiple regressions is better than the traditional credit scoring system based on logistic regression.

Common methods used for fraud detection include Naive Bayes (NB), Support Vector Machine (SVM), and K-Nearest Neighbor (KNN). They are adopted alone or integrated to build classifiers. Aggregate learning methods brought good predictive performance [9]. Kuo et al. [10] developed a genetic algorithm based fuzzy neural network to predict the stock market. Due to too many parameters and manual selection of parameters, the genetic algorithm is not effective. Malekipirbazari and Aksakalli [11] employed a random forest based classification method to predict the state of the borrower. Experiments on the popular social lending platform Lending Club show that the random forest based approach outperformed the credit score model.

Neural network models were studied for prediction and classification in the past. Google [12] used the information about the click rate and search volume collected by its search engine to establish a box office forecasting model. The prediction result was close to the actual box office. The accuracy rate reached 94%. Neural network models were also studied for credit risk management. They could construct complex nonlinear functions and were more flexible than traditional statistical modeling. Jin and Zhu predicted bank defaults by neural network [13]. They exploited the information about terms of loan, annual income, the amount of loan, debt-to-income ratio, credit grades, etc. Experimental results showed that neural network achieved the best prediction accuracy compared to linear classifiers, logistic regression and decision tree. Khashman found that when the training data and the testing data accounted for 4:6, the accuracy rate was the highest in their experiment [14].

The learning ability of convolutional neural networks is strong [15]. They can automatically learn the features which are most effective for classification. Geng et al. [16] used a simple linear iterative clustering algorithm to segment the original road image into super-pixels with uniform size. The convolutional neural network took these data as input and automatically classified images as road and non-road areas. They reported that this method worked well. In the task of satellite image scene classification, Zhang et al. [17] combined gradient acceleration and convolution network to generate a gradient boosting random convolutional network framework, which achieved good classification accuracy. Experiments were conducted on the UC Merced data set with 21 different categories of aerial scenes and the Sydney data set with 8 kinds of uses. They found that their model was successful. Duan et al. [18] designed a CNN and Extreme Learning Machine hybrid structure for gender classification. They integrated two classifiers to handle gender classification, and used unfiltered facial images in Adience Benchmark dataset to verify the hybrid structure. The experimental results showed that the hybrid architecture with CNN improved the performance on accuracy and efficiency. Suo et al. [19] used CNN algorithm to predict the risk of patients suffering from certain diseases. Experiments showed that the addition of CNN model present the longitudinal electric health records sequence better.

Selvin et al. [20] proposed a stock price forecast model based on deep learning. Based on stocks' day stamp, time stamp, transaction id, price and volume sold in each minute, the deep neural network model could capture hidden dynamics and give predictions. After comparing CNN, RNN, and LSTM models, they found that the CNN architecture performed better than the other two models. Xu et al. [21] proposed a psychologically inspired convolutional neural network to achieve automatic facial attractiveness prediction, using facial detail, illumination, and color to train the facial attractiveness predictor. Experiments show that PI-CNN prediction model has the highest correlation of 0.87 in the SCUT-FBP benchmark database, which is superior to manual design features and other deep learning methods. Krug et al. [22] employed CNN to predict miniaturized seizures. They investigate the generalization capability of the CNN-based method to measure the strength of generalized synchronization in EEG recordings from epilepsy patients. Experiments showed that the method worked well. Chris et al. [23] studied a CNN based framework, which was trained to classify small pieces of text into predefined font categories. They reported state-of-the-art performance in a challenging dataset of 40 Arabic computer fonts.

3 Methodology

3.1 Convolutional Neural Network

In our credit risk forecast model, we train a CNN model through a lot of data. The CNN model is divided into two parts in the learning process, i.e., forward propagation and backward propagation. In forward propagation, it mainly includes the following parts: the input layer forward propagation to the convolutional layer, the hidden layer forward propagation to the convolutional layer, the hidden layer forward propagation to the pooled layer, and the hidden layer forward propagation to the fully connected layer. For an input, the forward propagation of the convolutional layer can be expressed as the following equation.

$$h_l = \sigma(z_l) = \sigma(h_{l-1} * W_l + b_l) \tag{1}$$

Here l represents the number of layers. * represents convolution operation, and b represents the bias, and W is weight. n represents the number of feature maps, and σ is the activation function. In the convolution operation, two parameters exist, namely stride and padding. Stride is the pixel distance of each movement during the convolution process, and padding is for better recognition of the edge features.

The processing logic of the pooling layer is relatively simple. Our goal is to narrow down the input matrix. For example, if some of the input matrices are of N * N dimension, and our pooling size is of k * k dimension, then the output matrix is of N/k * N/k dimension. Common pooling methods mainly include max pooling, mean pooling, etc.

In backward propagation, two problems are mainly solved: the first one is δ^l of the known convolutional layer, which is used to derive δ^{l-1} of the previous hidden layer; the second one is δ^l of the known convolutional layer, which is used to derive the layer weights and bias.

The forward direction is from pooling to convolution. Assuming that the pooling layer is a 3 * 3 matrix at layer L−1 and the convolution kernel is 2 * 2, the convolution layer is 2 * 2 at L layer. The following equations can be derived.

$$\mathrm{h}_1 = p_1 * w_1 + p_2 * w_2 + p_4 * w_3 + p_5 * w_4 \tag{2}$$

$$\mathrm{h}_2 = p_2 * w_1 + p_3 * w_2 + p_5 * w_3 + p_6 * w_4 \tag{3}$$

$$\mathrm{h}_3 = p_4 * w_1 + p_5 * w_2 + p_7 * w_3 + p_8 * w_4 \tag{4}$$

$$\mathrm{h}_4 = p_5 * w_1 + p_6 * w_2 + p_8 * w_3 + p_9 * w_4 \tag{5}$$

It can be seen that p_1 is related to h_1; p_2 is related to h_1 and h_2; p_3 is related to h_2; p_4 is related to h_1 and h_3; p_5 is related to h_1, h_2, h_3 and h_4; p_6 is related to h_2 and h_4; p_7 is related to h_3; p_8 is related to h_3 and h_4; p_9 is related to h_4.

$$\Delta^l = \sigma'(h^l) \cdot (\Delta^{l+1} \circledcirc cir(W)) \tag{6}$$

Among them, Δ^l represents the output of the $l+l$th layer, ◎represents the convolution operation, and cir() function represents the rotation of the matrix by 180°. Forward propagation is from convolution to pooling. Since each neuron in the pooled layer corresponds to multiple neurons in the convolutional layer, the pooled layer needs to be up sampled to obtain a 4th-order matrix P1. Here $P1 = P \times E_{2*2}$. At this time, it becomes one-to-one sampling, so each node of the L−1th layer only corresponds to the only one node of the Lth layer after the expansion, and the gradient of W and b is required after the output is obtained, and the principle is similar to the back propagation in DNN.

$$\Delta^{l-1} = \lambda^l(\sigma'(h^{l-1}) \cdot P1 \tag{7}$$

λ_j^l indicates the weight of the L-th layer pooling layer. $\sigma'(Z_j^i)$ is the derivative of the activation function of the i + 1th layer (the independent variable is the upper layer). Therefore, we derive the gradient calculation of the weights:

$$\frac{\partial H}{\partial W^{l-1}} = \frac{\partial H}{\partial h^{l-1}} \cdot \frac{\partial h^{l-1}}{\partial W^{l-1}} = \Delta^{l-1} \cdot \frac{\partial h^{l-1}}{\partial W^{l-1}} \tag{8}$$

Here H is the standard deviation. For the b parameter, since b contributes to each neuron (feature), according to the derivation rule, the gradient of b is obtained:

$$\frac{\partial H}{\partial b^{l-1}} = \frac{\partial H}{\partial h^{l-1}} \cdot \frac{\partial h^{l-1}}{\partial b^{l-1}} = \sum(\Delta^{l-1}) \tag{9}$$

Among them, Δ^{l-1} is the result of the sensitivity of all feature maps of layer L−1.

3.2 Parameter Selection

In the process of full connection of neural networks, over-fitting often occurs. Dropout can solve this problem well. The principle is to set the probability P of each layer of neurons, reducing the connection between the individual neurons. To adjust the weight values in the propagation process, we use stochastic gradient descent methods. After comparing the performance of AdaGrad, RMSProp, Adam and SGD algorithms, we find that the first three methods perform well when training, but poorly when testing. When using the same number of hyper-parameters to adjust parameters, we find that SGD and SGD + momentum methods perform better than AdaGrad, RMSProp and Adam on test data. The equation for SGD is described as the following one. Here μ is the learning rate.

$$\Delta v^t = -\mu * g_t \tag{10}$$

The following equation is for SGD + Momentum.

$$\Delta v^t = \omega \Delta v^{t-1} - \mu g_t \tag{11}$$

Here ω is a momentum and its value is between 0 and 1, indicating the degree of dependence on the previous sample update. At the beginning of the training procedure, since the gradient may be large, the initial value is generally chosen to be 0.5. When the gradient is not so large, it is changed to 0.9. Momentum prevents falling into local optimum. Δv indicates the influence of the gradient direction of the historical moment on the gradient of the current moment. If the direction is the same, the gradient direction of the current time will be strengthened. If it is different, the gradient direction of the current time will be weakened. Therefore, the SGD + Momentum method can suppress the oscillation phenomenon and accelerate the convergence.

The Weight initialization method is described as follows. The ultimate goal of training CNN is to get the best parameters and make the objective function get the minimum value. Before adjusting the parameters, we need to initialize the parameters. Through a lot of experiments, the Xavier initialization method confirmed its validity [24–26]. Another initialization strategy is the He method, which was proposed by He et al. [27]. The basic idea is same as the Xavier method. The weight is scaled to complete the initialization. Unlike the Xavier method, this method only considers the forward process, So for normal distribution and uniform distribution. The equation is the input part of Xavier. There are two types of He and Xavier in Tensorflow. One is based on uniform distribution and the other one is based on normal distribution. It is assumed that both the weight W and the offset b obey the condition that the mean is 0.

$$\mathrm{uniform}_X(W) = \left[-\frac{2}{q_{in} + q_{out}}, \frac{2}{q_{in} + q_{out}}\right] \tag{12}$$

$$\mathrm{normal}_X(W) = \left[-\sqrt{\frac{6}{q_{in} + q_{out}}}, \sqrt{\frac{6}{q_{in} + q_{out}}}\right] \tag{13}$$

$$\mathrm{uniform}_H(W) = \left[0, \frac{2}{q_{in}}\right] \tag{14}$$

$$\mathrm{normal}_H(W) = \left[-\sqrt{\frac{6}{q_{in}}}, \sqrt{\frac{6}{q_{in}}}\right] \tag{15}$$

In the above equations, q_{in} and q_{out} represent the number of input neurons and output neurons, respectively. Uniform(W) is the uniform distribution initialization range, and normal(W) is the normal distribution initialization range.

3.3 Convolutional Neural Network for Loan Approval

The general framework for loan approval is shown in Fig. 1. After pre-processing of the raw data, feature vectors are fed into a convolutional neural network. Then multiple layers of convolution and pooling are designed. Finally, a fully connected layer is employed to give the decision on loan application.

Convolutional layers and pooling layers are unique to CNN models. The combination of convolutional layer and pooled layer can appear multiple times. We can also use the combination of convolutional layer and convolutional layer, or convolutional layer, convolutional layer and pooled layer, which is not limited when building the model.

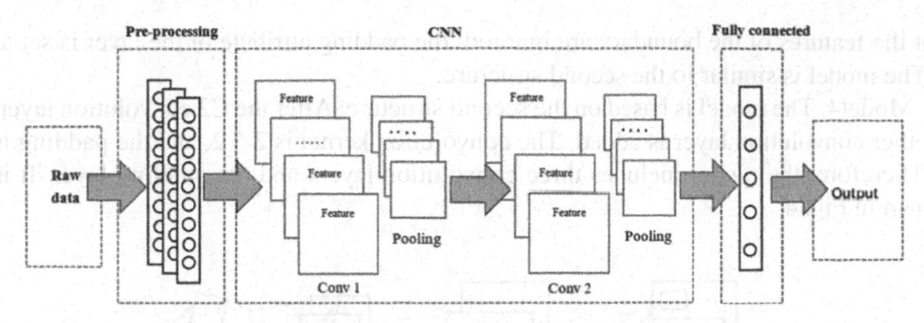

Fig. 1. The framework of loan application approval model

In Sect. 3.1, we describe the principles of convolutional and pooling layers in forward and backward propagation. We design different kinds of convolutional neural network structures here.

We study four different CNN models, which are named as model1, model2, model3 and model4. They are based on Structure 1, Structure 2, Structure 3, and Structure 4 respectively.

Model1. First, the feature maps of six 3 * 3 convolution kernels form the convolutional layer C1. Then pass through the 2 * 2 pooling layer P, and connect the convolutional layer C2 which is composed of the feature maps of 18 1 * 1 convolution kernels. A data set of 90 features is obtained. Finally the fully connected hidden layer is used as input to produce a final prediction result. It is shown in Fig. 2.

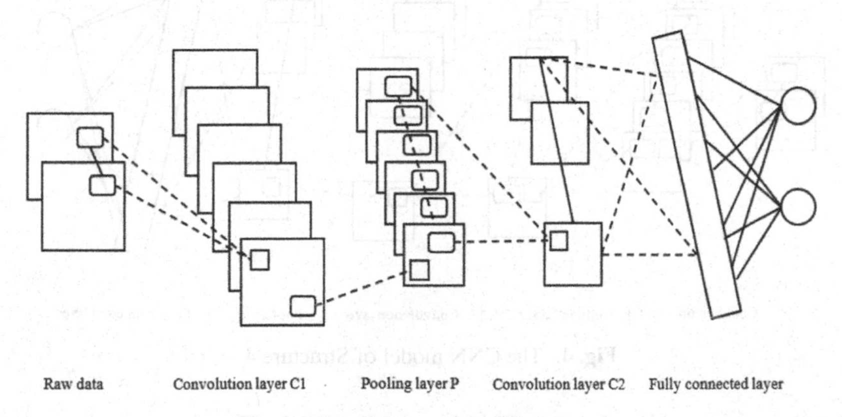

Fig. 2. The CNN model of Structure 1

Model2. First, the feature maps of six 2 * 2 convolution kernels form the convolutional layer C1. Then connect the convolutional layer C2 which is composed of the feature maps of 18 2 * 2 convolution kernels. Pass through the 2 * 2 pooling layer P. A data set of 90 features is obtained. Finally the fully connected hidden layer is used as input to produce a final prediction result. It is shown in Fig. 3.

Model3. On the basis of the first structure, a convolution layer is added after the C1 convolutional layer. The convolution kernel is 2 * 2. In order to avoid the phenomenon that the features of the boundary are ignored, the padding attribute of the layer is set to 1. The model is similar to the second structure.

Model4. The model is based on the second structure. After the C2 convolution layer, another convolution layer is added. The convolution kernel is 2 * 2, and the padding is 1. Therefore, the model includes three convolution layers and one pooling layer. It is shown in Fig. 4.

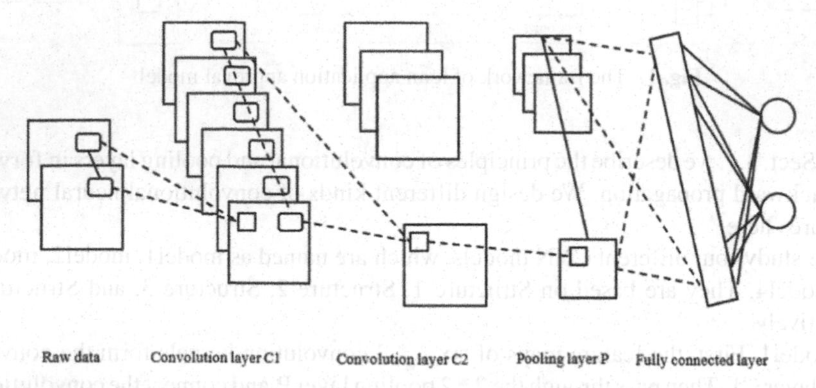

Fig. 3. The CNN model of Structure 2

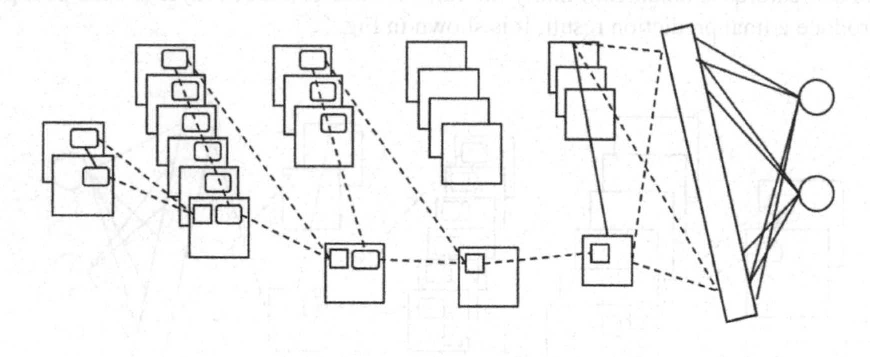

Fig. 4. The CNN model of Structure 4

4 Experiments

4.1 Data Set

The dataset we used in our experiments is from real loan applications. It is provided by Chinese Rong360 financial platform. The data set includes 55,596 customer records, 6070197 bank statement records, and 2338118 credit card consumption records. At the same time, the issuing times of the loans are also provided. For each customer, label 1 or 0 is given to denote whether he/she don't repay the loan on time. In this work, we classify users into two categories, i.e., "bad" customers and "good" customers.

Each customer's financial information includes his/her bank statement records and credit card consumption records. We calculate bill amount of current period, bill amount of last period, repayment amount of last period, the number of times, maximum of current period, minimum of current period for each customer every day, month, and three months. At the same time, we calculate these values for the periods before the loan issuing time and after the loan issuing time. We calculate feature values for customer credit card records and bank statement records. Finally, a feature map of 6 * 6 * 2 is built and fed into our CNN models.

4.2 Experimental Configuration

We employ a zero-center data preprocessing method to normalize the data. The offset is initialized to 0. According to our preliminary experiments, we set the ratio of training data and testing data as 8:2. The initial learning rate of the CNN model is set as 0.1. The learning rate is adjusted by the function of Tensorflow. Our models read 128 samples at a time. In addition, different experimental configurations are set, which is described as follows.

1. CNN. In the above four basic network structures, ReLu activation function is adopted and dropout technique improves the over-fit phenomenon. Weight values are initialized as 0.
2. CNN + W. In the above four basic network structures, different initialization methods are investigated, i.e., Xavier and He.
3. CNN + W + SGD. Besides weight initialization, we employ a stochastic gradient descent algorithm during back propagation.
4. CNN + W + SGD + momentum. In order to prevent the occurrence of local optimal solutions, a combination of stochastic gradient descent algorithm and momentum is used in the back propagation procedure.

4.3 Experimental Result

We show the accuracy scores of the above four CNN models with different configurations in Table 1. Here model1, model2, model3 and model 4 have different CNN structures. Xavier and He weight initialization methods are compared. SGD and momentum methods are also studied. In addition, accuracy, recall and f1 performance curves of these models with different configurations are shown in Fig. 5, Fig. 6 and Fig. 7, respectively. According to Table 1 and above figures, it can be seen that weight initialization, stochastic gradient descent and momentum are effective. He method performs better than Xavier method when the weight values are initialized. It confirms that He method works well with ReLu activation function. We find that model4 with SGD and momentum achieves the best accuracy and recall, i.e., 0.95 and 0.26 respectively.

Although the accuracy scores in Fig. 5 are satisfactory, the recall scores in Fig. 6 are not. To improve the recall performance, we adjust the threshold. Then we list the performance scores of four models under He + SGD + momentum configuration in

Table 2. We find that Model 4 achieves the best performance on accuracy, recall and f1-score. Model 1 and Model 3 get similar performance scores. We report that the best predictive model is the CNN Model 4 with He + SGD + momentum configuration. The accuracy, recall and F1-score are 0.75, 0.45 and 0.47 respectively.

Table 1. Accuracy of different CNN models

	CNN	CNN + W(Xavier)	CNN + W(He)	CNN + W(He) + SGD	CNN + W + SGD + momentum
Model1	0.90	0.91	0.92	0.92	0.92
Model2	0.91	0.92	0.93	0.93	0.94
Model3	0.91	0.92	0.92	0.93	0.93
Model4	0.93	0.93	0.93	0.94	0.95

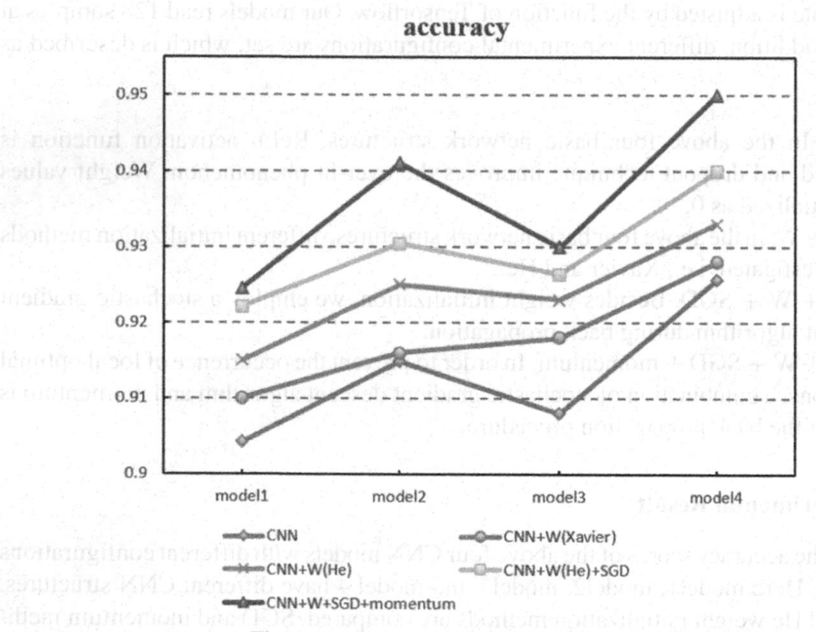

Fig. 5. Accuracy of different CNN models

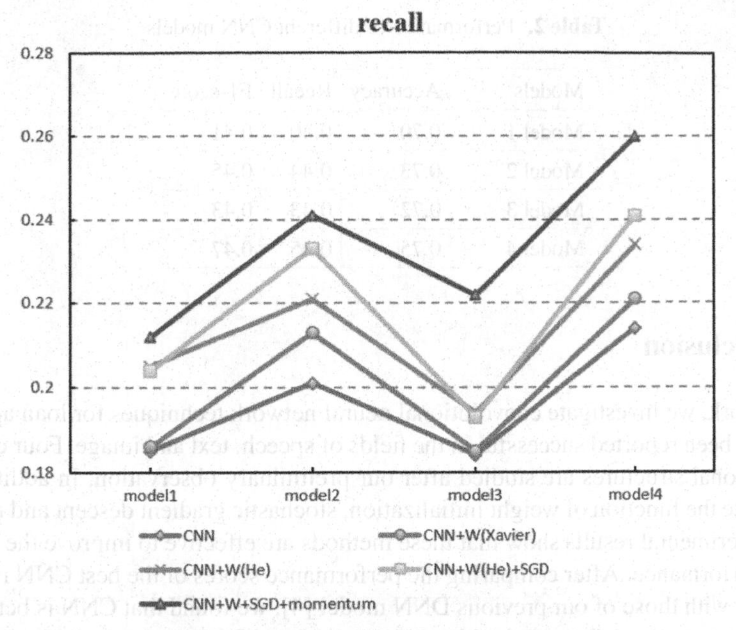

Fig. 6. Recall of different CNN models

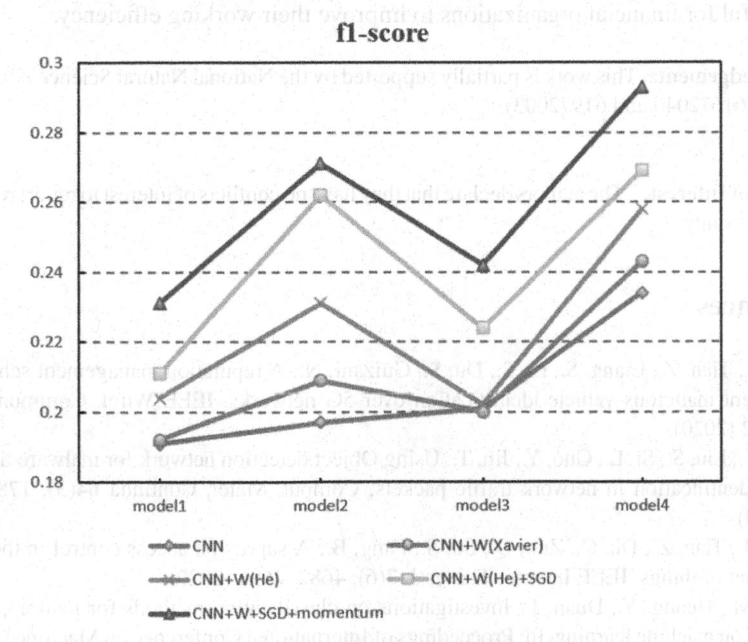

Fig. 7. F1-score of different CNN models

Table 2. Performance of different CNN models

Models	Accuracy	Recall	F1-score
Model 1	0.70	0.40	0.41
Model 2	0.73	0.44	0.45
Model 3	0.72	0.43	0.43
Model 4	0.75	0.45	0.47

5 Conclusion

In this work, we investigate convolutional neural network techniques for loan approval. CNN has been reported successful in the fields of speech, text and image. Four different convolutional structures are studied after our preliminary observation. In addition, we investigate the function of weight initialization, stochastic gradient descent and momentum. Experimental results show that these methods are effective to improve the classification performance. After comparing the performance scores of the best CNN model in this work with those of our previous DNN model [4], we found that CNN is better. The best accuracy and recall scores in this work are 0.95 and 0.26, respectively. It means that 26% of "bad" customers can be recognized, and the prediction accuracy is about 95%. It is helpful for financial organizations to improve their working efficiency.

Acknowledgement. This work is partially supported by the National Natural Science Foundation of China (61672040 and 61972003).

Conflicts of Interest. The authors declare that they have no conflicts of interest to report regarding the present study.

References

1. Su, S., Tian, Z., Liang, S., Li, S., Du, S., Guizani, N.: A reputation management scheme for efficient malicious vehicle identification over 5G networks. IEEE Wirel. Commun. **27**(3), 46–52 (2020)
2. Du, C., Liu, S., Si, L., Guo, Y., Jin, T.: Using Object detection network for malware detection and identification in network traffic packets. Comput. Mater. Continua **64**(3), 1785–1796 (2020)
3. Qiu, J., Tian, Z., Du, C., Zuo, Q., Su, S., Fang, B.: A survey on access control in the age of internet of things. IEEE Internet Things J. **7**(6), 4682–4696 (2020)
4. Wu, M., Huang, Y., Duan, J.: Investigations on classification methods for loan application based on machine learning. In: Proceedings of International Conference on Machine Learning and Cybernetics, pp. 541–546. Kobe, Japan (2019)
5. Arora, N., Kaur, P.: A Bolasso based consistent feature selection enabled random forest classification algorithm: an application to credit risk assessment. Appl. Soft Comput. J. **86**, 105936 (2020)

6. Melo, L., Nardini, F., Renso, C., Trani, R., Macedo, J.: A novel approach to define the local region of dynamic selection techniques in imbalanced credit scoring problems. Expert Syst. Appl. **152**, 115531 (2020)
7. Shih, J., Chen, W., Chang, Y.: Developing target marketing models for personal loans. In: Proceedings of IEEE International Conference on Industrial Engineering and Engineering Management, pp.1347–1351. Bandar Sunway, Malaysia (2014)
8. Serrano-Cinca, C., Gutiérrez-Nieto, B.: The use of profit scoring as an alternative to credit scoring systems in peer-to-peer (P2P) lending. Decis. Support Syst. **89**, 113–122 (2016)
9. Zareapoor, M., Shamsolmoali, P.: Application of credit card fraud detection based on bagging ensemble classifier. Procedia Comput. Sci. **48**, 679–685 (2015)
10. Kuo, R., Chen, C., Hwang, Y.: An intelligent stock trading decision support system through integration of genetic algorithm based fuzzy neural network and artificial neural network. Fuzzy Sets Syst. **118**(1), 21–45 (2001)
11. Malekipirbazari, M., Aksakalli, V.: Risk assessment in social lending via random forests. Expert Syst. Appl. **42**(10), 4621–4631 (2015)
12. Panaligan, R., Chen, A.: Quantifying Movie Magic with Google Search. Google Whitepaper (2013)
13. Jin, Y., Zhu, Y.: A data-driven approach to predict default risk of loan for online peer-to-peer (P2P) lending. In: Proceedings of International Conference on Communication Systems and Network Technologies, pp. 609–613. Gwalior, India (2015)
14. Khashman, A.: Neural networks for credit risk evaluation: investigation of different neural models and learning schemes. Expert Syst. Appl. **37**(9), 6233–6239 (2010)
15. Zeng, X., Ouyang, W., Wang, X.: Multi-stage contextual deep learning for pedestrian detection. In: Proceedings of IEEE International Conference on Computer Vision, pp. 121–128. Sydney, Australia (2013)
16. Geng, L., Sun, J., Xiao, Z., Zhang, F., Wu, J.: Combining CNN and MRF for road detection. In: Lu, H., Xu, X. (eds.) Artificial Intelligence and Robotics. SCI, vol. 752, pp. 103–113. Springer, Cham (2018). https://doi.org/10.1007/978-3-319-69877-9_12
17. Zhang, F., Du, B., Zhang, L.: Scene classification via a gradient boosting random convolutional network framework. IEEE Trans. Geosci. Remote Sens. **54**(3), 1793–1802 (2016)
18. Duan, M., Li, K., Yang, C., Li, K.: A hybrid deep learning CNN-ELM for age and gender classification. Neurocomputing **275**, 448–461 (2018)
19. Suo, Q., et al.: Personalized disease prediction using a CNN-based similarity learning method. In: Proceedings of IEEE International Conference on Bioinformatics and Biomedicine, pp. 811–816. Kansas City, USA (2017)
20. Selvin, S., Ravi, V., Gopalakrishnan, E., Menon, V., KP, S.: Stock price prediction using LSTM, RNN and CNN-sliding window model. In: Proceedings of International Conference on Advances in Computing, Communications and Informatics, pp. 1643–1647, Udupi, India (2017)
21. Xu, J., Jin, L., Liang, L., Feng, Z., Xie, D., Mao, H.: Facial attractiveness prediction using psychologically inspired convolutional neural network (PI-CNN). In: Proceedings of IEEE International Conference on Acoustics, Speech and Signal Processing, pp. 1657–1661, New Orleans, USA (2017)
22. Krug, D., Elger, C., Lehnertz, K.: A CNN-based synchronization analysis for epileptic seizure prediction: inter- and intra- individual generalization properties. In: Proceedings of International Workshop on Cellular Neural Networks and Their Applications, pp. 92–95, Santiago de Compostela, Spain (2008)
23. Tensmeyer, C., Saunders, D., Martinez, T.: convolutional neural networks for font classification. In: Proceedings of International Conference on Document Analysis and Recognition, pp. 985–990. Kyoto, Japan (2017)

24. Srivastava, N., Hinton, G., Krizhevsky, A., Sutskever, I., Salakhutdinov, R.: Dropout: a simple way to prevent neural networks from overfitting. J. Mach. Learn. Res. **15**(1), 1929–1958 (2014)
25. He, K., Zhang, X., Ren, S., Jian, S.: Deep residual learning for image recognition. In: Proceedings of IEEE Conference on Computer Vision and Pattern Recognition, pp. 770–778, Las Vegas, USA (2016)
26. Jia, Y., et al.: Caffe: convolutional architecture for fast feature embedding. In: Proceedings of ACM conference on Multimedia, pp. 675–678, Orlando, USA (2014)
27. He, K., Zhang, X., Ren, S., Sun, J.: Delving deep into rectifiers: surpassing human-level performance on ImageNet classification. In: Proceedings of IEEE International Conference on Computer Vision, pp. 1026–1034. Santiago, Chile (2015)

Chinese Verb-Object Collocation Knowledge Graph Construction and Application

Yuhang Zhao[1,2], Yi Li[1,2], and Yanqiu Shao[1,2](✉)

[1] School of Information Science, Beijing Language and Culture University, Beijing 100083, China

[2] National Language Resources Monitoring and Research Center (CNLR) Print Media Language Branch, Beijing 100083, China

Abstract. Verb is the core of a sentence. It can not only reflect the syntactic structure and semantic framework of the whole sentence, but also restrict the nominal elements which co-exist with them. They play a significant role in sentence. Verb-Object Collocation has received more and more attention owing to its high frequency, complexity and flexibility of using. Domestic researches on verb object collocation mainly focus on automatic recognition and construction of corresponding collocation knowledge base. Nevertheless, most of the existing Verb-Object Collocation knowledge base lacks semantic information and classification information. Thus, the application scenarios will be greatly limited. It cannot meet the semantic needs of natural language processing, either. In view of this situation, this paper deeply analyzes the semantic relationship of Verb-Object Collocation, and uses the related technology of knowledge graph to construct a Verb-Object Collocation Knowledge Graph, which obtains 40704 specific examples of Verb-Object Collocation and 20000 extraction templates of Verb-Object Collocation. At the same time, this paper constructs a Verb-Object Collocation Knowledge Graph to manage and maintain our knowledge through two parts: ontology layer and data layer. Finally, the effectiveness of the Verb-Object Collocation Knowledge Graph is verified on the task of entity recognition.

Keywords: Verb-Object Collocation Knowledge Graph · Semantic relational framework · Ontology construction · Verb-object collocation extraction

1 Introduction

In the 1950s, Firth [1] proposed the concept of collocation. He believed that collocation is the common co-occurrence of particular words. Benson et al. [2] considered that collocations are recurrent combinations or collocations. He also noticed the influence of grammatical rules on collocations. Collocation is an important manifestation of semantics and grammar. Among them, verb-object collocation has received extensive attention in academia because of its high frequency, complexity and flexibility of using. Verb-object collocation occupies the core position in sentence because verb is the core of a sentence, which can not only reflect the syntactic structure and semantic frame

X. Sun et al. (Eds.): ICAIS 2021, CCIS 1422, pp. 217–232, 2021.
https://doi.org/10.1007/978-3-030-78615-1_19

of the entire sentence, but also restrict the nominal components that co-occur with it. Compared with other parts of speech, verb has high complexity and flexibility, and it can effectively reflect the syntax and semantic information of sentence. A scientific and reasonable knowledge graph focused on the study of verb-object collocation can improve the depth of understanding of natural language and provide semantic information support for natural language processing. Therefore, it is of great significance to conduct an in-depth study on the semantic of verb-object collocations and construct a Verb-Object Collocation Knowledge Graph.

2 Related Work

2.1 Current Research Status of Collocation

The research on collocation originated early abroad. In 1963, Katz et al. [3] realized the importance of semantics when studying syntax. In corpus, they tried to introduce semantic tags to characterize words, and generate sentences by establishing projection rules. They described each argument of the verb with semantic selection restriction. Halliday et al. [4] proposed three closely related concepts in lexicon research: lexical items, collocations and word sets. He believes that collocation must be based on corpus and extract collocations from the corpus. Choueka et al. [5] extracted common English collocations, such as fried chicken, home run, Magic Johnson, etc., from the text of about 11 million words scale in New York Times. Church et al. [6] introduced statistical methods into the study of collocation. On this basis, the Xtract system developed by Smadja et al. [7] is a relatively complete system for quantitative analysis of collocation so far. Smadja et al. [7] puts forward formula for calculating the strength of word pairs, introduces the formula for dispersion of position information, and integrates the automatic part-of-speech tagging technology of corpus linguistics.

The domestic study on collocation has also undergone a series of developments, and the process has mainly experienced the transformation from studying various types of collocations in general to studying specific types of collocations. Sun Maosong et al. [8] firstly studied the automatic extraction of Chinese word collocations and used Smadja's algorithm to propose a quantitative evaluation system of collocation including strength, dispersion, and peak, and constructed a corresponding collocation judgment algorithm with a value of about 7100 million words-level Xinhua News Agency corpus as experimental platform, and the accuracy of the algorithm for automatic recognition of collocation reached 33.94%. Qu Weiguang et al. [9] proposed a frame-based method to extract Chinese collocations and their structure information simultaneously. It used ratio of relative word rank (RRWR) to pre-process the potential collocations, applied linguistic knowledge to restrict the part of speech of the candidate collocations. The average accuracy reached 84.73%. Wang Lu [10] combined relevant semantic knowledge, used typical sentence pattern rules and mathematical measures such as frequency and mutual information and extracted and extracted nine collocation word types which uses nouns, adjectives and verbs as central words from a large-scale corpus of the true text. And the accuracy rate is 90.9%. In recent years, in order to avoid the problem of inappropriate collocation caused by relying solely on statistics, some scholars have tried to apply more syntactic and semantic knowledge to the acquisition of word collocations, and at the same time, they gradually began to integrate natural language processing technology in the extraction of collocations.

2.2 Construction Status of Verb-Object Collocation Knowledge Graph

At present, the existing collocation resources are mainly based on collocation dictionaries compiled by experts, such as "Modern Chinese Collocation Dictionary" and "Modern Chinese Notional Word Collocation Dictionary". The former collocations are presented in the form of words, phrases and idioms, while the latter are presented in the form of grammatical constitutes, namely subject, object, attributive, adverbial, etc. However, these dictionaries require a lot of manual participation, and difficult to update and maintain. In addition, there are some wrong cases. In recent years, some scholars have tried to automatically extract collocations and build a large-scale collocation knowledge base. For example, Hu Renfen et al. [11] proposed a hybrid method integrating linguistic knowledge of surface and deeper layers as well as statistic information for automatic extraction of collocation with high proficiency and precision. 22298 collocations are extracted from a 2.4-million-word L2 textbook corpus, and 219451 collocations are extracted from 138-million-word Wikipedia corpus. On this basis, two large-scale collocation knowledge bases are built.

Among the various types of collocations, owing to the unique importance of verbs in sentences, verb-object collocations have become a key research object in academia. The current research on verb-object collocations is mainly concentrated in two fields: one is the recognition of verb-object collocations; the other is to build verb-object collocations knowledge base. Based on the semantic selection restriction of verbs on the object, Zhou Weihua [12] used analogy as the main principle to distinguish regular phenomena and irregular phenomena in the semantic collocation of verbs and objects, and adopted different treatments for them. He established rules for verb-object collocations that can be freely analogized, and built a corpus for verb-object collocations that cannot be freely analogized. Liu Zhichao [13] extracted common verb lists from HowNet, People's Daily, and Southern Weekend. Based on the collocation of the verb lists, a Chinese verb-object collocation bank with a scale of 110,000 collocations and containing semantic annotations and semantic information is constructed.

However, owing to the fact that only study the verb-object collocation, the constructed knowledge base will be greatly restricted in the application scenes and cannot meet the semantic needs of the natural language processing. Furthermore, it is also difficult to maintain and manage the knowledge base obtained from the common verb lists. Considering the above problems, this article uses the related technology of knowledge graph to construct the knowledge graph of verb-object collocation.

3 The Definition of Verb-Object Collocation Semantic Relations

The semantic relation between verbs and objects has always been the focus of research in Chinese. The classification of verb-object semantic relation is mainly based on the semantic role of the object in the structure. Different scholars' classification of object semantic roles is shown in the Table 1. It can be seen that linguists have not yet reached a consensus on the classification of objects. Different linguists have different object classifications, and linguists have inconsistent naming of the same object semantic roles. For example, the "使动宾语" defined by Xu Shu and Ma Qingzhu is the same as "致使宾语" defined by Meng Cong, which is not conductive to subsequent semantic analysis. In view of the above problems, this paper extracts verb-object collocations and analyzes semantic relations on the basis of SDP [14]. The data comes from different fields, including news, literature, Primary and secondary school textbooks, and teaching materials for Chinese as a foreign language. It should be noted that the objects can be nouns, verbs, adjectives, and pronouns in verb-object collocation. This article mainly focuses on verbs and nouns. The research object mainly refers to the verb-object collocation that the object is a noun.

Table 1. Semantic role classification of objects

Xu Shu [15]	Ma Qingzhu [16]	Meng Cong [17]	Ling Xingguang [18]	Huang Borong [19]
Patient	Patient	Patient	Patient	Patient
Agent	Agent	Agent		Agent
	Time	Time	Time	Time
Tool	Tool	Tool	Tool	Tool
	Manner	Manner	Manner	Manner
Reason		Reason	Reason	Reason
		Purpose	Purpose	Purpose
Causative	Causative			
	Quantity		Quantity	
…	…	…	…	…

This paper analyzes the extracted verb-object collocations according to the BH-SDP semantic dependency annotation system. Finally, we defined 21 kinds of verb-object collocation semantic relations. They are Patient, Agent, Content, Experiencer, Class, Dative, Belongings, Possessor, Product, Consequence, Tool, Material, Manner, Initial Location, Final Location, Location Through, Intention, Reason, Time, According, Direction.

1. Patient
   It indicates the object directly affected or dominated by the action.
   e.g. 1 他吃了三个苹果. (He ate three apples.)
   e.g. 2 倩碧在台推出男性保养品. (Clinique launched male skin care products in Taiwan.)
2. Experiencer
   It indicates the conscious subject of a specific action.
   e.g. 1 一个教室挤了快八十位同学. (The classroom was crowded with nearly 80 classmates.)

e.g. 2 这里来了大约一万五千人. (Here came about 15,000 people.)

3. Content

   It indicates the object involved in the action but has not changed.

   e.g. 1 育空新闻的资深记者莫理察也报告了这个好消息. (Moritza, a senior reporter from Yukon News, also reported the good news.)

   e.g. 2 至今许多画家、雕塑家仍在汉代文物中汲取创作元素. (So far, many painters and sculptors are still absorbing creative elements from the cultural relics of the Han Dynasty.)

4. Experiencer

   It indicates the issuer of non-behavior actions and the issuer has no subjective initiative.

   e.g. 1 例如屡次改革货币, 使民间经济根本发生动摇, 极为扰民. (For example, repeated currency reforms have fundamentally shaken the private economy, which greatly disturbs the people.)

   e.g. 2 孤悬外海的绿岛, 曾经是拘禁政治犯的重地之一, 留下无数的悲怆血泪. (The lonely green island off the coast was once one of the most important places for detaining political prisoners, leaving countless tears of sorrow and blood.)

5. Class

   It indicates the subject of judgment or implicit comparison, or the state of the final classification of the subject.

   e.g. 1 雪化成了水. (The snow turned into water)

   e.g. 2 业务导向已成为电视台首先考量的重点. (Business orientation has become the first focus of the TV station's consideration.)

6. Dative

   It indicates an inactive participant in action.

   e.g. 1 中共总理朱熔基到伦敦参加欧亚高峰会议, 行程中宴请侨领时说了什么? (Premier Zhu Rongji of the Chinese Communist Party went to London to attend the Eurasian Summit. What did he say when he hosted a banquet for overseas Chinese leaders during his trip?)

   e.g. 2 过去一些富人家的院子, 如今已变成了供人们参观的园林, 例如有名的沈厅和张厅. (The yards of the wealthy in the past have now become gardens for people to visit, such as the famous Shenting and Zhangting.)

7. Belongings

   It indicates the owner in the affiliation.

   e.g. 1 做一点有价值有意义的事并不难, 难的是不做那些不该做的事. (It is not difficult to do something valuable and meaningful, but it is difficult not to do things that should not be done.)

   e.g. 2 介绍这种理想社会的服务, 建筑, 室内装修, 包括图书馆. (Introduces the services, architecture, interior decoration of this ideal society, including libraries.)

8. Possessor

   It indicates the subject in the possession relationship.

   e.g. 1 这块蛋糕属于妹妹. (This cake belongs to sister)

   e.g. 2 面容不仅仅属于个人, 而且也属于社会, 成为人们文化符号体系中的一部分. (The face not only belongs to the individual, but also belongs to the society, becoming a part of people's cultural symbol system.)

9. Product
   It indicates new things created or produced by the subject.
   e.g. 1 因长年风化及海水的侵蚀, 形成曲折多变的海岸景观. (Due to several years of weathering and seawater erosion, a tortuous coastal landscape is formed.)
   e.g. 2 同一年他和一些朋友成立了一个革命组织. (In the same year, he and his friends established a revolutionary organization.)
10. Consequence
   It indicates the result of the change caused by the action.
   e.g. 1 所以吃素并不意味着营养不良, 相反, 吃肉却会给身体带来很多隐患. (So being vegetarian does not mean malnutrition. On the contrary, eating meat can bring many hidden problems.)
   e.g. 2 吸烟可能引起多种疾病. (Smoking may cause many diseases.)
11. Tool
   It indicates the tool that the action used.
   e.g. 1 他便吹响小号, 号声凄婉悲凉. (He blew the trumpet, it sounds sad.)
   e.g. 2 暑假他父亲归途阻塞, 到天津改乘轮船, 辗转回家. (During the summer vacation, his father was blocked on his way home, so he went to Tianjin to take a steamer and went home.)
12. Material
   It indicates the material used in the action.
   e.g. 1 大橹上过桐油, 天天被水冲洗, 非常干净. (The big scull has been put on tung oil, and it is washed with water every day and it is very clean.)
   e.g. 2 本来要在塑胶产品上镀铬以增加耐用度的, 无奈做不成功. (We were going to plate chrome on plastic products to increase durability, but it was unsuccessful.)
13. Tool
   It indicates the method used by the action.)
   e.g. 1 走运河比走陆路运费每吨便宜十五元. (It is 15 yuan cheaper per ton by the canal than by land.)
   e.g. 2 从吴江坐木船到苏州, 水程40余华里. (It is a journey of more than 40 Huali to take a wooden boat from Wujiang to Suzhou.)
14. Initial Location
   It indicates the initial position where the action occurred.
   e.g. 1 他不能离开他的办公室. (He cannot leave his office.)
   e.g. 2 没想到下飞机以后, 中国人跟我说汉语我听不懂. (I didn't expect that I couldn't understand when they spoke Chinese to me after getting off the plane.)
15. Final Location
   It indicates the final position of the action.
   e.g. 1 从吴江坐木船到苏州, 水程40余华里. (It is a journey of more than 40 Huali to take a wooden boat from Wujiang to Suzhou.)
   e.g. 2 只要在一分钟内赶到法国就可看到日落. (As long as you arrive in France within a minute, you can see the sunset.)
16. Location Through
   It indicates the spatial position of the action from the initial position to the final position.
   e.g. 1 小王子穿过沙漠. (The little prince walks through the desert.)
   e.g. 2 我记得曾坐小船经过山阴道. (I remember passing through the mountain vagina

17. Intention
    It indicates the purpose of the action.
    e.g. 1 为了考研究生，他放弃了和家人一起出国旅游. (In order to take the postgraduate entrance examination, he gave up traveling abroad with his family.)
    e.g. 2 都逼对方出去奔伙食费? (Are they forced to go out for food?)
18. Reason
    It indicates why the action occurred.
    e.g. 1 以音乐和舞蹈庆祝他的胜利. (Celebrate his victory with music and dance.)
    e.g. 2 他会责备你的过失. (He will blame you for your mistakes.)
19. Time
    It indicates the time when the action occurred.
    e.g. 1 中国人非常重视过春节. (Chinese people attach great importance to the Spring Festival.)
    e.g. 2 处于大明王朝鼎盛时代的朱权也是躲避过的. (Zhu Quan, who was in the heyday of the Ming Dynasty, also avoided it.
20. According
    It indicates the basis of the action.
    e.g. 1 父亲依了它的长相管它叫“小狮子”. (His father called it “little lion” according to its appearance.)
    e.g. 2 有时全凭运气加人气. (Sometimes it is all luck and popularity.)
21. Direction
    It indicates the direction the subject of the action is facing or moves towards to.
    e.g. 1 我打发这些运载旅客的列车, 一会儿发往右方, 一会儿发往左方.
    e.g. 2 人走向生命尽头时，音乐不再是保姆也不是恋人. (When a person approaches the end of life, music is no longer a nanny or a lover.)

4 The Construction of Verb-Object Collocation Knowledge Graph

In this part, based on the above analysis, we construct a knowledge graph. The main process is divided into the following two parts.

4.1 Ontology Layer Construction

The semantic classification of vocabulary is a significant part of the semantic system. It is the basis of the entire semantic system [20]. Ontology is an important part of the semantic knowledge graph. Whether the classification of ontology is reasonable and accurate directly affects the quality of the semantic knowledge graph. Thus, in this part, we are going to elaborate on the method of ontology construction. At present, in Natural Language Processing, there are many representative ontology classification projects, such as HowNet, Tongyici-cilin, Modern Chinese Semantic Dictionary, 973 Semantic Classification System and Modern Chinese Classification Dictionary. When constructing the ontology, considering that the verb-object relationship involves different parts of speech (nouns, verbs), we construct different ontologies (nouns, verbs). On this basis, each word can be classified into different categories.

When constructing ontology, we refer to the existing classification standards for nouns and verbs. When constructing the noun ontology, we refer to the method Zhu et al. [21] proposed, and obtain the ontology from HowNet [22], Tongyici-cilin [16], and Peking University Semantic Dictionary [23], use the vocabulary similarity calculation to merge the repeated ontology elements, and at the same time we construct the ontology mapping table. Finally, we obtain our own noun ontology by combining multiple ontology. The constructed noun ontology is continuously subdivided from entity downwards, and finally constitute an ontology with eight levels and 138 specific categories. (Due to the limitation of space, the whole classification won't be presented here.) (Table 2).

Table 2. Noun ontology classification

Ontology level	Ontology elements
Level one	Entity
Level two	Time, location, portion, attribute et al.
Level three	Substance, spirit, group, thing, et al.
Level four	Creature, emotions, willingness, appearance et al.
Level five	Animal, botany, microorganism, waste et al.
Level six	Human, tree, vegetable, fruit et al.
Level seven	Career, name, identity, insect et al.
Level eight	Rain, snow, road, car et al.

In the construction of verb ontology, we refer to the different classification methods in 905 Semantic System [24], 973 Semantic System [25], Peking University Semantic Dictionary [23] and Modern Chinese Classification Dictionary [26], we classify verbs into static and dynamic categories. Under this condition, dynamic verbs include action and activity, and static actions are divided into state and relationship. According to the tree structure, there are 25 major categories and 59 small categories. As shown in Table 3, verbs are subdivided in order according to the tree structure, the entire category system can be divided into 5 levels, including 2 word groups, 4 word categories, 25 major categories, 195 subcategories, etc. (Due to the limitation of space, the whole classification won't be presented here.)

4.2 Data Layer Construction

When constructing the data layer, we use two different construction methods. The data obtained by the two construction methods complement each other, making the knowledge graph more complete. The following sections will introduce the two methods in detail.

Data Layer Construction Based on Extraction Method. Based on the extraction method, we use a combination of syntactic dependency analysis [27] and semantic dependency analysis [28] to extract verb-object collocation triples from unstructured data. When extracting, we combine the two methods in order to analyze the sentence from different aspects. Firstly, we conduct word segmentation and part-of-speech tagging

Table 3. Verb ontology classification

Ontology level	Ontology elements
Level one	Dynamic, static
Level two	Movement, activity, state, relation
Level three	Head movement, economic activity, possession et al.
Level four	Exercise, play, put, fit et al.
Level five	Remain, exist, laugh, swim et al.

of the sentence, then carry out syntax dependency analysis and semantic dependency analysis.

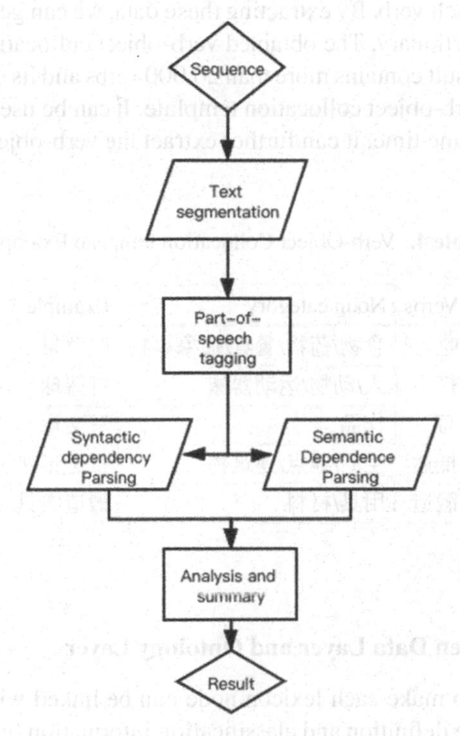

Fig. 1. The framework of extraction model

The overall extraction steps are as follows:

1. First of all, this article performs word segmentation and part-of-speech tagging on language sequences.
2. Secondly, we use the syntactic dependency parsing model to analyze the natural language and find the verb-object structure in it.

3. Thirdly, we further analyze the natural language by using the semantic dependency parsing model to find the semantic dependency relations between verb and object.
4. In the fourth step, the results of the second and third steps are analyzed to obtain the final verb-object collocation.

After the above steps in Fig. 1., we get different triples in the form of "verb, semantic dependency relation/syntactic dependency relation, noun" from the extraction. We use Neo4j database for management, establish vocabulary as nodes in the database, knowledge of parts of speech and other knowledge as attributes of nodes, and store the relations between verb-object collocation as the relations in the database.

Data Layer Supplement Based on Dictionary. We use the dictionary-based method to extract the object information from the knowledge base of other sources, and obtain the object of verbs. We supplement the data layer based on the Peking University Semantic Verb Dictionary, from which we can obtain the semantic category, subject, predicate and other information of each verb. By extracting these data, we can get the predefined verb-object triples of the dictionary. The obtained verb-object collocation example is shown in Table 4. The final result contains more than 20,000 verbs and its semantic information, which constitutes a verb-object collocation template. It can be used to connect data and ontology, and at the same time, it can further extract the verb-object relations.

Table 4. Verb-Object Collocation template Example

Verbs	Noun category	Example
吃	食物/药物/餐具/抽象事物	吃苹果
打	人/动物/运动器械	打篮球
写	作品	写文章
抵达	空间/地点/建筑物	抵达雅典
锻造	用具/材料	锻造农具

4.3 The Link Between Data Layer and Ontology Layer

In this part, in order to make each lexicon node can be linked with a certain ontology node and have accurate definition and classification information, we link data layer with ontology layer. Secondly, linking the verb with the noun ontology node using verb-object collocation template can also facilitate the usage of knowledge graph and data expansion. Therefore, at this stage, we mainly consider the following two types of knowledge.

(1) Knowledge of lexicon classification

Using the existing dictionary information, combined with the ontology constructed in the previous chapters, the data layer vocabulary can be linked with the ontology node, so that each word can correspond to specific category definition at the ontology level. And We link nouns and verbs with "isA" and "belongTo" respectively.

(2) Knowledge of verb-object collocation

Based on the verb-object collocation template obtained in the previous part, each verb and the corresponding noun ontology element are linked by using "objectOf" to further complement the knowledge graph.

Based on these two parts of knowledge, on the one hand, the data layer node can be linked with the ontology layer node, which can illustrate the data layer classification information; on the other hand, by adding verb-object collocation template knowledge, the relationship between the ontology layer node and the data layer node is supplemented, which also facilitates the further expansion of the knowledge graph.

4.4 Results and Analysis

Table 5. Example and analysis of the knowledge graph

Semantic classification	Example	Amounts
Patient	使用电脑	7586
Agent	跑过一只兔子	155
Content	看书	12792
Experiencer	留下一道水花	1405
Class	演花木兰	4616
Dative	赠予他	5514
Belongings	包括图书馆	2047
Possessor	属于中国	54
Product	写小说	2127
Consequence	造成污染	242
Tool	吃大碗	202
Material	烧柴油	20
Manner	写楷体	25
Initial location	离开学校	231
Final location	抵达雅典	2294
Location through	经过上海	178
Intention	考大学	12
Reason	愁工作	80
Time	过中秋节	525
According	依照说明书	51
Direction	奔向前方	985

After the above steps, we get the verb-object collocation triples as shown in Table 5. It can be seen that the obtained verb-object collocation triples have specific semantic relations. When constructing, we mainly use semantic dependency labels as the relationship labels of data layer nodes. We preliminary established a Verb-Object Collocation Knowledge Graph which contains 40704 specific examples of verb-object collocation and 20,000 verb-object collocation extraction templates. As shown in Fig. 2, the verb-object collocation examples are presented from perspective of the ontology layer and the data layer. The upper left part is the verb ontology, the upper right part is the noun entity, and the lower part is the data layer. The construction is displayed in the way that verb points to the object, and the detailed semantic dependency relations are marked between them. At the same time, the data nodes and the ontology nodes are connected by a dotted line, indicating the specific classification of the data layer elements. Taking the two data nodes of "belonging to" and "China" as an example, they form a verb-object collocation of "Possession". At the same time, the classification of "belonging" is "subordinate" and the classification of "China" is "country".

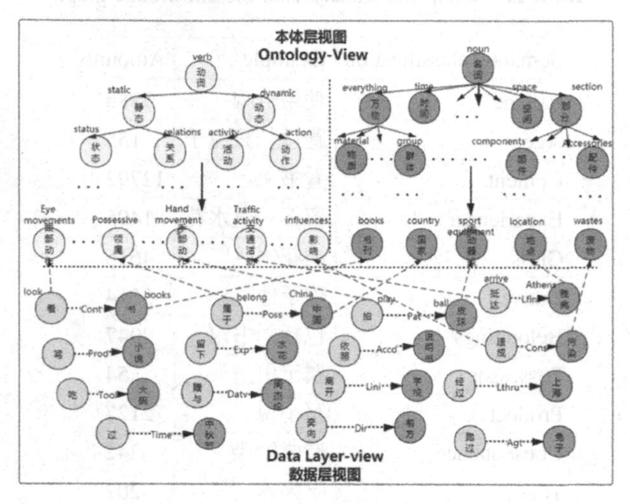

Fig. 2. Example of Verb-Object Collocation Knowledge Graph

5 Named Entity Recognition Adding Knowledge of Verb-object Collocation

When exploring the actual utility of the knowledge graph, this paper selects the name entity recognition task, and verifies the practicality of the verb-object collocation semantic knowledge graph by constructing a knowledge auxiliary module. The model architecture for name entity recognition is shown in Fig. 3. The main model framework is divided into two parts. The first part is the main module of entity recognition. The sequence is encoded by using BERT [29] as the embedding layer, the embeddings are further calculated by Bi-LSTM to model the semantic relations, then the results are feed into Conditions Random Field, extracting name entities from the text. The second part is the knowledge auxiliary module, which preprocesses the same text such as word segmentation and part-of-speech tagging, and then performs syntactic dependency analysis to find the verb-object structure in the text, and uses the verb-object collocation knowledge combined with the part-of-speech tagging results to determine whether the object is name entity. (For example, the collocation object type of the word "Abolition" is "institution", if thc part of speech of the object is ni, it will be marked as the name of the institution; the collocation object of the word confession is the country name nh or the person name nl, with the help of part of speech and The verb-object collocation rules can achieve the purpose of judging the object as a person's name or a place name) From the above steps, the rule-based name entity result is obtained in this part, and the two parts are finally combined into one. The result of the auxiliary module is used to corroborate the result of the main module, and the final recognition result is obtained.

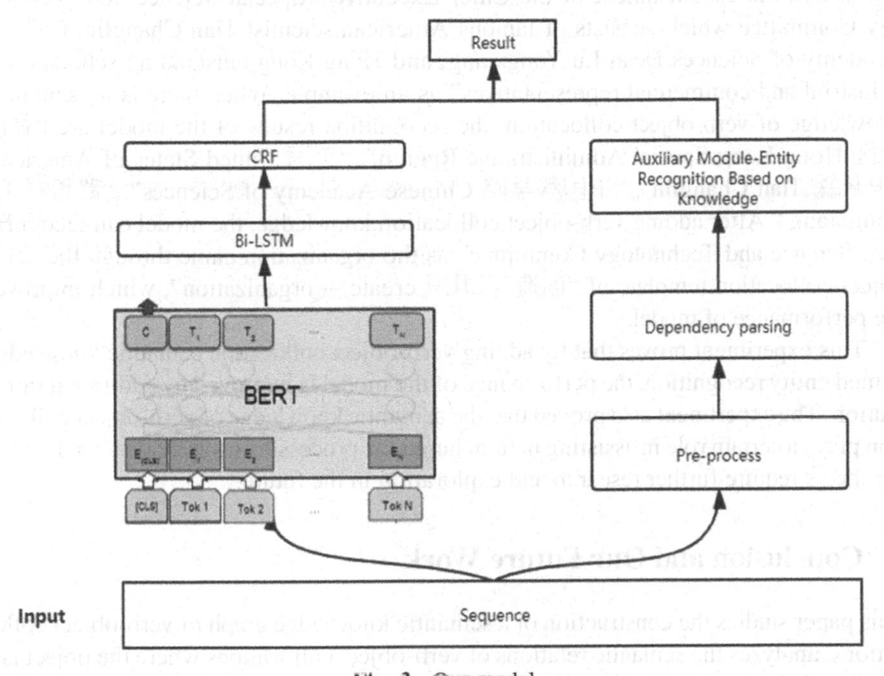

Fig. 3. Our model

Experimental setting: In the experiment, we use the public MSRA Chinese Name Entity Recognition dataset. The sentences in the dataset contain three types of entities: Person, Organization, and Location. The other characters are marked as Other. Table 6 is our experimental result.

MSRA dataset is a named entity recognition dataset developed and open-sourced by Microsoft Research Asia. It contains 45,000 sentences in training set and 3442 sentences in test set. This corpus annotates named entities in natural language texts. The data set is shown as following format: the entire corpus is labeled with BIO tags, 'B' represents the first Chinese character of the named entity, 'I' represents other Chinese characters of the named entity, and 'O' represents the Chinese characters of other non-named entities.

Table 6. Result of our NER model

Model	Precision (%)	Recall (%)	F1-score (%)
BERT + BiLSTM + CRF(no additional knowledge)	95.4	95.3	95.3
BERT + BiLSTM + CRF(add knowledge)	95.8	95.5	95.7

Example: 香港特区政府宣布成立集合美国著名科学家田长霖，中国科学院院长路甬祥及香港杰出学者，工商业巨子的行政长官特设科技委员会.

Take the sentence "The Hong Kong Special Administrative Region Government announced the establishment of the Chief Executive's Special Science and Technology Committee which consists of famous American scientist Tian Changlin, Chinese Academy of Sciences Dean Lu Yongxiang, and Hong Kong outstanding scholars and industrial and commercial representatives" as an example, When there is no semantic knowledge of verb-object collocation, the recognition results of the model are "香港特区 Hong Kong Special Administrative Region", "美国 United States of America", "田长霖 Tian Changlin", "中国科学院 Chinese Academy of Sciences", "路甬祥 Lu Yongxiang". After adding verb-object collocation knowledge, the model can recognize the "Science and Technology Committee" as the organization name through the verb-object collocation template of "创造 + 组织 create + organization", which improves the performance of model.

This experiment proves that by adding verb-object collocation semantic knowledge named entity recognition, the performance of the model is improved by additional information. The experiment also proved that the semantic knowledge of verb-object collocation plays a certain role in assisting natural language processing tasks, and its application prospects require further research and exploration in the future.

6 Conclusion and Our Future Work

This paper studies the construction of a semantic knowledge graph of verb-object collocations, analyzes the semantic relations of verb-object collocations where the object is a noun, and obtains specific examples of 40704 verb-object collocations and an extraction template for 20000 verb-object collocations. Meanwhile, this paper constructs a Verb-Object Collocation Knowledge Graph to manage and maintain knowledge through two parts: ontology layer and data layer. Secondly, in this paper, the prospect of Verb-Object Collocation Knowledge Graph is preliminary explored in the task of name entity recognition, which verifies the feasibility and effectiveness of the application of verb-object collocation semantic knowledge graph.

In the future, we will continue to expand the scale of the semantic knowledge graph and the scope of objects in the verb-object collocation structure, and conduct further research on the collocations that use adjectives, pronouns and other words as objects. The entire semantic knowledge graph can be expanded not only in terms of data scale, but also in overall quality to meet the needs of higher-level research. Furthermore, we select a suitable task for the application of verb-object collocation semantic knowledge graph in natural language processing, and further explore the practicality of this semantic knowledge graph.

Acknowledgement. This research project is supported by the National Natural Science Foundation of China (61872402), the Humanities and Social Science Project of the Ministry of Education (17YJAZH068), Science Foundation of Beijing Language and Culture University (supported by "the Fundamental Research Funds for the Central Universities") (18ZDJ03), the Open Project Program of the National Laboratory of Pattern Recognition (NLPR).

References

1. Firth, J., Rupert: Paper in Linguistics 1934–1951. Oxford University Press, London (1958)
2. BBI Homepage. https://www.amazon.com/BBI-Combinatory-Dictionary-English-Combinations/dp/0915027801. Accessed 28 Oct 2020
3. Katz, J.J., Fodor, J.A.: The structure of a semantic theory. Language **39**(2), 170–210 (1963)
4. Mak, H.: Lexis as a linguistic level. In: Bazell, C.E., Catford, C., Halliday, M.A.K., Robins, R.H. (eds.) Memory of JR Firth. Longmans, London (1966)
5. Choueka, Y., Klein, T.M., Neuwitz, E.: Automatic retrieval of frequent idiomatic and collocational expressions in a large corpus. J. Literary Linguist. Comput. **4**(1), 34–38 (1983)
6. Church, K., Hanks, P.: Word association norms, mutual information, and lexicography. Comput. Linguist. **16**(1), 22–29 (1990)
7. Smadja, F.: Retrieving collocations from text: Xtract. Comput. Linguist. **19**(1), 143–177 (1993)
8. Sun, M., Huang, C., Fang, J.: A probe into the quantitative analysis of Chinese collocation. Stud. Chin. Lang. **256**(1), 29–38 (1997)
9. Qu, W., Chen, X., Ji, G.: A frame-based approach to Chinese collocation automatic extracting. Comput. Eng. **30**(23), 22–38 (2004)
10. Wang, L.: The construction and application of binary collocation semantic knowledge base based on multiple knowledge sources. M.S. thesis, Beijing University of Science and Technology Information, Beijing (2011)
11. Renfen, H., Hang, X.: The construction of chinese collocation knowledge bases and their application in second language acquisition. Appl. Linguist. **2019**(1), 135–144 (2019)
12. Zhou, W.: Information processing-oriented research on the semantic collocation between verbs and objects in modern Chinese language. M.S. thesis, Huazhong Normal University, Wuhan (2007)
13. Liu, Z.: Research on construction technology of Chinese verb-object collocation bank. M.S. thesis, Shenyang Aerospace University, Shenyang (2011)
14. Cheng, X., Shao, Y.: The annotation scheme of sematic structure relations based on semantic dependency graph bank. In: Hong, J.-F., Su, Qi., Wu, J.-S. (eds.) CLSW 2018. LNCS (LNAI), vol. 11173, pp. 720–732. Springer, Cham (2018). https://doi.org/10.1007/978-3-030-04015-4_63
15. Xu, S.: Object and Complement. Heilongjiang Publishing House, Heilongjiang (1985)
16. Ma, J.: Tongyici-cilin. Shanghai Lexicographical Publishing House, Shanghai (1987)
17. Meng, C.: Dictionary of Verb Usage. Shanghai Lexicographical Publishing House, Shanghai (1987)
18. Lin, X., Wang, L., Sun, D.: A Dictionary of Modern Chinese Verbs. Beijing Language and Culture University Press, Beijing (1994)
19. Huang, B., Liao, X.: Chinese. Higher Education Press, Beijing (2017)

20. Lin, X.: Practice and understanding of word meaning classification. J. Renmin Univ. China **2**(1), 63–66 (1990)
21. Zhu, S., Li, Y., Shao, Y.: Research on construction and automatic expansion of multi-source lexical semantic knowledge base. In: Proceedings of the CCKS 2019, Beijing, pp. 74–85 (2020)
22. Dong, Z., Dong, Q.: HowNet - a hybrid language and knowledge resource. In: Proceedings of the International Conference on Natural Language Processing and Knowledge Engineering, Beijing, pp. 820–824 (2003)
23. Wang, H., Yu, S., Zhan, W.: New progress of the semantic knowledgebase of contemporary Chinese (SKCC). In: Proceedings of the 7th National Joint Conference on Computational Linguistics, Harbin, pp. 351–256 (2003)
24. Chinese Information Processing 905 Platform Project. Beijing Language and Culture University, Beijing (1990)
25. Project Semantic Classification System. Peking University, Beijing (1973)
26. Su, X.: A Thesaurus of Modern Chinese. Commercial Press, Beijing (2013)
27. Che, W., Li, Z., Liu, T.: LTP: a Chinese language technology platform. In: Proceedings the 23rd International Conference on Computational Linguistics: Demonstrations, Beijing, pp. 13–16 (2010)
28. Shen, Z., Li, H., Liu, D., Shao, Y.: Dependency-gated cascade Biaffine network for Chinese semantic dependency graph parsing. In: Tang, J., Kan, M.-Y., Zhao, D., Li, S., Zan, H. (eds.) NLPCC 2019. LNCS (LNAI), vol. 11838, pp. 840–851. Springer, Cham (2019). https://doi.org/10.1007/978-3-030-32233-5_65
29. Jacob, D., Wei, C.M., Lee, K., Kristina, T.: BERT: pre-training of deep bidirectional transformers for language understanding. arXiv preprint (2018)

Optimal Deployment of Underwater Sensor Networks Based on Hungarian Algorithm

Jinglin Liang[1,2,3], Qian Sun[1,2,3](✉), Xiaoyi Wang[1,2,3], Jiping Xu[1,2,3], Li Wang[1,2,3], Huiyan Zhang[1,2,3], Jiabin Yu[1,2,3], Gongxue Cheng[1,2,3], Fengbo Yang[1,2,3], and Ning Cao[4]

[1] School of Artificial Intelligence, Beijing Technology and Business University, Beijing 100048, China

[2] China Light Industry Laboratory of Industrial and Big Data, Beijing Technology and Business University, Beijing 100048, China

[3] Beijing Laboratory for Intelligent Environmental Protection, Beijing Technology and Business University, Beijing 100048, China

[4] Shandong Chengxiang Information Technology Co.Ltd, Dezhou 253000, China

Abstract. Underwater sensor networks are widely used in water quality monitoring and underwater resources detection. The deployment of sensor nodes greatly affects the monitoring performance of the network. After the initial deployment of underwater sensor nodes, due to the influence of water flow, the sensor nodes will deviate from the initial position, which will affect the coverage of the network. Therefore, an effective algorithm is needed to deploy the underwater sensor networks. In this paper, MCM flow movement model is introduced to simulate the influence of water movement on sensor node position. Then, the Hungarian algorithm is used to redeploy the sensor nodes. The simulation results show that, compared with direct movement, Hungarian algorithm can achieve the effective deployment of the network with less mobile distance and better prolong the network life cycle.

Keywords: Hungarian algorithm · Flow model · Energy · Coverage rate · Deployment

1 Introduction

The ocean is very important for human survival and social sustainable development. In recent years, the theoretical research and application of underwater sensor network technology have received more and more attention. On the one hand, coastal countries pay more attention to territorial sovereignty and the competition for marine resources is increasingly fierce. The technology of underwater sensor networks has important application in the fields of marine environmental monitoring, data collection, exploitation and utilization of marine resources, geological disaster prediction and marine national defense security. On the other hand, as the source of life, water is an indispensable part of our human life activities. With the acceleration of urbanization and industrialization, the

X. Sun et al. (Eds.): ICAIS 2021, CCIS 1422, pp. 233–244, 2021.
https://doi.org/10.1007/978-3-030-78615-1_20

continuous expansion of construction land and the increase of pollutants from multiple sources such as industry, human life, agricultural production, make water quality continue to deteriorate. The quality of water environment monitoring has a direct impact on our protection of water resources. Therefore, it has become one of the current research hotspots [1–3]. Underwater sensor deployment is the basis of underwater sensor network application and is of great research significance.

Compared with the land sensor node deployment, the node in the underwater sensor network has its particularity. The underwater environment is complex and changeable so the sensor nodes will deviate from the initial position due to the influence of water flow [4, 5]. In addition, the energy of the sensor node in the underwater is limited because it is mainly supplied by batteries. So, it is not easy to change the battery. As far as the current situation is concerned, many scholars at home and abroad have done in-depth research on the sensor deployment of underwater wireless sensor network (UWSNs). In reference [6], the influence of water flow model on node deployment in UWSNs is studied, but the specific method of redeploying node to monitor underwater environment is not given. Some researchers introduced the self-organizing mobile node deployment strategy based on node position dynamic adjustment according to the influence of water flow [7]. At every interval T, the coverage rate of sensor nodes in the monitoring area is detected. Redeployment is performed if the correlation index is not met. In reference [8], Cho et al. propose that uses the non-uniform rational B-spline function to design a novel distribution model to achieve more accurate placement of underwater wireless sensors.

In the deployment of underwater sensors, energy and the influence of water flow are the important factors that need to be considered [9–11]. Therefore, how to dynamically adjust the node position according to the influence of the underwater environment to ensure the minimum energy consumption and achieve a longer life cycle of the underwater sensor network is an urgent problem to be solved. At present, there is no perfect method to solve the problem of energy consumption in the research of underwater sensor deployment. In order to solve the problem of energy consumption, in this paper, firstly, the Meandering Current Mobility (MCM) flow movement model is introduced to simulate the influence of water movement on sensor node position, and then the Hungarian algorithm is used to redeploy the sensor nodes to monitor the underwater environment. Hungarian algorithm can achieve the effective deployment of the network with less mobile distance, lower energy consumption and better prolong the network life cycle.

2 The Influence of Water Movement on Sensor Node Position

In actual water environments, the sensor nodes will deviate after the initial deployment due to the influence of water flow. However, the underwater environment is so diverse and varied that there is no unified equation to describe it. In practice, a specific node movement model is usually established according to the underwater environment.

2.1 Types of Node Mobility Models

(1) The random walk model:

Some researchers presented a simple underwater sensor node movement model called the random walk model [12]. The model expression is as follows:

$$\begin{cases} x(t) = x(t-1) + v_{cx} \\ y(t) = y(t-1) + d_t v_{cx} \end{cases} \tag{1}$$

among them,

$$d_t \begin{cases} -1, d_{t-1} = 1 \cap y(t-1) > l_{cy} \\ 1, d_{t-1} = -1 \cap y(t-1) < -l_{cy} \end{cases} \tag{2}$$

l_{cy} is the threshold of node oscillation motion.

The random walk model is simple. The nodes are in a simple swing state and are easy to control. It is suitable for the research of UWSNs where the sensor is relatively static.

(2) The Eulerian model:

Some researchers proposed the Eulerian model: all sensor nodes are doing oscillating motion in a certain range, or doing rotating motion in other areas. The expression of the model is as follows:

$$\begin{cases} V_x = k_1 \lambda sin(k_2 y) cos(k_3 y) + k_1 \lambda cos(2k_1 t) + k_4 \\ V_y = -\lambda v cos(k_2 x) sin(k_3 y) + k_5 \end{cases} \tag{3}$$

Where V_x and V_y represent the speed of the node in the x-axis and y-axis directions respectively, k_1, k_2, k_3, λ, v are coefficients closely related to the environment such as water flow, k_4, k_5 are two random variables.

The nodes have a limited range of motion under the Eulerian model, and perform approximate periodic motion. The model ignores the vertical movement when analyzing the mobility of nodes.

(3) The Lagrangian model:

The Lagrangian model considers the influence of water flow and vortex on the nodes. This model is more suitable for the actual dynamic network. The model expression is as follows:

$$\begin{cases} u = -\dfrac{\partial \psi}{\partial y} \\ v = \dfrac{\partial \psi}{\partial x} \end{cases} \tag{4}$$

Where ψ represents an incompressible fluid in a two-dimensional plane. It is generally assumed that u is the velocity in the x-axis and v is the velocity in the y-axis. The trajectory of the node is formed with the fluid and can be described by the following Hamiltonian differential equation:

$$\begin{cases} \dot{x} = -\partial_x \psi(x, y, t) \\ \dot{y} = \partial_y \psi(x, y, t) \end{cases} \tag{5}$$

(4) The MCM model:

Based on the Lagrangian flow model mentioned above, a typical MCM model is derived [13]. The expression of the model is as follows:

$$\psi(x, y, t) = -tanh[\frac{y - B(t)sin(k(x - ct))}{\sqrt{1 + k^2 B(t)^2 cos^2(k(x - ct))}}] \tag{6}$$

$$B(t) = A + \varepsilon cos(\omega t) \tag{7}$$

Where $B(t)$ represents the width of the curve flow, A represents the average width of the bend of the entire flow field, ε represents the amplitude of the entire flow field, w represents the motion frequency of the flow field, k represents the number of bends per unit length of the flow in space, and c represents the phase velocity.

Nodes move relatively smoothly in this model. Combined with the actual underwater environment and the deployment of nodes, we analyze the impact of node movement on the deployment of underwater wireless sensor networks.

2.2 Sensor Node Location Variation Based on MCM Model

The underwater environment is complex and changeable, the monitoring target is highly dynamic and the sensor nodes are mobile. Combining the characteristics of the underwater sensor network and the underwater environment, after the initial deployment of underwater sensor nodes, due to the influence of water flow, the sensor nodes will deviate from the initial position, which will affect the coverage of the network. Through the water flow movement model, the influence of the water flow movement on the position of the sensor node is analyzed. This paper uses the MCM water flow model to simulate the influence of water movement on sensor node position and the coverage of the network.

In the two-dimensional plane of the monitoring water area, the sensors are deployed uniformly. The number of sensors is 54 and the radius is 1.5 km. The model is shown in Fig. 1.

The positions of the sensors change by the water flow, which are shown in Fig. 2.

The movement of the sensors will affect the coverage of the monitoring area, as shown in Fig. 3.

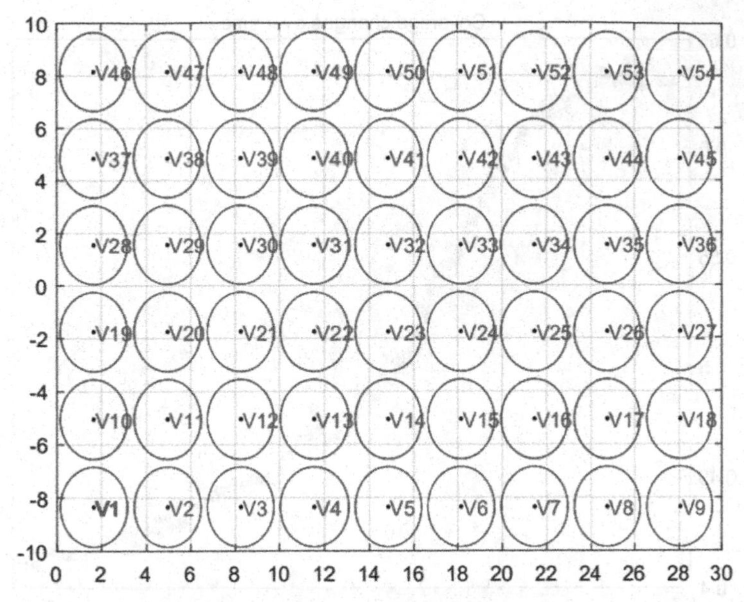

Fig. 1. Uniform deployment of sensors

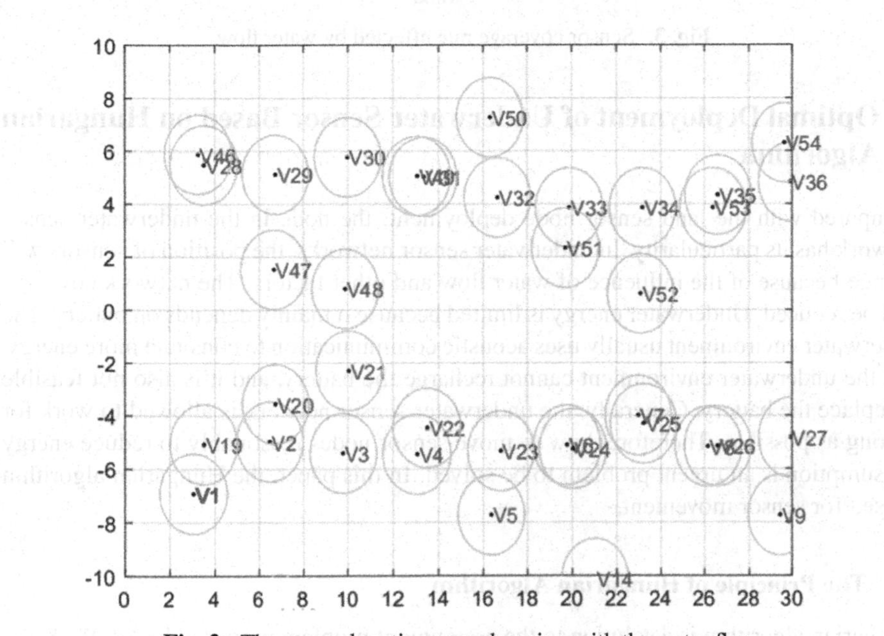

Fig. 2. The sensor locations are changing with the water flow

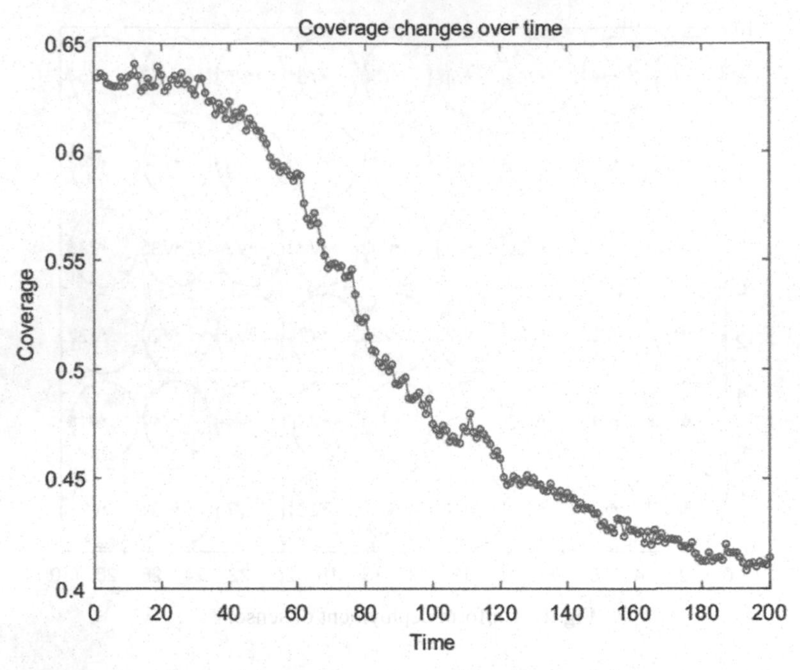

Fig. 3. Sensor coverage rate affected by water flow

3 Optimal Deployment of Underwater Sensor Based on Hungarian Algorithm

Compared with the land sensor node deployment, the node in the underwater sensor network has its particularity. In underwater sensor networks, the position of sensors will change because of the influence of water flow and other factors. The network coverage will be reduced. Underwater energy is limited because it mainly depends on battery. The underwater environment usually uses acoustic communication to consume more energy, and the underwater environment cannot recharge the battery, and it is also not feasible to replace the battery. Generally, the underwater sensor network is allowed to work for as long as possible. Therefore, how to move sensor nodes reasonably to reduce energy consumption is an urgent problem to be solved. In this paper, the Hungarian algorithm is used for sensor movement.

3.1 The Principle of Hungarian Algorithm

Hungarian algorithm is a solution to the assignment problem proposed by W· W· Kuhn in 1955. He cited a theorem of Hungarian mathematician D. Konig about zero elements in matrix: the maximum number of independent zero elements in coefficient matrix is equal to the minimum number of straight lines that can cover all zero elements.

The optimal solution of the assignment problem has this property: If the smallest element of the row (column) of coefficient matrix (C_{ij}) is subtracted from each element of the row (column), the new matrix b_{ij} is obtained. The optimal solution obtained by taking

b_{ij} as the coefficient matrix is the same as that obtained by using the original coefficient matrix. Using this property, the original coefficient matrix can be transformed into a new coefficient matrix with many zero elements, while the optimal solution remains unchanged. If n independent zero elements can be found in the coefficient matrix (b_{ij}) and substituted it into the objective function to obtain objective function is zero, it must be the smallest. This is the Hungarian algorithm.

3.2 Optimal Deployment Method Based on Hungarian Algorithm

Description of Sensor Movement Problem. In the underwater environment, after the initial deployment of underwater sensor nodes, due to the influence of water flow, the sensor nodes will deviate from the initial position, which will affect the coverage of the network. So how to move the sensor reasonably and optimize the network deployment is an urgent problem to be solved. The sensor movement problem is described as follows.

1. Assuming that n sensors need to be moved to n positions. The distance from each sensor to different position is different. Therefore, the energy consumed by the sensors is also different.
2. The distance from i-th sensor to j-th position is C_{ij}.
3. Try to determine a mobile scheme where n sensors are moved to n locations to minimize total energy consumption.

This problem is a linear nondifferentiable programming problem [14]. Let $x_{ij} = 1$ or 0, $x_{ij} = 1$ means to assign the i-th sensor to the j-th position, otherwise $x_{ij} = 1$. Then the mathematical model of multi-objective programming for sensor deployment is as follows.

$$z = \min \sum_{i=1}^{n} \sum_{j=1}^{n} C_{ij} X_{ij} \tag{8}$$

Meets:

$$\sum_{i=1}^{n} X_{ij} = 1, j = 1, \ldots, n \tag{9}$$

$$\sum_{j=1}^{n} X_{ij} = 1, i = 1, \ldots, n \tag{10}$$

$$X_{ij} = 1, 0 \tag{11}$$

Formula (10) means that the i-th sensor can only move to one position. Formula (9) means that the j-th position can only have one sensor. The matrix $C = (C_{ij})_{n \times n}$ consisting of C_{ij} is called the efficiency matrix of the assignment problem. Therefore, the problem of seeking the solution of the assignment model (9)–(11) is equivalent to: Select n distances from C to satisfy: ① Exactly one distance in each column is selected to ensure that each target position has only a sensor. ② Exactly one distance in each row is selected to ensure that each sensor only goes to one target position. ③ The sum of the selected n distances is the smallest. The Hungarian algorithm [15, 16] can be used to move the sensor to minimize the total energy consumption.

Specific Steps of Underwater Sensor Deployment Optimization. Underwater energy is limited because it mainly depends on battery. The underwater environment usually uses acoustic communication to consume more energy and it is not feasible to replace the battery. Generally, the underwater sensor network is allowed to work for as long as possible. Therefore, how to move sensor nodes reasonably to reduce energy consumption is an urgent problem to be solved. The Hungarian algorithm is used to move the sensor to the appropriate location, resulting in less total energy consumption and longer network life cycle. The steps to move the sensor based on the Hungarian algorithm are as follows:

Step 1: Mark each sensor and position with 1 to n.

Step 2: Calculate the distance from the i-th sensor to j-th position to form a coefficient matrix. The coefficient matrix is as follows:

$$\begin{pmatrix} C_{11} \; C_{12} \ldots C_{1n} \\ C_{21} \; C_{22} \ldots C_{2n} \\ \ldots \; \ldots \; \ldots \; \ldots \\ C_{n1} \; C_{n2} \ldots C_{nn} \end{pmatrix} \tag{12}$$

Step 3: The distance matrix from the sensor to the position is transformed, and the zero element appears in each row and column. Specific process is as follows: firstly, we can subtract the smallest distance from each row of the coefficient matrix; secondly, the smallest element of each column should be subtracted from the resulting coefficient matrix. There is no need to subtract the smallest element if any row (column) already has a zero element.

Step 4: We can try to assign the task to find the optimal solution according to the following steps.

After the transformation in step 3, there are zero elements in each row and column of the coefficient matrix, but we need to find several elements (the number of the elements is n) that are independent of each other. Specific process is as follows:

Firstly, we can start from a row with only one zero element, the zero element can be marked as ⊗. It indicates that the sensor represented by this row has only one target position. Then we can cross out the other zero elements of the column and mark it as H. Similarly, a column with only one zero element can be found and the zero element can be marked as ⊗, then we can cross out the other zero elements of the row and mark it as H. Repeat the above process until all zero elements are marked and crossed out.

Secondly, if there exists an unmarked zero element and there are at least two zero elements in the row (column) (indicating that the sensor represented by this row or column can choose any one of the two target locations),then we can start from the row (column) with the least number of zero elements left, comparing the number of zero elements in the column of each zero element in this row and marking the zero element of the column with the least zero elements, and then we can cross out the other zero elements at the same column and row. It can be repeated until all the zero elements have been marked and crossed out. If l(the number of $\otimes$) in the matrix is equal to n which represents of order of the matrix, the optimal solution of the assignment problem has been obtained. Otherwise, make the least straight lines to cover all zero elements to determine the most independent elements that can be found in the coefficient matrix if l in the matrix is less than n.

Underwater energy is limited because it mainly depends on battery. The underwater environment uses acoustic communication to consume more energy, and the underwater environment cannot recharge the battery, and it is not feasible to replace the battery. Generally, the underwater sensor network is allowed to work for as long as possible. Therefore, how to move sensor nodes reasonably to reduce energy consumption is an urgent problem to be solved. In underwater environment, the Hungarian algorithm can reasonably allocate the sensor for moving [17]. The total energy consumption of the sensor network is less.

4 Simulation Analysis

4.1 Experimental Design

Compared with the land sensor node deployment, the node in the underwater sensor network has its particularity. The underwater environment is complex and changeable, the sensor nodes are mobile. Combining the characteristics of the underwater sensor network and the underwater environment, after the initial deployment of underwater sensor nodes, due to the influence of water flow, the sensor nodes will deviate from the initial position, which will affect the coverage of the network. In this paper, the Hungarian algorithm is used to redeploy the sensor nodes to achieve the purpose of monitoring water area. In the water area of 20 km × 30 km, 54 sensors are deployed with a radius of 1.5 km. The initial energy is 2000 J, and the energy consumed is 50 J when the sensor moved 1 km.

4.2 Experimental Results and Analysis

In the underwater environment, the Hungarian algorithm is used to calculate the corresponding position that each sensor node should move to. Blue indicates that the sensor is moved using the Hungarian algorithm, and red indicates that the sensor is moved directly to its original position.

Figure 4 shows the residual energy of each sensor. As we can see from Fig. 4, compared with direct movement, the residual energy of most sensors is higher by using the Hungarian algorithm.

Figure 5 is a comparison diagram of the total moving distance between the two methods. It can be seen from the Fig. 5 that the total moving distance by using the Hungarian algorithm is less than that of moving directly.

Figure 6 is a comparison diagram of the total energy consumption of sensors by using direct movement and Hungarian algorithm. It can be seen from the Fig. 6 that the Hungarian algorithm saves nearly 500 J of energy. Using the Hungarian algorithm can effectively extend the network life cycle.

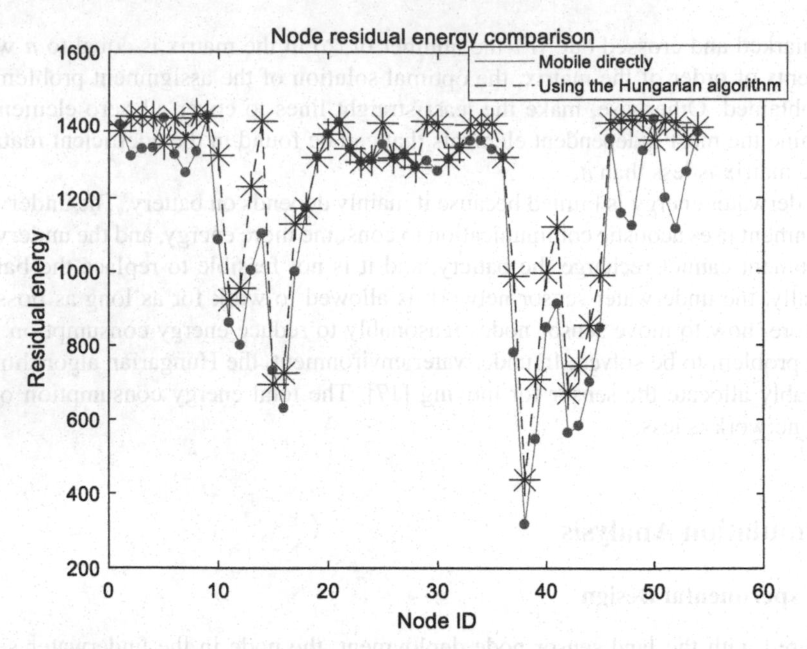

Fig. 4. Comparison diagram of node residual energy

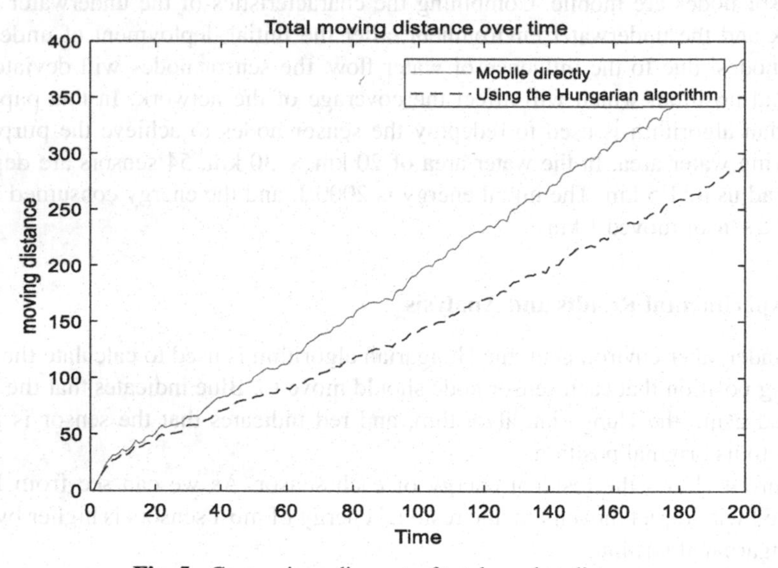

Fig. 5. Comparison diagram of total moving distance

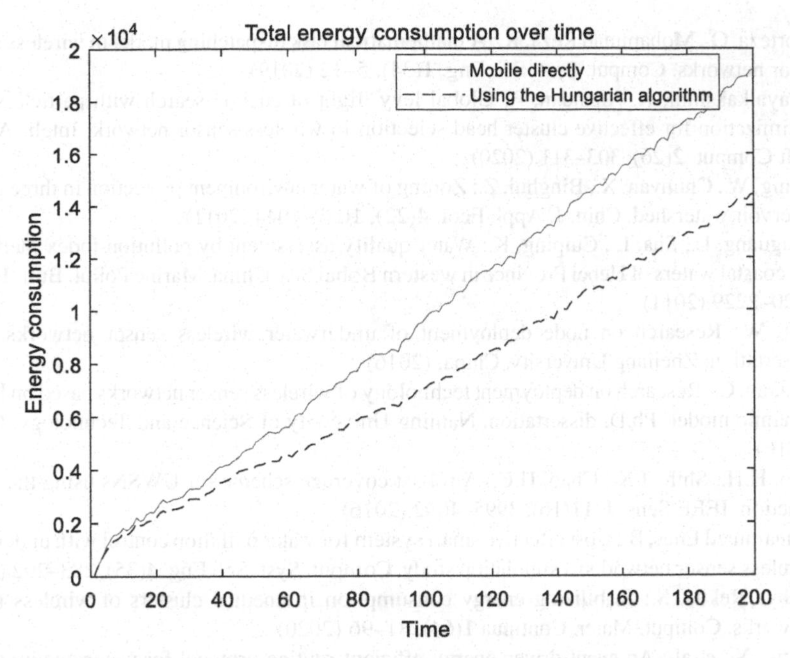

Fig. 6. Comparison diagram of total energy consumption

5 Conclusions

Water quality sensor networks are promising tools for the exploration of the oceans. The deployment of sensor nodes greatly affects the monitoring performance of the network. In the actual water area, the sensor nodes will deviate from the initial position due to the influence of water flow, which will affect the coverage of the network. Therefore, it is necessary to consider the mobility of the node in the process of node deployment. Energy consumption also needs to be considered because it is mainly supplied by batteries. This paper firstly introduces several flow models, then studies the influence of MCM flow model on sensor node position and the coverage of the network, and finally uses the Hungarian algorithm to redeploy sensor nodes. The simulation results show that, compared with direct movement, the Hungarian algorithm can achieve the effective deployment of the network with less mobile distance, lower energy consumption and better prolong the network life cycle.

Funding Statement:. This research was funded by the National Natural Science Foundation of China, No. 61802010; National Social Science Fund of China, No. 19BGL184; Beijing Excellent Talent Training Support Project for Young Top-Notch Team No. 2018000026833TD01; and Hundred-Thousand-Ten Thousand Talents Project of Beijing No. 2020A28.

References

1. Nianbin, W., Ming, H., Jianguo, S., et al.: IA-PNCC: noise processing method for underwater target recognition convolutional neural network. Comput. Materi. Continua 1(58), 169–181 (2019)

2. Morteza, O., Mohammad Reza, K.: A mathematical task dispatching model in wireless sensor actor networks. Comput. Syst. Sci. Eng. **1**(34), 5–12 (2019)
3. Vijayalkakshmi, K, Anandan, P.: Global levy flight of cuckoo search with particle swarm optimization for effective cluster head selection in wireless sensor network. Intell. Autom. Soft Comput. **2**(26), 303–311 (2020)
4. Lijing, W., Chunyan, X., Binghui, Z.: Zoning of water environment protection in three gorges reservoir watershed. Chin. J. Appl. Ecol. **4**(22), 1039–1044 (2011)
5. Shuguang, L., Sha, L., Cuiping, K.: Water quality assessment by pollution-index method in the coastal waters of Hebei Province in western Bohai Sea. China. Marine Pollut. Bull. **10**(62), 2220–2229 (2011)
6. Hui, W.: Research on node deployment of underwater wireless sensor networks. M.S. dissertation, Zhejiang University, China, (2016)
7. Jiguang, C.: Research on deployment technology of wireless sensor networks based on hydrodynamic model. Ph.D. dissertation, Nanjing University of Science and Technology, China, (2016)
8. Cho, H.H., Shih, T.K., Chao, H.C.: A robust coverage scheme for UWSNs using the spline function. IEEE Sens. J. **11**(16), 3995–4002 (2016)
9. Muhammed Enes, B.: Cost effective smart system for water pollution control with underwater wireless sensor networks: a simulation study. Comput. Syst. Sci. Eng. **4**(35), 283–292 (2020)
10. Nithya Rekha, S.: Stabilizing energy consumption in unequal clusters of wireless sensor networks. Comput. Mater. Continua **1**(64), 81–96 (2020)
11. Wang, X., et al.: An event-driven energy-efficient routing protocol for water quality sensor networks. Wireless Netw. **26**(8), 5855–5866 (2020). https://doi.org/10.1007/s11276-020-02320-4
12. Jules, J., Curt, S.: Sensor networks of freely drifting autonomous underwater explorers. In: Proceedings of the First ACM International Workshop on Underwater Networks (WUWNet 2006), pp. 93–96. Los Angeles, USA (2006)
13. Caruso, A., Paparella, F., Vieira, L.F.M., Erol, M., Gerla, M., et al.: The meandering current mobility model and its impact on underwater mobile sensor networks. In: IEEE INFOCOM 2008 - The 27th Conference on Computer Communications, pp. 221–225. Phoenix, AZ (2008)
14. Minghua, H., Tianyuan, Y., Yumeng, R.: Optimization of airport slot based on improved Hungarian algorithm. Appl. Res. Comput. **7**(36), 2040–2043 (2019)
15. Smriti, C., Giuseppe, N., Matthew, R., et al.: A distributed version of the Hungarian method for multirobot assignment. IEEE Trans. Rob. **4**(33), 932–947 (2017)
16. Yanpeng, L., Yanling, Q., Yue, L.: Multi-skilled personnel scheduling method based on improved Hungarian algorithm. J. National Univ. Defense Technol. **2**(38), 144–149 (2016)
17. Bin, C., Jiaxing, W., Jing, F., et al.: Querying similar process models based on the Hungarian algorithm. IEEE Trans. Serv. Comput. **1**(10), 121–135 (2017)

Discrete Multi-height Wind Farm Layout Optimization for Optimal Energy Output

Jianming Zhu[1], Shanshan Lin[1], Junzhe Wen[1], Jin Qin[1,2], Wenqing Yan[1], and Bin Xu[1,2,3](✉)

[1] Nanjing University of Posts and Telecommunications, Nanjing 210000, China
[2] NanJing Pharmaceutical Co., Ltd., Nanjing 210012, China
[3] Jiangsu Key Laboratory of Data Science and Smart Software, Jinling Institute of Technology, Nanjing 211169, China
xubin2013@njupt.edu.cn

Abstract. Optimization of wind farm layout has practical significance for the full utilization of wind energy. A novel wind farm layout optimization model is proposed in this paper. Both the continuous coordinate and the multi-hub height model are considered comprehensively. These two important factors are often discussed in isolation in previous studies. ADEG (Adaptive Differential Evolution with Greedy Method) is proposed to solve the above model. In ADEG, the differential evolution algorithm with parameter adaptive mechanism is used to improve the global search ability. The greedy algorithm is also introduced in ADEG to improve the local search ability. It balances the exploration and exploitation of the ADEG to improve the quality of the solution. The experimental results show that when the number of wind turbines is 5, 10, 20, 30 and 50 respectively, the output power of ADEG is 5.59%, 1.16%, 0.81%, 0.22% and 0.63% higher than that of DEEM (Differential Evolution with a new Encoding Mechanism) under the condition of multi-hub height model. When the number of wind turbines is 5 and 10 respectively, the output power of ADEG in the multi-hub height model is 1.96% and 0.07% higher than that in the fixed hub height model.

Keywords: Differential evolution algorithm · Self-adaptation parameter · Wind farm layout optimization

1 Introduction

Global warming has a far-reaching impact on human survival and development, and human beings are looking for alternative clean energy to deal with this problem. As a pollution-free and renewable characteristics of wind energy, it has become one of the most potential green energy.

Recently, wind turbines were mainly used to collect wind energy. In the wind farm, the wind power generation capacity of wind farm cannot be fully exerted since the influence of wake effect. To improve the power generation efficiency of wind farm and make rational use of wind energy resources, it needs to be realized through layout

X. Sun et al. (Eds.): ICAIS 2021, CCIS 1422, pp. 245–256, 2021.
https://doi.org/10.1007/978-3-030-78615-1_21

optimization in wind farms. Therefore, how to optimize the layout of wind farm has become a problem studied by many scholars.

Wind farm layout optimization is designed to solve the disturbance among wind turbines. Kusiak et al. [1] proposed a multi-hub height model in which two kinds of wind turbines with the height of 50 m and 78 m were used. Ning et al. [2] put two types of wind turbines at discrete multi-height in a line. The results show that the linear arrangement of wind turbines is beneficial to improve the output power of wind farms. Chowdhury et al. [3] used Particle Swarm Optimization to optimize the multi-height wind farm model. The results showed that adding turbines with different heights to the model could significantly increase the power output of wind farm. Wang et al. [4] designed a new coding mechanism to make the coordinates of wind turbines continuous. Many scholars [5–11] have studied the continuous coordinate model. It is proved that the continuous coordinate model is superior to the discrete coordinate model.

In this paper, a novel model is designed, which takes into account two factors: continuous coordinates and multi-hub height. ADEG is proposed to solve the above model. Differential evolution and greedy algorithms are introduced to improve the global search ability and local search ability of ADEG respectively. In order to further improve the quality of the solution, adaptive parameters are introduced into the ADEG. In order to verify the performance of the model and algorithm, the characteristics of fixed hub height and multi-hub height are analyzed in the experiment, and the convergence of the algorithm is explored.

2 Model Construction

2.1 Model Assumptions

In order to use Jensen wake model [12], it is assumed that the wind farm is built on a flat area. Meanwhile, in order to describe the wind speed more accurately, Weibull distribution is used to describe the wind speed distribution of a certain wind direction in this paper:

1. When the wind turbine is running, the wind speed and direction around it are relatively stable.
2. When the wind turbine blades are rotating, the rotation plane of the blades is always perpendicular to the wind direction.
3. In the layout process, the safe distance between any wind turbines is not less than five times the radius of the impeller [13].

2.2 Construction of Model

Speed Loss. The distance between the two rotor planes is $d_{j,i}$ [14]:

$$d_{j,i} = \left|(x_j - x_i)cos\varepsilon + (y_j - y_i)sin\varepsilon\right|, \tag{1}$$

where, ε is the wind direction, the coordinates of wind turbine i and j are (x_i, y_i) and (x_j, y_j), respectively.

Since the incoming wind speed of a certain wind turbine is not completely affected by the wake effect of the upstream wind turbine, $LOSS_{j,i}$ is set as the velocity loss in the wake area, which can be calculated from

$$LOSS_{j,i} = \frac{2t}{\left(1 + kd_{j,i}/R\right)^2}, \tag{2}$$

where t is the axial induction factor, R is the initial radius of wake area. The constant k is related to the height of the wind turbine.

Let *affected* be the area affected by the wake effect at the downstream wind turbine, as shown in Fig. 1. R_w is the radius of the wake zone. L is the distance between the wake flow center and the downstream wind turbine,

$$L = \sqrt{\left(x_j - x_i\right)^2 + \left(y_j - y_i\right)^2 + \left(h_j - h_i\right)^2 - d_{j,i}^2}. \tag{3}$$

The area affected [15] by the upstream wind turbine wake effect *affected* is calculated by

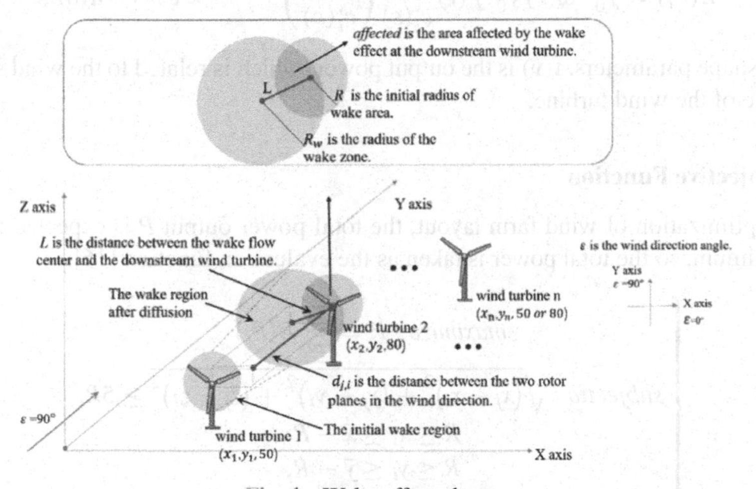

Fig. 1. Wake affected area.

$$\begin{aligned} affected &= S_1 + S_2 \\ &= R^2\left(\arcsin\left(\tfrac{x-L}{R}\right) + \tfrac{\pi}{2} + \sin\left(2\arcsin\left(\tfrac{x-L}{R}\right)\right)/2\right) \\ &+ R_w^2\left(\tfrac{\pi}{2} + \arcsin\left(\tfrac{x}{R_w}\right) - \sin\left(2\arcsin\left(\tfrac{x}{R_w}\right)\right)/2\right). \end{aligned} \tag{4}$$

Then the loss of wind speed at wind machine is $LI_{j,i}$,

$$LI_{j,i} = \frac{affected}{\pi R_w^2} \times LOSS_{j,i} \tag{5}$$

Assume that the wind speed of wind turbine i is affected by the wake effect of all other wind turbines in the wind power station. N is the number of wind turbines. The total speed loss is denoted as LI_i,

$$LI_i = \sqrt{\sum_{j=1, j \neq i}^{N} (LI_{j,i})^2}. \quad (6)$$

Average Output Power [6]. The scale parameter $c_i(\varepsilon)$ in Weibull distribution is modified to $c_i^{'}(\varepsilon)$ after being affected by Jensen wake:

$$c_i^{'}(\varepsilon) = c_i(\varepsilon) \times (1 - LI_i). \quad (7)$$

Let $\omega(\varepsilon)$ be the probability distribution of wind direction. The average output power $E(P_i)$ considering wind direction and wake effect is calculated by

$$E(P_i) = \int_0^{2\pi} \omega(\varepsilon) \int_0^{\infty} f(v) \frac{k_i(\varepsilon)}{c_i^{'}(\varepsilon)} \left(\frac{v}{c_i^{'}(\varepsilon)} \right)^{k_i(\varepsilon)-1} \times e^{-\frac{v^{k_i(\varepsilon)}}{c_i^{'}(\varepsilon)}} dv d\varepsilon, \quad (8)$$

$k_i(\varepsilon)$ is shape parameters. $f(v)$ is the output power, which is related to the wind speed at the blade of the wind turbine.

2.3 Objective Function

In the optimization of wind farm layout, the total power output P is expected to reach the maximum, so the total power is taken as the evaluation function [16].

$$\begin{cases} maximize : P = \sum_{i=1}^{N} E(P_i) \\ subjectto : \sqrt{(x_j - x_i)^2 + (y_j - y_i)^2 + (z_j - z_i)^2} \geq 5R, \\ R \leq x_i \leq \overline{x} - R, \\ R \leq y_i \leq \overline{y} - R, \\ i = 1, 2, \ldots, N, j = 1, 2, \ldots, N \; and \; j \neq i. \end{cases} \quad (9)$$

Among them, $\overline{x}$ and $\overline{y}$ are the upper bounds of the wind farm.

3 Method

ADEG is proposed to solve the multi-hub height model. In the algorithm, differential evolution and greedy algorithm are introduced. The global search ability of differential evolution algorithm is strong, but the local search ability is weak, and greedy algorithm can make up for its weak local search ability. ADEG balances the ability of local search and global search, and the whole algorithm has strong robustness. In addition, adaptive parameters are introduced into the algorithm to make the quality of the solution obtained by ADEG better.

3.1 ADEG Process

The specific flow of ADEG is as follows:

Step 1: Initialize the population and all parameters of the algorithm.
Step 2: The differential evolution algorithm with adaptive parameters is used for the parent population to generate candidate set.
Step 3: If the candidate set is empty, go to Step2. An individual is randomly selected from the candidate set to replace any individual in the parent population to generate the offspring population and remove the selected position from the candidate set.
Step 4: Compare the output power of the parent population and the offspring population. If the output power of the offspring population is greater than the output power of the parent population, let the current offspring population be the parent population of the next generation and go to Step 3.
Step 5: When the number of iterations reaches the maximum, output results and terminate the algorithm.

3.2 Parameter Self-adaptation

In order to further improve the quality of the solution, an adaptive mechanism [17–19] is introduced into ADEG. In the adaptive parameter mechanism, the mutation factor F and the crossover factor CR are used for adaptive control. When the values of F and CR are large, the global search ability of ADEG becomes stronger. When the values of these two parameters are small, the algorithm will enter the stage of local accurate search. The corresponding parameters of individuals who successfully evolve in the process of algorithm implementation need to be recorded. The corresponding mean value was selected according to these parameters. Then, according to the mean value, the probability distribution is constructed and the next generation parameters are selected. As the number of iterations increases, the parameters will adjust adaptively. The adaptive parameter variation strategy is

$$CR_i = randn_i(\Delta_{CR}, 0.1), F_i = randc_i(\Delta_F, 0.1), \tag{10}$$

where $\Delta_{CR} \leftarrow (1 - a) \times \Delta_{CR} + a \times mean_A(Success_{CR})$ and $\Delta_F \leftarrow (1 - a) \times \Delta_F + a \times mean_L(Success_F)$. Initializate parameters $\Delta_{CR} = 0.5$, $\Delta_F = 0.5$ and set a $\leftarrow$[0.05, 0.2].$Success_F$, $Success_{CR}$ are the set of success parameters, that is, the set of parameters F and CR corresponding to the better individuals of this generation of crossover and mutation. $mean_A$ is the standard arithmetic mean [20], the calculation $mean_L$ is as follows:

$$mean_L(Success_F) = \frac{\sum_{F \in Success_F} F^2}{\sum_{F \in Success_F} F}. \tag{11}$$

4 Experiments

4.1 Experimental Design

In order to verify the characteristics of the model, we compare the output power between the multi-hub height model and the fixed hub height model. In order to verify the performance of the algorithm, we select a state-of-the-art and excellent DEEM (Differential Evolution with a new Encoding Mechanism) for comparison. Not only the output power of the two algorithms is compared, but also their convergence is compared.

In the experiment, we define the wind direction as 0° from west to east, 90° from south to north, 180° from east to west and 270° from north to south. Specific parameters are shown in Table 1:

Table 1. Scenario parameters.

n	ε_n	$k_i(\varepsilon)$	$c_i(\varepsilon)$	δ_n	n	ε_n	$k_i(\varepsilon)$	$c_i(\varepsilon)$	δ_n
1	7.5°	2	13	0	13	187.5°	2	13	0.01
2	22.5°	2	13	0.01	14	222.5°	2	13	0.01
3	37.5°	2	13	0.01	15	217.5°	2	13	0.01
4	52.5°	2	13	0.01	16	232.5°	2	13	0.01
5	67.5°	2	13	0.01	17	247.5°	2	13	0.01
6	82.5°	2	13	0.2	18	262.5°	2	13	0.01
7	97.5°	2	13	0.6	19	277.5°	2	13	0.01
8	112.5°	2	13	0.01	20	292.5°	2	13	0.01
9	127.5°	2	13	0.01	21	307.5°	2	13	0.01
10	142.5°	2	13	0.01	22	322.5°	2	13	0.01
11	157.5°	2	13	0.01	23	337.5°	2	13	0.01
12	172.5°	2	13	0	24	352.5°	2	13	0

In this wind scene, the wind direction is divided into 24 units. The value of wind direction is approximately the center of each wind direction interval, where n is the index of wind direction interval. ε_n is the approximate wind direction. δ_n is the frequency of wind direction. Because the Weibull distribution is controlled by the shape parameter and the proportional parameter, when the proportional parameter is fixed, the larger the shape parameter is, the wind speed will concentrate on the larger value. When the shape parameter is unchanged, the smaller the proportional parameter is, the wind speed will concentrate on the smaller value. In this wind scenario, the wind direction is mainly concentrated in the 6th and 7th wind direction intervals. The wind speed of each wind has a similar probability distribution. Also, the wind speed is relatively large.

In order to explore the influence of hub height on output power, we set up fixed height model and multi-hub height model. Model 1: The height of the fixed hub is 50 m. Model 2: The height of the different heights of hubs are 50 and 80 m.

The scale of the wind farm will change according to the number of wind turbines. The corresponding relationship between the number of wind turbines and the scale of the wind farm is shown in Table 2:

Table 2. Scale of the wind farm.

N	5	10	20	30	50
X/Y (m)	500	800	1500	2000	2500

In the following statements, -S means that the algorithm is applied to fixed hub height model, -M means that the algorithm is applied to multi-hub height model. In the experiment, both DEEM and ADEG will run 30 times.

4.2 Experimental Results and Analysis

Table 3 records the average output power obtained by DEEM and ADEG. The experimental results show that when the number of wind turbines is 5, 10, 20, 30 and 50 respectively, the output power of ADEG-M is 5.59%, 1.16%, 0.81%, 0.22% and 0.63% higher than that of DEEM-M. In ADEG, when the number of wind turbines is 5 and 10 respectively, the output power of multi-hub height model is 1.96% and 0.07% higher than that in the fixed hub height model. We can see from the table that under the condition of the same number of wind turbines and model, the output power of ADEG is better than that of DEEM. In addition, it can be found that when the number of wind turbines is small, the multi-hub height model is better than the fixed hub height model. However, with the increase of the number of wind turbines, the advantages of the fixed hub height model will appear.

Table 3. Average output power comparison.

N(scale)	Algorithm			
	DEEM-S	DEEM-M	ADEG-S	ADEG-M
5 (500 m × 500 m)	4241.02	4157.74	4305.79	**4390.26**
10 (800m × 800m)	8387.80	8392.41	8483.60	**8489.77**
20 (1500 m × 1500 m)	17005.25	16964.46	**17109.97**	17101.85
30 (2000 m × 2000 m)	25472.21	25547.48	**25621.12**	25604.66
50 (2500 m × 2500 m)	41869.85	41884.05	**42218.69**	42146.46

Figure 2, 3 record the layout of ADEG and DEEM under the condition of a fixed height of wind turbine. This diagram can intuitively depict Layout of wind turbines in wind farm under different algorithms.

As can be seen from the figure, in this kind of wind scene, both DEEM and ADEG have a tendency to distribute the wind turbines side by side to reduce the wake effect, but it is clear that the ADEG arrangement effect is more ideal. Placing wind turbines in the main wind direction is more in line with the actual layout of the wind farm.

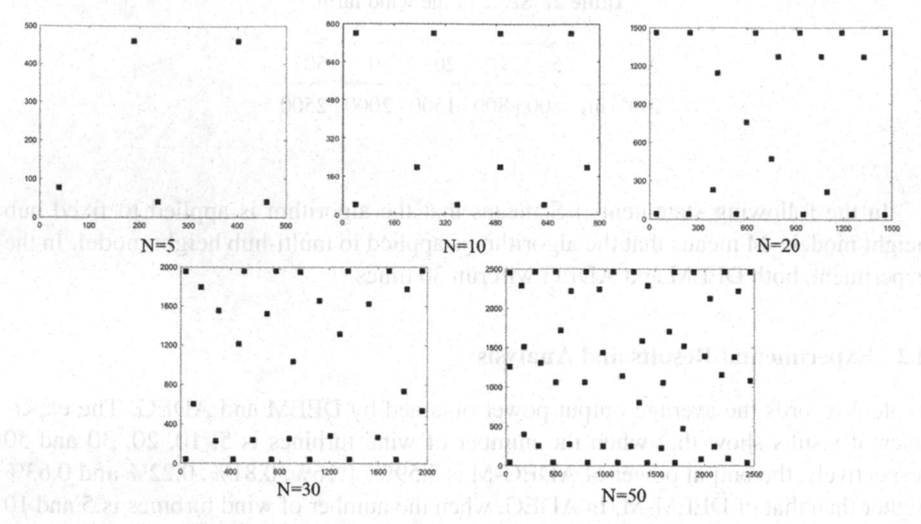

Fig. 2. The layout of wind turbines obtained by DEEM-S.

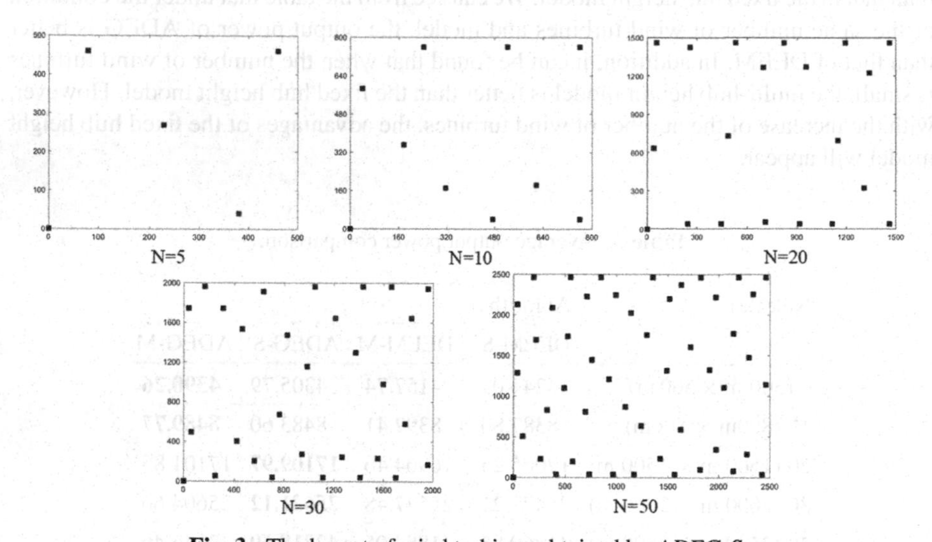

Fig. 3. The layout of wind turbines obtained by ADEG-S.

Figure 4 records the comparison of the output power of the two algorithms in the iterative process under the fixed hub height model.The solid line represents ADEG. We can clearly see that ADEG can get a better solution quickly and well. DEEM is inferior to ADEG in speed and quality.

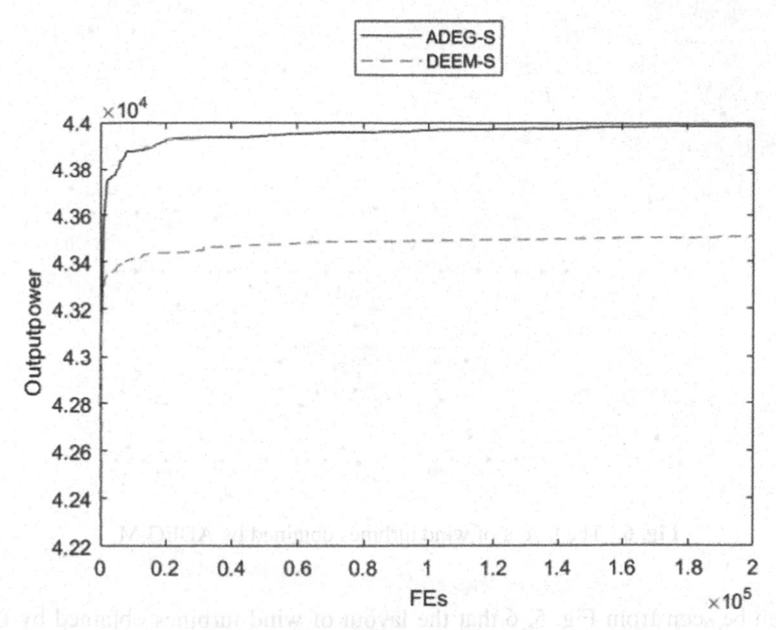

Fig. 4. The output power convergence comparison among the ADEG-S and DEEM-S.

Figure 5, 6 records the distribution of wind turbines obtained by ADEG and DEEM in the multi-hub height model. In order to show the difference, we take two different shapes of points to represent the two kinds of hub heights in the experiment. The point of the square indicates that the height of the hub is 80 m. The triangle point indicates that the height of the hub is 50 m.

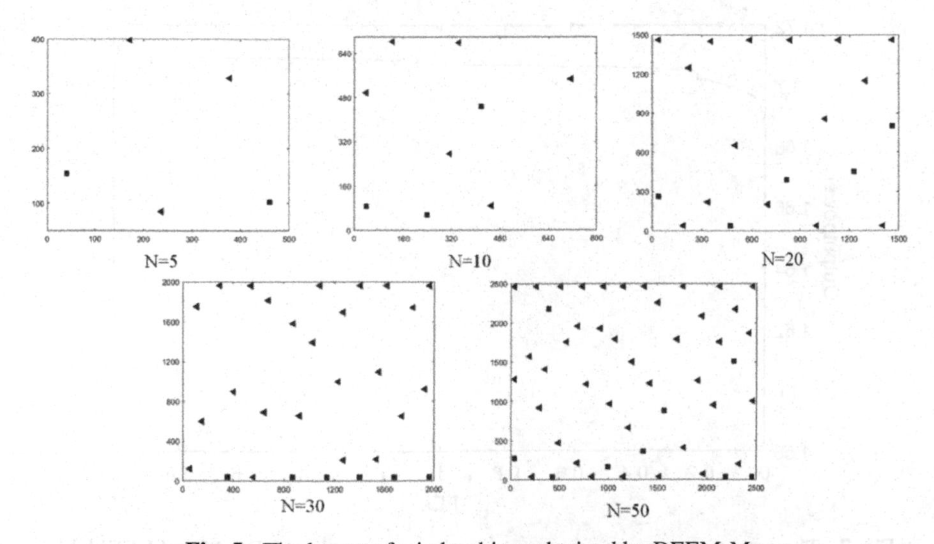

Fig. 5. The layout of wind turbines obtained by DEEM-M.

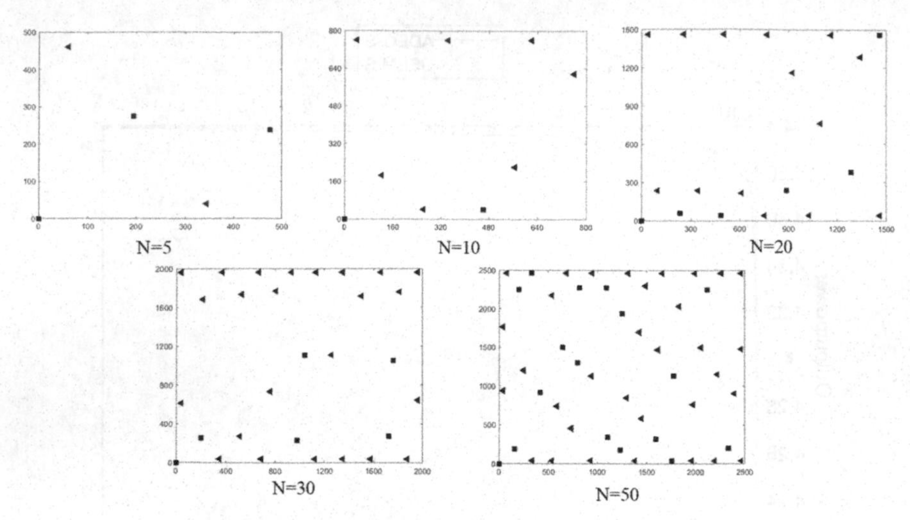

Fig. 6. The layout of wind turbines obtained by ADEG-M.

It can be seen from Fig. 5, 6 that the layout of wind turbines obtained by DEEM and ADEG is regular. The advantage of this layout is to reduce the influence of wake effect on power generation. The layout of ADEG is more regular than that of DEEM. Wind turbines of different heights are placed one after another in the one wind direction, which is more in line with the requirements of reducing the wake effect.

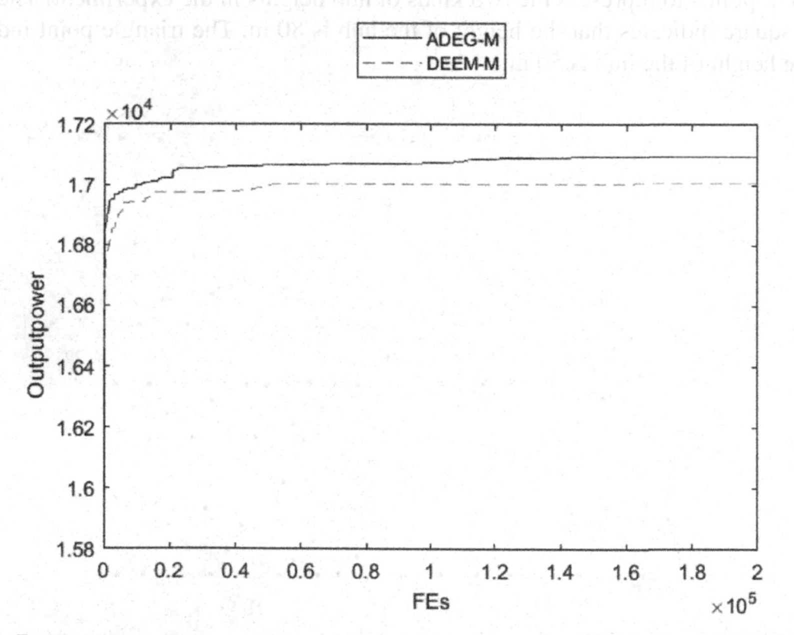

Fig. 7. The output power convergence comparison among the ADEG-M and DEEM-M.

Figure 7 records the comparison of the output power of the two algorithms in the iterative process under the multi-hub height model. We can see that the output power solved by ADEG is still higher than that of DEEM.

5 Conclusions

In this paper, both the continuous coordinate and the multi-hub height model are considered comprehensively. The experimental results show that the multi-hub height model can effectively reduce the influence of wake effect when the number of wind turbines is small. ADEG is proposed to solve the above model. Differential evolution and greedy algorithms are introduced into ADEG to balance the global search ability and local search ability. In order to further improve the quality of the solution, adaptive parameters are incorporated into the ADEG.

Acknowledgement. This paper was supported in part by Project funded by China Postdoctoral Science Foundation under Grant 2020M671552, in part by Jiangsu Planned Projects for Postdoctoral Research Funds under Grant 2019K223, in part by NUPTSF(NY220060), in part by the Opening Project of Jiangsu Key Laboratory of Data Science and Smart Software (No. 2020DS301) in part by Natural Science Foundation of Jiangsu Province of China under Grant BK20191381, in part by the National Natural Science Foundation of China under Grant No. 61802207.

Conflicts of Interest. No conflicts of interest exit in the submission of this manuscript, and manuscript is approved by all authors for publication.

References

1. Kusiak, A., Song, Z.: Design of wind farm layout for maximum wind energy capture. Renewable Energy **35**(3), 685–694 (2010)
2. Quan, N., Kim, H.M.: A mixed integer linear programing formulation for unrestricted wind farm layout optimization. J. Mech. Des. **138**(6), 061404 (2016)
3. Chowdhury, S., Zhang, J., Messac, A., Castillo, L.: Optimizing the arrangement and the selection of turbines for wind farms subject to varying wind conditions. Renewable Energy **52**(2), 273–282 (2013)
4. Wang, Y., Liu, H., Long, H., Zhang, Z., Yang, S.: Differential evolution with a new encoding mechanism for optimizing wind farm layout. IEEE Trans. Industr. Inf. **14**(3), 1040–1054 (2018)
5. Chen, K.: Wind turbine layout optimization with multiple hub height wind turbines using greedy algorithm. Renewable Energy **96**(5), 676–686 (2016)
6. Mora, J.C., Baron, J.M.C., Santos, J.M.R., Payan, M.B.: An evolutive algorithm for wind farm optimal design. Neurocomputing **70**(16), 2651–2658 (2007)
7. Wan, C., Wang, J., Yang, G., Zhang, X.: Optimal micro-siting of wind farms by particle swarm optimization. In: Tan, Y., Shi, Y., Tan, K.C. (eds.) ICSI 2010. LNCS, vol. 6145, pp. 198–205. Springer, Heidelberg (2010). https://doi.org/10.1007/978-3-642-13495-1_25
8. Saavedra-Moreno, B., Salcedo-Sanz, S.: Seeding evolutionary algorithms with heuristics for optimal wind turbines positioning in wind farms. Renewable Energy **36**(11), 2834–2844 (2011)

9. Eroglu, Y., Seckiner, S.U.: Design of wind farm layout using ant colony algorithm. Renewable Energy **30**(1), 53–62 (2012)
10. Wagner, M., Veeramachaneni, K., Neumnn, F.: Optimizing the layout of 1000 wind turbines. In: European Wind Energy Association Annual Event, UK, pp. 205–209 (2011)
11. Jiang, D., Peng, C., Fan, Z., Chen, Y.: Modified binary differential evolution for solving wind farm layout optimization problems. In: 2013 IEEE Symposium on Computational Intelligence for Engineering Solutions, Singapore, pp. 23–28. IEEE (2013)
12. Katic, I., Højstrup, J., Jensen, N.O.: A simple model for cluster efficiency. In: European Wind Energy Association Conference and Exhibition, pp. 407–410 (1986)
13. Long, H., Li, P.K., Gu, W.: A data-driven evolutionary algorithm for wind farm layout optimization. Energy **208**(6), 118310 (2020)
14. Lückehe, D.: Evolutionary Wind Turbine Placement Optimization with Geographical Constraints, pp. 3–12. Springer, Wiesbaden (2017). https://doi.org/10.1007/978-3-658-18465-0_1
15. Park, J., Law, K.H.: Cooperative wind turbine control for maximizing wind farm power using sequential convex programming. Energy Convers. Manage. **101**, 295–316 (2015)
16. Mosetti, G.: Optimization of wind turbine positioning in large windfarms by means of a genetic algorithm. Wind Eng. Ind. Aerodyn. **51**(1), 105–106 (1994)
17. Rainer, S., Price, K.: Differential evolution – a simple and efficient heuristic for global optimization over continuous spaces. Global Optim. **11**(4), 341–359 (1997)
18. Xu, B., Tao, L.L., Chen, X., Cheng, W.: Adaptive differential evolution with multipopulation-based mutation operators for constrained optimization. Soft Comput. **23**(10), 25–28 (2018)
19. Ramos, J., Leon, M., Xiong, N.: MPADE: an improved adaptive multipopulation differential evolution algorithm based on JADE. In: 2018 IEEE Congress on Evolutionary Computation (CEC), Rio de Janeiro, pp. 1–8. IEEE (2018)
20. Chen, Y., Li, H., Jin, K., Song, Q.: Wind farm layout optimization using genetic algorithm with different hub height wind turbines. Energy Convers. **70**(1), 56–65 (2013)

Multilevel Threshold Image Segmentation Based on Modified Moth-Flame Optimization Algorithm

Bin Xu[1,2,3](✉), Yunkai Zhao[1], Chong Guo[1], Yuxin Yin[1], and Jin Qi[1]

[1] Nanjing University of Posts and Telecommunications, Nanjing 210003, China
xubin2013@njupt.edu.cn

[2] NanJing Pharmaceutical Co., Ltd., Nanjing 210012, China

[3] Jiangsu Key Laboratory of Data Science and Smart Software, Jinling Institute of Technology, Nanjing 211169, China

Abstract. Among many image segmentation methods, multi-threshold segmentation is an effective method. However, the calculational complexity of multi-threshold segmentation is high, and the traditional calculation methods is difficult to obtain satisfactory results. In view of this situation, an Modified Moth-flame Optimization algorithm (MMFO) is proposed in this paper. Taking maximizing Kapur entropy as the objective function, MMFO algorithm is successfully applied to image multilevel threshold segmentation. The performance of the algorithm is evaluated by using several images which are widely used in the field of image segmentation. The experimental results show that, compared with other algorithms, MMFO algorithm can find a better solution more effectively.

Keywords: Image segmentation · Modified Moth-Flame Optimization algorithm · Multithreshold · Kapur's entropy

1 Introduction

Image is one of the main sources of information in the objective world. The acquisition of information in an image often depends on image segmentation. Image segmentation is a technology that helps people obtain information by dividing an image into several specific regions with unique attributes.

There are many kinds of image segmentation methods, including threshold-based segmentation [1], edge detection [2], region-based segmentation [3] and

This paper was supported in part by Project funded by China Postdoctoral Science Foundation under Grant 2020M671552, in part by Jiangsu Planned Projects for Postdoctoral Research Funds under Grant 2019K223, in part by NUPTSF (NY220060), in part by the Opening Project of Jiangsu Key Laboratory of Data Science and Smart Software (No.2020DS301), in part by Natural Science Foundation of Jiangsu Province of China under Grant BK20191381, in part by the Natural Science Foundation of Anhui (1908085MF207), in part by the National Natural Science Foundation of China under Grant No. 61802207.

X. Sun et al. (Eds.): ICAIS 2021, CCIS 1422, pp. 257–267, 2021.
https://doi.org/10.1007/978-3-030-78615-1_22

so on. Among them, the threshold-based segmentation method has gradually become the most commonly used method in image segmentation because of its high efficiency and stability.

Multilevel threshold segmentation of an image needs to find an optimal threshold set and divide pixels into several categories. The problem of Multilevel threshold Image Segmentation is essentially a constrained combinatorial optimization problem. Currently, many classical optimization algorithms and their improved algorithms are applied to multilevel threshold image segmentation to solve the problem of threshold selection. D. Shao et al. carried out the ultrasound image segmentation based on the improved differential search algorithm, and it achieved a preferable results [4]. Khairuzzaman et al. proposed an image segmentation method based on Masi entropy and particle swarm optimization algorithm (PSO), which greatly reduced the calculative complexity of multilevel threshold [5]. F. Huo et al. proposed an improved artificial bee colony algorithm (ABC), which improved the update strategy of ABC algorithm, and effectively applied it to image multilevel threshold segmentation [6]. O. Diego et al. combined the Electromagnetic Optimization (EMO) algorithm with the objective function proposed by Otsu and Kapur methods, then proposed an improved EMO algorithm, and applied it to multilevel threshold segmentation [7]. Moreover, many other optimization algorithms have been applied to this kind of problems [8,9].

Moth-flame optimization (MFO) algorithm is an intelligent optimization algorithm inspired by studying the flight strategy of natural moths. The related work shows that, compared with traditional intelligent optimization algorithms such as PSO algorithm and ABC algorithm, this algorithm has better optimization effect and higher stability to some extent [10]. In this paper, a Modified Moth-flame optimization (MMFO) algorithm is proposed by introducing a local search mechanism based on EMO algorithm into MFO algorithm, and applied to multilevel threshold image segmentation. Meanwhile, through experiments, the MMFO algorithm is compared with MFO algorithm and EMO algorithm, which indicates the effectiveness of MMFO algorithm.

2 Solution Overview

Kapur entropy method is a nonparametric threshold optimization method proposed by Kapur [11], and it is widely used in image segmentation. It is an image segmentation technique which can find the multi-level threshold producing the maximum entropy according to the probability distribution of the gray histogram.

Taking the L intensity levels from gray scale image or RGB (red, green, blue) image, Eq. (1) shows how to calculate the probability distribution of the intensity values:

$$Ph_i^b = \frac{h_i^b}{SP}, \sum_{i=1}^{SP} Ph_i^b = 1,$$
$$b = \begin{cases} 1,2,3 & if\, RGB\, image \\ 1 & if\, Gray\, scale\, image \end{cases}, \tag{1}$$

where i is a specific intensity level $(0 \leq i \leq L-1)$, b is used to judge whether the image is gray scale or RGB. SP is the total number of pixels in the image. h_i^b represents the number of pixels corresponding to i. Ph_i^b represents the probability distribution of i. The Eq. (2) gives the simplest segmentation (bi-level):

$$C_1 = \frac{Ph_1^b}{\omega_0^b(th)}, \ldots, \frac{Ph_{th}^b}{\omega_0^b(th)},$$
$$C_2 = \frac{Ph_{th+1}^b}{\omega_1^b(th)}, \ldots, \frac{Ph_L^C}{\omega_1^b(th)}, \tag{2}$$

where th is the segmentation threshold, $\omega_0(th)$ and $\omega_1(th)$ are probabilities distributions for C_1 and C_2 respectively, as it is defined as Eq. (3):

$$\omega_0^b(th) = \sum_{i=1}^{th} Ph_i^b, \omega_1^b(th) = \sum_{i=th+1}^{L} Ph_i^b. \tag{3}$$

For the bi-level example, the objective function of the Kapur's problem can be defined as Eq. (4):

$$\max_{\text{th}} H_1^b + H_2^b, \tag{4}$$

where the entropies H_1^b and H_2^b are calculated by the following Eq. (5):

$$H_1^b = \sum_{i=1}^{th} \frac{Ph_i^b}{\omega_0^b} \ln\left(\frac{Ph_i^b}{\omega_0^b}\right),$$
$$H_1^b = \sum_{i=th+1}^{L} \frac{Ph_i^b}{\omega_1^b} \ln\left(\frac{Ph_i^b}{\omega_1^b}\right), \tag{5}$$

Ph_i^b is the probability distribution of the intensity levels from Eq. (1). Equation (3) gives a description of how to calculate $\omega_0(th)$ and $\omega_1(th)$. The entropy-based approach of bi-level threshold values can be expanded for multiple threshold values. In this case, the new objective function is defined as Eq. (6):

$$\max_{\text{TH}} \sum_{i}^{k} H_i^b, \tag{6}$$

where $TH = th_1, th_2, ..., th_{k-1}$ is a vector that contains the multiple thresholds. Each entropy is calculated by its corresponding threshold, as shown in the Eq. (7)

$$H_1^b = \sum_{i=1}^{th_1} \frac{Ph_i^b}{\omega_0^b} \ln\left(\frac{Ph_i^b}{\omega_0^b}\right),$$
$$H_2^b = \sum_{i=th_1+1}^{th_2} \frac{Ph_i^b}{\omega_1^b} \ln\left(\frac{Ph_i^b}{\omega_1^b}\right),$$
$$\vdots$$
$$H_k^b = \sum_{i=th_{k-1}+1}^{L} \frac{Ph_i^b}{\omega_{k-1}^b} \ln\left(\frac{Ph_i^b}{\omega_{k-1}^b}\right), \quad (7)$$

where the values of the probability occurrence $\{\omega_0^b, \omega_1^b, ..., \omega_{k-1}^b\}$ of the k classes are obtained as Eq. (8):

$$\omega_0^b = \sum_{i=1}^{th_1} Ph_i^b,$$
$$\omega_1^b = \sum_{i=th_1+1}^{th_2} Ph_i^b,$$
$$\vdots$$
$$\omega_{k-1}^b = \sum_{i=th_{k-1}+1}^{L} Ph_i^b. \quad (8)$$

3 Multilevel Thresholding Using MMFO

In this paper, a MMFO algorithm is proposed by introducing a local search mechanism based on EMO algorithm into MFO algorithm. The relationship between MMFO and image segmentation is shown in Table 1.

Table 1. The relationship between MMFO and image segmentation.

MMFO	Image segmentation
Moth positions	Threshold segmentation solution
Flame positions	Threshold segmentation solution
Moth population	Threshold candidate set
Fitness function value	Entropy value of segmented image

3.1 MFO

Introduction of Related Concepts. The moth population M can be represented by the following matrix:

$$M = \begin{bmatrix} m_{1,1}, m_{1,2} & \cdots m_{1,d} \\ m_{2,1}, m_{2,2} & \cdots m_{2,d} \\ \vdots & \vdots \\ m_{n,1}, m_{n,2} & \cdots m_{n,d} \end{bmatrix}, \quad (9)$$

where n is the number of moths and d is the number of variables to be sought. The i th moth in the population is expressed as M_i.

The flame population F can be represented by the following matrix:

$$F = \begin{bmatrix} F_{1,1}, F_{1,2} \cdots F_{1,d} \\ F_{2,1}, F_{2,2} \cdots F_{2,d} \\ \vdots \qquad \vdots \\ F_{n,1}, F_{n,2} \cdots F_{n,d} \end{bmatrix}, \tag{10}$$

where n is the number of flames and d is the number of variables to be sought. The j th flame in the population is expressed as F_j.

In MFO algorithm, an adaptive mechanism about the number of flames is proposed for improving the exploitation to the best promising solutions. The number of flames is equal to the population of the moths in the first iteration of the algorithm and it will decrease according Eq. (11):

$$flame\_no = round\left(N - l * \frac{N-1}{T}\right), \tag{11}$$

where N is the maximum number of flames, l is the current number of iteration, and T indicates the maximum number of iterations.

Initialization of Population. The population size of moths is set to n. During MMFO initialization, M^0 is the initialization population and M_i^0 represents the i th moth in the initial population. The individual is initialized according to the following Eq. (12):

$$M_i^0 = \lfloor rand(1, d) \cdot (ub - lb) + lb \rfloor, i \in \{1, 2, \cdots, n\}, \tag{12}$$

where ub and lb are the upper and lower limits of the parameters respectively.

The Update of Position of the Moths. The position of the moth is updated by a logarithmic helix, as shown in Eqs. (13)–(15):

$$M_i = S\left(M_i, F_i\right), \tag{13}$$

$$S\left(M_i, F_j\right) = D_i * e^{bt} * \cos(2\pi t) + F_j, \tag{14}$$

$$D_i = \left|F_j - M_i\right|, \tag{15}$$

where D_i indicates the distance of M_i and F_j. In order to further emphasize exploitation, it is assumed that t is a random number in $[r, 1]$, where r is linearly decreased from −1 to −2 over the course of iteration. b is the shape parameter of logarithmic helix.

3.2 Local Search Operator for MMFO

In MMFO algorithm, a local search operator based on EMO is introduced to increase the effect of MMFO algorithm. The main steps are as follows:

Step1: Initialize the related parameters of local search.

Step2: Randomly select the dimensions that have not been selected in the current solution for linear search. The search step size of each dimension is shown in the following Eq. (16):

$$Length = rand \cdot \delta \cdot (ub - ul), \tag{16}$$

where δ is the local search parameter.

Step3: Choose the solution with larger fitness value between the two solutions before and after search.

Step4: Repeat steps 2 and 3 until the search for each dimension of the current solution is completed.

Step5: Repeat step 2, step 3, step 4 until all the solutions have completed the local search.

3.3 MMFO Algorithm Framework

The main steps of the MMFO algorithm are as follows:

Step1: Initialize population parameters, including population size, number of iterations, number of individual dimensions, etc.

Step2: Generate the initial population by randomly generating the position of each moth.

Step3: Sort the individuals in the population according to the fitness value, and then assign the value to the flame to form the initial position of the flame.

Step4: Update the position of the moth to obtain a new population.

Step5: Perform a local search on the new population and then update the population obtained in Step4.

Step6: Calculate and sort the fitness values of the moth population and the flame population, and use the position corresponding to the better fitness value as the position of the next-generation flame.

Step7: Adaptively reduce the number of flames.

Step8: Repeat step 4 to 8 until the maximum number of iterations is reached.

Step9: Output the optimal segmentation threshold, images before and after segmentation, and convergence curve of fitness value.

4 Experiment and Analysis

In this section, 9 images which are widely used in the field of image segmentation are selected for experiments. All the images have the same size (512 * 512 pixels) and they are in JPGE format. The original image and the corresponding histogram are shown in Fig. 1. The related work shows that, compared with GA, PSO and BF algorithm, EMO algorithm has better image segmentation effect

[7]. In order to illustrate the superiority of MMFO, the proposed algorithm is compared with EMO. During the experiment, MMFO is performed 35 times for each image. The numbers of thresholds for test are $th = 2, 3, 4, 5$ and the number of iterations is 150.

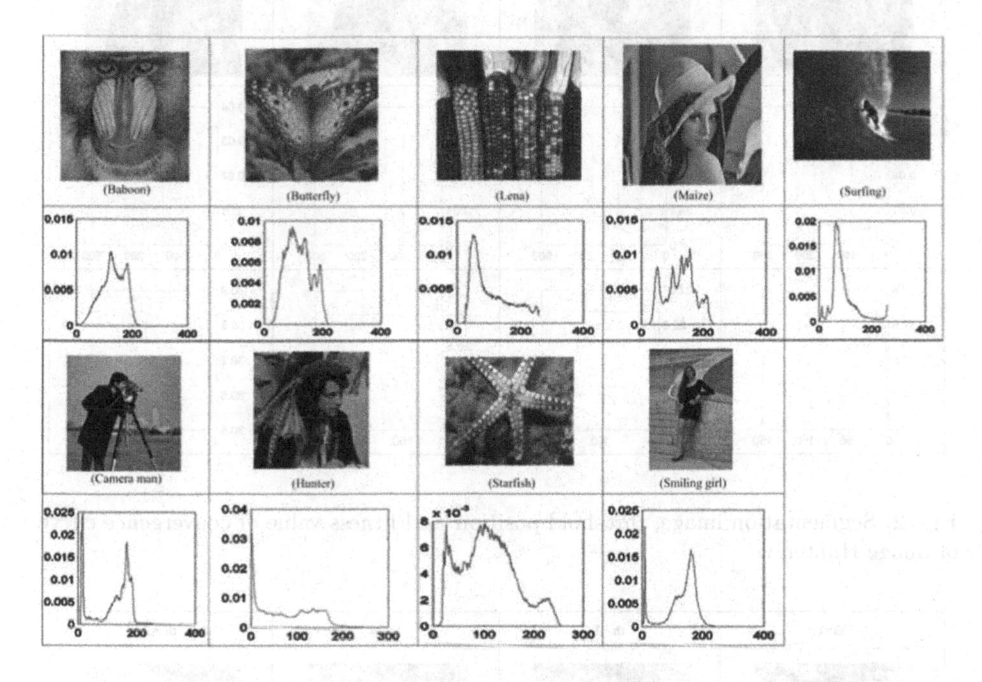

Fig. 1. Original image and the corresponding histogram.

4.1 Experimental Result

Figure 2, and 3 show the fitness, threshold position and convergence curve of the segmented image of Hunter and Butterfly in 9 images. It is difficult to directly judge the effect of image segmentation only visually, so the image will be analyzed by experimental data in the next section.

In order to compare the results of image segmentation, the peak signal-to-noise ratio (PSNR) is used to evaluate the quality of image segmentation. Since each image has run 35 times, it is possible to make PSNR fluctuate slightly in a certain range. The MEAN is the average value of 35 optimal fitness values, which implicates the overall quality of image segmentation and is positively correlated with the value of PSNR. STD indicates the stability of image segmentation and its mathematical meaning is the standard deviation of 35 optimal fitness values.

Table 2 shows the results of image segmentation by MMFO and MFO under different thresholds and the values of PSNR, Mean and STD indexes after segmentation. Each performance index has 36 values in Table 2. The results of experiment show that MMFO has 14 values better than the MFO in the index

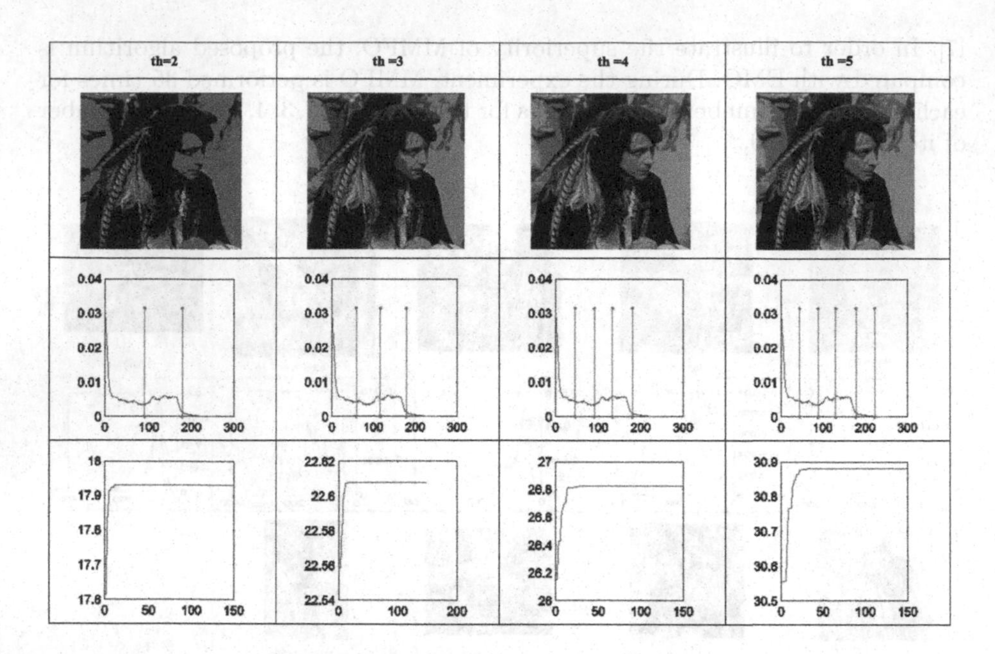

Fig. 2. Segmentation image, threshold position and fitness value of convergence curve of image Hunter.

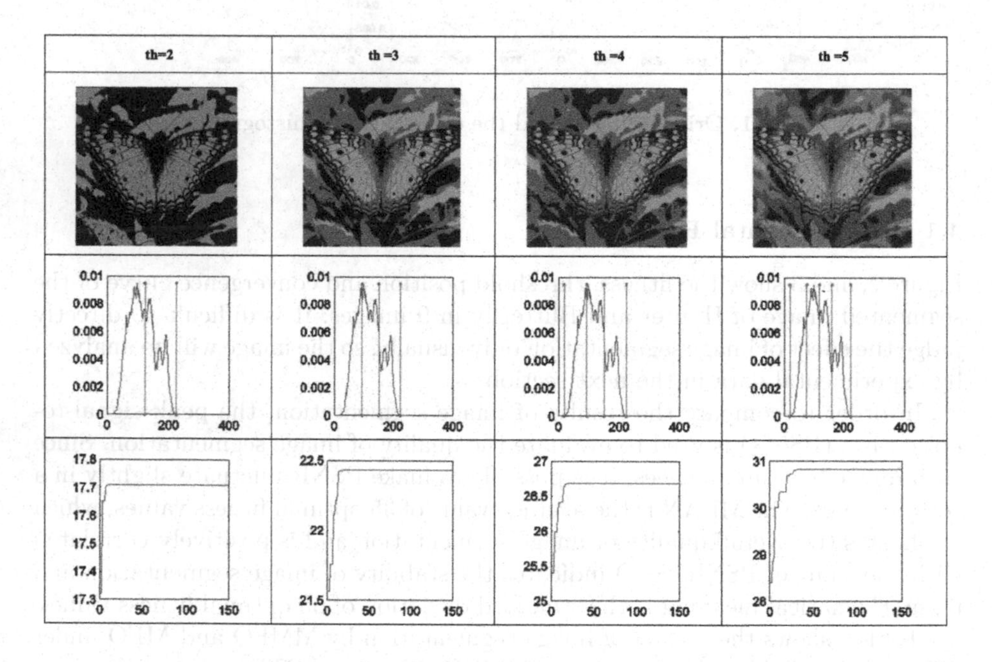

Fig. 3. Segmentation image, threshold position and fitness value of convergence curve of image Butterfly.

of MEAN, 20 values of MMFO are better than MFO in the index of STD and the index of PSNR between MMFO and MFO is roughly equal. Therefore, the local search in MMFO is not only increase the overall performance of the image segmentation, but also improves the stability of image segmentation. Experimental results obtained by MMFO, EMO, GA, PSO and BF methods under different thresholds are shown in Table 3. EMO algorithm is better in comparison with the GA, PSO and BF according to the related work. This paper compares MMFO with EMO in three performance indexes, containing PSNR, STD and MEAN. Table 3 shows that MMFO has 23 values better than the EMO in the index of PSNR, 27 values better than EMO in the index of STD and 22 values better than EMO in the index of MEAN. The results show that the proposed MMFO algorithm is better in quality and stability of image segmentation compared

Table 2. Comparisons between MMFO and MFO applied over the selected test images.

Image	k	Kapur's segmentation results by MMFO				Kapur's segmentation results by MFO			
		Threshold	PSNR	MEAN	STD	Threshold	PSNR	MEAN	STD
Baboon	2	88,144	16.007	17.680	1.08e–14	80,144	16.007	17.680	1.08e–14
	3	46,100,153	18.592	2.133	7.21e–15	46,100,153	18.592	22.133	7.21e–15
	4	34,75,115,160	20.523	26.392	1.39e–01	34,75,115,160	20.523	26.302	1.09e–01
	5	30,65,101,137,172	22.068	30.597	1.52e–01	30,65,101,137,172	22.068	30.538	2.15e–01
Butterfly	2	93,151	14.402	17.662	2.15e–01	93,151	14.402	17.612	1.35e–01
	3	93,151,224	14.402	22.430	0.00e+00	93,151,224	14.402	22.430	0
	4	75,117,160,224	17.681	26.623	4.43e–02	75,117,160,224	17.681	25.592	2.15e–02
	5	19,75,117,160,224	19.290	30.744	1.69e–01	19,75,117,160,224	19.290	30.710	1.87e–01
Lena	2	97,164	14.590	17.823	1.081e–14	97,164	14.590	17.823	1.08e–14
	3	82,127,177	17.178	22.111	7.209e–15	82,127,177	17.212	22.111	2.40e–04
	4	64,97,137,179	19.006	26.064	5.560e–02	64,97,137,179	19.030	26.030	4.96e–02
	5	16,64,97,137,179	20.325	29.940	1.001e–01	16,64,97,137,179	20.325	29.852	1.19e–01
Maize	2	99,177	13.590	18.634	1.081e–14	99,177	13.590	18.634	1.08e–14
	3	82,140,198	15.182	23.268	1.802e–14	82,140,198	15.182	23.268	1.30e–04
	4	73,118,164,211	16.346	27.503	1.442e–14	73,118,164,211	16.353	27.503	3.28e–04
	5	68,105,142,180,218	17.223	31.432	1.267e–04	68,105,142,180,218	17.666	31.430	3.10e–03
Camera man	2	44,106	14.463	17.584	2.85e–14	128,196	13.626	17.584	3.61e–15
	3	44,103,196	14.460	22.007	3.61e–15	44,103,149	14.460	22.007	3.61e–15
	4	44,96,146,196	20.153	26.586	1.08e–14	44,96,146,196	20.153	26.586	3.04e–13
	5	24,60,98,146,196	20.661	30.560	3.61e–15	39,84,119,154,196	22.306	30.530	6.37e–02
Hunter	2	90,178	15.190	17.929	3.61e–15	90,178	15.190	17.929	3.61e–15
	3	58,117,178	18.507	22.607	1.08e–14	58,117,178	18.507	22.608	1.08e–14
	4	46,91,134,180	21.039	26.822	2.16e–14	44,89,132,180	21.066	26.822	2.41e–04
	5	33,69,104,141,180	22.844	30.874	2.94e–02	33,69,104,141,180	22.844	30.859	5.77e–02
Sea star	2	90,170	14.395	18.752	7.21e–15	90,170	14.395	18.752	7.21e–15
	3	74,131,185	16.973	23.321	1.08e–14	73,129,185	17.066	23.321	7.19e–05
	4	66,114,161,205	18.404	27.570	0.00e+00	66,113,160,204	18.444	27.570	3.22e–04
	5	55,92,131,170,210	20.968	31.555	0.00e+00	53,90,129,170,210	20.358	31.555	6.79e–04
Smiling girl	2	108,197	13.514	17.320	3.61e–15	108,197	13.514	17.320	3.61e–15
	3	95,144,207	18.086	21.891	3.61e–14	95,143,203	18.133	21.891	2.82e–04
	4	36,84,140,201	18.826	26.016	1.80e–14	92,137,179,211	18.986	26.014	6.70e–03
	5	34,78,116,153,207	21.484	30.137	2.02e–02	34,78,116,153,207	21.484	30.130	3.46e–02
Surfer	2	103,171	11.878	18.342	7.21e–15	103,171	11.878	18.342	7.21e–15
	3	52,108,173	18.762	23.318	0.00e+00	52,108,173	18.762	23.316	8.30e–03
	4	52,102,152,201	19.719	27.835	1.80e–14	52,102,152,201	19.719	27.835	3.85e–04
	5	51,95,134,171,212	20.738	31.806	1.80e–14	51,91,126,165,208	21.229	31.801	1.77e–02

with EMO. Under the image of Butterfly, datas of three performance indexes obtained by MMFO are much better than EMO, so the MMFO achieved the greatest effect in image segmentation to the image of Butterfly.

Table 3. Comparisons between MMFO, EMO, GA, PSO and BF applied over the test images

Image	k	MMFO			EMO			GA			PSO			BF		
		PSNR	MEAN	STD	PSNR	MEAN	STD	PSNR	MEAN	STD	PSNR	MEAN	STD	PSNR	MEAN	STD
Baboon	2	16.007	17.680	1.08e–14	16.016	17.625	1.08e–14	12.184	16.425	0.0567	12.213	16.811	0.0077	12.216	16.889	0.0009
	3	18.592	22.133	7.209e–15	16.016	22.269	3.60e–15	14.745	21.069	0.1580	15.008	21.088	0.0816	15.211	21.630	0.0287
	4	20.523	26.392	1.392e–01	18.485	26.688	2.10e–03	16.935	25.489	0.1765	170574	24.375	0.0853	17.999	25.446	17.999
	5	22.068	30.597	1.392e–01	20.507	30.800	1.08e–14	19.662	29.601	0.2775	20.224	30.994	0.1899	20.720	30.887	20.720
Butterfly	2	14.402	17.662	2.15e–01	11.065	16.681	1.35e–01	10.470	15.481	0.0872	10.474	14.098	0.0025	10.474	15784	0.0014
	3	14.402	22.430	0.00e+00	14.176	21.242	3.56e–01	11.628	20.042	0.2021	12.313	19.340	0.1880	12.754	21.308	0.0118
	4	17.681	26.623	4.43e–02	16.725	25.179	3.45e–01	13.314	23.980	0.2596	14.231	25.190	0.2473	14.877	25.963	0.0166
	5	19.290	30.744	1.69e–01	19.026	28.611	2.32e–01	15.755	27.411	0.3977	16.337	27.004	0.2821	16.828	27.960	0.0877
Lena	2	14.590	17.823	1.081e–14	14.672	17.831	0.00e+00	12.334	16.122	0.0049	12.345	16.916	0.0033	12.345	16.605	0.0003
	3	17.178	22.111	7.209e–15	17.247	22.120	7.50e–04	14.995	20.920	0.1100	15.133	20.458	0.0390	15.133	20.812	0.0061
	4	19.006	26.064	5.560e–02	18.251	25.999	1.34e–02	17.089	23.569	0.2594	17.838	24.449	0.1810	17.089	26.214	0.0081
	5	20.325	29.940	1.001e–01	20.019	29.787	2.67e–02	19.549	27.213	0.3043	20.442	27.526	0.2181	19.549	28.046	0.0502
Maize	2	13.590	18.634	1.081e–14	13.633	18.604	2.026e–4	13.506	18.521	0.0725	13.466	18.631	0.0012	13.601	0.0022	13.601
	3	15.182	23.268	1.802e–14	15.229	22.941	5.92e–05	15.150	23.153	0.1582	15.018	23.259	0.0530	15.032	0.0068	15.032
	4	16.346	27.503	1.442e–14	16.280	26.936	2.97e–04	15.909	26.789	0.2697	15.834	27.470	0.1424	16.120	0.0128	16.120
	5	17.223	31.432	1.267e–04	17.211	31.023	3.27e–04	16.921	30.852	0.8971	16.319	31.255	0.4980	16.985	0.0978	16.985
Camera man	2	14.463	17.584	2.85e–14	13.626	17.584	3.60e–15	11.941	15.341	0.1270	12.259	16.071	0.1001	12.264	16.768	0.0041
	3	14.460	22.007	3.61e–15	18.803	21.976	4.91e–02	14.827	20.600	0.2136	15.211	21.125	0.1107	15.250	21.498	0.0075
	4	20.153	26.586	1.08e–14	20.586	26.586	1.08e–14	17.166	24.267	0.2857	18.000	25.050	0.2005	18.406	25.093	0.0081
	5	20.66	30.560	3.61e–15	20.66	30.506	6.35e–02	19.795	18.326	0.3528	20.963	28.365	0.2734	21.21	30.026	0.074
Hunter	2	15.190	17.929	3.605e–15	15.206	17.856	1.44e–14	12.349	16.150	0.0148	12.370	15.580	0.0068	12.373	16.795	0.0033
	3	18.507	22.607	1.081e–14	18.500	22.525	4.82e–04	14.838	21.026	0.1741	15.128	20.639	0.0936	15.553	21.860	0.1155
	4	21.039	26.822	2.163e–14	21.729	26.728	3.93e–04	17.218	25.509	0.2192	18.040	27.085	0.1560	18.381	26.230	0.0055
	5	22.844	30.874	2.940e–02	21.074	30.642	4.20e–02	19.563	29.042	0.3466	20.533	29.013	0.2720	21.256	28.856	0.0028
Sea star	2	14.395	18.752	7.21e–15	14.398	18.754	7.20e–15	14.282	18.754	0.0816	14.346	18.593	0.0002	14.280	18.753	0.0016
	3	16.973	23.321	1.08e–14	16.987	23.323	1.08e–14	8.2638	23.260	0.1987	16.949	23.289	0.1723	16.319	23.292	0.1813
	4	18.404	27.570	0.00e+00	18.304	27.582	5.02e–04	15.035	26.533	0.2691	18.389	27.407	0.2481	18.240	26.938	0.2092
	5	20.968	31.555	0.00e+00	20.165	31.562	7.51e–04	19.005	30.798	0.9740	19.849	31.288	0.6159	19.052	30.857	0.3553
Smiling girl	2	13.514	17.320	3.61e–15	13.420	17.334	0.00e+00	13.092	17.295	0.0178	13.352	17.321	0.0368	13.370	17.272	0.0038
	3	18.086	21.891	3.61e–15	18.254	21.904	6.06e–05	17.764	21.580	0.2179	18.201	21.887	0.0556	18.207	21.847	0.0178
	4	18.826	26.016	1.80e–14	18.860	26.040	1.96e–02	17.923	25.432	0.3024	18.063	25.815	0.2817	18.340	25.183	0.2119
	5	21.484	30.137	2.02e–02	19.840	30.089	5.42e–02	19.026	30.798	0.7128	19.200	29.700	0.5887	19.786	28.300	0.3813
Surfer	2	11.878	18.342	7.210e–15	11.744	18.339	1.02e–02	11.521	18.237	0.0219	11.698	18.194	0.1144	11.425	18.269	0.0489
	3	18.762	23.318	0.00E+00	18.584	23.231	7.49e–02	17.181	22.865	0.1715	18.413	22.214	0.2332	18.509	23.089	0.1369
	4	19.719	27.835	1.802e–14	19.478	27.863	3.06e–15	18.868	26.447	0.2093	19.125	26.676	0.4214	19.388	26.859	0.8240
	5	20.738	31.806	1.802e–14	20.468	31.823	6.50e–03	19.52	30.363	0.3182	19.49	30.587	0.4789	19.935	30.968	0.9684

5 Conclusions

In this paper, a MMFO algorithm is proposed by introducing a local search mechanism based on EMO algorithm into MMFO algorithm. Taking maximizing Kapur entropy as the objective function, the MMFO algorithm is successfully applied to image multi-level threshold segmentation. The experimental results show that, compared with the original MFO and EMO, MMFO has an advantage in segmentation quality, and effectively improves the stability of MMFO algorithm.

References

1. Yan, L., Feng, H., Chen, B., et al.: Adaptive local threshold segmentation for Fourier spatial filtering in automatic analysis of digital speckle interferogram. Opt. Eng. **59**(4), 046108–046108 (2020)

2. El-Sayed, M.A., Alib, A.A., Hussien, M.E.M.: A multi-level threshold method for edge detection and segmentation based on entropy. Comput. Mater. Continua **63**(1), 1–16 (2020)
3. Li, H., Pan, C., Chen, Wulamum, A., Yang, A.: Ore image segmentation method based on u-net and watershed. Comput. Mater. Continua **65**(1), 563–578 (2020)
4. Shao, D., Xu, C., Xiang, Y., et al.: Ultrasound image segmentation with multilevel threshold based on differential search algorithm. IET Image Proc. **13**(6), 998–1005 (2019)
5. Khairuzzaman, A.K.M., Chaudhury, S.: Masi entropy based multilevel thresholding for image segmentation. Multimedia Tools Appl. **78**(23), 33573–33591 (2019). https://doi.org/10.1007/s11042-019-08117-8
6. Huo, F., Sun, X., Ren, W.: Multilevel image threshold segmentation using an improved Bloch quantum artificial bee colony algorithm. Multimedia Tools Appl. (23), 2447–2471 (2019). https://doi.org/10.1007/s11042-019-08231-7
7. Diego, O., Erik, C., et al.: A multilevel thresholding algorithm using electromagnetism optimization. Neurocomputing **139**(2), 357–381 (2014)
8. Wang, S., Jia, H., Peng, X.: Modified salp swarm algorithm based multilevel thresholding for color image segmentation. Math. Biosci. Eng. **17**(1), 700–724 (2019)
9. Abdelkader, E.M., Moselhi, O., Marzouk, M., Zayed, T.: A multi-objective invasive weed optimization method for segmentation of distress images. Intell. Autom. Soft Comput. **26**(4), 643–661 (2020)
10. Mirjalili, S.: Moth-flame optimization algorithm: a novel nature-inspired heuristic paradigm. Knowl. Based Syst. **89**, 228–249 (2015)
11. Khehra, B.S., Pharwaha, A.S.: Image segmentation using teaching-learning-based optimization algorithm and fuzzy entropy. In: 15th International Conference on Computational Science and Its Applications, pp. 67–71. IEEE, Banff, AB, Canada (2015)

Research on Critical Illness Fundraising Methods Based on Blockchain and Smart Contracts

Huanrong Tang, Aoyun Jiang, and Jianquan Ouyang(✉)

Key Laboratory of Intelligent Computing Information Processing, Ministry of Education, School of Computer Science and School of Cyberspace Science, Xiangtan University, Xiangtan 411100, Hunan, China
oyjq@xtu.edu.com

Abstract. At present, with the development of science and medical level, more and more diseases can be cured. However, the cost of treating some diseases is high, and for many families with limited financial resources, it is difficult to spend so much money on treatment in a short period of time. As a result, more and more people are launching fundraising projects through third-party platforms such as Water Drops and Easy Fundraising, using the money raised to treat diseases. However, these third-party platforms have many problems, such as easy to disclose the private information of people who seeking help, the fundraising process is complex and slow, the use of project funds is difficult to timely monitor, the release of false fundraising projects for fundraising fraud, and so on. For these reasons, fewer members of the public will believe the authenticity of fundraising projects, so that it is not easy to donate to fundraising projects, which also makes it difficult for people who really need to raise funds to treat diseases. In view of this situation, this paper puts forward a blockchain and smart contract-based fundraising system for critical diseases, which can solve the problems existing in third-party platforms, and can help helpers quickly raise funds to cure diseases, it also can avoid the abuse of fundraising funds, false fundraising and other acts. Not only improve the transparency of fundraising behavior, but also enhance the enthusiasm of public participation in donations.

Keywords: Blockchain · Smart contract · IPFS · Serious illness fundraising

1 Introduction

In 2008, a person or a group of people going under the pseudonym Satoshi Nakamoto, published an article entitled Bitcoin: Peer-to-Peer Electronic Cash System [1]. The author implements the already familiar technologies in a revolutionary way, achieving a working model. As a result, blockchain technologies got a boost, cloud technologies [2], smart contracts emerged [3], and most innovative companies are now developing blockchain projects.

At present, with the development of technology and medical level, more and more diseases can be cured. However, the cost of treatment for some diseases is high. For many

X. Sun et al. (Eds.): ICAIS 2021, CCIS 1422, pp. 268–278, 2021.
https://doi.org/10.1007/978-3-030-78615-1_23

families with limited financial resources, it is difficult to spend so much money to treat the disease in a short period of time. Therefore, more and more people initiate fundraising projects through third-party platforms such as Water Drop Funding and Easy Funding, and use the raised funds to treat diseases. However, these third-party platforms have more problems, such as easy disclosure of private information of seekers, complicated and slow fundraising process, and difficulty in timely supervision of the use of project funds.

Many people tend to avoid sponsorship because of the lack of transparency in the sponsoring organization. In addition, many sponsoring organizations lack transparent and formal administration due to lack of working capital. For this reason, many private sponsors are only sponsored by large donation bodies. As a result, small organizations are not well managed for their long-term sponsors, and have problems with their lack of information power, human resources, and opacity [4].

In response to these situations, this article proposes a critical illness fundraising system based on blockchain and smart contracts. This system can be used for personal critical illness fundraising, which can solve these problems in third-party platforms, effectively raise funds for medical treatment, and can also avoid fraudulent donors from making false fundraising.

2 Related Work

In paper [4], the author uses blockchain to increase the transparency of funding organizations and encourage individuals to donate by protecting the privacy of donors. And to prevent this by providing audit services and certifying the ability of recipients of public administrative assistance.

In paper [5], the author proposes a blockchain-based decentralized donation tracking system, which is built on Ethereum and will provide complete transparency, accountability and direct donations to intended recipients. As a decentralized system, it is not subject to the jurisdiction of any third party, nor does it have any centralized entity like a database. Everything is stored in the blockchain network in the form of transactions. All transactions are executed or worked digitally on the basis of smart contracts similar to real world contracts.

In paper [6], the author proposes a crowdfunding platform based on Ethereum and smart contracts. The platform uses dual smart contracts: every time a new fundraising event is launched, the first smart contract deploys the second contract. The second contract contains all the logic needed for fundraising activities. The first contract is called the initiator, and the second contract is called the active contract. The administrator generates a payment request after collecting the required amount (in the form of Ether). This payment request needs to be approved by more than 50% of donors. If the payment request reaches the specified number of approvals, the administrator can finalize the request and transfer the required funds from the collected funds to the supplier.

In paper [7], the author proposes a privacy-protecting blockchain system in which all data is encrypted within a controllable time. Encrypt all data uploaded by participating users, and share the decryption key with all participants after the agreed encryption period. Within the agreed time, unless the user voluntarily discloses his data, the user's personal data will not be stolen. At the end of the encryption period, the decryption key disclosed by the system can verify whether the user cheated in the transaction. For users who commit fraud, the system will impose a penalty of forfeiting the deposit. At the same time, the system also compensates honest users who have been harmed by cheaters. This compensation comes from the punishment of the deceiver.

In paper [8], the system proposed by the author solves the problem of lack of transparency in the use of materials or funds donated by government departments or other donors. The author classified all users according to their roles: donors, organizations, retailers, and government officials.

In paper [9], the author proposes the blockchain based crowdfunding by using which the platform can give a private, secure and decentralized path for crowdfunding. The main objective of this paper is to let investors contribute to any project effectively by creating smart contracts through which the contributors can have a control over the invested money and also both the project creators and investors can effectively make and reserve funding for the project.

The method proposed in the article [4–9] is mainly concerned with the privacy protection of users and the supervision of the use of donated materials and funds, but can not prevent and identify false fundraising information, and there is the possibility that fundraisers and related organizations or individuals can jointly publish false information. At present, in the area of serious illness fundraising, people mainly rely on Water Drops or Easy Fundraising such third-party platforms to publish fundraising information for fundraising treatment, but this third-party platform sometimes occurs fundraising fraud, so many people lose trust in the third-party platform fundraising information, no longer donate to it.

Based on the technical characteristics of blockchain that cannot be tampered with, traceability, and decentralization, the application of blockchain technology to the field of public welfare donations can save the transaction costs and information disclosure costs, increase public's donation enthusiasm, and improve social welfare. The credibility of the organization can also solve the abuse of fundraising funds and fundraising fraud. Therefore, this article proposes a plan to apply blockchain and smart contract technology to the field of public welfare fundraising to help people in need raise funds for serious illnesses.

3 System Design and Implementation

3.1 Business Flow

The critical illness fundraising system proposed in this paper is a private chain distributed system composed of three parts: Helper, Hospital, and Donor based on the Ethereum environment. The business process of this program is shown in Fig. 1.

A. Helper: First, register an account on the platform and perform identity verification, and then submit the fund-raising application and diagnosis certificate, ID card information, description of illness, description of family situation and other supporting materials, and wait for the authenticity of the fundraising project by the hospital audit.

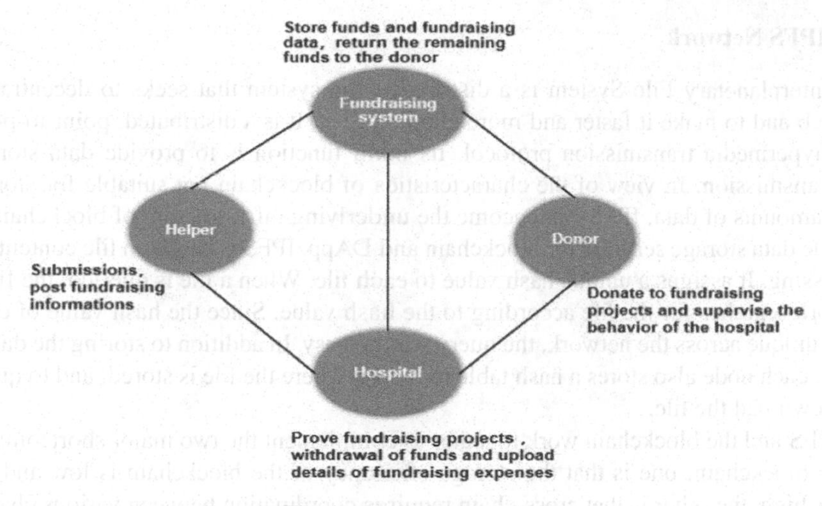

Fig. 1. Business flow chart.

B. Hospital: Verify the authenticity of the fundraising application and certification materials submitted by the seeker. If the fundraising project is authentic, it will be approved and fundraising can begin. If the authenticity certification materials of the fundraising project are missing, the review fails and the system informs the applicant to continue to submit additional certification materials. If the fundraising project is not true, the review fails and the fundraising application will be cancelled. In the course of disease treatment, the hospital has the right to withdraw funds from the smart contract account for payment of treatment costs, and upload the specific medication list and blockchain electronic invoices [10] to the system for donors and related organizations to view and supervise.
C. Donors: Register an account on the platform and perform identity verification. Select the fundraising project that has passed the authenticity verification on the platform to make donations. After donating, donors can check the use of funds at any time and supervise the blockchain invoices and medication list uploaded by the hospital to the platform.

The critical illness fundraising system proposed in this article consists of IPFS, smart contracts, and Ethereum virtual machine. The help seeker and the hospital upload the certification materials to the IPFS network, and then upload the hash value of the storage address of the mapping certification materials to the blockchain for storage. All funds of the fundraising project are temporarily stored in the account of the smart contract, and it is stipulated that the funds can only be withdrawn and used by the hospital, and the helper has no right to withdraw the funds. If the project funds are not used up due to the patient's sudden death or recovery and discharge, the remaining funds will be returned to the donor's account due to the traceability of the blockchain, the fundraising project will be terminated and no further donations will be accepted.

3.2 IPFS Network

The Interplanetary File System is a distributed file system that seeks to decentralize the web and to make it faster and more efficient [11]. It is a distributed, point-to-point new hypermedia transmission protocol. Its main function is to provide data storage and transmission. In view of the characteristics of blockchain not suitable for storing large amounts of data, IPFS can become the underlying infrastructure of blockchain to provide data storage services for blockchain and DApp. IPFS is based on file content for addressing. It assigns a unique hash value to each file. When a file is queried, the IPFS network searches for the file according to the hash value. Since the hash value of each file is unique across the network, the query will be easy. In addition to storing the data it needs, each node also stores a hash table to record where the file is stored, and to query and download the file.

IPFS and the blockchain work together to complement the two major shortcomings of the blockchain: one is that the storage efficiency of the blockchain is low and the cost is high, the other is that cross-chain requires coordination between various chains, which is difficult to coordinate. The help seeker and the hospital upload the relevant certification materials to the IPFS network. After the upload is completed, the user will receive the hash value from IPFS, and then can directly use the hash address to access the file, as shown in Fig. 2. Placing an unchanging and permanent IPFS link in the blockchain transaction, instead of putting the data itself in the blockchain, can solve the storage bottleneck problem of the blockchain itself and improve the performance of the blockchain.

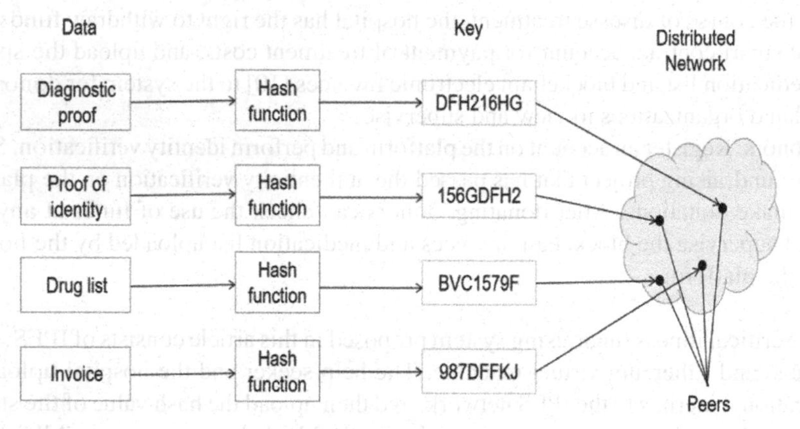

Fig. 2. How IPFS works.

3.3 Smart Contract

Smart Contracts on Ethereum are responsible for defining the logic of execution of the business process. One of the main characteristics of blockchain technology is immutability [12]. In this system, we use a smart contract to stipulate that the fundraising funds are stored in the account in the contract, and only the hospital is allowed to withdraw and use the fundraising funds, and the helper does not have the authority to withdraw funds. In this way, the earmarked funds can be used exclusively to prevent the abuse of funds. The transaction log of each withdrawal of funds by the hospital will also be uploaded to the blockchain for storage, so that subsequent donors and audit departments can monitor and audit the use of funds by the hospital. Due to the traceability of the blockchain, the unused funds in the contract will be returned to the donor's account after the fundraising project is terminated. The smart contract is written in solidity language, and the code of the donation part in the contract is shown in Fig. 3.

```
function _transfer(address _from, address _to, uint _value) internal {

    //The address to avoid transfers is 0x0
    require (_to != 0x0);

    //Check if the sender has enough balance
    require (balanceOf[_from] > _value);

    //Check for overflow
    require (balanceOf[_to] + _value > balanceOf[_to]);

    //Check frozen account
    require(!frozenAccount[_from]);
    require(!frozenAccount[_to]);

    //Subtract the amount sent from the sender
    balanceOf[_from] -= _value;

    //Add the same amount to the recipient
    balanceOf[_to] += _value;

    //Notify any clients that monitor the transaction
    Transfer(_from, _to, _value);

}
```

Fig. 3. The code of the donation part in the smart contract.

3.4 Blockchain Electronic Invoices

Blockchain invoices include not only traditional issuers and recipients, but also a range of agencies related to transactions and invoices, such as regulators, customs, industry and commerce, banks, third-party supply chains, making them a node on the blockchain. The tax authorities are introduced to implement the regulatory mechanism on the blockchain system, and the tax authorities uniformly formulate the operational standards and contract conditions of the blockchain, each node and entrant in the blockchain must participate and operate in accordance with pre-established trading rules [10].

In this system, the hospital has the tax digital certificate issued by the tax authority, and as the producer of the blockchain invoice, is responsible for the authenticity and validity of the invoice data. After the hospital draws funds from the smart contract, it will issue blockchain invoices for each expenditure and upload them to the system for tax authorities and donors to supervise the use of the funds, so as to make the fund-raising more open and transparent. In turn, improve the enthusiasm of the public to participate in donation activities.

3.5 Ethereum Virtual Machine

The Ethereum Virtual Machine provides an environment for all nodes to execute their own instructions. It converts a smart contract into a set of instructions to be executed by a node or a computer. For each transaction executed by the EVM, the related gas fee will be paid to the miner who validates and adds the transaction to the blockchain network.

3.6 Mist

Mist is the official online wallet management tool of Ethereum. Through Mist, we can easily connect to our private network to better develop, debug, and test our smart contracts. Mist needs the following components: Node.js, Meteor, Yarn, Electron, Gulp.

3.7 React

React is a declarative, efficient and flexible JavaScript library for building user interfaces. Using React, you can combine some short, independent code snippets into a complex UI interface. These code snippets are called "components".

3.8 Remix IDE

Remix IDE is an application, which provides the environment for scripting, compiling, debugging and executing the smart contract using Solidity language [13].

3.9 System Architecture

The blockchain platform as a whole can be divided into five layers: network layer, consensus layer, data layer, smart contract layer and application layer. According to the application scenarios of the public welfare field, this paper designs the system architecture as shown in Fig. 4.

The whole system is divided into four layers:

(1) Application service layer: account registration, donation, project release, project audit verification, capital use inquiry, fund withdrawal and return;
(2) Smart contract layer: query contract, donation contract, charging and returning contract;
(3) Blockchain service layer: transaction service, block service, distributed accounting service, data synchronization service;
(4) Storage layer: blockchain storage, local database storage, IPFS storage.

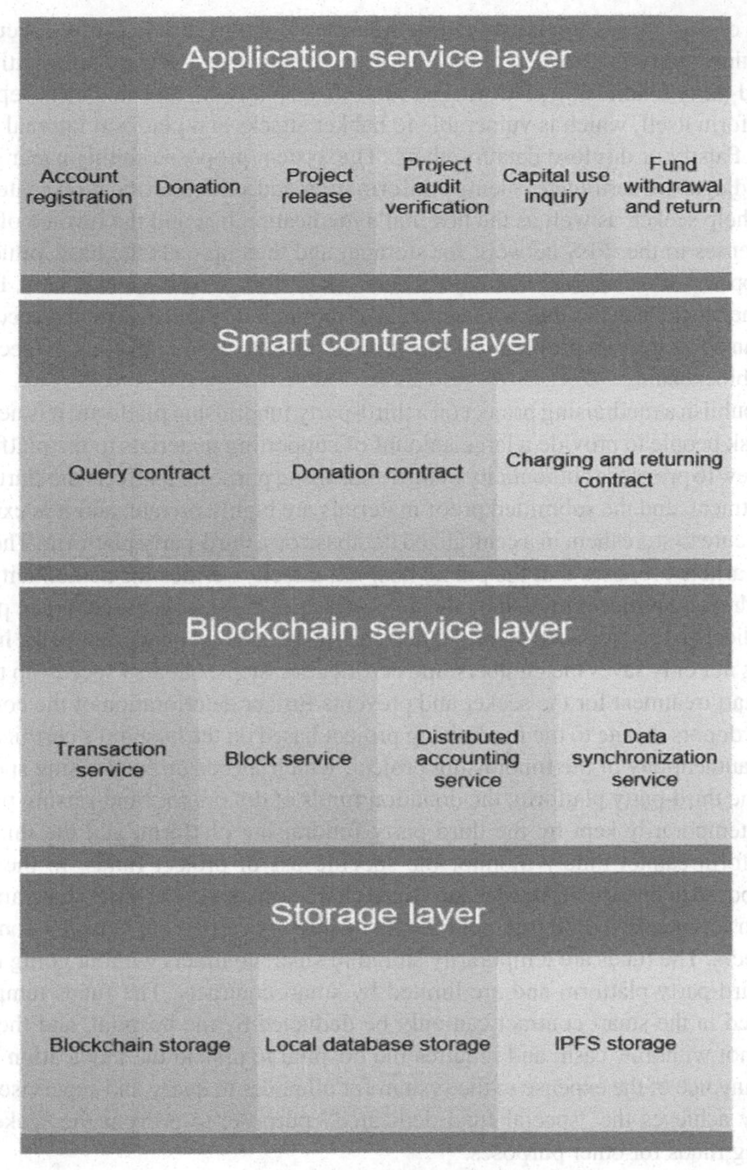

Fig. 4. System architecture diagram.

4 Evaluation

The critical illness fundraising system based on blockchain and smart contracts proposed in this article has more advantages than traditional third-party fundraising platforms, which are embodied in the following aspects:

(1) The critical illness fundraising system proposed in this article is more secure than the third-party fundraising platform for protecting users' private information. The third-party fundraising platform is a centralized platform, and all data is kept by the platform itself, which is vulnerable to hacker attacks and platform internal personnel. Falsify or disclose data to others. The system proposed in this paper uploads the diagnosis certificate, identity information and other certification materials of the help seeker, as well as the hospital's medication list, and the invoice of related expenses to the IPFS network for storage, and then uploads the hash value of the mapped storage address in the IPFS network to storage on the blockchain. Because of the blockchain has the characteristics of immutability and distributed accounting, it can solve the two problems of the storage bottleneck and information security of the blockchain.
(2) To publish a fundraising project on a third-party fundraising platform, it is necessary to ask people to provide a large amount of supporting materials to the platform for review to prove the authenticity of the fundraising purpose. It affects the start time of treatment, and the submitted proof materials are highly private, and it is extremely insecure to store them in a centralized database on a third-party platform. Therefore, this article proposes that the public hospital directly verifies the authenticity of the fundraising project initiated by the help-seeker in the system based on the patient's medical history, personal identification certificate and other materials in the hospital. This not only saves the cumbersome certification steps, but also speeds up the time to start treatment for the seeker and prevents further deterioration of the condition. The donors donate to the fundraising project based on the hospital's certification of the authenticity of the fundraising project, which speeds up fundraising speed.
(3) In the third-party platform, the donation funds of donors for fund-raising purposes are temporarily kept by the third-party fundraising platform, and the third-party platform cannot timely monitor the specific use of project funds. In the system proposed in this article, the decentralized characteristics of the blockchain and smart contracts are used. P2P transactions can be directly carried out between donors and helpers. The funds are temporarily stored in smart contracts without going through a third-party platform and are limited by smart contracts. The funds temporarily stored in the smart contract can only be deducted by the hospital, and the seeker cannot withdraw cash, and requires the hospital to upload the medication list and the invoice of the expense to the system for all nodes to query and supervise, which truly achieves the "special fund dedicated" purpose, to prevent the seeker from using funds for other purposes.
(4) Because of each generated block in the blockchain has a timestamp and a hash value that records the address information of the previous block, the source and destination of each fund can be traced. The unused funds of the project will be returned to the donor.

Compared with the methods proposed in the paper [4–9], they are mainly focus on the user's privacy protection, or the supervision of donated materials and funds. However, the methods proposed in this paper are mainly to solve the problem of fundraising fraud, because public hospitals are more convinced than third-party companies or individual organizations, make public hospitals attended by patients directly to prove the authenticity of fundraising behavior will be more convincing, and funds temporarily exist in intelligent contracts, only hospitals have the right to withdraw and use. Hospitals upload the details of the use of funds and blockchain invoices to blockchain for audit and oversight by the auditing department, which makes it easier to regulate the use of funds than to direct them to recipients or third-party organizations. This prevents helpers and third-party platforms or organizations from teaming up to post false fundraising information and reduce fundraising fraud.

5 Summary

The critical illness fundraising system based on blockchain and smart contracts proposed in this article uses the combination of blockchain's decentralization, immutability, distributed accounting, peer-to-peer transactions, traceability and smart contracts. It is more suitable for individuals to raise funds for serious illnesses. This system solves some of the problems common to several mainstream centralized third-party fundraising platforms. Generally speaking, this system is more convenient for people seeking help to raise funds for medical treatment. In the future, with the construction of smart cities and the popularization of citizens' personal credit points, it may also be considered to introduce personal credit points into this system. On the one hand, it can supervise the fundraising behavior of people seeking help and reduce the occurrence of fraud. On the other hand, it can improve the enthusiasm of citizens to participate in donation activities.

To sum up, the critical illness fundraising system based on blockchain and smart contracts proposed in this article is more convenient, safer and more reliable than centralized third-party fundraising platforms. Let us look forward to the vigorous development of blockchain technology in various fields in the future.

Funding Statement. This research was supported by NGII20180902.

Conflicts of Interest. The authors declare that they have no conflicts of interest to report regarding the present study.

References

1. Nakamoto, S.: Bitcoin: a peer-to-peer electronic cash system (2008). https://bitcoin.org/bitcoin.pdf
2. Minkovska, D., Ivanova, M., Rachev, M., Grosseck, G.: Cloud technologies for realization of an adaptive multimedia application. In: New Trends and Perspectives in Open Education International Conference Timisoara
3. Szabo, N.: Smart Contracts (1994). http://www.fon.hum.uva.nl/rob/Courses/InformationInSpeech/CDROM/Literature/LOTwinterschool2006/szabo.best.vwh.net/smart.contracts.html
4. Jeong, J., Kim, D., Lee, Y., Jung, J., Son, Y.: A study of private donation system based on blockchain for transparency and privacy. In: 2020 International Conference on Electronics, Information, and Communication (ICEIC), Barcelona, Spain, pp. 1–4 (2020). https://doi.org/10.1109/iceic49074.2020.9051328

5. Singh, A., Rajak, R., Mistry, H., Raut, P.: Aid, charity and donation tracking system using blockchain. In: 2020 4th International Conference on Trends in Electronics and Informatics (ICOEI) (48184), Tirunelveli, India, pp. 457–462 (2020). https://doi.org/10.1109/icoei48184.2020.9143001
6. Pandey, S., Goel, S., Bansla, S., Pandey, D.: Crowdfunding fraud prevention using blockchain. In: 2019 6th International Conference on Computing for Sustainable Global Development (INDIACom), New Delhi, India, pp. 1028–1034 (2019)
7. Zhong, P., Zhong, Q., Mi, H., Zhang, S., Xiang, Y.: Privacy-protected blockchain system. In: 2019 20th IEEE International Conference on Mobile Data Management (MDM), Hong Kong, Hong Kong, pp. 457–461 (2019). https://doi.org/10.1109/mdm.2019.000-2
8. Sirisha, N.S., Agarwal, T., Monde, R., Yadav, R., Hande, R.: Proposed solution for trackable donations using blockchain. In: 2019 International Conference on Nascent Technologies in Engineering (ICNTE), Navi Mumbai, India, pp. 1–5 (2019). https://doi.org/10.1109/icnte44896.2019.8946019
9. Yadav, N., Sarasvathi, V.: Venturing crowdfunding using smart contracts in blockchain. In: 2020 Third International Conference on Smart Systems and Inventive Technology (ICSSIT), Tirunelveli, India, pp. 192–197 (2020). https://doi.org/10.1109/icssit48917.2020.9214295
10. Liu, X.: Research and application of electronic invoice based on blockchain, p. 232 (2018)
11. Nyaletey, E., Parizi, R.M., Zhang, Q., Choo, K.R.: BlockIPFS - blockchain-enabled interplanetary file system for forensic and trusted data traceability. In: 2019 IEEE International Conference on Blockchain (Blockchain), Atlanta, GA, USA, pp. 18–25 (2019). https://doi.org/10.1109/blockchain.2019.00012
12. Kafeza, E., Ali, S.J., Kafeza, I., AlKatheeri, H.: Legal smart contracts in Ethereum Block chain: linking the dots. In: 2020 IEEE 36th International Conference on Data Engineering Workshops (ICDEW), Dallas, TX, USA, pp. 18–25 (2020). https://doi.org/10.1109/icdew49219.2020.00-12
13. Remix documentation. https://remix.readthedocs.io/en/latest/

Collaboration Energy Efficiency with Mobile Edge Computing for Data Collection in IoT

Guobin Zou[1,2,5,6], Jian Zhang[3](✉), Jian Tang[4], and Junwu Zhou[1,2,5,6]

[1] College of Information Science and Engineering, Northeastern University, Shenyang 110819, China

[2] State Key Laboratory of Process Automation in Mining & Metallurgy, Beijing 102600, China

[3] School of Computer and Software, Nanjing University of Information Science & Technology, Nanjing 210044, China

[4] Faculty of Information Technology, Beijing University of Technology, Beijing 100024, China

[5] Beijing Key Laboratory of Process Automation in Mining & Metallurgy, Beijing 102600, China

[6] BGRIMM Technology Group Co., Ltd., Beijing 102600, China

Abstract. In this paper, we investigated collaboration energy efficiency with mobile edge computing (MEC) mechanism in internet of things (IoT), which is a challenge issue. In order to prolong the lifetime of IoT, we adopt dynamic clustering methods to improve energy efficiency while guaranteeing energy balance. We deign the sensor selection scheme for collecting data according to Pareto optimality. Concretely, by considering the reality of random deployment and introducing the definition of node density for IoT, we develop a Pareto optimality for sensor nodes selection in terms of energy efficiency. Furthermore, we recruit voluntary mobile devices as mobile edge computing servers to offload the data from selected sensor nodes in the cluster and process them to the estimate the data state. Simulations demonstrate the efficiency for the performance on energy balance in terms of efficiently prolonging the IoT lifetime.

Keywords: IoT · Mobile edge computing · Pareto optimality · Data collection

1 Introduction

With the development of the micro-electro-mechanical system (MEMS) technology, internet of things (IoT) have been widely applied to many applications, such as building structure monitoring and control, data collection and localization, military defense and other fields [1, 2], where a large number of low cost and spatially dispersed position sensors are densely deployed as a general paradigm for information collection, transmission and processing data. Energy efficiency in IoT is one of the most important issues and often needs coverage of broad areas. However, the resources like bandwidth and energy are actually restricted, which has been investigated via the scheduled schemes of sensors to realize the goal of saving energy task while satisfying advanced performance of IoT.

The sensor selection problem has been developed to get the desired information gain or reduction in estimation error in [3–5]. Authors in [6] proposed distributed algorithms

X. Sun et al. (Eds.): ICAIS 2021, CCIS 1422, pp. 279–285, 2021.
https://doi.org/10.1007/978-3-030-78615-1_24

to select cluster members to gather data and formulated an optimization for data collection to maximize utilization of data quality through a rendezvous-based data collection algorithm which not only integrates above positive factors to track a target but also maintains WSNs' functions. In [7], the sensor selection problem with and without sensing range is demonstrated by utilizing a fixed area and a circle drawn with the help of communication range to select sensor nodes for two cases, respectively. In order to improve the performance of energy efficiency, the quantization technique has been adopted to decrease the transmission rates for energy saving [8], which would lead to additional errors on the available measurements in terms of quantization effects [9], which may degrade the filtering/estimation performance.

The problem of efficacy and effectiveness for saving and balancing energy consumption via collaborative mobile nodes is studied in [10], where the machine learning algorithm, namely MU-MAB, is employed to solve energy consumption problem, and the stable matching theory based on marginal utility for the allocation of the final one-to-one optimal combinations is used to achieve energy efficiency. Generally, mobility can improve the performance of IoT [11], which characterizes three node sets of the cluster as a theoretical foundation to enhance high performance of WSNs, and propose optimal solutions by introducing rendezvous and Mobile Elements (MEs) to optimize energy consumption for prolonging the lifetime of WSNs. Meanwhile, MEs can be used as mobile edge computing server to collect data from WSNs and process them for data collection due to their capacity of computing, storage, rich energy and so on. Considering offloading this type of deep learning (DL) tasks to a mobile edge computing (MEC) server, Bo Yang [12] proposed an optimization problem to minimize the weighted-sum cost including the tracking delay and energy consumption introduced by communication and computing of the UAVs, while taking into account the quality of data input to the DL model and the inference errors. In [13], the IoT devices can offload the intensive computing tasks to edge computing servers, while saving their own computing resources and reducing energy consumption. In [14], mobile cloud computing as an emerging and prospective computing paradigm, can significantly enhance computation capability and save energy for smart mobile devices (SMDs) by offloading computation-intensive tasks from resource constrained SMDs onto resource-rich cloud.

The main contribution of this paper is that we proposed a Pareto optimality for sensor nodes selection in terms of energy efficiency. Furthermore, we recruit voluntary mobile devices as mobile edge computing servers to collect the data from selected sensor nodes in the cluster. Unlike the most of previous data collection method, it is shown that the proposed algorithm exhibits excellent performances in terms of energy efficiency.

The layout of the paper is as follows: In Sect. 2, problem statements are described, including clustering. Section 3 is Energy statements. Quantization and sensor selection is demonstrated in Sects. 4 and 5. In Sect. 6, collaboration of sensor selection and data quantization scheme is given. Simulation results are discussed. We conclude the paper in Sect. 8.

2 Problem Statements

In order to improve the energy efficiency, we configure the cluster in grid for state estimation of certain data requirement, which is shown in Fig. 1. However, due to random deployment of sensor nodes in WSNs, the node density of the cluster is a random variable that could affect the performance of data collection. In order to balance the energy consumption of local or global WSNs, the sensor node selecting scheme is significant to improve the energy efficiency, which will be illustrated in the follow section.

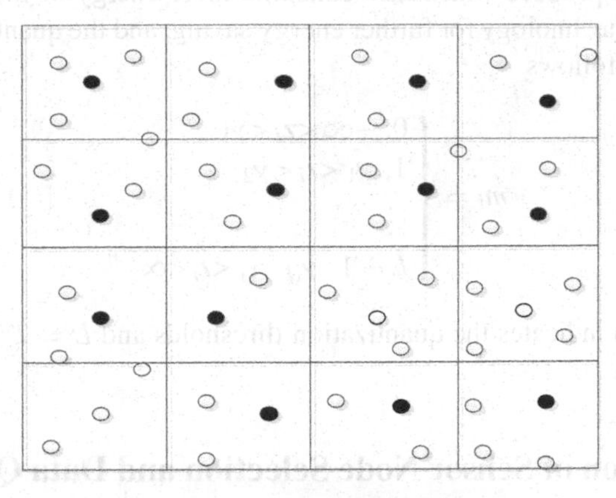

Fig. 1. Clustering for IoT

3 Energy Statements

The cost is expressed as energy consumption, and energy model in [11] is shown as follows. For the sake of convenience, we consider energy consumption at the cost of a b-bit message and a distance d between sensor s_i as a transmitter and sensor s_j as a receiver. The energy consumed in transmitting by the sensor s_i is

$$E_{Tx}(s_i, s_j) = b \cdot E_{Telec} + b \cdot \varepsilon_{amp} \cdot d_{i,j}^{\alpha} \tag{1}$$

and to receive this message, the radio expends:

$$E_{Rx}(s_j) = b \cdot E_{Relec} \tag{2}$$

In addition, the receiver s_j also spends energy in sensing or processing. The radio expends:

$$E_{Sx}(s_j) = b \cdot E_{Selec} \tag{3}$$

where E_{Telec} and ε_{amp} are determined by transmitter s_i, $d_{i,j}$ indicates the distance between sensor s_i and sensor s_j. E_{Relec} and E_{Selec} are decided by receiver s_j. α depends on the channel characteristics and is known as $\alpha = 2$.

Therefore, the total energy consumption between senor s_i and sensor s_j is demonstrated by

$$\begin{aligned} E(s_i, s_j) &= E_{Tx}(s_i, s_j) + E_{Rx}(s_j) + E_{Sx}(s_j)r \\ &= b \cdot E_{Relec} + b \cdot E_{Selec} + b \cdot E_{Telec} + b \cdot \varepsilon_{amp} \cdot d_{i,j}^2 \\ &= E_0 + e_0 d_{i,j}^2 \end{aligned} \tag{4}$$

4 Quantization

Considering the sensor node density, we deal with a more complicated scenario which includes redundant sensors. Although we introduce the wake-up mechanism, redundant information maybe produce which also consume more energy of IoT. As a result, we adopt quantization technology for further energy saving, and the quantization for sensor node i is shown as follows

$$m_i = \begin{cases} 0, -\infty < z_i < \gamma_1 \\ 1, \gamma_1 < z_i < \gamma_2 \\ :: \\ L-1, \gamma_{(L-1)} < z_i < \infty \end{cases} \tag{5}$$

where γ_i in Eq. (5) indicates the quantization thresholds and $L = 2^m$ is the number of quantization levels.

5 Collaboration of Sensor Node Selection and Data Quantization

In this section, we consider the optimality of sensor selection for energy efficiency in the process of data collection.

5.1 Pareto Optimality

We utilize Pareto optimality to select sensor nodes for data collection in terms of a given certain accuracy. In the grid cluster including a sensor set $S = \{s_1, s_2, \cdots, s_n\}$, we supposed that there exist a set $S' = \{s'_1, s'_2, \cdots, s'_r\}$ to accomplish the tracking task, where $S' \subseteq S$ holds. However, this is a permutation and combination problem and it is hard to solve this problem when the sensors are densely deployed in the area. Naturally, the permutation and combination problem of sensor selection for data collection can be evolved into the constraint optimization problem as follows.

$$\min \sum_{i=1}^{r} (E_{s'_i} - \bar{E})^2$$
$$s.t. \Phi_{S'} \le \Phi_0 \tag{6}$$

where $E_{s'_i}$ is the residual energy of sensor s'_i in set S', $\bar{E}$ denotes the average energy of the cluster, $\Phi_{S'}$ is the actual data accuracy obtained by the sensor combination of set S', and Φ_0 denotes the given data accuracy. According to expression (16), we design a Pareto optimality mechanism to get the approximate sensor combination set.

Step 1: arbitrarily select a sensor with a certain probability, and get $\{s'_1\}$;

Step 2: select another sensor in the set $S \backslash \{s'_1\}$ and get $\{s'_1, s'_2\}$. According to the concept of Pareto optimality, increasing sensor s'_2 is not reduce the revenues of s'_1;

Step 3: continue to select a set $S' = \{s'_1, s'_2, \cdots, s'_r\}$ until expression (16) is satisfied.

5.2 Collaboration of Sensor Selection and Data Quantization

Considering the energy consumption of data transmission, we collaborate the sensor selection and data quantization for energy efficiency in the data collection process. In this scenario, we deal with data collection of sensors being randomly deploy in the interest area. Actually, due to the approximate solution in terms of Sect. 5.1, we demonstrate four cases for collaboration of sensor selection and data quantization according to the density of the cluster, which are shown in Table 1.

Table 1. Collaboration of sensor selection and data quantization

	Quantization	No quantization
Selection	(1,1)	(1,0)
No selection	(0,1)	(0,0)

Case 1: for energy efficiency, we not only select sensors but also quantize the data, that is (1,1) in Table 1. In this scenario, the density of the cluster is higher.

Case 2: (1,0) in Table 1, that is, we plan to select sensors while not quantize the data;

Case 3: (0,1) denotes we are about to quantize the data of sensing data while not select sensors for data collection.

Case 4: (0,0) stands for neither selection nor quantization. In this scenario, the density of the cluster is lower.

6 MEC Server Placement

In the process of data collection, MEC server play roles of collecting data, processing data and transmitting data.

For further reducing the energy consumption of the cluster, we set the optimal position of MEC server according to Eq. (1). After selecting the sensing nodes, the position of each node is confirmed. Therefore, we can find a point to place the MEC server to minimize the expression (6). So we can get the following theorem.

Theorem 1. After selecting sensors, there exist a point in cluster for MEC server to make the energy consumption least.

Proof according to Eq. (1), it is easy to get the conclusion, so we omit it here.

7 Simulation Results

To illustrate advantages of the new method, some simulation results are discussed and some assumptions for simulations here. To illustrate advantages of the new method, some simulation results are discussed and some assumptions for simulations here. In these simulations, the number of sensors N = 400, sensor radius $R = 60$ m.They are randomly deployed in a square field 100 * 100 m^2. Simulation parameters about energy consumption model are listed in Table 2.

Table 2. Simulation parameters

Parameter	Value	Parameter	Value
α	2	E_0	0.5 J
ε	10 J/bit/m^2	N	400
d	60	E_{elec}	50 nJ/bit
E_{amp}	0.0013 PJ/bit m^4		

In Fig. 2, collaboration of sensor selection and data quantization as shown in Table 1 are considered with the number of sensors N = 400. Seen from Fig. 2, energy consumption changes with different scenarios. Especially, the scenario of (1,1) gets the best performance of IoT due to dense sensor for rich selection and data quantization. The scenario of (0,0) gets the worse performance of IoT when no selection and quantization. Apparently, this scenario consumes many times energy than the scenario of (1,1). Nevertheless, although different collaboration of sensor selection and data quantization plays the role of reducing different energy consumption, they can keep the performance of energy balance for IoT.

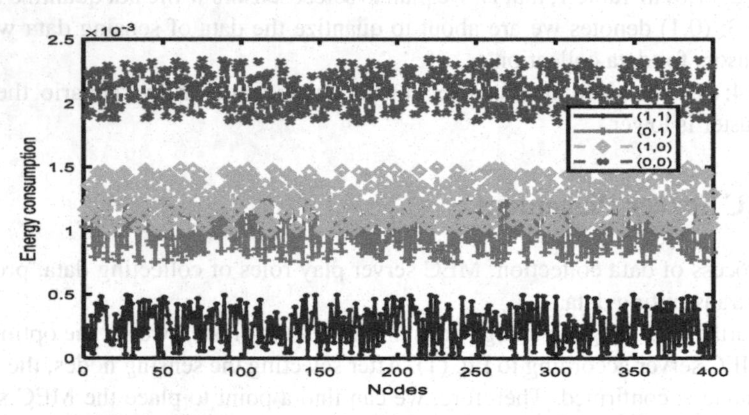

Fig. 2. Energy consumption with 50 rounds and N = 400

8 Conclusions

In this paper, dynamic clustering methods is adopted to improve energy efficiency while guaranteeing energy consumption balance. We deign the sensor selection scheme for carrying out the data collection tasks according to energy distribution of sensor nodes. Concretely, by considering the reality of random deployment and introducing the definition of node density for IoT, we develop a Pareto optimality for sensor nodes selection in terms of energy efficiency. Furthermore, we recruit voluntary mobile devices as mobile edge computing servers to collect the data from selected sensor nodes in the cluster, offloading and processing them to achieve energy efficiency.

Acknowledgments. This work is supported by the State Key Laboratory of Process Automation in Mining & Metallurgy and Beijing Key Laboratory of Process Automation in Mining & Metallurgy (BGRIMM-KZSKL-2020-02), and National Key Research and Development Project(2020YFE0201100,2019YFE0105000).

References

1. Akyildiz, F., Su, W., Sankarasubramaniam, Y., Cayirci, E.: Wireless sensor networks: a survey. Comput. Netw. **38**(4), 393–442 (2002)
2. Gezici, S., et al.: Localization via ultra-wideband radios: a look at positioning aspects for future sensor networks. IEEE Signal Process. Mag. **22**(4), 70–84 (2005)
3. Liu, S., Kar, S., Fardad, M., Varshney, P.K.: Sparsity-aware sensor collaboration for linear coherent estimation. IEEE Trans. Signal Process. **63**(10), 2582–2596 (2015)
4. Liu, S., Fardad, M., Masazade, E., Varshney, P.: Optimal periodic sensor scheduling in networks of dynamical systems. IEEE Trans. Signal Process. **62**(12), 3055–3068 (2014)
5. Cao, N., Sora, C., Engin, M., Pramod, K.V.: Sensor selection for target tracking in wireless sensor networks with uncertainty. IEEE Trans. Signal Process. **64**(20), 5191–5204 (2015)
6. Zhang, J., Wang, T., Tang, J.: Optimising rendezvous-based data collection for target tracking in WSNs with mobile elements. Int. J. Sensor Networks **30**(4), 218–230 (2019)
7. Liu, H.Q., So, H.C., Lui, K.W.K., Chan, F.K.W.: Sensor selection for target tracking in sensor networks. Progress Electromag. Res. **95**, 267–282 (2009)
8. Zhang, D., Yu, L., Zhang, W.A.: Energy efficient distributed filtering for a class of nonlinear systems in sensor networks. IEEE Sens. J. **15**(5), 3026–3036 (2015)
9. Hu, L., Wang, Z., Liu, X.: Dynamic state estimation of power system with quantization effects: arecursive filter approach. IEEE Trans. Neural Netw. Learn. Syst. **27**(8), 1604–1614 (2016)
10. Zhang, J., Tang, J., Wang, F.: Cooperative relay selection for load balancing with mobility in hierarchical WSNs: a multi-armed bandit approach. IEEE Access **8**, 18110–18122 (2020)
11. Zhang, J., Tang, J., Wang, Z.H., Wang, F., Yu, G.: Load-balancing rendezvous approach for mobility-enabled adaptive energy-efficient data collection in WSNs. KSII Trans. Internet Inf. Syst. **14**(3), 1204–1227 (2020)
12. Yang, B., Cao, X., Yuen, C., Qian, L.: Offloading optimization in edge computing for deep learning enabled target tracking by internet-of-UAVs. IEEE Internet Things J. https://doi.org/10.1109/jiot.2020.3016694
13. Cui, L., et al.: Joint optimization of energy consumption and latency in mobile edge computing for internet of things. IEEE Internet Things J. **6**(3), 4791–4803 (2019)
14. Guo, S., Liu, J., Yang, Y., Xiao, X., Li, Z.: Energy-efficient dynamic computation offloading and cooperative task scheduling in mobile cloud computing. IEEE Trans. Mobile Comput. **18**(2), 319–333 (2019)

Improving the Transformer Translation Model with Back-Translation

Hailiang Wang[1], Peng Jin[2], Jinrong Hu[1], Lujin Li[1], and Xingyuan Chen[2](✉)

[1] Department of Computer Science, Chengdu University of Information Technology, Chengdu 610225, China

[2] School of Artificial Intelligence, Leshan Normal University, Leshan 614000, China

Abstract. Back-translation has been proved to help improve the translation quality of statistical machine translation (SMT) systems and neural machine translation (NMT) systems. But in the previous work, Back-translation needed to use additional monolingual data to improve the quality of machine translation, translation models in these works are still independent. In this paper, we propose a novel approach using back-translation to improve the quality of NMT without additional monolingual data, and apply our approach to the Transformer model, which is the best NMT model in recent years. The specific method is to extend the Transformer model by using the same structure as Transformer model. The extended part is used for the backward translation of the predictive sentences obtained from forward translation to fit the source statement in each step of the training process. By optimizing both forward and back translation losses, the same effect can be achieved without the use of additional monolingual data. Experiments on the IWSLT2016 German-English datasets and show IWSLT2012 French-English datasets that our approach improves over Transformer significantly.

Keywords: Back-translation · Neural machine translation · Transformer

1 Introduction

With the success of deep learning in the field of image processing, this technology has also begun to be applied in the field of natural language processing. Neural machine translation (NMT) has developed rapidly in the past few years [1, 2]. Compared with traditional statistical machine translation (SMT), NMT directly regards the process of machine translation as an end-to-end process, and adopts the idea of sequence transformation to model the translation process. This is a new translation architecture that automatically learns hundreds of millions of parameters and makes it easier to use contextual information in the translation process. Nowadays, NMT has been considered as a new factual method and has been widely used in business machine translation systems. A variety of NMT models of different structures have been proposed such as RNNencdec [1–3], and Transformer [4]. A large number of experiments have shown that the ability of NMT in many languages greatly exceeds the traditional SMT models, among which the Transformer model has achieved the best performance.

X. Sun et al. (Eds.): ICAIS 2021, CCIS 1422, pp. 286–294, 2021.
https://doi.org/10.1007/978-3-030-78615-1_25

In recent years, the Transformer model has attracted much attention due to its excellent translation performance. However, reaching the level of human translation is still a huge challenge. Previous studies have shown that NMT models are likely to face the illness of inadequate translation [5], which is usually embodied in over- and under-translation problems [6]. Among them, over-translation means that some words will be repeatedly translated in the process of translation, under-translation means that some words may be omitted in translation. This issue may be attributed to the poor ability of NMT of recognizing the dynamic translated and untranslated contents. [6] proposes to add the coverage method in the SMT strategy to the traditional NMT model to track and judge whether the original sentence has been translated. [7] first demonstrate that explicitly tracking PAST and FUTURE contents helps NMT models alleviate this issue and generate better translation. But these methods are concentrated on the RNN model, and we cannot apply them directly into Transformer.

In our paper, combined the idea of back-translation and Transformer [4], we propose a model that can correct the over-and under-translation problems in the training process. The basic idea is to use translation, calculate loss from back translation x' and source sentence x, and then optimize the total loss after adding the loss obtained from forward translation module and back translation module. With such a structure, we can make the sentence y' in forward translation contain as much information as possible in the source statement x and improve the overall alignment between the source sentence and target sentence, so as to effectively alleviate the problem of inadequate translation.

Experiments show that our approach achieves an improvement of 1.01 and 3.29 BLEU points over Transformer on German-English and French-English translation. Compared with the Transformer model [4], our approach achieves substantial improvements.

2 Background and Related Work

2.1 Neural Machine Translation

Notation. Given a pair of sentences from source and target languages, e.g., ⟨x, y⟩, we denote x as a sentence of the "source" language, and y as a sentence of the "target" language. Additionally, we use the terms "source-side" and "target-side" of a translation direction to denote the input and the output sides of it, e.g., the source-side of the "tgt2src" translation is the target language.

We build upon recent work on neural machine translation which is typically a neural network with an encoder/decoder architecture. The encoder infers a continuous space representation of the source sentence, while the decoder is a neural language model conditioned on the encoder output. The parameters of both models are learned jointly to maximize the likelihood of the target sentences given the corresponding source sentences from a parallel corpus [8, 9]. At inference, a target sentence is generated by left-to-right decoding.NMT models aim to approximate the conditional distribution $\log p(y|x;\theta_{xy})$ over a target sentence $y = \langle y_1, \ldots, y_{L_y} \rangle$ given a source sentence $x = \langle x_1, \ldots, x_{L_x} \rangle$. Training criterion for a discriminative NMT model is to maximize the conditional log-likelihood $\log p(y|x;\theta_{xy})$ on abundant parallel bilingual data $\mathcal{D}_{xy} = \{x^{(n)}, y^{(n)} | n = 1 \ldots N\}$ of i.i.d observations.

Different neural architectures have been proposed with the goal of improving efficiency and/or effectiveness. This includes recurrent networks [2, 9], convolutional networks [3, 10] and transformer networks [4]. Recent work relies on attention mechanisms where the encoder produces a sequence of vectors and, for each target token, the decoder attends to the most relevant part of the source through a context-dependent weighted-sum of the encoder vectors [2, 9]. Attention has been refined with multi-hop attention [3], self-attention [4] and multi-head attention [4]. We use the Transformer architecture [4], which is shown in Fig. 1.

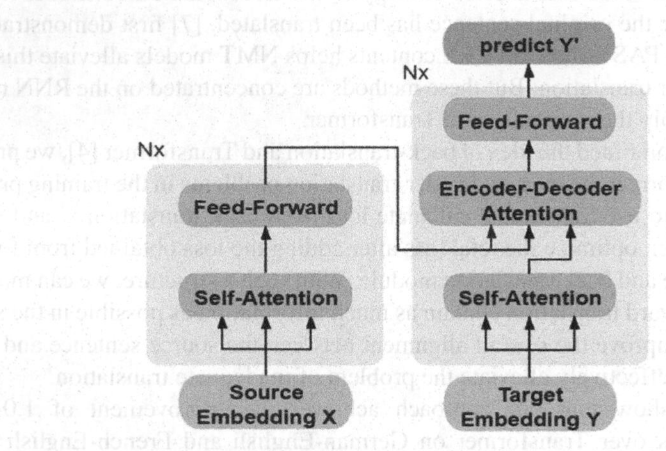

Fig. 1. The original Transformer translation model [4]

2.2 Back-Translation

Back-translation (BT) [11] is the most widely used approach for exploiting monolingual data. It requires training a target-to-source system in order to generate additional synthetic parallel data from the monolingual target data. Specifically, as shown in the Fig. 2, given parallel corpus pairs $\langle x, y \rangle$ (for source and target languages) and monolingual data z (for target language), First, train a model from target to source using parallel corpus $\langle x, y \rangle$, and BT utilizes z by applying tgt2src translation model ($TM_{y \to x}$) to get predicted translations x'. Thus, an extended parallel corpus pair $\langle x', z \rangle$ (for source and target languages) is obtained. This data complements human bitext to train the desired source-to-target system. BT has been applied earlier to phrasebase systems [12]. For these systems, BT has also been successful in leveraging monolingual data for domain adaptation [13, 14]. Recently, BT has been shown beneficial for NMT [15, 16]. It has been found to be particularly useful when parallel data is scarce [17]. In order to utilize non-Parallel Bilingual data, BT and its variants [11, 18] exploit non-parallel bilingual data by generating synthetic parallel data. Dual learning [19, 20] learns from non-parallel data in a round-trip game via reinforcement learning, with the help of pretrained language models.

However, in previous work, BT is mainly used to augment parallel bilingual data with non-parallel bilingual data. The BT method is not integrated into the machine translation

Fig. 2. The graphical model of BT

model. Some related work like joint BT [18] and dual learning [19] introduce iterative training to make $TM_{y \to x}$ and $TM_{x \to y}$ benefit from each other implicitly and iteratively. But translation models in these approaches are still independent.

3 Approach

3.1 Bi-Directional Translation Method

Motivation. We believe that if the predictive sentence y′ obtained after forward translation can know what information is added or missing from the source statement x during the training process, then forward translation can benefit from the dynamic overall context. To this purpose, We should design a mechanism to capture the defect in statement y′ to help forward prediction produce more sufficient information and correct sentences.

The bi-directional translation method based on deep learning Transformer model proposed in this paper is shown in Fig. 3. It consists mainly of two parts, and the structure of the two parts is actually the same, both using the transformer architecture [4]. The difference is that the first part is used for forward translation, then the results are passed to the second part for reverse translation, and the final optimization is combined with the loss gained by the two. In this process, the most important transformer model provides excellent parallel computing power [4] and the idea of optimizing with BT.

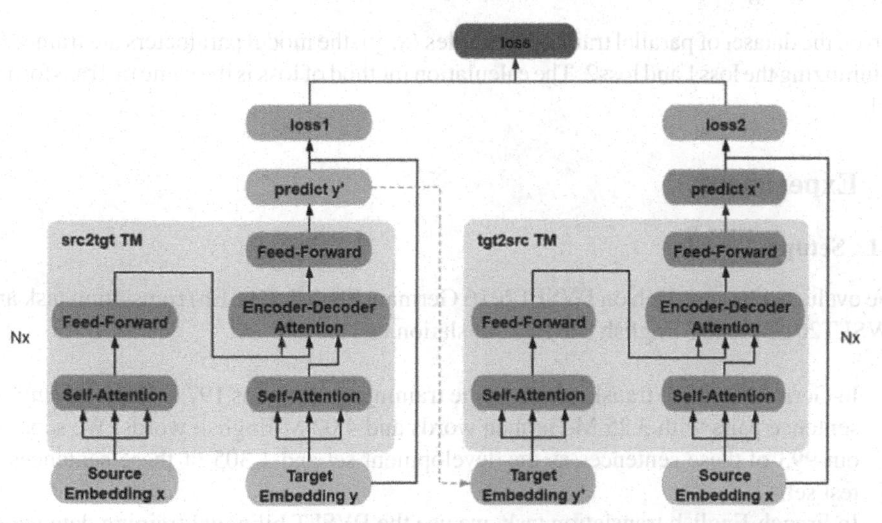

Fig. 3. The extended Transformer translation model. On the left is the original Transformer and on the right is what we introduced.

On the one hand, the Transformer model has demonstrated its ability to deal with machine translation problems, its attention mechanism [4] and positional embedding algorithm [4] make machine translation no longer limited by sentence length, which effectively improves the quality of translation. On the other hand, the related work based on BT also proves that BT is helpful to improve the quality of machine translation. Therefore, it was natural for us to think about combining the Transformer model with the BT approach.

3.2 Integrate the Transformer and BT

The way we've combined Transformer with the BT approach is different from what people have done before, which is to use the BT approach to turn monolingual data into parallel data and then improve the quality by expanding the corpus. We have added a Transformer module to Transformer, the first Transformer is used to train the source-to-target translation model (scr2tgt TM), the second Transformer is used to train the target-to-source translation model and is also used for BT (tgt2scr MT). We form a new parallel corpus pair $\langle y', x \rangle$ with the prediction statement y' generated in the training process of scr2tgt TM and the source sentence x, and then put $\langle y', x \rangle$ into tgt2scr TM for training at the same time. Loss1 generated by scr2tgt TM and loss2 generated by tgt2scr TM were optimized together to achieve higher quality y'. In this way, BT and the Transformer are combined. Scr2tgt TM and tgt2scr TM have the same structure as Transformer. For implementation details of Transformer, please refer to [4]. Note that we follow [4] to use residual connection and layer normalization in each sub-layer, which are omitted in the presentation for simplicity.

3.3 Training

Given the dataset of parallel training examples $\langle x, y \rangle$, the model parameters are trained by minimizing the loss1 and loss2. The calculation method of loss is the same in Transformer [4].

4 Experiments

4.1 Setup

We evaluate our approach on IWSLT2016 German-English (Ge-En) translation task and IWSLT2012 French-English (Fr-En) translation tasks:

1. In German-English translation task, the training set contains 197 K German-English sentence pairs with 3.25 M German words and 4.02 M English words. We separate out 993 of these sentences as the development set and 1,305 of these sentences as test set.
2. In French-English translation task, we use the IWSLT bilingual training data which contains 1,024 documents with 138 K sentence pairs as training set. For development and testing, we separate out 1,000 of these sentences as the development set and 1,000 of these sentences as test set.

In preprocessing, we use byte pair encoding [15] with 32 K merges to segment words into sub-word units for all languages, and source languages share the same vocabulary as target languages, and mapped all the out-of-vocabulary words in the parallel corpus to a special token UNK. For the original Transformer model and our extended model, the hidden size is set to 512 and the filter size is set to 2,048. The multi-head attention has 8 individual attention heads. We set N = Ns = Nt = 6. In training, we use Adam [21] for optimization. Each mini-batch contains approximately 2 K words. We use the learning rate decay policy described by [4]. We use two TITAN RTX GPUs for training. We implement our approach on top of the open-source code, which is obtained from GitHub.

4.2 Algorithm Evaluation Indicators

The evaluation indicator we used was BLEU(Bilingual Evaluation Understudy). BLEU is an algorithm for evaluating the quality of text which has been machine-translated from one natural language to another. Quality is considered to be the correspondence between a machine's output and that of a human: "the closer a machine translation is to a professional human translation, the better it is", this is the central idea behind BLEU. BLEU was one of the first metrics to achieve a high correlation with human judgements of quality, and remains one of the most popular automated and inexpensive metric. The evaluation metric for both tasks is case-insensitive BLEU score as calculated by the multi-bleu.perl script.

4.3 Comparison with Previous Work

We compare our approach with the Transformer [4], the state-of-the-art results of the Transformer model that cannot be duplicated using open-source code. We only compare the results repeated using open-source code. We use the same dataset to train Transformer and our model.

As shown in Table 1, using the same data, our approach achieves significant improvements over the original Transformer model [4]. In German-English translation task, our model outperforms Transformer by 1.01 BLEU points, and our model also outperforms Transformer by 3.29 BLEU points on French-English translation task. It should be noted that the improvement in BLEU scores in our model for German-English translation was much smaller than that for French-English translation. The underlying reasons lie in the following two aspects. First, the German-English dataset is 1.4 times larger than the French-English dataset, the amount of data has a big impact on performance for NMT. Second, compared with German, French is more distantly related to English, leading to the predominant advantage of utilizing target-side contexts in French-English translation. Since our approach mainly utilizes the predictive target sentences generated by forward translation in each step of training, it does not change the structure of the forward translation model itself. We believe that our work can be easily applied to other architectures in the same way.

Table 1. Comparison with Transformer on German-English translation task and French-English translation task. The evaluation metric is case insensitive BLEU score.

Method	Model	IWSLT2016 Ge-En		IWSLT2012 Fr-En	
		BLEU	ΔBLEU	BLEU	ΔBLEU
[4]	Transformer	28.06	–	22.64	–
This work	Bi-Transformer(ours)	29.07	+1.01	25.93	+3.29

4.4 Effect of Different Proportions of Loss

We combine the forward and back losses of the model in different proportions, can control the proportion of positive feedback provided to the whole model by the added reverse translation module, so as to obtain different degrees of improvement. As shown in Table 2, when the proportion of back loss is 0.8, the promotion obtained is much higher than that obtained when the proportion of reverse loss is 0.2. The performance of the model is positively correlated with the ratio of back loss to total loss. This shows that the BT module we added is extremely effective.

Table 2. Compare the effects of setting different proportion of loss on the performance of Bi-Transformer. The evaluation metric is case insensitive BLEU score.

Method	Model	Loss ratio	IWSLT2012 Fr-En
			BLEU
This work	Bi-Transformer(ours)	$0.2 * \text{Loss1} + 0.8 * \text{Loss2}$	25.93
This work	Bi-Transformer(ours)	$0.5 * \text{Loss1} + 0.5 * \text{Loss2}$	25.50
This work	Bi-Transformer(ours)	$0.8 * \text{Loss1} + 0.2 * \text{Loss2}$	25.33

4.5 Impact of Data Size

We compared the performance of Transformer and Bi-Transformer(ours) with different sample sizes, for fixed model and training procedure. We train different models using {25%, 50%, 75%, 100%} amount of data to mimic the "low-resource", "medium-resource" and "high-resource" regime. Figure 4 shows that the BLEU values of Transformer and Bi-Transformer at different percentages of the training data. As shown in Fig. 4, the performance of Bi-Transformer(Ours) is always better than that of Transformer with the same amount of training data. This suggests that our approach can still achieve the same improvement with less data.

4.6 Efficiency

To examine the efficiency of the proposed approach, we also list the relative speed of both training and testing. Our approach took twice as long to run one epoch as the

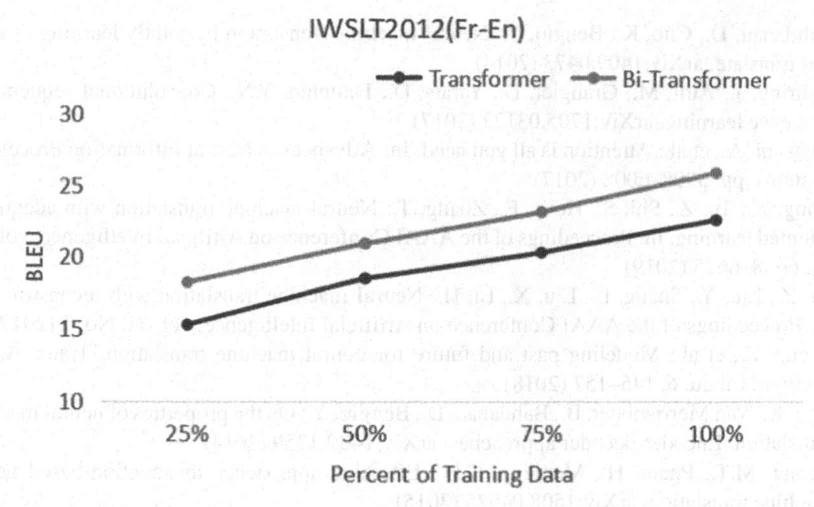

Fig. 4. BLEU Values of Transformer, Bi-Transformer trained on different amount of data

Transformer baseline in training phase. The reason is that the parameters of our model have doubled. however, it does not hurt the speed of testing too much. It is because only forward translations are performed during the test phase, the degradation of the testing efficiency is nonexistent.

Although our approach ran one epoch more slowly than the Transformer model, it converges faster than the Transformer model. It takes the original Transformer model about 9.2 h to converge during training. In contrast, it takes our model about 7.6 h to converge during training. This also shows that the added modules have improved translation performance.

5 Conclusion and Future Work

We have presented a method for exploiting BT information inside the state-of-the-art neural translation model Transformer. Experiments on German-English and French-English translation tasks show that our method is able to improve over Transformer significantly. At the same time, In future work, we plan to further validate the effectiveness of our approach on more language pairs.

Funding Statement. This work is supported in part by the Introduction Talent Research Start-up Program of Chengdu University of Information Technology (No. KYTZ202008), and in part by the Sichuan Science and Technology Program (2019YFH0085, 2019ZDZX0005, 2019YFG0196), and in part by the Science and Technology Project Affiliated to the Education Department of Chongqing Municipality (No. 19ZB0257).

References

1. Sutskever, I., Vinyals, O., Le, Q.V.: Sequence to sequence learning with neural networks. Adv. Neural. Inf. Process. Syst. **27**, 3104–3112 (2014)

2. Bahdanau, D., Cho, K., Bengio, Y.: Neural machine translation by jointly learning to align and translate. arXiv:1409.0473 (2014)
3. Gehring, J., Auli, M., Grangier, D., Yarats, D., Dauphin, Y.N.: Convolutional sequence to sequence learning. arXiv:1705.03122 (2017)
4. Vaswani, A., et al.: Attention is all you need. In: Advances in Neural Information Processing Systems, pp. 5998–6008 (2017)
5. Kong, X., Tu, Z., Shi, S., Hovy, E., Zhang, T.: Neural machine translation with adequacy-oriented learning. In: Proceedings of the AAAI Conference on Artificial Intelligence, vol. 33, pp. 6618–6625 (2019)
6. Tu, Z., Liu, Y., Shang, L., Liu, X., Li, H.: Neural machine translation with reconstruction. In: Proceedings of the AAAI Conference on Artificial Intelligence, vol. 31, No. 1 (2017)
7. Zheng, Z., et al.: Modeling past and future for neural machine translation. Trans. Assoc. Comput. Lingu. **6**, 145–157 (2018)
8. Cho, K., Van Merriënboer, B., Bahdanau, D., Bengio, Y.: On the properties of neural machine translation: Encoder-decoder approaches. arXiv:1409.1259 (2014)
9. Luong, M.T., Pham, H., Manning, C.D.: Effective approaches to attention-based neural machine translation. arXiv:1508.04025 (2015)
10. Kaiser, L., Gomez, A.N., Chollet, F.: Depthwise separable convolutions for neural machine translation. arXiv:1706.03059 (2017)
11. Sennrich, R., Haddow, B., Birch, A.: Improving neural machine translation models with monolingual data. arXiv:1511.06709 (2015)
12. Bojar, O., Tamchyna, A.: Improving translation model by monolingual data. In: Proceedings of the Sixth Workshop on Statistical Machine Translation, pp. 330–336 (2011)
13. Bertoldi, N., Federico, M.: Domain adaptation for statistical machine translation with monolingual resources. In: Proceedings of the Fourth Workshop on Statistical Machine Translation, pp. 182–189 (2009)
14. Lambert, P., Schwenk, H., Servan, C., Abdul-Rauf, S.: Investigations on translation model adaptation using monolingual data. 14. In: Workshop on Statistical Machine Translation, pp. 284–293 (2011)
15. Sennrich, R., Haddow, B., Birch, A.: Neural machine translation of rare words with subword units. arXiv:1508.07909 (2015)
16. Poncelas, A., Shterionov, D., Way, A., Wenniger, G.M.D.B., Passban, P.: Investigating backtranslation in neural machine translation. arXiv:1804.06189 (2018)
17. Karakanta, A., Dehdari, J., van Genabith, J.: Neural machine translation for low-resource languages without parallel corpora. Mach. Transl. **32**(1–2), 167–189 (2018)
18. Zhang, Z., Liu, S., Li, M., Zhou, M., Chen, E.: Joint training for neural machine translation models with monolingual data. In: Proceedings of the AAAI Conference on Artificial Intelligence, vol. 32, No. 1 (2018)
19. He, D., et al.: Dual learning for machine translation. Adv. Neural. Inf. Process. Syst. **29**, 820–828 (2016)
20. Xia, Y., Qin, T., Chen, W., Bian, J., Yu, N., Liu, T.Y.: Dual supervised learning. arXiv:1707.00415 (2017)
21. Kingma, D.P., Ba, J.: Adam: A method for stochastic optimization. arXiv:1412.6980 (2014)

Rigid Image Registration Based on Graph Matching

Zhiqin Lei, Jinrong Hu, and Bin Ye(✉)

Department of Computer Science,
Chengdu University of Information Technology, Chengdu 610225, China

Abstract. Medical image registration is an important technology in medical image processing and analysis. It has important application value in guiding radiotherapy, radiosurgery, minimally invasive surgery, endoscopy, and interventional radiology. Rigidly aligning two images is used to register rigid body structure, and it is also usually the first step for deformable registration with a large discrepancy. In the field of computer vision, one well-established method for image alignment is to find corresponding points from two images, and image alignment is based on identified corresponding points. Our approach lies in this category. Feature matching is crucial in finding corresponding points. However, conventional feature matching, such as SIFT, fails to consider the structural information between features. In this paper, a rigid image registration algorithm based on hypergraph matching is proposed. First, perform rough matching feature through SIFT (Scale invariant feature transform); then, based on coarse feature matching, filter out abnormal points through the hypergraph matching method that considers the structural information between features to obtain the correct feature matching pair; finally pass the obtained feature matching is registered to the calculated conversion matrix. Experimental results show that this method's registration results are better than those of SIFT and MIND methods regardless of whether it is rotated or scaled.

Keywords: Image registration · Rigid registration · Graph model · Hypergraph matching

1 Introduction

Medical image registration [1–7] is a critical step in medical image processing and analysis, and has important application value in guiding radiotherapy, radiosurgery, minimally invasive surgery, endoscopy, and interventional radiology. Medical image registration refers to seeking one or a series of spatial transformations for a medical image (image to be registered), so that the corresponding point on another medical image (reference image) reaches the spatial position and anatomical position. It is the same. The methods of medical image registration can be classified into rigid and deformable. Specifically, rigid registration is often used in the alignment of rigid body structures; and it is also

X. Sun et al. (Eds.): ICAIS 2021, CCIS 1422, pp. 295–304, 2021.
https://doi.org/10.1007/978-3-030-78615-1_26

usually the first step for deformable registration with a large displacement or discrepancy. Therefore, robust, accurate, and fast rigid alignment algorithms are still highly required.

Medical image rigid registration methods are mainly divided into two categories: intensity-based [8–10] and feature-based [11–13]. The registration method based on image intensity information directly establishes the similarity measurement function based on the intensity information, and finally uses the corresponding transformation to register the image in the case of maximum similarity. Although the implementation is relatively simple, the amount of calculation is large, and the noise and light Sensitive to the influence of changes, the result is poor for dealing with image registration problems with large differences. In the feature-based image registration method, the image is characterized by extracting certain features of the image. Compared with the intensity-based image registration method, it does not consider all image areas and only expresses the entire image in a characteristic way, it has a small amount of calculation, strong anti-interference ability, and relatively high noise and deformation resistance. In general, the feature-based image registration method has become a hot spot of current research because of its good cost performance and its ability to meet clinical speed requirements.

The image feature-based registration method often uses the corner points to extract image features because they are not sensitive to grayscale changes and image deformation. There are two types of commonly used corner extraction methods: direct detection methods, such as Harris, SUSAN, SIFT, Moravec, etc.; the other way is detection edge contour. Based on practical applications requirements, directly detecting corner points is used widely because of its fast and robust characteristics. Among them, the SIFT is used widely because of its good stability and invariance and its ability to adapt to features such as rotation and scale scaling used in feature-based image registration.

In feature-based image registration, the feature matching result's quality will directly affect the final registration result. A good feature matching result will be crucial to registration. In the traditional registration method based on SIFT [14] (Scale-invariant feature transform) features, the similarity between features achieves the initial matching of key points. Then the abnormal feature points are screened out by random sampling consensus algorithm [15] (RANSAC). However, this method only considers the feature information. It does not view the structural information between features, which leads to a great deal of mismatches in the result of feature matching. The final registration result is unsatisfactory.

Regarding the appealing problem, considering that Graph is often used to describe feature structure information, it can describe the correlation information between features. Graph Matching is often used in computer vision to solve the matching of key points to key points between images. It considers feature information and structural information (such as the similarity information of the edges formed between key points) between features. In the graph matching, the obtained feature matching result will have higher accuracy and stronger robustness because of considering the structure information.

2 Methods

The registration used in this paper is a registration method based on image features. The registration framework contains four main parts: key point detection, key point descriptor, key point matching, and transformation fitting. The quality of the feature point matching result will directly affect the quality of the registration result. In order to solve the problem of lots of mismatches caused by the feature matching part in the traditional feature-based registration that only depends on the feature information, we propose a graph-based model Feature matching image registration method. Next, we will describe how to implement feature matching in image registration through a graph model.

2.1 Graph Matching

Graph matching [16–20] has a wide range of multimedia and computer vision applications such as image registration [21–23], shape matching, and target tracking. Graph matching is to establish the correspondence between nodes and nodes between two or more graph structures. It is usually used to solve the matching from key points to key points in computer vision. Compared with the traditional computer vision method, graph matching considers the local features and the spatial geometric structure between the features simultaneously. It has stronger robustness and higher accuracy in the expression of image features. In graph matching, it is necessary to establish similarity information between nodes and spatial structure information, so the computational complexity is high, and the time complexity is high.

2.2 Rigid Image Registration Based on Image Matching

In image registration based on feature matching, a good feature matching result is a prerequisite for obtaining a more accurate fitting transformation. It is the key to obtaining a more precise registration result. Because the graph model also considers the feature information and the structural relationship between the features, it can get more accurate feature matching results. Still, there is a problem of high time complexity. To overcome this problem, in this paper, we pass the similarity screening between SIFT features. The feature point pair obtained after removing part of the abnormal points is used as the graph model's initial feature. The graph model is then constructed based on the initial feature, and the final feature matching result is obtained by solving the graph model. The method steps are as follows:

1. Feature extraction: feature extraction through SIFT;
2. Feature matching:
    a. Obtain initial matching features based on the similarity of SIFT features;
    b. Construct structural information between features according to the initial matching features, and establish a graph structure;
    c. Solve the graph matching and get the exact key point and key point matching result
3. Fitting transformation: Perform parameter fitting transformation according to the precise key point matching obtained by the graph matching to obtain the registration result.

The rigid image registration based on graph matching is shown in Fig. 1.

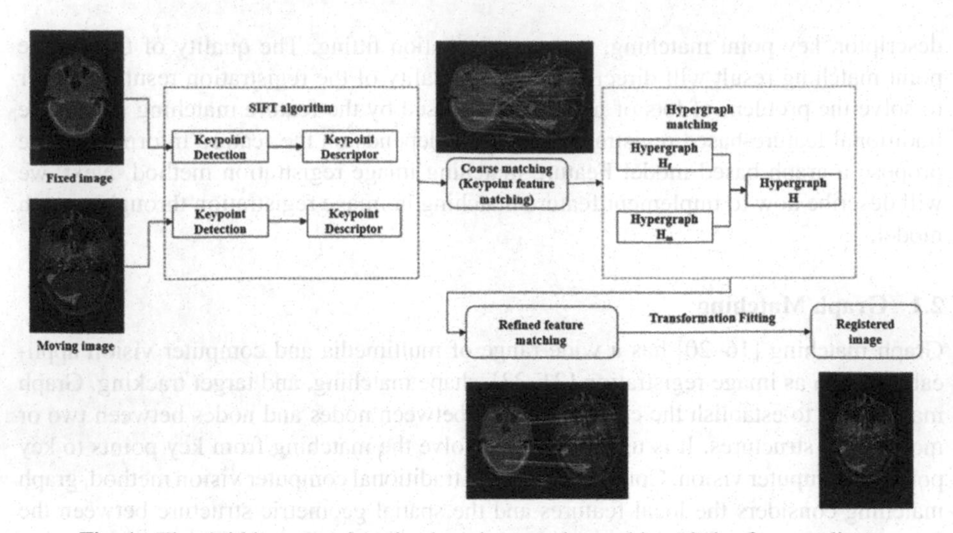

Fig. 1. The rigid image registration based on graph matching (Color figure online)

In the red box shown in the figure above, it is the part of the hypergraph matching. Let the features extracted by the SIFT algorithm undergo similarity measurement to obtain N pairs of rough matching features, namely $\{F_1, M_1\}$, $\{F_2, M_2\}$, …, $\{F_N, M_N\}$, $\{F_i, M_i\}$ represent the i-th rough matching pair, where $F_1, \ldots, F_N$ are the feature points in the reference image, and $M_1, \ldots, M_N$ are the feature points in the image to be registered.

After the rough matching features obtained by the SIFT algorithm, the extreme curvature function (as shown in formula 1) is used in $F_1, \ldots, F_N$ to measure the third-order similarity between the feature points F_i, F_j, and F_k to establish the fixed image Hypergraph Hf.

$$f_{i,j,k} = \left|F_i - F_k, F_j - F_k\right| \cdot \sum\nolimits_{i,j,k} \frac{1}{\sqrt{\|F_i - F_k\| \cdot \|F_j - F_k\|}} \tag{1}$$

Among them, $f(i, j, k)$ represents the weight of the super edge containing the feature points $\{F_i, F_j, F_k\}$. If the super edge is not included between $\{F_i, F_j, F_k\}$, then $f(i, j, k)$ is 0.

Similarly, in the moving image, the third-order similarity between the feature points M_i, M_j, and Mk is measured by using the polar curvature function to establish the hypermap Hm. As shown in formula (2):

$$\mathrm{m}_{i,j,k} = |M_i - M_k, M_j - M_k| \cdot \sum\nolimits_{i,j,k} \frac{1}{\sqrt{\|M_i - M_k\| \cdot \|M_j - M_k\|}} \tag{2}$$

Calculate the similarity of the hyper-edge between Hf and Hm by formula (3), and establish the hypergraph H, where $S(i, j, k)$ represents the similarity between the super edges formed between the i-th, j-th, and k-th rough matching $\{F_i, M_i\}$, $\{F_j, M_j\}$, $\{F_k, M_k\}$. σ is a scale control parameter used to scale the similarity between a pair of possible three-point matches, and it is proved by experiments that in order to be effective measure the similarity between super edges, σ is usually less than 0.01.

$$S_{i,j,k} = \exp[-\frac{\|f_{i,j,k} - m_{i,j,k}\|_2^2}{\sigma^2}] \quad (3)$$

In order to obtain accurate feature matching results based on rough matching pairs, evolutionary game theory is introduced into hypergraph matching, so that the problem of solving exact feature matching is transformed into a non-cooperative "matching game", to maximize Improve the overall profit of all players to get accurate feature matching results. Maximizing the average return is shown in formula (4):

$$G(\mathbf{x}_1, \mathbf{x}_2, \mathbf{x}_3) = \mathcal{S}\times_1\mathbf{x}_1^T\times_2\mathbf{x}_2^T\times_3\mathbf{x}_3^T \quad (4)$$

Among them, N rough matching feature pairs are represented as vertices in the hypergraph H, which are also called N existing strategies. In this paper, three vertices are calculated to be associated with three independent players in a non-cooperative game. Use x_1, x_2, x_3 to represent three independent players; the initial rough matching probability is represented by a vector x, where $\mathcal{X}_m$(m = 1, …, N) represents the probability that the Nth rough matching pair is a correct match. x satisfies the following conditions: $\forall m, xm \geq 0$ 且 $\sum_{m=1}^{N} x_m = 1, X_p$ (p = 1, 2, 3) represents a tensor notation to represent a tensor.

By finding the best choice probability for each rough matching pair, all players will get the maximum overall profit, as shown in formula (5):

$$\hat{\mathbf{x}} = \arg\max_{\mathbf{x}} G(\mathbf{x}, \mathbf{x}, \mathbf{x}) \quad (5)$$

The Nash equilibrium problem [24] in formula (5) is solved by calculating the local solution of nonlinear optimization, as shown in formula (6):

$$x_i(t) = x_i(t-1)\frac{\sum_{j=1}^{n}\sum_{k=1}^{n} S_{i,j,k}\prod_{n=j,k} x_n(t-1)}{\sum_{i=1}^{M}\sum_{j=1}^{M}\sum_{k=1}^{M} S_{i,j,k}\prod_{n=i,j,k} x_n(t-1)} \quad (6)$$

Where t represents the number of iterations, and x is updated through continuous loop iterations until the Eq. (6) converges to get the best $\hat{\mathbf{x}}$.

3 Experiments and Results

3.1 Data

This study has been conducted using CT and MR images of 100 Nasopharyngeal Cancer (NPC) patients (male/female: 52/48; mean age ± standard deviation: 50.3 ± 11.2 years; age range: 21 ± 76 years old) underwent chemoradiotherapy or radiotherapy at West China Hospital. The CT images were obtained by a Siemens SOMATOM Definition ASC system, with a voxel size ranges from 0.88 × 0.88 × 3.0 mm^3 to 0.97 ± 0.97 ± 3.0 mm^3. The MR images were captured by a Philips Achieva 3T scanner. We used in this study the T1-weighted images with contrast. The in-plane resolution is 0.61 × 0.61 mm^2 and a slice spacing of 0.8 mm. It is noted that every scan contains nasopharyngeal cancer.

3.2 Preprocess Data

We resampled the images to have a voxel size of 1 × 1 × 1 mm^3. Because of the different imaging r-anges covered by those images, we use ITK_snap to only kept the range from eyebrow and chin. Moreover, because the ground truth alignment of these two image modalities is unfortunately not easily obtainable, the standard of alignment is estimated using an off-the-shelf toolbox Elastix for the sake of efficiency. It is noted that the alignment is performed in 3D using Elastix with standard parameters.

Among them, we use CT as the reference image, and use the rotated or zoomed image of the paired MR as the image to be registered. The specific design is as follows: We randomly divide the images aligned by Elastix into two groups, each of which is 50 pairs; in one of the groups, fifty MR images were obtained by rotating the slice by a degree from −10 to 10 with a step of 5; In another group, fifty images were acquired by scaling the slice with a factor in [0.8, 1.2].

3.3 Evaluation Index

Color Overlay Map. This article directly uses the SimpleITK4 toolkit method for color overlay display to displays the registration result. The unaligned areas will display green artifacts. If there are no artifacts or fewer artifacts, then the registration effect is bet-ter. As shown in Fig. 2, after the reference image and the image to be registered are superimposed, the green or purple artifacts are the parts that are not aligned.

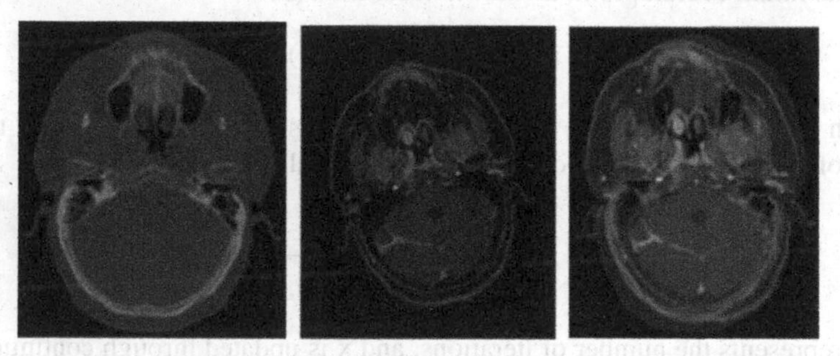

Fig. 2. The color overlapped image for the fixed and moving images. (a) fixed image; (b) moving image; (c) color overlapped image. (Color figure online)

Target Registration Error. Target registration error (TRE), It is a quantitative evaluation index for image registration performance measurement, which can effectively calculate the accuracy of registration. TRE can reflect the error bet-ween the reference image and the registered image, and its calculation formula (7):

$$TRE = \sqrt{\frac{1}{N}\sum\nolimits_{i=1}^{N} \left\| P_i - T \cdot P_i' \right\|_2^2} \tag{7}$$

Where P_i denotes a key point in the ground truth moving image, P_i' is the keypoint in the moving image, T is the estimated transformation matrix after transformation fitting, N is the number of key points.

3.4 Experiment Setup

We compared our proposed methods: affine image registration network AIRNet, SIFT, MIND [25]. AIRNet is a self-supervised and end-to-end deep learning registration method, which employs a CNN to predict the transformation parameters to register rigidly. It is considered a deep learning baseline for rigid registration. SIFT combined with RANSAC is a typical registration method based on traditional feature information. MIND is based on a similarity measure of image registration, which is a deep learning method.

The main parameter settings of the comparison algorithm in the experiment are as follows: The RANSAC parameter ω is used to determine if a matched point pair is an inlier or outlier in the transformation fitting procedure. We computed the local SIFT descriptors and MIND by using the open-source VLFeat toolbox on http://www.vlfeat.org/ and the MIND code shared by MP Heinrich on http://www.mpheinrich.de/software.html accordingly. In our method, the number of iterations t is set to 100, and the scale control parameter in graph matching σ is set to 0.005.

3.5 Experiment Results

In this section, we investigate the registration performance on the accuracy of three methods: AIRNet, SIFT, MIND. Randomly select 50 pairs of data in the data set for experimentation, Table 1 demonstrates TREs of the three methods for registration tasks where the ground truth moving images were rotated clockwise by a certain degree. Table 2 presents the TREs of these methods for registration tasks where the ground truth moving images were scaled by a specific factor.

Table 1. TREs of the rotated registration task on image pairs from the testing NPC

Rotation degree	0	5	10	15	20
AIRNet	1.08 ± 0.51	1.21 ± 0.69	1.32 ± 0.74	1.58 ± .081	1.86 ± 0.81
SIFT	7.28 ± 4.32	8.4 ± 5.16	8.14 ± 4.11	10.99 ± 9.95	12.121 ± 7.23
MsIND	5.67 ± 1.18	6.3 ± 1.36	6.95 ± 3.41	7.62 ± 4.54	9.98 ± 6.19
Our	3.52 ± 0.51	3.01 ± 0.41	6.25 ± 0.78	5.91 ± 0.52	6.45 ± 0.82

It can be seen from the data in Tables 1 and 2 that compared with the SIFT and MIND methods, the method proposed in this paper achieves the lowest TRE value regardless of whether it is rotated or scaled. At the same time, Fig. 3 shows the color overlay of AIRNet, SIFT, and MIND for NPC image registration: It can be seen from the figure that the registration result of this method is close to AIRNet, and is far superior to SIFT and MIND methods.

Table 2. TREs of the scaled registration task on image pairs from the testing NPC

Scaling factor	0.8	0.9	1	1.1	1.2
AIRNet	2.09 ± 0.62	1.61 ± 0.41	0.92 ± 0.42	1.92 ± 0.41	2.66 ± 0.1
SIFT	10.99 ± 6.56	8.01 ± 4.42	7.38 ± 4.46	8.45 ± 6.13	8.97 ± 3.56
MIND	6.55 ± 0.54	5.84 ± 0.41	5.28 ± 0.41	6.21 ± 0.42	7.27 ± 0.45
Our	3.31 ± 0.53	4.31 ± 0.41	3.12 ± 0.62	4.12 ± 0.68	5.12 ± 0.46

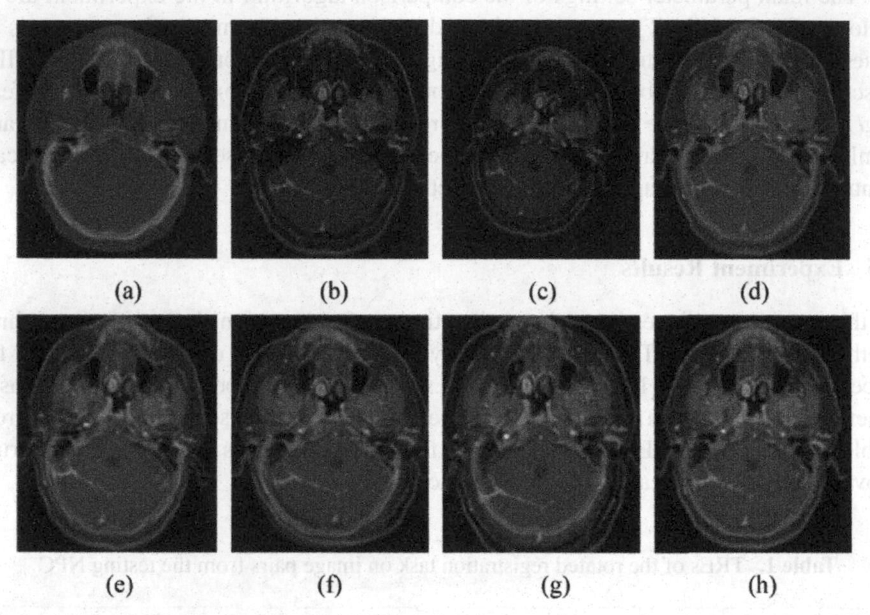

Fig. 3. (a) Fixed CT; (b) true MR; (c) initial moving image; (d) ground truth overlay; (e) AIRNet; (f) SIFT; (g) MIND; (h) our method

4 Conclusions

In this paper, we present a new image registration framework that considers both the image features and the image structure, which overcomes the problem of low registration accuracy caused by only considering the image features in the traditional feature-based matching problem. For our proposed method, through qualitative and quantitative analysis, the performance of the method was evaluated from the aspects of registration accuracy and image ghosting. The experimental results show that the registration effect of this method is better than that of SIFT method and MIND method. In future work, we plan to further validate the effectiveness of our approach on more data sets, and deep learning method is used to realize the problem.

Funding Statement. This work is supported in part by the Introduction Talent Research Start-up Pro-gram of Chengdu University of Information Technology (No. KYTZ202008), and in part by the Sichuan Science and Technology Program (2019YFH0085, 2019ZDZX0005, 2019YFG0196), and in part by the Science and Technology Project Affiliated to the Education Department of Chongqing Municipality (No. 19ZB0257).

References

1. Simonovsky, M., Gutiérrez-Becker, B., Mateus, D., Navab, N., Komodakis, N.: A deep metric for multimodal registration. In: Ourselin, S., Joskowicz, L., Sabuncu, M.R., Unal, G., Wells, W. (eds.) MICCAI 2016. LNCS, vol. 9902, pp. 10–18. Springer, Cham (2016). https://doi.org/10.1007/978-3-319-46726-9_2
2. Hu, J., Sun, S., Yang, X., Zhou, S.: Towards accurate and robust multi-modal medical image registration using contrastive metric learning. IEEE Access. **7**, 132816–132827 (2019)
3. Woo, J., Stone, M., Prince, J.L.: Multimodal registration via mutual information incorporating geometric and spatial context. IEEE Trans. Image Process. **24**(2), 757–769 (2014)
4. Padgett, K., et al.: SU-F-BRF-10: deformable MRI to CT validation employing same day planning MRI for surrogate analysis. Med. Phys. **41**(6Part23), 401 (2014)
5. Padgett, K.R., Stoyanova, R., Pirozzi, S., Johnson, P., Piper, J., Dogan, N.: Validation of a deformable MRI to CT registration algorithm employing same day planning MRI for surrogate analysis. J. Appl. Clin. Med. Phys. **19**(2), 258–264 (2018)
6. Rohé, M.-M., Datar, M., Heimann, T., Sermesant, M., Pennec, X.: SVF-net: learning deformable image registration using shape matching. In: Descoteaux, M., Maier-Hein, L., Franz, A., Jannin, P., Collins, D.L., Duchesne, S. (eds.) MICCAI 2017. LNCS, vol. 10433, pp. 266–274. Springer, Cham (2017). https://doi.org/10.1007/978-3-319-66182-7_31
7. Toews, M., Zöllei, L., Wells, W.M.: Feature-based alignment of volumetric multi-modal images. In: Gee, J.C., Joshi, S., Pohl, K.M., Wells, W.M., Zöllei, L. (eds.) IPMI 2013. LNCS, vol. 7917, pp. 25–36. Springer, Heidelberg (2013). https://doi.org/10.1007/978-3-642-38868-2_3
8. Klein, S., Staring, M., Murphy, K., Viergever, M.A., Pluim, J.P.: Elastix: a toolbox for intensity-based medical image registration. IEEE Trans. Med. Imaging **29**(1), 196–205 (2009)
9. Glodeck, D., Hesser, J., Zheng, L.: Potential of metric homotopy between intensity and geometry information for multi-modal 3D registration. Z. Med. Phys. **28**(4), 325–334 (2018)
10. Pluim, J.P.W., Maintz, J.B.A., Viergever, M.A.: Image registration by maximization of combined mutual information and gradient information. In: Delp, S.L., DiGoia, A.M., Jaramaz, B. (eds.) MICCAI 2000. LNCS, vol. 1935, pp. 452–461. Springer, Heidelberg (2000). https://doi.org/10.1007/978-3-540-40899-4_46
11. Zhao, D., Yang, Y., Ji, Z., Hu, X.: Rapid multimodality registration based on MM-SURF. Neurocomputing **131**, 87–97 (2014)
12. Lv, G., Teng, S.W., Lu, G.: Enhancing SIFT-based image registration performance by building and selecting highly discriminating descriptors. Pattern Recognit. Lett. **84**, 156–162 (2016)
13. Hossein-Nejad, Z., Nasri, M.: An adaptive image registration method based on SIFT features and RANSAC transform. Comput. Electr. Eng. **62**, 524–537 (2017)
14. Lowe, D.G.: Distinctive image features from scale-invariant keypoints. Int. J. Comput. Vis. **60**(2), 91–110 (2004)
15. Fischler, M.A., Bolles, R.: C: Random sample consensus: a paradigm for model fitting with applications to image analysis and automated cartography. Commun. ACM. **24**(6), 381–395 (1981)
16. Caetano, T.S., McAuley, J.J., Cheng, L., Le, Q.V.: Learning graph matching. IEEE Trans. Pattern Anal. Mach. Intell. **31**(6), 1048–1058 (2009)
17. Zhang, H., Ren, P.: Game theoretic hypergraph matching for multi-source image correspondences. Pattern Recognit. Lett. **87**, 87–95 (2017)

18. Umeyama, S.: An eigen decomposition approach to weighted graph matching problems. IEEE Trans. Pattern Anal. Mach. Intell. **10**(5), 695–703 (1988)
19. Wiskott, L., Fellous, J.-M., Krüger, N., von der Malsburg, C.: Face recognition by elastic bunch graph matching. In: Sommer, G., Daniilidis, K., Pauli, J. (eds.) CAIP 1997. LNCS, vol. 1296, pp. 456–463. Springer, Heidelberg (1997). https://doi.org/10.1007/3-540-63460-6_150
20. Zhou, F., De la Torre, F.: Factorized graph matching. In: IEEE Conference on Computer Vision and Pattern Recognition 2012, pp. 127–134. IEEE (2012)
21. Neemuchwala, H., Hero, A., Carson, P.: Image registration using entropic graph-matching criteria. In: Conference Record of the 36th Asilomar Conference on Signals, Systems and Computers, vol. 1, pp. 134–138. IEEE (2002)
22. Zeng, Y., Wang, C., Wang, Y., Gu, X., Samaras, D., Paragios, N.: Dense non-rigid surface registration using high-order graph matching. In: 2010 IEEE Computer Society Conference on Computer Vision and Pattern Recognition, pp. 382–389. IEEE (2010)
23. Deng, K., Tian, J., Zheng, J., Zhang, X., Dai, X., Xu, M.: Retinal fundus image registration via vascular structure graph matching. Int. J. Biomed. Imaging **2010** (2010)
24. Baum, L.E., Eagon, J.A.: An inequality with applications to statistical estimation for probabilistic functions of Markov processes and to a model for ecology. Bull. Am. Math. Soc. **73**(3), 360–363 (1967)
25. Heinrich, M.P., Jenkinson, M., Bhushan, M., Matin, T., Gleeson, F.V.: MIND: modality independent neighbourhood descriptor for multi-modal deformable registration. Med. Image Anal. **16**(7), 1423–1435 (2012)

A Multi-scale Fusion Method for Dense Crowd Counting

Liwen Shen, Zhao Qiu(✉), Ping Huang, Yu Jin, Chao Li, and Jinye Cai

Hainan University, Haikou 570228, Hainan, China
qiuzhao@hainanu.edu.cn

Abstract. We propose a dense crowd detection network, called MSFNet, which can deal with highly dense crowd scenes, make accurate counting estimation and generate high-quality density maps by deep learning. The network is mainly composed of two main parts: the front-end network uses VGG-16 as the 2D feature extraction module, and the back-end network uses convolution networks with different sizes of convolution kernels instead of linking operations. The network is composed of convolution layers, which is an easy training model. We verify our network on two representative data sets (ShanghaiTech Data Set, UCF CC 50 Data Set), and the performance has been improved.In the first data set of ShanghaiTech, the root mean square error (MSE) decreased by 10%, and the mean absolute error (MAE) and root mean square error (MSE) of the second data set both decreased by about 6%.

Keywords: Multi-scale fusion · Crowd counting · Dense crowd

1 Introduction

On December 31, 2014, a large number of people gathered on the Bund in Shanghai to hold a New Year's Eve event. Unfortunately, a stampede accident occurred and a major accident that caused 35 deaths. Unfortunately, large-scale stampede accidents have occurred all over the world, causing many casualties. Obtaining the number of people in a specific scene from video images has become one of the popular researches in computer vision technology in crowd control and public safety. In some specific scenarios, such as sports events, temple fairs, and scenic spots, timely access to important information such as the number of people and spatial distribution can enable relevant functional departments to wait for data support and make corresponding processing measures and personnel scheduling in a timely manner. The method of population counting can also be extended to other fields, such as vehicle counting [6, 11, 20], cell counting [3, 18, 19] and so on.

Because the number of people in a crowded crowd varies greatly, and the size of the person in the image is also different due to the position of the person in the image. It is difficult for a single-scale CNN network to deal with the crowd density of the entire image. [14, 21] complete crowd density estimation by using a multi-branch CNN network. This network structure is composed of three parallel CNN network branches, with

X. Sun et al. (Eds.): ICAIS 2021, CCIS 1422, pp. 305–314, 2021.
https://doi.org/10.1007/978-3-030-78615-1_27

different convolution kernel sizes, and different network structures handle different perception fields. Recently, [15] proposed a five-branch structure, in which three branches are similar to the previous multi-column CNN [21] branches, while the remaining two branches are used to obtain global and local context information. Some key points of these methods are: the use of parallel multi-column CNN network models with different convolution kernel sizes improves the performance of crowd density estimation.

In this work, we propose a CNN-based crowd density estimation method: a multi-scale fusion convolutional neural network (MSFNet). Similar to the previous method, in the previous method, a three-column parallel network structure is used to obtain more columns to obtain good performance. Our MSFNet is a relatively simpler architecture. Our model uses convolution as the backbone and can support input images of different resolutions. We use the first 10 layers of VGG-16 [7] with the fully connected layer removed as the front-end 2D feature extraction network, and use convolutional layers of different scales as the backend network to process different perception fields and advance deeper features without Loss of resolution. Our performance on the ShanghaiTech data set has improved, in which the root mean square error (MSE) in the first part is reduced by about 10%; in the second part, the mean absolute error (MAE) and the root mean square error (MSE) are both reduced Around 6%. Excellent results were also achieved on the UCF CC 50 data set.

2 Related Work

2.1 Detection-Based Methods

Most early research on population counting focused on detection methods, which used a sliding window detector to detect people and count their number [5]. These methods extract low-level features (such as edge features [17] and HOG [4]) from the entire human body to complete human detection and recognition. However, they are suitable for scenes where the number of people is relatively sparse. In scenes where people are relatively crowded, people block each other, making the detection effect poor. In order to solve the problem of crowding, some researchers use the characteristics of specific parts of the human body (such as head, torso, etc.) to complete the detection and recognition of people, and then obtain the number of people.

2.2 Regression-Based Approach

Since detection-based methods cannot solve the concealment problem caused by crowded crowds, some people study the mapping from a feature learned from an image to the number of people, and use regression algorithms to complete the number of people. This method fifirst extracts low-level features such as foreground features, edge features, texture features, and gradient features [2, 13]; then uses linear regression, ridge regression and Gaussian process regression to train the regression model to learn from the previously extracted the mapping relationship between low-level features and the number of people to complete the statistics of the number of people.

2.3 Density Map-Based Methods

Although the regression-based method can solve the problem of crowd occlusion to a certain extent, because the feature of the whole image is extracted for regression counting, the spatial information of the crowd in the image is ignored. In 2010 Carlos et al. [8] proposed a mapping relationship between learning the local features of an image and its corresponding density map, thereby adding spatial information of the crowd in the counting process. The paper first proposed the method of using density map to count the crowd, and then many image counts were based on this method.

2.4 Deep Learning-Based Methods

Compared with traditional methods, recognition, segmentation, and detection in computer vision use deep learning methods, which have made great progress, and crowd counting is no exception. In recent years, most of the methods are based on deep learning methods, using deep learning methods. Learning ability to get the image to its corresponding population density map or number [10, 15, 16, 21].

Zhang et al. [21] proposed a multi-column-based network structure (MCNN),which uses convolution kernels of different sizes to process people of different scales and count people. Onoro and Sastre [11] proposed a perceptible scale multicolumn counting model called Hydra CNN for object density estimation. Li et al. [9] proposed a single-column counting model using hole convolution for crowd density estimation. Sam et al. [14] completed crowd density estimation based on the multi-column counting model fed into different networks according to different density of images. Viresh et al. [12] proposed a two-branch structure that generates high resolution from low resolution for crowd counting. Sindgi et al. [15] used the context pyramid network to process the global and local context information of the crowd image to generate a crowd density map to achieve higher counting accuracy and generate a more accurate density map. Obviously, the performance of the CNN-based crowd counting method is better than the work mentioned in Sects. 2.1 to 2.3

2.5 Limitations of Existing Methods

Most of the existing solutions use multi-column networks, pyramid networks and hole convolution to generate density maps. However, these methods have some disadvantages: (1) Multi-column CNN network structure needs more time to train. (2) Multi-column CNN has a redundant structure. (3) To deal with people of different scales, it is necessary to classify the population density levels and send them to different networks. (4) A single network structure is different to deal with population densities of different scales. However, it is difficult to define the density level in real-time crowded scene analysis because the number of people will change in a large range. Similarly, increasing the density level of the classifier means that more columns need to be implemented, which will make the design more complicated and lead to more redundancy. Because the branching structure in [21] is inefficient, the lack of parameters to generate density map will reduce the final accuracy. In [9], a single column of hole convolution is used to expand the field of view and obtain a finer density map. However, the heads of images are different in size, so it is difficult to completely deal with different scales of people by using scale network. Considering all the above shortcomings, we propose a novel method to generate high-quality density maps.

3 A Multi-scale Fusion CNN

Our basic idea is to use an existing mature and deeper network as our front-end feature extraction network. The back-end uses a multi-scale network with different sizes of convolution kernels, and uses different-scale networks to capture head information of different sizes. Without excessively expanding the complexity of the network, it can receive a large range of advanced features and generate highquality density maps. In this section, we propose a multi-scale fusion network architecture, and then give the corresponding training method.

3.1 A Multi-scale Fusion CNN Architecture

We choose a VGG-16 network structure with strong migration learning capabilities and flexibility as the front-end network of MSFNet, because it is mostly composed of convolution operations and can easily connect back-end networks to generate density map. In CrowdNet [1], the author selects the first 13 layers of the VGG-16 network structure, and adds a 11 convolutional layer as the output layer to eliminate the fully connected layer of VGG-16 for crowd counting, but this method has poor performance. There are other network structures, such as [9], before sending the input image to the network, VGG-16 is used as a density level classifier to label the input image, and the back-end uses empty segment convolution to extract features to generate a density map. In this article, inspired by works [9, 14, 15], we try to use convolution kernels with different sizes as our back-end network. We use the same VGG-16 network structure as [9], and delete the classification part of the VGG-16 fully connected layer, select the first 10 layers as the front-end part of the network, and use different sizes of convolution kernels as the back-end network. Inspired by works [14, 15], we try to use convolutional networks with different sizes of kernels to extract more important information and maintain the output resolution.

3.2 Network Configuration

Images usually contain heads of different sizes, and filters of different sizes can capture crowd density characteristics of different scales. Inspired by multi-column neural networks [21] and single-column dilated convolution [9], we use filters of different sizes to learn the target density. We use the VGG-16 network as the front-end network (excluding the fully connected layer). According to [11], the VGG-16 network is more suitable for crowd counting, and the method based on the single-column network performs better than the method based on the multi-column network, that is, the deeper network is better than the wider network. According to [9], under the trade-offs of training time, memory consumption, and number of parameters, choosing the first 10 layers of VGG-16 to retain only 3 maximum pooling layers can achieve the best performance without affecting the output accuracy.

The overall structure of our MSFNet is shown in Fig. 1. It consists of two parts: The front-end network is the first 10 layers of VGG-16 and only three pooling layers are reserved. Back-end network uses filters of different sizes to feel the local visual field, and then connects and convolves it to the next layer. We use maximum pooling in 2 × 2 area and ReLU as activation function, because it has good performance for CNN network. We use 1 × 1 convolution to get the final output and map it to the density map. Then Euclidean distance is used to measure the difference between the estimated density map and the real value. The loss function is defined as follows:

$$\mathrm{L}(\theta) = \frac{1}{2N}\sum_{i=1}^{N} ||F(X_i;\theta) - F_i||_2^2 \tag{1}$$

Which θ is a parameter that can be learned in MSFNet. N is the number of training images, X_i is the input images and F_i is the true density map of the input images. $F(X_i; \theta)$ represents the estimated density map of X_i represented by θ generated by MSFNet. L is the loss between the estimated density map and the real density map.

3.3 MSFNet Optimization

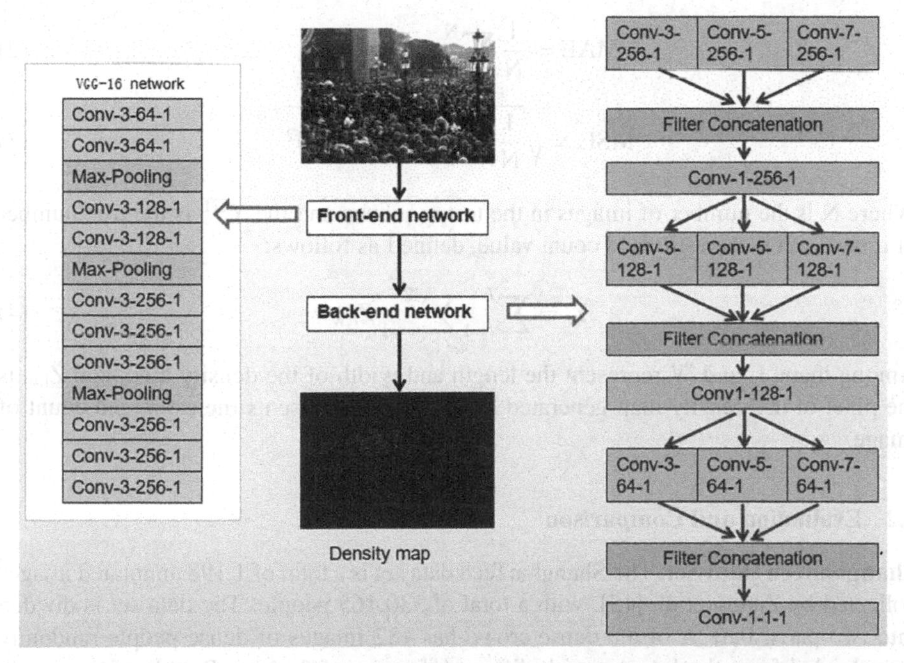

Fig. 1. Overall network structure of MSFNet.The front-end network uses a pre-trained VGG-16 network, and the back-end is connected to a multi-scale layer-by-layer fusion network

Using typical methods of training neural networks, the loss function (1) is optimized by batch-based random gradient descent and back propagation. However, in the actual training process, because the number of training samples is limited, and when the influence of the gradient on the deep neural network disappears, it is not easy to learn all the parameters at the same time. According to the successful experience of [10], we used the pre-trained VGG-16 in the Pytorch framework when training the model, and then connected to our multi-scale fusion back-end network to initialize our entire network. The trained network reduces the time for model training and improves the effect of the model to a certain extent.

4 Experiment

We do experiments on two challenging and representative data sets: Shanghaitech [15], UCF_CC_50 [8] and Compared with most of the methods mentioned in this paper, our MSFNet model is not complicated, its structure is relatively simple, and it is improved on all data sets with higher accuracy. Our model is based on pytorch framework, using pre-trained front-end network to initialize our network.

4.1 Evaluation Index

We use the usual evaluation indexes MAE and MSE to evaluate our model, which are defined as:

$$\mathrm{MAE} = \frac{1}{\mathrm{N}} \sum\nolimits_{\mathrm{i}=1}^{\mathrm{N}} |\mathrm{Y_i} - \mathrm{Y_i^{GT}}| \tag{2}$$

$$\mathrm{MSE} = \sqrt{\frac{1}{\mathrm{N}} \sum\nolimits_{\mathrm{i}=1}^{\mathrm{N}} |\mathrm{Y_i} - \mathrm{Y_i^{GT}}|^2} \tag{3}$$

Where N is the number of images in the test sequence and the $\mathrm{Y_i^{GT}}$ is the true number of images. Yi is the estimated count value, defined as follows:

$$Y_i = \sum\nolimits_{l=1}^{L} \sum\nolimits_{w=1}^{W} Z_{l,w} \tag{4}$$

Among them, L and W represent the length and width of the density map, and $\mathrm{Z_{l,w}}$ is the pixel of the density map generated by (l, w). $\mathrm{Y_i}$ represents the estimated count of image.

4.2 Evaluation and Comparison

Shanghaitech Data Set. The ShanghaiTech data set is a total of 1,198 annotated images collected by Zhang et al. [15], with a total of 330,165 people. The data set is divided into two parts: Part_A of the dense crowd has 482 images of dense people randomly downloaded from the Internet with different degrees of density; Part_B of the sparse population is composed of 716 crowd scene images taken from the streets of Shanghai. We follow the training method of Li et al. [10] to train and test our proposed model,

Table 1. Comparison results with other models. Part_A and Part_B indicate ShanghaiTech Part_A and Part_B, respectively.

	Part_A		Part_B	
Method	MAE	MSE	MAE	MSE
Zhang et al. [4]	181.8	277.7	32	49.8
MCNN [22]	110.2	173.2	26.4	41.3
Switching-CNN [15]	90.4	135.0	21.6	33.4
CP-CNN [13]	73.6	106.4	20.1	30.1
ic-CNN [13]	*68.5*	116.2	10.7	16.0
CSRNet [10]	*68.2*	115.0	10.6	16.0
MSFNet (ours)	**67.1**	**103.8**	**10.0**	**14.9**

and compare it with other six works.The results are shown in Table 1. Compared with the CSRNet method, our method has a 10% lower MSE in part_A; MAE and MSE in part_B are both reduced by about 6%. Our results in Part A are shown in Fig. 2, and the results in Part_B are shown in Fig. 3.

UCF_CC_50 Data Set. The UCF_CC_50 data set includes 50 images with different perspectives and resolutions [8]. The number of annotated people in each image is between 94 and 4543, with an average number of 1280. According to the standard setting in [8], we randomly divided 50 images into 10 groups, each group as the verification set, and the other 9 groups as the training set. The results areas follows (see Table 2):

Table 2. Comparison of results with other networks on the UCF_CC_50 dataset

Method	MAE	MSE
Zhang et al. [4]	419	541.5
MCNN [22]	377.6	509.1
Switching-CNN [15]	318.1	318.1
CP-CNN [13]	295.8	**320.9**
ic-CNN [13]	**260.9**	365.5
CSRNet [10]	266.1	397.5
MSFNet (ours)	289.2	350.4

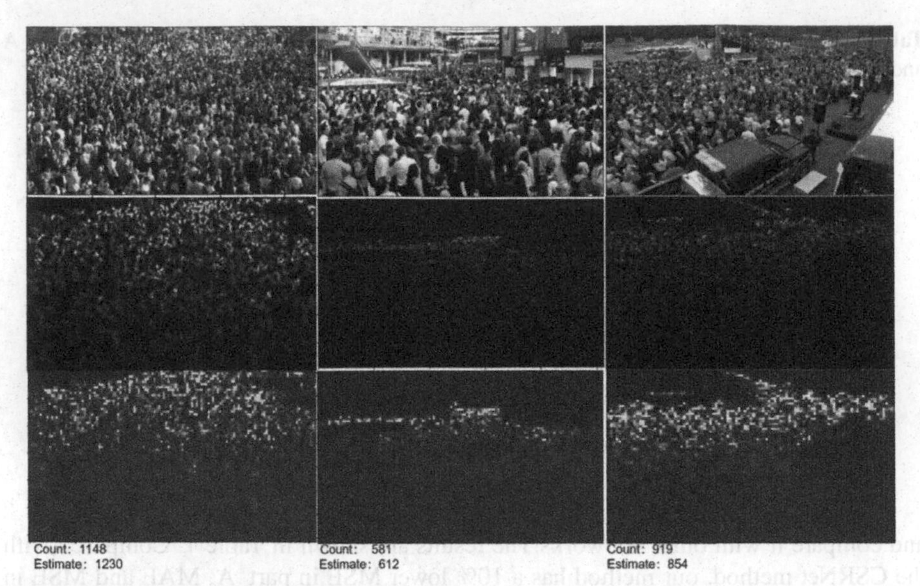

Fig. 2. Part_A partial result display. The top image is the input image, the middle image is the density map generated from the labeled data, and the bottom image is the predicted density map generated by the MSFNet network.

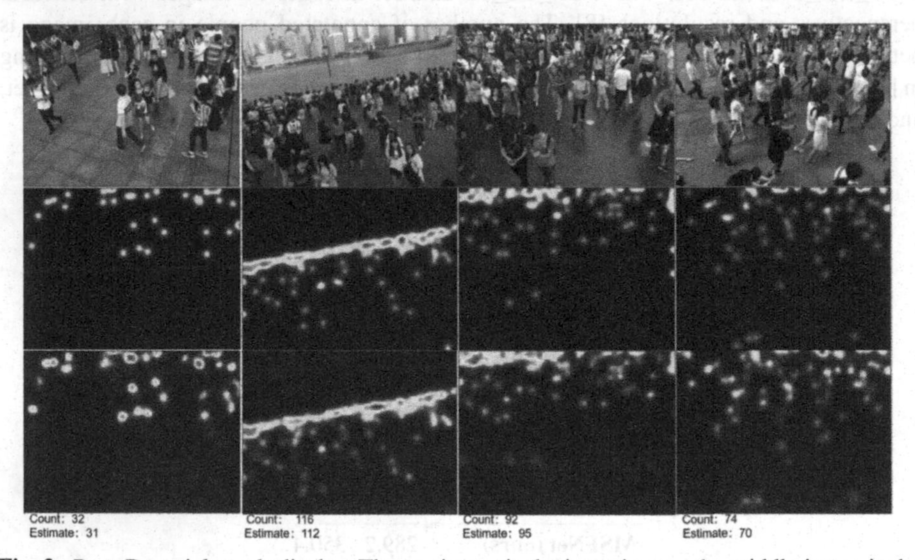

Fig. 3. Part_B partial result display. The top image is the input image, the middle image is the density map generated from the labeled data, and the bottom image is the predicted density map generated by the MSFNet network.

5 Summary

In this paper, a new architecture called MSFNet is proposed, which is used to generate crowd counting and estimated density map. It adopts a multi-scale fusion and easy to-train end-to-end network structure. We use convolution kernels of different sizes to extract different scale information of crowded people, and through convolution layers of convolution kernels of different sizes, we can deal with people of different scales without losing resolution. Our performance on two data sets has good results.

Acknowledgment. This work is Supported by Hainan Natural Science Foundation Project (No. 618MS028 and No. 2019RC100). Thanks to Professor Zhao Qiu, the corresponding author of the paper.

References

1. Author, F.: Article title. Journal **2**(5), 99–110 (2016). Boominathan, L., Kruthiventi, S., Venkatesh Babu, R.: CrowdNet: a deep convolutional network for dense crowd counting. In: Proceedings of the 24th ACM International Conference on Multimedia (2016)
2. Chan, A.B., Vasconcelos, N.: Bayesian Poisson regression for crowd counting. In: 2009 IEEE 12th International Conference on Computer Vision, pp. 545–551 (2009)
3. Chen, K., et al.: Feature mining for localised crowd counting. In: BMVC (2012)
4. Cong, Z., et al.: Cross-scene crowd counting via deep convolutional neural networks. In: IEEE Conference on Computer Vision Pattern Recognition (2015)
5. Dalal, N., Triggs, B.: Histograms of oriented gradients for human detection. In: 2005 IEEE Computer Society Conference on Computer Vision and Pattern Recognition (CVPR 2005), vol. 1, 886–893 (2005)
6. Doll'ar, P., et al.: Pedestrian detection: an evaluation of the state of the art. IEEE Trans. Pattern Anal. Mach. Intell. **34**, 743–761 (2012)
7. Hsieh, M.-R., Lin, Y., Hsu, W.: Drone-based object counting by spatially regularized regional proposal network. In: 2017 IEEE International Conference on Computer Vision (ICCV), pp. 4165–4173 (2017)
8. Idrees, H., et al.: Multi-source multi-scale counting in extremely dense crowd images. In: 2013 IEEE Conference on Computer Vision and Pattern Recognition, pp. 2547–2554 (2013)
9. Lempitsky, V., Zisserman, A.: Learning to count objects in images. In: NIPS (2010)
10. Li, Y., Zhang, X., Chen, D.: CSRNet: dilated convolutional neural networks for understanding the highly congested scenes. In: 2018 IEEE/CVF Conference on Computer Vision and Pattern Recognition, pp. 1091–1100 (2018)
11. Marsden, M., et al.: Holistic features for real-time crowd behaviour anomaly detection. In: 2016 IEEE International Conference on Image Processing (ICIP), pp. 918–922 (2016)
12. Oñoro-Rubio, D., López-Sastre, R.J.: Towards perspective-free object counting with deep learning. In: Leibe, B., Matas, J., Sebe, N., Welling, M. (eds.) ECCV 2016. LNCS, vol. 9911, pp. 615–629. Springer, Cham (2016). https://doi.org/10.1007/978-3-319-46478-7_38
13. Ranjan, V., Le, H., Hoai, M.: Iterative crowd counting. arXiv abs/1807.09959 (2018)
14. Ryan, D., et al.: Crowd counting using multiple local features. In: 2009 Digital Image Computing: Techniques and Applications, pp. 81–88 (2009)
15. Sam, D.B., Surya, S., Babu, R.V.: Switching convolutional neural network for crowd counting. In: 2017 IEEE Conference on Computer Vision and Pattern Recognition (CVPR), pp. 4031–4039 (2017)

16. Sindagi, V.A., Patel, V.: Generating high-quality crowd density maps using contextual pyramid CNNs. In: 2017 IEEE International Conference on Computer Vision (ICCV), pp. 1879–1888 (2017)
17. Szegedy, C., et al.: Going deeper with convolutions. In: 2015 IEEE Conference on Computer Vision and Pattern Recognition (CVPR), pp. 1–9 (2015)
18. Tuzel, O., Porikli, F., Meer, P.: Pedestrian detection via classification on riemannian manifolds. In: IEEE Transactions on Pattern Analysis and Machine Intelligence, vol. 30, pp. 1713–1727 (2008)
19. Walach, E., Wolf, L.: Learning to count with CNN boosting. In: Leibe, B., Matas, J., Sebe, N., Welling, M. (eds.) ECCV 2016. LNCS, vol. 9906, pp. 660–676. Springer, Cham (2016). https://doi.org/10.1007/978-3-319-46475-6_41
20. Wang, Y., Zou, Y.: Fast visual object counting via example-based density estimation. In: 2016 IEEE International Conference on Image Processing (ICIP), pp. 3653–3657 (2016)
21. Zhang, S., et al.: Understanding traffic density from large-scale web camera data. In: 2017 IEEE Conference on Computer Vision and Pattern Recognition (CVPR), pp. 4264–4273 (2017)
22. Zhang, Y., et al.: Single-image crowd counting via multi-column convolutional neural network. In: 2016 IEEE Conference on Computer Vision and Pattern Recognition (CVPR), pp. 589–597 (2016)

Chinese Text Classification Based on Adversarial Training

Jinye Cai[1], Zhao Qiu[1(✉)], XiaoRan Yang[2], Chao Li[1], Yu Jin[1], and LiWen Shen[1]

[1] Hainan University, Haikou 570228, Hainan, China
qiuzhao@hainanu.edu.cn
[2] Hefei University of Technology, Hefei 230009, Anhui, China

Abstract. As the most basic and important application in the field of natural language processing, text classification has always been a hot research topic in the field of natural language processing. So far, a large number of excellent models have emerged, including SVM, TextCNN, TextRNN, DPCNN, Transformer, etc. Higher accuracy has always been the goal pursued by text classification algorithms. Therefore, this paper constructs a text classification algorithm based on adversarial training. By adding the adversarial disturbance to the word embedding layer, and then using the adversarial samples and the original samples to train the model together, the model's ability to correctly classify adversarial samples is improved, and the robustness of the model is improved, so as to improve the classification accuracy and improve the classification effect of the classification model, and it is proved on three benchmark data sets that this method can effectively improve the accuracy of the text classification model.

Keywords: Natural language processing · Text classification Adversarial training · Adversarial disturbance · Adversarial samples

1 Introduction

With the advent of the big data era, massive amounts of data are being generated at all times. The mixing of a large number of irrelevant data increases the difficulty for us to extract effective information from it, and the cost of extracting effective information is also increasing. At this time, natural language processing plays an important role, and text classification, one of the important modules of natural language processing, has naturally received more attention.

Text classification is a very important module in natural language processing. It has a wide range of applications. It is widely used in spam filtering, news classification, part-of-speech tagging, etc., and it plays an important role. The process of text classification is roughly divided into four steps: (1) text preprocessing; (2) text representation and feature extraction; (3) construction of a classifier; (4) classification.

X. Sun et al. (Eds.): ICAIS 2021, CCIS 1422, pp. 315–327, 2021.
https://doi.org/10.1007/978-3-030-78615-1_28

From the early manual methods of rule-based feature matching, to machine learning, and then to the emergence of deep learning methods, a large number of excellent algorithms have emerged in the field of text classification, such as SVM [1], LSTM [2], BiLSTM+Attention [3] et al. Due to the advantages of RNN's time-series feature extraction and the advantages of CNN's spatial feature learning and capture, in recent years, improved models based on CNN and RNN have become the mainstream of text classification.

In order to solve the problem of the poor parallel computing ability of RNN and the difficulty of capturing long-distance features of CNN, Transformer (the Transformer mentioned in this article is not the complete model in the article "attention is all you need" [4], but the encoder part of it) came into being, and was proved by the research of Gongbo [5] et al. it has obvious advantages in semantic feature extraction, long-distance capture, task comprehensive feature extraction, and parallel computing power and efficiency.

The concept of adversarial training was first proposed by Goodfellow [6] and applied it in the field of computer vision to enhance the robustness of the model. Miyato [7] et al. first applied the adversarial training method to the text field. They defined the adversarial disturbance on continuous word embeddings as a regularization strategy to improve the generalization performance of the text classification model.

This paper adopts the neural network model based on Transformer, and adopts the method of adversarial training to counter disturbance in the word embedding layer, and adopts adversarial training to enhance the robustness of the classification model, so as to improve the robustness and accuracy of the text classification model.

2 The Research Process

The text classification technology was first proposed in the late 1950s. It was proposed by Luhn of IBM in the United States. They pioneered the application of word frequency statistics to automatic classification. Then in 1960, Maron published the first text classification paper, leading the research process of text classification technology. Initially, text classification technology was mainly based on knowledge engineering, which was classified by the definition of classification rules considered by experts in various fields. Although a high classification accuracy rate can be obtained, it is time-consuming and labor-intensive, and text classification research has also fallen into a bottleneck. It was not until the emergence of the Internet and the development of machine learning in the 1990s that text categorization re-entered people's vision. At this time, the main technique of text categorization was statistical learning, which was characterized by high accuracy without human intervention. Therefore, the performance has been significantly improved, and it became the mainstream classification method at that time. The development of text classification can be classified into the following four stages:

2.1 Theory Formation Stage

This stage ranged from the 1950s to the 1980s. During this period, Maron and others proposed a probabilistic indexing model, and Salton proposed the famous vector space model. These theories laid a solid foundation for the development of text classification.

2.2 Text Classification Technology Based on Knowledge Engineering

This stage mainly lasted from the 1980s to the 1990s, mainly based on knowledge-based engineering, using manual methods. Experts in the field write appropriate rules and manually build the classifier. Although a high accuracy rate can be obtained, it is time-consuming and labor-intensive and is not suitable for a large number of text tasks.

2.3 Text Classification Technology Based on Statistical Machine Learning

Because of the rapid development of network information technology, in the 1990s, the classification technology based on knowledge engineering has been rapidly developed, making machine learning gradually become the mainstream. In 1992, Lewis introduced the text classification system for the first time in his thesis, and conducted several implementation tests to guide future text classification research. Joachim applied support vector machine to text classification for the first time, which greatly improved the classification performance. Subsequently, Yiming Yang also used decision trees for text classification, and analyzed and compared various feature selection algorithms from an experimental perspective.

2.4 Deep Learning Development Stage

In recent years, deep learning has developed rapidly. BP neural network is used in the text classification field to make up for the shortcomings of linear and linear non-linear methods in expressing semantics. Subsequently, Johnson et al. proposed a semi-supervised, convolutional neural network learning framework using unidentified data. The C-LSTM model proposed by Zhou et al. uses CNN to extract text features, which are then classified by the LSTM long and short-term memory network, which greatly improves the efficiency of text classification. With the support of large amounts of data, deep learning models can improve the ability of features to express textual semantic information and have a good ability to learn and understand. With the continuous development of deep learning, text classification technology based on deep learning will become more mature.

3 Related Model

Currently, RNN and CNN are still the mainstream models for text categorization, but the Transformer model has gradually come into people's attention due to its excellent performance. So, the following will briefly introduce the RNN classification model, CNN classification model and Transformer classification model.

3.1 RNN Classification Model

Since RNN was introduced into the field of natural language processing, it has quickly attracted most of the attention and has been widely used in various tasks of NLP. The RNN text classification model proposed by Arevian [8] first splits the input sequence into word vectors one by one, and inputs them into the recurrent neural network. The entire sentence is regarded as a time series from front to back, and the linear sequence structure is adopted to continuously collect input information from front to back. The linear sequence structure of RNN itself not only has the problem of difficulty in optimization during backpropagation, but also limits the ability of RNN to perform parallel computing. Although the LSTM and GRU models were subsequently introduced, which solved the gradient problem well, they could not change the current situation of poor parallel computing capabilities of RNNs. The RNN classification model is shown in Fig. 1:

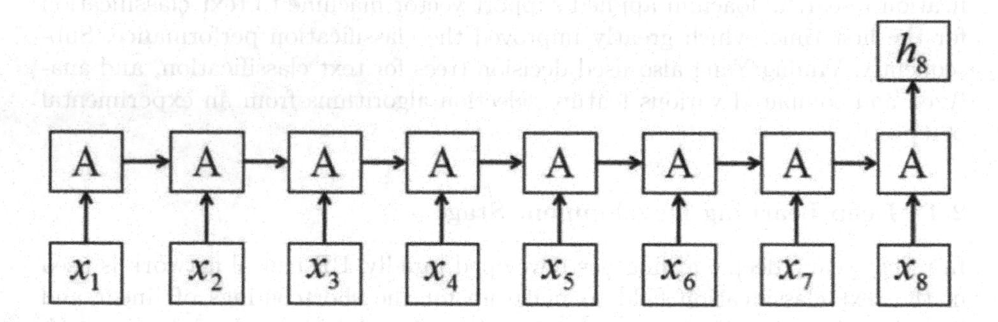

Fig. 1. RNN classification model.

3.2 CNN Classification Model

The earliest introduction of CNN into NLP was Kim's work [9] in 2014. The classification model is shown in Fig. 2. The CNN classification model first expresses the input in Word Embedding, converts one-dimensional text information input into a two-dimensional input structure, and then extracts features from each convolution kernel in the convolution layer to form feature vectors and feature sequences. Then the Pooling layer reduces the dimensionality of the features, and finally connects the fully connected layer neural network to form the final classification process. From the classification process of CNN, we can see that the features captured by CNN are largely determined by the window covered by the convolution kernel. The size of the convolution kernel window determines how far away features can be captured.

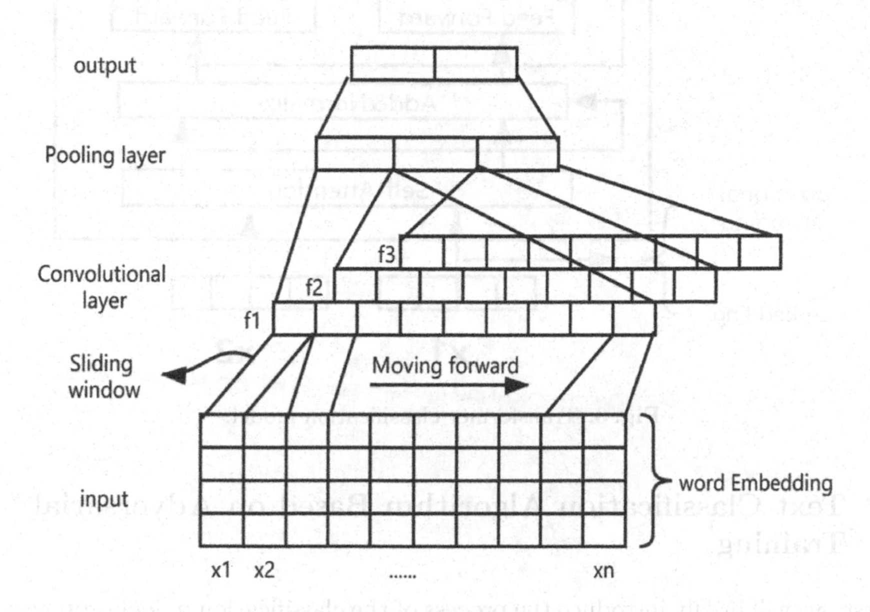

Fig. 2. CNN classification model.

3.3 Transformer Classification Model

Transformer was born in the paper "Attention is all you need" [4] on Google's translation task in 2017, which aroused considerable response. The Transformer model mentioned in this article is not what the original text refers to, but the Encoder part of it. The classification model is shown in Fig. 3, the input sequence is first converted into a fixed-length word vector, and then the processed word vector is position-encoded, and then input to the Encoder layer, and use the self-attention mechanism to score the similarity between words. Finally, the input sequence is classified through a series of regularization and residual connection operations. It has been proven in many subsequent experiments that Transformer has incomparable advantages in semantic feature extraction, long-distance feature capture, task comprehensive feature extraction, and parallel computing capabilities and efficiency. Refer to "Why Self-attention" [5] and "Improving Language Understanding by Generative Pre-Training" [10].

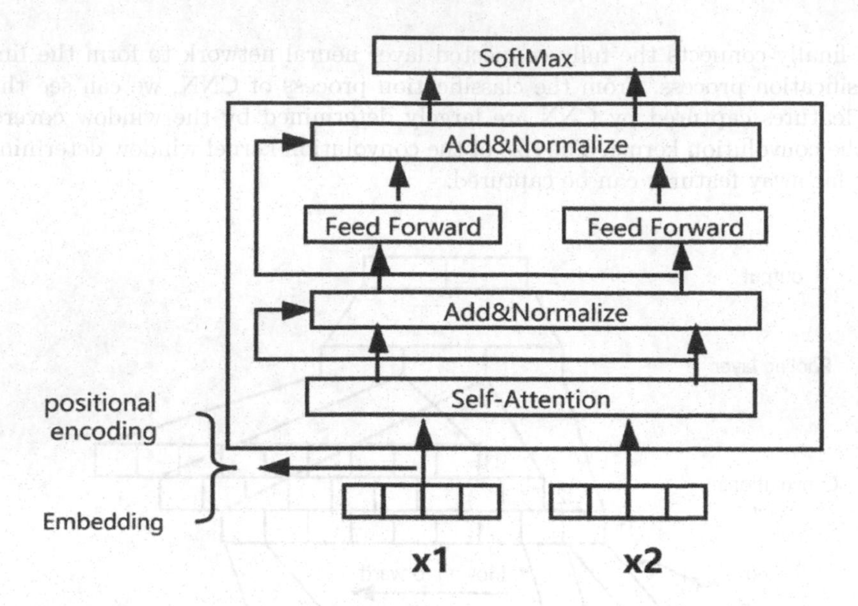

Fig. 3. Transformer classification model.

4 Text Classification Algorithm Based on Adversarial Training

First, we will briefly introduce the process of the classification model proposed in this paper. The text classification model based on adversarial training proposed in this paper is based on the traditional Transformer classification model, which classifies text through the change of word vector, position encoding and similarity calculation. The difference is that we add gradient attacks to the embedding layer of the Transformer classification model to form a counter sample and add the counter sample to normal training tasks. At this point, the whole classifier network forms an adversarial network. The embedding layer constantly generates adversarial samples with gradient attack, and the final classifier should correctly classify the adversarial samples and normal samples. In such a process of continuous confrontation training, the robustness of the classifier network can be increased, and then the accuracy of the classifier network can be improved.

4.1 Model Structure

The Transformer classification algorithm model structure based on adversarial training is shown in Fig. 4.

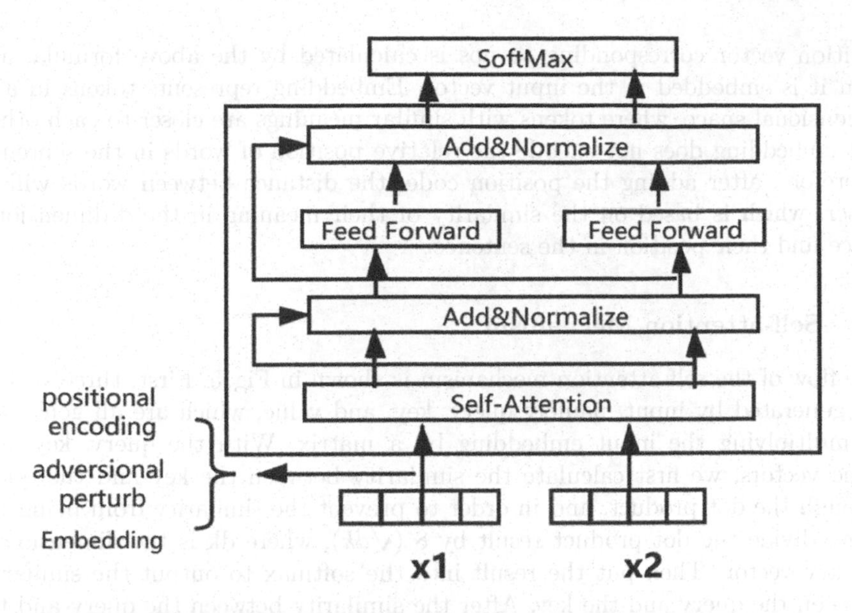

Fig. 4. Transformer_Adn classification model.

First, the model uses the word2vec word vector model and a custom vocab table to convert the input to the corresponding word vector, and then uses the FGM fast gradient attack algorithm to add the attack noise generated by the network gradient to the word vector, then add the corresponding position code to it. Finally enter the word vector to the encorder Module. Each encoder layer includes two sub-layers (Multi-head attention and Pointwise feed forward networks). Among them, the Multi-head attention layer performs a similarity calculation on the input vector through the self-attention mechanism. The Pointwise feed forward networks layer is a fully connected layer that contains two linear transformations and a nonlinear activation function. There is a residual connection around each sub-layer, and then the layer is standardized. Residual connections help avoid the vanishing gradient problem in deep networks.

4.2 Position Coding

Location information plays a very important role in extracting context. Therefore, position coding is added to provide the model with some information about the relative position of words in the sentence. The calculation formula of the position code is as follows:

$$PE(pos, 2i) = sin(pos/10000^{2i}/dmodel) \tag{1}$$

$$PE(pos, 2i+1) = cos(pos/10000^{2i}/dmodel) \tag{2}$$

Suppose the position of a word in the sentence is pos, and we need to generate a vector with dimension dmodel about pos. The value of each dimension of the position vector corresponding to pos is calculated by the above formula, and then it is embedded in the input vector. Embedding represents tokens in a d-dimensional space, where tokens with similar meanings are closer to each other. But embedding does not encode the relative position of words in the sentence. Therefore, after adding the position code, the distance between words will be closer, which is based on the similarity of their meaning in the d-dimensional space and their position in the sentence.

4.3 Self-attention Mechanism

The flow of the self-attention mechanism is shown in Fig. 5. First, three vectors are generated by input, namely query, key, and value, which are all generated by multiplying the input embedding by a matrix. With the query, key, and value vectors, we first calculate the similarity between the key and the query through the dot product, and in order to prevent the similarity from being too large, divide the dot product result by 8 ($\sqrt{dk}$), where dk is the dimension of the key vector. Then put the result into the softmax to output the similarity between the query and the key. After the similarity between the query and the key is obtained, use these similarities to perform a weighted sum on the value to obtain the output.

4.4 Adversarial Training

Adversarial training is a training method that introduces noise, which can regularize parameters and improve the robustness and generalization ability of the model. The hypothesis of adversarial training is: after adding disturbance to the input x, the output distribution is consistent with the original y distribution. When applied to a classifier, adversarial training adds the following terms to the cost function, as shown in Eq. 3.:

$$-\,log(y|x + r_{Adn};\theta) \tag{3}$$

Where x is the input and y is the parameter of the classifier, r_{Adn} is shown in Eq. 4:

$$r_{Adn} = argminlogp(y|x + r;\theta') \tag{4}$$

Where r is a perturbation on the input and θ' represents a constant, which means that although the gradient is calculated during the calculation of the anti-disturbance, the parameters are not updated, because the current anti-disturbance is optimal for the old parameters.

The earliest adversarial training was proposed by Goodfellow, and it was applied to the image field to enhance the robustness of the model. Then Miyato first applied the adversarial training method to the text field, and defined the adversarial disturbance on the continuous word embedding as a regularization strategy to improve the generalization performance of the text classification model.

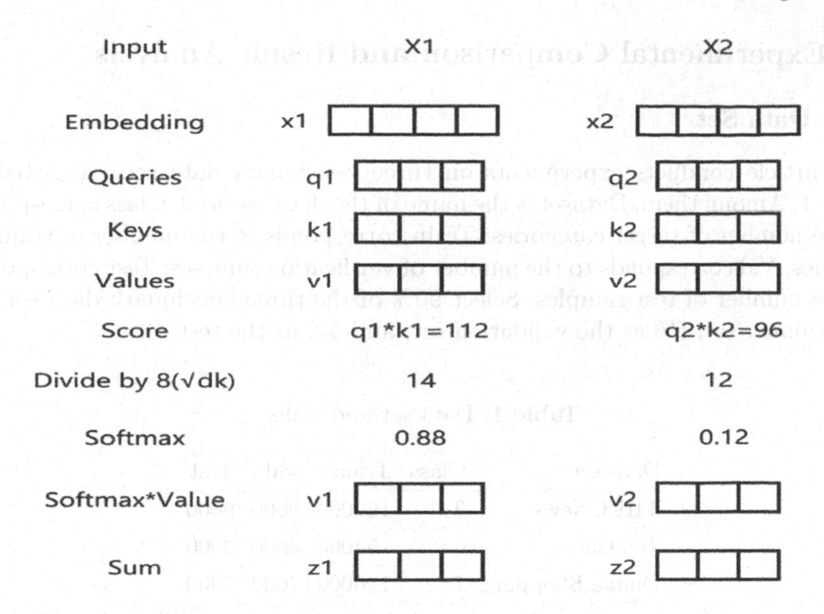

Fig. 5. Process of self-attention mechanism.

In this paper, in order to improve the accuracy of the basic model classification, the anti-perturbation is added to the classification model embedding layer through the FGM algorithm, thereby generating the anti-sample. And through adversarial training, the model is trained to correctly classify adversarial samples and original samples, thereby improving the classification accuracy of the model. The FGM algorithm [7] used is shown below, and its principle is to scale according to the specific gradient, so as to get a better confrontation sample, as shown in Eq. 5

$$r_{Adn} = \epsilon g/||g||_2 \ where \ g = \nabla_x logp(y|x;\theta'). \tag{5}$$

The specific steps of confrontation training are shown in the algorithm pseudo code:
Algorithm pseudo code:
For each x:

1. Calculate the forward loss of x and back propagation to get the gradient
2. Calculate r according to the gradient of the embedding matrix and add it to the current embedding, which is equivalent to x+r
3. Calculate the forward loss of x+r, backpropagate to get the gradient of the confrontation, and add it to the gradient of 1
4. Restore embedding to the value at 1 Update the parameters according to the gradient of 3.

5 Experimental Comparison and Result Analysis

5.1 Data Set

This article conducts experiments on three benchmark data sets, as listed in Table 1. Among them, Dataset is the name of the data set used; Class corresponds to the number of target categories; Train corresponds to the number of training samples; Val corresponds to the number of verification samples; Test corresponds to the number of test samples. Select 90% of the three benchmark data sets as the training set, 5% as the validation set, and 5% as the test set.

Table 1. Data set and scale.

Data set	Class	Train	Val	Test
THUCNews	9	162000	9000	9000
TouTiao	6	54000	3000	3000
Online_Shopping	7	126000	7000	7000

5.2 Experimental Parameters

On the three data sets, the common parameters used in the experiment are shown in Table 2.

5.3 Evaluation Index

The judgment of the quality of a method must be measured by public technical indicators. At present, the more common performance indicators mainly include precision, recall, F1 value (H-mean value), etc. The meaning and calculation formula of each performance index are shown in the Fig. 6. Among them, TP represents the true positive rate, which means that the prediction is 1, and the actual is 1, then the prediction is correct; TN means the true negative rate, which means that the prediction is 0, the actual is 0, and the prediction is correct; FP means the false positive rate, which means that the prediction is 1, and the actual If it is 0, the prediction is wrong; FN represents the false negative rate, which means that the prediction is 0, and the actual is 1, and the prediction is wrong.

Because text classifiers pay more attention to whether they can obtain valid data from a large amount of irrelevant data, this article chooses precision as the criterion for the classification model.

Table 2. Experimental public parameters.

Parameter name	Value	Parameter interpretation
dropout	0.5	Random inactivation
require_improvement	15000	If more than 15000 batches do not improve, the training ends
num_class	9/6/7	Number of categories on the three data sets
n_vocab	0	Vocabulary size, assignment at runtime
num_epochs	20	Number of epochs
batch_size	128	Batch data set size
pad_size	32	The length of each sentence
learning_rate	5e−4	Learning rate
embed	300	Word vector dimension
dim_model	300	Default parameter dimensions
num_head	6	Number of multi_head
num_encoder	6	Number of encoders

Index name	meaning	Calculation formula
Precision	The proportion of all predictions that are correct (positive and negative)	$\frac{TP}{TP+FP}$
Recall	The proportion of the correct prediction being positive to the total actual positive	$\frac{TP}{TP+FN}$
F1 value (H-mean value)	Arithmetic mean divided by geometric mean	$\frac{2}{\frac{1}{precision+1}+\frac{1}{recall+1}}$

Fig. 6. Evaluation index

5.4 Experimental Results and Analysis

According to the calculation formula of experimental data and performance indicators, we can get the accuracy of text classification on the three benchmark data sets (THUCNews, TouTiao, Online_shopping) and it is shown in the Table 3:

Table 3. Best accuracy on the three models.

Model	THUCNews	Online_Shopping	TouTiao
Transformer_Adn	93.25%	93.57%	94.02%

The experimental results are shown in Table 4, in which Transformer_Adn is the method proposed in this article. This table compares the accuracy between the current mainstream text classification methods and Transformer_Adn on 3 data sets (THUCNews, TouTiao, Online_shopping). The results obtained show that Transformer_Adn has achieved the best results on all three data sets. This proves that Transformer_Adn can effectively improve the accuracy of model classification.

Table 4. Comparison of different models on three data sets.

Model	THUCNews	Online_Shopping	TouTiao
Transformer_Adn	93.25%	93.57%	94.02%
Transformer [4]	92.42%	92.82%	93.75%
BiLSTM [11]	92.36%	92.62%	93.05%
BiLSTM+Attention [3]	92.51%	93.20%	93.25%
BiLSTM+Pooling [12]	92.62%	92.80%	93.18%
DPCNN [13]	92.93%	92.98%	93.86%

6 Concluding Remarks

The Transformer_Adn model proposed in this paper is based on the Transformer text classification model. Adversarial samples are added to the embedding layer for training, and adversarial training is used as a regularization method to achieve the purpose of improving the robustness of the model and the accuracy of text classification. The experimental results show that the model can effectively improve the effect of text classification. In the following research, I intend to continue to find better ways to add anti-disturbance, and apply the algorithm model to a wider range of fields such as sentiment analysis, and machine translation.

Acknowledgment. This work is Supported by Hainan Natural Science Foundation Project (No. 2019RC100) and Hainan Natural Science Foundation Project (No. 618MS028).

References

1. Wang, Z.-Q., Sun, X., Zhang, D.-X., Li, X.: An optimal SVM-based text classification algorithm. In: 2006 International Conference on Machine Learning and Cybernetics, pp. 1378–1381. IEEE (2006)
2. Hochreiter, S., Schmidhuber, J.: Long short-term memory. Neural Comput. **9**(8), 1735–1780 (1997)
3. Zhou, P., et al.: Attention-based bidirectional long short-term memory networks for relation classification. In: Proceedings of the 54th Annual Meeting of the Association for Computational Linguistics: Short Papers, vol. 2, pp. 207–212 (2016)

4. Vaswani, A., et al.: Attention is all you need. In: Advances in Neural Information Processing Systems, pp. 5998–6008 (2017)
5. Tang, G., Müller, M., Rios, A., Sennrich, R.: Why self-attention? A targeted evaluation of neural machine translation architectures. arXiv preprint arXiv:1808.08946 (2018)
6. Goodfellow, I.J., Shlens, J., Szegedy, C.: Explaining and harnessing adversarial examples. arXiv preprint arXiv:1412.6572 (2014)
7. Miyato, T., Dai, A.M., Goodfellow, I.: Adversarial training methods for semi-supervised text classification. arXiv preprint arXiv:1605.07725 (2016)
8. Arevian, G.: Recurrent neural networks for robust real-world text classification. In: IEEE/WIC/ACM International Conference on Web Intelligence (WI 2007), pp. 326–329. IEEE (2007)
9. Kim, Y.: Convolutional neural networks for sentence classification. arXiv preprint arXiv:1408.5882 (2014)
10. Radford, A., Narasimhan, K., Salimans, T., Sutskever, I.: Improving language understanding by generative pre-training (2018)
11. Schuster, M., Paliwal, K.K.: Bidirectional recurrent neural networks. IEEE Trans. Sig. Process. **45**(11), 2673–2681 (1997)
12. Lai, S., Xu, L., Liu, K., Zhao, J.: Recurrent convolutional neural networks for text classification. In: Twenty-Ninth AAAI Conference on Artificial Intelligence (2015)
13. Johnson, R., Zhang, T.: Deep pyramid convolutional neural networks for text categorization. In: Proceedings of the 55th Annual Meeting of the Association for Computational Linguistics: Long Papers, vol. 1, pp. 562–570 (2017)

Research on SVM Parameter Optimization Mechanism Based on Particle Swarm Optimization

Ning Cao[1] and Wenli Wang[2](✉)

[1] Sanming University College of Information Engineering, Sanming 365004, China
[2] Hebei University of Economics and Business, Shijiazhuang 050061, China

Abstract. As an common and effective pattern recognition method, support vector machine is used in many scenarios. The parameter selection of support vector machine directly affects its performance, determines its classification accuracy and generalization ability. This paper optimizes the parameter optimization mechanism of support vector machine (SVM) based on particle swarm optimization (PSO), adjusts the parameters of particle swarm optimization, and makes it have stronger global search ability. In this paper, support vector machine is firstly introduced, then the parameter optimization mechanism of support vector machine is elaborated, and the improved parameter selection model based on particle swarm optimization algorithm is proposed, then carries on the simulation experiment to the improved model on the open data set. Experimental results show that the improved model has higher classification accuracy. At last, the shortcomings of the study and other possible acting contents and optimization directions are described, and the focus and direction of the future research work are given.

Keywords: Support vector machine · Parameter optimization · Kernel function

1 Introduction

Support Vector Machine (SVM) [1] is a practical supervised learning classifier, and it has achieved very good results in practical applications [2]. Support vector machines are based on statistical learning theory. At present, the selection of support vector machine parameters does not have a well-established theory and unified standard, and the selection of kernel parameters is directly related to the classification accuracy of support vector machines. Literature [3] gives the traditional parameter selection methods and shortcomings.

At present, a large number of scholars have studied support vector machine [4–9]. The combination of support vector machine, artificial intelligence algorithm, neural network and so on has become a hot research topic, and has been applied in many scenes [10–13].

As a group search algorithm, the idea of particle swarm optimization originates from the predatory behavior of birds. The optimization is completed through the cooperation

X. Sun et al. (Eds.): ICAIS 2021, CCIS 1422, pp. 328–337, 2021.
https://doi.org/10.1007/978-3-030-78615-1_29

and iteration between particles. The fitness of particles is used to evaluate the quality of particles, and has an impact on the update of particles. The search process of particle swarm optimization is simple, the search speed is fast, and the parameters are less, it is easy to implement, and it has good combination. Therefore, it is applied well in many scenarios [14–17].

In Sect. 2, the basic concepts of support vector machine are introduced. In Sect. 3, the mechanism of parameter optimization is introduced. The application of particle swarm optimization algorithm in parameter optimization of support vector machine is explored. Then the optimization performance and classification performance of support vector machine are explored through experiments. In Sect. 4, the possible research directions of SVM in the future are explained.

In the optimization process, experiments prove that it has high classification accuracy and convergence speed. The experimental simulation environment used in this article is: Windows 7 × 64 PC, Matlab 2017 (a), libsvm-3.21 toolbox.

2 Support Vector Machine

Support vector machine is a learning system that uses a linear function hypothesis space in a high-dimensional feature space [3]. It is developed based on the VC dimension theory and the principle of structural risk minimization in the statistical learning theory, which solves the under-learning, over-learning and dimensionality disasters question [13]. The idea of the support vector machine is to map the input vector to a high-dimensional feature space by implementing a selected nonlinear transformation to construct a hyperplane with classification, as shown in Fig. 1. Suppose the training sample is:

$$\{x_{i,}y_i\}(_i = 1^N, y_i \in \{-1, 1\}$$

Set the hyperplane used for separation:

$$\omega \cdot x + b = 0 \tag{1}$$

Where ω is the normal vector of the hyperplane and b is the intercept. The goal of support vector machincs is to find an optimal hyperplane that maximizes the distance between positive and negative classes. That is, look for ω0 and b0 to make (1) hold. That is, the optimal hyperplane is:

$$\omega_0 \cdot x + b_0 = 0 \tag{2}$$

The support vector is the point on the hyperplane, which satisfies the following conditions:

$$\begin{cases} \omega \cdot x_i + b = -1 \\ y_i = 1 \end{cases} \text{or} \begin{cases} \omega \cdot x_i + b = 1 \\ y_i = -1 \end{cases} \tag{3}$$

Suppose the interval between positive and negative examples is:

$$dis = \frac{\omega}{||\omega||}(x_2 - x_1) \tag{4}$$

Maximize the interval between positive and negative examples, that is, maximize the value of (4). The problem can be transformed into (5).

$$\begin{cases} min\frac{||\omega||^2}{2} \\ S.t.\ y_i(\omega \cdot x_i + b) \geq 1\ \ \forall i = 1, 2, \ldots N \end{cases} \quad (5)$$

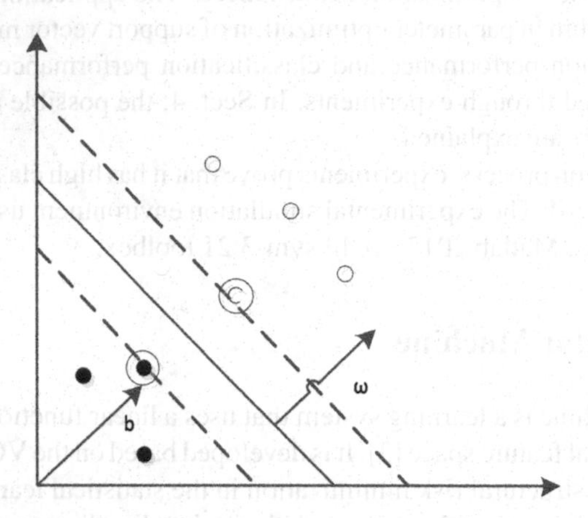

Fig. 1. Two dimensional classification hyperplane

When the sample is nonlinear and separable, the problem is transformed into (6).

$$\begin{cases} min\frac{1}{2}\omega^T\omega + c\sum_{i=1}^{l}\varepsilon_i \\ S.t\ y_i\left(\omega^T\varphi(\omega_i + b)\right) \geq 1 - \xi_i \end{cases} \quad (6)$$

Where C is the penalty factor and a is the Lagrange coefficient, the dual problem is transformed into (7).

$$\begin{cases} min\frac{1}{2}\alpha^T Q\alpha - e^T\alpha \\ S.t\ y^T\alpha = 0 \end{cases} \quad (7)$$

$Q_{ij} \equiv y_i y_j k(x_i, x_j)$, kernel function is $k(x_i, x_j) \equiv \varphi(x_i)^T\varphi(x_j)$.

3 Support Vector Machine Parameter Optimization Mechanism

The performance of the support vector machine largely depends on the choice of parameters [19]. The improper choice of parameters may lead to phenomena such as "over-fitting" and "under-fitting". Support vector machines use kernel functions to map data samples from low-dimensional to high-dimensional, and the main parameters are penalty factor C and kernel function parameters [21]. Many scholars have studied the parameter optimization mechanism of support vector machines. The literature [19] proposed based on the improved Drosophila algorithm to optimize the parameters of the support vector machine, and applied to the failure analysis of the analog circuit. Literature

[20] proposed a hybrid algorithm HPSOCS based on particle swarm optimization and genetic algorithm to optimize the parameters of support vector machines. Literature [21] proposed a support vector machine parameter selection mechanism based on social sentiment optimization algorithm. Literature [22] proposed a support vector machine parameter selection mechanism based on optimized bird flocks. Literature [23, 24] proposed a parameter optimization model based on artificial fish swarms, and literature [25] proposed a parameter optimization scheme based on particle swarms.

3.1 The Influence of Kernel Function on Classification Accuracy

Table 1. Familiar kernel function forms

Kernel function	Forms
RBF	$K(x_i, x_j) = \exp\left(-r\|x_i - x_j\|^2\right)$
Sigmoid	$K(x_i, x_j) = \tanh\left(rx_i^T x_j + r\right)$
Linear	$K(x_i, x_j) = x_i^T x_j$
Polynomial	$K(x_i, x_j) = \left(rx_i^T x_j + r\right)^\alpha$

The kernel function is one of the parameters of the support vector machine. The commonly used kernel functions include the radial basis kernel function (RBF), the Sigmoid kernel function, the linear kernel function (Linear), and the polynomial kernel function (Polynomial). The specific expressions are shown in Table 1. Show.

In order to explore the impact of different kernel functions on classification accuracy, the Heart, Iris and Wine data sets published by UCI and the C-SVC support vector machine model were used for testing. 50% of the data set is used for training and 50% is used for testing. The specific results are shown in Table 2.

Table 2. Classification accuracy

Data	RBF Accuracy /%	Sigmoid Accuracy /%	Linear Accuracy /%	Polynomial Accuracy /%
Heart	83.70	84.44	86.67	84.44
Iris	93.33	76.00	93.33	89.33
Wine	40.45	39.33	97.75	98.85

In order to ensure the accuracy of the experimental results and reduce the contingency of the results, the average of ten experimental results is taken as the experimental result. It can be seen from the experimental results that different kernel functions have a greater impact on the accuracy of the classification results.

3.2 Parameter Optimization Based on Particle Swarm Optimization

Introduction to Particle Swarm Algorithm. Particle swarm algorithm is a kind of swarm intelligence algorithm, a heuristic algorithm that simulates the interaction mechanism in the foraging behavior of group animals in nature to find the optimal solution to the problem to be solved. The particle swarm algorithm requires fewer parameters and has the characteristics of simple structure and easy implementation. It has received a lot of attention from researchers and is widely used in multi-objective optimization, nonlinear integer and mixed integer constrained optimization problems. In recent years, heuristic algorithms including particle swarm optimization have begun to be used for neural network training.

The particle swarm algorithm realizes the search in the optimization space through the interaction between different particles in the group to achieve the optimal position, that is, to obtain the optimal solution of the problem. The pseudo code of the particle swarm algorithm is as follows (Table 3).

Table 3. Pseudo code

Step	Action
Step 1	Population initialization
Step 2	Calculate fitness
Step 3	Update the optimal position of individual particle history
Step 4	Update PSO history optimal position
Step 5	Update particle speed and position
Step 6	Repeat 2–6 until the end condition is reached

The update rules of particle speed and position in particle swarm algorithm are as follows:

$$v_{i+1}(t+1) = \omega v_i(t) + c_1 r_1 (pbest_i(t) - x_i(t)) + c_2 v_2 (gbest - x_i(t)) \tag{8}$$

$$x_{i+1}(t+1) = x_i(t) + v_{i+1}(t+1) \tag{9}$$

Among them, pbest is the historical optimal position of the individual, gbest is the historical optimal position of the group, and ω is the inertia weight coefficient, which determines the degree of influence of the result of the previous iteration on this iteration, and is an important parameter of the particle swarm algorithm. At the same time, the adjustment of ω achieves the trade-off between the algorithm's global search and local search.

In the early stage of the algorithm, choosing a larger ω makes the algorithm obtain a stronger global search ability; in the later stage of the algorithm, choosing a smaller ω makes the particles converge to the global optimum.

SVM Parameter Optimization Model Based on Particle Swarm Optimization. Parameters have a greater impact on the classification accuracy of the support vector machine. Before training the training set, the penalty factor C and kernel function parameters need to be determined. This article aims to apply particle swarm optimization to the parameter optimization of support vector machines, and construct a support vector machine optimization model that focuses on the penalty factor C and the kernel function parameter g.

Remember the binary function F(C, g) about the parameters C and g, then this problem can be transformed into the following mathematical problem:

$$\begin{cases} max\ F(C, g) \\ S.t.\ C \in [C_{min}, C_{max}],\ g \in [g_{min}, g_{max}] \end{cases} \tag{10}$$

That is, the classification accuracy of the support vector machine is used as the objective function to obtain C and g in a given search area. The algorithm flow is shown in Fig. 2.

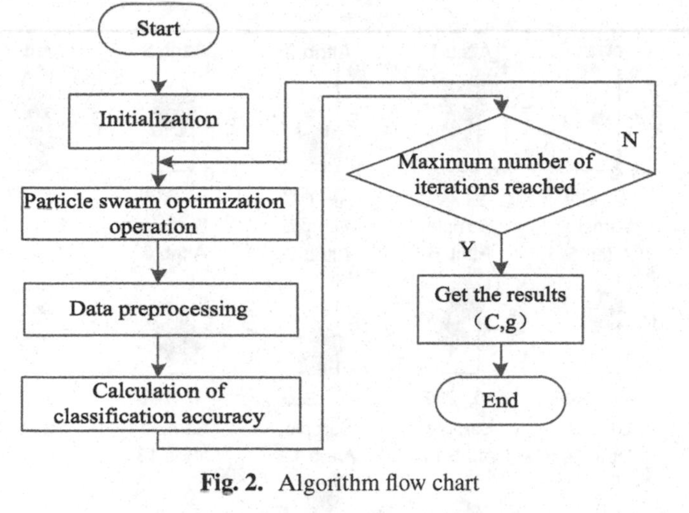

Fig. 2. Algorithm flow chart

4 Experiment and Simulation

In order to verify the validity of the model, the standard dataset wine published by UCI is tested in Windows 7 × 64 PC, matlab 2017 (a) and libsvm-3.21 toolbox.

The description of wine dataset information is shown in Table 4. The visualization of the dataset is shown in Fig. 3a and 3b.

The kernel function of support vector machine (SVM) in the experiment is RBF kernel. 50% data of all kinds of samples in wine dataset are used as training set, and 50% of data is used as test set. After 200 iterations, the optimal (C, g) number pairs

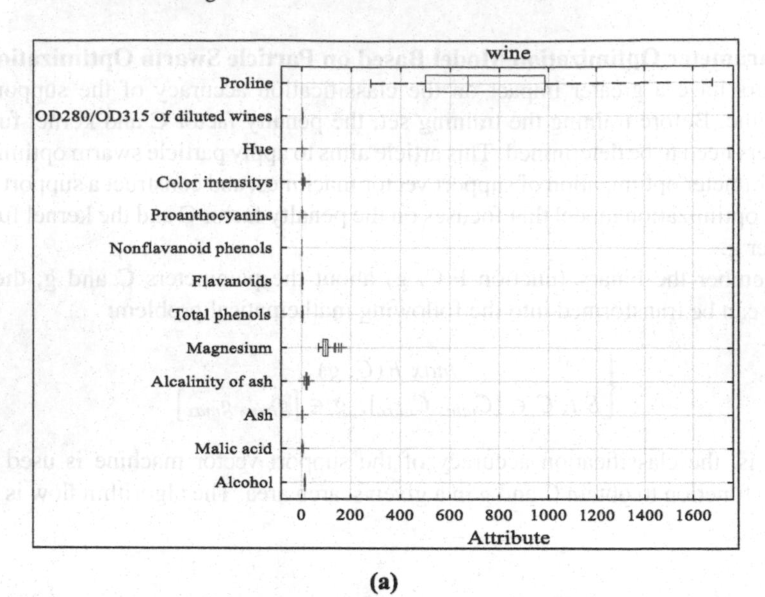

(a)

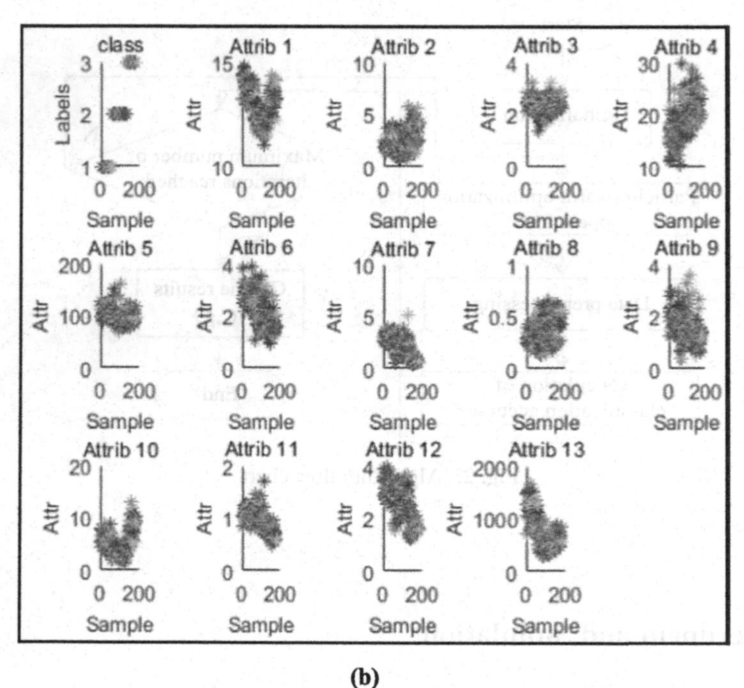

(b)

Fig. 3. (a) Dataset visualization. (b) Attribute visualization

obtained are (11.70, 4.91). The classification accuracy of support vector machine model based on this parameter is 97.75% for wine test set prediction. The relationship between evolution algebra and fitness is shown in Fig. 4.

Table 4. Dataset information

Dataset	Number of data	Attribute dimension	Labels
Wine	178	13	3

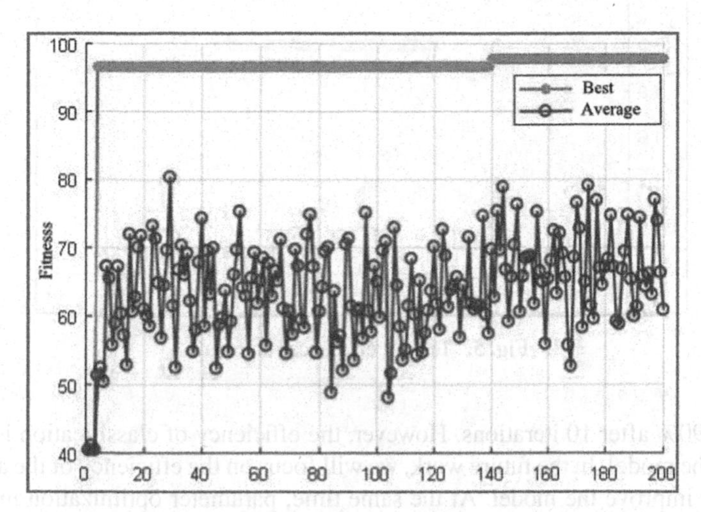

Fig. 4. Relationship between iteration times and fitness

In order to further explore the accuracy of SVM classification, the real classification of WINE data set was compared with the predicted situation.

The wine test set is tested by using the optimal training model obtained from the 200 generations evolution, and the results are summarized. The test classification effect is shown in Fig. 5.

Experimental results show that the algorithm can achieve high fitness when the number of iterations is 0–20. Moreover, the final iterative results of the algorithm are applied to SVM, and have good classification results.

5 Citations

Because of its superior performance, support vector machine has been widely concerned by the academic circle. SVM has good robustness and has been applied well in many fields. However, the classification performance of SVM is greatly affected by its parameters. Therefore, the exploration of SVM parameter mechanism is helpful to improve its classification performance, which has been explored in the industry and academia at present.

This paper introduces the principle of support vector machine, and then explores the influence of kernel function on its classification accuracy. In the paper, the SVM parameter optimization model based on particle swarm optimization algorithm has a certain improvement in classification accuracy, and the classification accuracy rate can reach

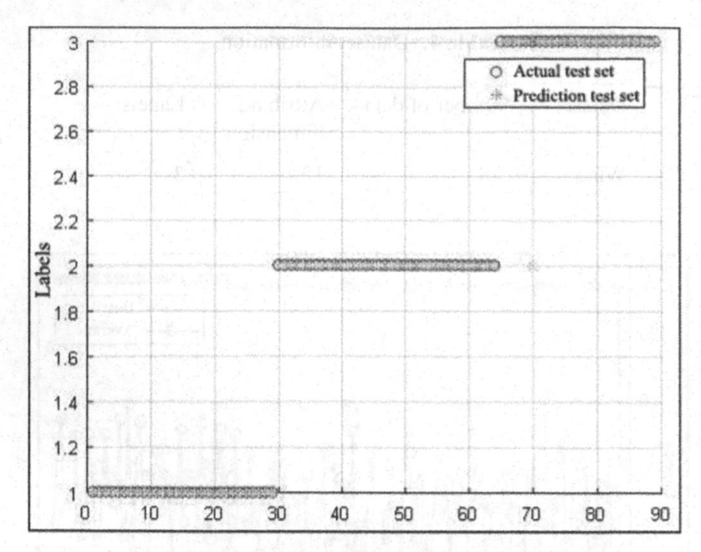

Fig. 5. Test set classification result

more than 90% after 10 iterations. However, the efficiency of classification is not considered in the model. In the future work, we will focus on the efficiency of the algorithm, and further improve the model. At the same time, parameter optimization mechanism of support vector machine based on group fusion intelligent algorithm and multi-stage indefinite sum vector machine are also possible optimization directions. Next, the focus of the research work is to use different kernel functions and intelligent optimization algorithms in different stages in order to achieve higher classification accuracy and efficiency.

Acknowledgement. This research is based upon works supported by the Grant 19YG02, Sanming University.

Conflicts of Interest. The authors declare that they have no conflicts of interest to report regarding the present study.

References

1. Cortes, C., Vapnik, V.: Support-vector networks. Mach. Learn. **20**(3), 273–297 (1995)
2. Shi, N., Xue, H., Wang, Y.Y.: Two-phase indefinite kernel support vector machine. J. Front. Comput. Sci. Technol. 1–9 (2019)
3. Zhang, X.Y., Li, Q.: The summary of text classification based on support vector machines. Sci. Technol. Inf. 344–345 (2008)
4. Zhang, X., Zhang, F., Xiao, X., Ren, X.H.: Fault diagnosis on strip snap of cold continuous rolling based on QPSO-SVM. Metall. Ind. Autom. **44**(06), 17–24 (2020)
5. Zhou, L.N., Lv, X.Y.: Research on DGA family classification method based on SVM. China Sciencepaper **15**(11), 1328–1333 (2020)

6. Yu, T.F., Zhang, H.J.: NOx generation prediction model of coal-fired boiler based on SVM and RBF neural network. Comput. Simul. **37**(09), 209–213+316 (2020)
7. Zhao, Y.K., Shen, L., Wang, X.L., Zhang, M.H.: Composite modulation recognition based on cascaded SVM and all-digital receiver. J. Hangzhou Dianzi Univ. (Nat. Sci.) **40**(05), 18–24 (2020)
8. Zhang, J.F., Wang, Z., Cui, W.S., Liu, M.: Research on an unbalanced data classification method based on SVM. J. Northeast Normal Univ. (Nat. Sci. Ed.) **52**(03), 96–104 (2020)
9. Ma, L.L., Guo, K.J., Wang, J.G.: A fault diagnosis method of vehicle transmission system based on improved SVM. Trans. Beijing Inst. Technol. **40**(08), 856–860 (2020)
10. Bemani, A., Baghban, A., Shamshirband, S., Mosavi, A., Csiba, P., et al.: Applying ANN, ANFIS, and LSSVM models for estimation of acid solvent solubility in supercritical CO_2. Comput. Mater. Continua **63**(3), 1175–1204 (2020)
11. El Mamoun, M., Mahmoud, Z., Kaddour, S.: SVM model selection using PSO for learning handwritten Arabic characters. Comput. Mater. Continua **61**(3), 995–1008 (2019)
12. Li, R., Liu, Y., Qiao, Y., Ma, T., Wang, B., et al.: Street-level landmarks acquisition based on SVM classifiers. Comput. Mater. Continua **59**(2), 591–606 (2019)
13. Wang, Z., Jiao, R., Jiang, H.: Emotion recognition using WT-SVM in human-computer interaction. J. New Media **2**(3), 121–130 (2020)
14. Mehiar, D.A.F., Azizul, Z.H., Loo, C.K.: QRDPSO: a new optimization method for swarm robot searching and obstacle avoidance in dynamic environments. Intell. Autom. Soft Comput. **26**(3), 447–454 (2020)
15. Lan, S.H., Han, T., Huang, Y.R., Xu, S.Y.: Research on dynamic window algorithm of mine mobile robot based on membrane computing and particle swarm optimization. Ind. Mine Autom. **46**(11), 46–53 (2020)
16. Cao, X.Y., Zhang, X.X.: Research of simplified particle swarm optimization algorithm with weight and learning factor. Comput. Technol. Dev. **30**(11), 30–36 (2020)
17. Chen, K,. Lei, G.M., Zhang, C.Y.: Design of optimization algorithm for building space utilization based on spatialized particle swarm. Electron. Des. Eng. **28**(21), 47–50+55 (2020)
18. Gao, L., Zhang, X., Wang, F.: Application of improved ant colony algorithm in SVM parameter optimization selection. Comput. Eng. Appl. **51**(13), 139–144 (2015)
19. Zhao, X., Liu, D.S.: Analog circuit fault diagnosis and comparison analysis based on SVM optimized by improved fruit fly optimization algorithm. J. Electron. Meas. Instrum. **33**(03), 78–84 (2019)
20. Nie, D.X., Wei, W.K., Zhuang, Z.H., Wu, H.T.: Water quality evaluation based on SVM optimized by the HPSOCS combined PSO and GA. Water Sci. Eng. Technol. 1–4 (2019)
21. Cheng, C.F., Sun, X.E.: Parameter selection of support vector machine based on social emotion optimization algorithm. Mod. Electron. Tech. **42**(12), 108–111+116 (2019)
22. Zhang, W.W., Liu, Y.J., Peng, J.J.: Improved bird swarm algorithm for parameters selection of SVM. Comput. Eng. Des. **38**(12), 3267–3271+3278 (2017)
23. Xu, M.: SVM parameter optimization based on artificial fish swarm algorithm. Shanxi Electron. Technol. (01), 30–33+61 (2019)
24. Qiu, Y.F., Li, Z.Y.: An improved artificial fish swarm algorithm and its application for SVM parameter optimization. Comput. Eng. Sci. **40**(11), 2074–2079 (2018)
25. Chen, J.Y., Xiong, H., Zheng, H.B.: Parameters optimization for SVM based on particle swarm algorithm. Comput. Sci. **45**(06), 197–203 (2018)

A Parameter Selection Method Based on Reinforcement Learning: The QoS-Privacy Trade-off for Adversarial Training Framework

Junping Wan[1], Zhiqiang Cao[1], Ying Ma[2], Gaohua Zhao[3], and Zhe Sun[1](✉)

[1] Guangzhou University, Guangzhou 510006, China
sunzhe@gzhu.edu.cn
[2] State Information Center, Beijing 100045, China
[3] Chinese Academy of Cyberspace Studies, Beijing 100010, China

Abstract. Deep learning is widely used in different fields. When service providers use deep learning models to provide image recognition services to clients, they need to collect data from them and also protect their privacy. An adversarial deep learning model is proposed to help service providers protect user's privacy when guaranteeing the quality of the service. However, how to decide on a threshold number to make an appropriate trade-off between Quality of Service (QoS) and privacy is worth researching. In this paper, we use deep reinforcement learning to automate the threshold searching process. In deep learning, the threshold number is also called hyperparameter. We first use a random hyperparameter to train an adversarial deep learning model. According to the performance of it, we train the hyperparameter optimizer to learn what kind of hyperparameter adjustment is good. Finally, we run the hyperparameter optimizer to get the best threshold. The best trade-off can be found automatically through our approach.

Keywords: Privacy-preserving · Trade-off · Deep reinforcement learning · Hyperparameter optimization

1 Introduction

In recent years, deep learning has achieved great success in the field of computer vision, such as face recognition [1]. Generally, service providers use a large number of images as training data to train deep learning models to provide various services. In order to improve service quality, service providers need to obtain high-quality image data from users as much as possible. However, for users, while obtaining high-quality services, the image data they provide may also leak their privacy. For example, the service provider can obtain private information that has nothing to do with the service from the image. This makes the contradiction between user privacy protection and data value mining inevitable.

To address this problem, some methods such as differential privacy and generalization [2, 3], appropriately blurring the privacy features by adding noises in the image, are used to protect the privacy of users. However, these kinds of methods inevitably

X. Sun et al. (Eds.): ICAIS 2021, CCIS 1422, pp. 338–350, 2021.
https://doi.org/10.1007/978-3-030-78615-1_30

make it more difficult to extract the feature information required for the service in the image and requires experts to spend a certain amount of work to find the appropriate noise addition location. Another method used more frequently is that users upload only the feature information of the image but not the raw pictures. This method usually uses an adversarial network to train a feature extractor, aiming to make the feature information contain more features required by the service(we call it service features below) and fewer features related to privacy(we call it privacy features below) [4]. However, in reality, there often exists a certain correlation between service features and privacy features [5]. In gender recognition, for example, information characteristics such as age will inevitably be extracted. If fewer privacy features are extracted, it may also result in less extraction of service features, which is detrimental to service quality. Therefore, it is necessary to make a trade-off between service features and privacy features when extracting features. How to slightly reduce the service quality and improve the privacy protection effect on the premise that the service demand is basically met is a trade-off issue that must be studied.

In adversarial training, a neural network is used as an image feature extractor, and then two similarly structured networks are set respectively, in which one network is used for service classification and the other is used for privacy classification. The features output by the image feature extractor are used as the input of the two classifiers. The iterative training of the feature extractor makes the image features output by the feature extractor easier to be correctly identified by the service classification network and more difficult to be identified by the privacy classification network [6]. Generally, when training a feature extractor, a parameter lambda is set for the loss function of the extractor to measure the trade-off between service and privacy. How to find the best lambda so as to improve the privacy protection effect as much as possible while ensuring the quality of service requires a lot of attempts. This is a very labor-intensive task, especially when the amount of training data is pretty large. Therefore, how to automatically find an appropriate lambda value, that is, to optimize the value of the lambda variable, is a problem of great research significance. We call this optimization process hyperparameter optimization, where hyperparameters are the parameters set before the neural network starts training.

Reinforcement learning provides a good solution to this problem. In reinforcement learning, an agent is trained to learn the reward value generated by each action so that it can find the best next action according to the current state so as to maximize the total expected return value of the action sequence. For example, Hu et al. [7] use reinforcement learning such as Monte Carlo strategy to enhance the performance of registration, Zheng et al. [8] introduce a reinforcement learning method in pattern recognition to predict future events. The traditional reinforcement learning method is generally to find the next best action by querying the known action value table when the agent is at a certain state. Its disadvantage is that it is not easy to build the value table in face of continuous state and action changes.

Since the lambda variable we want to optimize is continuous, the applicability of traditional reinforcement learning methods is not high, so we introduce deep reinforcement learning [9]. Wei et al. [10] proposed a multi-agent reinforcement learning framework to improve the Device-to-Device communication by creating several agents to conduct a game. In deep reinforcement learning, the neural network can be used as an agent. There is no need to store additional action value tables, and we want to enable the neural network to output specific values at any time according to the state input. Recently, Jia et al. [11] proposed a deep reinforcement learning method using LSTM and Dong et al. [12] proposed a dynamical hyperparameter optimization method using a continuous deep Q-learning action-forecast network to achieve their goals, respectively. Considering that our parameter optimization needs are not complicated, we use DDQN as our method. In addition, to enable the neural network to better learn the profit value under different hyperparameter changes, it is necessary to preprocess the state input data.

In this paper, we introduce an adversarial training framework that balances image service quality and privacy protection. Based on the theory of deep reinforcement learning, the DQN network is used to automatically adjust the hyperparameter lambda in the loss function of the image feature extractor to tune the trade-off between service quality and privacy protection and ultimately find the best hyperparameter lambda. Our main contributions are summarized as follows:

- We propose a hyperparameter optimization method based on deep reinforcement learning, which is used to balance image service quality and privacy protection. In our method, a neural network is trained as an agent to find the best trade-off threshold between service quality and privacy protection by adjusting the hyperparameter lambda.
- We try different preprocessing methods for DDQN data input and find the best data processing method.
- We conduct many experiments to verify that the neural network agent can converge to a certain level so that when a state value is fed into the network, it can output an optimal hyperparameter change plan. We also verify that no matter if the hyperparameter is optimized from any initial value, it can reach a certain optimal value in the end.

The rest of the paper is as follows. We describe related work such as reinforcement learning hyperparameter optimization methods in Sect. 2. Section 3 includes the basic concepts of privacy protection and reinforcement learning hyperparameter optimization. The detailed method proposed in our paper is shown in Sect. 4. In Sect. 5, the feasibility of the method and the evaluation of our hyperparameter optimizer are clearly described. Section 6 concludes this paper.

2 Related Work

2.1 Adversarial Training Network for Privacy Protection

To protect the privacy of raw data, besides adding noises into the data, some researchers use deep learning to achieve their privacy task. Osia et al. [4] use a client-server model to split one network into two parts called classifier and feature extractor, respectively. The feature extractor in one client receives raw data and outputs the extracted feature which is sent to the classifier to output the classification result. The classifier owner will not get the raw data. However, the image feature also can disclose the information of the picture. Liu et al. [13] introduce an adversarial training algorithm to the client-server model to resist the privacy classifier inferring privacy from the extracted features. Ding et al. [14]

proposed a privacy-preserving method in the reasoning stage of workflow. To reduce privacy leakage during feature extraction, they add privacy constraints when feature extracting to strengthen the privacy-preserving ability of the encoder. Wu et al. [15] proposed two novel strategies of model restarting and ensemble to protect privacy in action recognition. For data-driven (supervised) learning, they create a dataset with selected privacy attributes and target task labels. Li et al. [5] considered the privacy inferring and image reconstruction from image features and integrated them into one adversarial training framework with an adjustable trade-off hyperparameter. They adjusted the hyperparameter manually and discuss the effect of each hyperparameter. This approach is not practical enough in reality, because it will cost a lot of human effort to find the appropriate hyperparameter for privacy protection. So, it is essential to automate the hyperparameter searching process.

2.2 Deep Reinforcement Learning for Hyperparameter Optimization

In hyperparameter optimization, random search and grid search [16] are widely used. Xue et al. [17] adopted a parameter adaptive mechanism based on PSO and solved the feature problem. Zhang et al. [18] used EPLL to select parameter adaptively. More methods such as Bayesian optimization [19] works well. However, these methods generally work based on random trying on some structural hyperparameters such as training batches, number of network layers, etc. When addressing the problem of adjustment of coefficient in the loss function, reinforcement learning provides a good idea for the automation of parameter optimization.

Jia et al. [11] proposed a deep reinforcement learning architecture RPR-BP. They use a long short-term memory network as an agent that maximizes the expected accuracy of an ML model on a verification set. It selects a set of hyperparameters during each iteration. For updating its internal parameters, it uses the accuracy of the model on the verification set as the reward. Dong et al. [12] proposed a dynamical hyperparameter optimization method using a continuous deep Q-learning action-forecast network to achieve adaptive hyperparameter optimization for a given sequence. Because the observation space of visual object tracking is much complicated, to process high-dimensional state space and speed up the convergence, they bring in an efficient heuristic strategy. They use it to improve the correlation-filter-based tracker and the Siamese-based tracker. As can be seen above, reinforcement learning is good at dealing with continuous parameter optimization. Based on deep reinforcement learning, we can make a trade-off by training a hyperparameter optimizer.

3 Preliminaries

3.1 Motivations and Goals

Many well-known data service providers such as Google Inc. [20] and Apple Inc. [21] have taken some privacy to protect measures like differential privacy. Considering that experts need to make great efforts to find appropriate characteristics of the disturbance in differential privacy, researchers introduce adversarial learning [5] to build a multi-task recognition framework, in which the features related to service are tend to be recognized when the private features are not. Specifically, as for an image recognition task T, we denote the attribute needed to be identified for providing service as S and the private attribute that can be recognized by malicious attacker from the image as P. To prevent the private attribute P from being recognized correctly and preserve the attribute S to support the service and protect privacy as much as possible, we aim to build an image feature extractor to output image features and adjust the hyper-parameters in the extractor to control the feature ratio of S and P in the image features so as to find an appropriate trade-off point.

In general, the adjustment of hyper-parameter requires manpower to record the effect of the model as a reference. The process is constantly in random trials and it is time-consuming. The success in reinforcement learning has drawn researchers' great attention and inspire them to adapt it in hyper-parameter optimization. The basic idea of reinforcement learning is to create an agent to simulate the trying process of humans. We aim to use reinforcement learning to train a hyper-parameter optimizer to automate the parameter adjustment process.

3.2 Adversarial Training Framework

We firstly set a feature extractor to extract the image's feature which is needed for the service task. Then we set a task classifier to achieve the given service task using the extracted features. We also set an adversarial classifier and train it to infer the privacy to simulate the attacker. As shown in Fig. 1, the adversarial training framework used in our paper consists of three parts: feature extractor (FE), task classifier (TC), privacy classifier (PC). The concrete details are as follows.

Feature Extractor (FE) is used for image feature extraction. It is a raw picture input interface, and its output has no obvious meaning. However, when its output is used as the input of another classifier, the information contained in FE's output can be reflected according to the performance of the classifier. Since FE in adversarial training serves as a common part of the classification, it can be considered as parts of the deep neural network used in classification, in which there are multiple convolutional layers, max-pooling layers, and batch normalization layers. It is supposed to be trained with TC and PC together.

Target classifier (TC) is the essential part of application service, and it is also a neural network that receives input from the feature extractor. To improve the performance of TC, the service provider can adjust the structure of the neural network used according to classification task. The intermediate components of network are usually convolutional neural network (CNN), and the last layer of the network is the fully connected layer which outputs the probability of each possible classification result. We consider the classification result with the largest probability as the final classification output of the network. The difference between the output label of the network and the real label can reflect the performance of TC through the cross-entropy loss function Eq. (1).

$$L(TC) = -\sum_{j=1}^{N}\sum_{i=1}^{M} x_{ij}\log(x_{ij}^{'}) + (1 - x_{ij})\log(1 - x_{ij}^{'}) \tag{1}$$

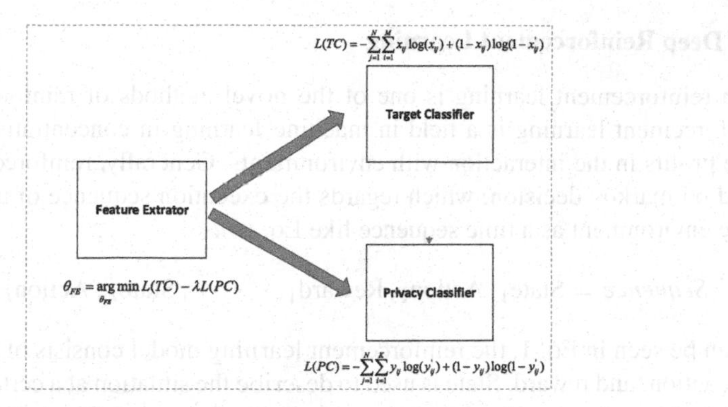

Fig. 1. The structure of the adversarial training framework

where N is the number of samples and M is the number of all classes. x_{ij} is the true label of the task, and $x_{ij}^{'}$ is the prediction given by TC. To make the classifier's prediction more accurate, the training process aims to minimize the above formula Eq. (2).

$$\theta_{TC} = \arg\min_{\theta_{TC}} L(TC) \tag{2}$$

Privacy classifier (PC) usually has the similar structure as TC, but different from TC's task is to complete the target classification task, the function of PC is to perform privacy classification from feature input, so as to achieve inferring privacy. The service provider can also amend the structure of PC according to the requirements. The performance of PC is evaluated by the similar cross-entropy loss function Eq. (3).

$$L(PC) = -\sum_{j=1}^{N}\sum_{i=1}^{K} y_{ij}\log(y_{ij}^{'}) + (1 - y_{ij})\log(1 - y_{ij}^{'}) \tag{3}$$

where y_{ij} is the true label of privacy, and $y_{ij}^{'}$ is the prediction given by PC. The optimization of PC is as the following formula Eq. (4).

$$\theta_{PC} = \arg\min_{\theta_{PC}} L(PC) \tag{4}$$

In the whole framework, we need to constantly improve the quality of service when preventing privacy leakage. We aim to limit the accuracy of PC to an acceptable value and enhance the accuracy of TC as much as possible. Consequently, we need to train FE using the following formula Eq. (5).

$$\theta_{FE} = \arg\min_{\theta_{FE}} L(TC) - \lambda L(PC) \tag{5}$$

where λ ranges from 0 to 1. When λ increases, the privacy protection is enhanced and QoS decreases.

In addition, the loss function is flexible, that is, the loss function can be not the same in different scenarios. This illustrates the trade-off parameter assignment is not strongly relevant to the concrete loss function used in the classifier's training. This adversarial can be widely used in trade-off assignment.

3.3 Deep Reinforcement Learning

Deep reinforcement learning is one of the novel methods of reinforcement learning. Reinforcement learning is a field in machine learning in concentrating on how to get more profits in the interaction with environments. Generally, reinforcement learning is based on markov decision, which regards the execution sequence of individual actions in the environment as a time sequence like Eq. (6).

$$Sequence = \text{State}_1, \text{Action}_1, \text{Reward}_1, \cdots\cdots, \text{State}_T, \text{Action}_T, \text{Reward}_T \quad (6)$$

As can be seen in Eq. 1, the reinforcement learning model consists of three main parts: state, action, and reward. State is used to describe the situation at a certain time t. Action is used to describe the action taken by an individual at time t. Reward is used to describe the reward obtained when an individual takes an action at time t along with that a state transition occurs. In the model, an action value is calculated for each action using the following equation Eq. (7).

$$Q(s, a) = R_s^a + \gamma \sum_{s' \in S} P_{ss'} \sum_{a' \in A} \pi(a'|s') Q(s', a') \quad (7)$$

The result is saved as a reference corresponding to each action. During the action selection process in a state, the next action with the greatest value will be selected and taken. The actions and states considered in traditional reinforcement learning are generally discrete values, however, the value of the action is difficult to be stored in a searchable list when it is a continuous value because the number of states and actions are infinite.

Deep reinforcement learning can address this problem well. Instead of creating a list to store the specific value of each action, reinforcement learning aims to train a neural network that can immediately output the value of all subsequent actions for a given state input, that is, train a neural network as an interface that outputs value results in real-time instead of pre-saving all value results for a query. To facilitate the neural network to learn to estimate the value of an action, deep reinforcement learning will set up an experience pool replay to record the value of several actions and update the neural network parameters in batches until the neural network converges. Mnih et al. [9] first proposed deep reinforcement learning. After that, double DQN is proposed to make the network converges quicker and behaves better. Some other methods such as prioritized experience replay and dueling DQN improve the training efficiency in different ways [22].

4 Framework Design

4.1 System Overview

Our system consists of two parts called the adversarial training framework and the hyperparameter optimizer. Among them, the adversarial training framework is used to train a feature extractor, a service classifier, and a privacy classifier under a given hyperparameter. We take the features extracted by the feature extractor as input and send them to the service classifier and privacy classifier respectively. The accuracy of the service classifier is trained to be higher and the accuracy of the privacy classifier is trained to be lower. Then, we train the hyperparameter optimizer according to the model effect of the adversarial training framework, so that it can take any state value as input and output the value of all optional actions after that state. The interaction between adversarial training framework and hyperparameter optimizer is described as Fig. 2.

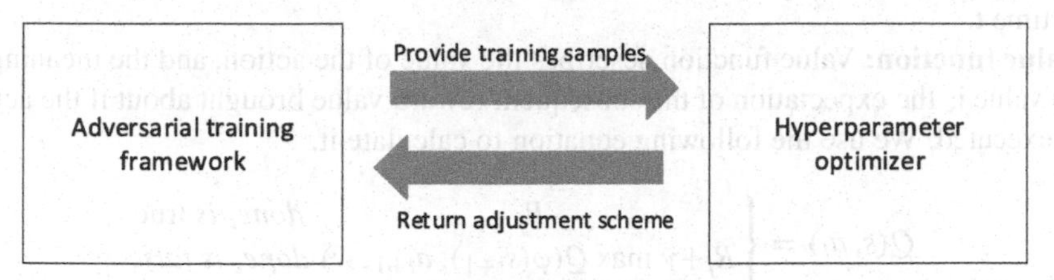

Fig. 2. The interaction between adversarial training framework and hyperparameter optimizer

After the training is done, for any initial hyperparameter, the best hyperparameter adjustment scheme can be obtained by inputting it into the hyperparameter optimizer. The final optimization result can be obtained through continuous adjustment of the hyperparameter.

4.2 Hyperparameter Optimization Process

The complete process consists of sample collection process, network optimization process, and hyperparameter optimization process.

Sample Collection Process. When the adversarial training model runs for the first time, we set a Long Short Term Memory network (LSTM) as the initial hyperparameter optimizer network. We use the current hyperparameter to train the adversarial training framework and obtain the service task and privacy prediction accuracy in the adversarial framework to be used for the subsequent calculation of action reward of hyperparameter changes. To make it clear, we list the terms of our framework in Table 1 and explain the detail of them.

Table 1. The detailed terms

Name	Notation	Description
State	S_i	The lambda set at time i
Action	a_i	The increase or decrease of lambda at time i
Reward	R_i	The reward corresponding to accuracy at time i
Value function	$Q(s_i, a_i)$	The score of action a_i at state S_i
Exploration rate	ε	The probability of choosing the best action
Discount rate	γ	The decay sum of subsequent value function

- **State:** State in our problem includes the lambda used in the previous T-1 moments. We denote it as $state_t = (\lambda_{t-T+1}, \cdots, \lambda_{t-1}, \lambda_t)$.
- **Action:** Action is the adjustment operation, specifically refers to the increase or decrease of a certain value of λ.
- **Reward:** Reward refers to the reward or penalty introduced according to the performance change of the adversarial training framework when the lambda value is adjusted at time t.
- **Value function:** Value function describes the value of the action, and the meaning of its value is the expectation of the subsequent reward value brought about if the action is executed. We use the following equation to calculate it.

$$Q(s, a_t) = \begin{cases} R_j & done_t \text{ is true} \\ R_j + \gamma \max_{a_{t+1}} Q(\varphi(s_{t+1}), a_{t+1}, w) & done_t \textit{ is false} \end{cases} \tag{8}$$

where R_j is the reward for the execution of action a_j, $done_t$ states that whether the action execution result meets the requirement of adversarial training framework.

- **Exploration rate:** exploration rate is a value range from 0 to 1. To execute more possible action, it is necessary to perform the best action with a probability of ε and randomly act with a probability of $1 - \varepsilon$ to avoid the local optimum.
- **Discount rate:** the calculation of value function is essentially the sum of all subsequent value function values. The discount rate reflects the importance of each subsequent reward when calculating the value. The lower the discount rate, the less is the rewards in the further times.

After the adversarial framework is trained, the service task accuracy rate and privacy estimation accuracy rate under the framework are again obtained, the current accuracy rate and the previous accuracy rate are compared, and whether the optimization is over is recorded. The reward value corresponding to the parameter changes is standardized after that.

Subsequently, the value of the preceding and following hyperparameters, the reward value, and the optimization end flag $(S_i, a_i, R_i, S_{i+1}, done_i)$ are stored as a five-tuple to the experience pool replay.

This process is repeated until the number of samples in the experience pool reaches N and enters the network optimization process.

Network Optimization Process. When the number of samples in the experience pool reaches N, we randomly select M samples like tuple $(S_i, a_i, R_i, S_{i+1}, done_i)$ from the experience pool. As can be seen in the tuple, R_i is the reward, which is used to calculate the value function in Eq. (8).

Now we can input the state S_i into the hyperparameter optimizer to obtain the estimated value of the best action. We use the difference between two values to update the network using the mean square error (MSE) function and backpropagation algorithm.

$$Loss = \frac{1}{m} \sum_{j=1}^{m} (y_j - Q(s, a_t))^2 \tag{9}$$

After the update is complete, the process returns to the sample collection process. The whole flow is shown in Algorithm 1.

Algorithm 1. The Training Process of Hyperparameter Optimizer

Input: initial state S_1

1: Initialize replay memory M
2: While optimizer is not convergent:
3: input S_i into the optimizer
4: choose the action a_i with the highest ouput using ε-greedy
5: execute the action, get S_i and calculate R_i
6: add $(S_i, a_i, R_i, S_{i+1}, done_i)$ into M
7: set S_{i+1} as S_i
8: sample a batch of tuples:
9: Update the optimizer using Eq. (8) and Eq. (9)

Hyperparameter Optimization Process. When the hyperparameter optimizer is trained to be converged, we randomly select an initial hyperparameter and apply it to the adversarial training framework to obtain its performance. If the performance does not meet the expected requirements, input the current hyperparameter into the hyperparameter optimizer to obtain the next adjustment scheme and apply it until the performance of the adversarial training framework reaches the target requirement, that is, the privacy accuracy rate is reduced to an acceptable value and the target accuracy rate is enough high to provide great service quality.

5 Experiment Evaluation

5.1 Experimental Settings

Adversarial Training Network. To evaluate the performance of our hyperparameter optimization method, we conduct an adversarial training framework. Our adversarial training is conducted with Python (version 3.83), Numpy (version 1.18.5), and Pytorch (version 1.6.0). Particularly, we conduct our experiment on a server with NVIDIA TELSA T4 GPU and 16 GB memory. We use the well-known CelebA dataset, which contains 202,599 photos of 10,177 different people, to conduct our smile recognition of photos based on CNN. We only use a subset of dataset to conduct the adversarial training network to speed up our experiment and explore the usability of our work. We conduct the experiment based on [5], which is also described in Sect. 3.2. We split the network into two parts: feature extractor, classifier. The structure of feature extractor and classifier can be seen in [5].

Hyperparameter Optimizer Network. We also implement a hyperparameter optimizer using a double deep q-network (DDQN). The network is based on LSTM which receives the λ state vector. The reward is calculated using the changes of accuracy and the action is to increase or decrease a small number. We vary the initial λ as $\lambda = \{0.1, 0.5, 1\}$

and observe the changing process of the parameters after the hyperparameter optimizer is converged. The changes of the parameter are shown in Fig. 3(a).

As can be seen, when we input $\lambda = 0.1$, the optimizer guides the process to increase λ, and then starts to decrease to a stable value, which means that when $\lambda = 0.6$, we get the nearly best optimization result. When we set $\lambda = 1$ as the initial value, it converges to a value near 0.6. This is because the accuracy of privacy is almost the same when λ ranges from 0.5 to 0.8, which means the reward is nearly the same. The accurate result is shown in Fig. 3(b).

The experiment result shows that the optimizer can help find the similar value of the best value but not the exact best value. This is because the learning effect is strongly related to the structure of the state vector and the formula for calculating reward. In addition, the training data size in our experiment is not large, which also may result in the inaccuracy of our experimental result.

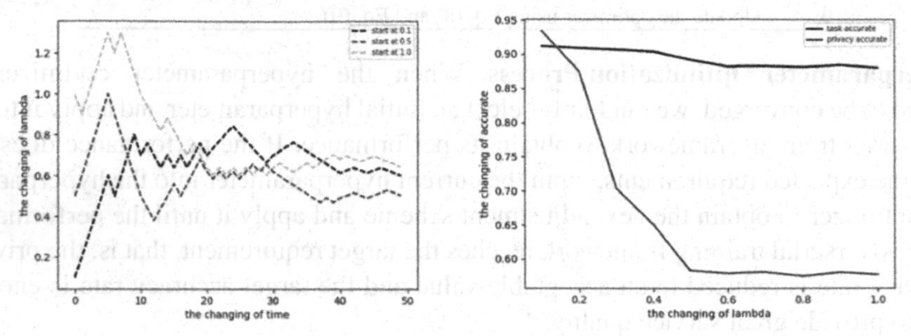

Fig. 3. (a) The changing of λ in optimization, (b) task accurate and privacy accurate corresponding to λ

6 Conclusion

In this paper, we address the trade-off between Quality of Service (QoS) and privacy protection problems using deep reinforcement learning to automate the threshold searching process. We use DDQN as a hyperparameter optimizer and train it, so that it can help us find the appropriate hyperparameter to meet the trade-off requirements. Specifically, our trained hyperparameter can provide a value function as a reference when adjusting the hyperparameter, and after several adjustments, we can get the final threshold we need. We prove that our method works well in our problem, and the future work is trying to use a small batch of data to train the optimizer and apply it in other situations with a huge amount of data, which makes our method more practical.

Funding Statement. This work was supported in part by the National Key R&D Program of China (No. 2018YFB2100400), in part by Zhejiang Lab (No. 2020NF0AB01), in part by the National Natural Science Foundation of China (No. 62002077, 61872100), in part by the China Postdoctoral Science Foundation (No. 2020M682657), in part by the Guangdong Basic and Applied Basic Research Foundation (No. 2020A1515110385), in part by the Key Research and Development Program for Guangdong Province (No. 2019B010136001).

References

1. Chan, T.H., Jia, K., Gao, S.: PCANet: a simple deep learning baseline for image classification? IEEE Trans. Image Process. **24**(12), 5017–5032 (2015)
2. Dwork, C., Roth, A.: The algorithmic foundations of differential privacy. Found. Trends Theor. Comput. Sci. **9**(3–4), 211–407 (2014)
3. Bhaskar, R., Bhowmick, A., Goyal, V., Laxman, S., Thakurta, A.: Noiseless database privacy. In: Lee, D.H., Wang, X. (eds.) Advances in Cryptology – ASIACRYPT 2011. LNCS, vol. 7073, pp. 215–232. Springer, Heidelberg (2011). https://doi.org/10.1007/978-3-642-25385-0_12
4. Osia, S.A., Taheri, A., Shamsabadi, A.S.: Deep private-feature extraction. IEEE Trans. Knowl. Data Eng. **32**(1), 54–66 (2018)
5. Li, A., Duan, Y., Yang, H.: TIPRDC: task-independent privacy-respecting data crowdsourcing framework for deep learning with anonymized intermediate representations. In: Proceedings of the 26th ACM SIGKDD International Conference on Knowledge Discovery and Data Mining, pp. 824–832. Association for Computing Machinery, New York (2020)
6. Sun, Z., Yin, L., Li, C.: The QoS and privacy trade-off of adversarial deep learning: an evolutionary game approach. Comput. Secur. **96**, 101876 (2020)
7. Hu, J., Luo, W., Wang, X.: End-to-end multimodal image registration via reinforcement learning. Med. Image Anal. **68**, 101878 (2021)
8. Zheng, H., Shi, D.: A multi-agent system for environmental monitoring using Boole-an networks and reinforcement learning. J. Cybersecurity **2**(2), 85 (2020)
9. Mnih, V., Kavukcuoglu, K., Silver, D.: Playing atari with deep reinforcement learning. arXiv preprint arXiv:1312.5602 (2013)
10. Wei, Y., Qu, Y., Zhao, M.: Resource allocation and power control policy for device-to-device communication using multi-agent reinforcement learning. Comput. Mater. Continua **63**(3), 1515–1532 (2020)
11. Jia, W.U., SenPeng, C., XiuYun, C.: RPR-BP: a deep reinforcement learning method for automatic hyperparameter optimization. In: 2019 International Joint Conference on Neural Networks (IJCNN), pp.1–8. IEEE, Piscataway (2019)
12. Dong, X., Shen, J., Wang, W.: Dynamical hyperparameter optimization via deep reinforcement learning in tracking. IEEE Trans. Pattern Anal. Mach. Intell. **43**(5), 1515–1529 (2019)
13. Liu, S., Shrivastava, A., Du, J.: Better accuracy with quantified privacy: representations learned via reconstructive adversarial network. arXiv preprint arXiv:1901.08730 (2019)
14. Ding, X., Fang, H., Zhang, Z.: Privacy-preserving feature extraction via adversarial training. IEEE Trans. Knowl. Data Eng. (2020). https://doi.org/10.1109/TKDE.2020.2997604
15. Wu, Z., Wang, H., Wang, Z.: Privacy-preserving deep action recognition: an adversarial learning framework and a new dataset. IEEE Trans. Pattern Anal. Mach. Intell. (2020). https://doi.org/10.1109/TPAMI.2020.3026709
16. Bergstra, J., Bengio, Y.: Random search for hyper-parameter optimization. J. Mach. Learn. Res. **13**(1), 281–305 (2012)
17. Xue, Y., Tang, T., Pang, W.: Self-adaptive parameter and strategy based particle swarm optimization for large-scale feature selection problems with multiple classifiers. Appl. Soft Comput. **88**, 106031 (2020)
18. Zhang, J., Qin, Z., Wang, S.: A new adaptive regularization parameter selection based on expected patch log likelihood. J. Cybersecurity **2**(1), 25 (2020)
19. Snoek, J., Larochelle, H., Adams, R.P.: Practical Bayesian optimization of machine learning algorithms. In: Proceedings of the 25th International Conference on Neural Information Processing Systems – vol. 2, pp. 2951–2959. Association for Computing Machinery, New York (2012)

20. Erlingsson, U., Pihur, V., Korolova, A.: Rappor: randomized aggregable privacy-preserving ordinal response. In: Proceedings of the 2014 ACM SIGSAC Conference on Computer and Communications Security, pp. 1054–1067. Association for Computing Machinery, New York (2014)
21. Fang, B., Jia, Y., Li, X.: Big search in cyberspace. IEEE Trans. Knowl. Data Eng. **29**(9), 1793–1805 (2017)
22. Fang, W., Pang, L., Yi, W.: Survey on the application of deep reinforcement learning in image processing. J. Artif. Intell. **2**(1), 39–58 (2020)

Word Research on Sentiment Analysis of Foreign News Reports Concerning China Based on the Hybrid Model of Opinion Sentence and BLSTM-Attention

Xiaohang Xu(✉) and Guowei Chen

School of New Media Studies, Communication University of China, Beijing 100024, China

Abstract. This paper proposes a sentiment analysis model based on a mixture of opinion sentences and BiLSTM-Attention mechanism, which is used to solve the problem of increasing foreign news reports concerning China in recent years that urgently need computers to assist in judging the sentiment polarity of China in reports. Firstly, text analysis is performed on the article to extract emotional sentences with emotional expression. Combined with the deep learning method for identification and judgment. Finally, use the foreign media reports on China in the actual project database as experimental data, use 70% of manual labeling and 30% for experimental verification. The experimental results show that compared with some existing methods, this method has a higher accuracy in the sentiment judgment of Chinese-related articles.

Keywords: Format natural · Language processing · Sentiment analysis · Attention mechanism · Opinion sentence

1 Introduction

With the increasing demand for machine to understand human language and use machine to process text on this basis, natural language processing (NLP) has become an increasingly hot field of artificial intelligence research in recent years. Research in the field of natural language processing includes text similarity research, machinc writing, machine translation, sentiment analysis and so on. Sentiment analysis is the process of identifying users' subjective emotions, opinions and attitudes from text data [1]. Using sentiment analysis can well grasp the trend of public opinion and user sentiment. At present, sentiment analysis research is widely used in various fields, such as: sentiment analysis of user comments on major e-commerce platforms [2] to obtain users' feelings for commodity to improve quality control; sentiment analysis of users' MicroBlog [3]; sentiment analysis for financial texts [4], sentiment analysis for film reviews [5], etc.

In recent years, foreign media reports and articles on China have increased day by day, and most of them have media emotions. For example, BBC reporters published insulting reports on China and misinterpreted reports on China about the new coronavirus. Extracting the emotions of foreign reports on China has a positive effect on

X. Sun et al. (Eds.): ICAIS 2021, CCIS 1422, pp. 351–360, 2021.
https://doi.org/10.1007/978-3-030-78615-1_31

public opinion control. For large-scale text data sets, how to extract the sentiment of the report quickly, how to select the suitable and efficient model to process the text, and how to ensure the accuracy of sentiment analysis are the urgent problems to be solved.

2 Research Status of Text Sentiment Analysis

In response to the current situation of the increasing number of foreign media reports on China and articles with special emotional color this year. Studying the sentiment of foreign media's articles on China is of great significance for public opinion guidance and targeted response to false statements. Up to now, the research methods of sentiment analysis are mainly divided into two categories: the method based on emotion dictionary and the method based on machine learning.

2.1 Based on Emotion Dictionary

The method based on emotion dictionary is a commonly used method in the early research of sentiment analysis. The main method is to use existing sentiment dictionaries, such as HowNet, WordNet, National Taiwan University Sentiment Dictionary (NTUSD), and General Inquirer (GI), emotion Dictionary of Harbin Institute of Technology, etc., or construct an emotional dictionary suitable for a specific field based on the existing basic dictionary to perform feature extraction, emotional word acquisition, emotional word matching and other operations for emotional analysis. Zhao [2] et al. constructed an emotional dictionary for capturing data subject areas to explore the emotional expression characteristics of different network platforms; Cao [6] et al. introduced an emotional dictionary to generate emotional labels to assist word vector generation, so that word vectors have emotional information; Hou [7] et al. used BosonNLP sentiment dictionary (Boson sentiment dictionary) matching feature point of view to complete the fine-grained sentiment analysis of film reviews; Chen [8] et al. used Hownet sentiment dictionary and Taiwan University simplified Chinese sentiment polarity dictionary (NTSUSD) to construct a hybrid dictionary for feature selection.

2.2 Based on Machine Learning

Due to the disadvantages of the affective dictionary method such as low efficiency and slow updating, coupled with the vigorous development of the deep learning field of machine learning, the method of using machine learning to construct neural network for emotion analysis has gradually become the mainstream. Cao [9] et al. proposed an attention-gated convolutional network model for target emotion analysis. At last, Softmax classifier was used to classify emotion features and output emotional polarity. Zeng [10] et al. proposed a hierarchical biattention neural network model aiming at the task of emotion classification at chapter level, which achieved good results in IMDB, YELP 2013 and YELP 2014 data sets. Zhao [11] et al. constructed a bi-directional long-duration memory neural network and a convolution neural network (BILSTM-CNN) serial hybrid model aiming at the problems of low accuracy, poor real-time performance and insufficient feature extraction of existing text emotion analysis methods. Peng [12] et al. proposed a network model combining two-way long and short memory network and aspect attention module (BLSTM-AAM), aiming at the difficulties of emotional polarity and ambiguous emotional polarity in many different aspects included in the sentence. Meng et al. [13] proposed a CRT mechanism mixed neural network for specific target emotion analysis. Although the current sentiment analysis research method based on machine learning is largely improve the training efficiency and discriminant accuracy, but the word did not express emotion words in highlighting the role, in order to solve this situation, given the Attention mechanism in image processing area for a specific part of the focus on care, will be introduced to the Attention mechanism analysis in the study of emotion.

According to the characteristics of foreign media reports related to China, considering the deficiencies of the emotion-based dictionary method and the redundancy of irrelevant information in the article, this paper USES viewpoint sentences and improves the machine-learning-based method to conduct the research on emotion analysis.

3 Emotion Analysis Algorithm of Foreign China-Related News Reports Based on Viewpoint Sentence and BLSTM-Attention Hybrid Model

3.1 Algorithm Idea and Process

The purpose of this paper is to study the foreign media reports released by the China emotion, found in manual analysis after reports a lot of marking report objective statement of paper length is larger, sentences and less evident in the easy identification of emotional expression, if training directly to the full text content analysis accuracy will increase the workload and influence the results, so in this paper, first of all to view other recognition, the article find articles sentences can express emotion; Secondly, the method of deep learning neural network is used for data processing. In view of the insufficient attention given to keywords by the existing deep learning method, the attention mechanism in image processing is introduced to the output processing of the neural network layer.

In the first step of the algorithm, the text is filtered through the viewpoint sentence rules to form the viewpoint sentence set. The second step is to embed the extracted viewpoint sentences with words. This step is applied to the Word2Vec model to transform the sentences into word vectors for further training.

Considering the importance of context in ee, the deformable bidirectional LSTM (BiLSTM) of LSTM (long and Short term memory neural network) is used to extract context information.

In order to fully reflect the different importance of different words in the expression of emotion, the attention mechanism is introduced after BiLSTM layer training, which can assign different weights to each word to form emotional characteristics.

Softmax classifier outputs the emotional polarity of each opinion sentence. See the next section for implementation details of each step of the model.

3.2 Algorithm Flow

Suppose the sentence extracted from the opinion sentence of article D is S1...... Sn, each sentence is made up of several words, Sk = {Wk1,Wk2... Wkp}. After embedding

Word2Vec word in each sentence and WkiSk for each word, VkiSkd can be obtained, where, I represents the ith word in the sentence and D represents the dimension of the vector. The embedded vector of the word is then thrown into BiLSTM, and the input sequence is read forward to LSTM (Vk1,Vk2… Vkp), then configure a reverse LSTM to read the sequence (Vkp… Vk1), the forward and the back are merged together in the way of connection as the Bi-LSTM output result hi in the hidden state of I at time. Finally, after the Attention mechanism assigns weights to the hidden sequence, softmax function is used to determine the final emotional tendency (Fig. 1).

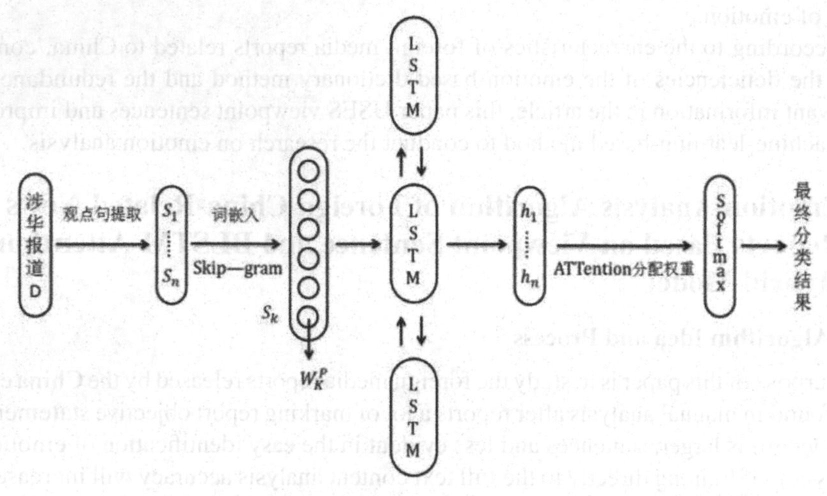

Fig. 1. Block diagram of the whole process

4 Model

4.1 Rules for Extracting Opinion Sentences

According to the characteristics of Chinese-related reports in foreign media, the author firstly extracts the viewpoint sentences from the collected texts. Firstly, a large number of sentence characteristics of Chinese-related reports are analyzed, and the extraction rules of viewpoint sentences are as follows:

Rule 1: First, set the targets that are often targeted by comments and emotional colors, like "China", "Xianye", "Sun", "Sun", "Osan", "Shanghai", "China virus", "wuhan" and "government", among others, and leave the sentences that contain them unfiltered.

Rule 2: Quote non-opinion sentences. Filter. Identify.

The Chinese Foreign Minister is visiting India this month. = The Chinese foreign Minister is visiting India this month.

Rule 3: Sentences with Numbers and dates are almost always objective sentences. Filter them.

The two sides have already held over 20 rounds of talks under the framework of SR Dialogue which was set up to find an early solution to The border dispute.

He declared, declared, asserted, concluded, etc. that one means not of opinion. He declared, said, he concluded…

Such as: The rush of new entrants and absence of regulatory oversight proved a toxic combination, as P2P lenders raced to offer ever higher interest rates to tempt investors, said Assistant Professor Gui Zhengqing of Wuhan University's Economcs and Management School.

Rule 4: Filter sentences that begin with research findings, data findings, claims, people's opinions, etc.

Overall, the U.S. Tech War with China has fallen short of the intensity implied by Ren's military metaphors.

Rule 5: After extraction from Rule 7, sentences beginning with personal pronouns other than the designated words in Rule 1 are filtered as non-viewpoint sentences.

For example, Doval and Wang are the designated Special Representatives (SRs) of the two countries for the Boundary Talks.

Rule 6: Sentences with special symbols are non-opinion sentences. Exclude.

Such as: China's once-mighty peer-to-peer (P2P) lending sector is on the ropes, with Beijing tightening regulations further in recent weeks in its multi-year campaign to reduce risks in its financial sector.

Due to the fact that the report only states objective facts and does not express opinions, if the number of extracted opinion sentences is 0, the emotional polarity of classified report is neutral.

4.2 Word Embedding and Two-Way BiLSTM

After the opinion sentence is determined, in order to better input model, each word of each opinion sentence is mapped into a low-dimensional vector for word embedding. First, NLTK English word segmentation is performed for each opinion sentence, and then Word2Vec is used for word vector representation. Word2Vec includes two training models: CBOW model, which input the prediction center word according to the given context, as shown in Fig. 2 below. Skip-gram model is the exact opposite of CBOW model, which makes predictions based on the output context of the central word, as shown in Fig. 3 below.

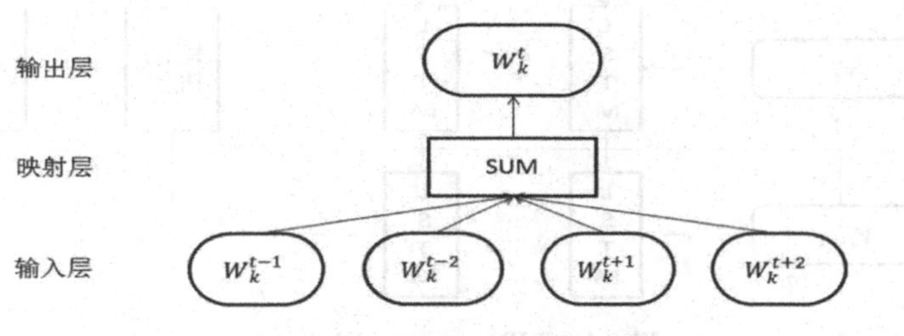

Fig. 2. CBOW model diagram

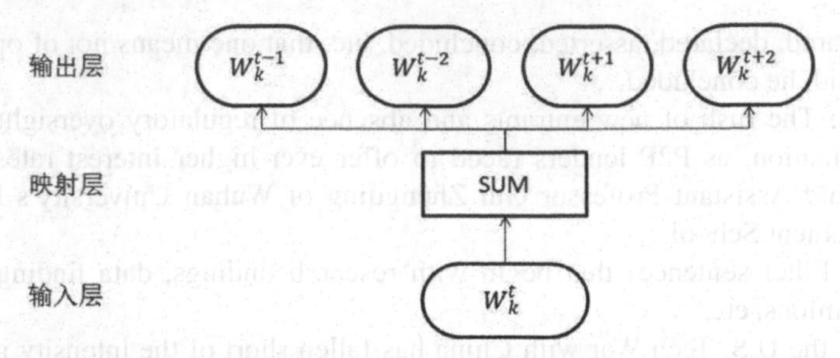

Fig. 3. SkIP-gram model diagram

Since skIP-gram can better generalize low-frequency words, this paper selects the skIP-gram model for word embedding, assuming that the opinion sentence of a report is S1…… Sn, each sentence is made up of several words: Sk = {Wk1,Wk2… Wkp}. After word embedding in each sentence, each word WkiSk can be obtained VkiSkd, where I represents the ith word in the sentence and D represents the dimension of the vector.

LSTM can solve the problem such as long-term dependence, gradient explosion and gradient disappeared, but it is read in sequence can only be read sequentially, not very good considering the context of the contact, therefore, this article consider to add a reverse LSTM to training input vector, good to extract context information, which in every time sequence configuration before and after a positive LSTM, reads input sequence (Vk1, Vk2,… Vkp), then configure a reverse LSTM to read the sequence (Vkp… Vk1). At time I, the input processing of forward LSTM and reverse LSTM are respectively:

In Formula 1, HI is the hidden state output result of forward LSTM at time I; In Formula 2, HI is the hidden state output of the backward LSTM at time I. The forward and backward direction are combined together in the way of connection as hi, the output result of Bi-LSTM's hidden state at time I. The BiLSTM model is shown in Fig. 4 below:

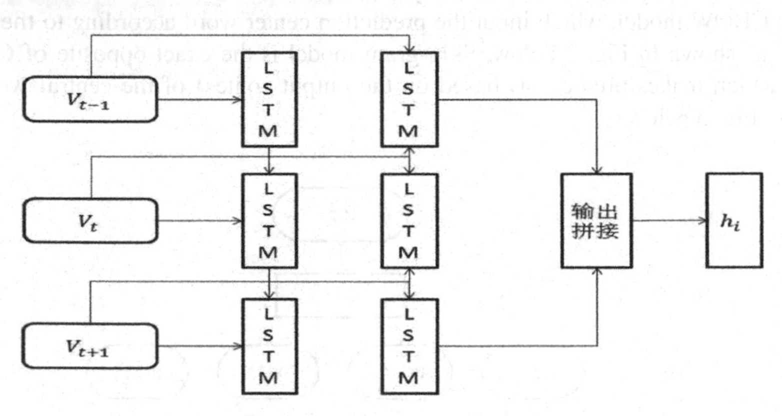

Fig. 4. SkIP-gram model diagram

After BiLSTM processing, the output is vector matrix, which is further weighted as the input of Attention.

4.3 Word Embedding and Two-Way BiLSTM

Attention mechanisms work by the principle that the human eye pays more attention to some parts of an image than to others. Attention mechanism was first applied in the field of image processing to quickly screen out the value of important parts in images. Later, attention mechanism was widely applied in various fields. Zhang Yangsen et al. [16] successfully extracted the words or sentences that were more important to the document by using the mechanism of double attention on the text, which improved the effect of text classification.

This paper assigns an appropriate weight of importance to each word in each opinion sentence according to the English vocabulary, so that the model can pay more attention to the vocabulary expressing emotion. Let the output sequence of the hidden layer be: (h1… Hi), the attention weight allocation rules formulated in this paper are as follows:

Where q is called query vector, when Q comes from H, it is called self-attention, and F is the matching function of Q and Hi, which can be implemented in various ways [17].

5 Experimental Design and Analysis

5.1 Experimental Data and Relevant Tools

In order to verify the practicality of the model in this paper, 13,200 articles based on keywords such as CHINA, Wuhan and Sunyang were adopted from December 2019. Among them, 65% of the extracted texts were selected as training sets and 35% as test sets for BiLSTM training. In this experiment, skip-gram in Word2Vec model is used to embed the word vector, and the 100-dimensional word vector is finally obtained. In this paper, Adam is used to update the rules for single LSTM model training. In order to avoid the occurrence of overfitting, DropOut mechanism is adopted, and parameter Settings are shown in the following Table 1.

Table 1. LSTM model parameters table

超参数	参数值
激活函数	tanh
优化方法	Adadelta
Dropout 比例	0.5
词向量维度	100
训练批次	10
Epoch	8

In this paper, the TensorFlow platform was used to train BiLSTM, and the number of LSTM model layers was set to 64, while the number of BiLSTM layers was set to 128. Pycharm tool was used for the experiment.

5.2 Evaluation Criteria

In order to verify the accuracy of the model proposed in this paper, a comparative experiment was conducted with the classical model:

(1) Match according to rules based on the emotion dictionary method.
(2) BLSTM: This model takes plain text as the unique input of the model, so that BLSTM can be used to train the word direction. This model includes a forward LSTM and a reverse LSTM, combining the output sequences of both as the representation vectors of text sequences, and also using the full connection layer for classification [18].

Accuracy rate, recall rate and F1 were used as accuracy indexes in this paper.

Table 2. Classification results of different models on data sets.

模型	指标	分类		
		正面	中立	负面
情感词典	P	89.1	95.0	92.1
	R	91.84	92.8	99
	F	90.61	92.5	77.68
BiLsTM	P	90.2	96.15	93.7
	R	91.3	97.55	82.3
	F	93.2	92.8	81.55
BiLsTM-ATTention	P	93.2	96.1	93.5
	R	94.8	97.85	89.7
	F	94.88	97.55	88.6

By analyzing the data in the Table 2, it can be seen that the average F1 value of the LSTM model without any features is 1.82% higher than that of the SVM model based on features. This shows that deep learning has more advantages than traditional machine learning methods in feature extraction and learning. By comparing the LSTM model and the BI-LSTM model, it can be seen that the recall rate of the BI-LSTM model under the negative classification is 3.19% higher than that of the LSTM model under the condition of little difference in accuracy rate, which indicates that the bidirectional cyclic neural network can better capture the information of negative emotions than the unidirectional cyclic neural network. By comparing LSTM, AE-LSTM and ATAE-LSTM, it can be concluded that adding aspect information or using attention mechanism can effectively judge the emotional polarity of different aspects, because the model can pay high attention to the characteristics of specific aspects in the training process. Finally, by comparing the BLSTM-AAM model and ATAE-LSTM model proposed by us, it can be seen that in terms of F1 value of positive classification and negative classification, the BLSTM-AAM model has been improved by 0.75% and 2.24%, respectively. This indicates that the use of multiple aspect attention modules can fully extract the hidden features of specific aspects, thus achieving a better performance in affective classification. At the same time, the validity of the proposed method is verified.

6 Conclusion

To foreign media reports released by emotional problem analysis, this paper puts forward a view words recognition rules and BiLSTM - Attention mechanism of hybrid model, to view other improved extraction sequence after work efficiency is also improved the accuracy of BiLSTM fully considering the context semantic information, and Attention mechanism is a key part of the focus on the emotional expression, compared with the traditional method, higher accuracy of sentiment analysis.

The shortcoming of this paper is that the accuracy of other fields needs to be verified by experiments, and the extraction rules of viewpoint sentences are not flexible enough.

References

1. Wenyu, Z., Jian, X.: Comparative analysis of the characteristics of emotional expression of users on different network platforms. The intelligence theory practice. http://kns.cnki.net/kcms/detail/11.1762.g3.20190917.0944.004.html
2. Haibin, L., Yuqing, M., Wanzhen, Z., Ming, Z., Jigang, W.: Multi-feature fusion of graphic weibo emotion analysis. Res. Comput. Appl. https://doi.org/10.19734/j.issn.1001-3695.2018.12.0929
3. Fupeng, L., Dongxiang, F.: An emotion analysis method of financial text based on transformer encoder. Electron. Sci. Technol. **31**(1), 31–41 (2008)
4. Huan, Z., Haonan, M., Jia, L.: Research on film and Television Recommendation Algorithm combining emotion analysis and probability language. The intelligence theory and practice. http://kns.cnki.net/kcms/detail/11.1762.G3.20191217.0830.002.html
5. Junbo, C., Xia, Y., Feixiang, X., Lidong, Y.: Improved CBOW emotional information acquisition study. Comput. Eng. Appl. http://kns.cnki.net/kcms/detail/11.2127.TP.20190325.1755.034.html
6. Yanhui, H., Huifang, D., Min, H., Xuelian, C.: Classification of fine-grained emotions in film reviews based on ontological characteristics. Comput. Appl. **12**(12), 1001–9081 (2019)
7. Xi, C., Xiaodong, Z., Guangwan, G., Fangxiong, X.: Chinese comment emotion analysis based on mixed vector model. Comput. Eng. https://doi.org/10.19678/j.issn.1000-3428.0053116
8. Weidong, C., Jiqi, L., Huaichao, W.: Objective emotion analysis using an attention-gated convolutional network model. J. Xidian Univ. 20190903. 1708–1714 (2019)
9. Biqing, Z., Xuli, H., Shengyu, W., Wu, Z., Heng, Y.: Study on emotion analysis of hierarchical biattentional Neural network model. J. Intell. Syst. 20191226. 1025–1102 (2019)
10. Hong, Z., Le, W., Weijie, W.: Text Emotion analysis based on BILSTM-CNN Serial Mixed model. Comput. Appl. **2019**(9), 1001–9081 (2019)
11. Zhuliang, P., Bowen, L., Chengan, F., Jie, W., Ming, X., Zeien, L.: Affective categorization based on BLSTM and aspect attention. Comput. Eng. pp. 1000–3428
12. Wei, M., Yongqing, Y., Wenfeng, L.: Specific target emotion analysis based on CRT mechanism mixed neural network. Res. Comput. Appl. https://doi.org/10.19734/j.issn.1001-3695.2018.08.0538
13. Hochreiter, S., Schmidhuber, J.: Long-term memory. Neural Comput. **9**(8), 1735–1780 (1997)

14. Bahdanau, D., Cho, K., Bengio, Y.: Neural machine translation by jointly learning to align and translate. In: Proceedings of International Conference on Learning Representations, pp. 1–15 (2015)
15. Yangsen, Z., Jia, Z., Gajuan, H., et al.: Journal of Tsinghua University: Natural Science. **58**(2), 122–130 (2015). (in Chinese)
16. Vaswani, A., Shazeer, N., Parmar, N., et al.: Attention is all you need. Long Beach. In: Proceedings of the Conference and Workshop on Neural Information Processing System (NIPS) (2017)
17. Kaixu, H., Weijian, R.: Application of support vector machine based on improved fisher kernel function in emotion analysis of Twitter database. Autom. Technol. Appl. **34**(11), 30–36 (2015)
18. Changyu, Z., Yaping, W., Jimin, W.: Twitter text topic mining and emotion analysis under the One Belt And One Road initiative. Books Inf. Work. **63**(19), 119–127

Cascaded Convolutional Neural Network for Image Super-Resolution

Jianwei Zhang, Zhenxing Wang, Yuhui Zheng, and Guoqing Zhang(✉)

Nanjing University of Information Science and Technology, Nanjing 210044, China

Abstract. Image super resolution is an important field of computer research. The current mainstream image super-resolution technology is to use deep learning to mine the deeper features of the image, and then use it for image restoration. However, most of these models mentioned above only trained the images in a specific scale and do not consider the relationships between different scales of images. In order to utilize the information of images at different scales, we design a cascade network structure and cascaded super-resolution convolutional neural networks. This network contains three cascaded FSRCNNs. Due to each sub FSR-CNN can process a specific scale image, our network can simultaneously exploit three scale images, and can also use the information of three different scales of images. Experiments on multiple datasets confirmed that the proposed network can achieve better performance for image SR.

Keywords: Super-resolution · Cascade structure · Convolutional neural network

1 Introduction

Single image super-resolution (SR) is an important issue in the field of computer vision research. Given a low resolution (LR) image, the purpose of single image SR is to recover a high resolution (HR) image from its corresponding LR image. Currently, single image SR based methods are generally divided into two categories: sparse coding-based methods [1–4] and deep learning-based methods [5–8].

Most sparse coding-based methods assume that each pair of patches from LR and HR images have similar coding coefficients in the patch space. Therefore, the HR image patches can be reconstructed as a linear combination of a few LR image patches. Yang et al. [1, 2] learned a pair of dictionaries D_L and D_H for LR and HR image patches respectively. The final reconstructed HR image can be obtained through the learned dictionary D_H and coding coefficient α, which encoded by the learned dictionary D_L [9–11]. At present, existing sparse coding-based methods mainly focus on how to learn better dictionaries [12, 13]. For example, Zhao et al. [12] designed a transfer robust sparse coding based on graph for image representation and Ahmed et al. [13] proposed a new multiple dictionary learning strategy.

Deep learning-based methods aim to directly learn an end-to-end mapping from LR images to HR images [5–8]. Dong et al. [7] first applied convolutional neural network (CNN) for the single image SR, and proposed a super-resolution convolutional neural

X. Sun et al. (Eds.): ICAIS 2021, CCIS 1422, pp. 361–373, 2021.
https://doi.org/10.1007/978-3-030-78615-1_32

network (SRCNN) and achieved better restoration performance. However, SRCNN only learns the mapping from bicubic interpolated LR images (not the original LR images) to HR images, the calculation cost of the network will be increased quadratically with the size of HR image increase [8]. In addition, SRCNN adopts a 5*5 convolution kernel to learn nonlinear mapping which obviously limited the learning ability of network in such a setting. To solve the above problems, researchers [14–16] proposed lots of approaches such as the large super-resolution convolutional neural network (SRCNN-ex), fast super-resolution convolutional neural network (FSRCNN), very deep super-resolution network (VDSR), Laplacian pyramid super-resolution network (LaPSRN) and enhanced deep super-resolution network (EDSR) to improve the structure of SRCNN. In addition, Christian et al. [17] exploited generative adversarial networks (GAN) [18] to handle out image super-resolution problem and proposed super-resolution generative adversarial network (SRGAN). GAN is different from convolutional neural network (CNN), and the main difference is that GAN aims to generate more realistic HR images that are more consistent with human eye, while the purpose of CNN is to faithfully restore the high-frequency information of images. Lots of improved networks based on GAN have been proposed in recent years [19, 20]. These networks have solved the above problems to some extent and achieved better restoration performance.

However, most of these models mentioned above only trained the images in a specific scale and do not consider the relationships between different scales of images. For example, if we set the scale factor is 4 during the training process, FSRCNN only utilizes the image information, where the image scale is 4, and does not use the complementary information resided in different scales. If we can simultaneously use the information of different scales of images in the training process, the recovered image qualities will be further improved.

Based on the above concerns, we design a cascaded super-resolution convolutional neural networks (CSRCNN), which consists of three cascaded FSRCNNs. Due to each sub FSRCNN can process a specific scale image, our network can simultaneously exploit three scale images, and can also use the information of three different scales of images in the training process. For each FSRCNN, we set its scaling factor is 2, which can double enlarge the size of input image. Since the images with different scales are trained together, the learned network can exploit more image information resided in different scales. In addition, we use L1 loss function to make the reconstructed image in texture and edge clearer. Experiments on multiple datasets show that the proposed our proposed network can achieve better performance for image SR.

2 Related Work

2.1 Deep Learning for Image Super-Resolution

The purpose of image super-resolution is to reconstruct high resolution image from a given low resolution one. Dong et al. [7] first applied convolutional neural network in the field of image super-resolution and proposed a super-resolution convolutional neural network (SRCNN) which achieved impressive performance for image restoration. At present, deep learning technology is widely used in the field of image super-resolution. Various network structures have been developed, such as deep network with residual learning [14], Laplace pyramid structure [15], residual block [6], and residual dense network [21]. Besides supervised learning and unsupervised learning [22–25], reinforcement learning [26–28] are also introduced to solve the problem of image super-resolution in recent years. Specifically, the literature [29] has a systematic description of image super resolution.

2.2 Fast Super-Resolution Convolutional Neural Networks

Dong et al. [8] proposed a fast super-resolution convolutional neural network (FSRCNN), aiming to accelerate the speed of SRCNN. Compared with SRCNN, FSRCNN can directly learn the mapping from the original LR image to HR image by introducing the deconvolution layer at the end of the network. In addition, FSRCNN adds a shrinking and an expanding layer at the beginning and the end of the mapping layer to enhance the presentation capability of non-linear mapping [30]. In the nonlinear layer, FSRCNN adopts a smaller filter (size = 3*3) to reduce the computational cost of the network. The whole network is designed as a compact hourglass-type CNN structure, which includes five parts: feature extraction, shrinking, non-linear mapping, expanding, deconvolution. The shrinking layer uses a filter of size 1*1 to reduce the LR feature dimension from d to s, where d is the feature dimension of LR image after feature extraction and s is the number of filters ($s \leq d$). The expanding layer uses d filters with a size of 1*1 to maintain consistency with the shrinking layer, which is the inverse operation of the shrinking layer. To avoid the "dead features" [31] caused by zero gradients in ReLU, the author uses $PReLU$ [32] as the activation function after each convolution layer. Thus, a complete FSRCNN can be represented as:

$$\begin{aligned} &Conv(5, d, 1) - PReLU - Conv(1, s, d) - PReLU - Conv(3, s, s) \\ &-PReLU - mConv(1, d, s) - PReLU - DeConv(9, 1, d). \end{aligned} \tag{1}$$

3 Cascaded Super-Resolution Convolutional Neural Network

3.1 Network Structure

We propose a cascaded super-resolution convolutional neural network (CSRCNN) framework in this section. Our proposed network consists of three cascaded FSRCNNs, where each FSRCNN can process a specific scale image. Figure 1 shows the structure of our CSRCNN.

In each sub FSRCNN, we set its scaling factor to 2 for double enlarging the size of input image. Suppose that all images have the same width-to-height ratio. We denote W and H respectively as the width and height of the original HR image. For the first sub FSRCNN, the image shape of the input is $W/8$*$H/8$ and the output will be as the input of next sub FSRCNN. Due to each sub FSRCNN can double enlarge the size of input image, the image shape of the input for the second subFSRCNN is $W/4$*$H/4$. Similarly, for the third sub FSRCNN, the image shape of the input is $W/2$*$H/2$. During the training process, images of different scales are trained simultaneously. Here, we use I^k to represent the input image of each sub FSRCNN, where $k = 0, 1, 2$ is the ID of each sub FSRCNN and the scale index of the input image. We set r_k to represent the scale ratio of LR image to HR image, the size of input image I^k is described as $r_k W^* r_k H$. The scale ratios are set as $r_0 = 1/8, r_1 = 1/4$ and $r_2 = 1/2$ respectively. $I^{k+1} = F_k(I^k)$ is the super-solved HR image obtained by sub FSRCNN-k. For each cascaded FSRCNN, the output image is twice the size of input image. We use $\tilde{I}^{k+1}$ represents the real HR image, which has the same size with the super-solved image $F_k(I^k)$. Our goal is to learn super-resolved functions that estimating the reconstructed HR images I^{k+1} and real HR images $\tilde{I}^{k+1}$.

3.2 Loss Function

Our network includes three cascaded FSRCNNs. For each subnetwork, the output of previous network will be the input of the next cascaded subnetwork. We use L1 loss functions as the sub-loss functions for all three subnetworks, and the sum of these sub-loss functions will eventually form the total loss function of the whole network. We use $L_1^{SR_0}, L_1^{SR_1}$ and $L_1^{SR_2}$ represent the loss functions of three subnetworks, respectively. The loss function of the whole network can be expressed as:

$$L_{total} = L_1^{SR_0} + L_1^{SR_1} + L_1^{SR_2}. \tag{2}$$

For each subnetwork, the loss function is defined as follows:

$$L_1^{SR_k} = \frac{1}{r_{k+1}^2 WH} \sum_{x=1}^{r_{k+1}W} \sum_{y=1}^{r_{k+1}H} \left\| \tilde{I}_{x,y}^{k+1} - F_k(I^k)_{x,y} \right\|_1, k = 0, 1, 2. \tag{3}$$

In Eq. (2), we select L1 loss function to replace MSE loss function. The main reason is that L1 loss function can make the reconstructed image clearer in texture and edge, while MSE loss function will lose high-frequency information in the reconstruction process, such as texture and edge [33].

3.3 Assessment Process

According to the scale ratios of LR images to HR images, we assign LR images to different stages of the cascaded networks during the evaluation process. When a LR image is input into FSRCNN-k, the input shape will be resized based on the corresponding network $r_k W^* r_k H$. For example, for a given test image with size of $W/3^*H/3$, we will resize the shape of the image as $W/2^*H/2$ by bicubic interpolation. The resized image will be arranged into FSRCNN-2. Here, each sub-network can double enlarge the size of input image. Finally, all images are enlarged to the uniform HR images [34]. Figure 2 shows the entire evaluation process of our network.

3.4 The Difference Between Our Model with FSRCNN

Our network contains three cascaded FSRCNNs, where each FSRCNN can process a specific scale image. For each sub FSRCNN, we use the L1 loss function, whereas

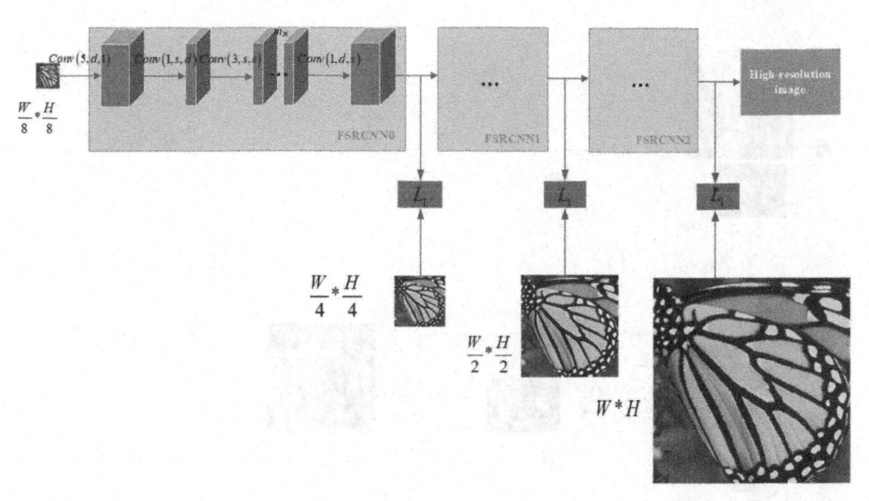

Fig. 1. The network structure of our CSRCNN. The network is composed of three cascaded FSRCNNs, in which the upscaling factor of each FSRCNN is 2. For the first sub FSRCNN, the input is an $W/8*H/8$ LR image and the output of previous layer will be the input of the next layer. Each sub FSRCNN can double enlarge the size of input image. We select L1 loss function as sub-loss functions of three subnetworks, and the sum of three sub-loss functions will eventually form the loss function of the whole network.

original FSRCNN uses the MSE loss function. In addition, FSRCNN only considers a specific scale image, while our network can simultaneously train different scales of images due to adopting the cascaded structure. In the test phase, our network can generate multiple intermediate results in a prediction process, while FSRCNN only obtains a single scale result.

4 Experiments

4.1 Dataset

Training and Testing Dataset. Following SRCNN and FSRCNN, we combine 91 images dataset and the general-100 dataset to train our network. In particular, general-100 dataset [8] contains 100 bmp-format images with ranges from 710 × 704 (large) to 131 × 112 (small) in size. All of these images are favorable quality with clear edges and less smooth regions. Following the setting used in [8], we adopt the data augmentation method [35], includes scaling and rotating images. 1) Scaling: each image is down-sampling with the factor 0.9, 0, 8, 0.7 and 0.6. 2) Rotation: each image is rotated with the degree of 90, 180 and 270. For testing dataset, we use Set5 [36], Set14 [37] and BSD200 [38] datasets. All testing images will be cropped according to the model structure which make the size of output image of the model as an integer.

Training Samples. Cascaded structure are adopted in our network. To form LR/HR sub-image pairs, we first down-sample the original training images by different scale factors n, then crop LR images and real HR images obtained in sampling process.

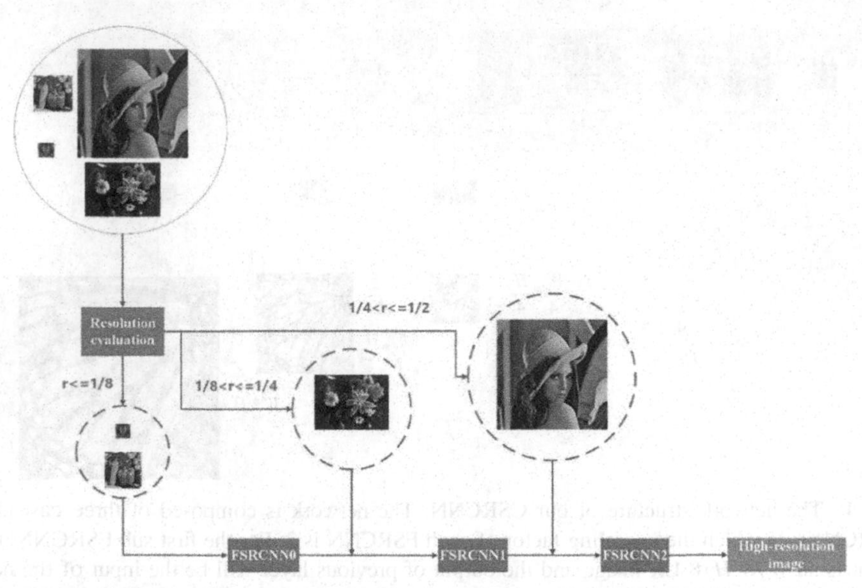

Fig. 2. LR images are assigned to different stages of the cascaded FSRCNNs according to their resolution scale ratios. Through the CSRCNN, all LR images are enlarged to the uniform HR ones.

4.2 Training Details

Training Strategy. In our experiment, we combine 91 images dataset and the general-100 dataset [8] for training. The 91-image dataset is used to train a network from scratch. Then, we add General-100 dataset to the network for fine-tuning when the training is saturated. Using this strategy, the training converges much earlier compare with training with both two datasets.

The Choice of Learning Rate. The learning rate plays a very import role for the performance of our network. For FSRCNN, the learning rates in the convolution layers and deconvolution layer are set to be 10^{-3} and 10^{-4}, the learning rates of all layers are reduced by half during fine - tuning. Obviously, the learning rates of convolution layer and deconvolutional layer are static in each iteration, while we adopt a dynamic strategy to update the learning rates for convolution layer and deconvolutional layer in our network. This strategy can make our network produce better learning rates in different iterations. We set $a_0 = 10^{-3}, 10^{-4}$ as the initial learning rates of the convolution layer and the deconvolution layer, and the total number of iterations in our network is set as n. When the number of iterations is m, the learning rates in convolution layer and the deconvolution layer are represented as:

$$a_m = a_0^* 0.1^{\rho\left(\frac{m}{0.8^* n}\right)} \tag{4}$$

Here, $\rho\left(\frac{m}{0.8^* n}\right)$ represents the integer part of the final result.

The Choice of Network Initialization. Network initialization has a great influence on network training, a good initialization can largely reduce the training time of the network. In our network, we select *PReLU* [32] as the activation function and use MSRA initialization [32]. MSRA is a gaussian with a mean of zero and variance of $n/2$, which make each layer neuron input/output variance is consistent.

Parameter Setting. Our proposed cascaded CSRCNN consists of three subnetworks. For each subnetwork, there are three sensitive variables governing the network's performance. The LR feature dimension d, the number of shrinking filters s, and the mapping of the depth m. Following the same setting in FSRCNN, we set $d = 56$, $s = 12$ and $m = 4$ for each subnetwork.

4.3 Experiments for Different Loss Functions

In this section, we conduct experiments to evaluate the effects of different loss functions on our network on three datasets when the upscaling factor is 3. We compare MSE-Loss, Huber-Loss, Charbonnier-loss and L1-loss functions and the results are listed in Table 1. The delta of Huber-Loss is 0.9 and the delta of Charbonnier-Loss is 0.0001. From Table 1, we can see that using L1-loss, our cascaded network can achieve better performance. Figure 3 shows the convergence curves of these four loss functions on Set5 datasets with upscaling factor 3. From the Fig. 3, we first observed that using L1-Loss, our cascaded network gets the best average test PSNR values, followed by Charbonnier-Loss, and Huber- Loss get the worst average test PSNR values. In addition, after 160 iterations, the PSNR value curves of these four loss functions tend to be stable. Among them, the network has a small fluctuation when we use Charbonnier-Loss and MSE-Loss as our network loss functions. We also find that the initial value of PSNR is low when we use L1 loss as the loss function of the network, with the increase of the number of iterations, the PSNR value will gradually increases and eventually tends to be stable. In this section, we evaluate the performance of our network versus the number of cascaded FSRCNNs. Since the scaling factor of each FSRCNN is set to 2, if we cascaded more than three FSRCNNs, the scale of the input image to the network will too small, which makes the effective information contained in these images are very limited. Therefore, in our model, we only cascaded three FSRCNNs. These three networks are respectively denoted as Net-1, Net-2 and Net-3. Net-1 contains one FSRCNN with upscaling factor is 4, and Net-2 contains two cascaded FSRCNNs, where upscaling factor is 2. We use PSNR as the evaluation criteria and test on three datasets for $4 \times$ SR. Table 2 lists the results our model versus different number of FSRCNNs and the number of FSRCNNs varies from 1 to 3. From Table 2, we can observe that Net-3 obtains the highest PSNR value on all testing datasets. Net-2 obtains higher results than Net-1. This indicates that cascading multiple basic models can achieve better performance than a single basic model. In addition, the increasing trend of PSNR is gradually decreasing with the increase of the number of cascaded FSRCNNs, one possible reason is that the extra information provided by LR images with small scale is limited.

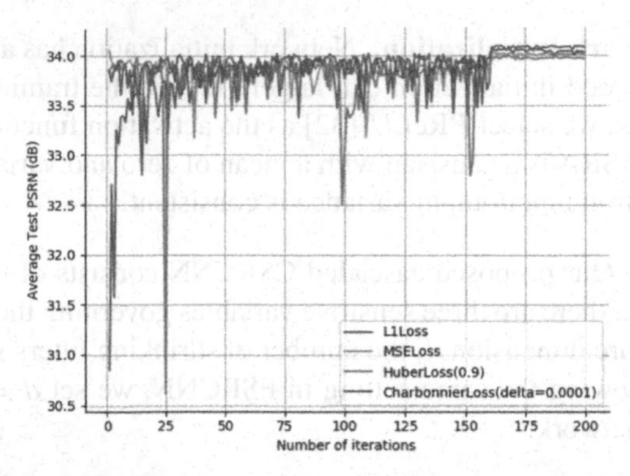

Fig. 3. Experiments for different loss functions on Set5 datasets when the upscaling factor is 3.

Table 1. The influence of different loss choices on CSRCNN with upscaling factor 3, where the test datasets are Set5, Set14 and BSD200.

Test-dataset	Upscaling factor	MSE-loss	Huber-loss(delta = 0.9)	Charbonnier-loss(delta = 0.0001)	L1-loss
		PSNR/SSIM	PSNR/SSIM	PSNR/SSIM	PSNR/SSIM
Set-5	3	34.02/0.9218	33.97/0.9213	34.04/0.9219	34.10/0.9233
Set-14	3	29.98/0.8321	28.86/0.8319	30.02/0.8332	30.02/0.8346
BSD200	3	29.69/0.8282	28.77/0.8279	29.68/0.8285	29.78/0.8302

Table 2. The influence of the number of cascaded FSRCNNs with upscaling factor 4, where the test datasets are Set5, Set14 and BSD200.

Test-dataset	Upscaling factor	Net-1	Net-2	Net-3
		PSNR	PSNR	PSNR
Set-5	4	30.55	30.82	**31.01**
Set-14	4	27.50	27.66	**27.71**
BSD200	4	26.92	27.42	**27.68**

4.4 Evaluation for the Number of Cascaded FSRCNNs

In this section, we evaluate the performance of our network versus the number of cascaded FSRCNNs. Since the scaling factor of each FSRCNN is set to 2, if we cascaded more than three FSRCNNs, the scale of the input image to the network will too small, which makes the effective information contained in these images are very limited. Therefore, in our model, we only cascaded three FSRCNNs. These three networks are respectively denoted as Net-1, Net-2 and Net-3. Net-1 contains one FSRCNN with upscaling factor is 4, and Net-2 contains two cascaded FSRCNNs, where upscaling factor is 2. We use PSNR as the evaluation criteria and test on three datasets for 4 $\times$ SR. Table 2

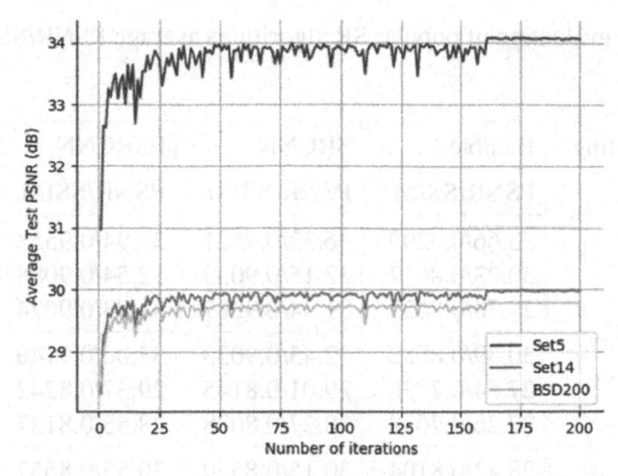

Fig. 4. Visual comparisons on Set5, Set14 and BSD200 with upscaling factor 3.

lists the results our model versus different number of FSRCNNs and the number of FSRCNNs varies from 1 to 3. From Table 2, we can observe that Net-3 obtains the highest PSNR value on all testing datasets. Net-2 obtains higher results than Net-1. This indicates that cascading multiple basic models can achieve better performance than a single basic model. In addition, the increasing trend of PSNR is gradually decreasing with the increase of the number of cascaded FSRCNNs, one possible reason is that the extra information provided by LR images with small scale is limited.

4.5 Comparison with Popular SR Algorithms

In order to verify the reconstruction performance of our proposed CSRCNN. We compare our model with three popular SR algorithms: Bicubic, SRCNN [7], and FSRCNN [8]. We carry out extensive experiments on three datasets: Set5 [36], Set14 [37] and BSD200 dataset [38]. We adopt two image quality indicators to evaluate the SR images i.e., PSNR and SSIM [39]. Table 3 reports quantitative comparisons for ×2, ×3 and ×4. From Table 3, we can observe that among all the compared methods,our proposed CSRCNN achieves the best results in three evaluation datasets for ×2, ×3 and ×4. Figure 4 shows the convergence curves of CSRCNN on three evaluation datasets with upscaling factor 3. Among them, the PSNR valuc of the proposed CSRCNN achieve 0.51 db, 1.04 db and 0.46 db improvement over FSRNCNN in Set5 datasets for ×2, ×3 and ×4. We observe that the PSNR value of CSRCNN obtains 1.04 db improvement over that of FSRCNN on factor 3. One possible reason is that we use bicubic interpolation to resize the size of the image before throwing the LR image into the network. Considering that our network is made up of three cascaded FSRCNNs, where each FSRCNN's upscaling factor is 2. We also made a quantitative comparison for ×8. This comparison result is shown in Table 4. Obviously, our network achieves higher PSNR and SSIM values than FSRCNN due to the cascaded design.

Figure 5 shows the visual comparisons on Set5 and Set14 datasets with upscaling factor 3. As you can see, our method can accurately reconstruct the details of the image, such as texture and edge.

Table 3. Quantitative evaluation of popular SR algorithms average PSNR/SSIM for scale factors 2, 3 and 4.

Test-dataset	Upscaling factor	Bicubic	SRCNN	FSRCNN	CSRCNN (our)
		PSNR/SSIM	PSNR/SSIM	PSNR/SSIM	PSNR/SSIM
Set-5	2	33.66/0.9299	36.33/0.9521	36.94/0.9558	**37.45/0.9570**
Set-14	2	30.23/0.8677	32.15/0.9039	32.54/0.9088	**32.82/0.9112**
BSD200	2	29.70/0.8625	31.34/0.9287	31.73/0.9074	**32.92/0.9122**
Set-5	3	30.39/0.8682	32.45/0.9033	33.06/0.9140	**34.10/0.9233**
Set-14	3	27.54/0.7736	29.01/0.8145	29.37/0.8242	**30.02/0.8346**
BSD200	3	27.26/0.7638	28.27/0.8038	28.55/0.8137	**29.78/0.8302**
Set-5	4	28.42/0.8104	30.15/0.8530	30.55/0.8657	**31.01/0.8702**
Set-14	4	26.00/0.7019	27.21/0.7413	27.50/0.7535	**27.71/0.7589**
BSD200	4	25.97/0.6949	26.72/0.7291	26.92/0.7398	**27.68/0.7552**

Table 4. Quantitative evaluation of popular SR algorithms average PSNR/SSIM for scale factor 8.

Test-dataset	Upscaling factor	Bicubic	SRCNN	FSRCNN	CSRCNN (our)
		PSNR/SSIM	PSNR/SSIM	PSNR/SSIM	PSNR/SSIM
Set-5	8	24.39/0.657	25.33/0.689	25.41/0.682	**25.74/0.715**
Set-14	8	23.19/0.568	23.85/0.593	23.93/0.592	**24.30/0.614**
BSD200	8	23.67/0.547	24.13/0.565	24.21/0.567	**24.50/0.581**

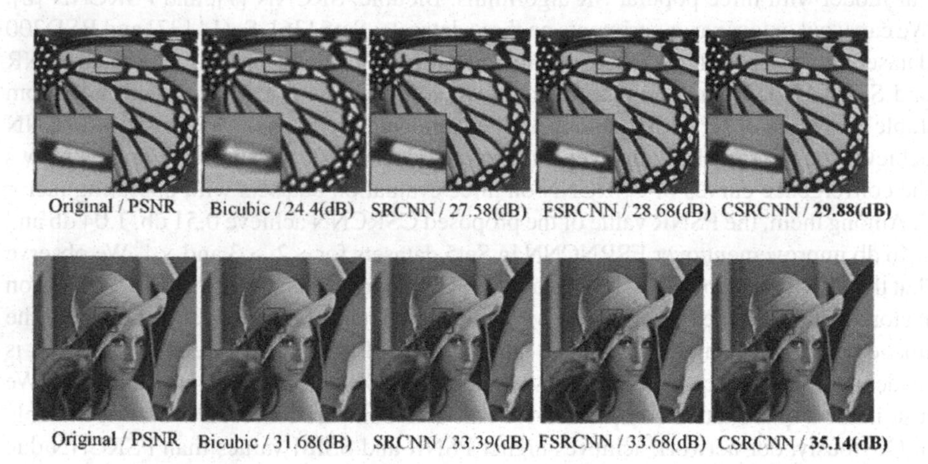

Fig. 5. Visual comparisons on Set5 and Set14 datasets with upscaling factor 3. The first row is "butterfly" image from Set5 dataset, and the second row is "lenna" image from Set14 dataset.

5 Conclusion

In this paper, we proposed a more efficient network for mining image information in different scales, named Cascaded super-resolution convolutional neural network (CSR-CNN). We adopted a cascaded network structure to mine the information of image in different scales. Furthermore, we adopted L1 loss function as the loss function of each subnetwork in the training process which make the reconstructed image clearer in texture and edge. Finally, we also explored the effects of the number of cascaded FSRCNNs for the performance of our network. Extensive experiments show that the proposed method achieves satisfactory SR performance.

Acknowledgement. This research is supported in part by the National Natural Science Foundation of China under Grant 61806099, in part by the Natural Science Foundation of Jiangsu Province of China under Grant BK20180790, in part by the Natural Science Research of Jiangsu Higher Education Institutions of China under Grant 8KJB520033, in part by Startup Foundation for Introducing Talent of Nanjing University of Information Science and Technology under Grant 2243141701077.

References

1. Farhadifard, F., Abar, E., Abar, E: Single image super resolution based on sparse representation via directionally structured dictionaries. In: Proceedings of Signal Processing and Communications Applications Conference, K.T.u, Mersin, Turkey, pp. 1718–1721(2014)
2. Yang, J., Wang, Z., Lin, Z.: Coupled dictionary training for image super-resolution. IEEE Trans. Image Process. **21**(2), 3467–3478 (2012)
3. Yuan, Z.L., Cai, S.U.: Image super-resolution via sparse representation. Computer engineering and design. IEEE Trans. Image Process. **19**(4), 2861–2873 (2016)
4. Ahmed, J., Shah, M.A.: Single image super-resolution by directionally structured coupled dictionary learning. EURASIP J. Image Video Process. **2016**(1), 1–12 (2016). https://doi.org/10.1186/s13640-016-0141-6
5. Cui, Z., Chang, H., Shan, S., Zhong, B., Chen, X.: Deep network cascade for image super-resolution. In: Fleet, D., Pajdla, T., Schiele, B., Tuytelaars, T. (eds.) Computer Vision – ECCV 2014. LNCS, vol. 8693, pp. 49–64. Springer, Cham (2014). https://doi.org/10.1007/978-3-319-10602-1_4
6. Choi, J.H., Kim, J.H.: Deep learning-based image super-resolution considering quantitative and perceptual quality. In: Proceedings of ECCV, Munich, Germany, pp. 347–359 (2018)
7. Dong, C., Loy, C.C., He, K., Tang, X.: Learning a deep convolutional network for image super-resolution. In: Fleet, D., Pajdla, T., Schiele, B., Tuytelaars, T. (eds.) Computer Vision – ECCV 2014. LNCS, vol. 8692, pp. 184–199. Springer, Cham (2014). https://doi.org/10.1007/978-3-319-10593-2_13
8. Dong, C., Loy, C.C., Tang, X.: Accelerating the super-resolution convolutional neural network. In: Leibe, B., Matas, J., Sebe, N., Welling, M. (eds.) Computer Vision – ECCV 2016. LNCS, vol. 9906, pp. 391–407. Springer, Cham (2016). https://doi.org/10.1007/978-3-319-46475-6_25
9. Zhang, G., Sun, H., Zheng, Y.: Optimal discriminative projection for sparse representation-based classification via bilevel optimization. IEEE Trans. Circuits Syst. Video Technol. **30**(2), 3065–3077 (2019)

10. Zhang, G., Zheng, Y., Xia, G.: Domain adaptive collaborative representation based classification. Multimedia Tools Appl. **78**(21), 30175–30196 (2018). https://doi.org/10.1007/s11042-018-7007-0
11. Zheng, Y., Wang, X., Zhang, G.: Multiple kernel coupled projections for domain adaptive dictionary learning. Multimedia Tools Appl. **21**(2), 2292–2304 (2019)
12. Zhao, P., Wang, W., Lu, Y.: Transfer robust sparse coding based on graph and joint distribution adaption for image representation. IEEE Trans. Knowl.-Based Syst. **147**(5), 135–167 (2018)
13. Ahmed, J., Memon, R., Waqas, M.: Selective sparse coding based coupled dictionary learning algorithm for single image super-resolution. In: Proceedings of ECCV, Munich, Germany, pp. 236–251 (2018)
14. Kim, J., Lee, J.K., Lee, K.M.: Accurate image super-resolution using very deep convolutional networks. In: Proceedings of CVPR, Las Vegas, America, pp. 1646–1654 (2016)
15. Lai, W.S., Huang, J.B., Ahuja, N.: Fast and accurate image super-resolution with deep Laplacian pyramid networks. IEEE Trans. Pattern Anal. Mach. Intell. **41**(2), 2599–2613 (2017)
16. Lim, B., Son, S., Kim, H.: Enhanced deep residual networks for single image super-resolution. In: Proceedings of CVPR, Hawaii, America, pp. 136–144 (2017)
17. Ledig, C., Theis, L., Huszar, F.: Photo-realistic single image super-resolution using a generative adversarial network. In: Proceedings of ICCV, Venice, Italy, pp. 4681–4690 (2017)
18. Wang, X., Yu, K., Wu, S.: ESRGAN: Enhanced super-resolution generative adversarial networks. In: Proceedings of CVPR, Salt Lake City, America, pp. 2560–2572 (2018)
19. Zhang, W., Liu, Y., Dong, C.: RankSRGAN: generative adversarial networks with ranker for image super-resolution. In: Proceedings of ICCV, Seoul, Korea, pp. 3096–3105 (2019)
20. Zhang, Y., Tian, Y., Kong, Y.: Residual dense network for image super-resolution. In: Proceedings of CVPR, Las Vegas, America, pp. 2472–2481 (2016)
21. Goodfellow, I.J., Abadie, J.P., Mirza, M.: Generative adversarial nets. In: Proceedings of ICCV, Barcelona, Spain, pp. 2672–2680 (2017)
22. Peng, Y.: Super-resolution reconstruction using multiconnection deep residual network combined an improved loss function for single-frame image. Multimedia Tools Appl. **79**(13–14), 9351–9362 (2019). https://doi.org/10.1007/s11042-019-7544-1
23. Timofte, R., De, V., Gool, L.V.: Anchored neighborhood regression for fast example-based super-resolution. In: Proceedings of ICCV, Sydney, Australia, pp. 1920–1927 (2013)
24. Wang, X., Gupta, A.: Unsupervised learning of visual representations using videos. In: Proceedings of ICCV, Santiago, Chile, pp. 2794–2802 (2015)
25. Meyer, K.D., Spratling, M.W.: Multiplicative gain modulation arises through unsupervised learning in a predictive coding model of cortical function. Neural Comput. **23**(6), 1536–1567 (2011)
26. Choi, J.J., Laibson, D.: Reinforcement learning and savings behavior. Finance **64**(6), 136–155 (2010)
27. Timofte, R., De, V., Smet, L., Gool, V.: A+: adjusted anchored neighborhood regression for fast super-resolution. In: Proceedings of CVPR, Columbus, America, pp. 111–126 (2014)
28. Yun, S.J., Choi, J., Yoo, Y., Yun, K.: Action decision networks for visual tracking with deep reinforcement learning. In: Proceedings of CVPR, Hawaii, America, pp. 2711–2720 (2017)
29. Wang, Z., Chen, J., Hoi, S.C.H.: Deep learning for image super-resolution: a survey. In: Proceedings of CVPR, Long Beach, America, pp. 2520–2538 (2019)
30. He, K., Zhang, X., Ren, S.: Deep residual learning for image recognition. In: Proceedings of CVPR, Las Vegas, America, pp. 770–778 (2016)
31. Zeiler, M.D., Fergus, R.: Visualizing and understanding convolutional networks. In: Fleet, D., Pajdla, T., Schiele, B., Tuytelaars, T. (eds.) Computer Vision – ECCV 2014. LNCS, vol. 8689, pp. 818–833. Springer, Cham (2014). https://doi.org/10.1007/978-3-319-10590-1_53

32. He, K., Zhang, X., Sun, J., Ren, S.: Delving deep into rectifiers: Surpassing human-level performance on imagenet classification. In Proceedings of ICCV, Santiago, Chile, pp. 1026–1034 (2015)
33. Yi, C., Huang, J.: Semismooth Newton coordinate descent algorithm for elastic-net penalized huber loss regression and quantile regression. Statistics **26**(2), 547–557 (2015)
34. Pang, J., Sun, W., Ren, S.: Cascade residual learning: a two-stage convolutional neural network for stereo matching. In: Proceedings of CVPR, Hawaii, America, pp. 887–895 (2017)
35. Wang, Z., Liu, D., Yang, J., Han, W., Huang, T.: Deeply improved sparse coding for image super-resolution. In: Proceedings of ICCV, Santiago, Chile, pp. 370–378 (2015)
36. Bevilacqua, M., Roumy, A., Guillemot, A., Morel, C., Alberi-Morel, M.L.: Low-complexity single-image super-resolution based on nonnegative neighbor embedding.In: Proc. BMVC, Surrey, British, pp. 135.1–135.10 (2012)
37. Zeyde, R., Elad, M., Protter, M.: On single image scale-up using sparse-representations. Curves and surfaces. In: Proceedings of Curves and Surfaces - 7th International Conference, Avignon, France, pp. 711–730 (2010)
38. Martin, D., Fowlkes, C., Tal, D.: A database of human segmented natural images and its application to evaluating segmentation algorithms and measuring ecological statistics. In: Proceedings Eighth IEEE International Conference on Computer Vision (ICCV), Vancouver, British Columbia, Canada, 2001, pp. 416–423. IEEE (2001)
39. Wang, Z., Bovik, A.C., Sheikh, H.R.: Image quality assessment: from error visibility to structural similarity. IEEE Trans. Image Process. **13**(2), 600–612 (2004)

Risk Probability Forecast Technology of the Gust in Beijing Winter Olympic Games Based on the Theory of Digital Image Processing

Yuxiao Chen[1], Jing Chen[1](✉), Hao Yang[2], Chuan Li[3], and Bo Pang[4]

[1] Numerical Weather Prediction Center/CMA, Beijing 100081, China
chenj@cma.gov.cn
[2] School of Computer Science, Chengdu University of Information Technology, Chengdu 610225, China
[3] National Satellite Meteorological Center/CMA, Beijing 100081, China
[4] School of Atmospheric Sciences, Chengdu University of Information Technology, Chengdu 610225, China

Abstract. Gust, which usually means a sudden, brief increase in the speed of the wind, has a great impact on some weather-sensitive activities, like Beijing Winter Olympic Games. Gust forecast has been a big challenge because of its uncertainty in time and spatial scale. The risk probability forecast has been often used to describe this uncertainty. Traditionally, the risk probability forecast of gust was obtained by neighborhood probability method. This method cannot accurately describe the uncertainty of time and spatial scale of gust. To solve this problem, based on the mean filtering algorithm in the digital image processing and the neighborhood probability method in meteorology, this study developed a new risk probability forecast technology of the gust, which converts the binary probability of the neighborhood probability method into the grayscale probability. And this new technology is used to post-process the two-week forecast results from the Global/Regional Assimilation and Prediction System (GRAPES) model of the China Meteorological Administration. The risk probability forecast obtained by the new technology, the observation, and the original forecast result were compared. The results show that the new risk probability products can better describe the characteristics of the gust. And compared with the gust peak speed from the original deterministic forecast results, the new risk probability products can better show some high-risk probabilities of the gust peak speed, which means it can provide forecasters with more useful information. It is concluded that the new technology is effective and significantly improves the skill of gust forecasting. This technology has good prospects for various applications and developments in numerical weather prediction.

Keywords: Digital image processing · Risk probability forecast of gust · Numerical weather prediction

X. Sun et al. (Eds.): ICAIS 2021, CCIS 1422, pp. 374–385, 2021.
https://doi.org/10.1007/978-3-030-78615-1_33

1 Introduction

The gust usually means a sudden, brief increase in the speed of the wind [1]. It has a more transient characteristic than a squall and its instantaneous wind speed is usually very high. The extreme gust has a great impact on some weather-sensitive activities, such as aviation, sailing and outdoor sports, which may cause lots of damages. The gust forecast has been a big challenge in weather forecast. Because of its uncertainty in time and spatial scale, it is very difficult to predict the accurate values of gust. And the risk probability forecast of gust is usually applied instead. The research of its technology has a great significance [1].

In 2022, the 24th Winter Olympic Games will be held in Beijing. Among them, the alpine skiing, the snowmobile, and the sled events will be held in Yanqing. The area of these events is located in the Haituo Mountain area, belonging to the Jundu Mountain System of the Yanshan Mountain Range. This area has vertical and horizontal ravines, high mountains and dense forests, complex terrain and large height drop [2], rapid and large wind speed changes. Especially in high-altitude alpine skiing tracks, the gust peak speed sometimes exceeds 30 m/s, and such extreme gust will bring a high risk to alpine skiing events. Therefore, for the weather prediction of the Winter Olympic Games, the risk probability forecast of gust will affect whether the relevant events can be held as scheduled, which is extremely important.

In recent years, with the development of numerical weather prediction (NWP) models, there are also more and more post-processing methods to producing risk probability forecast of gust by statistically processing the original forecast results of the model, such as the Bayesian model averaging (BMA) method [3–5], the probability matching (PM) method [6–8], and the neighborhood probability (NP) method [9–12]. Among these post-processing methods, the NP method has attracted more and more attention recently. Theis [9] proposed the NP method in 2005 firstly, which used the forecast results of one grid point and several points in its neighborhood space to calculate the probability of one grid point. This method can transform the traditional deterministic forecast result into the risk probability result. Later, Schwartz [10] extended the NP method to the post-processing of the ensemble forecast, which improved the reliability of the ensemble forecast. Now, researchers often use the NP method to obtain risk probability (RP) products about temperature, precipitation, gusts, etc. The RP product can better represent the uncertainty of the forecast and provide forecasters with more information, it seems to be suitable as a product of forecasting gust in the Winter Olympic Games. However, we noticed that there is a problem with the RP product obtained by the NP method. When calculating the probability by the NP method, for each grid point, the original forecast result is converted into binary probability (BP), that is, the probability that the forecast result is greater than the threshold is 1, otherwise it is 0. For example, when the threshold of the gust is 6 m/s, and the original forecast results of the two grid points are 5.9 m/s and 1 m/s, respectively. But two binary probabilities are all 0, which is obviously different from the original forecast results. This may cause the RP product of the gust obtained by the NP method cannot accurately describe the true characteristic of gust. It is a non ignorable problem for the traditional RP product obtained by the NP method.

The digital image processing technology can be uses to solve the problem for the traditional risk probability forecast of gust. The digital image processing technology mainly originated in the 1920s. Early image processing was used to improve image quality [13]. Later, people began to build digital human visual systems through computers. This technology is called image understanding or computer vision. The visual computing theory proposed by Marr in the late 1970s [14]. Mallat [15] applied wavelet analysis to image decomposition and reconstruction in 1988. It is considered to be a major breakthrough in mathematical methods of signal and image analysis, followed by rapid development of digital image processing technology. The theory of digital image processing is mainly to sample the discretized state of an image in a space, that is, to use the grayscale value of some points in the space to represent the image, and these points are called samples [16]. The differences between these samples are their grayscale values, and then rely on these grayscale values to segment images, extract characteristics and recognize images [17]. Over the years, with the maturity of the digital image processing technology, it has been applied in many industries, such as finance, aerospace, navigation, weather prediction, etc.

To solve this problem of traditional risk probability forecast of gust and improve its skill of gust forecasting, we were inspired by the theory of digital image processing from computer science, designed a new risk probability forecast technology. This technology is combined by the digital image processing technology and the NP method. And based on the forecast result of the Global/Regional Assimilation and Prediction System (GRAPES) model, two-week experiments were carried out for two different time periods (January 8–14, 2020, and February 3–9, 2020).

This paper is organized as follows. Section 2 introduces the method, experimental design and model data. Section 3 analyzes the experimental results, and Sect. 4 presents the conclusions.

2 Method, Experimental Design and Model Data

2.1 Method

According to the research of Theis [9] and Schwartz [10], the probability of the central grid is obtained by the original forecast of neighborhood grids. The calculation of the traditional NP method are as follows:

First, in order to obtain the BP of forecast results in threshold conditions, choosing a threshold q, the forecast results are processed as Eq. 1, the BP of a grid point (i, j) is 1 or 0.

$$BP_{f(i,j)} = \begin{cases} 1 \; F_{(i,j)} \geq q \\ 0 \; F_{(i,j)} < q \end{cases} \tag{1}$$

The $F(i, j)$ is the original forecast result of this grid point. We select the appropriate neighborhood range, the obtain the neighborhood probability of this grid point:

$$NP_{F(i,j)} = \frac{1}{N_b} \sum_{m=1}^{N_b} BP_{f(i,j)} \tag{2}$$

The N_b is the total number of grid points in the neighborhood range, the $BP_{f(i,j)}$ is the binary probability, and $NP_{F(i,\,j)}$ is the neighborhood probability of the grid point (i, j). Through the above calculations, we can transform original deterministic forecast results into spatial neighborhood probability, which can represent the uncertainty of the model to a certain extent.

In the digital image processing, the graphics reconstructed from real-world data is often corrupted by noise [18]. In order to avoid the influence of noise, the noise reduction processing is required. The mean filtering algorithm is a very effective method for noise reduction [19]. It replaces the grayscale value of the central pixel with the average of the grayscale values of all pixels in the local window [20], which is calculated as follows:

$$f_{m(i,j)} = \frac{1}{N \times N} \sum_{(i,j) \subset \Omega} f_{(i,j)} \tag{3}$$

The $f_{m(i,j)}$ is the grayscale value of the central pixel, the $f_{(i,j)}$ is the original value, the Ω is the sliding window, and the size is $N \times N$. Based on the traditional mean filtering algorithm, Chang [21] designed the adaptive weighted average filtering algorithm based on medium value in 2008. It is calculated as follows:

$$f_{m(i,j)} = \frac{1}{N \times N} \sum_{(i,j) \subset \Omega} w_{(i,j)} f_{(i,j)} \tag{4}$$

$$w_{(i,j)} = \frac{1}{1 + \max\{T,\ (f_{(i,j)} - M)^2\}} \tag{5}$$

The T is the mean value of the square of the difference between each pixel and the median and the M is the median grayscale value in the filter window. And the $w_{(i,j)}$ is weighted factor, which changes with different grayscale values of each point. This algorithm considered that the influence of each pixel in the window is different when calculating the average grayscale value of the center pixel [20].

We are inspired by the above NP method in meteorology and the mean filtering algorithms in the digital image processing technology, and consider to change the BP of the NP method into the grayscale probability which is a linear probability between 0 and 1. Then we design a new risk probability forecast technology. It is calculated as follows:

$$\left\langle \begin{array}{c} GP_{q_1(i,j)} = \begin{cases} 1 & F_{(i,j)} \geq q_1 \\ F_{(i,j)}/q_1 & F_{(i,j)} < q_1 \end{cases} \\ ,\ldots\ldots, \\ GP_{q_n(i,j)} = \begin{cases} 1 & F_{(i,j)} \geq q_n \\ (F_{(i,j)} - q_{n-1})/(q_n - q_{n-1}) & F_{(i,j)} < q_n \end{cases} \end{array} \right\rangle \tag{6}$$

In the above equations, q_1 is the minimum threshold value starting from 0, q_n is the next threshold value that increases sequentially. When calculating the grayscale probability at the minimum threshold value (q_1), equation ($GP_{q1(i,j)}$) should be used, when calculating the next threshold value, equation ($GP_{qn(i,j)}$) should be used. $GP_{qn(i,j)}$ is the grayscale probability. According to Eq. 2, the spatial grayscale probability of a grid point is calculated by Eq. 7:

$$SGP_{F(i,j)} = \frac{1}{N_b}\sum_{m=1}^{N_b} GP_{q_n(i,j)} \tag{7}$$

In addition, according to the research of Bouallègue [12], we considered the lag phenomenon of forecasting. Therefore, we extend the current time t to lag time $t + l$ as the temporal neighborhood range, then obtain the space-time grayscale probability of a grid point, as shown in the following equation:

$$RP_{F(i,j)} = \frac{1}{l+1}\sum_{t}^{t+l} SGP_{F(i,j)} \tag{8}$$

The $RP_{F(i,j)}$ represents the space-time grayscale probability of a grid point. After the above calculation, the new RP is finally obtained.

2.2 Experimental Design and Data

The GRAPES is a numerical weather prediction system independently developed by China Meteorological Administration (CMA) [22]. It includes the global forecast system (GRAPES_GFS) [23], the mesoscale regional model (GRAPES_Meso) [24, 25] and the regional ensemble prediction system (GRAPES_REPS) [26, 27]. The forecasted gust data are from the 3 km GRAPES_Meso model (3 km horizontal resolution, 70.00–145.00° E, 10.00–60.10° N). The model runs once a day (initialized at 0000 UTC) out to a 36 h forecast length. The lateral boundary conditions and initial conditions of 3 km GRAPES_Meso model are provided (directly downscaled) from the Global Forecasting System developed by the National Centers for Environmental Prediction, the National Oceanic and Atmospheric Administration (NCEP-GFS/NOAA).

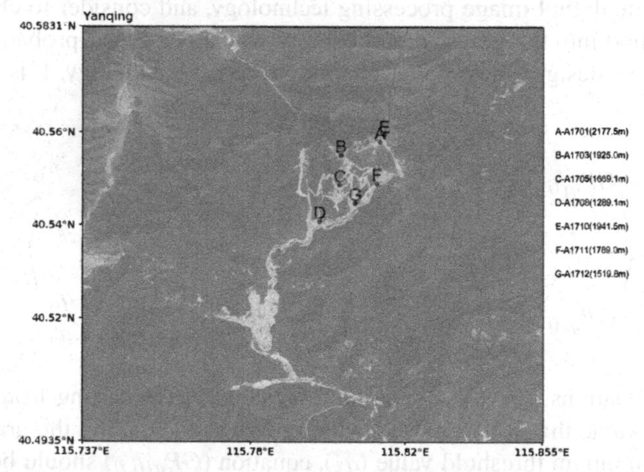

Fig. 1. The location distribution and height of the stations on alpine skiing track in Yanqing

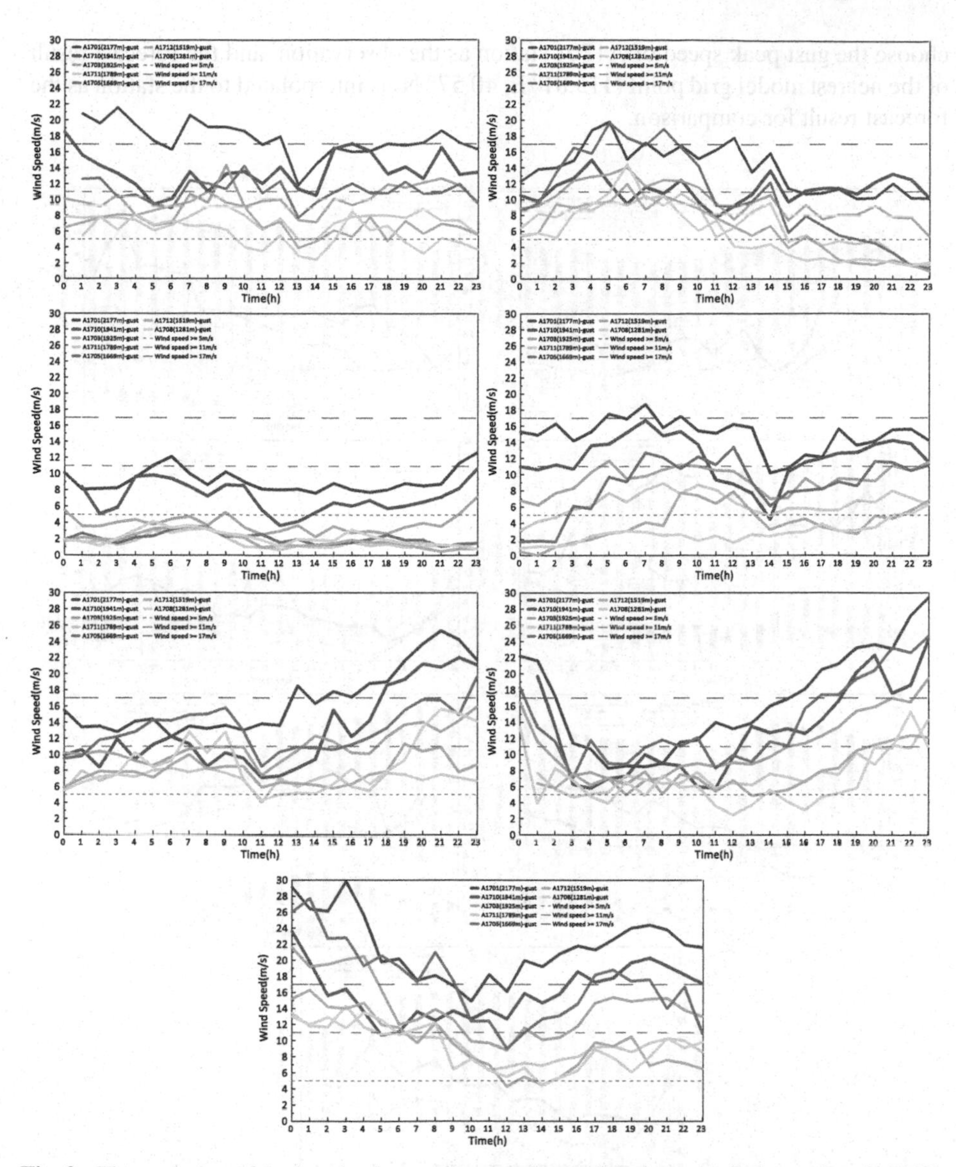

Fig. 2. The evolution of the hourly observed gust peak speed (units: m/s) (February 3 to 9, 2020)

The observed gust is obtained from several automatic meteorological stations on alpine skiing track in Yanqing. The location distribution and height of these stations are shown in Fig. 1. Through analyzing the observed gusts for a week (Fig. 2. February 3 to 9, 2020), it is found that the gust peak speed increases with the height of these stations. At A1701 station (crimson line in Fig. 2), which is located at the highest point of the alpine skiing track, we can find that the gust peak speed is very high in this week, and it is the highest in all stations at most of the time. It means that the area where A1701 station located is high-risk for the alpine skiing event. Therefore, in this experiment, we choose the gust peak speed of A1701 station as the observation, and the forecast result of the nearest model grid point (115.81° E, 40.57° N) is interpolated to the station as the forecast result for comparison.

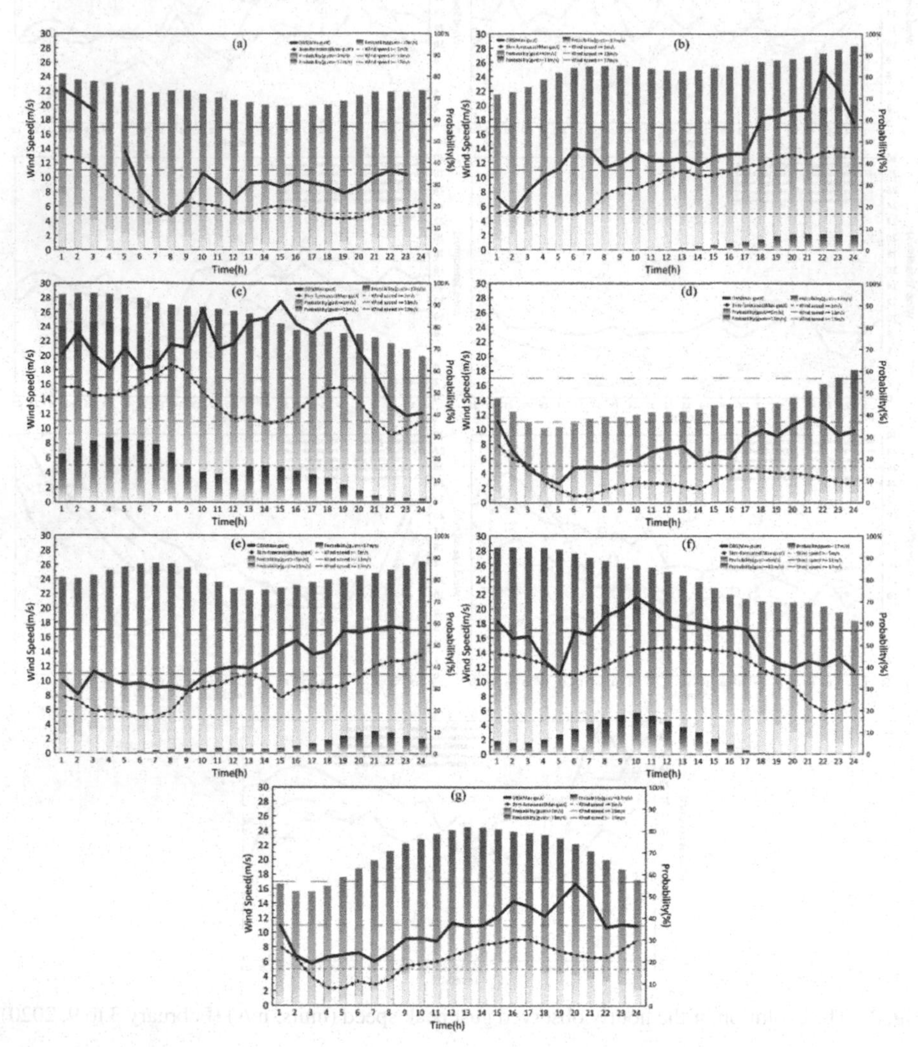

Fig. 3. The evolution of the hourly observed gust peak speed, the forecasted gust peak speed (units: m/s) and the RPs (units: %) (January 8 (a)–14 (g), 2020)

According to some previous research [14–17] and the actual situation of this study, we select 3 times the horizontal grid distance (9 km) as the spatial neighborhood radius and 6 h as the temporal neighborhood radius of the RP. Moreover, according to the date and requirements of the Winter Olympic Games, two-week experiments are selected for two different time periods (January 8–14, 2020, and February 3–9, 2020). And if the gust peak speed is greater than 11 m/s, the forecaster will focus on the impact of gusts on the event. If the gust peak speed is greater than 17 m/s, the organizers will consider the suspension of the event. Therefore, we choose 5 m/s, 11 m/s and 17 m/s as the gust threshold of RP. All experimental information can be found in Table 1.

Table 1. Experimental information.

Name	Detail
Observed gust	A1701 station
Forecasted gust	Model grid point (115.81° E, 40.57° N) From 3 km GRAPES_Meso model
Neighborhood radius	Grid distance: 9 km, time: 6 h
Date	January 8–14, 2020, and February 3–9, 2020
Gust threshold of RP	5 m/s, 11 m/s and 17 m/s

3 Results

3.1 Experimental Results for January

The first experimental period is from January 8 to 14, 2020. The average temperature is low and the average wind speed is high in this high mountain area. Figure 3 shows the evolution of the hourly gust peak speed for the observation, the original forecasted result and the RP. In the figure, the blue line of 'OBS' represents the observed gust peak speed; the blue dotted line of '3 km-forecasted' represents the forecasted gust peak speed; and gray, green and orange bar charts of 'Probability' represent the RP when the gust peak speed is greater than 5 m/s, 11 m/s and 17 m/s, respectively. The y-axis on the left-hand side of the figure indicates the wind speed value, and the y-axis on the right-hand side of the figure indicates the probability value. It can be found that the original forecasted gust peak speed is obviously lower than the observed gust peak speed in Fig. 3, especially for some moments with the extreme gust, the difference is even greater. For example, from 1800UTC on January 9th to 0400UTC on January 10th (Fig. 3b, 3c), the observed gust peak speed reaches 24 m/s, which means a high risk for the alpine skiing event. But the original forecast gust peak speed is only 15 m/s, it cannot reflect the actual high-risk situation. Fortunately, in this case, there is the 20 to 30% high-risk probability (orange bar chart), indicating that a strong gust is likely to occur during this period, which can provide the forecaster with a risk warning for the event. And in general, the evolution of the RP is closer to the observed gust peak speed than that of the original forecast. For example, from 0100 UTC on January 13 to 1800 UTC on January 14 (Fig. 3f, 3g), the evolution of the three RPs (gray, green and orange bar charts) is more similar with the observation, it is obviously better than the original forecast results.

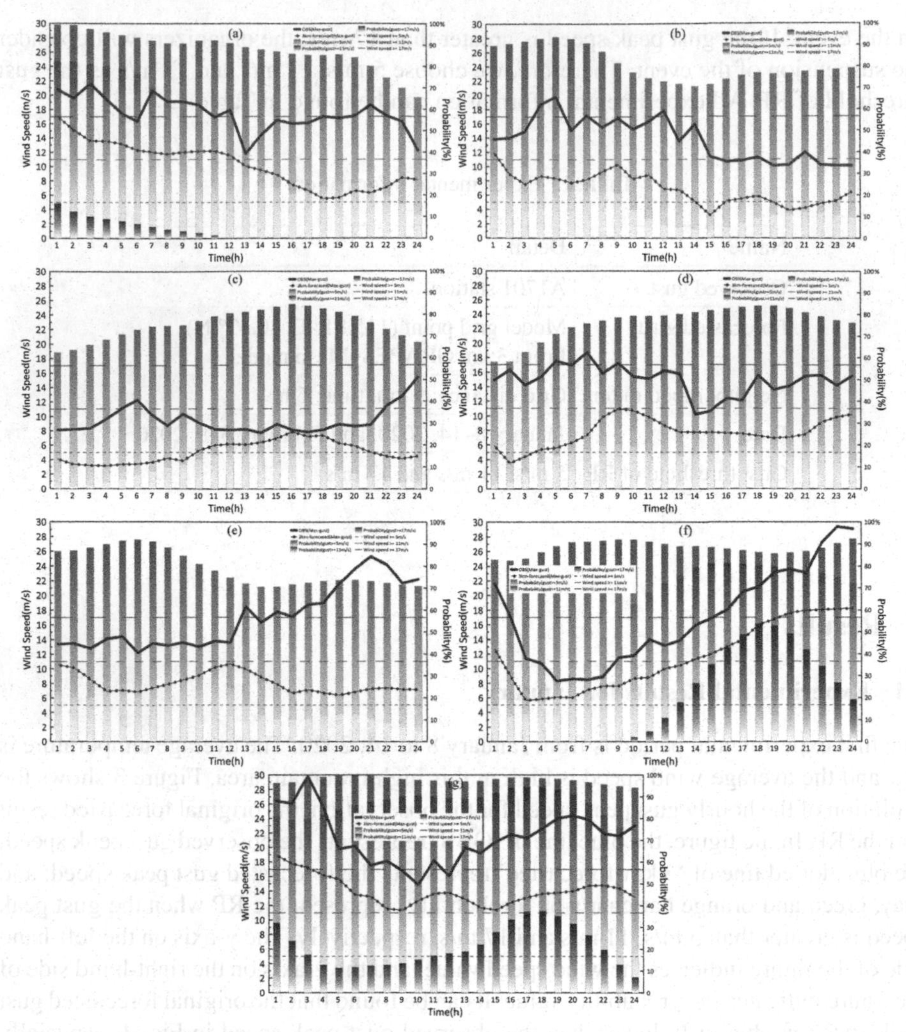

Fig. 4. The evolution of the hourly observed gust peak speed, the forecasted gust peak speed (units: m/s) and the RPs (units: %) (February 3 (a)–9 (g), 2020)

3.2 Experimental Results for February

The second experimental period is from February 3 to 9, 2020. This experimental period is similar to the date of the Beijing Winter Olympic Games (February 4 to 20, 2020). Figure 4 show the evolution of the hourly gust peak speed for the observation, the original forecast result and the RP. It can be found that the original forecasted gust peak speed is also obviously lower than the observed gust peak speed. For some cases of the extreme gust, such as the period from 0100UTC to 1200UTC on February 3rd (Fig. 4a). The observed gust peak speed reaches 20 m/s, the high-risk probability is about 5 to 15% (orange bar chart), which indicates that the strong gust is likely to occur during this period. Another example is from 0100UTC to 2300UTC on February 9th (Fig. 4g), the average of the observed gust peak speed reaches 20m/s and the maximum is about 30 m/s, but the average of the original forecasted gust peak speed is much lower than the observation. It means that if the alpine skiing event is held on that day, there will be high-risk. However, there are some high-risk probabilities about 20 to 40%, that mean a risk warning for holding alpine skiing event. It is an important information that original forecasted result cannot provide.

In summary, the RP obtained by the new technology shows many advantages by analyzing the results of two-week experiments, it provides forecasters with more information for references.

4 Summary and Discussion

A new risk probability forecast technology is designed based on the digital image processing technology and the neighborhood probability method. Two-week forecast experiments are carried out for two different time periods (January 8–14, 2020, and February 3–9, 2020), which based on the forecast result of 3 km GRAPES_Meso model. And the observed gust peak speed, the original forecasted gust peak speed, and the RPs are compared. The results and conclusions are as follows:

(1) The analysis of several cases shows that the original forecasted gust peak speed is obviously lower than the observed gust peak speed, but the RP can show some high-risk probabilities which might occur a strong gust. The RP can provide forecasters with more information that original forecasted result cannot provides.

(2) The analysis of two-week experiments shows that the evolution of the RP is more similar with the observation, it is obviously better than that of the original forecast, which means that the RP product obtained by this new technology can improve the skill of forecasting gust.

This study verified the effectiveness of the risk probability forecast technology and showed some advantages of the RP obtained by the new technology. However, there are problems and challenges in the application also. First, due to the coarse resolution of the model, the RP obtained is interpolated to each station in a simple way, but the significant difference of height between each station was not considered. Second, in this study, the calculation method of the RP is a linear calculation process. The detailed influence of thc calculation method on the result needs further exploration.

Acknowledgement. A Special thanks and appreciation are extended to Dr. Guo Deng for his assistance in Numerical Weather Prediction Center (NWPC) of CMA. We are grateful to Dr. Hua Tong and Yutao Zhang from NWPC for their exceptional efforts to provide related data. And we also thanks Anqi Xie from Chengdu University of Information Technology for her ideas about the technology.

Funding Statement. This work is sponsored by the National Science and Technology Major Project of the Ministry of Science and Technology of China (2018YFF0300103 and 2018YFC1507405) and Sichuan Science and Technology Program (2020YFS0355 and 2020YFG0479).

References

1. Cheng, X., Quan, L., Hu, F., Wang, B.: The fractal and chaotic characteristic of gustwind. Climatic and Environ. Res. **12**(3), 256–266 (2007). https://doi.org/10.3969/j.issn.1006-9585.2007.03.006
2. Lü, Y., Zhou, Q., Hou, Q., Li, B.: 3D GIS application on winter Olympics planning: a case study of Yanqing zone of 2022 winter Olympics. Bull. Surveying Mapp. (10), 133–137 (2019). https://doi.org/10.13474/j.cnki.11-2246.2019.0334
3. Raftery, A.E., Gneiting, T., Balabdaoui, F., Polakowski, M.: Using Bayesian model averaging to calibrate forecast ensembles. Mon. Weather Rev. **133**(5), 1155–1174 (2005). https://doi.org/10.1175/mwr2906.1
4. Schmeits, M.J., Kok, K.J.: A Comparison between raw ensemble output, (Modified) Bayesian model averaging, and extended logistic regression using ECMWF ensemble precipitation reforecasts. Mon. Weather Rev. **138**(11), 4199–4211 (2010). https://doi.org/10.1175/2010mwr3285.1
5. Sloughter, J.M., Gneiting, T., Raftery, A.E.: Probabilistic wind speed forecasting using ensembles and Bayesian model averaging. J. Am. Stat. Assoc. **105**(489), 25–35 (2010). https://doi.org/10.1198/jasa.2009.ap08615
6. Ebert, E.E.: Ability of a poor man's ensemble to predict the probability and distribution of precipitation. Mon. Weather Rev. **129**(10), 2461–2480 (2000). https://doi.org/10.1175/1520-0493(2001)129%3c2461:AOAPMS%3e2.0.CO;2
7. Jun, L., Du, J., Chen, C.: Introduction an analysis to frequency or area matching method applied to precipitation forecast bias correction. Meteorol. Monthly **40**(5), 580–588 (2014). https://doi.org/10.7519/j.issn.1000-0526.2014.05.008
8. Zhu, Y., Luo, Y.: Precipitation calibration based on the frequency-matching method. Weather Forecast. **30**(5), 1109–1124 (2015). https://doi.org/10.1175/WAF-D-13-00049.1
9. Theis, S.E., Hence, A., Damrath, U.: Probabilistic precipitation forecasts from a deterministic model: a pragmatic approach. Meteorol. Appl. **12**(3), 257–268 (2005). https://doi.org/10.1017/S1350482705001763
10. Schwartz, C.S., et al.: Toward improved convection-allowing ensembles: model physics sensitivities and optimizing probabilistic guidance with small ensemble membership. Weather Forecast. **25**(1), 263–280 (2010). https://doi.org/10.1175/2009waf2222267.1
11. Liu, X., Chen, J., Chen, F., Xia, Y., Fan, Y.: Scale sensitivity experiments of precipitation neighborhood ensemble probability method. Chin. J. Atmos. Sci. **44**(2), 282–296 (2020). https://doi.org/10.3878/j.issn.1006-9895.1903.18228
12. Bouallègue, Z.B., Theis, S.E., Gebhardt, C.: Enhancing COSMO-DE ensemble forecasts by inexpensive techniques. Meteorol. Z. **22**(1), 49–59 (2013). https://doi.org/10.1127/0941-2948/2013/0374
13. Gonzalez, R.C., Woods, R.E., Masters, B.R.: Digital Image Processing, 2nd edn. John Wiley and Sons Inc, New York (1991)
14. Marr, D., Poggio, T.: A computational theory of human stereo vision. Proc. R. Soc. Lond. **204**(1156), 301–328 (1979). https://doi.org/10.1098/rspb.1979.0029
15. Mallat, S.G.: A theory for multiresolution signal decomposition: the wavelet representation. IEEE Trans. Pattern Anal. Mach. Intell. **11**(7), 674–693 (1989). https://doi.org/10.1109/34.192463
16. Lu, T., Yang, J., Deng, M., Du, C.: A visibility estimation method based on digital total-sky images. J. Appl. Meteorol. Sci. **29**(6), 701–709 (2018). https://doi.org/10.11898/1001-7313.20180606
17. Chen, H., Wan, Y., Wang, G.: Progress of digital image technology processing research. Ind. Control Comput. **26**(1), 72–74 (2013). https://doi.org/10.3969/j.issn.1001-182X.2013.01.034

18. Yagou, H., Ohtake, Y., Belyaev, A.: Mesh smoothing via mean and median filtering applied to face normal. In: GMP 2002, Proceedings, pp. 124–131. Proc. IEEE, Wako, Saitama, Japan (2002).: https://doi.org/10.1109/GMAP.2002.1027503
19. Lee, Y., Kassam, S.: Generalized median filtering and related nonlinear filtering techniques. IEEE Trans. Acoust. Speech Signal Process. **33**(3), 672–683 (1985). https://doi.org/10.1109/TASSP.1985.1164591
20. Zhu, S., You, C.: A modified average filtering algorithm. Comput. Appl. Softw. **30**(12), 97–116 (2013). https://doi.org/10.3969/j.issn.1000-386x.2013.12.025
21. Chang, R., Mu, X., Yang, S., Qi, L.: Adaptive weighted average filtering algorithm based on medium value. Comput. Eng. Des. **29**(16), 4257–4259 (2008). https://doi.org/10.16208/j.issn1000-7024.2008.16.036
22. Xue, J., Chen, D.: Scientific Design and Application of Numerical Prediction System GRAPES, 1st edn. Science Press Inc, Beijing (2008)
23. Shen, X., et al.: Development and operation transformation of GRAPES global middle-range forecast system. J. Appl. Meteorol. Sci. **28**(1), 1 (2017). https://doi.org/10.11898/1001-7313.20170101
24. Huang, L., et al.: Main technical improvements of GRAPES_Meso V4.0 and verification. J. Appl. Meteorol. Sci. **21**(5), 524–534 (2010). https://doi.org/10.11898/1001-7313.20170103
25. Wang, Y., Li, L.: Verification of GRAPES_Meso V3.0 model forecast results. J. Appl. Meteorol. Sci. 28(1), 25–37 (2017). https://doi.org/10.11898/1001-7313.20100502
26. Zhang, H., Chen, J., Zhi, X., Li, Y., Sun, Y.: Study on the application of GRAPES regional ensemble prediction system. Meteorol. Monthly **40**(9), 1076–1087 (2014). https://doi.org/10.7519/j.issn.1000-0526.2014.09.005
27. Chen, J., Wang, J., Du, J., Xia, Y., Chen, F., Li, H.: Forecast bias correction through model integration: a dynamical wholesale approach. Q. J. R. Meteorol. Soc. **146**(728), 1149–1168 (2020). https://doi.org/10.1002/qj.3730

Depth Estimation Based on Optical Flow and Depth Prediction

Pengyang Shen, Xinrui Jia(✉), and Liguo Zhang

School of Computer Science and Technology, Harbin Engineering University, Harbin, China
{shenpengyang,jiaxinrui,zhangliguo}@hrbeu.edu.cn

Abstract. Depth sensing is essential for 3D reconstruction and scene understanding. Depth sensors can support dense depth results. But often face the difficult of environmental interference such as the restricted operating range, low spatial resolution, interference from other sensor, and the high cost of power. In this paper, we try to estimate per-pixel depth and its uncertainty continuously from video stream and the optical flow based on deep learning (DL) method. This can turn the RGB video stream to RGB-D video stream. It is different from normal deep learning methods as we try to get the depth probability distribution for each pixel rather than a single depth value of the graph. These depth probability volumes are amassed over time under a Bayesian filtering framework as more incoming frames and optical flow graph are processed sequentially. This effectively reduces depth uncertainty and improves accuracy, robustness, and temporal stability. Compared to prior work, the proposed approach achieves more accurate and stable results, and generalizes better to new datasets.

Keywords: Depth estimation · Optical flow · Depth prediction · Depth sensing

1 Introduction

3D models are used in a lot of areas, such as trajectory planning [26], terrain debris recognition [27], Video source identification [28] and fatigue investigations [29]. Depth sensing is crucial for 3D reconstruction [1–4,25]. Active depth sensors measure dense metric depth, but often have limited operating range and spatial resolution [6], consume too much power, and suffer from multi-path reflection and interference between sensors [7]. On the contrary, get the depth from the images does not face these questions, but will accept other long-standing challenges such as scale ambiguity and drift for monocular methods [8], as well as the correspondence problem and high computational cost for stereo [9] and multi-view methods [10].

Inspired by recent success of deep learning in 3D vision [1–4,11,12], in this work, we try to estimate depth and its uncertainty continuously from a monocular video stream by DL-based method, with the goal of effectively turning an RGB video into an RGB-D video. We have two key ideas:

X. Sun et al. (Eds.): ICAIS 2021, CCIS 1422, pp. 386–396, 2021.
https://doi.org/10.1007/978-3-030-78615-1_34

We introduce the optical flow into the depth prediction. Moreover, this can enhance the Depth Probability Volume (DPV), the DPV provides both a Maximum-Likelihood-Estimate (MLE) of the depth map, as well as the corresponding per-pixel uncertainty measure.

The influence of the camera's unstable speed can be reduced by the optical flow. As more incoming frames and optical flow are processed sequentially, the DPVs across different frames are accumulated over time. The Bayesian filtering theory guide us to the accumulation steps and implemented as a learnable deep network. It effectively reduces depth uncertainty and improves accuracy, robustness, and temporal stability.

We raise that the DL-based depth estimation methods can give the depth distributions as result but not depth values, and these statistical distributions should be integrated over time. Moreover, the input of optical flow can make the depth distributions integrate faster and more accurate. The reason we choose the depth distributions is there are a lot of uncertainly of methods to estimate the dense depth from single-view images, such as the lack of texture, occlusion, scale drift, and specular/transparent material. While some recent work started focusing on uncertainty estimation [13] for certain computer vision tasks, and we are one of them try to find a method to solve these problems.

We test our method on KTTIT datasets and compare with recent state-of-the-art, DL-based, depth estimation methods [1,2,12]. We also perform the "cross-dataset" evaluation task. It tests models trained on a different dataset without fine-tuning. The robustness and generalization ability can be well evaluated by these cross-dataset tasks. Our method use the input data with reasonably good camera pose estimation, outperforms these prior methods on depth estimation with better accuracy, robustness, and temporal stability.

2 Related Work

2.1 Depth from Depth Camera and Radar

Active sensors such as LiDAR [5] are widely used at measure dense metric depth and sensor-specific confidence measure. However, there are still many problem to face such as limited operating range and spatial resolution [6], consume too much power, and suffer from multi-path reflection and interference between sensors [7]. In this paper, we try to make the RGB camera to RGB-D camera and try to use optical flow to enhance the accuracy.

2.2 Depth from Images

Get depth from images is the core problem of the computer vision [14]. Traditional single-view method often make strong assumptions on scene structures [15]. Stereo and multi-view methods [16] rely on triangulation too much, and not good at textureless regions, transparent/specular materials, and occlusion field. What's more, these methods are too computationally expensive for real-time applications from global bundle adjustment. For depth estimation from sing-view video, there is still a problem of scale ambiguity and drifting [17]. To sidestep these questions, many computer vision system do not use RGB images for dense 3D reconstruction [16] but for camera pose [17]. In this paper, we do our work by learning-based method.

2.3 Application of Optical Flow Method in Detecting Moving Objects

Optical flow method detect moving objects by estimate and divide the optical field. Some methods detect moving objects by dense optic flow field [18]. Nevertheless, the performance is greatly affected by too many pixels used to get dense optic flow. Recently more and more methods try to use learning-based method to get the optical flow and get great result [11]. In this paper, we try to use the optical flow to find the move of camera and enhance the change field of depth.

2.4 Learning-Based Depth Estimation

Recently researches have shown greatly progress in depth sensing directly from images, including single-view methods [3,11], multi-view stereos [1], depth and motion from two views [2], and some of them try to set their methods into the visual SLAM systems [20]. With the promising performance, these DL-based methods are still not suitable for applications because their robustness and generalization ability is yet to be thoroughly tested. In our work, we try to use Bayesian filtering theory to solve the problems and use optical flow as extra input data to get result better.

3 Our Method

Our work consists of three part. The first part is the DO-NET. It estimates the Depth Probability Volume (DPV) for each input frame with the input frame and optical flow. The second part is the BP-NET. It is used to integrate DPVs with the Bayesian network over time. The third part is the RD-NET. We set the work of resize the output depth images with input images in this part. The structure is shown in Fig. 1. In this paper, we denote the DPV as $p = (d; u, v)$, which represents the probability of pixel (u, v) having a depth value d, where $d \in [d_{min}, d_{max}]$. The DPV is defined on the 3D view frustum attached to the camera due to perspective projection. d_{min} and d_{max} is the nearest and farest planes of the 3D frustum. We set the 64 planes over the inverse of depth. The DPV contains the complete statistical distribution of depth for a given scene. In this paper, we directly use the non-parametric volume to represent DPV. Given the DPV, we can compute the Maximum-Likelihood Estimates (MLE) for depth and its confidence:

$$Depth : d(u, v) = \sum_{d=d_{min}}^{d=d_{max}} p(f; u, v) \cdot d \tag{1}$$

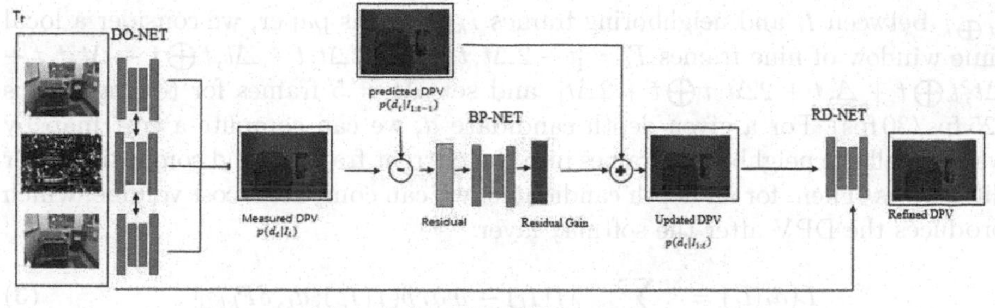

Fig. 1. The structure of the network, Our method takes the frames in a local time window in the video as input and outputs a Depth Probability Volume (DPV) that is updated over time. The update procedure is in a Bayesian filter fashion: we first take the difference between the local DPV estimated using the DO-Net and the predicted DPV from previous frames and optical flow between them to get the residual; then the residual is modified by the BP-Net and added back to the predicted DPV; at last the DPV is refined and upsampled by the RD-Net, which can be used to compute the depth map and its confidence measure.

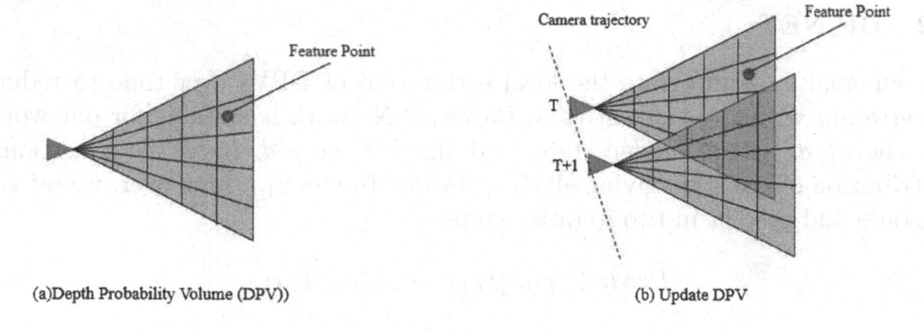

Fig. 2. Representation and update for DPV: (a) The DPV is defined over a 3D frustum defined by the pinhole camera model (b) The DPV gets updated over time as the camera moves.

$$confidence : \hat{c}(u, v) = p(\hat{d}, (u, v)) \tag{2}$$

We often treat DPV as a hidden state of the condition when processing the video stream. As the camera moves, the DPV $p = (d)$ will change, especially the overlapping volumes will change more than other places, as shown in Fig. 2. What' more, if we know the camera motion, we can predict the DPV of the next frame through the Bayesian Network. In addition, with the more frames input into the system, the predict DPV will get better accuracy.

3.1 DO-NET

We use a CNN named DO-Net to estimate the conditional DPV with the input of frame I_t, its temporally neighboring frames I_k and the optical flow image

$I_{t \bigoplus k}$ between I_t and neighboring frames I_k . In this paper, we consider a local time window of nine frames $T_t = [t - 2\Delta t, t \bigoplus t - 2\Delta t, t - \Delta t, t \bigoplus t - \Delta t, t, t + \Delta t, t \bigoplus t + \Delta, t + 2\Delta t, t \bigoplus t + 2\Delta t]$, and set $\Delta t = 5$ frames for testing videos (25 fps/30 fps). For a given depth candidate d, we can compute a cost map by warping all the neighboring frames into the current frame I_t and computing their differences. Then, for all depth candidates, we can compute a cost volume, which produces the DPV after the softmax layer:

$$L(d_t|I_t) = \sum_{k \in T_t, k \neq t} \| f(I_t) - warp(f(I_k); d_t, \delta P_{kt}) \| \tag{3}$$

$$p(d_t|I_t) = softmax(L(d_t|I_t)) \tag{4}$$

Where $f(\cdot)$ is a feature extractor, δP_{kt} is the relative camera pose from frame I_k to frame I_t, d_t is the depth candidate of I_t, $warp(\cdot)$ is an operator that warps the image features from frame I_k to the reference frame I_t, which is implemented as 2D grid sampling. In this paper, we use PSM-Net [19] as the feature extractor $f(\cdot)$, which outputs a feature map of 1/4 size of the input image.

3.2 BP-NET

As our goal is to integrate the local estimation of DPVs over time to reduce uncertainty with the video stream, Bayesian Network is suitable for our work. We choose d_t as the hidden state, and the volume $p(d_t|I_t)$ is the conditional distribution of the state giving all the previous frames $I_{1:t}$. Moreover, we set the upgrade and predict in two iterative steps:

$$Predict : p(d_t|I_{1:t}) \rightarrow p(d_{t+1}|I_{1:t}) \tag{5}$$

$$Update : p(d_{t+1}|I_{1:t}) \rightarrow p(d_{t+1}|I_{1:t}) \tag{6}$$

In addition, the prediction step is to warp the current DPV from the camera coordinate at t to the camera coordinate at t+1:

$$p(d_{t+1}|I_{1:t}) = warp(p(d_{t+1}|I_{1:t}), \delta P_{t,t+1}) \tag{7}$$

The $\delta P_{t,t+1}$ is the relative camera pose from time t to time t + 1, and $warp(\cdot)$ here is a warping operator implemented as 3D grid sampling. At time $\mathrm{t}+1$, we can compute the local DPV $p(d_{t+1}|I_{1+t})$ from the new measurement I_t using the D-Net. This local estimate is used to update the hidden state d_t:

$$p(d_{t+1}|I_{1:t+1}) = p(d_{t+1}|I_{1:t}) \cdot p(d_{t+1}|I_{1+t}) \tag{8}$$

Note that we always normalize the DPV in the above equations and ensure $\int_{d_{min}}^{d_{max}} p(d) = 1$. However, one problem is it integrates both correct and incorrect information in the prediction step. For example, when there are occlusions or disocclusions, the depth values near the occlusion boundaries change abruptly. This will bring the mistake into the next predict regions. To solve this problem, we set $E_d(d) = -\log p(d)$, and to reduce the weight of the prediction, we multiply a weight $\lambda \in [0, 1]$ with the predict term:

$$E(d_{t+1}|I_{1:t+1}) = \lambda \cdot E(d_{t+1}|I_{1:t}) + E(d_{t+1}|I_{t+1}) \tag{9}$$

We call this as "global damping". However, global damping may also prevent some correct depth information to be integrated to the next frames, since it reduces the weights equally for all voxels in the DPV. Therefore, we propose an "adaptive damping" scheme to update the DPV:

$$E(d_{t+1}|I_{1:t+1}) = E(d_{t+1}|I_{1:t}) + g(\Delta E_{t+1}, I_{t+1}) \tag{10}$$

The ΔE_{t+1} is the difference between the measurement and the prediction, $g(\cdot)$ is a CNN named BP-NET. This is work for transform ΔE_{t+1} into the correction term to the prediction. For regions with correct depth predictions, the values in the volume of DPVs are consistent and the residual in ΔE_{t+1} is small as the DPV will not be updated in Eq. 10. On the other hand, for regions with incorrect depth probability, the residual would be large and the DPV will be corrected by $g(\Delta E_{t+1}, I_{t+1})$. This way, the weight for prediction will be changed adaptively for different DPV voxels.

3.3 RD-NET

Finally, since the DPV $p(d_t|I_{1:t})$ is estimated with 1/4 spatial resolution of the input image, we select a CNN, named RD-Net, to upsample and refine the DPV back to the original image resolution. The RD-Net is essentially an U-Net with skip connections, which takes input the low-res DPV from the BP-Net $g(\cdot)$ and the image features extracted from the feature extractor $f(\cdot)$, and outputs a high-resolution DPV.

4 Experimental Results

During training, we use ground truth camera poses. For all our experiments, we use the ADAM optimizer [21] with a learning rate of 10^{-6}, $\beta_1 = 0.9$ and $\beta_2 = 0.999$. The whole framework, including DO-NET, BP-NET and RD-NET, is trained together in an end-to-end fashion for 25 epochs.

To emphasis on accuracy and robustness, we test our methods on the datasets KITTI [22]. And for accuracy evaluation, we do choose the statistical metrics [3] as it provide overall estimate of the entire depth map. We carefully select a few recent DL-based depth estimation methods and try our best for a fair comparison. For two-view methods, we compare with the DeMoN [2], which shows high quality depth prediction from a pair of images. For multi-view methods, we compared with MVSNet [1]. And we also choose the Neural RGBD Sensing(RGBD) [12] to show our change at the depth predict work. As shown in Table 1 our

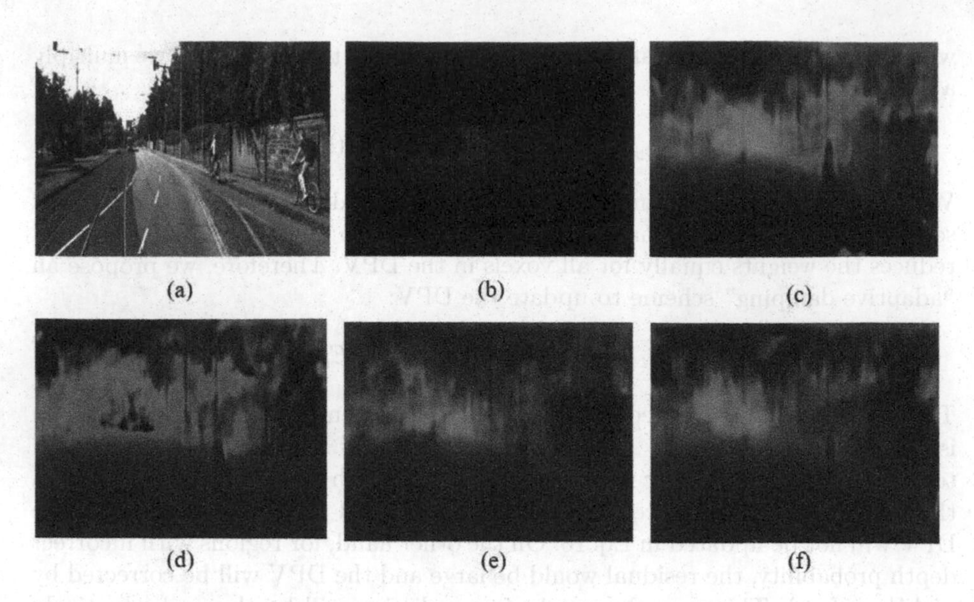

Fig. 3. The depth on the KITTI datasets of different method: (a) original image (b) ground truth image (c) DeMoN (d) MVSNet (e) RGBD (f) Ours

method significantly outperforms DeMoN and MVSNet in this datasets with use of statistical metrics [3]. Moreover, have some promote compare with RGBD method. The result is shown on Fig. 3 and Table 1.

Table 1. Comparison of depth estimation over the KITTI [22] datasets.

	$\sigma < 1.25$	abs	rmse	scale inv
DeMoN	31.88	0.3888	0.8549	0.4473
MVSNet	54.87	0.3481	0.8305	0.3743
RGBD	69.26	0.1758	0.4408	0.1899
Ours	70.03	0.1379	0.4195	0.3045

In Table 2, we choose the ScanNet [23] as the test data cause of it has dense, accurate metric depth as ground truth. The result is shown in Fig. 4 and Table 2. In addition, we try to use the model train on the KITTI test on the ScanNet as Cross-dataset tests for depth estimation in Table 3, 4. Moreover, we choose the DORN [24] method as the compared method. As shown, our method achieves better robustness and generalization ability.

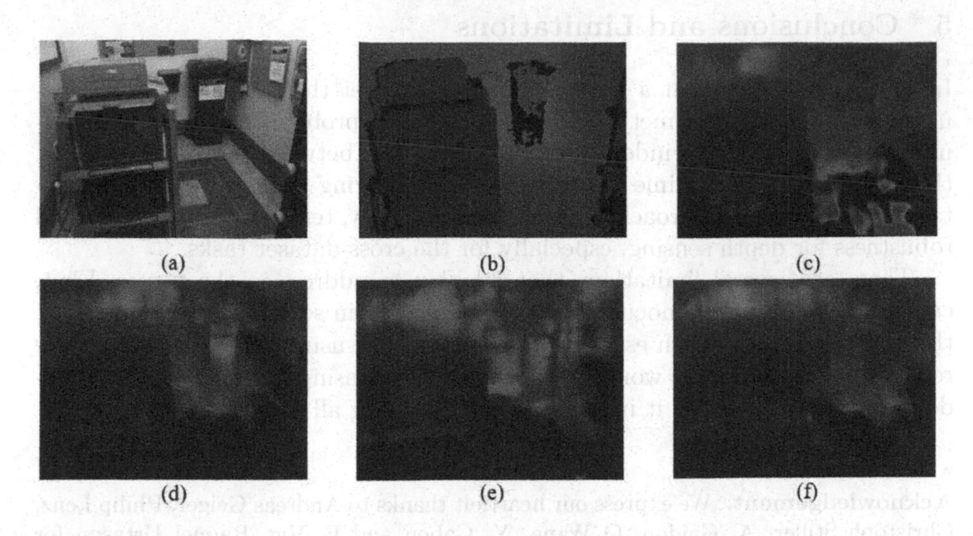

Fig. 4. The depth on the ScanNet datasets of different method train on KITTI: (a) original image (b) ground truth image (c) DeMoN (d) MVSNet (e) RGBD (f) Ours

Table 2. Comparison of depth estimation over the ScanNet [23] datasets

	$\sigma < 1.25$	abs	rmse	scale inv
DeMoN	67.80	0.1904	5.114	0.2628
MVSNet	86.43	0.1238	2.8684	0.1635
RGBD	93.12	0.0998	2.8294	0.1070
Ours	94.25	0.0890	2.7063	0.0960

Table 3. Cross-dataset tests for depth estimation (KITTI to ScanNet)

	$\sigma < 1.25$	abs	rmse	scale inv
DORN	69.61	0.2256	9.618	0.3986
Ours	74.38	0.3765	1.195	0.3576

Table 4. Cross-dataset tests for depth estimation (ScanNet to KITTI)

	$\sigma < 1.25$	abs	rmse	scale inv
DORN	25.44	0.6352	8.603	0.4432
Ours	73.35	0.2546	5.360	0.2048

5 Conclusions and Limitations

In this paper, we present a DL-based method to get the depth from a monocular video camera. Our method estimates a depth probability distribution volume from a local time window and the optical flow between the nearby frames, then integrates it over time through Bayesian filtering framework. Experimental results show our approach achieves high accuracy, temporal consistency, and robustness for depth sensing, especially for the cross-dataset tasks.

There are several limitations that we plan to address in the future. First, camera poses from a monocular video still suffer from scale drifting. This affect the accuracy of our depth estimation although with using the optical flow have reduce it. Second, in this work we focus on depth sensing from a local time window, rather than solving it in a global context using all the frames.

Acknowledgement. We express our heartfelt thanks to Andreas Geiger, Philip Lenz, Christoph Stiller, A. Gaidon, Q Wang, Y. Cabon and E. Vig, Raquel Urtasun for providing the data.
Funding Statement. Our research fund is funded by Fundamental Research Fund for the Central Universities (3072020CFQ0602, 3072020CF0604, 3072020CFP0601) and 2019 Industrial Internet Innovation and Development Engineering (KY1060020002, KY1060020008).

References

1. Yao, Y., Luo, Z., Li, S., Fang, T., Quan, L.: MVSNet: depth inference for unstructured multi-view stereo. In: Ferrari, V., Hebert, M., Sminchisescu, C., Weiss, Y. (eds.) ECCV 2018. LNCS, vol. 11212, pp. 785–801. Springer, Cham (2018). https://doi.org/10.1007/978-3-030-01237-3_47
2. Ummenhofer, B., et al.: Demon: Depth and motion network for learning monocular stereo. In: Proceedings of the IEEE Conference on Computer Vision and Pattern Recognition (ICCV), pp. 5038–5047. IEEE, Honolulu Hawaii (2017)
3. Eigen, D., Puhrsch, C., Fergus, R.: Depth map prediction from a single image using a multi-scale deep network. In: Advances in Neural Information Processing Systems, pp. 2366–2374. Morgan Kaufmann (2014)
4. Liu, F., Shen, C., Lin, G., Reid, I.: learning depth from single monocular images using deep convolutional neural fields. IEEE Trans. Pattern Anal. Mach. Intell. **38**(10), 2024–2039 (2016)
5. Broxton, P.D., Leeuwen, W.J.D., Biederman, J.A.: Improving snow water equivalent maps with machine learning of snow survey and lidar measurements. Water Resour. Res. **55**(5), 3739–3757 (2019)
6. Chan, D., Buisman, H., Theobalt, C., Thrun, S.: A Noise-Aware Filter for Real-Time Depth Upsampling. 1nd edn. (2008). Andrea Cavallaro and Hamid Aghajan
7. Maimone, A., Fuchs, H.: Reducing interference between multiple structured light depth sensors using motion. In: 2012 IEEE Virtual Reality Workshops (VRW), pp. 51–54. IEEE, Costa Mesa (2012)
8. Saxena, A., Chung, S.H., Ng, A.Y.: 3-d depth reconstruction from a single still image. Int. J. Comput. Vision **76**(1), 53–69 (2008)

9. Tippetts, B., Lee, D.J., Lillywhite, K., Archibald, J.: Review of stereo vision algorithms and their suitability for resource-limited systems. J. Real-Time Image Proc. **11**(1), 5–25 (2013). https://doi.org/10.1007/s11554-012-0313-2
10. Snavely, N.: Scene reconstruction and visualization from internet photo collections: a survey. IPSJ Trans. Comput. Vis. Appl. **3**(1), 44–66 (2011)
11. Godard, C., Aodha, O.M., Brostow, G.J.: Unsupervised monocular depth estimation with left-right consistency. In: 2017 IEEE Conference on Computer Vision and Pattern Recognition (CVPR), pp. 6602–6611. IEEE, Honolulu (2017)
12. Liu, C., Gu, J., Kim, K., Narasimhan, S.G., Kautz, J.: Neural RGBD sensing: depth and uncertainty from a video camera. In: Conference on Computer Vision and Pattern Recognition (CVPR), pp. 10978–10987. IEEE, Long Beach (2019)
13. Cipolla, R., Gal, Y., Kendall, A.: Multi-task learning using uncertainty to weigh losses for scene geometry and semantics. In: Conference on Computer Vision and Pattern Recognition(CVPR), pp. 7482–7491. IEEE, Salt Lake City (2018)
14. Seitz, S.M., Curless, B., Diebel, J., Scharstein, D., Szeliski, R.: A Comparison and evaluation of multi-view stereo reconstruction algorithms. In: Computer Society Conference on Computer Vision and Pattern Recognition, pp. 519–528. IEEE, New York (2006)
15. Saxena, A., Chung, S.H., Ng, A.Y.: 3-d depth reconstruction from a single still image. Int. J. Comput. Vision **76**(1), 53–69 (2007)
16. Schönberger, J.L., Zheng, E., Frahm, J.-M., Pollefeys, M.: Pixelwise view selection for unstructured multi-view stereo. In: Leibe, B., Matas, J., Sebe, N., Welling, M. (eds.) ECCV 2016. LNCS, vol. 9907, pp. 501–518. Springer, Cham (2016). https://doi.org/10.1007/978-3-319-46487-9_31
17. Mur-Artal, R., Tardos, J.D.: ORB-SLAM2: an open-source SLAM system for monocular, stereo, and RGB-D cameras. IEEE Trans. Robot. **33**(5), 1255–1262 (2017)
18. Kurnianggoro, L., Shahbaz, A., Jo, K.: Dense optical flow in stabilized scenes for moving object detection from a moving camera. In: International Conference on Control, Automation and Systems (ICCAS), pp. 704–708. IEEE, Gyeongju (2016)
19. Chang, J.R., Chen, Y.S.: Pyramid stereo matching network. In: Conference on Computer Vision and Pattern Recognition (CVPR), pp. 5410–5418. IEEE, Salt Lake City (2018)
20. Tateno, K., Tombari, F., Laina, I., Navab, N.: CNN-SLAM: Real-time dense monocular slam with learned depth prediction. In: 2017 IEEE Conference on Computer Vision and Pattern Recognition (CVPR), pp. 6243–6252. IEEE, Honolulu (2017)
21. Kingma, D.P., Ba, J.: Adam: a method for stochastic optimization. In: 3rd International Conference on Learning Representations, ICLR 2015, San Diego, CA, USA, pp. 1–15 (2015)
22. Geiger, A., Lenz, P., Urtasun, R.: Are we ready for autonomous driving? The KITTI vision benchmark suite. In: IEEE Conference on Computer Vision and Pattern Recognition, pp. 3354–3361. IEEE, Providence (2012)
23. Dai, A., et al.: ScanNet: richly-annotated 3D reconstructions of indoor scenes. In: IEEE Conference on Computer Vision and Pattern Recognition (CVPR), pp. 5828–5839. IEEE, Honolulu (2017)
24. Engel, J., Koltun, V., Cremers, D.: Direct sparse odometry. IEEE Trans. Pattern Anal. Mach. Intell. **40**(3), 611–625 (2018)
25. Shi, Q., Bin, L., Keyang, C., Xiao, Z., Guifang, D., Feng, L.: Multi-directional reconstruction algorithm for panoramic camera. Comput. Mater. Continua **65**(1), 433–443 (2020)

26. Xing, H., Zhao, Y., Zhang, Y., Chen, Y.: 3D trajectory planning of positioning error correction based on PSO-A* algorithm. Comput. Mater. Continua **65**(3), 2295–2308 (2020)
27. Xu, H., Yang, H., Shen, Q., Yang, J., Liang, H.: Automatic terrain debris recognition network based on 3d remote sensing data. Comput. Mater. Continua **65**(1), 579–596 (2020)
28. Li, J., Lv, Y., Ma, B., Yang, M., Wang, C., et al.: Video source identification algorithm based on 3d geometric transformation. Comput. Syst. Sci. Eng. **35**(6), 513–521 (2020)
29. Xiao, Z., Zhang, W., Zhang, Y., Fan, M.: Fatigue investigations on steel pipeline containing 3d coplanar and non-coplanar cracks. Comput. Mater. Continua **62**(1), 267–280 (2020)

The Algorithms for Word Segmentation and Named Entity Recognition of Chinese Medical Records

Yuan-Nong Ye[1,2,3(✉)], Liu-Feng Zheng[1,2], Meng-Ya Huang[1,2], Tao Liu[1,2(✉)], and Zhu Zeng[3(✉)]

[1] Bioinformatics and Biomedical Big Data Mining Laboratory, Department of Medical Informatics, School of Big Health, Guizhou Medical University, Guiyang 550025, China
{yyn,liutao}@gmc.edu.cn

[2] Key Laboratory of Environmental Pollution Monitoring and Disease Control, Ministry of Education, Guizhou Medical University, Guiyang 550025, China

[3] Cells and Antibody Engineering Research Center of Guizhou Province, Key Laboratory of Biology and Medical Engineering, School of Biology and Engineering, Guizhou Medical University, Guiyang 550025, China
zengzhu@gmc.edu.cn

Abstract. A complete inpatient electronic medical record contains a lot of information. In recent years, numerous researchers have carried out research on word segmentation of medical texts. Since the medical record text is written by free text, structuring the medical record text is an important part of the medical record intelligent analysis when these medical named entities are recognized. At present, named entity recognition methods are mainly divided into lexicographical and rule-based methods and machine learning-based methods. The machine learn-based approach takes the named entity recognition task as the annotation problem of sequence data, mainly considering the context information. The features commonly used in feature construction are contextual feature and dictionary features. BERT+LSTM+CRF were used to train the named entity recognition model. Open source CRF++ was adopted as the tool we relied on. We trained the LSTM+F model using the results of the original word segmentation and the information in the context as features. We carried out a 5-fold cross validation. The results showed that the overall F-1 score (MICRO-F) of named entity recognition reached 0.92, which confirmed that the model could accurately complete the task of medical named entity recognition.

Keywords: Word segmentation · Named entity recognition · Natural language processing

1 Introductions

A complete inpatient electronic medical record contains a lot of information [1], such as the index page of the inpatient medical record, summary of the medical record, admissions records, inspection reports, and discharge records. The discharge record is a

X. Sun et al. (Eds.): ICAIS 2021, CCIS 1422, pp. 397–405, 2021.
https://doi.org/10.1007/978-3-030-78615-1_35

high-level summary of the patient's inpatient diagnosis and treatment process, including the patient's hospital condition summary, admission diagnosis, condition changes in the hospitalization and the entire diagnosis and treatment processes. It has both structured data and unstructured data [2, 3]. The data onto discharge summary is mostly narrative text, which is urgently needed to develop excellent intelligent algorithms for semantic analysis and knowledge discovery [4].

In recent years, numerous of researchers have carried out research on word segmentation of medical texts [5, 6]. Since Chinese character does not have a natural separation mark, and most abroad electronic medical records have been achieved structured benefited by the excellent clinical terminology systems, such as SNOMED CT [7], UMLS[8], and RxNorm[9]. Hence, the oversea studies of medical record data are mainly focused on database establishment of medical terminology and ontology knowledge, medical natural language processing, medical data modeling and software development [10]. There are few researches on Chinese medical record segmentation in foreign countries. Lu used statistical methods to extract 470 key medical terms from one million cases and converted them into English for cluster detection to realize the monitoring of clinical symptoms [11]. Dong constructed a data-driven framework by using unsupervised iterative algorithm and SVM, and converted the free text in the electronic medical record into structured text in the form of time event description [12]. However, clinical activities are described in free text in Chinese medical records. For example, the clinical words are depending on doctor preferences without restrained, which brings unlisted word, (near-)synonym and abbreviation written in the medical records. Therefore, the segmentation of Chinese medical records and the extraction of clinical terms are the foundation of electronic medical records data analysis. Chen et al. made a research on word segmentation of the text of current history in electronic medical record, which provide a reference to realizing the structure of electronic medical record [13]. Li proposed medical named entity recognition algorithm by combining CRF and rule [14]. Wang et al. constructed a medical literature relevance database with the method of recurrence-free word segmentation in order to discovering new medical term [15]. Deng used CRF, maximum entropy and MIRA models and make use of the basic features of entity recognition [16]. Recently, Based on improved mutual information, Zhu and his colleagues proposes an algorithm for extracting new login sentiment words from Chinese micro-blog, which is good performance in the extraction of new login sentiment words [17]. The Baseline system was established to identify relevant concepts in electronic medical records. Although the existing Chinese medical record text segmentation research has made good progress in China, it still cannot meet the needs of medical record text processing.

The BERT model, recently proposed, takes advantage of the self-attention mechanism to integrate contextual information [18]. It has been pre-trained in English biomedical texts and, but without Chinese biomedical content [19]. We apply the BERT technology to Chinese medical record segmentation and entity recognition. We trained BERT in the biomedical corpus, including the general introduction, treatment, prevention and dietary care of each disease. We obtained 1.05G of clinical text by crawling on the Internet. In order to improve the accuracy of word segmentation, the medical vocabulary and commonly drug name are integrated into the dictionary for supplement segmentation tools. We combine the dictionary with BERT+LSTM+CRF to training corpus by using deep learning. And then, we do word segmentation for Chinese electronic medical records. Ultimately, we build a Chinese medical record segmentation model with fully unsupervised.

2 Materials and Methods

2.1 Materials

Because of the privacy implications of electronic medical records, it is difficult to acquire clinical textual data with permission from hospitals. Therefore, we crawled the clinical data that had been published online as corpus. About 1.05 GB clinical textual data with multiple medical fields was downloaded as corpus, such as dermatology, nephrology, dentistry, *etc.*

Supervised machine learning can extract medical knowledge and patient information from medical records. Manual annotation of text is the critical step of supervised learning for training named entity recognition (NER) model. We manually annotated two types of information: medically named entities (including disease diagnosis, clinical symptom examination and treatment, etc.), and semantic associations between entities (relationship between treatment and disease, relationship between treatment and symptoms, relationship between disease and symptoms, and relationship between disease and symptoms).

2.2 Methods

Linear Layers. After getting the words embedded from the BERT model, we can simply model the logarithmic probability using a fully connected linear layer.

$$z = W^T x + b \tag{1}$$

Where x denotes embedded word, z_i is the probability i. To get the tag with the highest probability, we can embed the predicted tag based on a series of words.

Conditional Random Field (CRF) Layers. CRF is a popular method of tag decoding [20]. Adding only linear layers may ignore some of the strong dependencies among the output labels in NER, which could be solved by adding the CRF layer. For a input sentence, $X=(x1; x2;; xn)$ and its tag, $y=(y1; y2;; yn)$, the *CRF* score is calculated by the following formula:

$$S(X, y) = \sum_{t=1}^{n} A_{t,yt} + \sum_{t=0}^{n} B_{yt,yt+1}. \tag{2}$$

The *CRF* score is consist of two matrixes, A and B: A is an emission fraction matrix calculated by formula (1). B is a transition fraction matrix between y_t and. y_{t+1}. And the probability that the softmax on all possible tag sequences produces sequence Y is:

$$p(y|\mathrm{X}) = \frac{e^{S(X,y)}}{\sum_{\overline{y} \in Yx} e^{S(X,\overline{y})}}. \tag{3}$$

In the training phase, We maximize the logarithmic probability of the correct tag sequence:

$$\log(p(y|X)) = s(X, y) - \log(\sum_{\overline{y} \in Yx} e^{S(X, \overline{y})}), \tag{4}$$

Where, *Yx* is all possible sequences of tags of sentence *X*. In the decoding phase, we only predict the output sequence to obtain the maximum score derived from the following formula:

$$y^* = \arg\max_{\overline{y} \in Yx} S(X, \overline{y}). \tag{5}$$

The Veterbi algorithm could be used to calculate the decoding process.

Bidirectional Long Short-Term Memory (LSTM). LSTM is a neural network that operates on sequential data [21]. It takes the vector of (x1; x2;; xn) as input. These sequences correspond to the embedding of each word in a sentence. Then another sequence of (h1; h2;; hn) that represents some information are returned. We use the followings to implement the order of each step in the input:

$$\begin{aligned}
i_t &= \sigma(W_{xi}x_t + b_{ii} + W_{hi}h_{t-1} + b_{hi}), \\
f_t &= \sigma(W_{xf}x_t + b_{if} + W_{hf}h_{t-1} + b_{hf}), \\
o_t &= \sigma(W_{xo}x_t + b_{io} + W_{ho}h_{t-1} + b_{ho}), \\
\widetilde{c}_t &= \tanh(W_{xc}x_t + b_{ic} + W_{hc}h_{t-1} + b_{hc}), \\
c_t &= f_t c_{t-1} + i_t \widetilde{c}_t, \\
h_t &= o_t \tanh(c_t),
\end{aligned} \tag{6}$$

Bi-LSTMA combines of past and future information, which can calculate both the left-sided sequences $\overrightarrow{h_t}$ and the right-sided sequences $\overleftarrow{h_t}$ of each input word *Xt*. We concatenate the final representation of the word by using the following model:

$$h_t = \left[\overrightarrow{h_t}; \overleftarrow{h_t}\right]. \tag{7}$$

A simple markup model is to use ht as a feature for making independent markup decisions of each output with y_t. It can be achieved by adding a linear layer output. However, the most common approach is adding a CRF layer at the top of the BI-LSTM output. The architecture of our model with BERT+Bi-LSTM+CRF is shown below (Fig. 1).

Dictionary and Radical. We propose a post-processing method using dictionary information. We can find the corresponding terms in the text through the bidirectional matching method based on the terminological dictionary (dictionary of medicine, dictionary of surgery et al.). Hence, We modify the probability of the corresponding label for the term by adding constants to the input in the CRF layer.

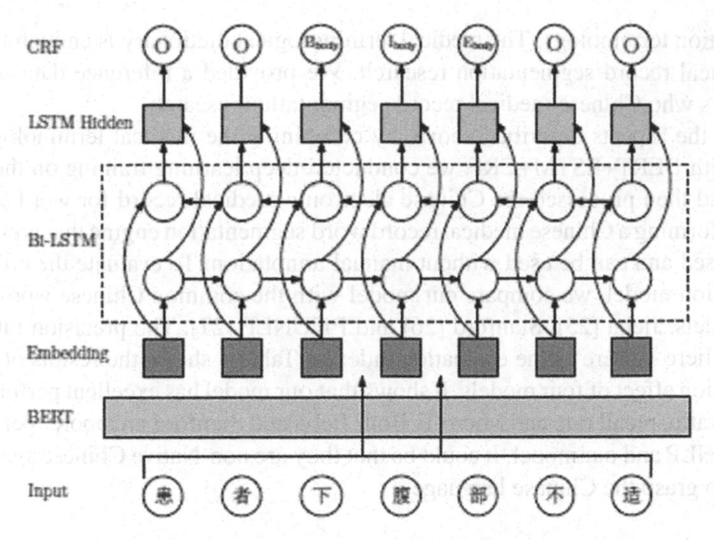

Fig. 1. The model architecture of with BERT+Bi-LSTM+CRF.

In recent years, the basic features of Chinese characters have been widely used to enhance different Chinese natural language processing [22–24]. Here, the model is also adopted the basic features of Chinese characters (radical). We studied the radical distribution of different entities in two entities. Figure 2 shows the distribution of several radicals. Radical of "口"(means mouth) appear more in symptom entities, and radical of "月" (which is a simplified form of group meat) appear more in body entities. Anatomical entities and radical of "疒" (Meaning of pain) are frequently identified in the description of symptoms.

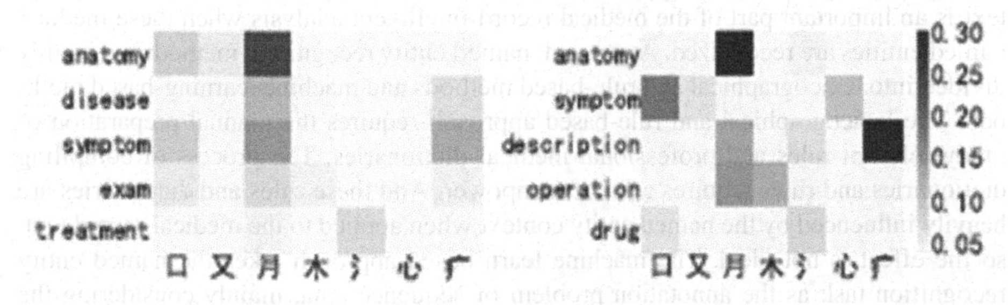

Fig. 2. The distribution of several radicals.

3 Results and Discussions

A medical terminological dictionary with 68,222 items was created in this study (Appendix A–F), which contains disease terminology, independent symptom terminology, symptom descriptions terminology, anatomic site terminology, drug terminology and operation terminology. The medical terminological dictionary is enough to do Chinese medical record segmentation research. We provided a reference data set for the researchers who Chinese medical record segmentation research.

Using the aspects described above, by combining the medical terminological dictionary with BERT+LSTM+CRF, we conducted deep learning training on the training corpus, and then processed the Chinese electronic medical record for word segmentation, thus forming a Chinese medical record word segmentation engine that is completely unsupervised and can be used without manual annotation. To evaluate the effect of the segmentation model, we compare our model with the common Chinese word segmentation models, Jieba [25], Stanford [26] and PKUNLP [27]. The precision ratio, recall rate and Micro-F were as the evaluation indexes. Table 1 shows the results of the word segmentation effect of four models. It shows that our model has excellent performance in precision ratio, recall rate and Micro-F. Both Jieba and Stanford are poorer performance than PKUNLP and our model. It could be that they are non-Native Chinese speakers and not easy to grasp the Chinese language.

Table 1. The compared results of various segmentation algorithms.

Model	Precision ratio	Recall rate	Micro-F
Jieba	84.6%	73.6%	78.7%
Stanford	75.5%	71.0%	73.2%
PKUNLP	94.3%	94.3%	94.2%
Our model	97.5%	98.1%	97.1%

Since the medical record text is written by free text, structuring the medical record text is an important part of the medical record intelligent analysis when these medical named entities are recognized. At present, named entity recognition methods are mainly divided into lexicographical and rule-based methods and machine learning-based methods. The lexicographical and rule-based approach requires the manual preparation of many relevant rules and professional medical dictionaries. The process of compiling dictionaries and rules requires a lot of manpower. And these rules and dictionaries are heavily influenced by the named entity context when applied to the medical record text, so the effect is not ideal. The machine learn-based approach takes the named entity recognition task as the annotation problem of sequence data, mainly considering the context information.

The features commonly used in feature construction are contextual feature and dictionary features. BERT+LSTM+CRF were used to train the named entity recognition model. Open source CRF++ was adopted as the tool we relied on. We trained the LSTM+CRF model using the results of the original word segmentation and the information in the context (window size is 5) as features. We carried out a 5-fold cross validation. The results showed that the overall F-1 score (MICRO-F) of named entity recognition reached 0.92, which confirmed that the model could accurately complete the task of medical named entity recognition (Table 2).

Table 2. Comparison of named entity extraction results.

Model	Precision ratio	Recall rate	Micro-F
FT-BERT	88.05%	90.47%	89.24%
FT-BERT+CRF	88.42%	90.76%	89.57%
FT-BERT+BiLSTM+CRF	88.37%	90.54%	89.44%
FT-BERT+BiLSTM+CRF+DIC+AdaBoost	89.55%	91.15%	90.35%

Acknowledgement. We thank Mr. Xin-Ting Huang for some useful suggesting. This study was jointly funded by the Science and Technology Foundation of Guizhou Province (2018-2820, 2018-1133, 2019-2811), the Science and Technology Fund project of Guizhou Health Commission (gzwjkj2019-1-40), the "Tian Cheng Hui Zhi" Foundation of Science and Technology Development Center of ministry of Science and Technology (2017A01044), and the Cell and Gene Engineering Innovative Research Groups of Guizhou Province (KY-2016-031).

Author Contributions. Conceived and designed the experiments: YNY and TL. Performed the experiments: YNY, LFZ and MYH. Analyzed the data: ZZ. Wrote the paper: YNY and ZZ.

Conflicts of Interest. The authors declare that they have no conflicts of interest to report regarding the present study.

Appendix

Appendix A: Disease Terminology Dictionary. (http://47.108.55.55/file/AppendixA.pdf)

Appendix B: Independent Symptom Terminology Dictionary. (http://47.108.55.55/file/AppendixB.pdf)

Appendix C: Symptom Descriptions Terminology Dictionary. (http://47.108.55.55/file/AppendixC.pdf)

Appendix D: Anatomic Site Terminology Dictionary. (http://47.108.55.55/file/AppendixD.pdf)

Appendix E: Drug Terminology Dictionary. (http://47.108.55.55/file/AppendixE.pdf)

Appendix F: Operation Terminology Dictionary. (http://47.108.55.55/file/AppendixF.pdf)

References

1. Garies, S., Birtwhistle, R., Drummond, N., Queenan, J., Williamson, T.: Data Resource Profile: national electronic medical record data from the Canadian Primary Care Sentinel Surveillance Network (CPCSSN). Int. J. Epidemiol. **46**, 1091–1092f (2017)
2. Vest, J.R., Grannis, S.J., Haut, D.P., Halverson, P.K., Menachemi, N.: Using structured and unstructured data to identify patients' need for services that address the social determinants of health. Int. J. Med. Inform. **107**, 101–106 (2017)

3. Fang, W., Pang, L., Yi, W.: Survey on the application of deep reinforcement learning in image processing. J. Artif. Intell. **2**, 39–58 (2020)
4. Menychtas, A., Tsanakas, P., Maglogiannis, I.: Knowledge discovery on IoT-enabled mHealth applications. In: Vlamos, P. (ed.) GeNeDis 2018. AEMB, vol. 1194, pp. 181–191. Springer, Cham (2020). https://doi.org/10.1007/978-3-030-32622-7_16
5. Qiao, X., et al.: A method of text extremum region extraction based on JointChannels. J. Artif. Intell. **2**, 29–37 (2020)
6. Niu, S., Li, X., Wang, M., Li, Y.: A modified method for scene text detection by ResNet. Comput. Mater. Con. **65**, 2233–2245 (2020)
7. Donnelly, K.: SNOMED-CT: the advanced terminology and coding system for eHealth. Stud. Health Technol. Inform. **121**, 279–290 (2006)
8. Gabetta, M., Larizza, C., Bellazzi, R.: A Unified Medical Language System (UMLS) based system for literature-based discovery in medicine. Stud. Health Technol. Inform. **192**, 412–416 (2013)
9. Liu, S., Ma, W., Moore, R., Ganesan, V., Nelson, S.: RxNorm: prescription for electronic drug information exchange. IT Prof. **7**, 17–23 (2005)
10. Niu, H., Yao, J., Zhao, J., Wang, J.: SERVQUAL model based evaluation analysis of railway passenger transport service quality in China. J. Big Data **1**, 17–24 (2019)
11. Lu, H.M., et al.: Multilingual chief complaint classification for syndromic surveillance: an experiment with Chinese chief complaints. Int. J. Med. Inform. **78**, 308–320 (2009)
12. Xu, D., et al.: Data-driven information extraction from chinese electronic medical records. PLoS One **10**, e0136270 (2015)
13. Chen, H., Huang, K.: The overview of structuring electronic medical record (IN CHINESE). China Digital Med. **006**, 36–39 (2011)
14. Li, W., Zhao, D., Li, B., Peng, X., Liu, J.: Combining CRF and rule based medical named entity recognition (IN CHINESE). Appl. Res. Comput. **032** , 1082–1086 (2015)
15. Wang, J., Hu, T.-J., Li, D.: The applications of a method for chinese word segmentation without thesaurus based on recurrence in the text mining of chinese biomedical literature (IN CHINESE). J. Med. Intell. **30**, 21–25 (2009)
16. Deng, B.-Y.: Research on concept extraction in electronic medical record. M.S. Research on concept extraction in electronic medical record, China (2014)
17. Zhu, G., Liu, W., Zhang, S., Chen, X., Yin, C.: The method for extracting new login sentiment words from chinese micro-blog basedf on improved mutual information. Comput. Syst. Sci. Eng. **35**, 223–232 (2020)
18. Zhang, A., Li, B., Wang, W., Wan, S., Chen, W.: MII: a novel text classification model combining deep active learning with BERT. Comput. Mater. Con. **63**, 1499–1514 (2020)
19. Devlin, J., Chang, M.-W., Lee, K., Toutanova, K.J.: BERT: pre-training of deep bidirectional transformers for language understanding. arXiv preprint arXiv 04805 (2018)
20. Lafferty, J., McCallum, A., Pereira, F.C.: Conditional random fields: probabilistic models for segmenting and labeling sequence data (2001)
21. Fang, W., Zhang, F.H., Ding, Y.W., Sheng, J.: A new Sequential image prediction method based on LSTM and DCGAN. CMC-Comput. Mater. Con. **64**, 217–231 (2020)
22. Chen, L., Song, L., Shao, Y., Li, D., Ding, K.: Using natural language processing to extract clinically useful information from Chinese electronic medical records. Int. J. Med. Inform. **124**, 6–12 (2019)
23. Peng, H., Cambria, E., Hussain, A.: A review of sentiment analysis research in Chinese language. Cogn. Comput. **9**, 423–435 (2017)
24. Shi, X., Zhai, J., Yang, X., Xie, Z., Liu, C.: Radical embedding: delving deeper to chinese radicals. In: Proceedings of the 53rd Annual Meeting of the Association for Computational Linguistics and the 7th International Joint Conference on Natural Language Processing, vol. 2, Short Papers, pp. 594–598 (2015)

25. Sun, J.: Jieba. Chinese Word Segmentation Tool (2012)
26. Manning, C.D., Surdeanu, M., Bauer, J., Finkel, J.R., Bethard, S., McClosky, D.: The Stanford CoreNLP natural language processing toolkit. In: Proceedings of 52nd Annual Meeting of the Association for Computational Linguistics: System Demonstrations, pp. 55–60 (2014)
27. Zhao, Y., Jiang, N., Sun, W., Wan, X.: Overview of the NLPCC 2018 shared task: grammatical error correction. In: Zhang, M., Ng, V., Zhao, D., Li, S., Zan, H. (eds.) Natural Language Processing and Chinese Computing. NLPCC 2018. Lecture Notes in Computer Science, vol. 11109, pp. 439–445. Springer, Cham (2018)

Temperature Forecasting Based on LSTM in Coastal City

Jinhui He[1], Hao Yang[1](✉), Bin Zhao[2], Da Li[2], and Fajing Chen[2]

[1] Chengdu University of Information Technology, Chengdu 610225, China
haoyang@cuit.edu.cn

[2] Numerical Weather Prediction Center, CMA, Beijing 100081, China

Abstract. Temperature is the main index that affects human comfort, and it is easily affected by air flow, rainfall, ocean currents and other factors to produce large fluctuations. In this paper, LSTM (Long Short-Term Memory) neural network algorithm is used to predict the long-term temperature of coastal cities. The training and testing data included eight measurements (every three hours) a day, spanning 10 years (2009–2019). Shanghai was chosen as the coastal city to be predicted. The structure of the neural network is composed of two kinds of neural networks, two LSTM layers and one full connection layer. The function of LSTM layer is to deal with the correlation of temperature time series, while the full connection layer maps the output layer of LSTM to the final prediction result. By comparing Back Propagation (BP), Recurrent Neural Network (RNN) and LSTM, the results show that LSTM is more suitable for temperature time series prediction.

Keywords: Temperature forecasting · Machine learning · Artificial intelligence · Deep learning · Neural network

1 Introduction

Weather prediction is the use of science and technology to predict the state of the atmosphere in a period of time in the future. As early as in ancient China, people began to try to predict the weather by "cloud observation" and other methods. The accuracy of modern weather forecasts has been greatly improved as people have gradually grasped the laws of weather change. Among many weather elements, air temperature is an important factor that people are concerned about [1]. For example, in the field of geothermal power generation, knowing in advance the temperature change in the future can reasonably control the power generation and make maintenance plans in advance. In the field of weather forecasting, Accurate and timely weather forecast has great social and economic benefits for economic construction, national defense construction and protection of people's life and property safety. Therefore, air temperature prediction is particularly important.

The prediction of atmospheric meteorological elements can be divided into short-term, medium-term and long-term according to the length of time scale, and can be divided into physical model, statistical model, artificial neural network model according

X. Sun et al. (Eds.): ICAIS 2021, CCIS 1422, pp. 406–417, 2021.
https://doi.org/10.1007/978-3-030-78615-1_36

to the method of prediction. Long-term forecasting [2–4] has a large time span, generally taking the year as the unit. It can be used to predict the annual large generating capacity of power plants, formulate power generation strategies and conduct wind energy resource assessment in the early stage of geothermal energy construction, etc. The correlation prediction method of measurement is the long-term temperature prediction model, and the common methods can include linear and nonlinear methods. The mid-term forecasting is mainly based on the month [5, 6], which is mainly used to help enterprises rationally arrange thermal energy production capacity. In addition, accurate mid-term forecasting is also a scientific reference basis for wind power plants to make overhaul and maintenance plans for units. The short-term forecasting is generally considered as the temperature prediction in the next 3 days [7–9]. It is mainly used to plan the economic dispatch of geothermal power plants in cooperation with the power department, optimize the power generation network, and ensure the safe and stable operation of the power system. The main feature of the physical model is that it can be used for temperature Prediction based on the analysis and transformation of external meteorological elements such as wind direction, wind speed and air pressure, etc. Numerical Weather Prediction (NWP) [10, 11] is the most common physical model for temperature prediction. Physical model can well describe geothermal off-site temperature prediction under the premise of factors, but the temperature prediction results based on NWP largely depend on the accuracy of numerical weather prediction. In addition, it is difficult to collect meteorological elements comprehensively in a period of time, and the data analysis also needs to rely on computers. Therefore, compared with the short-term temperature prediction, the physical model is more suitable for the mid-term and long-term prediction of the preliminary evaluation of geothermal power plant construction.

Different from physical models, statistical models are more widely used in short-term temperature prediction. Without considering the external meteorological factors, the statistical model focuses more on the analysis of the temperature series itself, and builds the model to realize the prediction by analyzing and mining the correlation hidden in the temperature historical data. Common time forecasting models include Autoregression Model (AR) [12, 13], Moving Average Model (MA) [14] and Autoregressive Integrated Moving Average model (ARIMA) [15–19]. The operation of time series model is simple and fast, but the statistical model has some limitations in temperature prediction. On the one hand, the time series model is constructed on the premise of a certain statistical hypothesis, and the actual temperature time series may not be in line with this hypothesis. On the other hand, the process of model identification and parameter estimation does not have a specific standard and is considered to be very subjective, which leads to large fluctuations in the accuracy of model prediction.

In recent years, with the development of computer technology, the concept of artificial neural network began to emerge and shine in more and more fields. As the name suggests, artificial neural networks are designed to mimic the structure and function of human brain large neural networks through self-learning through a large number of large node connections. It is very suitable for solving high dimensional, nonlinear, large and complex problems. Among them, Back Propagation (BP) neural network is the most common artificial neural network model. Through empirical analysis, compared with the traditional BP neural network, the large BP neural network with optimized parameters can further improve the accuracy of wind speed prediction [20]. And other models have been used in temperature prediction. Recurrent Neural Network (RNN) [21] is a kind of recursive neural network that takes sequence data as input, conducts recursion in the evolution direction of the sequence and all nodes (cyclic units) are connected by chain. Cyclic neural network has memorability, parameter sharing and complete Turing, so it has certain advantages in learning the nonlinear characteristics of the sequence. However, gradient explosion and gradient disappearance occurs when RNN processes long time series data. To solve this problem, Long Short-Term Memory(LSTM) added three gating forgetting doors [22], input doors and output doors. The improved LSTM can deal with long sequence well [23]. Based on the advantages and disadvantages of different models, this paper chooses LSTM which is more suitable for the long time series data to design the model, and different neural network models were set up as control experiments to analyze the prediction of each model.

Shanghai is located in the Yangtze River Delta region, in the east of China, the mouth of the Yangtze River, east of the East China Sea, north and west of Jiangsu and Zhejiang provinces, bounded by east longitude 120°52′–122°12′, north latitude 30°40′–31°53′. It is a typical coastal city. Among the current temperature prediction related papers, there are few papers on the temperature prediction of coastal cities. Shanghai is surrounded by water on both sides and has more precipitation, besides, due to the ocean current, the warm current plays a role in increasing the temperature and humidity of the coastal region climate, the cold current has a cooling and dehumidifying effect on the coastal climate. By accurately predicting the temperature of coastal cities, it plays an important role in protecting people's lives and property and promoting economic development.

2 Methodology

On the basis of the common RNN, the LSTM neural network has the long-term memory function by adding the memory element in the hidden layer. RNN increases the horizontal connection between each unit in the hidden layer on the basis of the human spirit network, transfers the value of a time step on the neural network to the current time step, so that the neural network has the memory function, and is applied to deal with the machine learning problems of NLP [24] and intertemporal sequence with context. The structure of RNN expansion is shown in Fig. 1.

The hidden layer in RNN is mainly used to capture the information of time series and carry out the most important iterative computation work. The hidden layer of the original RNN has only one shape S, which is sensitive to transient input. The LSTM adds a cell state to the RNN to hold the long-term state. See Fig. 2 for the structure of LSTM neural network.

LSTM optimizes and obtains the optimal model by adjusting the information of the whole history and the weight of the current information through three controllable doors. The LSTM control door is shown in Fig. 3. On the basis of RNN, LSTM calculates the single state before when according to the output value of the previous temperature and the input value of the current temperature.

$$\acute{c}_t = \tanh\left[W_c \cdot (S_{t-1}, X_t) + b_c\right] \quad (1)$$

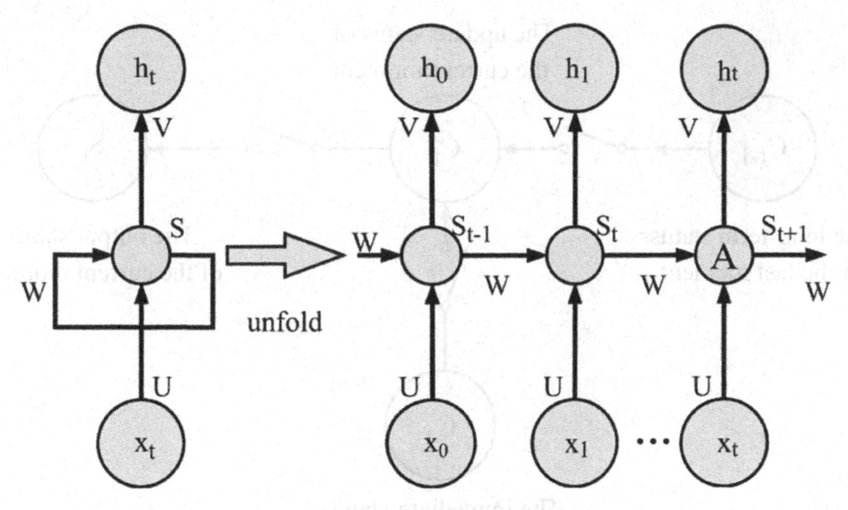

Fig. 1. Structure of RNN

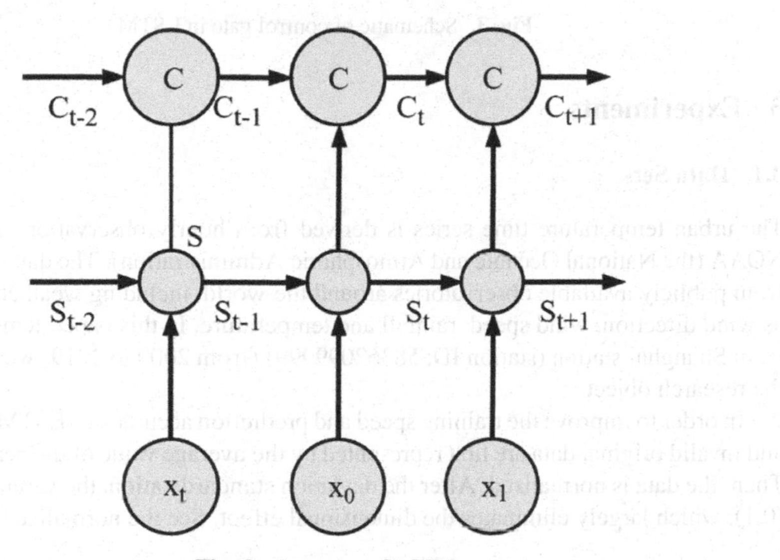

Fig. 2. Structure of LSTM

In the formula (1), $\acute{c}_t$is the singlet state of memory when the preceding time is engraved; (S_{t-1}, X_t) denotes the connection of two directional quantities; W_c and b_c are the weights and bias parameters of the forgotten gate; S_{t-1} is the output value of the sutra element at the last moment.

By the control of the forgetting gate, the new unit state c_t can preserve the information long ago, and by the control of the input gate, it can avoid the unimportant characteristic of the current time step into the single element of memory. The final temperature output value of LSTM is determined jointly by the output gate and the unit state.

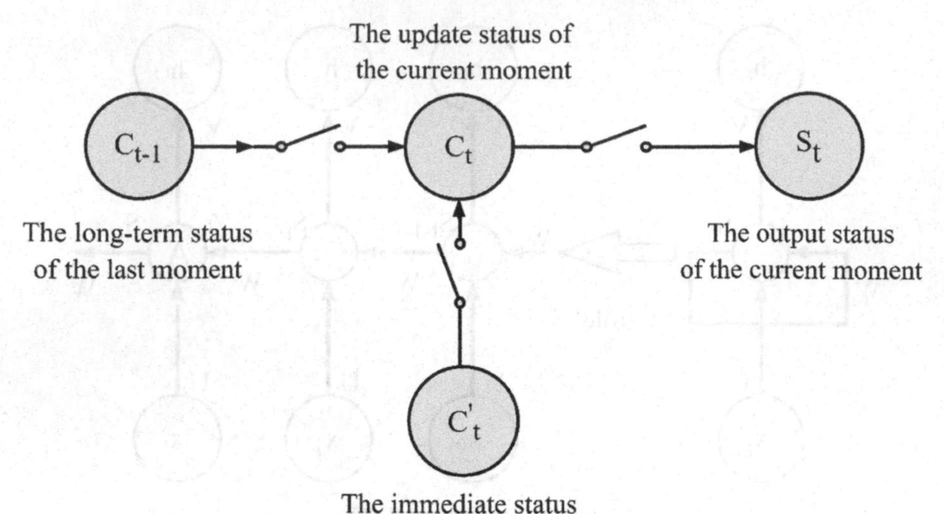

Fig. 3. Schematic of control gate in LSTM

3 Experiments

3.1 Data Sets

The urban temperature time series is derived from hourly observations published by NOAA (the National Oceanic and Atmospheric Administration). The data includes data from publicly available observatories around the world, including weather factors such as wind direction, wind speed, rainfall and temperature. In this paper, temperature data from Shanghai station (station ID: 58362099999) (from 2009 to 2019) were selected as the research object.

In order to improve the training speed and prediction accuracy of LSTM, the missing and invalid original data are first represented by the average value of the near time point. Then, the data is normalized. After the deviation standardization, the variable interval is (0,1), which largely eliminates the dimensional effect. See the normalized formula (1).

$$x_{norm} = \frac{x - x_{min}}{x_{max} - x_{min}} \tag{2}$$

Where x_{max} is the maximum value of the sample data, and x_{min} is the minimum value of the sample data.

The normalized temperature sequence is shown in the Fig. 4.

In order to facilitate the use of subsequent data, the processed data should be divided into training set, verification set and test set. For the selected temperature time series, the interval data from 2009 to 2017 were taken as training set, the interval data from 2017 to 2018 were divided into validation sets, and the interval data from 2018 to 2019 were taken as test sets.

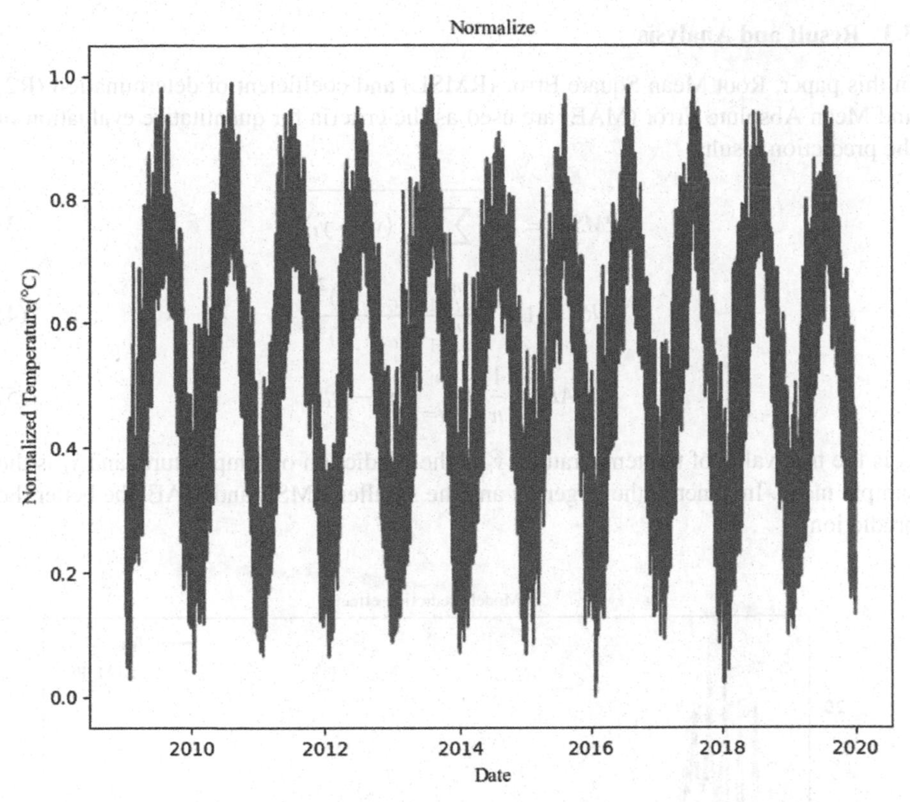

Fig. 4. Normalized temperature sequence

3.2 Experiment Setup

We divided the training data obtained above into 60 steps, namely (X_1,X_2...X_{60}) is input to the LSTM layer as input. After two layers of LSTM and one full connection layer, the model is self-trained according to the input. After the training is completed, the loss value and error are checked on the validation set, and the model parameters are modified according to the feedback on the validation set. In order to compare the effect of the model, two groups of comparative experiments were designed according to BP neural network and RNN, and the same data set, verification set, test set and time step were also set to 60. In the training process of the model, the weight of LSTM network is updated through Adam optimization method to minimize the network loss.

Python language and Keras, the deep learning library based on Tensorflow, the Google open source framework, were used as the back-end test basis to build and train the above three models, and the trained models were used to predict the test set respectively. We run the experiments under the environment of Intel Core(TM) CPU 10900F @2.80-GHz, 64G RAM, Ubuntu 18.04 64-b operating system, and Python 3.6.

3.3 Result and Analysis

In this paper, Root Mean Square Error (RMSE) and coefficient of determination (R2) and Mean Absolute Error (MAE) are used as the criteria for quantitative evaluation of the prediction results

$$RMSE = \sqrt{\frac{1}{n}\sum_{t=1}^{n}(y_t - \widehat{y_t})^2} \tag{3}$$

$$R^2 = 1 - \frac{\sum_{t=1}^{n}(y_t - \widehat{y_t})^2}{\sum_{t=1}^{n}(y_t - \overline{y_t})^2} \tag{4}$$

$$MAE = \frac{1}{n}\sum_{t=1}^{n}|y_t - \widehat{y_t}| \tag{5}$$

y_t is the true value of the temperature, $\widehat{y_t}$ is the prediction of temperature, and $\overline{y_t}$ is the sample mean. In general, the larger R^2 and the smaller RMSE and MAE, the better the prediction.

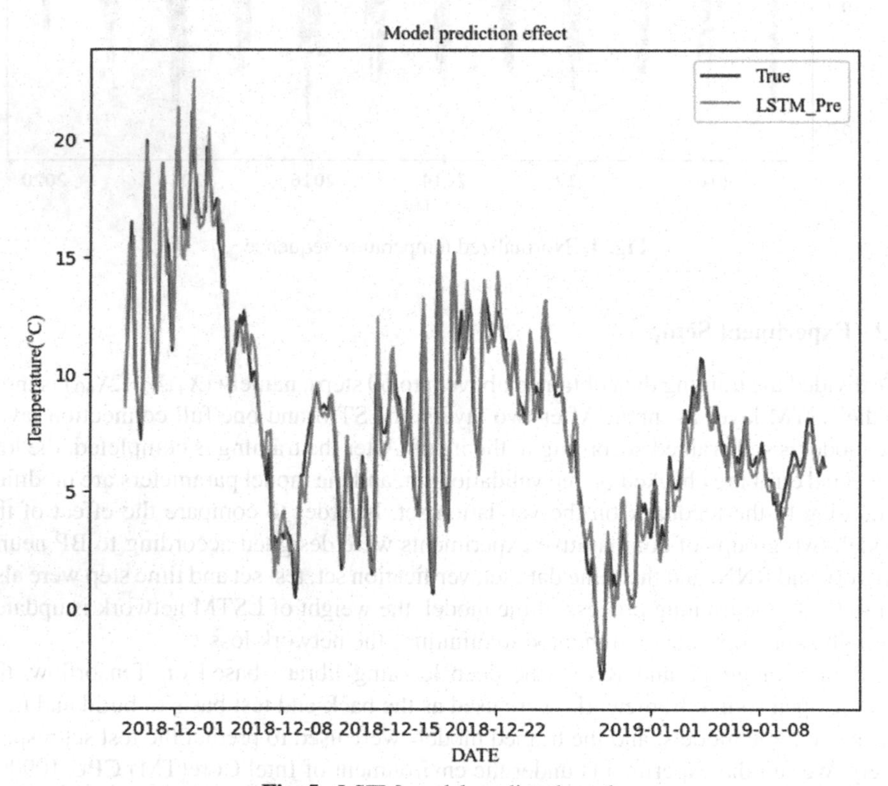

Fig. 5. LSTM model predicted results

As shown in Fig. 5, 6, and 7, the LSTM model has the best fitting effect for the simulation and prediction of Shanghai in December 2018, and can well predict the temperature in a certain time. The predicted value of the RNN model is generally lower than the real value, which may be related to the gradient disappearance and gradient explosion occurred in the long-term time series processed by RNN model. The gradient disappearance refers to that in the process of back propagation and calculation of the gradient of the loss function to the weight, the gradient becomes smaller and smaller as the propagation goes back, which means that the neurons in the front layer of the network will be much slower than the training in the later layer, or even will not change. The result is not accurate, the training time is very long. However, the BP model can only approximate the variation trend of the temperature, and the prediction effect of the three models is the worst. From the mathematical point of view, the traditional BP neural network as a local search optimization method, it is a complex nonlinear problem to solve, network weights is through the partial improvement direction gradually adjusted, so can make the algorithm into local extremum, weights of convergence to the local minimum points, thus cause the failure of network training.

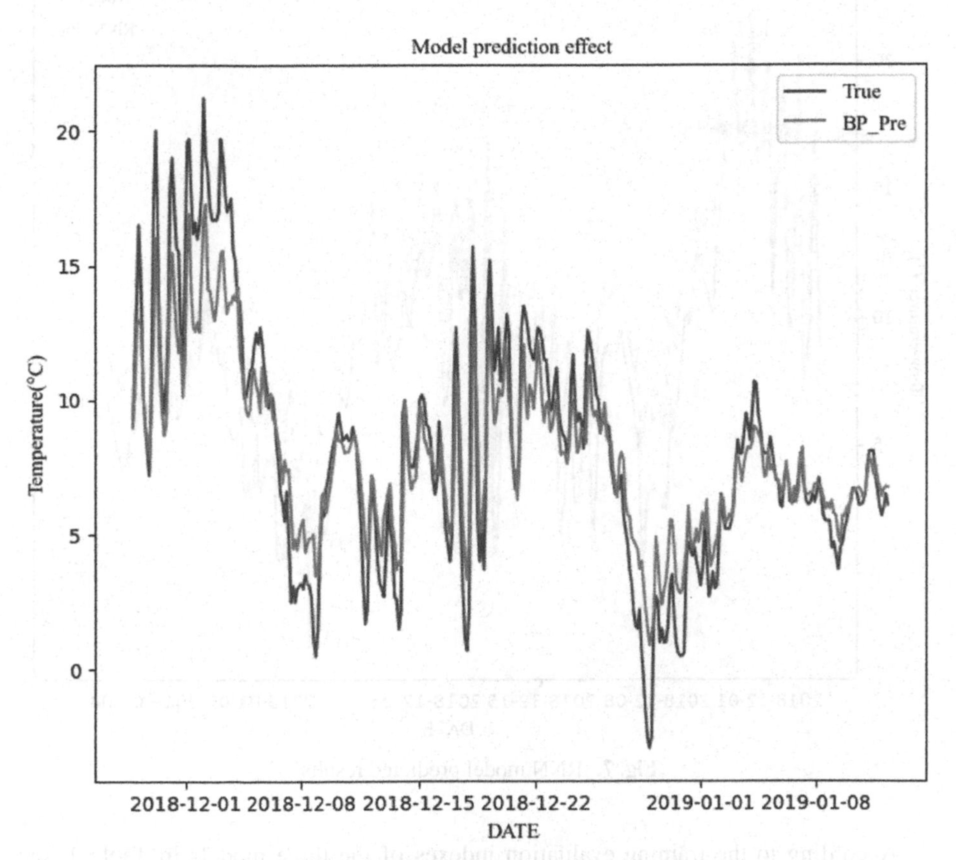

Fig. 6. BP model predicted results

It can be seen from Table 1 that, in the training process, the loss value, root mean square error and average absolute value error of LSTM model are the best among the three models, followed by RNN model, and BP model is the worst. From a numerical perspective, there is not much difference between RNN and LSTM, while the error of BP model is the largest among the three models. From the model training error, it can be concluded that RNN and LSTM have certain fitting effect for time series data prediction, while BP is not suitable for time series data prediction. As can be seen from Table 2, among the three models, LSTM and RNN had relatively good prediction effects, with R^2 at over 90%, RMSE and MAE at around 0.002 and 0.04, while BP model R^2 was less than 80%, while RMSE and MAE were as high as 0.009 and 0.078. Combined with the results in Table 1 and 2 and Fig. 5, 6 and 7, it can be found that in the prediction of time series data, LSTM can match the true value of temperature, while RNN has a relatively lagged performance, while BP can hardly be used in the prediction of temperature.

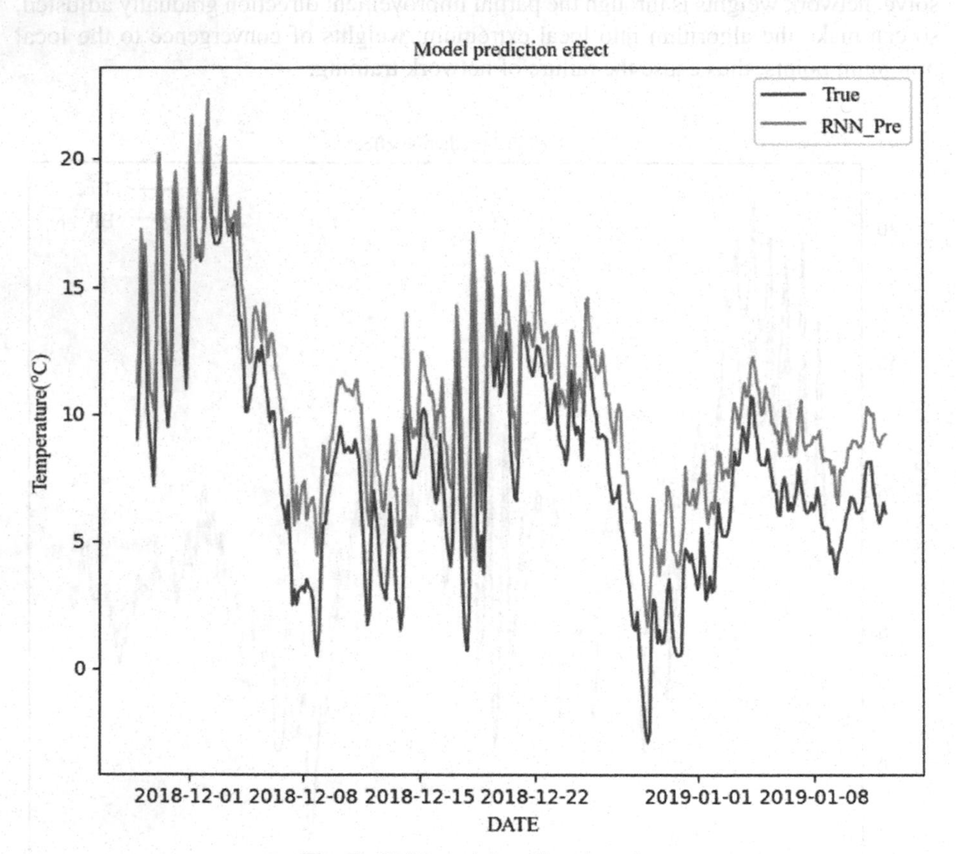

Fig. 7. RNN model predicted results

According to the training evaluation indexes of the three models in Table 1, the traditional BP neural network, RNN and LSTM have certain effects on temperature prediction, among which LSTM has the best effect. Similarly, according to the three prediction and assessment indicators in Table 2, LSTM is more suitable for temperature time series prediction whether in terms of training accuracy or prediction accuracy, which indicates that LSTM is effective in simulating and predicting temperature in coastal cities.

Table 1. Evaluation of training results

Model	Loss	Val_loss	RMSE	Val_RMSE	MAE	Val_MAE
BP	8.5226e−04	0.0127	0.0298	0.1122	0.0219	0.1011
RNN	7.4413e−04	7.7633e−04	0.0273	0.0279	0.0195	0.0202
LSTM	7.1903e−04	7.8187e−04	0.0268	0.028	0.0192	0.0203

Table 2. Evaluation of prediction results

Model	R^2	RMSE	MAE
BP	0.79925	0.0094	0.07882
RNN	0.93857	0.0028	0.0428
LSTM	0.94409	0.0026	0.0408

4 Conclusion

In this paper, the long-term and short-term memory neural network is used to predict the temperature of coastal cities, and the long-term memory ability of LSTM is utilized to improve the memory deficiency of RNN. By predicting the real coastal city temperature and comparing with the traditional BP and RNN models, the following conclusions can be drawn: LSTM model can effectively solve the prediction lag problem caused by the insufficient memory capacity of RNN; Compared with BP and RNN models, LSTM model has higher fitting accuracy and overall prediction.

In general, the LSTM model adopted in this paper has certain applicability in the field of temperature prediction in coastal cities. However, through observation, it can be found that the prediction accuracy of LSTM is still inadequate. In future work, new models can be tried to be integrated to improve the prediction accuracy.

Acknowledgement. A Special thanks and appreciation are extended to Anqi Xie for her assistance from Chengdu University of Information Technology.

Funding Statement. This work was sponsored by the Sichuan Science and Technology Program (2020YFS0355 and 2020YFG0479).

Conflicts of Interest. The authors declare that they have no conflicts of interest to report regarding the present study.

References

1. Dessai, S., Hulme, M., Lempert, R., Pielke Jr., R.: Climate prediction: a limit to adaptation. Adapting to climate change: thresholds, values, governance, pp. 64–78 (2009)
2. Woldesellasse, H., Marpu, P.R., Ouarda, T.B.M.J.: Long-term forecasting of wind speed in the UAE using nonlinear canonical correlation analysis (NLCCA). Arab. J. Geosci. **13**(18), 1–9 (2020). https://doi.org/10.1007/s12517-020-05981-9
3. Azad, H.B., Mekhilef, S., Ganapathy, V.G.: Long-term wind speed forecasting and general pattern recognition using neural networks. IEEE Trans. Sustain. Energy **5**(2), 546–553 (2014)
4. Aghelpour, P., Mohammadi, B., Biazar, S.M.: Long-term monthly average temperature forecasting in some climate types of Iran, using the models SARIMA, SVR, and SVR-FA. Theoret. Appl. Climatol. **138**(3–4), 1471–1480 (2019). https://doi.org/10.1007/s00704-019-02905-w
5. Mirasgedis, S., et al.: Models for mid-term electricity demand forecasting incorporating weather influences. Energy **31**(2–3), 208–227 (2006)
6. Hu, Z., Bao, Y., Chiong, R., Xiong, T.: Mid-term interval load forecasting using multi-output support vector regression with a memetic algorithm for feature selection. Energy **84**, 419–431 (2015)
7. White, M.A., Nemani, R.R.: Real-time monitoring and short-term forecasting of land surface phenology. Remote Sens. Environ. **104**(1), 43–49 (2006)
8. Szkuta, B., Sanabria, L.A., Dillon, T.S.: Electricity price short-term forecasting using artificial neural networks. IEEE Trans. Power Syst. **14**(3), 851–857 (1999)
9. Ke, J., Zheng, H., Yang, H., Chen, X.M.: Short-term forecasting of passenger demand under on-demand ride services: a spatio-temporal deep learning approach. Transp. Res. Part C Emerg. Technol. **85**, 591–608 (2017)
10. Chen, N., Qian, Z., Nabney, I.T., Meng, X.: Wind power forecasts using Gaussian processes and numerical weather prediction. IEEE Trans. Power Syst. **29**(2), 656–665 (2013)
11. Bauer, P., Thorpe, A., Brunet, G.: The quiet revolution of numerical weather prediction. Nature **525**(7567), 47–55 (2015)
12. Zhao, Y., Ye, L., Pinson, P., Tang, Y., Lu, P.: Correlation-constrained and sparsity-controlled vector autoregressive model for spatio-temporal wind power forecasting. IEEE Trans. Power Syst. **33**(5), 5029–5040 (2018)
13. Poggi, P., Muselli, M., Notton, G., Cristofari, C., Louche, A.: Forecasting and simulating wind speed in Corsica by using an autoregressive model. Energy Convers. Manage. **44**(20), 3177–3196 (2003)
14. Akrami, S.A., El-Shafie, A., Naseri, M., Santos, C.A.G.: Rainfall data analyzing using moving average (MA) model and wavelet multi-resolution intelligent model for noise evaluation to improve the forecasting accuracy. Neural Comput. Appl. **25**(7–8), 1853–1861 (2014). https://doi.org/10.1007/s00521-014-1675-0
15. Lee, Y.-S., Tong, L.-I.: Forecasting time series using a methodology based on autoregressive integrated moving average and genetic programming. Knowl.-Based Syst. **24**(1), 66–72 (2011)
16. Khashei, M., Bijari, M.: A novel hybridization of artificial neural networks and ARIMA models for time series forecasting. Appl. Soft Comput. **11**(2), 2664–2675 (2011)
17. Kavasseri, R.G., Seetharaman, K.: Day-ahead wind speed forecasting using f-ARIMA models. Renew. Energy **34**(5), 1388–1393 (2009)
18. Ediger, V.Ş, Akar, S.: ARIMA forecasting of primary energy demand by fuel in Turkey. Energy Policy **35**(3), 1701–1708 (2007)
19. Box, G.E., Pierce, D.A.: Distribution of residual autocorrelations in autoregressive-integrated moving average time series models. J. Am. Stat. Assoc. **65**(332), 1509–1526 (1970)

20. Yu, F., Xu, X.: A short-term load forecasting model of natural gas based on optimized genetic algorithm and improved BP neural network. Appl. Energy **134**, 102–113 (2014)
21. Shi, H., Xu, M., Li, R.: Deep learning for household load forecasting—A novel pooling deep RNN. IEEE Trans. Smart Grid **9**(5), 5271–5280 (2017)
22. Gers, F.A., Schmidhuber, J., Cummins, F.: Learning to forget: continual prediction with LSTM (1999)
23. Zhang, Q., Wang, H., Dong, J., Zhong, G., Sun, X.: Prediction of sea surface temperature using long short-term memory. IEEE Geosci. Remote Sens. Lett. **14**(10), 1745–1749 (2017)
24. Chowdhary, K.: Natural language processing. In: Fundamentals of Artificial Intelligence, pp. 603–64. Springer (2020)

Recent Progress of Medical CT Image Processing Based on Deep Learning

Yun Tan[1,2], Jiaohua Qin[2(✉)], Lixia Huang[1], Ling Tan[3], Xuyu Xiang[2], Hao Tang[3], Haikuo Peng[2], and Jiang Wu[4]

[1] Hunan Applied Technology University, Changde 415100, Hunan, China
[2] Central South University of Forestry and Technology, Changsha 410004, Hunan, China
[3] Second Xiangya Hospital of Central South University, Changsha 410011, Hunan, China
[4] Zhejiang Sci-Tech University, Hangzhou 310018, Zhejiang, China

Abstract. Computed tomography (CT) can provide comprehensive and high-resolution sectional anatomy images of human tissues and has been widely used in medical examination and diagnosis. With the rapid development of machine learning, medical CT image processing based on deep learning has become a hotspot and achieved great progress. In this paper, the characteristics of CT image are analyzed. Then the research status and progress of medical CT image processing based on deep learning are investigated. The common deep learning models and typical applications are reviewed, which include segmentation, classification and detection. Medical CT image processing based on deep learning has shown higher performance compared with traditional methods, which can reduce the burden of doctors and assist doctors to make more accurate decisions. However, it still faces the challenges of insufficient large-scale datasets, difficult dataset annotation and interpretability of the models that applications need. In the conclusion, the future development trend is analyzed.

Keywords: Medical CT image · Deep learning · Segmentation · Classification · Detection

1 Introduction

Computed tomography (CT) is a type of diagnostic imaging method in medical radiology. It can provide comprehensive and high-resolution sectional anatomy images of human tissues and has been widely used in medical examination and diagnosis since its appearance. However, the results of CT examination largely depend on the subjective judgment of imaging doctors, which are easy to be affected by doctors' experience and subjective factors. In the clinical process, imaging doctors need to read a large number of images, and it is possible to miss the subtle features due to visual fatigue. Therefore, there are usually some differences in the diagnosis results of CT images among different radiologists.

With the development of computer vision and the integration of interdisciplinary fields in recent years, medical CT image processing based on computer technology

X. Sun et al. (Eds.): ICAIS 2021, CCIS 1422, pp. 418–428, 2021.
https://doi.org/10.1007/978-3-030-78615-1_37

has become a hot spot and achieved great progress. In recent years, deep learning has achieved rapid development. It has powerful ability of automatic feature extraction, complex model construction and efficient feature expression by the hierarchical model similar to the human brain, and can extract features from the pixel level raw data from the bottom to the high level, which provides a new idea for solving the problem of medical CT image processing [1, 2]. Since the beginning of 2020, due to the shortage of global medical resources, the research results of medical CT image processing based on deep learning have grown explosively.

In this paper, we introduce the characteristics of CT images and investigate the research status and challenges of deep learning in the field of medical CT image processing, which includes CT image segmentation, classification and detection. Through the review, we introduce the common models of deep learning, typical applications of medical CT image processing and its research progress. Finally, we summarize the challenges and possible countermeasures.

2 Characteristics of CT Image

CT usually uses X-ray beam to scan a certain part of the human body with a certain thickness. The detector receives the X-ray through the plane. The measured signal is converted into digital information through analog-to-digital conversion (ADC), and then processed by the computer, so as to obtain the X-ray absorption value of each unit volume of this layer, namely CT value, and arrange it into a digital matrix. Through the digital-to-analog converter (DAC), each digit in the digital matrix is transformed into pixels with different gray levels from black to white, and arranged according to the matrix to form a CT image, also known as a cross-sectional image. Therefore, CT image is a type of reconstructed image. The gray level of CT image reflects the absorption of X-ray in organs and tissues. For example, the white pixels mean high-density areas with high absorption, which represent bone or calcified tissue with high calcium content. The gray pixels represent medium density area with moderate X-ray absorption, such as muscles and organs. While the black pixels represent low-density area with low X-ray absorption, such as lung tissue.

According to the principle of CT imaging, CT images usually have the following characteristics [3, 4]:

1) High density resolution. In CT images, the X-ray absorption coefficient of each voxel can be calculated by mathematical methods, which can achieve high measurement accuracy. Therefore, CT images can distinguish the tissue with small density difference. It can clearly show the anatomical structure of some human organs and some pathological tissues with density change.
2) Existence of artifacts. The artifact refers to the abnormal area that has nothing to do with the tissue structure being scanned in CT images. There are three main reasons for artifacts. First of all, equipment damage or unstable working state may lead to artifacts, which is generally circular or round. The second is the patient's reason, such as the artifacts caused by movement, which generally present strip low-density shadow. Finally, improper scanning conditions may also lead to artifacts, such as setting unsuitable scanning fields and display fields.

3) Interference of noise. Noise refers to the variation of CT values relative to its average value in a CT image of homogeneous matter, which usually leads to uneven CT values and grainy images. The noise level is related to the amount of X-ray and the sensitivity of detector, which will affect the image quality.
4) Correlation of adjacent images. Compared with other medical images such as ultrasound and MRI (Magnetic Resonance Imaging), there is obvious temporal and spatial correlation between adjacent images since CT images are generated by a serials of interval scanning.

During medical CT image processing, the above characteristics should be fully considered, which will directly affect the accuracy of CT image diagnosis. The high resolution and the adjacent correlation can help to realize location and detection efficiently. While the presence of noise and artifacts may lead to detection errors with high possibility.

3 Segmentation of Medical CT Image

Segmentation is very important during the process of medical CT images. The basic concept is to segment different feature regions according to different requirements. In recent years, deep learning network provides a promising solution for automatic image segmentation [5–7]. At present, the mainstream network frameworks being used for segmentation of medical CT images are convolutional neural network (CNN) [5, 6], full convolutional network (FCN) [7, 8] and U-Net [9].

FCN is the initial model of deep learning semantic segmentation with full convolution neural network and up sampling operation. In order to improve the segmentation precision, jump join is used to combine the low-level spatial information with the high-level semantic information [7]. Christ et al. used cascaded FCN to implement segmentation on CT abdomen images, which includes two steps based on the consideration of spatial correlation [8].

The most typical network used for segmentation is the U-Net proposed by O. ronneberger et al. in 2015 [9]. It is a U-shaped structure generated by adding a large number of characteristic channels in the upper sampling part under the architecture of fully convolution network, which allows the network to transfer the context information to the higher resolution layer, and makes the expansion path and contraction path roughly symmetrical. Only pixels with complete context obtained from the input image are included in the segmentation graph. This overlap-tile strategy allows the seamless segmentation of any large image. At the same time, the strategy of data augmentation makes it needn't a lot of training data, so it is especially suitable for small-scale medical data. The U-Net has achieved good performance in different biomedical segmentation applications and has become the most commonly used network architecture in the field of medical image processing.

Hiasa et al. used Bayesian convolutional neural networks with the U-Net architecture to implement automatic segmentation of individual muscles from CT images, in which uncertainty metric was applied and has been further proved to be effective [10]. An improved U-net model was applied to liver segmentation combined with context information by Liu, Z. et al. [11], which achieved good performance. U-net was also used for metal segmentation in dental CT and was expected to improve the performance of existing metal artifact reduction methods [12]. Zhou et al. introduced residual structure and deep separable convolutions layer to improve U-net network and applied it to the segmentation of CT image with hepatocellular carcinoma [13].

In order to further improve the accuracy of medical image segmentation, Z. Zhou et al. proposed the U-net++ structure in 2018, which redesigned the jump path and depth monitoring based on the U-Net [14]. It improved skip connection and reduced the semantic gap. The results showed good performance of segmentation on CT scans of both chest and abdominal, which significantly exceeded U-Net. The context relation between slices was also considered to further improve the accuracy of segmentation [15].

To solve the problem of 3D segmentation, O. Cicek et al. proposed a 3D U-Net architecture by replacing all 2D operations with 3D operations and achieved good performance [16]. Liu et al. applied dilated convolution and separable convolution in 3D U-net, which considered multi-scale feature and correlation information of channels to achieve highly accurate segmentation of liver [17]. Wang et al. proposed a two-stage framework based on 3D U-Net for accurate segmentation of organ at risk [18], in which the segmentation work has been divided to two sub-tasks. Recently, Radiuk used modified 3D U-net to solve the problem of multi-organ segmentation on CT images and achieved an average Sorensen-Dice similarity coefficient (SDSC) score of 84.8% [19].

The DRINet proposed by L. Chen et al. focused on the problem of distinguishing the categories with small differences on shape, location and size [20]. The Dense-Res-Inception Net means the convolution block with dense connections and the residual inception module in deconvolutional block separately, which can improve the gradient propagation. Moreover, the architecture is very flexible, and it is easy to integrate these modules into other CNN architectures. Compared with the standard U-net network architecture, the performance of DRINet has been significantly improved in multi class segmentation of cerebrospinal fluid on CT images and multi organ segmentation on abdominal CT images.

Recently, generative adversarial network (GAN) has been gradually used for segmentation. The main idea of GAN based segmentation is to use the discriminator to refine the initial segmentation result achieved by the generator [21, 22]. At the same time, neural architecture search (NAS) has also been extended for segmentation and has achieved even better performance with fewer parameters than U-net on CT dataset [23]. The proposed NAS-Unet used U-shaped architecture as the backbone network, which is followed by two parallel search units of UpSC and DownSC. Yu et al. proposed the coarse-to-fine neural architecture search (C2FNAS) method to solve the problem of inconsistency between search stage and deployment stage due to the limitation of memory and search space [24].

Actually, the labeling of medical images can only be completed by professionals with corresponding medical knowledge. Therefore, in most cases, annotation data is relatively scarce. Y. Zhang et al. tried to train effective segmentation models using unlabeled images in 2017 [25]. A new deep adversarial network model (Dan) was proposed, which was composed of segmentation network and evaluation network used to evaluate segmentation quality. It can be effectively applied to obtain fairly good segmentation results for unlabeled image data. Attention based semi-supervised depth network (ASDnet) proposed in 2018 also used unlabeled medical image data during network training [26]. The network structure consists of a segmented network and a confidence network. In addition, by considering the importance of each sample, a multi-level dice loss function was proposed to improve the network training effect. Dice similarity coefficient (DSC) and average surface distance (ASD) were used to measure the consistency between manual segmentation and automatic segmentation. The experimental results showed that ASDnet was better than U-net on the same data set.

4 Classification of Medical CT Images

As a basic task of medical CT image processing, classification has also introduced deep learning and achieved remarkable improvements. Based on input images, the trained CNN model will give binary or multivariate output. Pre training of CNN on natural images has shown amazing and powerful effect, which has even exceeded the accuracy of human experts or other learning models in some tasks [27]. Compared with the images of ordinary scenarios, the number of medical images is far less. Therefore, the classification of medical images usually adopts the algorithm of transfer learning. The strategy of using the pre trained network as the feature extractor is effective. Medical images can be used to fine tune the pre trained network.

Classification of lung CT images has always been the difficulty and focus of medical CT image processing. A multiscale convolutional neural network (MCNN) with hierarchical learning framework was proposed by Shen et al. [28]. To capture the heterogeneity of pulmonary nodules, discriminant features were extracted from alternating layers. The framework has shown good classification results on CT image sets of the lung image database. Setio et al. proposed a computer-aided detection system for pulmonary nodules [29]. It can automatically learn discriminant features from training data. There were three candidate detectors corresponding to entity, sub entity and large nodule detection respectively, which were used to obtain the candidate nodules and input into the network. Then the final classification result was obtained by fusion method. Data enhancement and discarding technology were introduced to avoid over fitting. The experimental results on LIDC-IDRI dataset showed that the multi view convolution neural network was very suitable for reducing false positives.

Kim et al. used a deep neural network to extract abstract feature, which was combined with traditional image feature [30]. Through effective feature selection, the classification of pulmonary nodules was realized based on CT images of lung. To implement the tuberculosis (TB) disease classification from pulmonary CT images, patch-based CNN was combined with support vector machine (SVM) [31], which has been proved to be effective.

A fully automatic lung CT cancer diagnosis system (DeepLung) was proposed in 2018 [32]. It consisted of two parts: node detection and classification. 3D dual channel module and U-net structure were used to detect nodules and learn their features. On the common data set LIDC-IDRI, it showed equivalent performance as an experienced doctor. Bhatt et al. used morphological methods for the lung nodule segmentation and then CNN was applied for malignant and benign classification [33]. Ren et al. investigated the manifold representation of lung nodule images and introduced the idea of enforcing regularization to network training [34]. Nardelli et al. used CNN to solve the problem of artery-vein classification from chest CT images [35]. Three steps were included: At first, segmentation was implemented in scale space. Then 3D CNN was used for vessels classification. Finally, the results were optimized by graph-cuts. The accuracy outperformed other methods. At the same time, scale characteristics have proved to be meaningful. Therefore, multi-scale information was integrated to deep CNN to classify emphysema by Peng et al. [36].

Recently many methods based on deep learning have emerged for coronavirus disease classification. In [37], five keras-related deep learning models were applied for COVID-19 classification. A novel multitask deep learning model was proposed by Amyar et al., which made full use of the common information of several tasks including reconstruction, segmentation and classification [38]. The multi-layer perceptron can realize classification. The results showed good performance for multi tasks.

For the classification of CT images of other parts, Gao et al. integrated 2D and 3D CNN networks to solve the problem of CT brain images classification [39], which can help for early diagnosis of Alzheimer's disease. Yuan et al. applied ResNet50 to the classification of parotid tumor CT images, which introduced desensitization process based on the consideration of privacy protection [40]. Le et al. applied deep learning on carotid CT scans to differentiate symptomatic and asymptomatic stroke, which was expected to identify high-risk patients [41].

In another hand, Lee et al. proposed a novel artificial neural network (ANN) for classification of intracranial haemorrhage, which was more efficient than CNN method [42]. It provided a new idea for future research.

5 Detection of Medical CT Image

Fast and robust detection of anatomical structure or pathology is a another important task of medical CT image analysis. It usually needs to recognize all the objects in the image and determine their physical location and category. In fact, the task of detection is a complex process, which usually includes the process of segmentation and classification.

An efficient and robust deep learning algorithm was proposed for carotid artery bifurcation detection in [43]. It used a shallow network to initially test all the data, and then got the final detection results through a deep network. At the same time, the separation filter and network sparseness were used to accelerate the network. Small 3D patches were extracted from multi-resolution image pyramid to improve robustness. Combining the hierarchical features of deep learning and Haar wavelet features, the detection accuracy on head-neck CT images was further improved.

In [44], three CNNs were used to locate ROIs in three image planes separately, which can be combined to form a 3D bounding box. So as to realize the detection of different parts such as heart, aortic arch and descending aorta. Cai et al. combined multi-layer deep learning network with feature fusion, which can realize the identification of vertebra locations [45]. Kalinovsky et al. focused on the detection of tuberculosis lesion on 3D lung CT images. Sliding-window technique, slice-wise segmentation and 3D semantic segmentation were applied for lesion detection based on deep convolutional network [46].

Zreik et al. proposed a detection method for coronary artery stenosis based on deep learning analysis on coronary CT angiography (CTA) images [47]. At first, multiscale CNN was applied to implement the segmentation of myocardium. Then SVM was used to identify the patients with coronary artery stenosis. Mobiny et al. [48] proposed the capsule network (CapsNet) as an alternative of CNN. It was shown that when the number of training samples was small, the performance of CapsNet was significantly better than CNN. The proposed dynamic routing mechanism can effectively improve the speed of CapsNet. Finally, a convolutional decoder was proposed, which can reduce reconstruction error and achieve higher classification accuracy. Tan et al. proposed an automatic detection method of aortic dissection on CTA images [49]. Morphology operations were used for region of interest (ROI) segmentation and then several deep learning models were applied for aortic dissection detection. Among them, DenseNet121 achieved best performance. Sori et al. proposed a two-path CNN to realize the detection of lung cancer [50]. In order to improve the detection accuracy, the residual learning denoising model (DR-Net) was used to eliminate the interference of noise. In the two paths of CNN, both local and global features have been integrated by employing different receptive field sizes.

Undoubtedly, the detection of covid-19 based on lung CT scans has been a hot issue since the beginning of 2020. Purohit et al. used CNN for covid-19 detection combined with image augmentation technique, which achieved sensitivity of 94.78% on CT scan images [51]. Mohamad et al. used EfficientNet architecture as backbone [52]. The abstracted multiscale features were convoluted at different rates, so as to achieved better representation. The test results demonstrated good robustness and good accuracy of 96.16%.

Aiming at the problem of insufficient annotation data, Wang et al. Proposed a focalmix method [53]. It is the first to apply weak supervised learning to 3D image detection. Great progress has been made in luna16 and NLST data sets. ResNext+ was proposed by Mohammed et al. for weakly-supervised COVID-19 detection [54]. Long short term memory (LSTM) was used to extract the axial correlation between adjacent slices and attention module was introduced to generate the slice prediction.

6 Conclusion

Through the survey, it can be seen that medical CT image processing based on deep learning has made great progress in recent years. However, there are still some challenges and limitations as follows:

1) The lack of large-scale datasets of CT images. For deep learning network, the problem of over fitting or poor robustness is easy to occur when the test set is small. Therefore, deep learning models need a large and public CT image dataset in order to find more general features and improve training performance. At present, although there are some public datasets, but for complex medical diagnosis, to achieve the highest accuracy, the amount of medical data is still far from enough and sometimes the number of positive and negative samples is too different. Transfer learning can be used to pre train the deep learning network model of natural image data set for

medical image processing and analysis. Image augmentation, image reconstruction techniques and generative adversarial networks (GAN) can be used for high quality medical CT image database construction.

2) The difficulty of dataset annotation. The existing deep learning methods largely rely on large labeled datasets to a great extent and these annotations need the participation of professionals, which are difficult to obtain, especially in the case of covid-9 epidemic. Therefore, the methods based on weak supervision and unsupervised will be the research hotspot of medical CT image processing based on deep learning in the future.
3) The problem of unexplainability of deep learning model. In the medical field, the interpretability of the model is required to be higher. However, the features learned by deep learning model cannot be explained effectively at present. It can be predicted that the research on interpretable methods of deep learning model and the formation of interpretable imaging markers will be the research hotspot in the field of medical image processing. This is also the problem that should be solved before the CT image processing methods based on deep learning are widely used.

Funding Statement. This work was supported in part by the National Natural Science Foundation of China under Grant 61772561 and 62002392; in part by the Key Research and Development Plan of Hunan Province under Grant 2019SK2022; in part by the Science Research Projects of Hunan Provincial Education Department under Grant 18A174 and 19B584; in part by the National Natural Science Foundation of Hunan under Grant under Grant 2019JJ50866, 2020JJ4140 and 2020JJ4141; in part by Zhejiang Provincial Natural Science Foundation of China under Grant LGF19F010008; in part by Key Laboratory of Universal Wireless Communications (BUPT) (No.KFKT-2018101, Ministry of Education, P. R. China).

References

1. Razzak, M.I., Naz, S., Zaib, A.: Deep learning for medical image processing: Overview, challenges and the future. Classif. BioApps **26**(1), 323–350 (2018)
2. Litjens, G., et al.: A survey on deep learning in medical image analysis. Med. Image Anal. **42**(1), 60–88 (2017)
3. Chen, Y., et al.: Artifact suppressed dictionary learning for low-dose CT image processing. IEEE Trans. Med. Imaging **33**(12), 2271–2292 (2014)
4. Cui, X.Y., Gui, Z.G., Zhang, Q., Shangguan, H., Wang, A.H.: Learning-based artifact removal via image decomposition for low-dose CT image processing. IEEE Trans. Nucl. Sci. **63**(3), 1–12 (2016)
5. Moeskops, P., et al.: Deep learning for multi-task medical image segmentation in multiple modalities. In: Ourselin, S., Joskowicz, L., Sabuncu, M.R., Unal, G., Wells, W. (eds.) MICCAI 2016. LNCS, vol. 9901, pp. 478–486. Springer, Cham (2016). https://doi.org/10.1007/978-3-319-46723-8_55
6. Badrinarayanan, V., Kendall, A., Cipolla, R.: Segnet: a deep convolutional encoder-decoder architecture for image segmentation. IEEE Trans. Pattern Anal. Mach. Intell. **39**(12), 2481–2495 (2017)
7. Long, J., Shelhamer, E., Darrell, T.: Fully convolutional networks for semantic segmentation. In: Proceedings of the IEEE Conference on Computer Vision and Pattern Recognition, pp. 3431–3440 (2015)

8. Christ, P.F., et al.: Automatic liver and lesion segmentation in CT using cascaded fully convolutional neural networks and 3D conditional random fields. In: Ourselin, S., Joskowicz, L., Sabuncu, M., Unal, G., Wells, W. (eds.) Medical Image Computing and Computer-Assisted Intervention – MICCAI 2016. Lecture Notes in Computer Science, vol. 9901. Springer, Cham (2016)
9. Ronneberger, O., Fischer, P., Brox, T.: U-Net: convolutional networks for biomedical image segmentation. In: International Conference on Medical image computing and computer-assisted intervention, pp. 234–241 (2015)
10. Hiasa, Y., Otake, Y., Takao, M., Ogawa, T., Sato, Y.: Automated muscle segmentation from clinical CT using bayesian U-Net for personalized musculoskeletal modeling. IEEE Trans. Med. Imaging **99**(1), 1–15 (2019)
11. Liu, Z., et al.: Liver CT sequence segmentation based with improved U-Net and graph cut. Expert Syst. Appl. **126**(7), 54–63 (2019)
12. Hegazy, M.A.A., Cho, M.H., Cho, M.H., Lee, S.Y.: U-Net based metal segmentation on projection domain for metal artifact reduction in dental CT. Biomed. Eng. Lett. **9**(4), 375–385 (2019)
13. Zhou, S., Li, W.Y., Lin, R.: Design of the segmentation algorithm of HCC based on the improved U-Net network. J. Phys. Conf. Ser. **1631**(1), 12–60 (2020)
14. Zhou, Z., Rahman Siddiquee, M.M., Tajbakhsh, N., Liang, J.: UNet++: a nested U-Net architecture for medical image segmentation. In: Stoyanov, D., et al. (eds.) DLMIA/ML-CDS -2018. LNCS, vol. 11045, pp. 3–11. Springer, Cham (2018). https://doi.org/10.1007/978-3-030-00889-5_1
15. Azad, R., et al.: Bi-directional ConvLSTM U-Net with densely connected convolutions. In: IEEE/CVF International Conference on Computer Vision Workshop, Piscataway, pp. 406–415. IEEE Press (2019)
16. Çiçek, Ö., et al.: 3D U-Net: learning dense volumetric segmentation from sparse annotation. In: International Conference on Medical Image Computing and Computer-Assisted Intervention, pp. 424–432 (2016)
17. Liu, C., Cui, D., Shi, D., Hu, Z., Lang, J.: Automatic liver segmentation in CT volumes with improved 3D U-Net. In: the 2nd International Symposium, ISICDM 2018: Proceedings of the 2nd International Symposium on Image Computing and Digital Medicine, pp. 78–82 (2018)
18. Wang, Y., Zhao, L., Song, Z., Wang, M.: Organ at risk segmentation in head and neck CT images by using a two-stage segmentation framework based on 3d U-Net. IEEE Access **99**(1), 1–18 (2019)
19. Radiuk, P.: Applying 3d U-Net architecture to the task of multi-organ segmentation in computed tomography. Appl. Comput. Syst. (2020). https://doi.org/10.2478/acss-2020-0005
20. Chen, L., Bentley, P., Mori, K., Misawa, K., Fujiwara, M., Rueckert, D.: DRINet for medical image segmentation. IEEE Trans. Med. Imaging **37**(11), 2453–2462 (2018)
21. Dong, X., Lei, Y., Wang, T., Thomas, M., Yang, X.: Automatic multi-organ segmentation in thorax CT images using U-Net-GAN. Med. Phys. **46**(5), 2157–2168 (2019)
22. Singh, V.K., et al.: Breast tumor segmentation and shape classification in mammograms using generative adversarial and convolutional neural network. Expert Syst. Appl. **139**(1), 112855.1–112855.14 (2020)
23. Weng, Y., et al.: NAS-Unet: Neural architecture search for medical image segmentation. IEEE Access **7**(1), 44247–44257 (2019)
24. Yu, Q., et al.: C2FNAS: coarse-to-fine neural architecture search for 3D medical image segmentation. In: Proceedings of the IEEE/CVF Conference on Computer Vision and Pattern Recognition, pp. 4126–4135(2020)
25. Zhang, Y., Yang, L., Chen, J., Fredericksen, M., Hughes, D.P., Chen, D.Z.: Deep adversarial networks for biomedical image segmentation utilizing unannotated images. In: Descoteaux,

M., Maier-Hein, L., Franz, A., Jannin, P., Collins, D.L., Duchesne, S. (eds.) MICCAI 2017. LNCS, vol. 10435, pp. 408–416. Springer, Cham (2017). https://doi.org/10.1007/978-3-319-66179-7_47
26. Nie, D., Gao, Y., Wang, L., Shen, D.: ASDNet: attention based semi-supervised deep networks for medical image segmentation. In: Frangi, A.F., Schnabel, J.A., Davatzikos, C., Alberola-López, C., Fichtinger, G. (eds.) MICCAI 2018. LNCS, vol. 11073, pp. 370–378. Springer, Cham (2018). https://doi.org/10.1007/978-3-030-00937-3_43
27. Dutta, S., Manideep, B.C.S., Rai, S., Vijayarajan, V.: A comparative study of deep learning models for medical image classification. Iop Conf. **263**(4), 1–9 (2017)
28. Shen, W., Zhou, M., Yang, F., Yang, C., Tian, J.: Multi-scale convolutional neural networks for lung nodule classification. In: Ourselin, S., Alexander, D.C., Westin, C.-F., Cardoso, M.J. (eds.) IPMI 2015. LNCS, vol. 9123, pp. 588–599. Springer, Cham (2015). https://doi.org/10.1007/978-3-319-19992-4_46
29. Setio, A., et al.: Pulmonary nodule detection in CT images: false positive reduction using multi-view convolutional networks. IEEE Trans. Med. Imaging **35**(5), 1160–1169 (2016)
30. Kim, B.C., Sung, Y.S., Suk, H.I.: Deep feature learning for pulmonary nodule classification in a lung CT. In: International Winter Conference on Brain-Computer Interface. IEEE (2016)
31. Gao, X., Qian, Y.: Prediction of multi-drug resistant tb from CT pulmonary images based on deep learning techniques. Mol. Pharm. **15**(10), 4326–4335 (2017)
32. Zhu, W., et al.: Deeplung: Deep 3d dual path nets for automated pulmonary nodule detection and classification. In: IEEE Winter Conference on Applications of Computer Vision (WACV), pp. 673–681 (2018)
33. Bhatt, J., Joshi, M., Sharma, M.: Early detection of lung cancer from CT images: nodule segmentation and classification using deep learning. In: Tenth International Conference on Machine Vision, pp. 1–15 (2018)
34. Ren, Y., Tsai, M.Y., Chen, L., Wang, J., Shen, C.: A manifold learning regularization approach to enhance 3D CT image-based lung nodule classification. Int. J. Comput. Assist. Radiol. Surg. **15**(2), 287–295 (2019)
35. Nardelli, P., et al.: Pulmonary artery-vein classification in CT images using deep learning. IEEE Trans. Med. Imaging **37**(11), 2428–2440 (2018)
36. Peng, L., Lin, L., Hu, H., Li, H., Chen, Y.W.: Classification of pulmonary emphysema in CT images based on multi-scale deep convolutional neural networks. In: 25th IEEE International Conference on Image Processing (ICIP), pp. 3119–3123. IEEE (2018)
37. Deng, X., Shao, H., Shi, L., Wang, X., Xie, T.: A classification–detection approach of covid-19 based on chest x-ray and CT by using keras pre-trained deep learning models. Comput. Model. Eng. Sci. **125**(2), 579–596 (2020)
38. Amyar, A., Modzelewski, R., Li, H., Ruan, S.: Multi-task deep learning based CT imaging analysis for covid-19 pneumonia: classification and segmentation. Comput. Biol. Med. **126**(1), (2020)
39. Gao, X.W., Hui, R., Tian, Z.: Classification of CT brain images based on deep learning networks. Comput. Methods Programs Biomed. **138**(1), 49–56 (2017)
40. Yuan, J., Fan, Y., Lv, X., Chen, C., Wang, Y.: Research on the practical classification and privacy protection of CT images of parotid tumors based on ResNet50 model. J. Phys. Conf. Ser. **1576**(1), (2020)
41. Le, E.P.V., et al.: Contrast CT classification of asymptomatic and symptomatic carotids in stroke and transient ischaemic attack with deep learning and interpretability. Eur. Heart J. **41**(2), 2418 (2020)
42. Lee, J.Y., Kim, J.S., Kim, T.Y., Kim, Y.S.: Detection and classification of intracranial haemorrhage on CT images using a novel deep-learning algorithm. Sci. Rep. **10**(1), 1–7 (2020)

43. Zheng, Y., Liu, D., Georgescu, B., Nguyen, H., Comaniciu, D.: 3D deep learning for efficient and robust landmark detection in volumetric data. In: Navab, N., Hornegger, J., Wells, W.M., Frangi, A.F. (eds.) MICCAI 2015. LNCS, vol. 9349, pp. 565–572. Springer, Cham (2015). https://doi.org/10.1007/978-3-319-24553-9_69
44. Vos, B.D.D., et al.: 2D image classification for 3D anatomy localization: employing deep convolutional neural networks. In: SPIE Medical Imaging, Proceedings of SPIE, vol. 9784, pp. 97841 (2016)
45. Cai, Y., Landis, M., Laidley, D.T., Kornecki, A., Lum, A., Li, S.: Multi-modal vertebrae recognition using transformed deep convolution network. Comput. Med. Imaging Graph. **51**(1), 11–19 (2016)
46. Kalinovsky, A., Liauchuk, V., Tarasau, A.: Lesion detection in CT images using deep learning semantic segmentation technique. In: The International Archives of the Photogrammetry, pp. 13–19 (2017)
47. Zreik, M., et al.: Deep learning analysis of the myocardium in coronary CT angiography for identification of patients with functionally significant coronary artery stenosis. Med. Image Anal. **44**(1), 72–85 (2018)
48. Mobiny, A., Van Nguyen, H.: Fast CapsNet for lung cancer screening. In: Frangi, A.F., Schnabel, J.A., Davatzikos, C., Alberola-López, C., Fichtinger, G. (eds.) MICCAI 2018. LNCS, vol. 11071, pp. 741–749. Springer, Cham (2018). https://doi.org/10.1007/978-3-030-00934-2_82
49. Tan, Y., et al.: Automatic detection of aortic dissection based on morphology and deep learning. CMC-Comput. Mater. Con. **62**(3), 1201–1215 (2020)
50. Sori, W.J., Feng, J., Godana, A.W., Liu, S., Gelmecha, D.J.: DFD-Net: lung cancer detection from denoised CT scan image using deep learning. Front. Comput. Sci. **15**(1), 152701 (2021)
51. Purohit, K., Kesarwani, A., Kisku, D.R., Dalui, M.: COVID-19 Detection on chest X-Ray and CT scan images using multi-image augmented deep learning model. Preprints, pp. 1–8 (2020). https://doi.org/10.1101/2020.07.15.205567
52. Alrahhal, M.M., Bazi, Y., Jomaa, R.M., Zuair, M., Alajlan, N.: Deep learning approach for covid-19 detection in computed tomography images. Comput. Mater. Con. **65**(3), 1–15 (2020)
53. Wang, D., et al.: FocalMix: Semi-supervised Learning for 3D medical image detection. In: Proceedings of the IEEE/CVF Conference on Computer Vision and Pattern Recognition, pp. 3951–3960 (2020)
54. Mohammed, A., Wang, C., Zhao, M., Ullah, M., Cheikh, F.A.: Semi-supervised network for detection of covid-19 in chest CT scans. IEEE Access **8**(1), 155987–156000 (2020)

Road Damage Detection and Classification Based on M2det

Xinxin Gan[1,2], Jiantao Qu[3], Junru Yin[3], Wei Huang[3], Qiqiang Chen[3], and Wenxia Gan[1](✉)

[1] Wuhan Institute of Technology, Wuhan 430205, China
charlottegan@whu.edu.cn
[2] SIPPR Engineering Group Co., Ltd., Zhengzhou 450000, China
[3] School of Computer and Communication Engineering, Zhengzhou University of Light Industry, Zhengzhou 450000, China

Abstract. Road damage detection is very important for road maintenance. Deep learning is one of the popular method in road damage detection. Deep learning road damage detection methods include Fast R-CNN, Faster R-CNN, Mask R-CNN, RetinaNet. These methods always have some problems, shallow and deep features can not be extracted simultaneously, the detection accuracy of big targets or small targets is low, the speed of detection is slow. etc. This paper proposes one method based on M2det which can extract the shallow and deep feature. For its multi-scale and multi-level, it belongs to one-stage. We use M2det for road damage detection, training on a large number of images photographed by a vehicle-mounted smartphone, then comparing it with the other one-stage methods. Finally, experimental results show that our method is better than the-state-of-art methods in road damage detection.

Keywords: Road damage detection · One-stage detector · M2det · Feature pyramid structure

1 Introduction

Roads play a significant role in the human lives. Since road construction directly determines a city's economic development, especially in big cities with more population and trade. More or less, the road affected by some unpredictable factors such as weather, traffic, human activities, etc. As a result, the relevant staffs need to check the road damage from time to time and take different measures according to the degree of damage. Nevertheless, this tedious work will consume vast human, material and financial resources, especially in large countries which have the biggest territories, like China or US which has over 100,000 km highway to be tested and maintained periodically. A low-cost and based on machine automation method that can detect and classify the damaged areas in images which are captured by the vehicle mounted smartphones is urgently needed to find out [1]. Given the GPS information of images, this method can output the result of road damage detection correctly and rapidly. In particular, it can recognize the damage area of the image and annotate it in the bounding box and attach the type of damage.

X. Sun et al. (Eds.): ICAIS 2021, CCIS 1422, pp. 429–440, 2021.
https://doi.org/10.1007/978-3-030-78615-1_38

Recently, an increasing number of computer vision problem (object detection, image classification, image segmentation, etc.) have been solved by DCNN (deep convolutional neural network) [2] that demonstrated state-of-the-art performance. For detecting the damage in the road image, various methods have been proposed to save time and cost.

Fast R-CNN integrates the numerous steps of R-CNN which not only greatly improves the speed but also increases the accuracy of detection. Convolution of the whole image instead of convolution of each region proposal, ROI (Region of Interest) Pooling, classification and regression are trained together on the network multi-task loss are the three cores of the algorithm. Seungbo Shim et al. [3] proposed Fast R-CNN to use in road surface damage detection which shows variation and power in terms of parameters, and the overall performance is excellent.

Faster R-CNN is a unified, deep-learning-based, and end-to-end object detection framework running at near real-time frame rates. Faster R-CNN is extended the Fast R-CNN, while Fast R-CNN is evolved from R-CNN. The difference with Fast R-CNN is Faster R-CNN use the Region Proposal Network (RPN) instead of Selective Search (SS). Wang et al. [4] proposed a method for road damage detection and classification based Faster R-CNN, and they used the dataset which is same as this paper. There is a major imbalance in the number of different classes depending on the frequency of occurrence and the degree of urgency to maintain. They collected and evaluated the number of each type in the training set, and then used data augmentation techniques for the insufficient types. As a result, they achieved a great Mean F1-Score of 0.6255 in the competition. Despite the accuracy of using the Faster R-CNN, its execution speed is less than desirable.

Singh et al. [5] discussed Mask R-CNN which can be used to simultaneously detect and classify different types of road damage. Mask R-CNN was an extension of the Faster R-CNN model, and used for object detection, localization and instance segmentation in natural images. In their approach, Mask R-CNN was trained on road images that came from real-world, with bounding box information for the damage and its type. They handled the detection and classification problem end-to-end by adopting bounding box regression loss and classification loss together for the model. The experiments show that a high level of accuracy can be obtained using Mask R-CNN, at the same time, the speed of it is slower.

Though the accuracy of one-stage is lower than two-stage which is caused by positive and negative class imbalance, Laha Ale et al. [6] considered one-stage detection models since they were faster than two-stage detection models. RetinaNet used Feature Pyramid Networks (FPN) [7] to extract multiple-scale information and performed classification and box regression in one stage. It is universal for one-stage detection models to improve the performance of prediction by prior anchors. An image may generate thousands of candidate bounds, of these, there is a small fraction of them contain target that named positive samples, the others which do not contain target that named negative samples. This phenomenon lead to a problem, positive and negative class imbalance, and it also cause the performance of detection by one-stage less effective than the two-stage. Focal Loss is a new Loss method for balancing positive and negative samples of one-stage detection model.

Of these, two-stage detectors such as Fast R-CNN and Faster R-CNN can achieve high accuracy can achieve high accuracy, but they are poorly efficient. We aim to explore light and efficient one-stage detectors that possibly achieve high accuracy prediction. Current challenges to road damage detection include: shadows affecting damage brightness, obscurants, different states of the same type of damage in different places which have had a significant impact on the extraction of features. The traditional pyramid feature extraction network only focuses on the features extracted at this layer, so that they can only extract significant features for large or small targets. Therefore, we put forward Road Damage Detection based on M2det which is a multi-scale, multi-level detection model [8], detects features at different levels and fuses them. It focuses on shallow features as well as deep features, and is friendly to both large and small targets for detection. We use M2det as the basic framework and VGG16 is chosen as the feature extraction network. Since M2det is depending on input size, backbone, and whether or not multi-scale is utilized, M2det can achieve both high speed (and relatively excellent accuracy) and high accuracy (and relatively excellent speed), it is a speed-accuracy-compatible detector.

The specific architecture of the paper is as follows. Section 2 introduces the proposed method. Section 3 analyzes the experiments and results. Finally, Sect. 4 concludes this paper.

2 The Proposed Method

M2det is a model that use backbone network and MLFPN to extract image features [9], then adopt an SSD-like manner to predicts dense bounding-box and category scores, finally, it will get the result of detection by NMS (Non-Maximum suppression). M2det architecture for road detection and classification is show in Fig. 1.

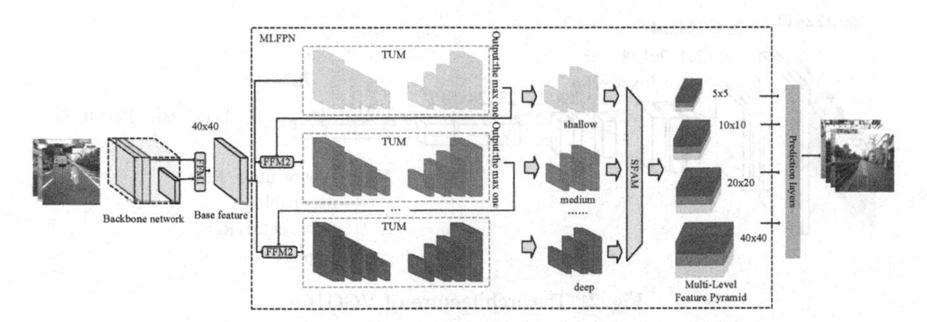

Fig. 1. M2det architecture for road detection and classification.

VGG16 is used as the backbone network, removing all the fully connected layers while keeping the convolutional layer and the maximum pooling layer. FFM1(Feature Fusion Module) takes out two different level shapes from VGG for preliminary fusion. The two different FFMs are used for preliminary fusion and feature-enhanced fusion respectively. Using TUM (Thinned U-shape Module) first is used to compress the feature layer and then up-sampling for feature fusion to obtain multiple valid feature layers.

SFAM (Scale Feature Aggregation Module) adjusts the attentive mechanism of each channel for the valid feature layer, then determines the weight of each channel. Finally, we predict the result by the effective feature layers with the attention mechanism. Specifically, Subsect. 2.1 introduces the architecture of VGG, Subsect. 2.2 introduces the models of MLFPN (Multi-level Feature Pyramid Network), Subsect. 2.3 introduces the detail of FFM, Subsect. 2.4 introduces the architecture of TUM, Subsect. 2.5 introduces the architecture of SFAM. Additionally, Subsect. 2.6 introduces the loss function.

2.1 VGG

VGG was proposed by the group of Visual Geometry Group in Oxford [10]. The network was related at ILSVRC (ImageNet Large Scale Visual Recongnition Challenge) 2014 and the major work was demonstrated that increasing the depth of the network can affect the final performance of the network to some extent. There are two types of VGG structures, VGG16 and VGG19, which are not different fundamentally except different network depths, VGG16 [11] is introduced in this paper. An improvement of VGG16 over AlexNet is that VGG16 adopt several consecutive 3 × 3 convolutional kernels instead of the larger convolutional kernels in AlexNet (11 × 11, 7 × 7, 5 × 5).For the given RF(receptive field), the use of stacked small convolutional kernels are preferable to the large convolutional kernels, since multiple nonlinear layers increase the network depth to ensure learning of more complex patterns and it has a lower cost. In detail, three 3 × 3 convolutional kernels are instead of 7 × 7 convolutional kernels and two 3 × 3 convolutional kernels are instead of 5 × 5 convolutional kernels in VGG which aim at enhancing the depth of the network and in some degree the effectiveness of the neural network, while ensuring that it has the same perceptual field. The architecture of VGG16 is shown in Fig. 2.

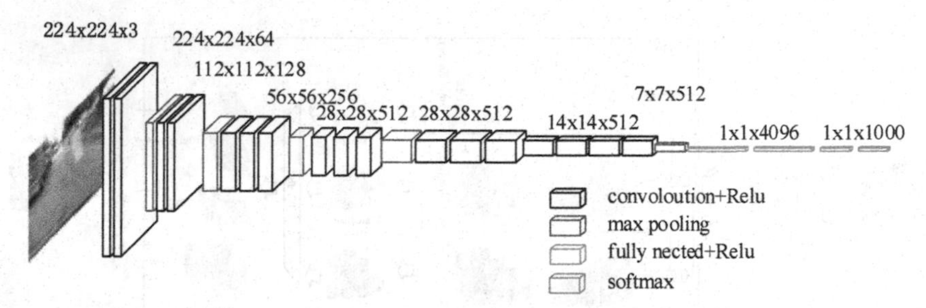

Fig. 2. The architecture of VGG16.

2.2 MLFPN

As shown in Fig. 1 the MLFPN consists of three main parts: FFM, TUM and SFAM.

First, FFMv1 fuses the shallow layer and deep layer features that extracts from the backbone network to obtain the base feature. Second, each TUM can generate multiple feature maps that have different scales, and each FFMv2 fuses the base feature and the output of the previous TUM, then inputs the result to the next TUM as input. The output of each level is described in the following formula.

$$\left[x_1^l, x_2^l, \cdots, x_3^l\right] = \begin{cases} T_l(X_{base}), l = 1 \\ T_l\left(F\left(X_{base}, x_i^{l-1}\right)\right), \; l = 2 \ldots L \end{cases} \tag{1}$$

where X_{base} denotes base features, x_i^l denotes features that with the *i-th* scale in the *l-th* TUM, L indicates the number of TUMs, T_l denotes the *l-th* TUM processing, and F denotes the FFMv1 processing. Finally, SFAM aggregates multi-level and multi-scale features via scale-wise splicing and channel-wise attention.

2.3 FFM

FFM is used to fuse features of different levels in M2Det, FFMv1 used for preliminary fusion and FFMv2 used for feature-enhanced fusion. The details of the two FFMs are shown in Fig. 3.

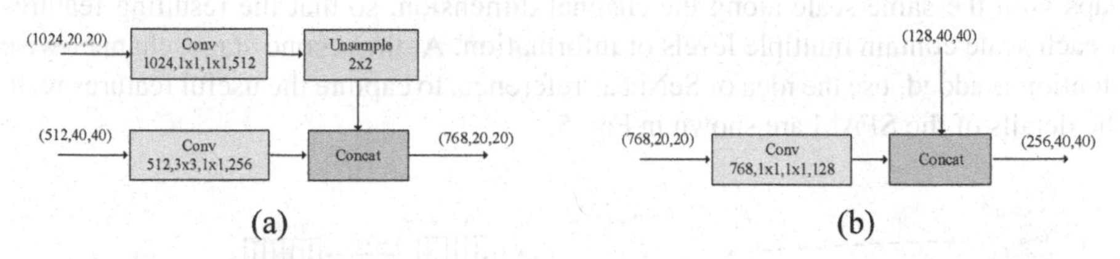

Fig. 3. The detail of FFMs, (a) represent FFMv1, (b) represent FFMv2.

In FFMv1, the (20,20,1024) feature layer are convoluted and up-samples, the shape is changed to (40,40,512); the (40,40,512) feature layer are convoluted, the shape is changed to (40,40,256). Then the two convolution results are stacked to a (40,40,768) preliminary fused feature layer. In FFMv2, the (40,40,128) feature layer of the six valid feature layers from the TUM, is enhanced fused with the initial fused feature layer extracted by FFM1, then output a (40,40,256) enhanced fused feature layer. This feature layer can be passed into the TUM as input again.

2.4 TUM

The TUM uses a U-shaped network which is thinner than FPN and RetinaNet. A 1x1 convolution was added after the up-sampling and point-wise add to enhance learning and maintain feature smoothness. The outputs of each decoder in a TUM constitute the multi-scale output of the TUM jointly. The outputs of each TUM together constitute the multi-level and multi-scale features, the front TUM providing the low level feature and the back TUM providing the high level feature. The details of the TUM are shown in Fig. 4.

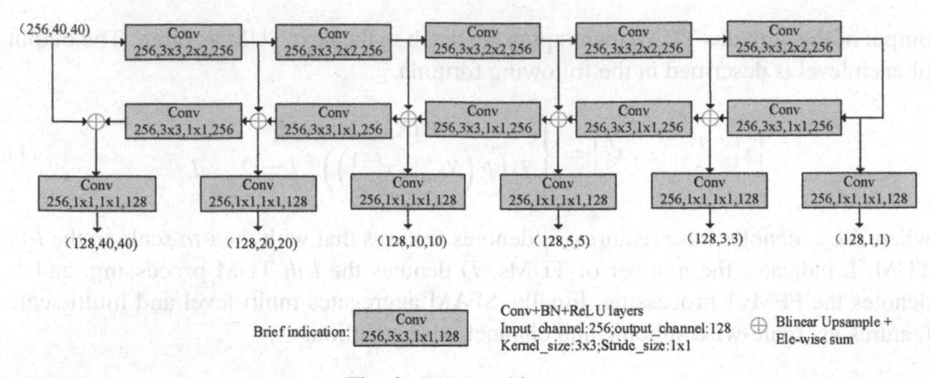

Fig. 4. TUM architecture

2.5 SFAM

SFAM aims to aggregate the multi-level multi-scale features that generated by TUMs to construct a multi-level feature pyramid. At the first stage, SFAM splices together feature maps with the same scale along the channel dimension, so that the resulting features of each scale contain multiple levels of information. At the second stage, channel-wise attention is added, use the idea of SeNet as reference, to capture the useful features well. The details of the SFAM are shown in Fig. 5.

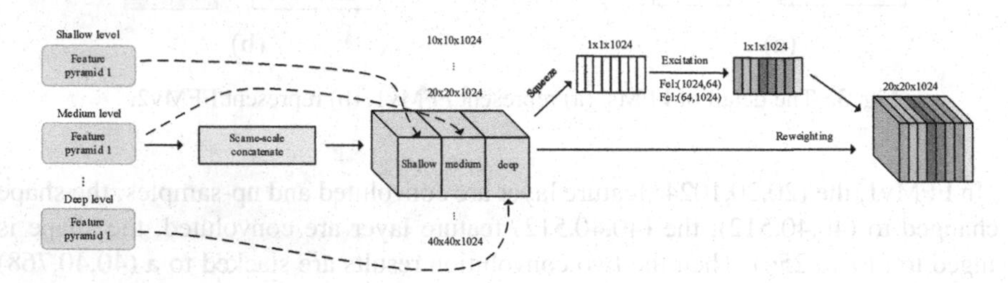

Fig. 5. SFAM architecture

2.6 Loss Function

Input a set of images, we first extract two different feature layers from the backbone network for fusion, then we input the fused feature to multiple TUM modules for feature compression. In order to train the proposed model, we use a multitasking loss function, defined as follows:

$$\mathrm{L}(\{p_j\}, \{t_j\}) = \frac{1}{N_{cls}} \sum\nolimits_j L_{cls}\left(p_j, p_j^*\right) + \lambda \frac{1}{N_{reg}} \sum\nolimits_j p_j^* L_{reg}\left(t_j, t_j^*\right) \tag{2}$$

$$L_{cls}(p, l) = -\log pl \tag{3}$$

$$L_{reg}\left(t_j, t_j^*\right) = smoothL1\left(t_j - t_j^*\right) \tag{4}$$

where j is the index of an anchor in a mini-batch, p_j represent the predicted probability of anchor j being an object region. p_j^* is the ground-truth label, of these, 0 for negative and 1 for positive. In [12] defined the predicted box t_j and the ground-truth box t_j^*. In Eq. (2), the two terms $L_{cls}\left(p_j, p_j^*\right)$ and $p_j^* L_{reg}\left(t_j, t_j^*\right)$ are normalized with N_{cls}, N_{reg}, and a balancing weight λ, respectively. And, the function of smooth L1(x) is defined as follows:

$$\text{smoothL1}(x) = \begin{cases} 0.5x^2, & if\,|x| < 1 \\ |x| - 0.5, & others \end{cases} \tag{5}$$

3 Experiments and Results

In this section, we show the implementations details, and get a good result after training with a large dataset. We also trained on the same dataset with two other methods and compared the results of the three experiments, analyzing their strengths and weaknesses according to different evaluation criteria. Specifically, our experiments consisted of five parts: 3.1 introducing implement details about the experiment; 3.2 introducing the datasets distribution ratio; 3.3 showing the detecting result of M2det; 3.4 introducing the Evaluation criteria and compare the mAP in three different method; 3.5 compare the results of detection and show the detecting image of result.

3.1 Implementations Details

Our experiment based on M2det which is running with keras v2.1.5, the hardware environment is CUDA 10.0.13 and cuDNN 7.4.1. GPU is NVIDIA GeForce RTX 2070, on windows 10 which has 16 GB memory.

In M2det model, the original image are resized to (320,320,3), and the resized image are input to backbone network that is VGG16. In M2Det, we remove all the full connection layer of VGG16, only retaining the convolutional layer and the maximum pooling layer, which means that Conv1 to Conv5. FFM1 will perform the initial fusion of the features extracted by VGG. In the feature extraction using VGG, we will take out the feature layer with the shape of (40,40,512) and (20,20,1024) for the next operation. In FFM1, it will perform a convolution of (20,20,1024) with 512 channels, a convolution kernel size of 3×3, and a step size of 1×1, and then up-sample it to change its shape to (40,40,512); it will also perform a convolution of (40,40,512) with 256 channels, a convolution kernel size of A convolution of 1×1 with a step size of (40,40,256) changes its Shape to (40,40,256); then the two convolutions are stacked to become a (40,40,768) initial fused feature layer. After we input a (40,40,256) valid feature layer, TUM will extract the features in U-shape, the structure here is similar to a feature pyramid, feature compression is done first, and then up-sampling is done for feature fusion. we can obtain six valid feature layers with sizes (40,40,128), (20,20,128), (10,10,128), (5,5,128), (3,3,128), and (1,1,128) by using TUM, we can obtain six valid feature layers, and to further strengthen the feature extraction capability of the network, M2det takes out the (40,40,128) feature layer out of the six valid feature layers and performs enhanced fusion with the initial fused feature layer extracted by FFM1, and again outputs a (40,40,256) enhanced fused feature layer. At this time, the enhanced fused feature layer from FFM2 can be passed into the TUM again for U-shaped feature extraction. The attentive mechanism of each channel is adjusted to determine the weight that should be given to each channel number.

3.2 Datasets

For detecting the damage of road, we selected 7240 images of roads with damage as the datasets, these images photographed by a vehicle-mounted smartphone [13]. Road damaged type are manually labeled as nine types in the dataset. Eight types among of them are common extremely, in Fig. 6. We show some examples of each road damage type.

Fig. 6. Examples of each damage type in the dataset.(a)is 'D00', (b)is 'D01', (c)is 'D10', (d)is 'D11', (e)is 'D20', (f)is 'D40', (g)is 'D43', (h)is 'D44'.

In details, 'D10' represent equal interval, 'D11' represent construction joint part, 'D20' represent alligator crack, 'D40' represent rutting, bump, pothole and separation, 'D43' represent cross walk blur. We divided the datasets into training set and testing set in a ratio of 7:3, i.e. 5068 images for training and 2172 images for testing, and took 10% from the training set as the validation set.

3.3 Results of M2det

Our method yields a good detection results after training a large amount of dataset, and the result of detection is showing that the damage is included by the bounding box and labeled with the type of damage in the road image, as shown in Fig. 7 most of the damage are detected by our method and the bounding box of the road damage are accurate.

Fig. 7. The result of detection.

3.4 Evaluation Criteria

mAP (mean Average Precision) as a measure of detection accuracy in object detection which is expressed by the following formula:

$$\text{mAP} = \frac{\sum_{i=1}^{n} (AP)_i}{n} \tag{6}$$

$$\text{AP} = \frac{\sum_{i=1}^{m} \left(\sum_{j=1}^{n} P_i \right)}{m} \tag{7}$$

$$P_i = \frac{T_i}{S_i} \tag{8}$$

In addition, AP (Average Precision) means the average precision of one type, P_i stands for the precision of one image. T_i represents the number of correctly detected in one category in a picture, and S_i represents the total number of categories in a picture.

We used M2det, RetinaNet, and YOLO-v3 [14] respectively, which are trained on a large dataset, and finally compared the AP values and mAP values obtained from the detection results, which are presented in the Fig. 8 and Fig. 9.

The value of mAP can be used as an overall evaluation criterion, but the accuracy of detection of different types of road damage in the pictures varies greatly, so we also compared the AP values obtained for different damage types, which detail are shown in Table 1.

As we can see from the Fig. 8, 9 and Table 1 the 'D43' crack has the highest AP value among all the methods, similarly, 'D44', 'D01', 'D20' also get good performance. Because these four types of cracks have more samples in the dataset. In addition, it is not difficult to find that we use M2det to detect 'D30' cracks which get 47.62% AP value, while RetinaNet and YOLO-v3 can not detect 'D30' cracks, the reason is that the number of "D30" cracks in the dataset are so few that the two methods could not detect well, but our method also works better for such small sample detection.

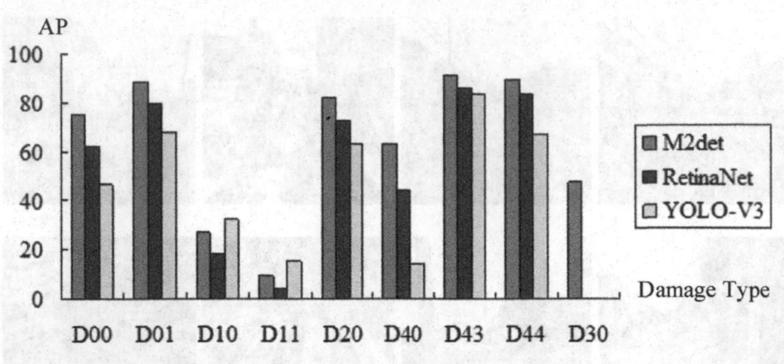

Fig. 8. AP value of different Damage Type.

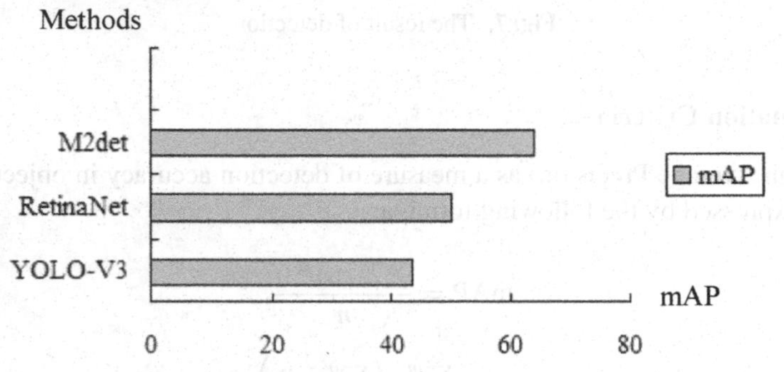

Fig. 9. mAP value of different Damage Type.

Table 1. The value of AP and mAP on three methods.

Method	AP								mAP
	D00	D01	D10	D11	D20	D40	D43	D44	
M2det	**75.44**	**88.88**	26.80	9.24	**82.15**	**63.66**	**91.78**	**90.04**	**63.96**
RetinaNet	62.47	80.06	17.96	4.19	72.88	44.21	86.51	84.09	50.26
YOLO-v3	46.19	68.51	**32.11**	**15.37**	63.76	14.14	84.08	67.69	43.54

3.5 Compare the Result of Detection

Not only the values of AP and mAP could be used as a standard for evaluating the detecting results, but also we can analyze the image of detecting results. The same dataset often be detected as different performance in different methods. For example, Fig. 10 shows the detection results obtained by using three different detection methods in the same scene.

By comparing Fig. 10 (a) (b) (c) three images, it is obviously to see that there is only one road damage in the image. These three methods have detected the correct results. By contrast, the effect of using YOLO-v3 is better than RetinaNet because the number of red

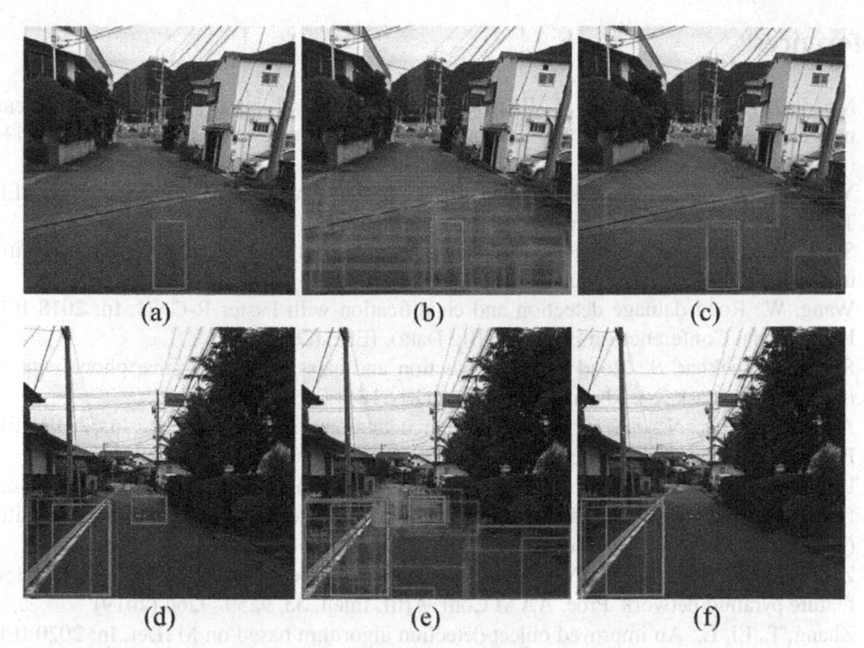

Fig. 10. The detecting result of M2det, RetinaNet, YOLO-v3 for the same scenario where (a) (d) represents M2det, (b) (e) represents RetinaNet, (c) (f) represents YOLO-v3.

bounding box which is use RetinaNet is more than YOLO-v3, these red bounding boxes means the wrong targets. And we use M2det to get the best performance. We do not deal with the wrong target while detecting the correct target. Therefore, our method has better generalization ability. In addition, by comparing Fig. 10 (d) (e) (f) three pictures, it is not difficult to find another phenomenon. We all know that the detection speed of YOLO-v3 is relatively fast, but the accuracy of the performance in this experiment with YOLO-v3 is slightly inferior than M2det. The "D44" is a small sample in this example, YOLO-v3 has missed the detection target, but M2det adopts a multi-scale and multi-level structure, which can extract the shallow and deep features of object, so that all targets are correctly detected. Although RetinaNet also detected all targets correctly, as we said earlier, it produces too many red boxes, so its performance in the detection process is not as good as M2det.

4 Conclusion

In this paper, we propose a method based on M2det, it uses a feature pyramid extraction network which is a multi-scale multi-level detection model. We select 70% of the dataset for training and the remaining 30% for testing, in order to evaluate the detection results, we compare the AP and mAP values of the various methods. The proposed method yields a mAP of 63.96%, which is the maximum value among the three methods, which indicating that the proposed method can obtain high accuracy in road damage detection, and the experimental results also show that the method gives good results in detecting both large and small targets.

References

1. Maeda, H., Sekimoto, Y., Seto, T.: Road damage detection and classification using deep neural networks with smartphone images. Comput. Aid. Civ. Infrastruct. Eng. **33**(12), 1127–1141 (2018)
2. Yang, Y.: Recomputation of dense layers for the performance improvement of DCNN. IEEE Trans. Softw. Eng. **PP**(99), 1 (2019)
3. Seungbo, S., Chanjun, C., Ki, R.S.: Road surface damage detection based on object recognition using Fast R-CNN. J. Korea Inst. Intell. Transp. Syst. **18**(2), 104–113 (2019)
4. Wang, W.: Road damage detection and classification with Faster R-CNN. In: 2018 IEEE International Conference on Big Data (Big Data). IEEE (2018)
5. Singh, J., Shekhar, S.: Road damage detection and classification in smartphone captured images using mask R-CNN. arXiv preprint arXiv:1811.04535 (2018)
6. Ale, L., Zhang, N., Li, L.: Road damage detection using RetinaNet. In: 2018 IEEE International Conference on Big Data (Big Data). IEEE (2018)
7. Lin, T., Dollár, P., Girshick, R., He, K., Hariharan, B., Belongie, S.: Feature pyramid networks for object detection. In: 2017 IEEE Conference on Computer Vision and Pattern Recognition (CVPR), pp. 936–944, Honolulu (2017)
8. Zhao, Q., Sheng, T., Wang, Y.: M2Det: a single-shot object detector based on multi-level feature pyramid network. Proc. AAAI Conf. Artif. Intell. **33**, 9259–9266 (2019)
9. Zhang, T., Li, L.: An improved object detection algorithm based on M2Det. In: 2020 IEEE International Conference on Artificial Intelligence and Computer Applications (ICAICA), pp. 582–585, Dalian (2020)
10. Pan, Y., Zhang, G., Zhang, L.: A spatial-channel hierarchical deep learning network for pixel-level automated crack detection. In: Automation in Construction, p. 119 (2020)
11. Theckedath, D., Sedamkar, R.: Detecting affect states using VGG16, ResNet50 and SE-ResNet50 networks. SN Comput. Sci. **1**(2), 1–7 (2020)
12. Girshick, R: Fast R-CNN. In: Proceedings of the IEEE International Conference on Computer Vision, Santiago, Chile, 11–18 December 2015, pp. 1440–1448 (2015)
13. Maeda, H., Sekimoto, Y., Seto, T., et al.: Road damage detection and classification using deep neural networks with smartphone images. Comput. Aid. Civ. Infrastruct. Eng. **33**(12), 1127–1141 (2018)
14. Choi, J., Chun, D., Kim, H.: Gaussian YOLOV3: an accurate and fast object detector using localization uncertainty for autonomous driving. In: Proceedings of the IEEE International Conference on Computer Vision, pp. 502–511 (2019)

Weather Temperature Prediction Based on LSTM-Bayesian Optimization

JingRong Wu, DingCheng Wang(✉), ZhuoYing Huang, JiaLe Qi, and Rui Wang

College of Computer Science and Software, Nanjing University of Information Science and Technology, Nanjing 210044, China

Abstract. With the continuous improvement of observation technology, the complexity of meteorological data elements has increased sharply, and the volume of the model has expanded, which brings inconvenience to conventional weather forecasting and conventional weather forecasting methods based on traditional statistical forecasting. This paper proposes a LSTM weather forecast method based on Bayesian optimization. Through the constructed sample data, the Bayesian optimization method is used to select the optimal parameters of the LSTM, and then the sample is reconstructed through the optimal LSTM, which has achieved better results in terms of accuracy. This study can explore more reasonable sample construction methods for weather attribute characteristics, and LSTM optimal parameter selection methods, and provide a simple, easy-to-use, high-precision weather prediction method for meteorological experts.

Keywords: LSTM · Temperature · Prediction · Bayesian optimization

1 Introduction

In meteorology, the physical quantity that measures the warmth or coldness of air is called air temperature. Temperature has a great impact on human production and life. As people's demand for temperature prediction and observation accuracy increase, the number density of automatic weather stations has also increased significantly, therefore the volume of meteorological data has expanded rapidly in the past ten years [1]. The advent of the era of meteorological big data has greatly improved the forecast accuracy of weather phenomena. Singh S et al. [2, 3] proposed a comprehensive back-propagation temperature prediction model based on the combination of time series genetic algorithm and neural network by studying the temperature attributes in meteorology and its dependence on specific data sequences. From the perspective of time series, Yang Han et al. [4] selected meteorological and temperature attributes to conduct refined forecasting research 24 h in advance, compared and improved traditional machine learning models, and applied cutting-edge technologies in deep learning to meteorological temperature prediction. Liu Xinda et al. compared the shallow BP neural network and support vector machine methods with the study of meteorological temperature prediction based on the DBN model and the SAE model, and proposed the advantages of the deep learning

X. Sun et al. (Eds.): ICAIS 2021, CCIS 1422, pp. 441–450, 2021.
https://doi.org/10.1007/978-3-030-78615-1_39

model in meteorological prediction, and used the support vector machine to perform deep learning related models Improvement [5].

Deep learning has made significant progress in computer vision [6], speech recognition, natural language processing and other fields, providing new ideas for weather forecasting [7]. This paper studies a Bayesian optimized LSTM deep learning method for temperature prediction research. For specific problems, when building a neural network model for temperature prediction, samples are often constructed based on past experience, such as setting the size of the sliding window. This method of relying on experience is probably not the optimal choice, resulting in the inability to train an accurate network and making larger temperature prediction errors. This paper studies the influence of different experimental sample structures and the selection of hyperparameters in the Long Short-term Memory network on atmospheric temperature prediction. The Bayesian optimization method is used to make the operation of the experiment easier and the experimental results clear and easy to understand.

2 Long Short-Term Memory Networks Based on Bayesian Optimization

Long Short-term Memory network is an improved recurrent neural network, which can solve the problem that ordinary recurrent neural networks cannot handle long-term dependence on information. It was developed by Hochreiter and Schmidhuber in 1997 [8]. LSTM adds a method of carrying information across multiple time steps. It saves the information for later use, thereby preventing the earlier signals from gradually disappearing during processing.

The control flow of LSTM is similar to that of RNN, which processes the flow of cells and information during the propagation process. LSTM's core concept is the cell flow and "gate" structure. Cell flow is the path of information transmission, allowing information to be continuously transmitted in the sequence. The "gate" structure contains many functions, which can normalize data information. The most important thing in LSTM is the calculation of the cell state. The cell state of the previous layer is multiplied by the forgetting vector point by point. When the value is close to 0, the information should not exist in the new cell state, and the information will also be discard. Then add this value to the output value of the input gate point by point, so as to update and replace the information. Figure 1 shows the basic structure of LSTM.

The optimization problem is actually a problem of seeking extreme values. In mathematics, the problem of seeking extreme values is closely connected with the problem of seeking derivation. However, many functions are not suitable for derivation in practical problems, so it is difficult to use derivation methods to explore the extreme values of functions. The advantage of Bayesian optimization is to infer the extreme value of the function through continuous guessing and exploration sampling [9]. Suppose it is desired to find the minimum value of the function f(X), X is a set of restricted parameters. The Bayesian optimization method is different from other methods in that it builds a probability model of f(X), and then explores this model to determine where to evaluate X next. The Bayesian optimization function uses all the previously evaluated information, not only the current gradient. Compared with the mainstream tuning methods Grid Search

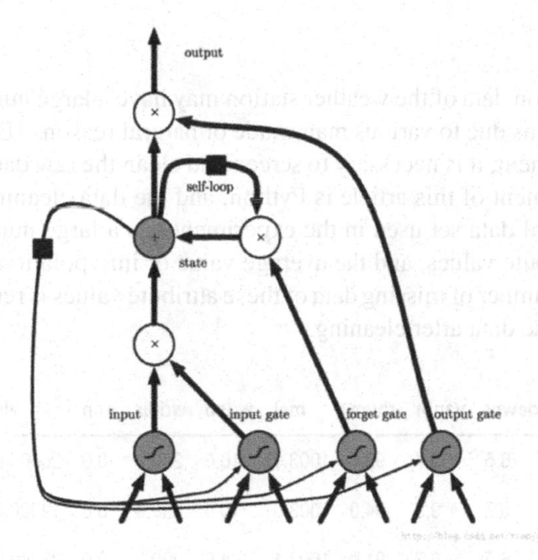

Fig. 1. LSTM structure

and Random Search, Bayesian optimization is less expensive, but it may not find the global optimal solution. Figure 2 shows the workflow of Bayesian optimization function [10].

Algorithm 1 Sequential Model-Based Optimization

Input: $f, \mathcal{X}, S, \mathcal{M}$
$\mathcal{D} \leftarrow \text{INITSAMPLES}(f, \mathcal{X})$
for $i \leftarrow |\mathcal{D}|$ **to** T **do**
  $p(y \mid \mathbf{x}, \mathcal{D}) \leftarrow \text{FITMODEL}(\mathcal{M}, \mathcal{D})$
  $\mathbf{x}_i \leftarrow \arg\max_{\mathbf{x} \in \mathcal{X}} S(\mathbf{x}, p(y \mid \mathbf{x}, \mathcal{D}))$
  $y_i \leftarrow f(\mathbf{x}_i)$ ▷ Expensive step
  $\mathcal{D} \leftarrow \mathcal{D} \cup (\mathbf{x}_i, y_i)$
end for

Fig. 2. Algorithm workflow

3 Weather Data Processing

The dataset used is hrly_Irish_weather. The original data set includes 18 items of data such as country, station, date, and rainfall. Relying on common sense, delete the data items of country, station, date that are obviously not related to the temperature prediction model. The attributes in the filtered model include 13 items of data such as rainfall, temperature, humidity, steam, relative humidity, air pressure, wind speed, sunshine, visibility, and cloud height. The data is collected every 1 h, and 24 pieces of data are obtained 24 h a day, totaling about 190,000 pieces of data. The data set is divided into 3 parts, namely training set, validation set and test set.

3.1 Data Cleaning

The original observation data of the weather station may have a large number of missing and duplicated problems due to various man-made or natural reasons. Therefore, before the start of the experiment, it is necessary to screen and clean the raw data obtained. The experimental environment of this article is Python, and the data cleaning work is done in Python. The original data set used in the experiment has a large number of missing values in certain attribute values, and the average value or interpolation method cannot be used, so the large number of missing data of these attribute values is removed. Figure 3 shows the header of the data after cleaning.

	rain	temp	wetb	dewpt	vappr	rhum	msl	wdsp	wddir	sun	vis	clht	clamt
0	0.0	7.6	7.1	6.5	9.7	93.0	1003.0	10.0	200.0	0.0	26000.0	16.0	8.0
1	0.0	7.6	7.2	6.7	9.8	94.0	1003.0	10.0	190.0	0.0	19000.0	18.0	8.0
2	0.0	7.8	7.2	6.5	9.7	91.0	1003.1	8.0	200.0	0.0	22000.0	18.0	7.0
3	0.0	7.7	6.8	5.7	9.2	87.0	1003.4	12.0	220.0	0.0	26000.0	18.0	6.0
4	0.5	7.5	6.5	5.3	8.9	86.0	1003.2	13.0	210.0	0.0	30000.0	999.0	3.0

Fig. 3. The header of the data after cleaning

3.2 Data Standarlization

In order to improve the convergence speed and accuracy of the model and reduce the computational burden during training and the influence of the dimensions of the data, the data in the training set is standardized. Through centralization and standardization, data with a mean value of 0 and a standard deviation of 1 obeys the standard normal distribution. In subsequent experiments, both the recurrent neural network and the LSTM model used data standardization in the experiment. Figure 4 shows the header of the standardized data.

	rain	temp	wetb	dewpt	vappr	rhum	msl	wdsp	wddir	sun	vis	clht	clamt
0	-0.281805	-0.718894	-0.555807	-0.322670	-0.422925	0.871038	-0.723031	-0.450617	0.036265	-0.469706	0.004913	-0.599120	0.936813
1	-0.281805	-0.718894	-0.529669	-0.274977	-0.390204	0.971956	-0.723031	-0.450617	-0.069453	-0.469706	-0.482612	-0.594146	0.936813
2	-0.281805	-0.668876	-0.529669	-0.322670	-0.422925	0.669200	-0.715458	-0.731363	0.036265	-0.469706	-0.273673	-0.594146	0.488066
3	-0.281805	-0.693885	-0.634219	-0.513440	-0.586526	0.265525	-0.692737	-0.169871	0.247699	-0.469706	0.004913	-0.594146	0.039319
4	0.700920	-0.743903	-0.712632	-0.608826	-0.684687	0.164607	-0.707884	-0.029498	0.141982	-0.469706	0.283499	1.845763	-1.306922

Fig. 4. The header of the data after standarlization

4 Experiment

4.1 Experimental Environment

The experiment uses Keras as the basic tool for building deep learning models. Keras is a tensorflow-based deep learning framework written by pure python, which is highly modular, simple and fast. The Bayesian optimization function uses the open source Bayesian optimization package fmfn/BayesionOptimization on GitHub. The method of using this function is: take the Keras model and parameter range as parameters, and return the maximum value of the loss function. Since what is sought is the minimum value of the loss function, the loss function becomes a negative number during the experiment.

4.2 Sample Structure

For time series data, the tensor dimension accepted by the network is (batch_size, timesteps, features). There are three parameters delay, lookback and step. Taking a certain time point as the standard, in the lookback time past this time point, one data is sampled at each step time point to predict the temperature at the time point after this time point delay data points. Due to the complexity of the temperature formation theory, it is impossible to accurately know the optimal size of the sliding window and the measurement of the time point in advance, so the Bayesian optimization function is used to search for the optimal time series sliding window size [11], input data amount and batch size. Based on the idea of control variables, an LSTM network with only 16 hidden units is constructed. Considering the large value range of the parameters, each parameter is represented by an integer multiplied by a fixed unit value. Expressed as follows:

$$\text{Starting point of sliding window: lookback} = \text{lookback\_n} * 24 \tag{1}$$

$$\text{End of sliding window: delay} = \text{delay\_n} * 2 \tag{2}$$

$$\text{Step length: step} = \text{step\_n} * 1 \tag{3}$$

$$\text{Batch size: batch\_size} = \text{batch\_size\_n} * 64 \tag{4}$$

The optimal parameter combination found after 20 iterations of the Bayesian optimization function is shown in Table 1. The overall distribution of parameters is shown in Fig. 3.

Table 1. Optimal parameter combination

delay_n	lookback_n	step_n	batch_size_n	Loss
0	15	1	5	0.1205

Analyzing Table 1 and Fig. 5, the distribution of delay and step is relatively fixed. When the delay is 0, the input sample date is closest to the date that needs to be predicted.

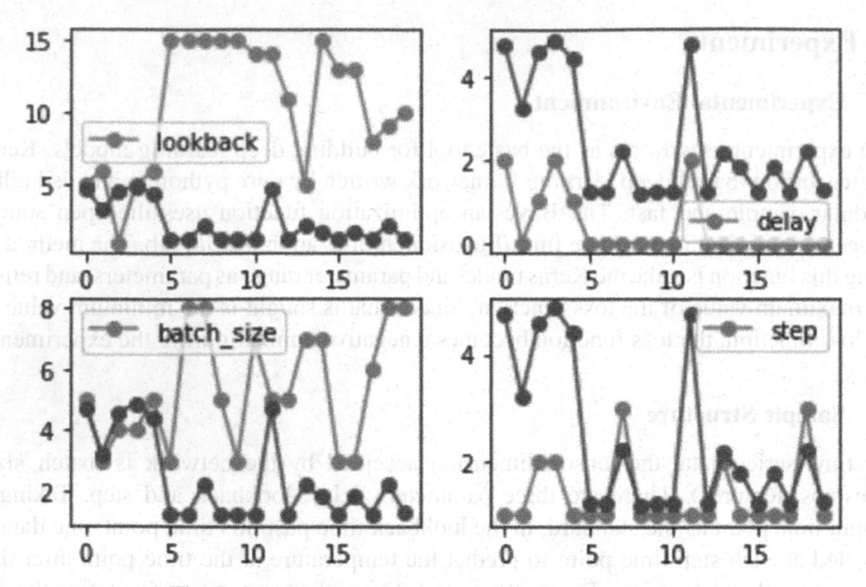

Fig. 5. Distribution of parameters (Color figure online)

For the network, the weight of the subsequent input sequence is larger, that is, the network can clearly "remember" the information of the input sequence, so it can be better reflected in the prediction results. A step of 1 means that each point of the current sliding window is sampled, that is, all data in the sliding window is the input data, so the amount of input data is the largest of all possibilities. Lookback is the size of the sliding window. The larger the sliding window is, the larger the amount of input data is. As far as the trend of the graph is concerned, the value of the sliding window can continue to increase, but in practical applications, if the amount of data is too large, the consumption of training time often increases significantly, so it is necessary to find a balance between the amount of data and the training time.

4.3 Training Network

After constructing the samples suitable for training, the Bayesian optimization function is used to search for the best combination of LSTM hyperparameters. Using the learning rate adaptive optimization function Adam, Adam will automatically adjust the learning rate according to the loss. Overfitting will occur if the network is too large, so the dropout method is used to regularize the neural network. Set dropout = 0.4. As far as the performance of the neural network is concerned, the more hidden units there are and the larger the network is, the better the prediction effect will be. However, too large network will increase traing time and lead to the occurrence of overfitting. Set the upper limit of the number of hidden units to 64, and the upper limit of the number of iterations to 80. This model uses MAE (Mean Absolute Error) as the loss function and evaluation standard. The calculation formula of MAE is:

$$\frac{1}{m}\sum_{i=1}^{m}\left|(y_i - \hat{y}_i)\right| \tag{5}$$

The search results of Bayesian function show the optimal number of hyperparameter groups as shown in Table 2:

Table 2. Optimal hyperparameter combination

nodeNums	Epochs	Loss
64	47	0.1047

As shown in Fig. 4, the network formed by this group of hyperparameters is very stable, and there is basically no fluctuation in the loss or over-fitting phenomenon since the fifth iteration. The result of using the test set prediction on this network is shown in Fig. 5. The red line represents the true value, and the blue line represents the predicted value. It can be seen from the figure that the true value is basically the same as the predicted value, the degree of fitting is very high, though some errors are large on several individual points (Figs. 6 and 7).

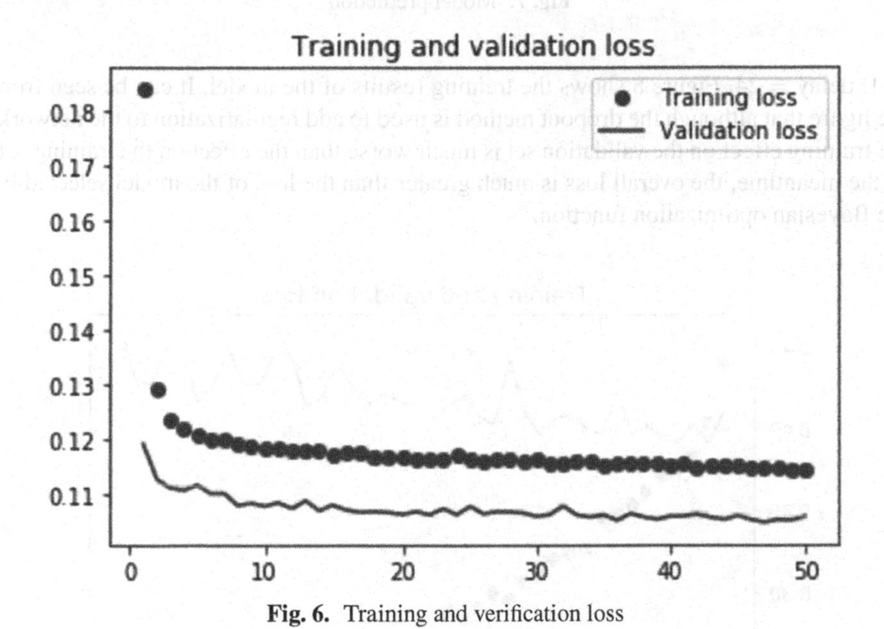

Fig. 6. Training and verification loss

4.4 Model Comparison

The temperature prediction model in Deep Learning with Python is selected for comparison. The sample constructed in the book is [12]: Given the observation data in the past 5 days, the sampling frequency of the observation data is one data point per hour, and the goal data is 24 h later. For the data of this experiment, set lookback = 120, step

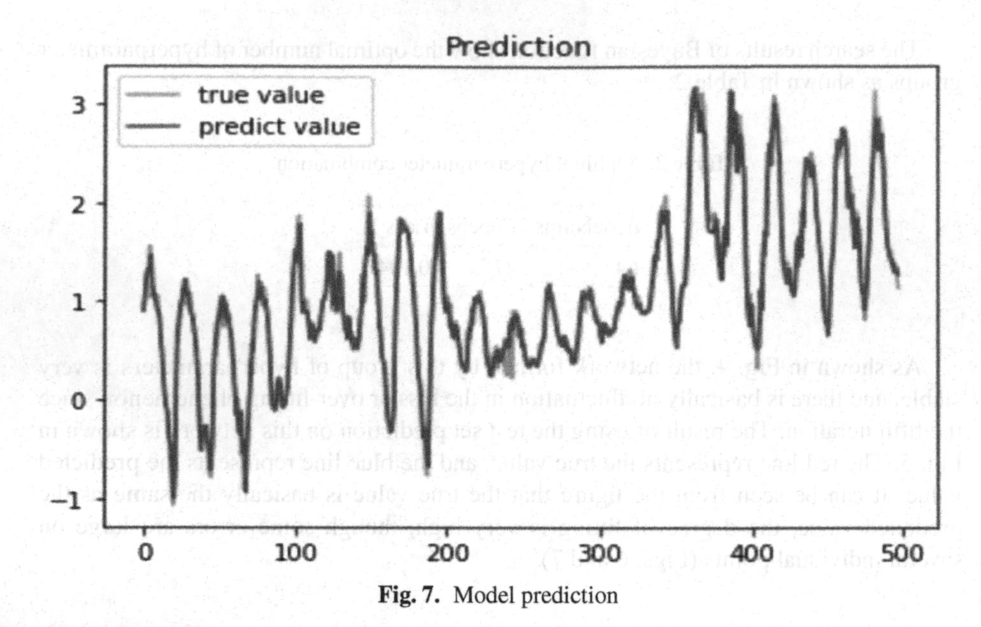

Fig. 7. Model prediction

= 1, delay = 24. Figure 8 shows the training results of the model. It can be seen from the figure that although the dropout method is used to add regularization to the network, the training effect on the validation set is much worse than the effect on the training set. In the meantime, the overall loss is much greater than the loss of the model selected by the Bayesian optimization function.

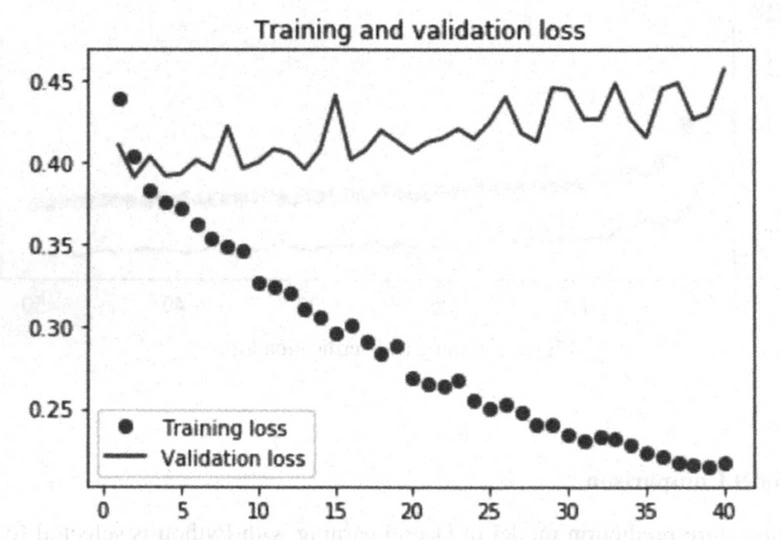

Fig. 8. Training and verification loss of the control model

5 Conclusion

The change of atmospheric temperature is affected by a variety of meteorological factors, therefore temperature prediction is a complicated problem for both deep learning and meteorological fields. Constructing a suitable sample is very important to the accuracy of temperature prediction. However, it is very difficult to construct samples because of the complexity of related theories involved and the size of temperature data samples. In order to solve some of these problems, the Bayesian optimized LSTM temperature prediction model proposed in this paper uses the Bayesian optimization function to find the optimal parameter combination. Experiments show that the method of constructing samples using Bayesian optimization functions is practical and feasible. The advantage is that it can reduce a lot of research on theories in specific fields and directly use functions to search for the best samples. Using Bayesian to adjust the hyperparameters of LSTM is more convenient than manual adjustment. It only needs to determine the approximate range of the parameters and does not need to be precised to a certain data point. In comparison with other sample construction models, the results show that the Bayesian optimized LSTM model is much better than another model in aspects of accuracy and suppression of overfitting. However, this study only used the meteorological data of a fixed region, and did not consider the different geographic characteristics of each region. The generalization of the model needs further verification. Therefore, in further research, a larger amount of data should be obtained to prove that the model can be used in extensive applications.

Acknowledgement. This paper can not be completed without the teacher's guidance and the support of our school. Thank our school for giving us an opportunity to do this research.

Funding Statment: This work was supported in full by NUIST Students' Platform for Innovation and Entreprneurship Training Program.

Conflicts of Interest: We have no conflicts of interest to report regarding the present study.

References

1. Shen, W.: Analysis of the "Big Data Application" of meteorological data——discussion on the applicability of thinking reform in "Big Data Era". In: China Information Technology. No. 11, pp. 20–21 (2014)
2. Singh, S., Bhambri, P., Gill, J.: Time series based temperature prediction using back propagation with genetic algorithm technique. Int. J. Comput. Sci. Issues **8**(5), 28–31 (2011)
3. Singh, S., Gill, J.: Temporal weather prediction using back propagation based genetic algorithm technique. Int. J. Intell. Syst. **6**(12), 55–61 (2014)
4. Han, Y.: Research on weather prediction based on deep learning. Master's thesis of Harbin Institute of Technology (2017)
5. Liu, X.: Research on meteorological temperature prediction based on deep learning. Master's thesis of Ningxia University (2016)
6. Siwei, C., Kebin, J., Congcong, W., Jun, L.: Research and application of deep learning in multi-weather classification algorithm. High Technol. News **30**(10), 1010–1017 (2020)
7. Zheng, N., Ping, L.: Preliminary study on refined air temperature forecast based on LSTM deep neural network. Comput. Appl. Softw. **35**(11), 233–236 (2018)

8. Greff, K., Srivastava, R.K.: LSTM: a search space odyssey. IEEE Trans. Neural Netw. Learn. Syst. **28**(10), 2222–2232 (2017)
9. Hochreiter, S., Schmidhuber, J.: Long short-term memory. Neural Comput. **9**(8), 1735–1780 (1997)
10. Snoek, J., Larochelle, H., Adams, R.P.: Practical Bayesian optimization of machine learning algorithms. arXiv:1206.2944 [stat.ML]
11. Frazier, P.I.: A tutorial on Bayesian optimization. arXiv:1807.02811 [stat.ML]
12. Chollet, F.: Deep Learning with Python, pp. 211–212. Manning Publications Co., New York (2017)

Research Situation and Development Trends of Deep Learning Application in Meteorology

Rui Wang, Dingcheng Wang(✉), Jialc Qi, Jingrong Wu, Shuo Liang, and Zhuoying Huang

College of Computer Science and Software, Nanjing University of Information Science and Technology, Nanjing 210044, China

Abstract. With the development of artificial intelligence, deep learning as an important branch has been paid attention to for its important achievements in machine learning, which has been developed rapidly and widely used in different fields. The complexity and uncertainty of meteorological environment, as well as the exponential and incremental index of meteorological data and the introduction of deep learning, can help to improve the accuracy of weather forecasting and related work. This paper summarizes the current situation and problems of the application of deep learning in the field of meteorology, and provides reference for scholars who further improve the accuracy of deep learning and weather forecasting.

Keywords: Deep learning · Meteorological forecast · Neural network

1 Introduction

Due to the complexity and uncertainty of the meteorological environment, as well as the massive and incremental exponential of meteorological data, it is necessary to study intelligent weather forecasting and forecasting which adapts to the modern information technology environment on the basis of the development of traditional meteorological forecasting and forecasting. Deep learning, with its characteristics of data-based cyclic iterative learning, discovers the law of things change from massive data, can effectively mine the temporal and spatial features from the spatio-temporal data. Meteorological data is a typical large geospatial data [1]. Deep learning-based weather forecasting is expected to be an important supplement to the research of modern weather prediction methods.

Deep learning originates from neural network, in a sense, it is the development of neural network. By extending the depth of network to improve the ability of machine learning, it has achieved successful cases in image processing, pattern recognition and other fields. At present, there are many models of deep learning, such as deep feedforward network, convolutional neural network, restricted Boltzmann machine, deep belief network, recurrent neural network and long short-term memory network. These models have a lot of potential research value for the complex weather forecast research, and have been engaged in related research at home and abroad. This paper summarizes and

X. Sun et al. (Eds.): ICAIS 2021, CCIS 1422, pp. 451–462, 2021.
https://doi.org/10.1007/978-3-030-78615-1_40

divides the related research status at home and abroad, aiming to provide reference for further research.

2 Deep Learning Model

2.1 Deep Feed Forward Network (DFN)

Deep feedforward network, also known as feedforward network or multilayer perceptron, is a typical deep learning model. In the feedforward neural network, different neurons belong to different layers, each layer of neurons can receive the signal from the previous layer and produce the signal output to the next layer. It can be represented by the directed acyclic graph of Fig. 1.

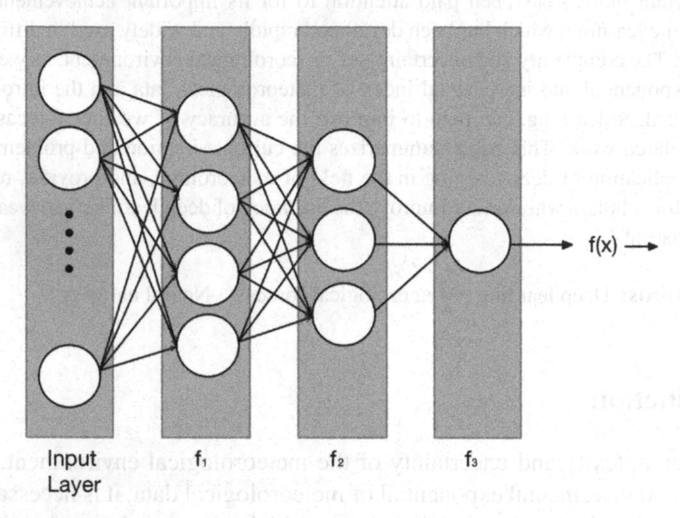

Fig. 1. Directed acyclic graph of feedforward network

Deep feedforward network is usually composed of many different functions. Assuming that there are three functions f1, f2 and f3 forming a chain compound structure, then: f(x) = f3(f2(f1(x))). f1 is called the first layer of the network, f2 is the second layer of the network, and f3 is the third layer of the network. The full length of the chain is called the depth of the model. Each layer can receive the signal from the previous layer, which is generated by the mapping function fn(x) and input to the next layer, This chain structure is also the most commonly used structure of neural networks [2]. In this process, the signal propagates unidirectionally from the input layer to the output layer, and there is no feedback in the whole network, resulting in a value close to the target.

2.2 Convolutional Neural Network (CNN)

Convolutional neural network is a kind of feedforward neural network with convolution calculation and depth structure. Convolution neural network includes convolution layer, pooling layer and fully connected layer. Convolution layer and fully connected layer have parameters, but pooling layer has no parameters. The basic structure of convolution neural network is shown in Fig. 2 below. The convolution layer is a feature extraction layer, and each convolution layer in the convolution layer consists of several convolution units. After the image is input, the convolution operation is performed through the convolution template to extract local features, and the features extracted from the current layer are transmitted to the higher layer to continue the convolution operation. Pooling layer divides the features extracted by convolution layer into several regions, and takes the average value or maximum value to get the features with smaller dimensions. All local features are combined into global features by full connection layer. Compared with other neural network models, convolutional neural network needs less training parameters because of its local receptive field and weight sharing, and its structure becomes simpler and more adaptable. Convolutional neural network has excellent performance in processing machine learning problems, especially in processing image data, such as computer vision and the largest image classification data set [3].

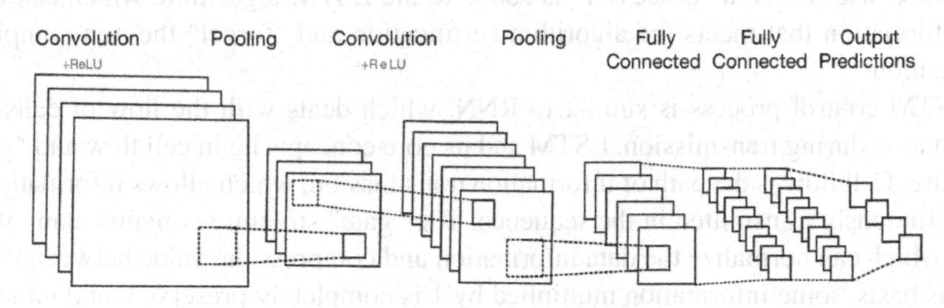

Fig. 2. CNN structure diagram

2.3 Recurrent Neural Network (RNN)

Recurrent neural network is a kind of recursive neural network with chain structure, which takes sequence data as input and recurs in the evolution direction of sequence. In structure, compared with the common multilayer neural network, each neuron in the hidden layer contains a feedback input, and the output of the neuron can act on itself at the next moment. The following Fig. 3 is the RNN structure diagram, where x is the input unit, y is the output unit and R is the neural network unit. In the figure below, the module R of neural network is reading the input value Xt and outputting Yt, transferring the information from the current module to the next module. This structure enhances the association between nodes. In practical application, RNN can use its internal memory to process input sequences with arbitrary time series. When building the model, it is not necessary to assign a set of individual parameters to each input feature set, which greatly simplifies the training overhead. Recurrent neural network has certain advantages in learning nonlinear features of sequences. However, when learning long sequences, the phenomena of gradient disappearance and gradient explosion will occur, and the nonlinear relationship with long time span cannot be grasped. This shortcoming also paves the way for the appearance of LSTM in the later period [4].

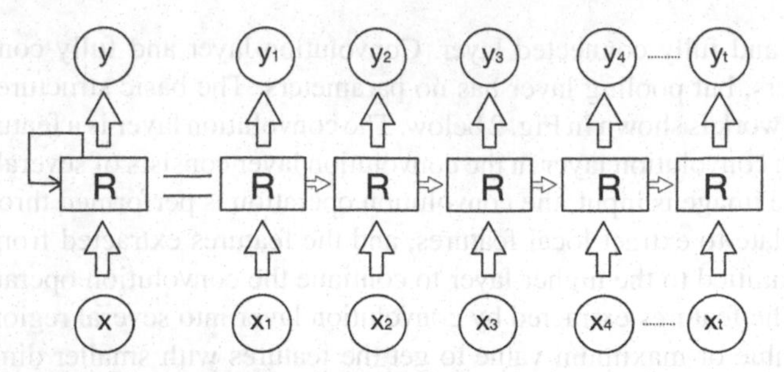

Fig. 3. RNN structure diagram

2.4 Long Short-Term Memory Network (LSTM)

The long short-term memory network is a special kind of recurrent neural network, which is to solve the problem of gradient disappearance and gradient descent in general RNN. Compared with RNN, a "processor" is added to the LSTM algorithm, which can filter out information that meets the algorithm certification and "forget" the non-compliant information.

LSTM control process is similar to RNN, which deals with the flow of cells and information during transmission. LSTM and its core concepts lie in cell flow and "gate" structure. Cell flow is the path of information transmission, which allows information to be continuously transmitted in the sequence. The "gate" structure contains many functions, which can normalize the data information and compress the value between [0, 1]. On this basis, some information multiplied by 1 is completely preserved, and unnecessary information is forgotten. The most important thing in LSTM is the calculation of cell state. The cell state of the previous layer is multiplied by forgetting vector point by point. When it is close to 0, the information should not exist in the new cell state, and the information will be forgotten and discarded. After that, the value and the output value of the input gate are added point by point, so as to realize the update and substitution of information. LSTM can learn the long-term dependence and grasp the long-term rules and long-term characteristics according to the time correlation. It is suitable for processing and predicting the events with very long interval and delay in time series [5]. At present, LSTM is widely used in language recognition, machine translation and text prediction [6, 7].

Figure 4 below shows the LSTM structure diagram. Where xt is the input at time t; ht is the information reservation at time t; tanh is a double tangent function, which compresses the range of input values to $[-1, 1]$, and expands the feature effect in the cycle process when the feature difference is obvious [8].

2.5 Deep Belief Network (DBN)

Deep belief network is a deep learning model structure composed of multiple restricted Boltzmann machines (RBM). Restricted Boltzmann machine is similar to Boltzmann machine in structure, which has a visible layer and a hidden layer. The visual layer contains multiple visual nodes for data input, and the hidden layer contains multiple

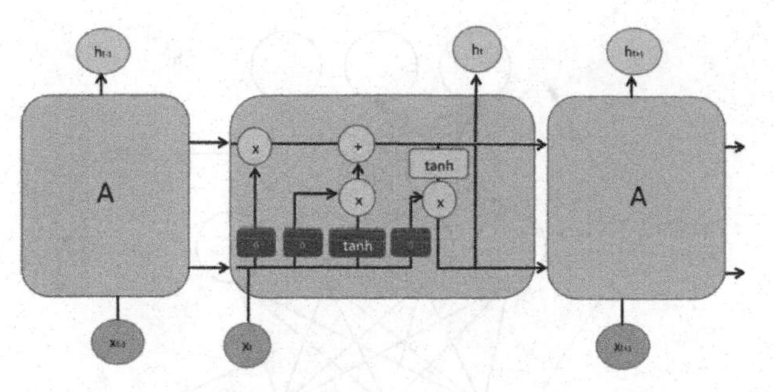

Fig. 4. LSTM structure diagram

hidden nodes for data output. As shown in Fig. 5 below, the network model of restricted Boltzmann machine is shown.

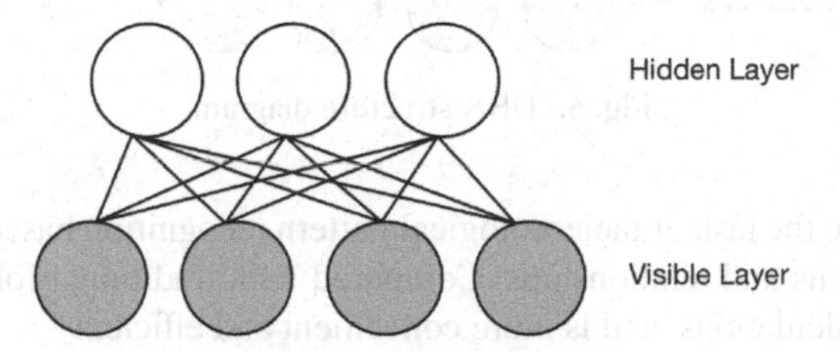

Fig. 5. RBM structure diagram

Because the deep belief network is composed of multiple restricted Boltzmann machines, the deep belief network model has multiple hidden layers. As shown in Fig. 6 below, each RBM has a hidden layer and a visible layer. The hidden layer of the lower layer can be used as the visible layer of the upper layer, that is, the output of the lower layer can be used as the input of the upper layer. In some application scenarios, such as temperature prediction scenario, because the deep belief network uses supervised fine-tuning, compared with the unsupervised fine-tuning neural network model, with the increase of the number of iterations, the prediction accuracy continues to improve, and the classification effect is better. At present, deep belief network is widely used in face recognition, emotion analysis and other fields.

3 Deep Learning Meteorological Application

Meteorology is a complex data analysis discipline, which has the characteristics of short time and constant change. Weather forecast is an important branch of meteorology, which involves massive data in different time and space. Accurate meteorological forecasting needs to collect a large amount of data, and the collected data are converted and analyzed. There is a certain delay in the artificial forecasting mode, and it is hard to realize the real-time meteorological forecasting. Deep learning has advantages in dealing with nonlinear complex data. It can automatically learn the complex relationships between

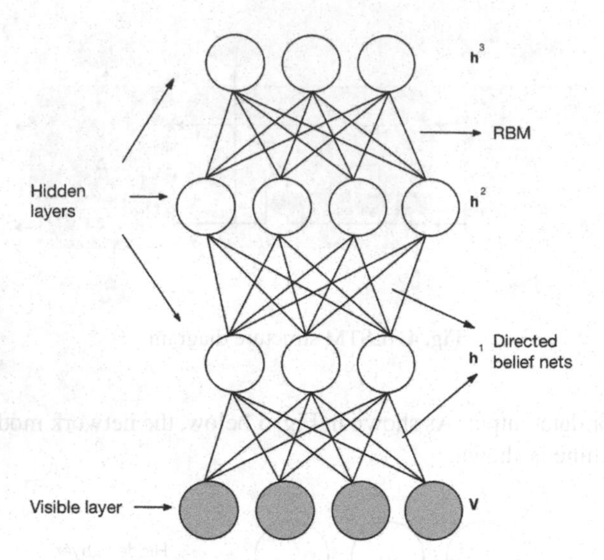

Fig. 6. DBN structure diagram

data, and complete the task of meteorological pattern recognition based on these knowledge representations and relationships. Compared with traditional forecasting models, it saves a lot of calculations, and is more convenient and efficient.

The application of deep learning in meteorological industry has just started, and there are many application scenarios. The main R&D and application scenarios include weather system intelligent identification, proximity forecasting (radar image extrapolation, precipitation proximity forecasting, cumulative precipitation forecasting, strong convection classification forecasting, etc.), quantitative precipitation objective forecasting, typhoon ensemble forecasting, numerical forecast revision and forecast intelligent editing and generation [9, 10]. Here are some common applications.

3.1 Deep Learning of Meteorological Disasters and Numerical Prediction of Conventional Meteorological Factors

Traditional weather forecasting mainly relies on meteorological monitoring facilities, monitoring equipment to observe the image analysis, to obtain prediction results. For example, a ground weather station is established to measure wind and wind speed, precipitation is measured in real time by radar, and multispectral images are obtained by meteorological satellites.

However, these measurement methods are not suitable for all parts of the world. To achieve global monitoring requires very high radar coverage and satellite coverage, and it is difficult to establish observation stations on the ocean. In addition, the real-time nature of traditional forecasting methods cannot meet today's weather forecasting needs. For example, in numerical forecasting, weather forecasting takes several hours to calculate, which makes each prediction based on data from a few hours ago, which limits the understanding of the current weather situation. In this context, the method based on deep learning came into being.

The early research of deep learning in meteorological field mainly focused on meteorological prediction, especially in weather disaster forecast and precipitation forecast. For decades, many achievements have been made at home and abroad. For example, in the aspect of disastrous weather prediction, foreign scholars use feedforward neural network to predict the probability of tornado in a single storm unit, and use gradient lifting decision tree, elastic network and random forest to predict the probability of hail weather [11].

In rainfall prediction, related scholars have applied the related technologies of convolution neural network in combination with meteorological radar. Meteorological radar map is an echo image of weather information detected by microwave signals emitted by meteorological radar on radar display. Rainfall intensity, rain area range, future rainfall intensity and movement can be judged from different colors, color block sizes and changes. Each layer of neural network can calculate the color depth, color area coverage and regional variation of the image, and predict the possibility, intensity, coverage and duration of future meteorological conditions. Through research and comparison, it is found that compared with the traditional method, the deep learning method is more real-time and effective in massive and various data, and it shows outstanding ability in improving the delay.

3.2 Deep Learning Improves Prediction Accuracy and Reduces Prediction Error

In these meteorological application scenarios of artificial intelligence, the research of proximity forecasting based on deep learning has been carried out more, and some achievements have been made. In the aspect of near prediction, the traditional prediction models are mostly mathematical methods and shallow neural networks. When faced with a large amount of data and more kinds of data, it is easy to have problems such as modeling difficulties and lower prediction accuracy. Traditional statistical forecasting lacks physical basis and has poor stability, and the empirical extrapolation method adopted in the early stage has low accuracy. At present, in order to achieve the goal of punctuality and high accuracy in weather forecasting, dynamic numerical models are becoming more and more complex. However, the deep learning model has certain advantages in dealing with a large number of nonlinear data, and its model has a multi-layer network structure. By training a large number of data, representative eigenvalues can be extracted, and hidden relationships among data can be found, so as to achieve the purpose of accurate classification and prediction [12].

In view of the characteristics of large amount of data processed by deep learning model and high precision, relevant scholars have carried out research in different branches of meteorological field, trying to improve prediction accuracy, recognition accuracy and reduce errors through deep learning model.

Xulin Liu [13] and others put forward a deep learning prediction model based on convolutional neural network and sequence to sequence, which uses convolutional neural network and Seq2Seq network to obtain spatio-temporal characteristics, and uses the generated model to predict PM2.5, which improves the prediction accuracy of PM2.5 concentration in the next hour. The experimental results show the potential of the improved depth model in solving the problem of short-term prediction of spatiotemporal series data;

Guosong Wang [14] and others built two LSTM deep learning models to forecast short-term wind speed. Through analysis, it is found that the LSTM model has greatly reduced the prediction error compared with WRF model. LSTM model has successfully learned the characteristics of wind speed prediction in different directions, and has certain anti-interference ability, which can reduce the error of terrain and other factors on numerical prediction results;

In the prediction of air pollution index, the existing prediction methods have some problems, such as strong dependence on samples, input too much redundant data, low prediction accuracy and so on. Moreover, due to the fixed learning rate of most neural networks, the training process of neural networks is long. Guangtao Wei and Jian Zhou [15] of Nanjing University of Posts and Telecommunications put forward an air pollution index prediction method based on deep belief network, the method takes the air pollution index data of c hours before t hour as the input of the deep belief network model, outputs the prediction value of $t+1$ hour, and decides whether to update the network model according to the error between the prediction value and the actual value. In addition, the prediction results of air pollution indicators are compared under different prediction windows, and the best prediction window is determined, which further improves the accuracy of prediction.

Hong Huang [16] and others put forward a method of SO_2 concentration detection based on deep belief network and extreme learning machine, On the basis of feature extraction of high-dimensional data of SO_2 UV absorption spectrum, an inversion model was established to solve the problem of overlapping feature information extraction difficulty and extraction accuracy. Compared with the original differential absorption spectrum technology, the accuracy is higher.

3.3 Using Deep Learning to Revise Numerical Forecast

In modern weather forecasting, numerical weather forecasting still plays an important role. Because the numerical weather forecast depends on theoretical calculation, some factors such as system deviation and local influence may be received in the actual forecast process, which leads to a certain deviation between the forecast result and the actual result. Although the forecast result can be corrected manually, a large number of observation data need to be collected in the correction process, which consumes a lot of manpower and material resources and has a small correction area. However, some deep learning models, such as LSTM model, have strong anti-interference ability, are faster in processing large amounts of data than manual methods, and have a larger correction area. Relevant scholars use this characteristic of deep learning to carry out research in this field. For example, in the research on wind speed prediction, Quande Sun [17] and others revised the wind speed near the ground in North China predicted by ECMWF based on three machine learning methods: LASSO regression, random forest and deep learning. The results show that the correction effect of the three machine learning algorithms is better than that of MOS method, which shows the potential of machine learning method in numerical forecast correction.

3.4 Use Deep Learning to Recognize, Detect and Classify Meteorological Images

Meteorological prediction requires not only collecting a large amount of data, but also analyzing related images. Accurate image recognition can provide valuable information for meteorological prediction. Traditional recognition methods require manual design of features, and rely strongly on recognition experience, and the recognition rate is low in complex recognition scenarios [18]. In the current meteorological field, in the face of massive image data, because scientific and intelligent identification methods are not fully applied, most of them still rely on manual reading and analysis of images to complete the identification task, which has low accuracy, objectivity, and stability. The application of deep learning image recognition technology to the meteorological industry can fully mine the information of weather radar and satellite cloud images, thereby improving the efficiency of recognition and reducing meteorological disasters [19]. Some deep learning models, such as convolution neural network, show advantages in image processing, and have better performance than traditional artificial recognition methods. They can complete image recognition and classification in a very short time, so as to realize real-time weather forecast.

For example, in cloud recognition, the color, shape, structure, brightness and texture of cloud images determine the cloud species, genus and precipitation. Relevant scholars use convolution neural network to determine the location and intensity of weather system according to the meteorological characteristics in the image, and judge their future movement and evolution trend [20].

China Meteorological Administration and Tianjin University jointly developed the national severe convection service product processing system. As far as hail warning is concerned, through deep learning, the system can clearly analyze the hail characteristics on the radar echo, so as to better identify the hail cloud, calculate its moving speed and direction, and give the possible influence range of kilometer level hail. This indicates that artificial intelligence technology can effectively improve the disaster warning capability of the meteorological system [21].

4 Problems Faced

With the rapid development of deep learning technology in meteorological field, there are still some problems and deficiencies. For example, in the related research of short-term and impending precipitation forecasting, the research of short-term and impending precipitation forecasting is mostly a single algorithm research, focusing on how to improve the forecasting accuracy. Although the timeliness and accuracy are improved compared with the traditional methods, the neural network model is highly dependent on historical data, and when the historical data is abnormal, the output data is often inconsistent with reality, and the fault tolerance is poor. Weather system is a high-order nonlinear system with multiple unstable sources. Its complex internal interaction and random changes lead to the variability of weather and climate, so its initial values and inputs are not completely determined, which leads to the prediction results using neural network technology can't achieve a high accuracy. Secondly, deep learning often requires a large number of samples, and in the prediction of disastrous weather, such as hail, typhoon meteorological observation data and climate observation data set is far less than the number of samples required for deep learning of meteorological forecast. Thirdly, the output variables of the weather forecast field are more than the input variables of the initial field, and they are not the same variables, and the temporal and spatial resolution of the forecast field often needs to be more precise than that of the initial field, which is difficult to achieve in the current deep learning technology.

Some questions remain open with respect to the input data: How many volumes of data shall be input to deep learning system to maximize training accuracy with minimal computational cost? Is older data or recent data more suitable for the training process? In weather forecast scenario, these questions are crucial to consider when time is a constraining factor or when usable data is scarce. In addition, the neural network model has many shortcomings, such as complex network structure, easy to fall into local minima and slow convergence speed [22, 23, 24, 25, 26]. From an academic point of view, we can optimize and improve this aspect.

5 Conclusion

This paper first summarizes the current application of deep learning in various fields and the needs of new weather forecasting models. Next, the typical deep learning model: deep feedforward network, convolutional neural network, recurrent neural network, long short-term memory network are analyzed theoretically. The abstract concepts are embodied, and the advantages and disadvantages of each neural network are compared and analyzed. Then, several common applications of deep learning in meteorology are analyzed, and the current research results of each kind of application are summarized and analyzed The reason why neural network model can be successfully applied in the field of meteorology. Finally, it points out the difficulties of neural network in the field of Meteorology and some problems to be solved in the future.

To sum up, although the application of deep learning technology in the meteorological field is at the exploratory stage, there are some basic problems, but the advantages of deep learning in dealing with a large number of nonlinear data determine that it can adapt to the current complex meteorological model. In the current and future meteorological forecasting field, deep learning technology will become a powerful supplement to the current meteorological forecasting model. In meteorological business, deep learning technology will also promote meteorological business to be intelligent and automatic.

Acknowledgement. The results of the study can't be separated from the teacher's guidance and the support of the school, at the end of this paper, here thank Mr. Wang has been guiding us to write and modify the paper, for our research content pointed out the direction, thank the school gave us a platform and financial support, let us have such an opportunity to realize our own value.

Funding Statement. This work was supported by NUIST Students' Platform for Innovation and Entrepreneurship Training Program (No. 202010300202) and the National Natural Science Foundation of China under Grant 62072251.

Conflicts of Interest. We have no conflicts of interest to report regarding the present study.

References

1. Ren, X.L., Li, X.Y.: Deep learning-based weather prediction: a survey. Big Data Research, vol. 23 (2020)
2. Goodfellow, I., Bengio, Y., Courville, A.: Deep Learning: The MIT Press, 800 pp. (2016). ISBN: 0262035618.
3. Albawi, S., Mohammed, T.A., Al-Zawi, S.: Understanding of a convolutional neural network. In: Proceedings of the IEEE, Antalya, AYT, Turkey, pp. 1–6 (2017)
4. Yang, H.: Research on meteorological prediction based on deep learning. M.S. dissertation, Harbin Institute of technology, China (2017)
5. Tao, Y.: Research on meteorological prediction based on long term and short term memory network. M.S. dissertation, Nanjing University of Information Science &technology, China (2019)
6. Gers, F.A., Schmidhuber, J., Cummins, F.: Learning to forget: continual prediction with LSTM. Neural Computation, vol. 12, no. 10 (2000)
7. Hochreiter, S., Schmidhuber, J.: Long short-term memory. Neural Comput. **9**(8), 1735–1780 (1997)
8. Cui, C.G., Zou, Y.H.: LSTM photovoltaic prediction based on deep learning. J. Shanghai Univ. Electric Power **26**(6), 544–552+579 (2019)
9. Tang, W., Zhou, Y., Dong, H., Zhang, D.Y., Zhao, W.F.: Current situation and international comparison of artificial intelligence technology in meteorological field in China. Adv. Meteorological Sci. Technol. **9**(5), 55–56+62 (2019)
10. Tang, W., Zhou, Y., Dong, H., Zhang, D.Y., Zhao, W.F.: Analysis on the current situation and influence of artificial intelligence in meteorological forecast. China informatization **4**(11), 69–72 (2017)
11. He, J.Y., Tang, W., Zhou, Y., Shen, W.H.: Opportunities and challenges of the application of artificial intelligence in meteorological science. China informatization **6**(12), 79–81 (2019)
12. Sun, L.Y., Liu, M.L., Zhou, L.X., Yu, Y.: Forest fire prediction method based on deep learning of meteorological factors. J. Forestry Eng. **33**(3), 132–136 (2019)
13. Liu, X.L., Zhao, W.F., Tang, W.: Using the prediction model of PM2.5 in the future hour by CNN-Seq2seq. Mini-Micro Syst. **41**(5), 1000–1006 (2020)
14. Wang, G.S., et al.: Application of LSTM deep neural network to coastal wind speed prediction based on observation and reanalysis data. Acta Oceanographica **42**(1), 67–77 (2020)
15. Wei, G.T.: Research and implementation of air index prediction method based on deep belief network. M.S. dissertation, Nanjing University of Posts and Telecommunications, China (2019)
16. Huang, H., Lan, H.Y., Huang, Y.X.: Concentration detection methods based on deep belief network and extreme learning machine SO2. J. Atmospheric Environ. Optics **27**(3) (2020)
17. Sun, Q.D., et al.: Machine learning-based numerical weather forecast wind speed revision study. Meteorological Monthly **45**(3), 426–436 (2019)
18. Wen, K.: Research and application of image recognition method based on deep learning, M.S. dissertation, Central China Normal University, China (2017)
19. Wang, S.H., Liu, Q., Ma, X.X.: Research on weather modification business based on image recognition. Comput. Technol. Dev. **29**(5), 172–177 (2019)
20. Liu, J.Q.: Analysis of the application of machine learning in climate research. Telecom World **25**(9), 218–219 (2018)
21. Tang, W., Zhou, Y., Wang, Z., Gong, J.L., Shen, W.H.: Analysis of the current situation of the application of artificial intelligence in weather forecast and its influence. China Informatization **4**(11), 69–72 (2017)

22. Liu, Y.C.: A summary of the application of deep learning under artificial intelligence in weather forecasting. Comput. Products Circul. **4**(11), 121+135 (2020)
23. Liu, X.D.: Research on meteorological temperature prediction based on deep learning, M.S. dissertation, Ningxia University, China (2016)
24. Han, Y.: The application of artificial intelligence in the field of weather forecasting. Telecom World **26**(4), 265–266 (2019)
25. Booz, J., Wei, Y., Guo, B.X., Dtiffith, D., Golmie, N.: A deep learning-based weather forecast system for data volume and recency analysis. In: Proceedings of the ICNC, Honolulu, HI, USA, pp. 697–701 (2019)
26. Li, Z.D.: Meteorological element grid based on deep learning. M.S. dissertation, Hunan Normal University, China (2019)

Exploiting Spatial-Spectral Feature for Hyperspectral Image Classification Based on 3-D CNN and Bi-LSTM

Junru Yin, Changsheng Qi(✉), Wei Huang, and Qiqiang Chen

School of Computer and Communication Engineering, Zhengzhou University of Light Industry, Zhengzhou 450000, China

Abstract. Hyperspectral remote sensing has been gaining more and more attention in recent years because of the rich spectral and spatial information contained in hyperspectral image (HSI). With the rapid development of deep learning, many deep learning methods have been applied to classify HSI. In the existing 3-D convolution methods, a widely used method is to project the original data into a low-dimensional subspace, so a small amount of the useful spectral information can be lost. To solve this problem, this paper propose a unified network framework using band grouping-based bidirectional long short-term memory (Bi-LSTM) network and 3-D convolutional neural network for HSI classification. In this framework, the issue of spectral feature extraction is considered as a sequence learning problem, and the Bi-LSTM as a spectral feature extractor is adopted to address it. To evaluate the performance of the proposed method, the Indian Pines remote sensing data sets are used for HSI classification experiments. The results demonstrate that the performance of proposed method is better than the state-of-the-art HSI classification methods.

Keywords: 3-D convolutional neural network · Bidirectional long short-term memory (Bi-LSTM) · HSI classification

1 Introduction

The research in HSI analysis is more and more important due to its potential applications in real life [1]. HSIs are obtained by high-altitude hyperspectral imaging spectrometers that collect spectral data of different bands reflected by various substances on the ground [2]. Image information can reflect the size, shape, defect, and other external quality characteristics of the sample. Since different components have different spectral absorption, the image will reflect a certain defect significantly at a certain wavelength, and the spectral information can fully reflect the difference in the physical structure and chemical composition of the sample. These characteristics determine the unique advantages of HSI technology in image classification [3] and target detection [4].

The classification of HSIs has been one of the hot topics in the past decades. While the abundant spectral information is very useful in improving the accuracy of classification, but its high dimensionality poses new challenges. Compared to natural images,

X. Sun et al. (Eds.): ICAIS 2021, CCIS 1422, pp. 463–475, 2021.
https://doi.org/10.1007/978-3-030-78615-1_41

HSI contains hundreds of spectral bands, including a wide range of the electromagnetic spectrum in the visible, near-infrared, mid-infrared, and thermal infrared. As the number of spectral bands increases, the number of calculations increases significantly and the classifier performance improves to some extent. When the number of feature dimensions exceeds the threshold, continuing to increase will instead lead to the poor performance of the classifier, which is called the Hughes phenomenon. The presence of homo-spectral and hetero-spectral species in HSI results in increased intra-class variability and decreased inter-class similarity, making feature extraction more difficult. In addition, because HSI labels are quite difficult to obtain, data are rarely publicly available, and labeling each pixel in the HSI is very time-consuming, there are few existing HSI data tagging samples and a very limited number of training samples available.

To solve these problems, many researchers have been proposing various methods. HSI classification techniques are constantly evolving, from hand-designed feature extraction techniques based on traditional methods to end-to-end feature extraction techniques based on deep learning in recent years. The traditional approach mainly lies in feature extraction and classifier training, in which researchers design feature extractors by imposing constraints on the model to extract useful information. Such as maximum likelihood methods [5], K-Means Clustering [6], K-Nearest Neighbor (KNN) [7] and logistic regression. In recent years, Support Vector Machine (SVM) [8] has been widely used in the classification of HSIs due to its powerful advantages, such as in processing high-dimensional, small sample data. SVM is mainly focused on transforming linearly inseparable problems into linearly separable problems by searching for the optimal hyperplane, which finally achieves the classification task. Hand-designed feature extractors usually only extract the superficial features of HSIs, and do not do well in extracting the deeper information of HSIs. Deep learning methods can effectively extract deep details of HSI and automatically learn deep features from the data set, so these methods have become inevitable to use deep learning to solve the problem of HSI classification. Classical neural network models based on deep learning are Autoencoder (AE) [9], Stacked Autoencoder (SAE) [10], Deep Belief Network (DBN) [11], and so on. Although these methods have relatively good classification performance because the model has more parameters and can only extract shallow image features. To address these problems, a deep neural network model based on CNN was designed for HSI classification, which exhibited better feature extraction and higher classification performance. The CNN architecture consists of two parts, one is the feature extraction network, which learns the high-level features of the input data, and the other is the classifier, which computes the probability that each input sample belongs to a certain category. In the development of CNN applications for HSI classification, Hu et al. [12] proposed deep CNNs with five 1-D layers that received pixel vectors as input data and classified HSI data cubes only in the spectral domain. Zhao et al. [13] proposed a CNN2D architecture for extracting deep spatial features using multi-scale convolutional AEs based on Laplace pyramids, meanwhile, the principal component analysis (PCA) to extract three principal components. Then, logistic regression is used as a classifier that connects the extracted spatial features with spectral information. Chen et al. [14] proposed three convolutional models for creating input blocks of their CNN3D model using full-pixel vectors from the original HSI.

Although deep learning-based methods have achieved greater success in HSI classification, deep learning-based methods still have many drawbacks. The 2-D CNN alone is not able to extract good discrimination feature maps from the spectral dimensions. Similarly, a deep 3-D CNN is more computationally complex and this alone seems to perform worse for classes having similar textures over many spectral bands. To address these issues, this paper proposes a joint unified network of spatial-spectral for HSI classification. The network uses 3-D CNN to extract the spatial-spectral information of HSI, adds 2-D CNN to 3-D CNN to further extract spatial features, enhances the extraction of spectral information by bidirectional long-short term memory (Bi-LSTM), and finally uses softmax for classification to form an end-to-end unified neural network. In this network, 3-D CNN can extract the spatial-spectral features of the HSI, and consider the continuity of the hyperspectral data between neighboring bands, Bi-LSTM can well process the information from the continuous data. Besides, Bi-LSTM and CNN share a loss function that allows both spectral and spatial features to be trained simultaneously, and all parameters in the network can be optimized simultaneously, allowing for good coordination of spatial-spectral information.

The rest of this paper is organized as follows. Section 2 introduces the framework of proposed method. Section 3 presents the HSI data sets and experimental results. Conclusions and other discussions are summarized in Sect. 4.

2 The Framework of Proposed Method

In this paper, a method for combining Bi-LSTM and 3-D CNN is proposed for HSI classification. The method is divided into two main parts, one is the extraction of spectral features by Bi-LSTM on the raw data and the other is the extraction of spatial-spectral features using 3-D CNN after the PCA dimension reduction data. In order to build a unified network so that the parameters can be optimized simultaneously, we concatenate the last layer of FC layers of Bi-LSTM and 3-D CNN to form a new FC layer, which is followed by another FC layer. The framework diagram of the proposed method is shown in Fig. 1.

2.1 Bi-LSTM

Whether it is a CNN or an artificial neural network, they can only process one input individually, and the previous and subsequent inputs are completely unrelated. However, some tasks need to be able to better process sequential information, the preceding input is related to the following input. RNNs can solve these problems, and RNNs can successfully deal with the spectral contextual information of HSIs.

Figure 2 architecture the structure of RNN. Given a sequence of values $x^{(1)}, x^{(2)}, \ldots, x^{(t)}$ as input data, the formula for each cell structure in the RNN network is shown as follows:

$$h^{(t)} = \varnothing(Wh^{(t-1)} + Ux^{(t)} + b_a) \quad (1)$$

$$O^{(t)} = Vh^{(t)} + b_o \quad (2)$$

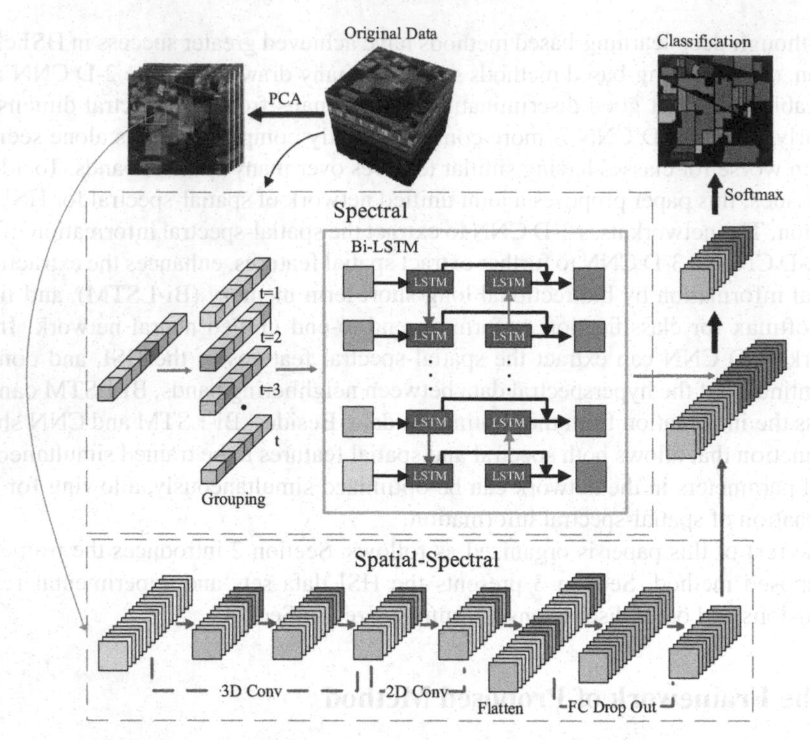

Fig. 1. The framework diagram of the proposed method.

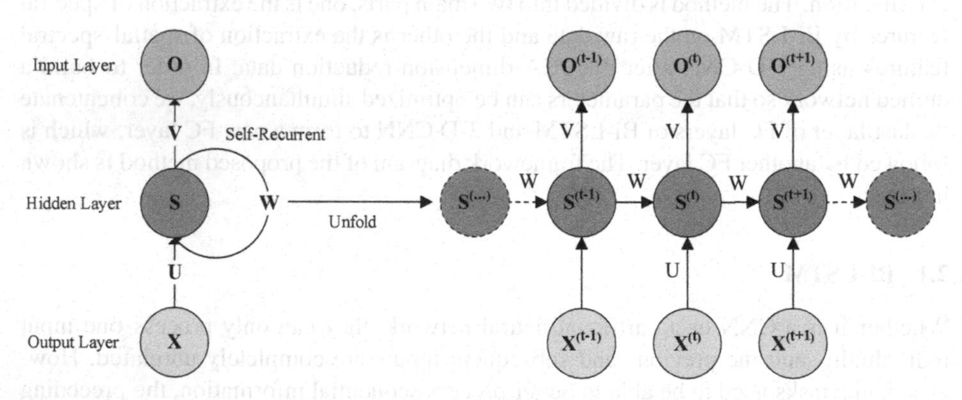

Fig. 2. The architecture of RNN

Where W denotes the weight matrices between the hidden node and itself from the previous time step, U denotes the weight matrices between the input node and the hidden node, V denotes the weight matrices between the hidden node and the output node. b_a and b_o are bias vectors. $x^{(t)}$, $h^{(t)}$, and $O^{(t)}$ are the input value, hidden value, output value of time t, respectively. $\varnothing$ is a nonlinear activation function, which generally uses the tanh function. The initialization value of $h^{(0)}$ in Eq. (1) are set to zero. As can be seen from the Eq. (1), the output is jointly determined by the $h^{(t-1)}$ at time $t-1$ and the input $x^{(t)}$ at time t. As $|W| < 1$ or $|W| > 1$, $h^{(t)}$ will be closer to infinity or zero as time increases. This will cause the gradient vanishing or exploding problem.

LSTM is a special kind of RNN that is designed to address the problem of gradient vanishing or exploding of RNNs during the backpropagation stage. Compared to regular RNNs, LSTM is better able to solve the problem of long dependencies in sequences and reduce the loss of information.

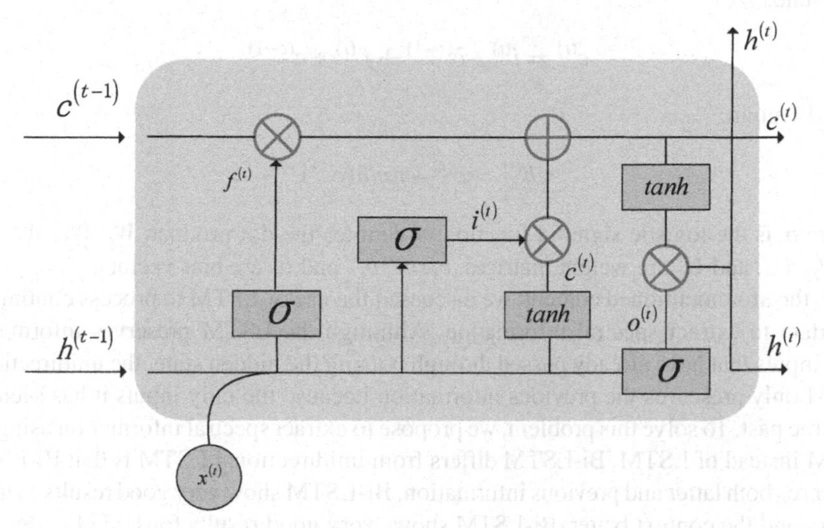

Fig. 3. The architecture of LSTM

The architecture of LSTM is shown in Fig. 3. The memory cell which replaces the hidden unit in traditional RNNs is the key to LSTM, represented by the horizontal line running through the cell. The cell state runs throughout the cell but has few branches, which ensures that information flows unchanged throughout the RNNs. The LSTM network can delete or add information about the cell state through a structure called a gate. The gate is a combination of a sigmoid function and a dot product operation that can selectively decide what information to let through. They are an input gate, a forget gate, an output gate, and a candidate cell value. The input gate can either allow the input information to update the state of the memory cell or block it. The output gate will control whether the cell state will affect other neurons at the next time step. The forget gate will regulate the cell state, making the cell remember or forget its previous state. The candidate cell value stores updated information from the output of the input gate operation. The forward propagation of the LSTM for time t is defined as Eq. (3–8):
Input Gate:

$$i^{(t)} = \sigma(W_i x^{(t)} + U_i h^{(t-1)} + b_i) \tag{3}$$

Forget Gate:

$$f^{(t)} = \sigma(W_f x^{(t)} + U_f h^{(t-1)} + b_f) \tag{4}$$

Output Gate:

$$o^{(t)} = \sigma(W_o x^{(t)} + U_O h^{(t-1)} + b_o) \tag{5}$$

Candidate value:

$$\tilde{c}^{(t)} = \tanh(W_c x^{(t)} + U_c h^{(t-1)} + b_c) \tag{6}$$

Cell State:

$$c^{(t)} = i^{(t)} * \tilde{c}^{(t-1)} + f^{(t)} * c^{(t-1)} \tag{7}$$

LSTM Output:

$$h^{(t)} = o^{(t)} * tanh(c^{(t)}) \tag{8}$$

Where σ is the logistic sigmoid function, $*$ denotes the dot product. W_i, W_f, W_o, W_c, U_i, U_f, U_o, and U_c are weight matrices. b_i, b_f, b_o, and b_c are bias vectors.

In the aforementioned content, we discussed the use of LSTM to process continuous HSI data to extract spectral information. Although the LSTM preserves information from inputs that have already passed through it using the hidden state, the unidirectional LSTM only preserves the previous information because the only inputs it has seen are from the past. To solve this problem, we propose to extract spectral information using Bi-LSTM instead of LSTM. Bi-LSTM differs from unidirectional LSTM is that Bi-LSTM preserves both latter and previous information. Bi-LSTM show very good results as it can understand the context better. Bi-LSTM shows very good results for HSI classification as it can understand the context better.

Let n be the number of bands and t be the number of time steps in the Bi-LSTM. Then, the sequence length of each time step is defined as $m = \mathit{flloor}(n/t)$, where $\mathit{flloor}(n/x)$ denotes rounding down x. For each pixel in the HSI, let $Z = [Z_1, Z_2, \ldots Z_i, \ldots Z_n]$ be the spectral vector, where Z_i is the reflectance of the i-th band. The grouping strategy is shown as:

$$\begin{aligned} x^{(1)} &= \left[Z_1, Z_{1+t}, \ldots Z_{1+t(m-1)}\right] \\ x^{(2)} &= \left[Z_2, Z_{2+t}, \ldots Z_{2+t(m-1)}\right] \\ &\cdots \\ x^{(i)} &= \left[Z_i, Z_{i+t}, \ldots Z_{i+t(m-1)}\right] \\ &\cdots \\ x^{(t)} &= [Z_t, Z_{2t}, \ldots Z_{tm}] \end{aligned} \tag{9}$$

Where $x^{(i)}$ is the sequence at time i. Each group covers most of the spectral range. The spectral distance between different time steps will be relatively shorter under such circumstances.

2.2 3-D CNN

CNNs include one-dimensional convolutional neural networks (1-D CNN), two-dimensional convolutional neural networks (2-D CNN), and three-dimensional convolutional neural networks (3-D CNN). 1-D CNN is mainly used for sequence data processing, 2-D CNN is often used for image recognition, and 3-D CNN is mainly used for medical image and video data recognition. CNN consists of three structures: convolution layer, activation function, and pooling layer. The purpose of the convolutional layer is to extract the different features of the input, with more multi-layered convolutional layers being able to iteratively extract more complex features from lower-level features. The activation function is designed to increase the nonlinearity of the neural network model. The pooling layer preserves the main features while reducing the number of parameters and calculations, preventing overfitting and improving model generalization. The weight-sharing network structure of CNNs makes them more similar to biological neural networks, reducing the complexity of the network model, and reducing the number of weights. The schematic diagrams of the 2-D convolution and 3-D convolution are shown in Fig. 4.

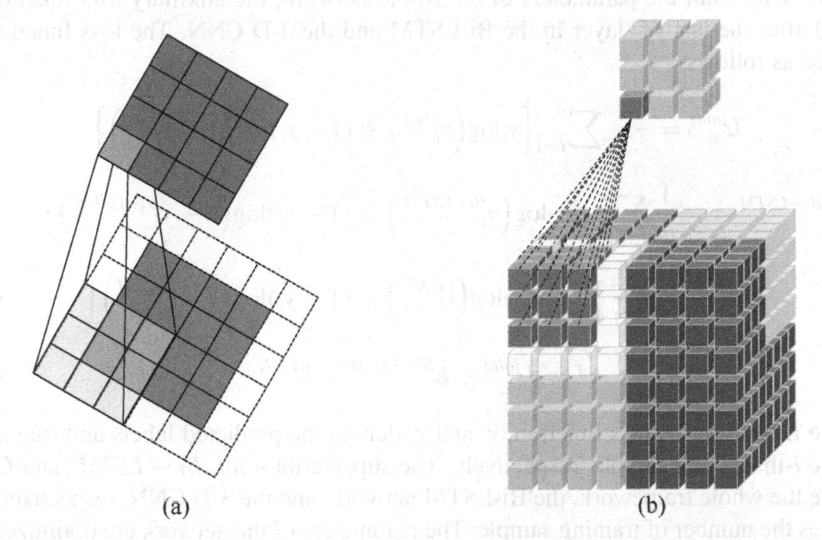

Fig. 4. The schematic diagrams of 2-D convolution and 3-D convolution (a) 2-D convolution; (b) 3-D convolution.

Let the HSI be denoted by $X \in R^{W \times H \times B}$, where X is the original data, W is the width, H is the height, and B is the number of spectral bands. Each HSI pixel in X contains a one-hot label vector $Y = (y^1, y^2, y^3, \ldots, y^C) \in R^{1 \times 1 \times C}$ and a value in each of the B spectral bands, where C represents the land-cover categories. To remove the spectral correlation and reduce the computational costs during the convolution operation, the number of spectral bands is reduced from B to P by the PCA, while keeping the W and H of the spatial dimension constant. We not only reduce the spectral band, but also preserve the spatial information that is important for feature extraction. We create the

3-D patches centered on each pixel, the neighbor region is extracted with the size of $w \times w \times P$, where $w \times w$ is the size of the window, P is the number of first principle components that have been reserved by the PCA. The truth labels of the patches are decided by the label of the central pixel. Assume that we have M samples, the data set can then be denoted by $XP \in R^{M \times w \times w \times P}$.

To achieve the spatial-spectral feature maps in the 3-D CNN, we repeated 3-D convolution three times. Consider the important role of 2-D convolution in extracting spatial information, we apply 2-D convolution to increase the number of spatial feature maps before the flatten layer. To improve the generalization of the model and prevent overfitting, the dropout is applied once after each fully connected (FC) layer.

2.3 Loss Function

In this framework, we use the softmax function after the last FC layer to get the probability of which class the pixel is. Besides, to increase the nonlinearity and speed up the convergence of the network, we use the rectified linear units (Relu) function after each layer.

To better train the parameters of the whole network, the auxiliary loss function is added after the last FC layer in the Bi-LSTM and the 3-D CNN. The loss function is defined as follows:

$$L^{joint} = -\frac{1}{m}\sum\nolimits_{i=1}^{m}\left[y_i \log\left(\hat{y}_i^{joint}\right) + (1-y_i)\log\left(1-\hat{y}_i^{joint}\right)\right] \tag{10}$$

$$L^{Bi-LSTM} = -\frac{1}{m}\sum\nolimits_{i=1}^{m}\left[y_i \log\left(\hat{y}_i^{Bi-LSTM}\right) + (1-y_i)\log\left(1-\hat{y}_i^{Bi-LSTM}\right)\right] \tag{11}$$

$$L^{CNN} = -\frac{1}{m}\sum\nolimits_{i=1}^{m}\left[y_i \log\left(\hat{y}_i^{CNN}\right) + (1-y_i)\log\left(1-\hat{y}_i^{CNN}\right)\right] \tag{12}$$

$$L = L^{joint} + L^{Bi-LSTM} + L^{CNN} \tag{13}$$

Where L denotes the loss function, $\hat{y}_i$ and y_i denote the predicted labels and true label for the i-th training sample, respectively. The superscript $joint$, $Bi - LSTM$, and CNN denote the whole framework, the Bi-LSTM network, and the 3-D CNN, respectively. m denotes the number of training sample. The parameters of the network are optimized by the mini-batch stochastic gradient descent algorithm.

3 Experiment and Analysis

In this section, we introduced a public benchmark HSI data set and evaluated the performance of the proposed method using three evaluation metrics, such as overall accuracy (OA), average accuracy (AA), and kappa coefficient (Kappa).

3.1 Data Set Description and Training Details

The Indian Pines data set, gathered by Airborne Visible/Infrared Imaging Spectrometer (AVIRIS) in 1992 from Northwest Indiana, an agricultural area characterized by its crops of regular geometry and also irregular forest regions. The data set is composed of 145 × 145 pixels with a spatial resolution of 20 m by pixel and 224 spectral reflectance bands, out of which 24 spectral bands covering the region of water absorption have been discarded, the remaining 200 bands range in the wavelength range of 0.4–2.5 μm. The available samples are divided into sixteen classes and about half of the data (10249 pixels from a total of 21025) contains labeled samples. The false-color composite image and the ground-truth map are shown in Fig. 5. We randomly select 10% pixels from each class as the training set, and use the remaining pixels as the testing set. The detailed numbers of training and testing samples are listed from Table 1.

(a)

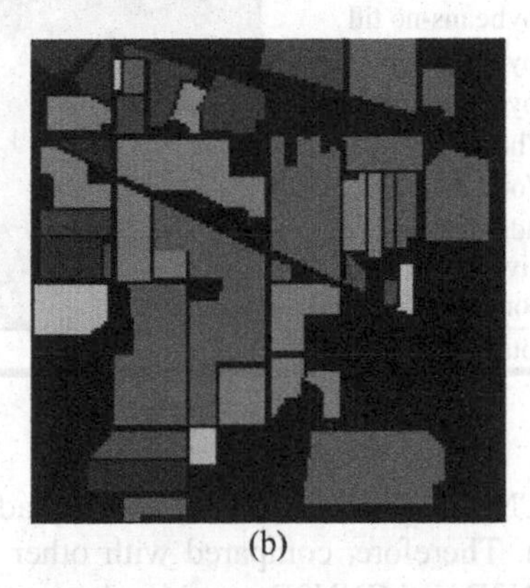
(b)

Fig. 5. The Indian Pines data set: (a).False-color image; (b).Ground-truth map.

All the experiments are run on a computer with the NVIDIA GTX 1070 graphical processing unit (GPU) and an Inter i7- 6700 3.40-GHz processor with 32 GB of RAM. All the experiments in this paper are randomly repeated 10 times with random training and testing data. We have chosen the optimal learning rate of 0.0001 based on the classification outcomes. In order to optimize the network, we adopt the mini-batch stochastic gradient descent algorithm, which batch size of 128. In the 3-D CNN, the input window size of the 3-D patches is 25 × 25 × 30.

3.2 Classification Results

In this paper, compare our proposed method with state-of-the-art methods, such as CNN1D, CNN2D, CNN3D [15], HybridSN [16], and SSUN [17]. Table 2 shows the results acquired by six methods on the Indian Pines data set. From these results, we can observe that our method is significantly better than other methods. The input data

Table 1. The details of Indian pines data set

Class name	Corresponding color	Number of training samples	Number of testing samples	Total number of samples
Alfalfa		5	41	46
Corn-no till		143	1285	1428
Corn-min till		83	747	830
Corn		24	213	237
Grass/pasture		48	435	483
Grass/tree		73	657	730
Grass/pasture-mowed		3	25	28
Hay-windrowed		48	430	478
Oats		2	18	20
Soybeans-no till		97	875	972
Soybeans-nin till		245	2210	2455
Soybeans-clean till		59	534	593
Wheat		20	185	205
Woods		126	1139	1265
Bldg-grass-tree-drives		39	347	386
Stone-steel towers		9	84	93
Total		1024	9225	10249

of CNN1D is a 1-D vector, which leads to the loss of spatial information of the input data. Therefore, compared with other methods, the classification effect is the worst. CNN2D and CNN3D consider the spatial and spatial-spectral information respectively, in which the classification effect is significantly improved compared to CNN1D. Therefore, spatial and spatial-spectral information is very important for HSI classification. In our method, the lack of 3-D CNN processing spectral information is made up, and the experimental results after adding Bi-LSTM are significantly better than other methods. The classification maps of different methods are shown in Fig. 6.

Table 2. Classification results for the Indian Pines data set using 10% of the available labeled data.

Class name	CNN1D	CNN2D	CNN3D	HbridSN	SSUN	Proposed method
Alfalfa	32.44	69.51	80.73	85.12	99.51	97.80
Corn-no till	69.84	91.18	97.68	94.59	96.93	95.06
Corn-min till	62.82	90.98	98.70	97.68	98.34	99.59
Corn	43.05	87.18	94.08	92.11	97.51	97.09
Grass/pasture	88.48	89.98	96.25	97.45	96.99	99.08
Grass/tree	96.59	97.41	99.53	98.60	98.87	99.30
Grass/pasture-mowed	40.40	87.20	88.00	97.60	97.20	100.00
Hay-windrowed	98.28	98.84	99.98	99.58	99.88	100.00
Oats	39.44	57.22	85.56	92.22	78.33	99.44
Soybeans-no till	67.34	93.34	97.12	98.11	95.94	98.99
Soybeans-nin till	82.92	95.36	98.90	98.77	98.48	99.60
Soybeans-clean till	73.15	88.31	94.40	93.95	97.08	98.56
Wheat	97.95	97.24	99.78	95.35	98.43	99.78
Woods	94.04	98.64	99.72	99.62	98.90	99.96
Bldg-grass-tree-drives	60.98	91.59	95.76	96.43	97.58	99.83
Stone-steel towers	83.45	90.12	96.55	88.45	93.81	96.79
OA (%)	78.89	93.73	97.99	97.34	97.89	98.78
AA (%)	71.58	89.01	95.17	95.35	96.53	98.80
Kappa × 1 00	74.88	92.85	97.70	96.96	97.60	98.60
Testing time(s)	0.36	4.91	3.31	4.54	1.19	5.34

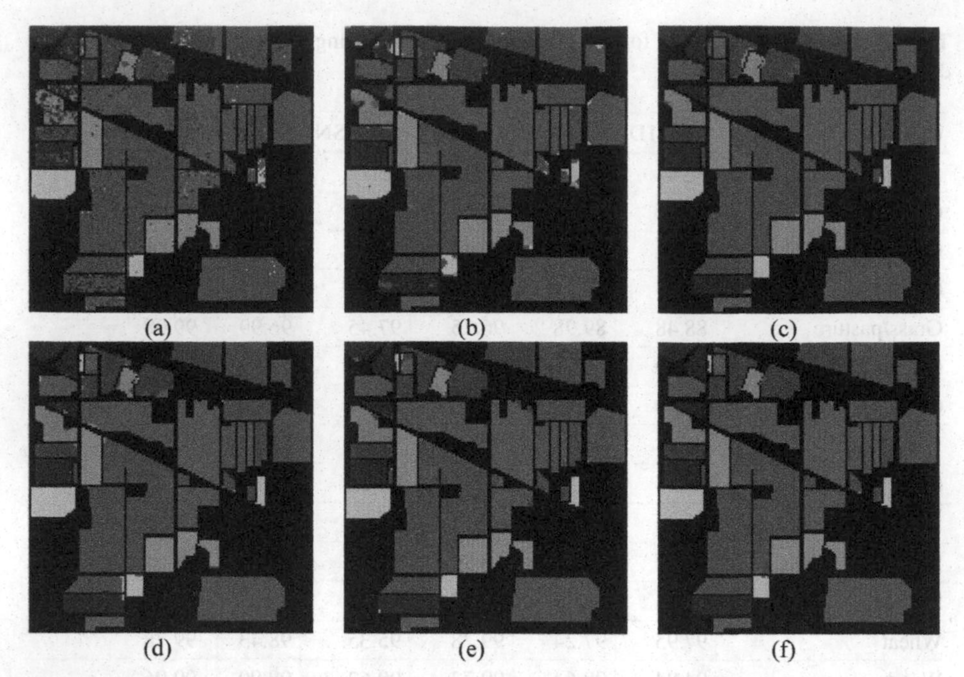

Fig. 6. Classification maps for the Indian Pines data set. (a) CNN1D. (b) CNN2D. (c) CNN3D. (d) HybridSN. (e) SSUN. (f) Proposed method.

4 Conclusion

The paper proposed a unified network framework for extracting the spatial-spectral features. In this network, Bi-LSTM can extract high-quality spectral features considering the complete spectral contextual information, which compensates for the shortcomings of 3-D CNN in extracting spectral features. The experiment results show the proposed method has significantly improved the effectiveness of HSI classification. While the classification effect improved, the computation time increased only slightly. In the future work, we will either replace the LSTM with the Gate Recurrent Unit to improve the speed of the network or use the optimized 3-D CNN to further improve the HSI classification results.

References

1. Chang, C.I.: Hyperspectral Imaging: techniques for spectral detection and classification. In the Library of Congress, vol. 1. Springer Science & Business Media, pp.1–63 (2003)
2. Hosseini, A., Ghassemian, H.: Classification of hyperspectral and multi-spectral images by using fractal dimension of spectral response curve, In 20th Iranian Conference on Electrical Engineering, pp. 1452–1457 (2012)
3. Roy, S.K., Manna, S., Dubey, S.R., Chaudhuri, B.: LiSHT: non-parametric linearly scaled hyperbolic tangent activation function for neural networks (2019), arXiv:1901.05894. https://arxiv.org/abs/1901.05894

4. Ren, S., He, K., Girshick, R., Sun, J.: Faster R-CNN: towards real-time object detection with region proposal networks. Institute of Electrical and Electronics Engineers transactions on pattern analysis and machine intelligence, pp. 91–99 (2016)
5. Moiane, A., Machado, A.M.L.: Class-based affinity propagation for hyperspectral image dimensionality reduction and improvement of maximum likelihood classification accuracy. Boletim de Ciências Geodésicas, vol. 25, no. 1 (2019)
6. Chatterjee, A., Yuen, P.W.T.: Endmember learning with K-Means through SCD model in hyperspectral scene reconstructions. J. Imaging **5**(11) (2019)
7. Zhang, L., Du, B.: Deep learning for remote sensing data: a technical tutorial on the state of the art. Inst. Electr. Electron. Eng. Geosci. Remote Sens. Magaz. **4**(2), 22–40 (2019)
8. Wang, K., Cheng, L., Yong, B.: Spectral-similarity-based kernel of SVM for hyperspectral image classification. Remote Sensing **12**(13), 2154 (2020)
9. Chen, X., Kingma, D.P., Salimans, T.: Variational lossy autoencoder. arXiv preprint arXiv:1611.02731 (2016)
10. Suk, H.-I., Lee, S.-W., Shen, D.: Latent feature representation with stacked auto-encoder for AD/MCI diagnosis. Brain Struct. Funct. **220**(2), 841–859 (2013). https://doi.org/10.1007/s00429-013-0687-3
11. Chen, Y., Zhao, X., Jia, X.: Spectral–spatial classification of hyperspectral data based on deep belief network. Inst. Electr. Electron. Eng. J. Select. Top. Appl. Earth Observ. Remote Sens. **8**(6), 2381–2392 (2015)
12. Hu, W., Huang, Y., Wei, L., Zhang, F., Li, H.: Deep convolutional neural networks for hyperspectral image classification. J. Sens. (2015)
13. Zhao, W., Guo, Z., Yue, J.: On combining multiscale deep learning features for the classification of hyperspectral remote sensing imagery. Int. J. Remote Sens. **36**(13), 3368–3379 (2015)
14. Chen, Y., Jiang, H., Li, C., Jia, X.P., Ghamisi, P.: Deep feature extraction and classification of hyperspectral images based on convolutional neural networks. Inst. Electr. Electron. Eng. Trans. Geosci. Remote Sens. **54**(10), 6232–6251 (2016)
15. Paoletti, M.E., Haut, J.M., Plaza, J.: Deep learning classifiers for hyperspectral imaging: a review. ISPRS J. Photogramm. Remote. Sens. **158**, 279–317 (2019)
16. Roy, S.K., Krishna, G., Dubey, S.R.: HybridSN: Exploring 3-D–2-D CNN feature hierarchy for hyperspectral image classification. Inst. Electr. Electron. Eng. Geosci. Remote Sens. Lett. **17**(2), 277–281 (2019)
17. Xu, Y., Zhang, L., Du, B.: Spectral–spatial unified networks for hyperspectral image classification. IEEE Trans. Geosci. Remote Sens. **56**(10), 5893–5909 (2018)

Automated Methods for Symptom Normalization in Traditional Chinese Medicine Records

Yuxiao Zhan[1,2], Dezheng Zhang[1,2], Qi Jia[1,2], Haifeng Xu[1,2], and Yonghong Xie[1,2](✉)

[1] School of Computer and Communication Engineering, University of Science and Technology Beijing, Beijing 100083, China
xieyh@ustb.edu.cn

[2] Beijing Key Laboratory of Knowledge Engineering for Materials Science, Beijing 100083, China

Abstract. The normative symptoms in traditional Chinese medicine (TCM) records is the basis for syndrome differentiation and accurate diagnosis. To further improve the accuracy of TCM diagnosis, this paper proposes a series of automated methods to normalize the four types of symptoms in TCM records: tongue, tongue coating, pulse and current symptoms (刻下症). We define a normative symptom system of traditional Chinese and Western medicine recognized by TCM experts, then a retrieval architecture for normative symptom words, including tongue, tongue coating, and pulse was constructed by the suffix tree and dictionaries. And the backtracking algorithm is applied to map the nonstandard symptom words of the three types of symptoms to this architecture to achieve the normalization of the tongue, tongue coating and pulse in TCM records. For the current symptoms, this research designs a classification model referred to an automated machine learning (AutoML) architecture, Autostacker. The model could classify the colloquial symptom words and predict the first three normative symptom words. Then we separately calculate the semantic similarity by the trained Directional Skip-Gram (DSG) model to achieve the normalization of the colloquial symptom words in the current symptoms. This series of automated methods was applied to 480,000 medical records. Results indicate that the accuracy of normalization of tongue, tongue coating, and pulse is about 96%, and 90% of colloquial words in the current symptoms successfully converted into normative symptoms. The experimental results demonstrate the effectiveness of the normalization of symptoms in TCM records.

Keywords: Traditional Chinese medicine · Symptom normalization · Suffix tree · Autostacker · Directional Skip-Gram

1 Introduction

The traditional Chinese medicine (TCM) record is the clinical symptom text recorded in the whole process of diagnosis and treatment by TCM doctors. Symptoms in TCM records are the main basis for doctors to diagnose the illness condition. Therefore, the normalization of symptoms is significant for the medical informationization. Since most

X. Sun et al. (Eds.): ICAIS 2021, CCIS 1422, pp. 476–487, 2021.
https://doi.org/10.1007/978-3-030-78615-1_42

of the symptoms in TCM records are derived from TCM literature or the experience of famous TCM doctors, there are phenomena such as irregular description of symptoms, complex names of symptoms, and unclear meanings of symptoms, etc. To achieve syndrome differentiation and accurate diagnosis, and to promote the conversion of clinical data to scientific research data, TCM symptoms must be normalized. The normalized process of symptoms in TCM records is to convert thc subjective and colloquial symptom words written by doctors into normalized symptom terms, which are recognized by experts and based on certain papers. The normalization of TCM symptoms is not only conducive to improving the accuracy of diagnosis, but also a significant link in TCM clinical research.

This paper mainly normalizes the tongue, tongue coating, pulse, and current symptoms (刻下症) in TCM records. The representation structures and problems of the first three types of symptoms are relatively similar, which are generally the number of synonyms, the improper way of expression of words and the guidance of the original data without a normalized system. For example, "string-like pulse" (弦脉) is subjectively recorded by doctors as "string" (弦); "Red tongue" (舌红) and "light red tongue" (舌淡红) are a pair of synonyms. The two words are similar from the perspective of computer semantics, but they actually correspond to different human states of "red tongue" (舌质红绛) and "light red tongue" (舌质淡红) in TCM. Compared with other symptom categories, words of current symptoms are numerous and abundant, which belong to semi-structured data. There are mainly problems of mixing symptomatic phrases and non-symptomatic phrases as well as mixing symptomatic phrases in traditional Chinese and Western medicine. For example, "The upper part of the esophagus has a burning sensation", which belongs to the category of esophagitis in the modern Chinese and Western medicine system, but there is no such description in the normative system of TCM.

In order to address the above problems of irregular symptoms, this paper presents a series of automated methods to achieve the normalization of tongue, tongue coating, pulse and current symptoms. First, we define a normative symptom system of traditional Chinese and Western medicine recognized by TCM experts to determine the normative guidance for the subsequent normalization of the four types of symptoms. Then we focus on constructing a retrieval architecture based on the suffix tree [1] and dictionaries for normalized symptom words of tongue, tongue coating and pulse. And we discuss how to use the backtracking algorithm to map thc irregular and ambiguous symptom words in the TCM records to this retrieval architecture, so as to realize the normalization of tongue, tongue coating and pulse. This paper designs a classification model referred to Autostacker [2], an automated machine learning (AutoML) architecture when normalizing irregular current symptoms. Finally, we predict which type of normative symptom words the colloquial symptom word in the information belong to, and then calculate the semantic similarity between this word and the predicted normative symptom words.

The purpose of proposing AutoML methods in classification tasks is to combine a variety of classification models and more complex model fusion architectures to quickly learn the mapping relationship between symptom words and normalized symptom words in the normative symptom system in various aspect. It could more accurately locate the colloquial symptom words under the normative symptom words. In addition, this research involves the task of calculating semantic similarity in natural language processing because of the symptom words in the language of TCM usually have richer semantic information. The semantic relationship and semantic information within Chinese characters cannot be accurately distinguished if traditional calculation approaches of semantic similarity are considered. The semantic similarity of each symptom word in 480,000 medical records and 25,339 normative symptom words were calculated separately, which is a huge amount of calculation and high time complexity. Considering that this word is in the context of the rich and complex semantic relations of TCM symptom terms, we train the Directional Skip-Gram (DSG) model [3] with a large amount of normalized symptom text data of TCM in the principle of reducing complexity. We measure the semantic similarity between the colloquial symptom word and the top three normative symptom words with the highest classification probability from the prediction results of this model. Then, we will transform this colloquial symptom word into the most normalized symptom word. We applied the series of automated methods to 480,000 medical records and the experimental results prove that these methods could obtain the normative tongue, tongue coating, pulse and current symptoms terms recognized by TCM.

2 Related Work

Various studies have recently proposed many normalized methods for TCM symptoms. Liang et al. [4] discussed the normalization symptoms of TCM based on the maximum probability method by comparing two extraction schemes of symptom. Bai et al. [5] applied the Delphi method to achieve the normalization of the diagnosis items for the internal damage and cough of the Miao medicine. A multivariate statistical method which combines cluster analysis and logistic regression analysis to screen out the main and minor symptoms of the basic TCM syndromes of liver cirrhosis was proposed [6]. The method provides a new approach for TCM syndrome differentiation. There are also some studies that specifically focus on normalized research on the terminology of certain diseases. For example, Ni et al. [7] used bibliometrics to normalize the symptom terminology of the spleen and stomach of TCM, established common symptom terms in this field and developed a TCM terminology database of spleen and stomach. Zhang. et al. [8] applied four methods such as core symptom extraction and logical relationship retention to normalize epilepsy symptoms, and provided reference technical terms for the mining of epilepsy diagnosis and treatment data.

However, there is no similar method in the normalization of these four types of symptoms: tongue, tongue coating, pulse, and current symptoms, but they are beneficial to TCM syndrome differentiation and diagnosis. Compared with previous works, our research focuses more on the use of computer and AutoML methods to address the four types of nonstandard symptoms. This work could provide basic support for TCM big data, which is of great significance to the TCM auxiliary diagnosis and treatment system. We discussed the semantic features contained in the symptoms of tongue, tongue coating and pulse, and then referred to the suffix tree [1] to design a retrieval architecture containing normalized symptoms of these three types of normative symptom words. Suffix tree is a tree structure that describes all suffixes of a given string and it is widely applied for efficient string matching and query. The suffix tree has been frequently proven to have an important role in processing text information tasks [9–11].

The proposal and application of AutoML are in a stage of rapid development in recent years, and it has identified obvious effects in several mature tools [12–16]. Chen [17] et al. reviewed the current developments of AutoML in terms of three categories and introduced both main-stream AutoML techniques and frameworks, the survey could give insight into the progress of AutoML. We refer to this survey and evaluate the TPOT [18] algorithm and the DarwinML [19] algorithm, but there are respectively insufficient flexibility and complex algorithm design. Therefore, when normalizing current symptoms, we refer to Autostacker [2], an AutoML modeling architecture using Evolutionary Algorithm (EA) to quickly evolves candidate pipelines with high predictive accuracy, to design a classification model to make symptom words in current symptoms of TCM records more normative. The architecture of Autostacker combines multiple stacked layers and nodes in each layer, which represents a primitive model in the machine learning sklearn [20] library.

In order to complete the semantic similarity task of symptoms in TCM records, we review the famous literature and methods. Continuous Bag-Of-Words (CBOW) and Skip-Gram [21] are very effective in English text. However, there is a big difference between Chinese and English semantic similarity tasks, especially for the TCM records which is composed of multiple words. There is abundant internal information in a word and every character could play a great role. However, CBOW and Skip-Gram cannot extract information of Chinese internal characters by modeling directly from the word level. Character-enhanced Word Embedding (CWE) [22], an improved word combination model based on CBOW, adds Chinese character information as a context to the training process. CWE believes that two symptom words with similar semantics are actually not related. For example, "dry eyes" (眼干燥) and "dry mouth" (口干燥). Directional Skip-Gram (DSG) [3] is a simple but effective model improved from Skip-Gram, which could clearly distinguish context in word prediction. Therefore, we train the DSG by large quantities of data that meet the normalization symptom. By adding the direction vector to each of the words, DSG could learn the vector representation of words from the specific direction of the context. Comparing the complexity and actual effect between DSG and the original Skip-Gram models indicates that DSG is equal to other Skip-Gram models in terms of training speed, but outperforms others in semantic similarity calculation and part-of-speech tagging tasks.

3 Materials and Methods

3.1 Normative Symptom System

The definitions of 21449 normalized TCM symptom words refer to the symptom normative defined by Zhang et al. [23]. We further expanded 3890 common symptom words based on the study of Yu et al. [24] and incorporated them into the normative symptom system of TCM. The naive TCM normative symptom system cannot solve the mixed condition of traditional Chinese and Western medicine symptoms. Therefore, a normative symptom system of western medicine was established referring the body parts commonly used and the description of symptoms in western medicine books. There are 25339 normative symptom words of TCM and 7704 of Western medicine in the normative symptom system of traditional Chinese and Western medicine (Table 1).

Table 1. Examples of normative symptom system of traditional Chinese and Western medicine.

TCM	Western medicine
Irritable (急躁易怒)	Double red eyes (双眼红)
Lack of vitality (少神)	Red eyelid (眼睑红)
Tongue body numb (舌体麻木)	Nystagmus (眼球震颤)
Red crimson tongue (舌质红绛)	Submandibular lymph nodes (双颌下淋巴结)

3.2 The Retrieval Architecture

Tongue, tongue coating and pulse in the normative symptom system of TCM could be summarized as the form: a body location with symptom description. This research reworks the suffix tree [1] and adds each symptom word to the suffix tree based on this feature. And we use an additional dictionary to indicate which normative symptom word it belongs. A retrieval architecture constructed by a suffix tree and dictionaries is designed to realize the normalization of tongue, tongue coating and pulse. (This paper takes tongue as an example to illustrate the normalization methods since the attributes of tongue, tongue coating and pulse have similar characteristics.)

The root of this suffix tree which does not contain any Chinese characters is empty (see Fig. 1). First, the location of tongue body in the tongue symptom words was removed. Then, we sort the words of the tongue symptom description by number of characters in ascending order and construct the suffix tree from back to front according to the number of words. Finally, the normative symptom word corresponding to the symptom description word are put into the suffix tree as a form of dictionaries. For example, "Partial Red" (a description of symptom), "Red" has appeared in the first level of the suffix tree, then put the word "Partial" directly into the next layer and add the corresponding normative symptom word "Red Crimson Tongue" (舌质红绛) in the suffix tree as dictionaries.

3.3 Matching Query Process

By using the depth-first traversal and backtracking algorithm on the above-mentioned architecture can complete that nonstandard and ambiguous tongue, tongue coating and pulse in TCM medical records are mapped to the normative symptom system on the condition of lower time complexity. This paper introduces a relatively complex example: "Tongue Dark Red Dim" (舌深红黯) to show the flow of the algorithm design. In order to explain the robustness of the algorithm more convincingly, this example sets an assumption that "Dark Red" (深红) does not exist in the normative symptom words.

The matching query steps for "Tongue Dark Red Dim" (舌深红黯) are as follows (see Fig. 2):

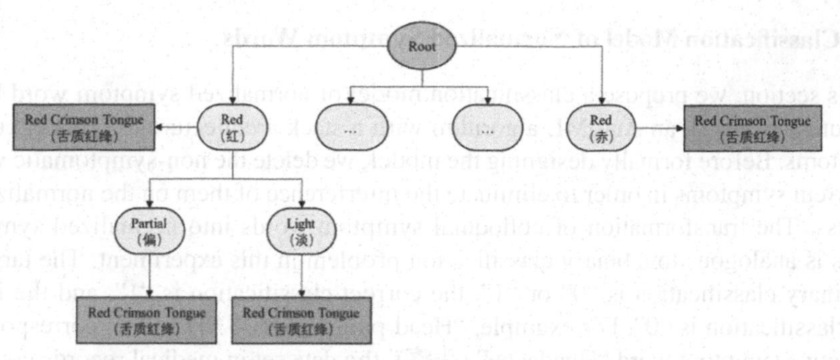

Fig. 1. The retrieval architecture contains normative symptom words of the "Red Crimson Tongue" (舌质红绛).

- First, deleting it if there is a description of the part and invert the word. The first character of the depth-first query algorithm is "Dim" (黯), it will continue to query "Red" if it is successful;
- At the second query, "Red" could not be found then returned to the previous state. The program obtains the normative symptom words corresponding to "Dim" (黯) and deletes the word "Dim" (黯) from the requested words;
- When querying the third time, only "Dark Red" (深红) is left in the word. After query "Red" successfully, continuing to query "Dark" (深) according to the "Red";
- At the fourth query, since there is no "Dark" (深) under "Red", take the normative symptom word corresponding to "Red", and delete the word "Red" from the requested words;
- The program returns directly and deletes the word "Dark" (深) when the fourth query "Dark" (深) is unsuccessful.

Finally, the query program ends with the requested string empty, and the corresponding two words in normative symptom system of traditional Chinese and Western medicine are "Red Crimson Tongue" (舌质红绛) and "Dim Purple Tongue" (舌质紫黯).

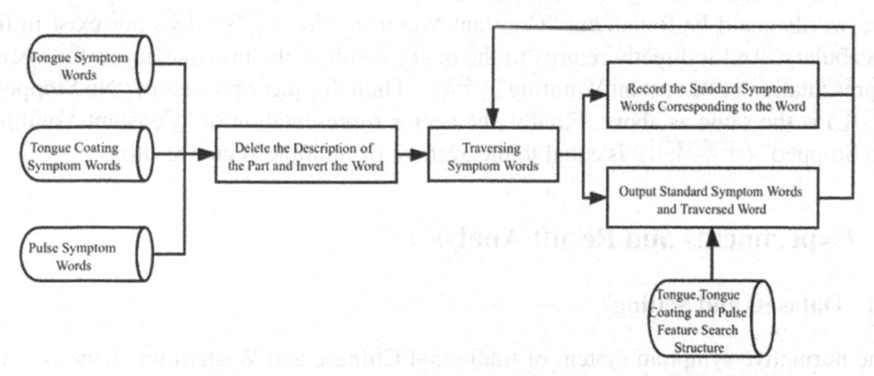

Fig. 2. Normalization flow chart of tongue, tongue coating and pulse.

3.4 Classification Model of Normalized Symptom Words

In this section, we propose a classification model of normalized symptom word based on Autostacker [2], an AutoML algorithm with a stack architecture to process current symptoms. Before formally designing the model, we delete the non-symptomatic words in current symptoms in order to eliminate the interference of them on the normalization process. The transformation of colloquial symptom words into normalized symptom words is analogous to a binary classification problem in this experiment. The target in the binary classification is "0" or "1", the correct classification is "1", and the incorrect classification is "0". For example, "Head pain" (头部疼痛) and its correspondent normative symptom word "Headache" (头疼), the data set in medical records could be represented as "Backache-Headache-0" (后背疼痛-头疼-0) and "Head pain-Headache-1" (头部疼痛-头疼-1). Based on this classification definition, this paper sets the operator of model in Autostacker to the currently more commonly effective classification algorithms, namely RandomForest, GBDT, SVM, LR etc., then trains a classification model of normative symptom words that includes the normative system of TCM symptoms.

3.5 Semantic Similarity Calculation with DSG

The objective words are defined as the first three words with the highest probability predicted by the above-mentioned classification model corresponding to the colloquial symptom words. The semantic similarity between the three objective words and the colloquial symptom words is calculated separately based on the classification results. We train the DSG model [3] through a large number of text data sets of normalized symptom words. However, the semantics of TCM symptoms are usually more complicated. In order to avoid the phenomenon of Out Of Vocabulary (OOV) due to the semantic complexity of symptom phrases, this paper applies the greedy matching algorithm to treat a symptom word as composed of multiple "factors". Then this problem is abstracted as cyclically finding the largest continuous subsequence in words. This approach of symptom phrases disassembly to obtain the more accurate vector representation is currently the most suitable. The process of above approach can be explained by one example, "Constant Vomiting No Stopped" (干呕不止) (symptom). This method orderly queries "Constant" (干), "Constant Vomiting" (干呕) and "Constant Vomiting No" (干呕不), if the first two words could be found, but "Constant Vomiting No" (干呕) does not exist in the vocabulary. And it directly returns to the query result of the previous layer, the vector representation of "Constant Vomiting" (干呕). Then the query process of "No Stopped" (不止) is the same as above. Finally, the vector representation of "Constant Vomiting No Stopped" (干呕不止) is equal to the average of multiple vector results.

4 Experiments and Result Analysis

4.1 Datasets and Settings

The normative symptom system of traditional Chinese and Western medicine is composed of 25,339 normative symptom words of TCM and 7704 normative symptom words of western medicine. The positive and negative samples of our training set are composed of 25339 normative symptom words and the same number of nonstandard words.

This paper uses the commonly used ensemble learning classification algorithms Catboost [25], Xgboost [26] and GBDT [27] to test the effect of the model mentioned in Sect. 3.4, whether the model might reach the level of the commonly used classification algorithm.

The hyperparameter settings of the classification algorithm in the experiment are shown in Table 2.

Table 2. Hyperparameter settings of classification algorithm.

Algorithm	Learning_rate	Max_depth	Subsample	N_estimators
Catboost	0.08	5	0.7	1000
Xgboost	0.1	5	0.7	1000
GBDT	0.05	7	0.8	1200

4.2 Comparison of Classification Algorithm

Table 3 shows the comparison of training time and accuracy between Autostacker and other classification algorithms. We introduced the classifier indicator Receiver Operating Characteristic (ROC), Area Under Curve (AUC) and four parameters (TP, FN, FP, TN) to determine accuracy.

$$Accuracy = \frac{(TP + TN)}{(TP + FN + FP + TN)} \quad (1)$$

Table 3. Comparison of accuracy and time of classification algorithms.

Algorithm	Accuracy	Training time
Catboost	0.939	101.7 s
Xgboost	0.852	50.2 s
GBDT	0.875	63.8 s
H2O AutoML	0.960	3.5 h
Autostacker	0.958	2 h

Since Autostacker needs a period of time to query the optimal machine learning model pipeline, the training time has a disadvantage compared with the above classification algorithms. The accuracy rate is much higher than the commonly used classification algorithms except that it is 2 thousand points lower than H2O AutoML model. However, the training time of Autostacker is shorter than H2O AutoML model, which generally shows that the effect of Autostacker is generally better.

4.3 Results

The results of the mapping of tongue, tongue coating and pulse from 480,000 medical records to the normative symptom system were sampled, and the accuracy rate of symptom normalization verified by TCM experts is about 96%. There are 26,038 colloquial symptoms of current symptoms among the 480,000 medical records need to be normalized. The results of the normalization are shown in Fig. 3.

After removing 1,060 non-symptomatic words (4%), 18533 colloquial symptom words could be transformed into the normative system of TCM symptoms, accounting for about 71%. There are 4,815 colloquial symptom words corresponding to the normative system of Western medical symptoms, which accounts for about 19%. Therefore, 90% of the colloquial symptom words are successfully converted into words in the normative symptom system. In the normalized results of current symptoms, 6% of the words were not successfully transformed into normalized symptom words, most of which were non-symptom description words and only a few were missing words.

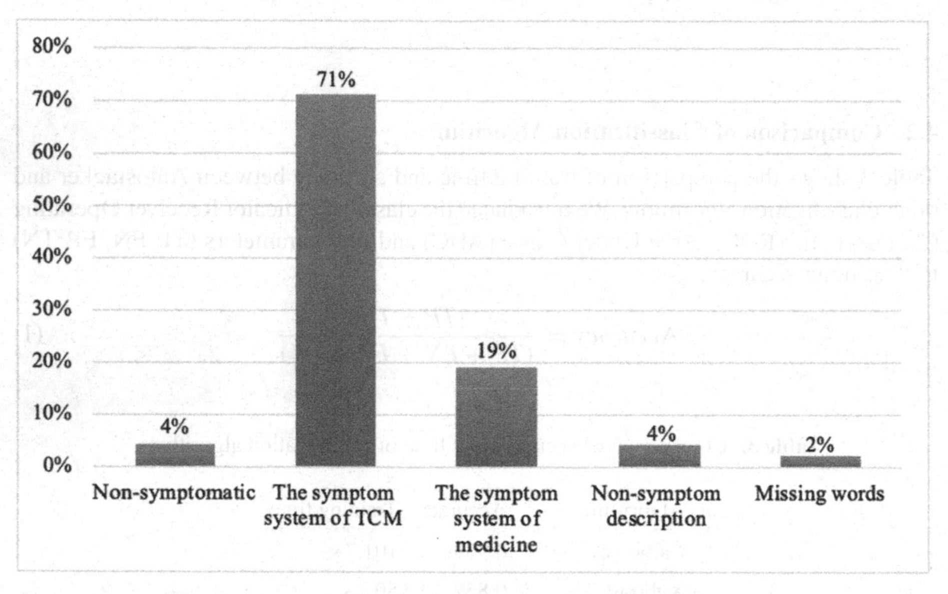

Fig. 3. The result of normalization of words in medical records.

4.4 Analysis of Incorrect Results

The incorrect conditions and proportions of tongue, tongue coating and pulse are shown in Fig. 4. Incomplete normalization means that there should be two or more normalized symptom words but only one predicted by the model, accounting for 64.0%. The percentage of missing normalization that the normalization result should be predicted but not is 29.6%. 6.4% of cases cannot find symptom words in the normative symptom system of traditional Chinese and Western medicine. We analyze the reasons for the errors of incomplete normalization and missing normalization are mainly the incomplete Symptom words in the normative symptom system of traditional Chinese and Western medicine. Although most symptom words might be normalized through this model, there will still be certain errors. We will continue to improve the normative symptom words in the normative symptom system of traditional Chinese and Western medicine to effectively reduce the error rate in the next work.

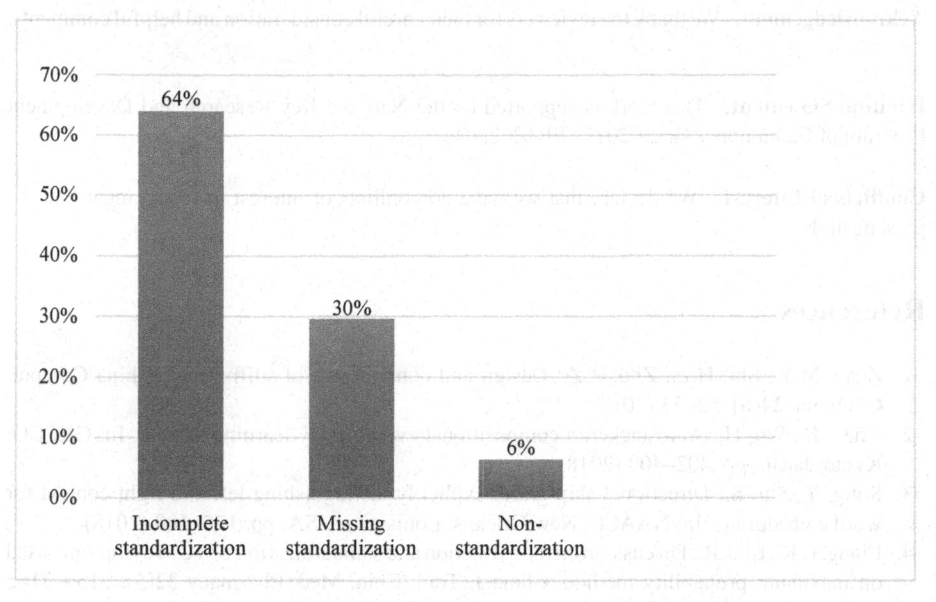

Fig. 4. Incorrect conditions of symptom normalization.

5 Conclusion

Normalization of symptom data in TCM medical records could better help TCM clinical diagnosis and improve the accuracy of TCM syndrome differentiation. This paper focuses on the four types of symptoms in medical records: tongue, tongue coating, pulse, and current symptoms. A retrieval architecture constructed by the suffix tree and dictionaries including normative symptom words of aforementioned the first three symptom is designed. The backtracking algorithm is applied to convert irregular or ambiguous words into normalized symptom words recognized by TCM. The final experimental results indicate that this approach is effective. For the current symptoms, a classification model is designed referred to Autostacker, an AutoML algorithm, to complete the classification operation of the colloquial symptom words. Then we calculate the semantic similarity between these colloquial symptom words and the normative symptom words in the normative symptom system of traditional Chinese and Western medicine to directly transform the colloquial symptom words into normalized symptom words. This method reduces the time consumption of the symptom normalization process and obtains normative and meaningful symptom information data.

However, our research still has shortcomings, which need to be resolved in future work. we intend to further improve the normative symptom system of traditional Chinese and Western medicine to cover all normalized symptom words in the field of TCM. In addition, the method we use might be further improved in the future, so that it can be applied to the normalization of more TCM records in the Chinese medicine data center. This is also an important direction for our future research.

Acknowledgement. We thank the reviewers for their careful consideration and helpful comments.

Funding Statement. This work is supported by the National Key Research and Development Program of China under Grant 2017YFB1002304.

Conflicts of Interest. We declare that we have no conflicts of interest to report regarding the present study.

References

1. Zhao, M.Y., Shi, H.Z., Zhu, Z.Z.: Design and construction of suffix trees. China Comput. Commun. **24**(6), 52–53 (2019)
2. Chen, B., Wu, H.: Autostacker: a compositional evolutionary learning system. In: GECCO, Kyoto, Japan, pp. 402–409 (2018)
3. Song, Y., Shi, S.: Directional skip-gram: explicitly distinguishing left and right context for word embeddings. In: NAACL, New Orleans, Louisiana, USA, pp. 175–180 (2018)
4. Liang, L.K., Li, J.B.: Discussion on the extraction and standardization of TCM symptom based on maximum probability method. China J. Trad. Chin. Med. Pharmacy **32**(5), 2159–2162 (2017)
5. Bai, Y., Zhou, X., Li, X.R., Ge, Z.X., Fu, L.H., et al.: Based on the delphi method for the diagnosis of cough caused by endotoxin in miao medicine standardization study. Chin. J. Ethnomed. Ethnopharmacy **28**(12), 19–22 (2019)
6. Li, Y., Liu, Y., Xu, Y.P., Wang, X.P., Kou, X.N.: Study on the combination of traditional Chinese medicine symptoms of hepatitis B cirrhosis based on cluster analysis and logistic regression analysis. J. Liaoning Univ. TCM **19**(9), 164–167 (2017)
7. Ni, F., Qu, J.Q., Wang, C.X., Cui, J.P., et al.: A study on the standardization of symptoms of spleen and stomach syndrome based on bibliometrics. J. Liaoning University TCM **45**(10), 2086–2087 (2018)
8. Zhang, N.N., Cao, X.Y., Lin, R.F., Wang, B., Shi, H.X., et al.: Thematic Discussion II: The Discussion and Application of Traditional Chinese Medicine Theory. Modernization of Traditional Chinese Medicine and Materia Medica-World Sci. Technol. **22**(05), 1386–1391 (2020)
9. Gobonamang, T.K., Mpoeleng, D.: Counter based suffix tree for DNA pattern repeats. Theoret. Comput. Sci. **814**, 1–12 (2020)
10. Zhao, J., Feng, H., Zhu, D.: DTA-SiST: de novo transcriptome assembly by using simplified suffix trees. BMC Bioinform. **20**(25), 1–12 (2019)
11. Zhu, G.H., Chen, G., Lu, L.: DGST: efficient and scalable suffix tree construction on distributed data-parallel platforms. Parallel Comput. **87**, 87–102 (2019)
12. Klein, A., Falkner:, S.: Fast bayesian optimization of machine learning hyperparameters on large datasets. In: AISTATS, FL, USA, pp. 528–536 (2017)

13. Falkner, S., Klein, A.: Bohb: robust and efficient hyperparameter optimization at scale. In: ICML, Stockholmsmässan, Stockholm SWEDEN (2018)
14. Khurana, U., Samulowitz, H.: Automating predictive modeling process using reinforcement learning. arXiv:1903.00743 (2019)
15. Falkner, S., Klein, A.: Bandit-based automated machine learning. In: BRACIS, São Paulo, SP, Brazil, pp. 121–126 (2018)
16. Hutter, F., Hoos, H.H.: Sequential model-based optimization f or general algorithm configuration. In: International Conference on Learning and Intelligent Optimization, pp. 507–523. Springer, Heidelberg (2011)
17. Chen, Y.W., Song, Q.Q., Hu, X.: Techniques for automated machine learning. In: arXiv:1907. 08908 (2019)
18. Olson, R.S., Bartley, N.: Evaluation of a tree-based pipeline optimization tool for automating data science. In: GECCO, Denver, Colorado, USA, pp. 507–523 (2016)
19. Qi, F., Xia, Z.H., Zhu, D.: A graph-based evolutionary algorithm for automated machine learning. BMC Bioinformatics **20**(25), 1–12 (2018)
20. Pedregosa, F., Varoquaux, G., Gramfort, A.: Scikit-learn: machine learning in Python. J. Mach. Learn. Res. **12**(25), 2825–2830 (2011)
21. Mikolov, T., Chen, T., Corrado, K.: Efficient estimation of word representations in vector space. J. Mach. Learn. Res. **12**(25), 2825–2830 (2013)
22. Chen, X., Xu, L., Liu, Z.: Joint learning of character and word embeddings. In: AAAI, Austin, Texas, USA, pp. 485–492 (2015)
23. Zhang, Q.M., Liu, B.Y.: Study on symptoms of Traditional Chinese medicine. In: Chinese Medicine Ancient books Publishing House, Beijing, China (2013)
24. Yu, D.L.: Study on the dialectical significance of the Functional characteristics of TCM symptoms and Parts. In: Ph.D. dissertation, China Academy of Chinese Medical Sciences, China (2014)
25. Prokhorenkova, L., Gusev, G.: CatBoost: unbiased boosting with categorical features. In: NIPS, Montreal, Canada, pp. 6638–6648 (2018)
26. Chen, T., Guestrin, C.: Xgboost: a scalable tree boosting system. In: ACM, New York, NY, USA, pp. 785–794 (2016)
27. Ke, G., Meng, Q.: Lightgbm: a highly efficient gradient boosting decision tree. In: NIPS, Long Beach, CA, USA, pp. 3146–31544 (2017)

A Nested Named Entity Recognition Method for Traditional Chinese Medicine Records

Haifeng Xu[1,2], Honglan Liu[1,2], Qi Jia[1,2], Yuxiao Zhan[1,2], Yan Zhang[1,2](✉), and Yonghong Xie[1,2](✉)

[1] School of Computer & Communication Engineering, University of Science and Technology Beijing, Beijing 100083, China
xieyh@ustb.edu.cn

[2] Beijing Key Laboratory of Knowledge Engineering for Materials Science, Beijing 100083, China

Abstract. Recently, with the development of deep neural networks, the named entity recognition (NER) task has been well studied in many domains. Among these domains, the information extraction and structuring of Traditional Chinese medicine (TCM) literature is a popular application of NER approach. TCM records are the summary of TCM knowledge and experience, but there are some obstacles imposed by using common machine learning methods, TCM corpus contains a large number of nested entities. And due to entity boundary problems caused by the difference between words and characters in TCM corpus, many methods that have great performance on the English datasets are not suitable for the NER task in TCM field. In order to solve such problems, we propose a nested NER model for TCM records. First, we use word-character-level embedding to enable the model to achieve more accurate extraction results on TCM records corpus. Then, referring to the main entity categories that need to be recognized from the TCM records, we designed a two-layer labelling strategy. This allows our nested NER model to extract more fine-grained results, and has better support for follow-up work such as knowledge base construction. Finally, we conduct some experiments to verify that our model can effectively achieve the NER task for TCM records.

Keywords: Nested named entity recognition · TCM records · Fine-gained entity extraction

1 Introduction

Traditional Chinese medicine (TCM) is an important part of the modern medical research. With the development of the modern medical system, the science of TCM is constantly improving and progressing. A majority of the knowledge of TCM is contained in literatures. Ancient TCM books are a summary of the precious experience left by famous Chinese doctors over thousands of years, and contain the wisdom of traditional Chinese medicine science. The ancient TCM medical records integrate the diagnosis mechanism and prescriptions, and systematically restores how famous Chinese doctors made diagnosis and gave treatment methods based on the patient's condition, which has

X. Sun et al. (Eds.): ICAIS 2021, CCIS 1422, pp. 488–497, 2021.
https://doi.org/10.1007/978-3-030-78615-1_43

great clinical reference value. In recent years, how to extract useful knowledge from unstructured TCM medical records and form a knowledge base has received extensive attention and research.

Named entity recognition (NER) [1] has been extensively studied as the first step in the construction of a knowledge base. Because grammatical structure of ancient TCM medical records is different from those of modern Chinese, a single character usually contains a lot of information. And there are entity boundary problems caused by the difference between words and characters in NER tasks of TCM corpus. Therefore, many NER methods that get good results on the English datasets and modern Chinese datasets are not applicable to the TCM medical records corpus. The current research on the extraction of TCM medical record entities mainly focuses on the extraction of simple entities such as traditional Chinese medicine and prescriptions [2]. However, the most important and complex entity type in medical records is symptom entity, which usually contains a large number of nested entities. As shown in the Fig. 1, according to the traditional sequence labeling model, the "Severe pain in the lower abdomen several times" is labeled as a "symptom" entity, but it is impossible to extract inner entities such as disease location, frequency, and severity at one time. The coarse-grained results of traditional entity recognition methods obviously cannot fully express the knowledge contained in the medical record text.

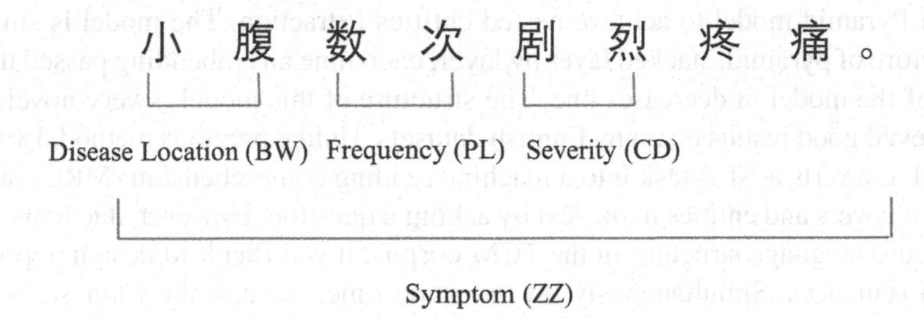

Fig. 1. Extraction example of symptom entity.

The symptom entity is crucial to the establishment of the knowledge base of TCM medical records. Only extracting the symptom entity will lose a lot of key information. If all entity types are divided into two rounds of extraction, it will greatly increase the time consumed by the model, and it will not be possible to use the correlation between symptoms and intrinsic attributes. Therefore, according to the characteristics of the TCM medical record corpus, this paper designs a nested NER model that can extract the outer coarse-grained entity and the inner fine-grained entity.《中华历代名医医案全库》[3] (Chinese medical records of all famous doctors) is an authoritative book in the field of TCM. We use it as a dataset for experiments, and our method proposed in this paper has achieved an F_1 value of 85.1%, which has reached state-of-the-art in the dataset. We summarize the mainly contributions of our paper as follows:

- In order to make the model achieve better performance in TCM corpus, we use the final input word embeddings that combine word embeddings and character embeddings, which could make better use of semantic information. And this strategy is in line with the feature that one single character in the corpus of TCM contains a large amount of information.

- We design a two-layer labeling strategy for TCM medical records. The fine-grained entity types are labeled in the inner layer and the coarse-grained entity types are labeled in the outer layer. This kind of labeling strategy could better realize the extraction of symptom entities and other common entities on the TCM medical records dataset.

2 Related Work

Named entity recognition has always been a research hotspot in the field of Natural language processing (NLP). With the development of neural networks and deep learning in recent years, more and more NER models based on deep learning have been proposed [4]. Most deep learning methods treat the NER problem as a sequence labeling task [5–7], which has achieved better performance than previous methods in various metrics, but in some complex datasets, the nested entity recognition problem cannot be solved well.

Recently, there have been attempts to perform nested NER tasks, for example, Muis and Lu [8] use the method based on mention hypergraph model to recognize overlapping mention in the corpus. At the same time, they designed a mention separator to encode the mentions, which makes the model effective for nested NER task. Wang, Jue et al. [9] design a Pyramid model to achieve nested entities extraction. The model is similar to the structure of pyramid, stacked layer by layer, each time an embedding passed through a layer of the model, it decreases one. The structure of this model is very novel, and it has achieved good results on many English datasets. Unlike previous method, Li Xiaoya [10] et al. converts a NER task into a machine reading comprehension (MRC) task and requires answers and entities in the text by asking a question. However, due to its unique and obscure language structure in the TCM corpus, it is difficult to design appropriate question sentences. Simultaneously. At the same time, because the Chinese medicine corpus is similar to the grammar of ancient Chinese, many effective methods on the English datasets are not effective in the TCM datasets.

As deep learning technology matures and various NER models have achieved great performance on English datasets, NER in the field of TCM has been widely studied in recent years. Wang et al. [11] used Conditional Random Fields (CRFs) to extract entities from Free-Text Clinical Records of TCM. Because they mainly relied on manually features, the results of the extraction were not as good as deep learning methods. With the practicability of the supervised method gradually accepted by researchers, Wang and Lu et al. [12] try to use the surprised method to recognize the symptom entities in the clinical records of Chinese medicine. With the increasing importance of the NER task of TCM corpus, more and more researchers have begun to be interested in the named entity extraction of TCM literatures. Zhang et al. [1] designed a bidirectional LSTM-CRF–based model to extract entities from Chinese electronic health records. Recently, the distant supervised method [13, 14] has also been used in the NER task of TCM corpus. Although the model's dependence on manually labeled samples is reduced, the problem of excessive noisy labels has a great negative impact on the NER results.

With the extensive research of NER tasks in the field of TCM, more fine-grained extraction of targets has gradually become an important research direction in the next step [15]. Therefore, this paper proposes a nested named entity recognize model to obtain a more fine-grained NER method suitable for TCM.

3 Methodology

In this chapter, we first introduce the combination of word-level and character-level vector representation, and then demonstrate the two-layer labeling strategy used in this paper. Finally, we will illustrate the structure of our nested TCM NER model, and give a specific example to show how the model works.

3.1 Input Representation

At present, most of the input vector representation of Chinese NER models based on word embedding. But in TCM corpus, one single character usually contains complete semantic information, only using word vector is not suitable as input representation. Therefore, we firstly map each word into a word vector through the word vector table that pre-trained by word2vec model. Then we use original text to train word2vec model and get character embeddings to map each character into a character vector. The final representation vector that fed into the model are concatenated by word vector and character vector. Through this method, we can utilize semantic information and syntactic information better, so the model could get preferable results on the TCM dataset.

3.2 Two-Layer Labeling Strategy

In order to solve the problem of nested entities extraction represented by symptom entities in TCM medical records, this paper designs a two-layer labeling strategy for TCM corpus. Firstly, based on the most representative entity types of TCM medical records, six outer entities including symptom, pulse, and tongue et al. are designed. Then, in order to satisfy the requirements of fine-grained extraction, we designed six fine-grained entities including disease location, frequency, and color et al. as inner entities. As the most important information in medical records corpus, symptom entity usually contains most of the inner entities. Therefore, such a labeling strategy could more completely extract the effective information in the TCM medical records, and at the same time, it has a crucial improvement in the extraction of symptom entities. And we found through experiments that the inner entity has a heuristic effect on the outer entity, which can further improve the accuracy of the recognition result of the outer entity.

3.3 Model Structure

Our model is designed based on the stacked Flat NER layer [1, 16]. As shown in the Fig. 2, we segment the input sentence "哮喘且喉有痰声。" (Asthma and phlegm in the throat.), get the word vector and the character vector by word2vec model. How many characters are contained in each word, the word vector of this word is repeated how many times. And we combine each character-level vector with the corresponding word-level vector. Then we get the character embeddings that are finally fed to the model.

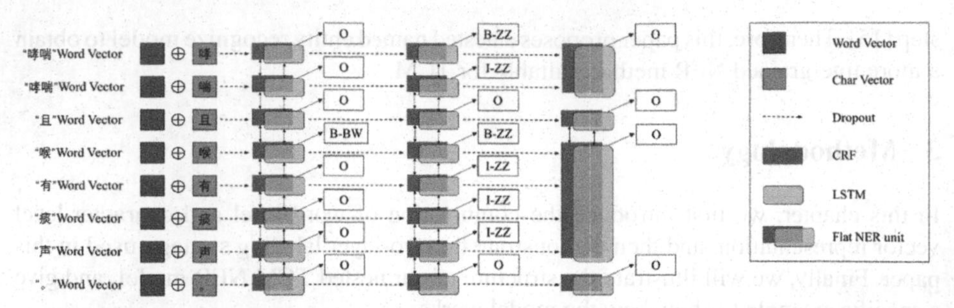

Fig. 2. The structure of TCM nested NER model.

When the fusion representation vector of each character is fed to the first layer of the model, the Flat NER layer composed of LSTM and CRF is used to extract the inner entity types. If an inner entity is identified, merge and average the context representation vector of the currently discovered inner entity area:

$$P_i = \frac{1}{end - start + 1} \sum_{i=start}^{end} v_i \tag{1}$$

Where v_i represents the representation vector of the $i-th$ character, *start* represents the starting position of the recognized entity, and *end* is the end position of the entity. After the vector of each character contained in the recognized entity word is averaged, each recognized entity is treated as one word and its representation vector is passed to the next layer. For characters that have not recognized the entity type, the original representation vector remains, pass to the second Flat NER layer with the representation vector of the recognized entity for the extraction of the outer entities. The recognition result undergoes the same merging and averaging operation as the first layer, and passes to the next layer for recognition. The recognition results of next layer are all "O", which means all the entities have been recognized. Then, the model stops stacking and completes the final entity detection.

4 Experiments

4.1 Datasets

The datasets used in our experiments is the famous TCM book entitled《中华历代名医医案全库》[2] (Chinese medical records of all famous doctors). This book was written by the expert team of Professor Lu Zhaolin of Beijing University of Chinese Medicine. The book records in detail the consultation process and the treatment plan given by ancient Chinese medicine experts, including more than 15,000 famous TCM practitioners and more than 18,000 cases.

In order to objectively evaluate the effectiveness of our model, we randomly sampled 4096 sentences from the book and asked experts in the field of TCM to manually annotate them. The number of various entities in the final dataset is shown in Table 1 and Table 2. The abbreviation of the entity name is designed according to the Chinese pinyin of the entity, and we will give a complete explanation in the Appendix.

Table 1. Number distribution of outer entities.

Entity	Abbreviation	Amount of entity
中药	ZY	8205
症状	ZZ	6405
脉象	MX	976
舌象	SX	630
方剂	FJ	541
剂量	JL	3282

Table 2. Number distribution of inner entities.

Entity	Abbreviation	Amount of entity
病位	BW	2435
时间	SJ	513
程度	CD	426
颜色	YS	355
量	L	218
频率	PL	106

4.2 Model Evaluation

In order to comprehensively evaluate the performance of our model, the evaluation method is crucial. Therefore, we choose the three commonly used evaluation metrics in the NLP field: Precision, Recall, F_1. Generally speaking, we judge the performance of a model based on the F_1 value of the output result. F_1 is calculated by the following formula:

$$F_1 = \frac{2 * P * R}{P + R} \tag{2}$$

P stands for Precision and *R* stands for Recall. In this way, it can not only evaluate whether the model recognition result is accurate, but also reflect whether the model recognition result is comprehensive.

4.3 Compare with Other Method

Compared Methods. LSTM-CRF [1] is a widely used sequence labeling model. Long Short-term Memory Networks (LSTM) has a good ability to capture context information, and Conditional Random Field (CRF) could optimize the results of sequence labeling.

Therefore, the LSTM-CRF model can achieve great performance on NER tasks and has good practicability. But this model can only extract flat entities, and cannot solve the problem of extracting nested entities.

Mention Hypergraph [8] is different from the common sequence labeling methods, uses nodes and directed hyperedges to encode mentions. The mention separator introduced by Muis and Lu et al. can better encode nested mentions. Although this novel method is very suitable for extracting nested mentions, it is found in follow-up experiments that the method is not effective in TCM datasets.

Span-based Model [17] can solve the problem of entity nesting by identifying all possible mentions in the span. This method uses a pre-trained model and get fine-grained recognition results of nested entities, achieved state-of-the-art performance on many datasets.

In order to effectively evaluate the performance of our proposed model, we used the above three methods to conduct experiments on the TCM dataset, and compared with our method from the three metrics of P, R and F_1. Table 3 shows the specific results of the four methods mentioned above. From precision scores, recall scores and F_1 values, we can see that our methods have achieved the best results, and the overall state-of-the-art model has been achieved. From the experimental results, we observed that our method has achieved the best results of precision scores, recall scores and F_1 values. It is the state-of-the-art model of the dataset. Experiments demonstrate that our method is more suitable for entity extraction of TCM corpus than mainstream methods.

Table 3. Comparative experimental results of NER methods.

Method	Precision	Recall	F_1
LSTM-CRF	80.91	73.22	76.87
Mention hypergraph	77.81	66.34	71.62
Span-based model	81.26	85.72	83.43
Our method	85.64	86.18	85.91

Validity of the Character-Word-Level Embedding. In order to make the model better adapt to the semantic characteristics of TCM corpus, we chose word-character-level embedding as the input embedding of the model. In Table 5, we compared the experimental results of using character-level embedding and word-character-level embedding. Through experiments, we can observe that using word-character-level-embedding has achieved better results in precision score, recall score and f1 value than character-level embedding.

Error Analysis. Through experiments, we found some existing problems in our model by analyzing the experimental results of each entity. As shown in Table 4, we observed that due to the relatively standardized grammar of the pulse and tongue entities, the extraction results of the pulse and tongue entities were very good, reaching f1 values of 96.18 and 96.66, respectively. The extraction of symptom entities in TCM corpus has

Table 4. Specific results of our model.

Category	Precision	Recall	F_1
症状 (ZZ)	82.11	82.99	82.55
病位 (BW)	93.05	91.63	92.33
中药 (ZY)	87.05	87.05	87.05
脉象 (MX)	95.45	96.92	96.18
时间 (SJ)	78.84	87.23	82.82
程度 (CD)	77.14	71.05	73.97
舌象 (SX)	96.66	96.66	96.66
颜色 (YS)	94.11	84.21	88.88
剂量 (JL)	97.09	98.52	97.80
量 (L)	55.00	78.57	64.70
方剂 (FJ)	92.30	80.00	85.71
频率 (PL)	50.00	42.85	46.15
Overall	85.64	86.18	85.91

Table 5. Comparative experimental results of the different embedding.

Method	Precision	Recall	F_1
Character-level embedding	82.76	85.97	84.33
Word-Character-level embedding	85.64	86.18	85.91

always been difficult due to its variable length and complex structure. The experimental results prove that our model has a good performance in the extraction of symptom entities, and achieved an f1 value of 82.55.

But due to the uncertainty of entity boundaries in TCM corpus, the extraction results of some symptom entity such as frequency entity (PL) and quantity entity (L) are not very satisfied. Although experiments have proved that the use of word-character-level embedding can solve the problem of the uncertainty of entity boundaries to a certain extent, we observed that the extraction results of some entities composed of single characters still need to be improved.

By using the double-layer labeling mode, the inner entity can inspire the recognition of the outer entity and improve the recognition effect of the outer entity, but how to further improve the recognition ability of the inner entity is our next research direction.

5 Conclusion and Future Work

In this paper, we proposed nested NER model for TCM text. We use word-character-level embedding to enable the model to better learn semantic and syntactic information. And two-layer labeling strategy we designed allows the model to better extract fine-grained results. Our work provides a good nested NER method for TCM records, which can extract valuable knowledge from previous TCM books and records.

Our future work includes further investigations on how to apply the correlation between entities to increase the precision of inner entities and improve the speed of the model.

Acknowledgement. We would like to thank the anonymous reviewers for their valuable comments.

Funding Statement. This work is supported by the National Key Research and Development Program of China under Grant 2017YFB1002304.

Conflicts of Interest. We declare that we have no conflicts of interest to report regarding the present study.

Appendix

See Table 6

Table 6. English-Chinese comparison table of entity name.

English	Chinese	Pinyin	Pinyin abbreviations
Medicine	中药	ZhongYao	ZY
Symptom	症状	ZhengZhuang	ZZ
Pulse	脉象	MaiXiang	MX
Tongue	舌象	SheXiang	SX
Medicine prescriptions	方剂	FangJi	FJ
Dose	剂量	JiLiang	JL
Disease location	病位	BingWei	BW
Onset time and duration	时间	ShiJian	SJ
Severity	程度	ChengDu	CD
Color	颜色	YanSe	YS
Quantity	量	Liang	L
Frequency	频率	PinLu	PL

References

1. Lample, G., Ballesteros, M., Subramanian, S., et al.: Neural architectures for named entity recognition. In: NAACL, San Diego, California, USA, pp. 260–270 (2016)

2. Zhang, Y., Wang, X.W., Hou, Z., Li, J.: Clinical named entity recognition from Chinese electronic health records via machine learning methods. JMIR medical informatics, vol. 6, no. 4 (2018)
3. Lu, Z.L.: Chinese Medical Records of all Famous doctors. Beijing Science and Technology Press, Beijing, China (2015)
4. Character-level neural network for biomedical named entity recognition: Gridach. Mourad. J. Biomed. Inform. **70**, 85–91 (2017)
5. Ma, X.Z., Hovy, E.: End-to-end Sequence Labeling via Bi-directional LSTM-CNNs-CRF. In: ACL, pp. 1064–1074, Berlin, Germany (2016)
6. Qin, Y., Zeng, Y.: Research of clinical named entity recognition based on BI-LSTM-CRF. J. Shanghai Jiaotong Univ. (Sci.) **23**(3), 392–397 (2018)
7. Zhang, Y., Yang, J.: Chinese ner using lattice lstm. In: ACL, pp. 260–270, Melbourne, Australia (2018)
8. Aldrian Obaja, M., Lu, W.: Labeling gaps between words: recognizing overlapping mentions with mention separators. In: EMNLP, pp. 2608–2618, Copenhagen, Denmark (2017)
9. Jue, W., Shou, L., Chen, K., Chen, G.: Pyramid: a layered model for nested named entity recognition. In: ACL, pp. 5918–5928 (2020)
10. Li, X., Feng, J., Meng, Y., Han, Q., Wu, F., Li, J.: A unified MRC Framework for named entity recognition. In: ACL, pp. 5849–5859 (2020)
11. Wang, Y.Q., Liu, Y.G., Yu, Z.H., Chen, L., Jiang, Y.G.: A preliminary work on symptom name recognition from free-text clinical records of traditional Chinese Medicine using conditional random fields and reasonable features. In: BioNLP, pp. 223–230, Montréal, Canada (2020)
12. Wang, Y.Q., et al.: Supervised methods for symptom name recognition in free-text clinical records of traditional Chinese medicine: an empirical study. J. Biomed. Inform. **47**, 91–104 (2014)
13. Wang, Y., Tang, D., Shu, H., Su, C.: An empirical investigation on fine-grained syndrome segmentation in TCM by learning a CRF from a noisy labeled data. J. Adv. Inf. Technol. **9**, 45–50 (2018)
14. Zhang, D.Z., et al.: Improving distantly-supervised named entity recognition for traditional Chinese medicine text via a novel back-labeling approach. IEEE Access **8**, 145413–145421 (2020)
15. Zhang, T., Wang, Y., Wang, X., Yang, Y., Ye, Y.: Constructing fine-grained entity recognition corpora based on clinical records of traditional Chinese medicine. BMC Medical Informatics and Decision Making (2020)
16. Ju, M., Miwa, M., Ananiadou, S.: A neural layered model for nested named entity recognition. In: NAACL-HLT, pp. 1446–1459, New Orleans, Louisiana (2018)
17. Eberts, M., Ulges, A.: Span-based joint entity and relation extraction with transformer pre-training. In: ECAI, Spain, Santiago de Compostela (2020)

Face Feature Extraction and Recognition Based on Lighted Deep Convolutional Neural Network

Ying Tian(✉), Yang Wang, and Longlei Cui

School of Computer Science and Software Engineering, University of Science and Technology Liaoning, Anshan 114051, China

Abstract. Face recognition is one of the most challenging research fields in image processing, pattern recognition and artificial intelligence, and its practical application has a bright future. Face is vulnerable to illumination and expression factors in the process of image acquisition. The traditional machine learning method for extracting face features usually requires high clarity of face images to achieve ideal results. Face feature extraction and recognition based on deep learning method can effectively solve this kind of problem, but the recognition rate is greatly improved, at the same time, the size of the network and the difficulty of training are also sharply increased. It does not meet the increasingly lightweight requirements of the network. This paper studies the current mainstream DCNN architecture, especially MobileNet and ShuffleNet at light-weight mobile level. Some high-efficiency and practical structural features of convolution are summarized; A Nonlinear separable deep convolutional neural network X-Net is designed; Focusing on the loss of the sample in the real face dataset, focus loss is introduced to reduce the effect of sample deviation on network fitting. The experimental results show that the feature extraction based on deep convolutional neural network in this paper can use dynamic library DLL to play an effective role in practical commercial projects.

Keywords: Deep convolutional neural network · Face feature extraction · Face recognition · Non-linear separation

1 Introduction

Face recognition is a classic biometric technology, which is widely used in video surveillance, entrance guard system and so on. As the core method in face recognition, face feature extraction is the main research direction of artificial intelligence. Face recognition mainly use cameras to collect face images or videos, based on various facial feature information, then automatically detect and track faces in images or videos, and implement recognition and classification at last. Face images are more complex and accurate than other images, the face images are easily effected by illumination, expression and other factors, which makes the changes in the class large and the spacing between classes smaller. It is extremely difficult to make a good distinction between classification changes and how to remove the influence of environment, which greatly effects the accuracy of face recognition.

X. Sun et al. (Eds.): ICAIS 2021, CCIS 1422, pp. 498–511, 2021.
https://doi.org/10.1007/978-3-030-78615-1_44

In recent years, many scholars use deep learning to solve the pattern recognition problem [1]. The deep structure neural network simulates cerebral cortex and adopts hierarchical processing method. Each layer can extract different level features of input data effectively, at the same time, it reduces the computational complexity and improves the network generalization ability. Convolutional neural network is a new artificial neural network method which combines artificial neural network and deep learning technology. It has the characteristics of global training which combines local receptive region, hierarchical structure, feature extraction and classification processing [2]. Convolutional neural networks use the last fully connected layer value as the feature extracted from the input sample, through external data to supervised learning. It can ensure that the obtained features are highly invariant to intra-class variations.

The application of convolutional neural network to face recognition shows satisfactory results. In 2014, DeepFace neural network mode [3] of Facebook AI lab achieved 97.35% accuracy on LFW face dataset; In the same year, DeepID network model [4, 5] proposed by Tang et al. also achieved extremely high recognition accuracy. In 2015, FaceNet [6] proposed by Google reached 99.63% recognition accuracy on the LFW dataset. However, with the improvement of face recognition accuracy, the network size and the difficulty of training also increase sharply [7]. He et al. [8] proposed the ResNet network has depth layers number, and there are some defects in the parameters control, which lead to the network model larger. But none of this is in line with the network's increasingly lightweight needs.

This paper studies the mainstream deep convolutional networks and summarizes some efficient and practical depth convolutional structural features. Especially like lightweight mobile networks, such as MobileNet [9] and ShuffleNet [10], According to these classical DCNN structures, a nonlinearly separated network X-net is proposed. It is a lightweight convolutional neural network, under the premise of retaining the detailed feature and the spatial association feature, can increased the spatial correlation of the convolution, and remove complex environmental impacts effectively.

2 Related Work

2.1 Convolutional Neural Network

The basic structure of a convolutional neural network mainly consists of three parts: the input layer, the hidden layer and the output layer. Hidden layer is the basic structure of the convolutional neural network, including the convolutional layer, the pooling layer, the fully connected layer, and the regularized layer [11, 12].

The main purpose of convolutional layer design is to extract feature from input image data. Convolution operation is a good way to determine the spatial position relationship between features. Each output feature map may be obtained by multiple input feature maps convolution. The implementation is shown in Eq. (1):

$$X_j^1 = f(\sum_{i \in M_j} x_i^{l-1} \times k_{ij}^l + b_j^l) \tag{1}$$

Where M_j represents the set of input feature maps, k is the definition of the convolutional kernel, l is the network layers number, b is the bias value size, each output feature map will has a bias value. Different convolutional kernel structure and size will directly effect the result of convolution calculation and obtain different convolutional feature maps.

The pooling layer output is actually downsampling the feature map of the previous layer output. Each network computing layer consists of multiple feature maps, each of which is mapped to a plane, and the weights of all neurons on the plane are equal [11], which reducing the network free parameters number and reducing the complexity of network parameter selection. The pooling layer is shown in Eq. (2):

$$X_j^l = f(\beta_j^l down(x_j^{l-1}) + b_j^l) \tag{2}$$

Where β_j^l and b_j^l are the weights and bias used for the jth output feature map. down(.) is the pooling function, which uses the average value of the continuous $n \times n$ size region of the input image in this layer. This not only increases the generalization ability and maintains the displacement invariance, but also improves the speed.

The fully connected layer is usually the last layer of the convolutional neural network. Before that, the network gets a series of feature maps with lower dimensions. The forward calculation process of the fully connected layer is to multiply each neuron in the previous layer by the weight W, and then add it to a bias value b to obtain the output value of fully connected layer, which is completely inherited from the weighted sum process of the traditional neural network. The back propagation starts with the derivative of the previous layer output. In general, the partial derivative of loss to the expected x can be obtained by the chain rule.

The first step is to obtain the partial derivative of the previous layer output a_i to the input x_j, as shown in Eq. (3):

$$\frac{\partial a_i}{\partial x_i} = \frac{\sum_j^n w_{ij} \times x_j}{\partial x_j} = w_{ij} \tag{3}$$

The second step is to obtain the partial derivative of loss to x_j through the chain rule. The specific implementation is shown in Eq. (4):

$$\frac{\partial loss}{\partial x_k} = \sum_j^n \frac{\partial loss}{\partial a_j}\frac{\partial a_j}{\partial a_k} = \sum_j^n \frac{\partial loss}{\partial a_j} \times w_{jk} \tag{4}$$

As can be seen from Eq. (4), in the forward calculation, the node of the previous layer can exert influence on the next layer node through the weight w. Similarly, in the process of back propagation, the gradient of the next layer node can be propagated to the previous layer node through the weight w, and then need to derive the weight coefficient w, the derivation is shown in Eq. (5):

$$\frac{\partial loss}{\partial w_{kj}} = \frac{\partial loss}{\partial a_k}\frac{\partial a_k}{\partial x_{kj}} = \frac{\partial loss}{\partial a_k} \times x_j \tag{5}$$

It can be seen from Eq. (5) that the partial derivative of the convolutional layer output is equal to the derivative of the next layer loss versus to the bias coefficient.

2.2 Deep Convolutional Neural Network

Professor Hinton proposed the Deep Convolutional Neural Network (DCNN) in 2012 [13]. The initial input of a two-dimensional image is a pixel array, local similar pixel blocks constitute points and various lines, and points and various lines are randomly combined to form various local textures. Multiple textures constitute local images, and finally local images constitute a complete image. The initial image input is mainly shallow features and the relations. After multi-layer processing, it is gradually combined into high-level features which can be connected.

According to the different layers of the network, the extracted image features can be divided into shallow, middle and high features. The shallow network model can not effectively represent hierarchical features because of the limited number of layers and single network structure, and it is also easy to lose many useful features in the process of feature extraction. The feature extraction method based on deep model can effectively remedy the deficiency of shallow model. Deep models generally have dozens to hundreds of layers. As the number of layers increases, it has higher feature expression ability, and learning ability is stronger than that of shallow models. Therefore, this paper chooses the deep convolutional neural network for face feature extraction.

2.3 Lightweight Deep Convolutional Network Architecture

Lightweight deep convolutional network refers to a fast, miniaturized and mobile network model. Generally it has efficient network structure, not only ensure a certain level of recognition rate, but also reflect the characteristics of lightweight [13]. After 2016, lightweight mobile DCNN such as MobileNet and ShuffleNet appeared one after another, FaceBook introduces the improved mobile lightweight deep learning framework (Caffe2) based on the Caffe framework and the Coffee2go with cross-platform computing power, which greatly improves the mobile devices computing speed. Caffe2go has a lightweight core computing structure, and the additional modules increases its computationally intensive power on mobile devices.

The MobileNet network model is based on a deep separable convolutional network structure [9]. The basic idea of separating convolution is to linearly decompose the deep standard convolutional structure into a combination of deep convolutional and 1 * 1 convolution. Deep convolutional is used to perform single convolution operation on each input channel. 1 * 1 convolution is the linear superposition operation used to create the deep convolutional layer. In MobileNet, depthwise separable convolution introduces a single convolutional kernel in each input channel, and then use 1 * 1 pointwise convolutional kernel to combine different depth convolution, and finally outputs the feature matrix. This decomposition method is linear, which reducing the amount of deep convolution calculations and reducing the size of the model.

ShuffleNet is a more efficient deep convolutional neural network based on MobileNet [10], which aims to make up for the limited computing capacity of mobile devices and reduce computing costs to a certain extent. The calculation is based on two operations: pointwise group convolutional and channel shuffle. It has achieved good results in the MS COCO public test dataset, and has better performance than other standard convolutional networks, and has a lower error rate than the MobileNet network.

3 Face Feature Extraction and Recognition Based on Lighted Deep Convolutional Neural Network

3.1 Nonlinear Separable DCNN Architecture

Combining the advantages of MobileNet and ShuffleNet, this paper designs a convolutional network structure that can be nonlinear separated. Continue to add the structure of the incremental feature construction and amplification path, increase the deep convolutional spatial correlation, ensure the detail features and spatial correlation features, introduce the conv3*3/1*1 composite structure, so that the calculation amount and parameters number can be controlled.

Based on the depthwise separable convolutional structure, this paper proposed the network structure is similar to the Conv/Bn/Scale/Relu in MobileNet, as shown in Fig. 1.

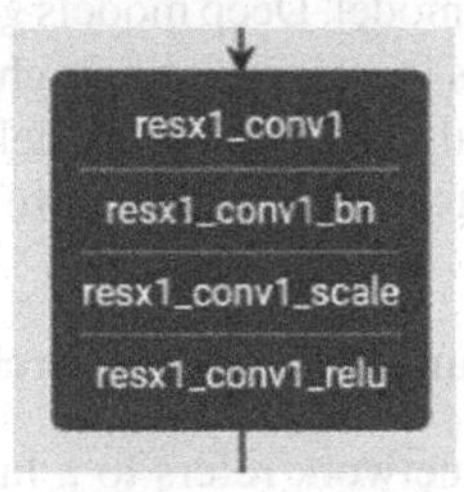

Fig. 1. 1*1 convolutional structure

And then introduce a structure containing the convolutional kernel of 1×1, which exists as a branching structure or a nonlinear convolutional structure. The nonlinear branching structure with 1×1 convolutional structure is shown in Fig. 2:

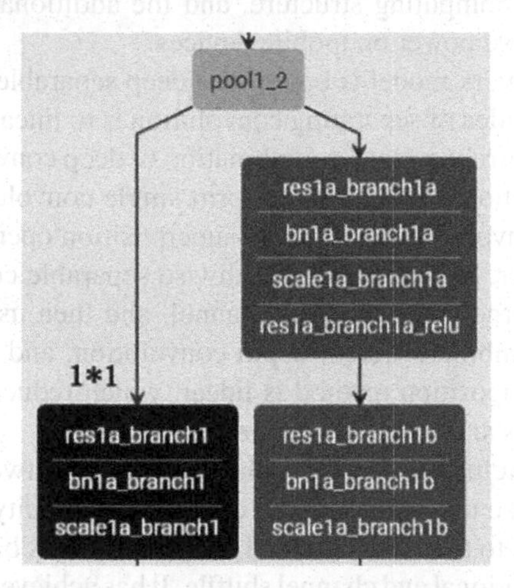

Fig. 2. Nonlinear bifurcation convolution

In Fig. 2, on the basis of the incremental feature construction and the amplification path, the 1 * 1 convolutional structure is mainly introduced where the nonlinear calculation needs to be added. The 1 * 1 convolutional kernel processing the two-dimensional signal, which essentially scales the input data. The 1 * 1 convolutional kernel has only a single kernel value. The convolutional kernel mobile computing on the image data, and the function on the single channel is to add a weight coefficient to the data itself.In the case of multi-channel data such as RGB, multiple 1 * 1 convolutional kernels usually have multiple effects.

3.2 Nonlinear Separable Structure X-Net

Based on the study of some Light DCNNs, this paper proposes a new nonlinear separable structure Res + Inception block. The basic structure is shown in Fig. 3, and details of each part are shown in Table 1:

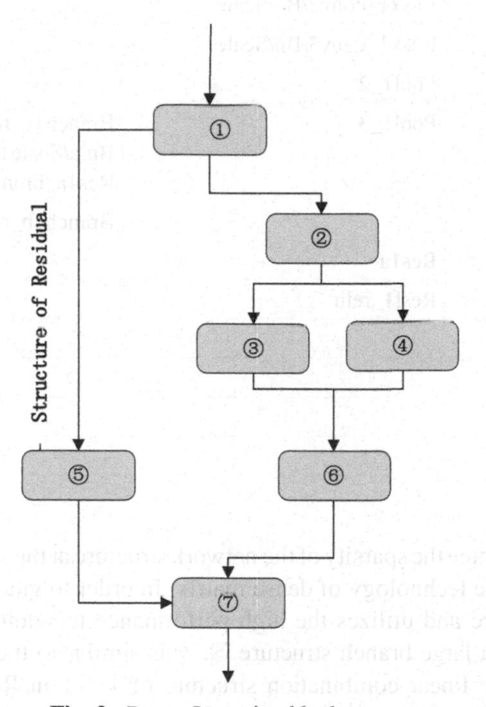

Fig. 3. Res + Inception block structure

In Table 1: from Conv1/Bn/Scale/Relu structure to Resx1_concat, corresponding to the ① ~ ⑦ in Fig. 3 is the detailed information of the basic structure that nonlinearly separable Res + Inception block designed in this paper.

This paper proposed X-Net structure can be seen by combining Fig. 3 and Table 1. Linearly adding a unique small branch structure ②-③-④ similar to Inception under basic structure such as Conv/Bn/Scale/Relu and Concat/Relu in ShuffleNet. The nonlinear branch convolution designed in Fig. 2.

Part of this small branching structure is based on the Conv3*3/1*1 convolutional units like MobileNet, combined channel numbers through⑥reduces the amount of network computation and parameters.⑥is similar to the Resx1_concat structure in ShuffleNet and can randomly combine channel numbers. This Conv3*3/1*1 convolutional structure similar to MobileNet is the 1*1 convolutional structure designed in Fig. 1. The aim is to effectively extract more subdivided image features while ensuring computing amount and the number of parameters are relative decrease.

Table 1. X-net basic structure information

Conv1/Bn/Scale/Relu		
Pool1_1		
Conv1_2	Resx1_conv1/Bn/Scale/Relu	
	Resx1_conv2/Bn/Scale	
	Resx1_conv3/Bn/Scale	
Resx1_match_conv	Pool1_2	
	Pool1_3	Branch1a_res1a/ Bnla/Scale1a/ Res1a_branch1a_relu
		Branch1b_res1a/Bn1a/Scale1a
	Res1a	
	Resl1_relu	
Resx1_concat		
Resx1_concat_relu		
Fn		
Dropout		

In order to guarantee the sparsity of the network structure at the same time and utilizes the high performance technology of dense matrix. In order to guarantee the sparsity of the network structure and utilizes the high performance technology of dense matrix, the algorithm adds a large branch structure ⑤, ⑤ is similar to the incremental feature construction and the linear combination structure of 1 * 1 in ResNet. The structure design is based on the linear combination of Residual structure and 1 * 1 convolution in the Residual network. On a relatively larger level, it constitutes a special inception incremental structure.

Finally through ⑦ to combine, on a larger level to constitute an Inception composite incremental structure, ⑦ is a similar to the Resx1_concat structure in ShuffleNet, which increased channel numbers of network. Therefore, the proposed structure has a strong nonlinear feature extraction ability, and the spatial correlation is enhanced. It can extract some missing details and spatial correlation features.

Therefore, this paper proposed the depth convolutional structure X-Net which can be nonlinearly separated. This structure increases the spatial correlation of deep convolutional, and has a better extraction effect on detailed features and spatial correlation features. Because of the characteristics of the Conv3*3/1*1 structure, the calculation amount and parameter numbers are also well controlled. Then add the Relu layer, the fully connected layer, and the focus loss layer. Adding the Dropout regularization layer also has a very good effect on preventing overfitting. In this way, through repeated reducing dimension and increasing dimension, under the premise of ensuring feature extraction, the calculation amount and the number of parameters are also controlled.

3.3 Focus Loss Function

As the structure of deep convolutional becomes deeper and more complex, the dataset size also increase geometrically. The problem of unbalanced sample categories is very serious, so a new loss function, Focus loss, appears [14]. The focus loss is modified by the standard cross entropy loss.

Taking the case of the two classifications as an example, the classic cross entropy loss refer to Eq. (6):

$$CE(p, y) = \begin{cases} -\log(p) \ \ if\,(y = 1) \\ -\log(1-p)\ otherwise \end{cases} \tag{6}$$

Since it is binary classification, the value of y is 1 or −1, and p represents probability, and the range from 0 to 1 in (6),. When the real label is 1, that is y = 1, if the probability p of class 1 predicted by some sample x is 0.6, then the loss is −log(0.6), and the loss ≥ 0. If p = 0.9, then the loss is −log(0.9), so the loss of p = 0.6 is greater than the loss of p = 0.9.

The focus loss algorithm is inspired by the cross entropy loss and has a good effect on the offset data. The meaning of focus loss can be obtained from Eq. (7) and Eq. (8)

$$CE(p_t) = -\log(p_t) \tag{7}$$

$$FL(p_t) = -(1-p_t)^{\gamma}\log(p_t) \tag{8}$$

Equation (7) is the standard cross entropy calculation formula. On the basis of Eq. (7), Eq. (8) introduces a new variable parameter as the coefficient, which can be used in the weight of sample selection, to increase samples weight that are difficult to be classified, to reduce the weight of samples that are easy to be classified. The loss value function allows the network model to focus on aspects that difficult to classify. The range of weight coefficient pt is [0, 1], The number of classified samples is no longer completely determined by the classified sample numbers, By changing pt, the sample number eventually learned from the network can be changed. If $\gamma = 0$, it becomes the standard cross entropy loss.

4 Experiments

4.1 Preprocessing of Face Dataset

The training dataset CASIA-WebFace used in the experiment was obtained through the processing of Li ziqing group from the Chinese academy of sciences. The original dataset contains 494,414 face data of 10,575 people, after preprocessing and alignment cutting, can get 422,515 face dataset with size of 224 * 224 of 10,307 people. The training set and verification set were generated in the ratio of 4:1, with 333,872 and 88643 respectively, then generates the LMDB dataset. The test dataset are the public LFW standard face recognition dataset.

Face alignment, denoising and offset correction are the common methods for data preprocessing. The focus is correction offset of the dataset. The correction offset method selected peoposed is Zero-centered first, and then normalization.

Zero-centered operation means that in the data, to solve the mean value of each dimension is solved first, and then subtracted it from each of the dimensions data. The RGB three-channel image should be subtracted its mean value. The mean value is generally the common feature to the whole image. Subtracting the mean value, for noise and background processing, also has a good effect.

Normalization processing is to normalize the valid data to the same degree space, so that different pixel scales can achieve consistency. Normalized data, has repair effect on noise and interference, and sample offset can also be corrected. In addition, the obtained data, because the scale is highly uniform, can effectively ensure the synchronization of data in the process of feature extraction and matrix calculation.

In the experiment, key point technology is used to locate five key points of face, including eyes, two lips, and nose. Faces located by key points can be used for face alignment and clipping, and the results are shown in Fig. 4 and Fig. 5. Figure 4 is the initial face image, and Fig. 5 is the aligned face image, which can remove face background and other factors.

4.2 Feature Extraction and Processing

The feature extractor training is a kind of DCNN training. The commonly optimization training methods are decreasing the learning rate from 0.01 to decline, parameter initialization, changing the activation function, using Zero-center for data preprocessing, gradient clipping, regularization, using SGD + Momentum instead of ADAM and so on. The trained feature extractor uses deconvolution visualization to extracts features in the fully connected layer, and the extracted feature vectors are compared with the corresponding feature vectors stored in the feature database. Setting a classification threshold to determine whether the feature vector belongs to the same person.

This paper proposes a cosine-based cosine similarity calculation [15], that similar to the space vector coordinate system, cosine of the angle between the two vectors, the cosine values is positively correlated with individual differences. The larger the cosine, the angle is closer to 0, the smaller the differences between individuals. J. Ma [16] analyzed that Cosine has great advantages in calculating similarity in high dimensional

Fig. 4. Initial face dataset

feature digital vectors. The proposed method also proves that using Cosine to calculate the similarity of text vectors is effective.

In the process of deconvolution visualization, the input is the feature image that has been calculated and extracted from each layer. After the reverse operation of deconvolution, can obtains the relevant features extracted from the specific layer. Deconvolution is a kind of computational processing with unsupervised learning, without learning training and optimization processing. Only gets the visualized feature results, it is reversible with ordinary convolution operations.

Face deconvolution can extract useful key features in the case of shielding part of the background interference. Generally, at the bottom of the network, it learns the global basic characteristics in the broad sense, such as color information, object background and edge orientation. Further to the next layer network, above the basic global features, it begin to extract the basic texture features, or grid, trend and so on. The higher layers features begin to become distinctive, such as angle, ray offsets and so on. The last few layers learn the complete, non-linear combination of basic features.

4.3 Comparison and Analysis of Experimental Results

Comparing the method this paper proposed with the traditional feature extraction method, under the same conditions, it can be seen that the proposed method not only improves the recognition rate, but also takes less time to extract features, the specific situation shown in Table 2. It can be seen from the experimental data in Table 2 that the

Fig. 5. Aligned face dataset

combination of Wavelet and SVM [17] takes less than 158 ms, the recognition rate of 0.86 has not reached the current commercial level.

Table 2. The results compared with traditional feature extraction methods

Method	Average recognition rate (%)	Average time (ms)
PCA + Gabor	0.91	895
Wavelet + SVM	0.86	158
Adaboost + LBP + SVM	0.95	586
X-Net	0.97	130

The combination of PCA and Gabor [18] currently has a general time consumption and recognition rate, which cannot meet real-time and efficient face recognition. PCA and Gabor combined method [18] currently has a general time consumption and recognition rate, which cannot meet the needs of real-time and efficient face recognition. Compared with Adaboost + LBP + SVM [19], the proposed method improves the recognition rate by 0.02, and the feature extraction speed is more than 3 times faster. The effect is obvious relatively.

Comparing our method with current two classical deep convolutional feature extraction methods, under the same condition, the comparison results are shown in Table 3:

Table 3. The results compared with other convolutional methods

Method	Average recognition rate	Extraction rate (ms)	Model size (M)
AlexNet	0.65	80	31
VGG-16	0.99	80	673
X-Net	0.97	130	230

Through the comprehensive comparison of the experimental data in Table 3, compared with AlexNet, the proposed method can improve the recognition rate from 0.65 to above 0.97. With the improvement of recognition rate, feature extraction can meet the needs of most face recognition scenes. Compared with VGG and other networks, under the premise of recognition rate decreases little, the feature extraction network model is smaller, and the extraction speed is increased by more than 6 times, which greatly enhances the feature extraction network adaptability.

4.4 Mobile Level Feasibility

The method proposed in this paper absorbs methods and structures of the lightweight mobile convolutional neural network such as MobileNet and ShuffleNet, and it can be seen that it is superior to the ordinary convolutional neural network in the research of recognition rate and model size. With the emergence of sparse network pruning compression method, similar to DSD at present, after pruning, the X-net will be more miniaturized and loaded on the mobile learning framework of Caffe2 and Caffe2go to make the network more feasible.

5 Conclusion

Based on the classical deep convolutional neural network, this paper proposed a nonlinearly separation network X-Net. It is a lighted convolutional neural network. Under the premise of retaining the detail features and spatial association features, which increases the spatial association of convolution, makes the network feature extraction ability higher, the calculation speed faster and generalization ability stronger. Considering that the face dataset samples obtained from the life scene is generally unbalanced, the focus loss is introduced from the object detection field, the over-fitting phenomenon caused by sample deviation is reduced. Based on the feature extraction method of deep convolutional neural network, Dynamic Link Library plays an efficient role in actual commercial projects.

But it still has a lot of shortcomings, such as glasses, bangs, heavy makeup and so on, it brings much troubles to face recognition and detection. It is still necessary to continue to find and study the network structural design optimization also need to be further strengthened or discretize the parameters to reduce. In addition, the neural network engineering of the time of forward calculation.

Conflicts of Interest. The authors declare that they have no conflicts of interest to report regarding the present study.

References

1. Lecun, Y., Bottou, L., Bengio, Y., Haffner, P.: Gradient-based learning applied to document recognition. Proc. IEEE **86**(11), 2278–2324 (1998)
2. Zhou, F.Y., Jin, L., Dong, P.J.: Review of convolutional neural network. Chin. J. Comput. **40**(06), 1229–1251 (2017)
3. Taigman, Y., Yang, M., Ranzato, M.A., Wolf, L.: Deepface: closing the gap to human-level performance in face verification. In: Proceedings of the IEEE Conference on Computer Vision and Pattern Recognition, pp.1701–1708 (2014)
4. Sun, Y., Wang, X., Tang, X.. Deep learning face representation from predicting 10,000 classes. In: Proceedings of the 2014 IEEE Conference on Computer Vision and Pattern Recognition (CVPR), pp.1891–1898 (2014)
5. Sun, Y., Chen, Y., Wang, X., Tang, X.. Deep learning face representation by joint identification-verification. In: Advances in Neural Information Processing Systems, pp.1988–1996 (2014)
6. Schroff, F., Kalenichenko, D., Phibin, J.. Facenet: a unified umbedding for face recognition and clustering. In: Proceedings the IEEE Conference on Computer Vision and Pattern Recognition, pp.815–823 (2015)
7. Szegedy, C., Liu, W., Jia, Y., Sermanet, P., Reed, S., Anguelov, D.: Going deeper with convolutions. In: Proceedings of the IEEE Counference on Computer Vision and Pattern Recognition, pp.1–9 (2015)
8. He, K., Zhang, X., Ren, S., Sun, J.. Delving deep into rectifiers: surpassing human-level performance on imagenet classification. In: Proceedings of the IEEE International Conference on Computer Vision, pp.1026–1034 (2015)
9. Andrew, G., Howard, M.L., Zhu, B., Chen, et al.: Mobilenet: efficient convolutional neural networks for mobile vision applications, arXiv:1704.04861 (2017)
10. Zhang, X.Y., Zhou, X.Y., Lin, M., Sun, J.: huffleNet: an extremely efficient convolutional neural network for mobile devices, arXiv:1707.01083, Aprli (2017)
11. Hinton, G.E., Osindero, S., The, Y.W.: A fast learning algorithm for deep belief nets. Neural Comput. **18**(7), 1527–1554 (2006)
12. Lv, G.H., Luo, S.W., Huang, Y.P., et al.: A novel regularization method based on convolution neural network. J. Comput. Res. Dev. **51**(09), 1891–1900 (2014)
13. Zeiler, M.D., Taylor, G.W., Fergus,R.: Adaptive deconvolutional networks for mid and high level feature learning. In: Proceedings of the International Conference on Computer Vision, pp. 2018–2025 (2011)
14. Lin, T.Y., P. Girshick, G.R., He, K., Dollar, P.: Focal loss for dense object detection, arXiv: 1708.02002v1 [cs.CV], 7 August 2017
15. Yuan, X.Q., Huang, F.G., Zhang, J.P.: Face recognition system based on the discrete cosine transform. Appl. Sci. Technol. **04**, 32–33 (2003)
16. Ma, J.: Research on Mult-Pose Face Location and Recognition Method, Ph.D. dissertation. North University of China, China (2006)

17. Xie, S.F., Shan, S.G., Chen, X.L., Meng, X., Gao, W.: Learned local gabor patterns for face representation and recognition, Signal Processing, vol. 12 (2009)
18. Xin, G., Kate, S.M., Wang, L., Li, M., Wu, Q.: Context-aware fusion: a case study on fusion of gait and face for human identification in video. Pattern Recognition **10** (2010)
19. Zhu, B.L., Wang, D.W., Li, M.J., et al.: Study on 3D face recognition system against expression change. Chin. J. Sci. Instr. **02** (2014)

Non-orthogonal Multiple Access is Used for Power Allocation to MaximizeThroughput in Cognitive Radio Network Systems

Yongming Huang(✉), Xiaoli He(✉), Weijian Yang, Yuxin Du, and Xinwen Cheng

School of Computer Science, Sichuan University of Science and Engineering, Zigong 643000, China

Abstract. In order to improve the resource utilization rate of spectrum, this paper performs power allocation based on the combination of Cognitive Radio Network (CRN) and Non-Orthogonal Multiple Access (NOMA). The primary user (PU) and secondary user (SU) in the cell of the paper model exist at the same time. The SU uses the NOMA method to access the system. Under the premise of ensuring the quality of service (QoS) of the PU and the SU, and the SU The cumulative interference to the PU cannot exceed the "interference temperature threshold" threshold for the normal operation of the PU. This paper proposes a power allocation method under the premise of the maximum available transmit power of the secondary base station (ST) and the QoS requirement constraints, which can maximize the throughput of the SU in the system, and also consider the channel conditions, transmit power and normal communication of the SU The signal-to-interference-to-noise ratio threshold and other situations. Aiming at nonlinear problems, a power allocation algorithm is proposed using convex optimization. By solving the power allocation factor, the system throughput of SU is improved. Finally, the simulation results show that the algorithm can effectively improve the system throughput of multiple SU models, and the research results lay a theoretical foundation for the application of CRN and NOMA technology.

Keywords: Cognitive radio networks · Non-orthogonal multiple access · Power allocation · Reinforcement learning

1 Introduction

In recent years, with continuous breakthroughs made on the road of global technological innovation, wireless communication technology has developed rapidly. Although the rapid development of wireless communication technology has greatly improved the quality of communication and greatly increased the data transmission rate, with the rapid development of smart mobile devices and the corresponding services of the Internet of Things (IoT) With the coming exponential data traffic growth, the existing mobile communication technology still cannot meet the increasing communication needs of mankind. Therefore, countries will focus on the fifth generation of mobile communications (5G) [1–3].

X. Sun et al. (Eds.): ICAIS 2021, CCIS 1422, pp. 512–523, 2021.
https://doi.org/10.1007/978-3-030-78615-1_45

Among them, multiple access technology is an important factor affecting system capacity. Non-orthogonal multiple access technology (NOMA) is a new multiple access solution to meet the requirements of 5G wireless communication for system capacity. NOMA technology is different from traditional orthogonal transmission. It adopts non-orthogonal transmission at the sending end, actively introduces interference information, and realizes correct demodulation through serial interference cancellation technology at the receiving end [4–8]. The basic idea of non-orthogonal transmission is to use complex receiver design in exchange for higher spectral efficiency. At the same time, Cognitive Radio Network (CRN) makes full use of the spectrum of wireless networks through its excellent characteristics of spectrum sensing. CRN is also a promising key technology in 5G.

Wireless spectrum is the lifeblood of mobile communication systems, and all network construction and mobile application services are indispensable. Therefore, the scientific allocation of spectrum resources is a prerequisite for the success of mobile communication systems, and it is related to global roaming and the scale effect of the industrial chain [9–12].

However, both CRN and NOMA are technologies aimed at effective utilization of spectrum resources. If the CRN and NOMA technology are organically combined, the shortage of spectrum resources and insufficient throughput can be solved. Therefore, this is one of the research hotspots in the communication field in recent years. Nowadays, most of the independent research on NOMA or CRN. People are also constantly trying to combine the two technologies for research, mainly focusing on the maximization of the number of access users and resource allocation. There are some researches on power allocation, but it is still not perfect. Therefore, the power allocation problem in NOMA-CRN faces a big risk challenge, which is the focus of this article.

1.1 Related Work

In recent years, as one of the key technologies of 5G, NOMA has been extensively studied in a large number of literatures. At the same time, the CRN technology and the NOMA technology are constantly being combined to study the performance of the NOMA-based CRN hybrid network. Literature [1] studied a new NOMA-assisted spectrum sharing framework, which can enhance spectrum utilization under multi-user conditions. It provides a reliability-oriented scheduling scheme for SU, and the goal is to obtain the minimum interruption probability of PU and SU. And a fair-oriented secondary user scheduling scheme is proposed, the goal is for all candidate SUs to have an equal opportunity to be arranged for cooperation. The closed expressions of PU and SU interruption probability are obtained. The literature [2] studies the security in the NOMA-CRN network, and enhances the transmission confidentiality by deliberately introducing interference in the network system. By combining NOMA and CRN, the confidentiality is greatly enhanced. The literature [6] studied the throughput in the NOMA-CRN hybrid network. For the fixed power allocation of NOMA, it can be proved that the fixed power allocation of NOMA technology obtains greater system throughput than the traditional OMA technology network. Users with good channel conditions choose NOMA access in fixed power allocation, which can obtain more channel gains than choosing OMA access. In the NOMA-CRN network, QoS with poor channel conditions can be guaranteed. The literature [9] considered the power allocation in the NOMA-CRN hybrid network model and maximized the number of slave users in the access system. The literature [13] studied in the NOMA-CRN hybrid network, under the premise of ensuring normal communication between PU and SU, with the goal of maximizing the number of access slave users, a power allocation algorithm was proposed. Literature [14] For the relay cooperative transmission system based on NOMA, based on the perception of base station power limitation, with the throughput of the secondary user as the goal, the deployment of expanded forwarding relays and the use of NOMA technology to broadcast user information are proposed. A power distribution algorithm.

1.2 Contribution

In this study, unlike previous research work, our goal is to maximize the throughput of slave users through power allocation. Therefore, we adjust the power allocation under multiple constraints (i.e., interference temperature, QOS guarantee, power limit and power factor). More specifically, our main contributions to this work are summarized as follows:

- First, this article considers the issue of NOMA-CRN. Therefore, in the optimization problem, multiple SUs are allowed to reuse the spectrum resources of the PU.
- Subsequently, the identified optimization problem is a nonlinear programming problem. Therefore, a power distribution algorithm is proposed using convex optimization to obtain the best solution and improve the overall performance of the network.
- Finally, we provide numerical results to evaluate the effectiveness of our proposed PA-RL algorithm. Numerical simulation results show that the algorithm can obtain better theoretical optimal performance when the signal-to-noise ratio is high. In addition, these tasks can effectively extend the life of the network and provide some insights for CRN research.

1.3 Organization of the Paper

The rest of this paper is organized as follows. Section 2 establishes system and network model, including system model and mathematical modeling. Section 3 presents a solution to the problem. Numerical results are addressed. in Sect. 4. Finally, Sect. 5 concludes this paper.

2 System and Network Model

2.1 System Model

This article mainly considers the NOMA-CRN downlink model. The system adopts the Underlay spectrum sharing mode, that is, SU and PU can access the spectrum at the same time, thereby improving the utilization of spectrum resources. In addition, the secondary user uses MONA to access the system.

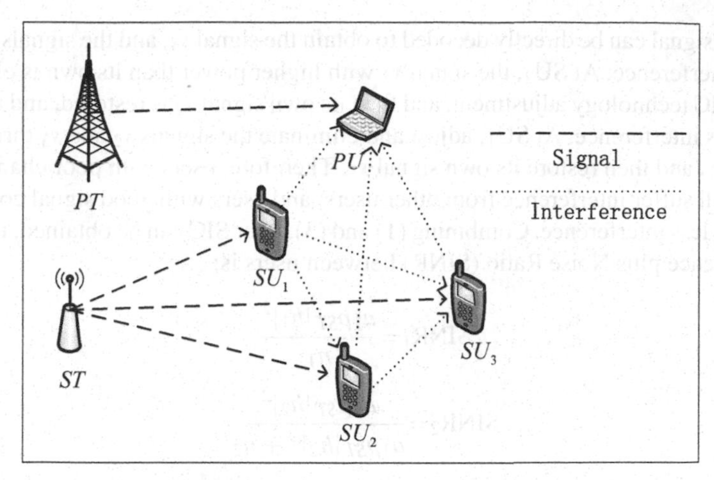

Fig. 1. System model of underlay CRN: PU and SUs share the downlink link scenario

In Fig. 1, the system includes a primary base station (PT) and primary user (PU), a NOMA-based secondary base station (ST) and several secondary users $SU_i(i = 1,2,\ldots,n)$. In order to simplify the model, this article only makes modeling analysis for three secondary users (SU_1, SU_2 and SU_3), and does not consider the interference caused by primary users to secondary users. In the downlink, the channel gain coefficient from ST to $SU_i(i = 1,2,3)$ is expressed as $h_i(i = 1, 2, 3)$. Suppose SU_3 is far away from ST,SU_1 is close to ST, and SU_2 is between SU_1 and SU_3, then we can get

$$|h_1|^2 > |h_2|^2 > |h_3|^2 \tag{1}$$

According to the principle of NOMA, ST allocates more power to SU_3 with poor channel conditions and less power to SU_1 with good channel, and linearly superimposes multiple secondary signals to form a composite signal, which is then transmitted to the target user (SU_1, SU_2 and SU_3). The signal s sent by the sender can be expressed as:

$$\mathrm{s} = \sum_{i=1}^{3} \sqrt{a_i p_{ST}} s_i \tag{2}$$

s_i is the information sent by ST to secondary user i, a_i is the power allocation factor of SU_i, $a_1 < a_2 < a_3$ and $a_1 + a_2 + a_3 = 1$, p_{ST} is the transmission power of ST. The signal received at $SU_i(i = 1,2,\ldots,n)$ can be expressed as:

$$\mathrm{y}_{\mathrm{i}}^{\mathrm{SU}} = \mathrm{h}_{\mathrm{i}} \sum_{i=1}^{3} \sqrt{a_i p_{ST}} s_i + n_i, i = 1, 2, 3 \tag{3}$$

n_i represents the mean value is 0, and the variance σ^2 is additive white Gaussian noise.

In the downlink, since the NOMA method is used to send superimposed information between ST and SU, in order to avoid mutual interference between SUs, the Successful Interference Cancelation (SIC) technology is used to eliminate Multiple Access Interference when SU receives signals. At SU_3, since the power obtained by itself is large, the mixed signal can be directly decoded to obtain the signal s_3, and the signals s_1 and s_2 exist as interference. At SU_2, the signal s_3 with higher power than its own is eliminated through SIC technology adjustment, and then its own signal s_2 is restored, and the signal s_1 exists as interference. At SU_1, adjust and eliminate the signals s_2 and s_3 through SIC technology, and then restore its own signal s_1. Therefore, users with poor channel coefficients will suffer interference from other users, and users with good signal coefficients will suffer less interference. Combining (1) and (3), after SIC can be obtained, the Signal to Interference plus Noise Ratio (SINR) between users is:

$$\mathrm{SINR}_1 = \frac{a_1 p_{ST} |h_1|^2}{n_1} \tag{4}$$

$$\mathrm{SINR}_2 = \frac{a_2 p_{ST} |h_2|^2}{a_1 p_{ST} |h_2|^2 + n_2} \tag{5}$$

$$\mathrm{SINR}_3 = \frac{(1 - a_1 - a_2) p_{ST} |h_3|^2}{(a_1 + a_2) p_{ST} |h_3|^2 + n_3} \tag{6}$$

$\mathrm{SINR_i}$ represents the SINR of the i-th slave user. In order to ensure the QoS of the slave user and enable it to communicate normally, there are:

$$\mathrm{SINR}_i \geq \mathrm{SINR}_i^{\mathrm{th}}, i = 1, 2, 3 \tag{7}$$

$\mathrm{SINR_i^{th}}$ represents the SINR threshold that the i-th slave user must meet for normal communication. At the same time, the power of the secondary user cannot exceed its threshold:

$$\mathrm{a_i p_{ST}} < \mathrm{p_i^{th}}, \mathrm{i} = 1,2,3 \tag{8}$$

$\mathrm{p_i^{th}}$ represents the maximum available power threshold for the i-th secondary user. Since the secondary users use the NOMA method to transmit superimposed codes, for the primary user, we regard the superimposed signals of the secondary users and the power integration allocated between them as interference. The received signal can be expressed as:

$$\mathrm{y^{PU}} = h_{PT-PU}\sqrt{p_{PT}}x + \underbrace{h_{ST-PU}\sqrt{p_{PT}}s}_{\text{interference}} + \sigma^2 \tag{9}$$

Where h_{PT-PU} and h_{ST-PU} are the channel gain coefficients from PT and ST to PU, respectively. x is the useful signal sent by the base station to the primary user, and s is the interference signal from the secondary base station. The SINR of the primary user can be expressed as follows:

$$\gamma = \frac{|h_{PT-PU}|^2 p_{PT}}{|h_{ST-PU}|^2 p_{ST} + \sigma^2} \tag{10}$$

In order to ensure that the QoS of the primary user will not be affected, the following constraints are imposed:

$$\gamma \geq \gamma^{th} \tag{11}$$

Interference temperature, a model recommended by the FCC at the end of 2003 to quantify and manage interference sources. The expression is as follows:

$$T_I(f_c, B) = \frac{P_I(f_c, B)}{\kappa B} \tag{12}$$

Where $P_I(f_c, B)$ is the average interference power with frequency f_c and bandwidth B. κ is Boltzmann's constant with a value of 1.38×10^{-23}J/k. As the interference of the PU as the SU, the cumulative interference cannot exceed the PU interference temperature threshold Γ^{th}, otherwise the SU cannot share the spectrum resource. Where B = 1Hz, the interference temperature constraint is expressed as:

$$\frac{\sum_{i=1}^{3} a_i p_{ST} |h_i|^2}{\kappa} \leq \Gamma^{\mathrm{th}} \tag{13}$$

2.2 Mathematical Modeling

Based on the above discussion, a power allocation optimization problem with the goal of maximizing sub-user throughput is proposed. The optimization target expression is as follows:

$$\max_{(a_1, a_3)} R = \sum_{i=1}^{3} \log_2(1 + SINR_i) \tag{14a}$$

$$s.t. \ \gamma \geq \gamma^{th} \tag{14b}$$

$$SINR_i \geq SINR_i^{th}, i = 1, 2, 3 \tag{14c}$$

$$a_1 + a_2 + a_3 = 1 \tag{14d}$$

$$0 < a_1 < a_2 < a_3 \tag{14e}$$

$$\Gamma \leq \Gamma^{\mathrm{th}} \tag{14f}$$

$$\mathrm{a_i p_{ST}} \leq P_i^{th} \tag{14g}$$

Among them, (14b) and (14c) respectively represent the QoS requirements that the primary and secondary users must meet. (14d) and (14e) represent the constraints of the power allocation factor. (14f) indicates that the cumulative interference of the secondary user to the primary user cannot exceed the interference temperature threshold of the primary user. (14g) indicates that the power of the secondary user cannot exceed the upper limit of the maximum power threshold of the secondary user.

3 Problem Solution Method

For the convenience of representation, this article defines the additive white Gaussian noise received by the secondary users as $n(\sigma^2)$, let $A_i = p_{ST}|h_i|^2$. From (4), (5), (6):

$$\mathrm{SINR}_1 = \frac{a_1 A_1}{n} \tag{15a}$$

$$\mathrm{SINR}_2 = \frac{a_2 A_2}{a_1 A_2 + n} \tag{15b}$$

$$\mathrm{SINR}_1 = \frac{a_1 A_1}{n} \tag{15c}$$

In order to facilitate the calculation, according to the constraint condition (14d), the objective function can be obtained:

$$R = \sum_{i=1}^{3} \log_2(1 + SINR_i) = \log_2(a_1 A_1 + n) - \log_2(a_1 A_2 + n)$$
$$+ \log_2[(1 - a_3)A_2 + n] - \log_2[(1 - a_3)A_3 + n] + \log_2(A_3 + n) - \log_2 n \tag{16}$$

3.1 Convexity Judgment

In order to prove that the objective function is concave and convex, let:

$$\begin{aligned} f(x, y) = {} & \log(A_1 x + n) - \log(A_2 x + n) + \log\big[(1 - y)A_2 + n\big] \\ & - \log\big[(1 - y)A_3 + n\big] + \log(A_3 + n) - \log(n) \end{aligned} \tag{17}$$

In order to facilitate the calculation, this paper considers the additive white Gaussian noise that SU receives as n. We can get:

$$\frac{\partial f}{\partial x} = \frac{A_1}{(A_1 x + n)\ln 2} - \frac{A_2}{(A_2 x + n)\ln 2} \tag{18a}$$

$$\frac{\partial f}{\partial y} = \frac{A_3}{\big[(1 - y)A_3 + n\big]\ln 2} - \frac{A_2}{\big[(1 - y)A_2 + n\big]\ln 2} \tag{18b}$$

$$\frac{\partial^2 f}{\partial x^2} = \frac{[A_2(A_1 x + n) + A_1(A_2 x + n)](A_2 - A_1)n}{(A_2 x + n)^2 (A_1 x + n)^2 \ln 2} < 0 \tag{18c}$$

$$\frac{\partial^2 f}{\partial x \partial y} = 0 \tag{18d}$$

$$\frac{\partial^2 f}{\partial y^2} = \frac{\big[2(1 - y)A_2 A_3 + A_3 n + A_2 n\big](A_3 - A_2)n}{\big[(1 - y)A_3 + n\big]^2 \big[(1 - y)A_2 + n\big]^2 \ln 2} < 0 \tag{18e}$$

In summary, it is not difficult to conclude that $f(x, y)$ is a concave function.

3.2 Maximum Throughput Optimization Based on Convex Optimization Theory

It can be seen from the above that the problem to be optimized (14a) is a concave function, which can be solved by the Lagrange dual method. Before constructing the Lagrange dual function, this paper transforms the constraint condition (14c) as follows

$$SINR_1 \geq SINR_1^{th} \;=>\; \frac{nSINR_1^{th}}{A_1} - a_1 \leq 0 \tag{19a}$$

$$SINR_2 \geq SINR_2^{th} \;=>\; \frac{nSINR_2^{th}}{A_2} + a_3 + a_1\left(SINR_2^{th} + 1\right) - 1 \leq 0 \tag{19b}$$

$$SINR_3 \geq SINR_3^{th} \;=>\; SINR_3^{th}\left(\frac{n}{A_3} + 1\right) - a_3\left(SINR_3^{th} + 1\right) \leq 0 \tag{19c}$$

The Lagrange dual function of the optimization problem (14a) can be expressed as:

$$\begin{aligned}
\mathrm{L}(a_1, a_3, \lambda, v, \varsigma) &= \mathrm{R} + \lambda_1\left(\frac{nSINR_1^{th}}{A_1} - a_1\right) + \lambda_2\left[\frac{nSINR_2^{th}}{A_2} + a_3 + a_1\left(SINR_2^{th} + 1\right) - 1\right] \\
&+ \lambda_3\left[SINR_3^{th}\left(\frac{n}{A_3} + 1\right) - a_3\left(SINR_3^{th} + 1\right)\right] + \varsigma_1\left(a_1 p_{\mathrm{st}} - P_1^{th}\right) + \varsigma_2\left[(1 - a_1 - a_3)p_{\mathrm{st}} - P_2^{th}\right] \\
&+ \varsigma_3\left[a_3 p_{\mathrm{st}} - P_3^{th}\right] + v\left[\frac{a_1 A_1 + (1 - a_1 - a_3)A_2 + a_3 A_3}{\kappa} - \Gamma^{\mathrm{th}}\right]
\end{aligned} \tag{20}$$

Among them, λ_i, v and ς_i are the non-negative Lagrange operators corresponding to constraint (14c), constraint (14f) and constraint (14g) respectively. Correspondingly, the Lagrange dual function corresponding to the function (20) can be expressed as:

$$g(\lambda, v, \varsigma) = \max_{\{a_1, a_3\}} L(\{a_1, a_3\}, \lambda, v, \varsigma) \tag{21}$$

At the same time, the dual problem corresponding to optimization problem (14a) can be expressed as:

$$\min_{\lambda, v, \varsigma \geq 0} g(\lambda, v, \varsigma) \tag{22}$$

For (19a), according to the KKT condition, the derivation of each Lagrangian operator can be obtained:

$$\frac{\partial L}{\partial a_1} = \frac{\partial R}{\partial a_1} - \lambda_1 + \left(SINR_2^{th} + 1\right)\lambda_2 + p_{\mathrm{st}}\varsigma_1 - p_{\mathrm{st}}\varsigma_2 + \frac{A_1 - A_2}{\kappa}v = 0 \tag{23a}$$

$$\frac{\partial L}{\partial a_3} = \frac{\partial R}{\partial a_3} + \lambda_2 - \left(SINR_3^{th} + 1\right)\lambda_3 - p_{\mathrm{st}}\varsigma_2 + p_{\mathrm{st}}\varsigma_3 + v\frac{A_3 - A_2}{\kappa} = 0 \tag{23b}$$

$$\frac{\partial L}{\partial \lambda_1} = \frac{nSINR_1^{th}}{A_1} - a_1 = 0 \tag{23c}$$

$$\frac{\partial L}{\partial \lambda_2} = \frac{nSINR_2^{th}}{A_2} + a_3 + a_1\left(SINR_2^{th} + 1\right) - 1 = 0 \tag{23d}$$

$$\frac{\partial L}{\partial \lambda_3} = SINR_3^{th}\left(\frac{n}{A_3} + 1\right) - a_3\left(SINR_3^{th} + 1\right) = 0 \tag{23e}$$

$$\frac{\partial L}{\partial \varsigma_i} = \begin{cases} a_i p_{\text{st}} - P_1^{th} = 0, i = 1, 3 \\ (1 - a_1 - a_3)p_{\text{st}} - P_2^{th} = 0, i = 2 \end{cases} \tag{23f}$$

$$\frac{\partial L}{\partial v} = \frac{a_1 A_1 + (1 - a_1 - a_3)A_2 + a_3 A_3}{\kappa} - \Gamma^{\text{th}} = 0 \tag{23g}$$

According to the KKT condition, the closed expression of the optimal power distribution of SU_1 and SU_3 can be solved

$$a_1 = \frac{\sqrt{(A_1 + A_2)^2 n^2 - 4A_1 A_2 \left\{ n^2 - \frac{(A_1 - A_2)n}{\ln 2 \left[\begin{array}{c} \lambda_1 - \left(SINR_2^{th} + 1\right)\lambda_2 \\ -p_{\text{st}}\varsigma_1 + p_{\text{st}}\varsigma_2 - \frac{A_1 - A_2}{\kappa} v \end{array} \right]} \right\}} - (A_1 + A_2)n}{2A_1 A_2} \tag{24}$$

$$a_3 = 1 - \frac{\sqrt{\begin{array}{c}(A_2 + A_3)^2 n^2 \\ -4A_2 A_3 \left\{ n^2 - \frac{(A_3 - A_2)n}{\ln 2 \left[\begin{array}{c} -\lambda_2 + \left(SINR_3^{th} + 1\right)\lambda_3 \\ +p_{\text{st}}\varsigma_2 - p_{\text{st}}\varsigma_3 - v\frac{A_3 - A_2}{k} \end{array} \right]} \right\}\end{array}} - (A_2 + A_3)n}{} \tag{25}$$

Among them $(x)^+ = \max(0, x)$. For the dual function (21) to be differentiable, a subgradient is introduced to solve the dual problem (22), and the solution is as follows:

$$\lambda_i^{(l+1)} = \left[\lambda_i^{(l)} + \theta_i^{(l)} \frac{\partial L}{\partial \lambda_i}\right]^+, i = 1, 2, 3 \tag{26a}$$

$$\varsigma_i^{(l+1)} = \left[\varsigma_i^{(l)} + \vartheta_i^{(l)} \frac{\partial L}{\partial \varsigma_i}\right]^+, i = 1, 2, 3 \tag{26b}$$

$$v^{(l+1)} = \left[v^{(l)} + \eta^{(l)} \frac{\partial L}{\partial v}\right]^+ \tag{26c}$$

Where $\theta_i^{(l)}$, $\vartheta_i^{(l)}$, $\eta^{(l)}$ are the step size of the iteration number l.

4 Simulation Result

This article is mainly in the NOMA-CRN hybrid network, realize simulation result using python. The noise power spectral density in the system.

As shown in Fig. 2, the transmission power of ST fluctuates between 20 dbm–40 dbm. Algorithm 1 in Fig. 2 is used in NOMA. In order to ensure that edge users get higher power allocation, simulation analysis is performed under the conditions of SU_1 and SU_2 meeting the minimum threshold of 2bit/Hz. It can be seen from Fig. 2 that as the transmission power of ST increases, the performance of the power allocation method proposed in this article in the NOMA environment is much better than that of OFDMA because it can achieve a higher sum rate.

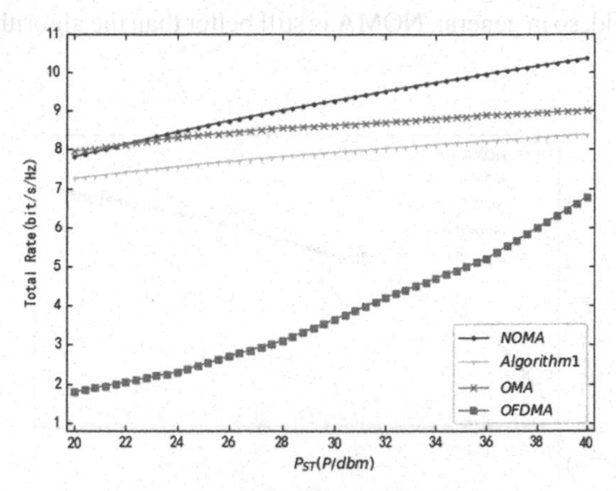

Fig. 2. The sum of the maximum transmission power and the transmission rate.

Next, compare the individual user rates in the simulation results. The simulation parameter settings are the same as the previous one. In Fig. 3, it is not difficult to see that when the ST transmit power reaches 28 dbm or more, the power allocation method under NOMA is higher than the OMA allocation algorithm. This is because the system allocates more power. In general, NOMA Power allocation for a single user is better than that under NOMA.

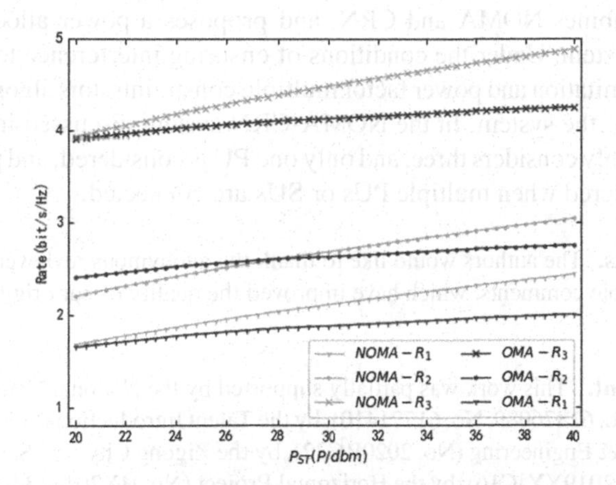

Fig. 3. NOMA and OMA secondary user rate comparison.

For the aforementioned sub-users with better channel conditions, the SINR meets the minimum threshold of 2bit/Hz, and the edge users adopt a larger power allocation factor. We can analyze it through Fig. 4. In Fig. 4, and SNR maintain the lowest threshold of 2bit/Hz, so R1 and R2 remain unchanged. Under this condition, we can find that although the overall trend of R3 in algorithm1 is better than R3 in NOMA, the degradation is smaller. However, the magnitudes of NOMA-R1 and NOMA-R2 are larger than the minimum threshold, so in general, NOMA is still better than the algorithm1 optimization solution.

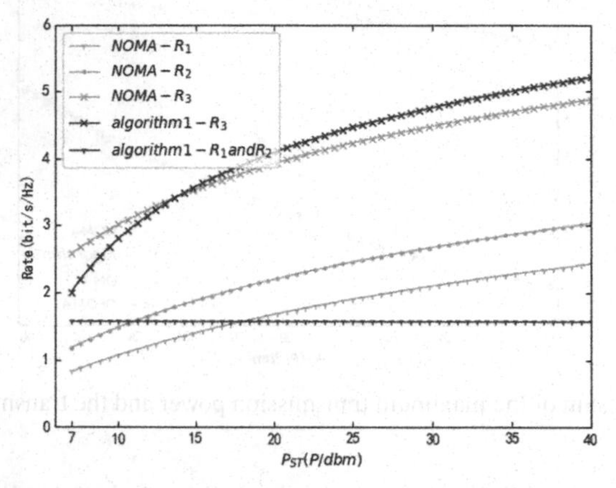

Fig. 4. NOMA and algorithm1 secondary user rate comparison.

5 Conclusions

This article combines NOMA and CRN, and proposes a power allocation algorithm for the hybrid system. Under the conditions of ensuring interference temperature, QoS quality, power limitation and power factor multiple constraints, this algorithm maximizes the throughput of the system. In the NOMA-CRN system discussed in this article, the number of SUs only considers three, and only one PU is considered, and power allocation should be considered when multiple PUs or SUs are connected.

Acknowledgments. The authors would like to thank the anonymous reviewers for their selfless reviews and valuable comments, which have improved the quality of our original manuscript.

Funding Statement. This work was partially supported by the National Natural Science Foundation of China (No. 61876089, No. 61771410), by the Talent Introduction Project of Sichuan University of Science & Engineering (No. 2020RC22), by the Zigong City Key Science and Technology Program (No. 2019YYJC16), by the Horizontal Project (No. HX2017134, No. HX2018264), by the Enterprise Informatization and Internet of Things Measurement and Control Technology Sichuan Provincial Key Laboratory of universities (No. 2020WZJ02, No. 2014WYJ08), and by Artificial Intelligence Key Laboratory of Sichuan Province (No. 2015RYJ04).

Conflicts of Interest. The authors declare that they have no conflicts of interest to report regarding the present study.

References

1. Lv, L., Yang, L., Jiang, H., Luan, T.H., Chen, J.: When NOMA meets multiuser cognitive radio: opportunistic cooperation and user scheduling. IEEE Trans. Veh. Technol. **67**(7), 6679–6684 (2018)
2. Wei, L., Jing, T., Fan, X., Wen, Y., Huo, Y.: The secrecy analysis over physical layer in NOMA-enabled cognitive radio networks. In: 2018 IEEE International Conference on Communications (ICC), Kansas City, MO, pp. 1–6 (2018)
3. Liu, G., Chen, X., Ding, Z., et al.: Hybrid half-duplex/full-duplex cooperative non-orthogonal multiple access with transmit power adaptation. IEEE Trans. Wirel. Commun. **17**(1), 506–519 (2018)
4. Xu, L., Zhou, Y., Wang, P., Liu, W.: Max-min resource allocation for video transmission in NOMA-based cognitive wireless networks. IEEE Trans. Commun. **66**(11), 5804–5813 (2018)
5. Xu, W.P.: Research on resource allocation strategy of multi-user cognitive radio network. Ph.D. dissertation, Donghua University, China (2019)
6. Yue, X., Liu, Y., Kang, S., et al.: Exploiting full/half-duplex user relaying in NOMA systems. IEEE Trans. Commun. **66**(2), 560–575 (2018)
7. Zhou, M.Y.: Research on non-orthogonal multiple access resource allocation algorithm in cognitive radio networks. M.S. dissertation, Chongqing University of Posts and Telecommunications, China (2019)
8. Dong, Y.F.: Research on non-orthogonal multiple access technology based on relay selection in cognitive radio networks. M.S. dissertation, Yantai University, China (2019)
9. Liu, X., Wang, Y., Liu, S., Meng, J.: Spectrum resource optimization for NOMA-based cognitive radio in 5G communications. IEEE Access **6**, 24904–24911 (2018)
10. Zhou, F., Wu, Y., Liang, Y., et al.: State of the art, taxonomy, and open issues on cognitive radio networks with NOMA. IEEE Wirel. Commun. **25**(2), 100–108 (2018)
11. Yu, Z.Y.: Research on spectrum sharing and non-orthogonal multiple access technology in cognitive wireless networks. Ph.D. dissertation, Shan Dong University, China (2018)
12. Chen, X., Jia, R., Ng, D.W.K.: On the design of massive non-orthogonal multiple access with imperfect successive interference cancellation. IEEE Commun. **67**(3), 2539–2551 (2019)
13. Nandan, N., Majhi, S., Wu, H.: Secure beamforming for MIMO-NOMA-based cognitive radio network. IEEE Commun. Lett. **22**(8), 1708–1711 (2018)
14. Zhang, Y.H.: Research on key technologies of non orthogonal cooperative transmission in wireless communication networks. Ph.D. dissertation, Beijing University of science and technology, China (2020)

Research on Drug Response Prediction Model Based on Big Data

Guijin Li and Minzhu Xie(✉)

College of Information Science and Engineering, Hunan Normal University, Changsha 410081, China

xieminzhu@hunnu.edu.com

Abstract. Personalized medicine, also known as precision medicine, refers to a medical model of providing the best treatment plan for a patient according to his or her personal genomic information. The research and practice of personalized medicine have become a hot topic in current medical research, and predicting the response of cell lines to specific drugs is one of the core problems. Using computer algorithms to predict the responses of cell lines to drugs based on huge amounts of existing omics information is currently one focus of bioinformatics. A variety of predictive methods have been proposed. The paper introduces the baseline analysis data, surveys some classical prediction methods and models, and details on the application of matrix decomposition, heterozygous network and deep learning at the drug response prediction. At last, some existing problems and future development trend and prospects are discussed.

Keywords: Drug response prediction · Multi-omics data · Personalized medicine · Computer algorithm · Bioinformatics

1 Introduction

The results of a large number of cancer genome studies have shown that the genetic mutations of patients with the same cancer have large individual differences, and the use of patient-specific genome profiles to design personalized treatment plans is essential for effective control of disease progression [1, 2]. Personalized treatment for each person's physiological characteristics (or precision medicine [3]) is a new medical concept and medical model developed with the rapid advancement of genome sequencing technology and the cross application of biological information and big data science. The same drug has different therapeutic effects on different patients suffering from the same disease. Predicting the different responses of specific cell lines of different individuals to drugs is the core problem of precision medicine.

There are two commonly used measures of cell line response to drugs: IC50 and AUC. IC50 [4] is called the half inhibitory concentration, which represents the concentration of the drug required to halve cell activity. The AUC [5] value represents the area under the drug concentration-cell activity response curve. The lower the value of IC50 and AUC, the more effective the drug. Traditional drug response prediction methods

X. Sun et al. (Eds.): ICAIS 2021, CCIS 1422, pp. 524–537, 2021.
https://doi.org/10.1007/978-3-030-78615-1_46

based on functional experiments are time-consuming, expensive, and require high experimental equipment [6]. The continuous development of high-throughput sequencing and other technologies has produced massive amounts of biological data. For example, the National Cancer Institute (NCI) has published the largest tumor-related mutation gene database NCI60. The database provides data on the response of 60 tumor cell lines to thousands of drugs, as well as cell line mRNA, microRNA expression, DNA methylation and mutation profile data [7, 8]. Abundant biological data enables researchers to better understand individual human cell lines at the molecular level. Driven by multiple omics data, methods for predicting drug response using computational models have emerged one after another.

Different from the experimental methods and statistical methods commonly used in clinical research, the computational prediction model based on multi-omics data emphasizes the integrity of the association between drugs and cell lines. The drug response prediction method based on the computational model is shown in Fig. 1. These models generally use cell line genotype, gene expression, proteomics, and drug molecular structure as input information for the model. Then build and train an optimized prediction model based on the response data of known cell lines to drugs, Finally, the specific cell line information that needs to be predicted is input into the optimized model, and the output obtained is generally a drug sensitivity index or a drug response category.

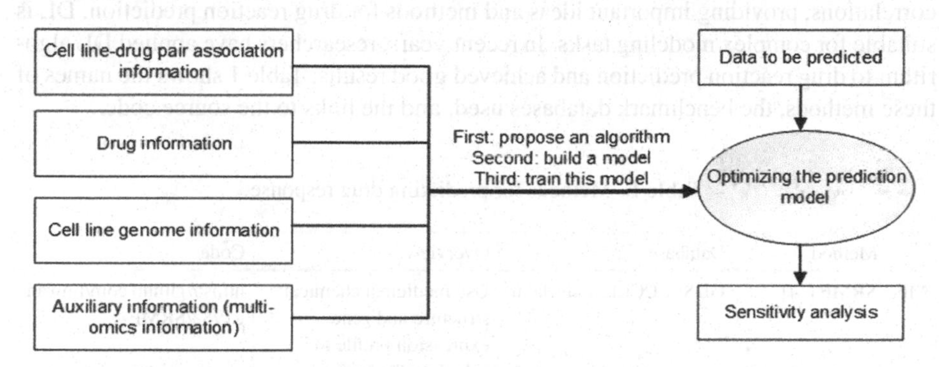

Fig. 1. Prediction method of drug response based on computational model

The initial drug response prediction methods are mostly based on calculation models such as matrix decomposition or linear regression due to the relatively few available biological data information. With the development of high-throughput sequencing and other technologies, the number of open multi-omics database data has increased exponentially, and people have begun to use the correlation between various biological data to construct multi-layer heterogeneous networks and analyze them in combination with classification models. In recent years, deep learning technology has begun to be applied to massive multi-omics data, and predictive models combining multi-omics information and deep learning technology have been continuously proposed. Models based on matrix decomposition and heterogeneous networks are called "traditional models", and prediction models based on deep learning are called "new models". There have been some reviews on drug response prediction [9–13] mainly introducing typical "traditional models" and data integration methods. This article not only analyzes the latest methods and typical methods based on the "traditional model", but also elaborates the application of deep learning in drug prediction, hoping to help readers understand the latest research progress in this field. At the same time, it provides links to related methods and databases to facilitate researchers to quickly obtain relevant information. This review first elaborates the drug prediction method based on the calculation model, then introduces the benchmark database used by the drug response prediction model, and finally looks forward to the problems and future development trends in the field.

2 Method

Drug response prediction methods based on computational models can be divided into three categories: methods based on matrix factorization (MF), methods based on networks (Net), and methods based on deep learning (DL). MF is an unsupervised learning algorithm that can handle interactions from different data sources. It has attracted wide attention because of its effective clustering and prediction of missing values. The main advantage of the heterogeneous network-based method is that it can fuse a large amount of data, construct a complex heterogeneous network and then infer direct and indirect correlations, providing important ideas and methods for drug reaction prediction. DL is suitable for complex modeling tasks. In recent years, researchers have applied DL algorithm to drug reaction prediction and achieved good results. Table 1 shows the names of these methods, the benchmark databases used, and the links to the source code.

Table 1. Methods for predicting drug response

	Method	Database	Overview	Code
MF	SRMF [14]	GDSC, CCLE, PubChem	Use medicinal chemical structure and gene expression profile to calculate the known similarity and use it as a regularization term	https://github.com/linwang1982/SRMF
	DSPLMF [15]	GDSC, CCLE	Based on gene expression profile, copy number change, somatic mutation data, and chemical structure information, logistic matrix factorization is used to calculate the probability	https://github.com/emdadi/DSPLMF

(*continued*)

Table 1. (*continued*)

	Method	Database	Overview	Code
	WGRMF [16]	PubChem, GDSC, CCLE	The similarity matrix is obtained from the medicinal chemical structure and gene expression profile data, and the p-nearest neighbor graph is constructed to sparse it	https://doi.org/10.1016/j.omtn.2019.05.017
Net	MSDRP [17]	GDSC, PubChem	By integrating drug similarity network, cell line similarity network and cell line-drug response network information, a modular approach is proposed	https://github.com/shiminwang1994/MSDRP.git
	HNMDRP [18]	STRING, KEGG, GDSC, PubChem	A three-layer heterogeneous network is constructed based on the heterogeneous relationship of cell lines, drugs, and targets, and information flow algorithms are used on this network	https://github.com/USTC-HIlab/HNMDRP
	NRL2DRP [19]	GDSC, PPI	Construct heterogeneous networks of cell lines, drugs and proteins, support vector machines as classification models	https://github.com/USTC-HIlab/NRL2DRP
DL	DNNs [20]	GDSC, TCGA	Based on the cell line-drug pair response and gene expression information, a feedforward deep neural network model is established	https://github.com/TeoSakel/deepdrug-response
	MOLI [21]	GDSC, PDX, TCGA	Based on the multi-omics information, a coding sub-network is established to obtain low-dimensional features, which are used as the input of the classification sub-network for model training	https://github.com/hosseinshn/MOLI

(*continued*)

Table 1. (*continued*)

Method	Database	Overview	Code
RefDNN [22]	CCLE, GDSC	The Elastic Net model is used to extract new low-dimensional features of cells, and then put them into the classifier for training	https://github.com/mathcom/RefDNN
tCNNs [23]	GDSC, PubChem, LINCS	Establish two convolutional neural sub-network models to extract low-dimensional features of cell lines and drugs respectively	https://github.com/Lowpassfilter/tCNNS-Project
Deep DR [24]	CCLE, GDSC, TCGA	The model includes mutation encoders and gene expression encoders pre-trained on the data, and an integrated prediction network	https://bmcmedgenomics.biomedcentral.com/articles/supplements

2.1 Matrix Factorization Method (MF)

The basic idea of matrix decomposition is to decompose a large matrix into the product of low-rank matrices. This type of method is widely used for high-dimensional data dimensionality reduction, and can handle any number of heterogeneous data [14, 25]. In the drug response prediction problem, the main input data is the drug response matrix $R \cdot R$ is a real matrix with N rows and M columns, where N is the number of cell lines and M is the number of drugs. The matrix decomposition method decomposes the matrix R into two low-rank matrices U and $V \cdot U$ and V can be regarded as the potential characteristic matrices representing cell lines and drugs, respectively. Matrix decomposition maps the drug response matrix R to a low-dimensional latent space, which is regarded as the product of the potential feature matrix of the cell line and the drug. This article introduces three MF-based drug response prediction methods: SRMF, DSPLMF and WGRMF.

Based on similar cell lines and similar drugs with similar drug responses, Wang et al. [14] used a similar regularized matrix factorization (SRMF) method to predict the response of cell lines to anticancer drugs. SRMF assumed that the drug response matrix R is approximately equal to the product of two low-rank matrices U and V. Obtained the drug similarity matrix and the cell line similarity matrix through the medicinal chemical structure data and gene expression profile data. In order to minimize the difference between the similarity of the drug in the potential space and the similarity of the chemical structure, as well as the difference between the similarity of the cell line in the potential space and the similarity of the gene expression profile, two constraints are introduced to avoid overfitting of the training data.

Akram et al. [15] introduced a method based on a recommendation system to predict the sensitivity of cancer drugs (DSPLMF), using the logistic matrix factorization method to calculate the probability of cell lines being sensitive to drugs. The design of the algorithm is as follows: at first, constructs the similarity between each pair of cell lines based on multi-omics information, and defines the similarity between each pair of drugs according to the chemical substructure, and use IC50 as the cell line-drug pair classification (sensitive/Resistance) threshold. They use the DSPLMF model to obtain the potential feature vectors of each cell line and drug. For each new cell line, they used a decision tree classifier to find its k-nearest neighbor, and estimated the probability that the cell line is sensitive to drugs based on the potential feature vector of its neighbors. Finally, the author compares the probability with the threshold to get the response classification of each new cell line-drug pair. The advantage of this method is that potential cell lines and drug characteristics can be obtained, thereby obtaining better predictive performance.

Guan et al. [16] used a weighted graph regularization matrix factorization method to predict cell line anticancer drug response (WGRMF). Research results showed the effectiveness of this method in predicting drug response in cell lines. This method constructs a p-nearest neighbor graph to sparse the drug similarity matrix and the cell line similarity matrix respectively. Then the sparse matrix is subjected to matrix decomposition to obtain the potential matrix of drugs and cell lines. This method retains effective cell line and drug neighborhood information, which helps to improve prediction accuracy.

2.2 Network-Based Methods (Net)

With the development of high-throughput sequencing technology, massive amounts of biological data have been obtained. In the existing research results, the researchers constructed the cell line-drug into a heterogeneous network and then combined multiple omics data for analysis. The nodes in the heterogeneous network represent drugs and cell lines, and the edges represent the connections between these nodes. Compared with other algorithms, network-based methods can merge large amounts of data, are relatively more interpretable, and are more in line with the current understanding of biological data. This review mainly introduces three classic drug reaction prediction methods based on heterogeneous networks: MSDRP, HNMDRP, and NRL2DRP.

Wang et al. [17] proposed a modular score prediction model based on the constructed two-layer heterogeneous network (MSDRP). The MSDRP framework consists of four parts: (a) integrate multiple similarity data information and response information between drugs and cell lines; (b) construct drug similarity network, cell line similarity network and drug cell line response network; (c) according to the known response information of the target cell line (or drug) and all drugs (or cell lines) in the data set, divide all drugs (or cell lines) into different sub-modules; (d) calculate the intra-module score and the inter-module score of the target cell line (or drug) to obtain the correlation score. According to the literature, the model has good predictive performance even in the case of unknown cell line-drug response.

Zhang et al. [18] proposed an information flow method based on heterogeneous networks (HNMDRP). The method includes four steps: (a) obtain the known sensitive or drug-resistant association information between the cell line and the drug; (b) introduce similarity measures and integrate all heterogeneity information (including cell line gene expression profiles, PPIs, drug targets, and medicinal chemical structure information); (c) integrate and construct a comprehensive three-layer heterogeneous network model Fig. 2 shows all sub-networks, nodes and interactions); (d) execute an algorithm based on information flow on a heterogeneous network to obtain the scores of all cell line-drug pairs, where the score is the predicted score of the drug response. According to experiments and reliable data and literature verification, the model also has a good effect in predicting the response of unknown cell lines-drugs.

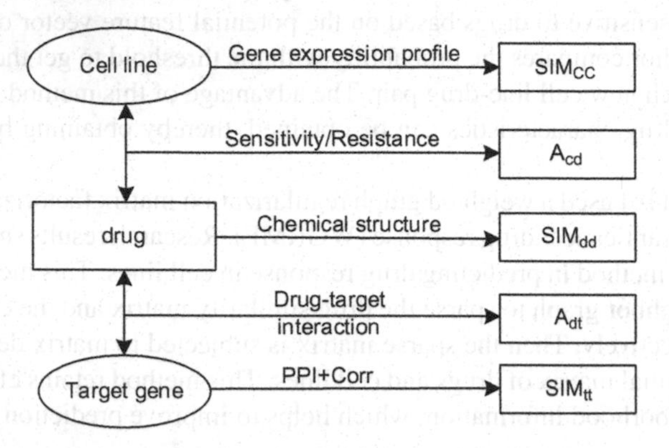

Fig. 2. All sub-networks, nodes and interactions

Yang et al. [19] proposed a prediction method based on network representation learning (NRL2DRP). The method is divided into three stages: (a) construct heterogeneous networks by integrating genome mutation information, drug response information and protein-protein interaction information; (b) extract the representation vector of the cell line through the network representation learning method, which can maintain the similarity of the vertex neighborhood and serve as a predictor of feature construction drug response; (c) input the representation vector of the cell line as a feature into the drug classification model, where the response index is used as the classification category. Due to the problem of data imbalance, the author chose a support vector machine (SVM) model for classification tasks.

2.3 Deep Learning-Based Methods (DL)

Deep Learning (DL) [26–28] is a new research direction in the field of Machine Learning (ML) [29] technology. DL can also be understood as a deep neural network. For example, a multilayer perceptron with multiple hidden layers is a deep learning structure. In traditional ML models, researchers need to design different methods to describe the characteristics of samples. The quality of these extracted sample features has a critical impact on the performance of traditional ML models. The DL method can directly use the original data as training samples, allowing the network to automatically extract features, skipping the traditional feature selection step, and avoiding the complexity of manually selecting features and the dimensional disaster of high-dimensional data. Through multi-layer processing, the initial "low-level" feature representation is gradually transformed into an abstract "high-level" feature representation, and then a "simple model" can be used to complete complex classification and other learning tasks. In recent years, DL technology has made great breakthroughs in many fields, making artificial intelligence related technologies have made great progress. In the DL model, starting from the input layer, the output of the previous layer is used as the input of the next layer until the final result is output. The training network is actually an optimization problem, and its goal is to optimize the gap between the predicted value of the output and the true value. Table 2 introduces several basic models in the DL architecture and their application scenarios in drug response problems.

Table 2. Basic models and applications of deep learning

	Models	Description	Note
Basic models	Deep Neural Networks (DNN) [30]	Suitable for most classification tasks	DNN is often used as the final classification sub-network for prediction (sensitive/resistant)
	Auto encoders (AE) [31]	Encoder and decoder are composed of two parts	AE is often used to reduce dimensionality and automatically obtain the characteristics of potential cell lines (drugs)
	Convolutional Neural Networks (CNN) [32]	It is formed by cross-stacking convolutional layer and pooling layer. The convolutional layer extracts features, and the function of the pooling layer is dimensionality reduction	CNN is often used to extract features of potential cell lines (drugs)
	Recurrent Neural Netwosrk (RNN) [33]	The formation of a network structure with loops is more in line with the structure of biological neural networks	It is better to retain the correlation information between different data in heterogeneous network

DL is also a powerful technical support in the field of bioinformatics [28]. For example, DL technology has been proven to be superior to traditional compounds in predicting compound activity [34–36], compound toxicity [37] and other compound characteristics [38] ML method. In the research on the prediction of drug effects, different types of DL structures are being tried and used by researchers, and some research results have been obtained. These results indicate that the application of DL technology can play an important role in predicting the effect of treatment and studying the relationship between genomic characteristics and drug resistance. This review mainly introduces five classic drug reaction prediction methods based on deep learning: **DNNs, MOLI, RefDNN, tCNNS** and **Deep DR**.

Sakellaropoulos et al. [20] proposed a method to predict the response of cancer treatment drugs. This method predicts drug response based on feedforward deep neural networks (DNNs). The author first obtained the response data (IC50 value) between 1001 cell lines and 251 drugs from the GDSC database, as well as the gene expression information of the cell lines; then built DNNs models to predict patient drug response, and used two widely used learning algorithms Compare. Experiments show that, compared with the current state-of-the-art machine learning framework, DNNs can more effectively capture the interaction between complex biological data.

Hossein et al. [21] proposed a multi-omics post-integration model based on deep neural network to predict drug response (MOLI). The author introduces three different types of omics information, somatic mutation, copy number aberration, and gene expression data, and establishes three different deep learning coding sub-networks, and learns different omics data to obtain abstract features after dimensionality reduction, and then abstract the features are connected into a representation vector, and these data are integrated to predict the drug response. Then, the representation vector is used as the input of the classification sub-network to train the network. At this stage, a unique combined loss function composed of binary cross-entropy loss and ternary loss is used to improve the performance of the model. MOLI's high predictive ability shows that it has practical value in the research of precision medicine.

Jonghwa et al. [22] proposed a deep neural network prediction model based on reference drugs (RefDNN). The model includes multiple Elastic Net models and a deep feedforward neural network classifier (DNN). Elastic Net generates low-dimensional features of cell line gene expression data by predicting drug resistance labels of input cell lines to multiple reference drugs. Then, splice the data features of the known drug structure and the obtained low-dimensional features of the cell line. Finally, put the combined features after stitching into DNN for training. DNN uses structural similarity maps (SSPs) based on user-defined reference drug [39], using the SSP of its target drug (SSP is a vector, which is the Tanimoto coefficient value obtained by comparing the binary fingerprint of the target drug and the reference drug) Integrate the results of these elastic network models, and then determine whether the cell line is resistant to the target drug. This model is not only superior to the existing computational prediction models in most comparative experiments, but also more robust than other deep learning models in predicting the response of unknown drug-cell line pairs.

Liu et al. [23] proposed a dual convolutional neural network model for clinical phenotype screening (tCNNS). The model first established two models of CNN branches to extract the characteristics of drugs and cell lines respectively. One of the convolutional networks is used to extract the characteristics of the drug from its simplified molecular input line input specification (SMILES) format [40], and the other convolutional network is used to extract the characteristics of cancer cell lines from the genetic feature vector. Then, a fully connected network (FCN) is established, the purpose of which is to perform a regression analysis between the output of the two branches and the IC50 value. This model can accurately predict the effect of drugs on cancer cell lines, with stable performance, less data but high quality, and the model can also solve the problem of outliers in other feature spaces.

Recent studies screened the response of multiple human cancer cell lines to a series of anti-cancer drugs and clarified the link between cell genotype and susceptibility. Due to the essential differences between cell lines and tumors, it is still challenging to translate into predicting drug response in tumors. Chiu et al. [24] developed a model for predicting drug response based on cancer cell or tumor mutations and expression profiles (Deep DR). It can extract abstract features of mutations and gene expression, and apply the information obtained from cancer cell lines to tumors. The model contains three sub-networks (DNNs). The first two sub-networks are a mutation encoder and a gene expression encoder, both of which are pre-trained using data retrieved from the GDSC database. The latter is the drug response prediction network that integrates the first two subnets. This method has several basic limitations: model complexity is high, and there is an overfitting problem; in addition, in order to enrich the complexity of tumor mutations used for model training, more genome mutation information can be added.

3 Database Resources

With the continuous in-depth study of the problem of drug response prediction, multi-omics data information has attracted the attention of researchers. Common multi-omics data include: gene expression profiles, MicroRNA expression profiles, copy number alteration, mutation profiles, DNA methylation profiles, etc. Various database resources published and continuously updated in recent years provide richer and more effective materials for computational method experiments. For example, in the gene expression relationship data commonly used to study cancer problems, the GEO database [41] issued by the National Center for Biotechnology Information is often used; the STRING [42, 43] and BIOGRID [44] containing protein interaction relationships; the TCGA [45] database with clinical patient samples as the core contains multi-sample, multi-type and multi-angle related data. And the KEGG [46] is a practical database resource for genome sequencing and other high-throughput experimental techniques generated from large molecular data sets.

In this review, we will introduce two of the most commonly used public benchmark databases (Table 3): Cancer Cell Line Encyclopedia (CCLE) and Genomics of Drug Sensitivity in Cancer (GDSC). CCLE is a tumor genomics research project initiated by the Broad Institute, collecting and collating omics data of more than one thousand tumor cell lines. Researchers can use this large public tumor genome database for in-depth data mining, and can browse the data after simple registration. The GDSC database was developed by the Sanger Institute in the United Kingdom and is the largest public resource for information on drug sensitivity and molecular markers of cancer cells. Contains a large number of drug sensitivity and genome data sets. CCLE contains drug sensitivity data of more than 11,000 experiments on 24 drugs and 1457 cell lines, as well as gene mutations and expression profiles of approximately 1,000 cell lines. GDSC includes data from approximately 200,000 drug response experiments, which tested 266 drugs on nearly 1,000 human cancer cell lines.

Table 3. The benchmark database mentioned in this review

	GDSC	CCLE
Number of experiments	200,000	11,000
Total number of cell lines	1,001	1457
Number of drugs tested on drug screening	266	24
Gene expression	√	√
DNA methylation	√	√
Copy number alteration	√	√
Mutational profiles	√	√
Website	www.cancerRxgene.org	https://portals.broadinstitute.org/ccle

4 Discussion

In this review, the types of drug response prediction models and the main data sources that appear in the context of the rapid increase of multi-omics information are introduced. Research analysis found that: On the one hand, massive omics information data is publicly available, and the development of predictive models that can effectively integrate data from many different biomedical sources and understand their impact on predicting drug response is still a research hotspot for researchers. Existing research results show that the predictive performance of the model that integrates different types of information for drug sensitivity analysis has relatively efficient performance. However, the effects of drugs are usually related to many factors, including genotype, phenotype, drug and chemical molecular structure, environmental factors and so on. For the research of drug response prediction, a hot direction in the future may be how to predict the response of drug combination by integrating more different types of biological data information. At the same time, with the increase of data, more challenges will arise, such as data redundancy that a large amount of data may generate, and the inconsistent importance of different types of biological data for drug response.

On the other hand, in recent years, the use of deep learning methods to predict drug response has become more and more popular, which may be a good direction to solve high-dimensional problems. Using different types of molecular and genomic data, more accurate and personalized drug treatment choices can be made. Although some calculation models based on deep learning perform well, there is still room for improvement in data mining in this field, and deep learning models lack interpretability and versatility. Researchers cannot use the computer theory of their algorithms to explain cancer to guide them. The application of DL technology in the prediction of drug response has not yet been fully explored. How to use DL technology to develop efficient and accurate prediction models is still worthy of in-depth exploration by researchers. In summary, the calculation method has certain limitations, but it has a good performance and predictable improvement in the prediction of drug effects. How to better use the calculation method is still one of the focuses of the future development of precision medicine.

5 Conclusion

The development of funny preclinical drug response prediction models based on biological data and computer technology has a very important role and significance for the realization of precision medicine. Although there is still a long way to go before clinical application, these computer science-based preclinical models can help people get closer to the goal of precision medicine in the clinic, and the results of these studies will help guide future development directions. As omics data and computer technology play an increasingly critical role in this topic, efficient and accurate preclinical drug prediction methods will have a wider range of needs and applications. Therefore, the strategy of developing an efficient drug response prediction model based on bioinformatics-related computing technology and massive biological auxiliary information will be an important development direction for drug response prediction in the future.

Acknowledgement. This work has been supported by the National Natural Science Foundation of China (Grant No. 61772197).

References

1. Roses, A.D.: Pharmacogenetics in drug discovery and development: a translational per spective. Nat. Rev. Drug Discov. **7**(10), 807–817 (2008)
2. Dugger, S.A., Platt, A., Goldstein, D.B.: Drug development in the era of precision medicine. Nat. Rev. Drug Discov. **17**(3), 183–196 (2018)
3. Bo, H., Yongqian, S., Ping, L., et al.: Precision medicine for oncology: concepts, techniques, and prospects. Tech. Rev. **33**(15), 14–21 (2015)
4. Sebaugh, J.L.: Guidelines for accurate EC50/IC50 estimation. Pharm. Stat. **10**(2), 128–134 (2011)
5. Neubig, R.R., Spedding, M., Kenakin, T., et al.: International union of pharmacology committee on receptor nomenclature and drug classification. XXXVIII. Update on terms and symbols in quantitative pharmacology. Pharm. Rev. **55**(4), 597–606 (2003)
6. Liu, D.Y., Ye, K.Q., Wang, H.Z., et al.: Research progress in the prediction of antitumor drug sensitivity based on functional test. Electron. J. Trans. Med. **004**(001), 1–6 (2017)
7. Alley, M.C., Scudiero, D.A., Monks, A., et al.: Feasibility of drug screening with panels of human tumor cell lines using a microculture tetrazolium assay. Can. Res. **48**(3), 589–601 (1988)
8. Shoemaker, R.H.: The NCI60 human tumour cell line anticancer drug screen. Nat. Rev. Cancer **6**(10), 813–823 (2006)

9. Ali, M., Aittokallio, T.: Machine learning and feature selection for drug response prediction in precision oncology applications. Biophys. Rev. **11**(1), 31–39 (2018). https://doi.org/10.1007/s12551-018-0446-z
10. DeNiz, C., Rahman, R., Zhao, X., et al.: Algorithms for drug sensitivity prediction. Algorithms **9**(4), 77 (2016)
11. Huang, S., Chaudhary, K., Garmire, L.X.: More is better: recent progress in multi-omics data integration methods. Front. Genet. **8**, 84 (2017)
12. Güvenç, P.B., Mamitsuka, H., Kaski, S.: Improving drug response prediction by integrating multiple data sources: matrix factorization, kernel and network-based approaches. Brief. Bioinform. **22**(1), 346–359 (2021)
13. Azuaje, F.: Computational models for predicting drug responses in cancer research. Brief. Bioinform. **18**(5), 820–829 (2017)
14. Wang, L., Li, X., Zhang, L., et al.: Improved anticancer drug response prediction in cell lines using matrix factorization with similarity regularization. BMC Cancer **17**(1), 1–12 (2017)
15. Emdadi, A., Eslahchi, C.: Dsplmf: a method for cancer drug sensitivity prediction using a novel regularization approach in logistic matrix factorization. Front. Genet. **11**, 75 (2020)
16. Guan, N.N., Zhao, Y., Wang, C.C., et al.: Anticancer drug response prediction in cell lines using weighted graph regularized matrix factorization. Molecular Therapy-Nucleic Acids **17**, 164–174 (2019)
17. Wang, S., Li, J.: Modular within and between score for drug response prediction in cancer cell lines. Mol. Omics **16**(1), 31–38 (2020)
18. Zhang, F., Wang, M., Xi, J., et al.: A novel heterogeneous network-based method for drug response prediction in cancer cell lines. Sci. Rep. **8**(1), 1–9 (2018)
19. Yang, J., Li, A., Li, Y., et al.: A novel approach for drug response prediction in cancer cell lines via network representation learning. Bioinformatics **35**(9), 1527–1535 (2019)
20. Sakellaropoulos, T., Vougas, K., Narang, S., et al.: A deep learning framework for predicting response to therapy in cancer. Cell Rep. **29**(11), 3367–3373 (2019)
21. Sharifi-Noghabi, H., Zolotareva, O., Collins, C.C., et al.: MOLI: multi-omics late integration with deep neural networks for drug response prediction. Bioinformatics **35**(14), 501–509 (2019)
22. Choi, J., Park, S., Ahn, J.: RefDNN: a reference drug based neural network for more accurate prediction of anticancer drug resistance. Sci. Rep. **7**(10), 1–11 (2020)
23. Liu, P., Li, H., Li, S., et al.: Improving prediction of phenotypic drug response on cancer cell lines using deep convolutional network. BMC Bioinform. **20**(1), 1–14 (2019)
24. Chiu, Y.C., Chen, H.I.H., Zhang, T., et al.: Predicting drug response of tumors from integrated genomic profiles by deep neural networks. BMC Med. Genomics **12**(1), 143–155 (2019)
25. Suphavilai, C., Bertrand, D., Nagarajan, N.: Predicting cancer drug response using a recommender system. Bioinformatics **34**(22), 3907–3914 (2018)
26. LeCun, Y., Bengio, Y., Hinton, G.: Deep learning. Nature **521**(7553), 436–444 (2015)
27. Baptista, D., Ferreira, P.G., Rocha, M.: Deep learning for drug response prediction in cancer. Brief. Bioinform. **22**(1), 360–379 (2021)
28. Wainberg, M., Merico, D., Delong, A., et al.: Deep learning in biomedicine. Nat. Biotechnol. **36**(9), 829–838 (2018)
29. Menden, M.P., Iorio, F., Garnett, M., et al.: Machine learning prediction of cancer cell sensitivity to drugs based on genomic and chemical properties. PLoS One **8**(4), e61318 (2013)
30. Kim, K.G.: Book review: deep learning. Healthcare Inform. Res. **22**(4), 351 (2016)
31. Kingma, D.P., Welling, M.: Auto-encoding variational bayes. ArXiv Preprint ArXiv **1312**, 6114 (2013)
32. LeCun, Y., Jackel, L.D., Boser, B., et al.: Handwritten digit recognition: applications of neural network chips and automatic learning. IEEE Commun. Mag. **27**(11), 41–46 (1989)

33. Lipton, Z.C., Berkowitz, J., Elkan, C.: A critical review of recurrent neural networks for sequence learning. ArXiv Preprint ArXiv **1506**, 00019 (2015)
34. Lenselink, E.B., TenDijke, N., Bongers, B., et al.: Beyond the hype: deep neural networks outperform established methods using a ChEMBL bioactivity benchmark set. J. Cheminformatics **9**(1), 1–14 (2017)
35. Koutsoukas, A., Monaghan, K.J., Li, X., et al.: Deep-learning: investigating deep neural networks hyper-parameters and comparison of performance to shallow methods for modeling bioactivity data. J. Cheminformatics **9**(1), 1–13 (2017)
36. Ma, J., Sheridan, R.P., Liaw, A., et al.: Deep neural nets as a method for quantitative structure–activity relationships. J. Chem. Inf. Model. **55**(2), 263–274 (2015)
37. Mayr, A., Klambauer, G., Unterthiner, T., et al.: DeepTox: toxicity prediction using deep learning. Front. Environ. Sci. **3**, 80 (2016)
38. Korotcov, A., Tkachenko, V., Russo, D.P., et al.: Comparison of deep learning with multiple machine learning methods and metrics using diverse drug discovery data sets. Mol. Pharm. **14**(12), 4462–4475 (2017)
39. Eckert, H., Bajorath, J.: Molecular similarity analysis in virtual screening: foundations, limitations and novel approaches. Drug Discov. Today **12**(5–6), 225–233 (2007)
40. Iorio, F., Knijnenburg, T.A., Vis, D.J., et al.: A landscape of pharmacogenomic interactions in cancer. Cell **166**(3), 740–754 (2016)
41. Barrett, T., Wilhite, S.E., Ledoux, P., et al.: NCBI GEO: archive for functional genomics data sets—update. Nucleic Acids Res. **41**(D1), D991–D995 (2012)
42. Szklarczyk, D., Franceschini, A., Kuhn, M., et al.: The STRING database in 2011: functional interaction networks of proteins, globally integrated and scored. Nucleic Acids Res. **39**(suppl_1), D561–D568 (2010)
43. Szklarczyk, D., Gable, A.L., Lyon, D., et al.: STRING v11: protein–protein association networks with increased coverage, supporting functional discovery in genome-wide experimental datasets. Nucleic Acids Res. **47**(D1), D607–D613 (2019)
44. Oughtred, R., Chatr-aryamontri, A., Breitkreutz, B.J., et al.: BioGRID: a resource for studying biological interactions in yeast. Cold Spring Harbor Protocols 2016(1), pdb. top080754 (2016)
45. Tomczak, K., Czerwińska, P., Wiznerowicz, M.: The Cancer Genome Atlas (TCGA): an immeasurable source of knowledge. Contem. Oncol. **19**(1A), A68 (2015)
46. Kanehisa, M., Furumichi, M., Tanabe, M., et al.: KEGG: new perspectives on genomes, pathways, diseases and drugs. Nucleic Acids Res. **45**(D1), D353–D361 (2017)

A Cross-Modal Image-Text Retrieval System with Deep Learning

Shuang Liu, Han Qiao, and Qingzhen Xu(✉)

School of Computer Science, South China Normal University, Guangzhou 510631, China

Abstract. With the generation of massive data, cross-modal image and text retrieval has attracted more and more attention. However, there is a heterogeneity gap between the data of different modalities, which cannot be measured directly. To solve this problem, the researchers project the data of different modalities into the common representation space to compensate for the differences between the data of different modalities. These studies focus on improving the mean average retrieval accuracy. Few research methods have been directly applied to the design of cross-modal retrieval systems. Therefore, this paper proposes a cross-modal image-text retrieval system based on deep learning. First, we design a cross-modal image and text retrieval system to realize the application of the retrieval system. Then, we apply deep learning to the cross-modal retrieval system to make full use of the research results of cross-modal retrieval. Finally, the designed system can carry out the bidirectional retrieval of images and texts to meet the requirements of images and texts as queries. The results show that the proposed cross-modal image and text retrieval system is capable of bidirectional retrieval of images and text, which is helpful to use different modalities images and text data.

Keywords: Cross-modal retrieval · Metric learning · Retrieval system

1 Introduction

Cross-modal image-text retrieval is the retrieval of image and text in two different modalities. With the rapid growth of data, the same semantics are described using data different modalities. Cross-modal retrieval can use different modality data to meet the needs of people's retrieval [1–3]. In recent years, it has attracted the attention of researchers in academia and industry. However, data in different modalities have different characteristics and cannot be measured directly.

To solve the data difference between different modalities, the common method is representation learning. That is, data in different modalities are transformed into a common space where different types of data can be measured directly. Researchers have proposed many approaches to representational learning to find a common space. Traditional methods use statistical correlation analysis to learn a linear transformation, for example, Canonical Correlation Analysis (CCA) [4] to learn the common subspace by maximizing pairwise correlation for heterogeneous data. However, linear transformation is difficult to represent relationships between complex heterogeneous data.

X. Sun et al. (Eds.): ICAIS 2021, CCIS 1422, pp. 538–548, 2021.
https://doi.org/10.1007/978-3-030-78615-1_47

Inspired by the successful application of deep learning in cross-modal retrieval, in recent years, researchers have proposed a number of deep learning-based methods to learn discriminative common representation space [5–10]. These methods usually use deep neural networks to perform nonlinear fitting and thus enhance the correlation between modalities. For example, Mikolajczyk et al. [6] proposed Deep Canonical Correlation Analysis (Deep CCA), which is mainly used to learn the nonlinear transformation between different modalities and make the data of different modalities highly linearly correlated. It is a nonlinear extension of CCA. Wang et al. [9] proposed a multimodal deep neural network, which is used to extract features of images and text. This method uses the label information of the sample to learn the internal semantics of the image and text, and uses the European distance to measure the distance between the image and text pairs, so as to supervise the cross-modal representation learning. Wang et al. [10] proposed Adversarial Cross-Modal Retrieval (ACMR) based on the ability of generating adversary network with strong modeling data distribution. The algorithm models the optimization process of common space as a form of counter training and effectively promotes the alignment of heterogeneous modal representation distribution. However, the existing methods usually focus on improving the retrieval accuracy and pay little attention to the application of cross-modal retrieval. Based on this, we propose a cross-modal image-text retrieval system based on deep learning.

The contributions of this paper are as follows:

- A cross-modal image-text retrieval system is proposed. Cross-modal retrieval usually focuses on the improvement of the retrieval accuracy, and seldom focuses on the application of cross-modal retrieval.
- Cross-modal retrieval system introduces deep learning. The introduction of deep learning facilitates retrieval.
- The bidirectional retrieval of images and text is introduced. The system not only supports image retrieval of text, but also supports text retrieval of image.

The rest of this paper is organized as follows: Sect. 2 introduces the related work of cross-modal image-text retrieval. Section 3 presents the design of the cross-modal image-text retrieval system. The principle of cross-modal retrieval module is introduced in Sect. 4. Section 5 implements the cross-modal image-text retrieval system.

2 Related Work

In order to bridge the modality gap, it is common practice to transform the data of different modalities into the same common representational space, that is, representation learning. There are two types of representational learning: real value based representational learning and binary based representational learning. The binary-based method, namely the cross-modal hash algorithm, has a high computational efficiency. This method maps heterogeneous data to the common hamming space. Because heterogeneous data is represented as binary information in hamming space, the cross-modal retrieval rate in this space is high. Real-value algorithms usually extract the features of different modalities data and use the rich information across-modal samples to establish associations.

Real-value representation learning approaches can be divided into four kinds: unsupervised approaches, pairwise approaches, rank-based approaches and supervised approaches. Unsupervised methods learn common representations of different types of data by associating information. Canonical Correlation Analysis (CCA) [11] and its extensions [12–15] are typical unsupervised approaches. Pair-based methods use sample pairs to learn the associations of different modalities data. Typical methods [16–18] include correspondence autoencoder (Corr-AE), semi-paired discrete hashing (SPDH) and the Modality-Specific Deep Structure (MSDS) method. The rank-based methods use sorting to realize the discrimination of the same modality data and the invariance of different modalities data. Typical methods [19, 20] include Learning Deep Structure-Preserving Image-Text Embeddings and Ranking Canonical Correlation analysis (RCCA). The supervised method uses category label information to shorten the distance between samples of the same category and to lengthen the distance between samples of different categories. Typical methods [21–25] include the Adversarial Learning, Cross-Modal Generative Adversarial Networks for Common Representation Learning (CM-Gans), Adversarial Cross-Modal Retrieval and Deep Adversarial Metric Learning for Cross-Modal Retrieval (DAML).

3 Design of a Cross-Modal Image-Text Retrieval System with Deep Learning

In order to complete a cross-modal image-text retrieval system, this paper designs the system and analyzes its module composition based on the requirement analysis of the system.

3.1 Problem Formulation

Cross-modal image-text retrieval is a mutual retrieval of image and text modalities, which is beneficial to image and text matching. However, the existing research usually focuses on the average precision of retrieval and seldom focuses on the image-text retrieval system directly. Therefore, this paper proposes a cross-modal retrieval system with deep learning.

3.2 The Structural Design of the System

To realize the application of image-text retrieval, we design a cross-modal image-text retrieval system [26, 27]. Based on the requirement analysis of the system, we propose the architecture design of the system.

The requirements of cross-modal retrieval system include image and text retrieval, page display and user management, see Fig. 1 below. The main requirement of the system is image retrieval of text and text retrieval of image. Two modalities data, image and text, are retrieved across the modality gap. In order to meet the needs of users, the system provides a page display function, to facilitate the interaction between users and the system. To facilitate the user management, the system provides user management functions.

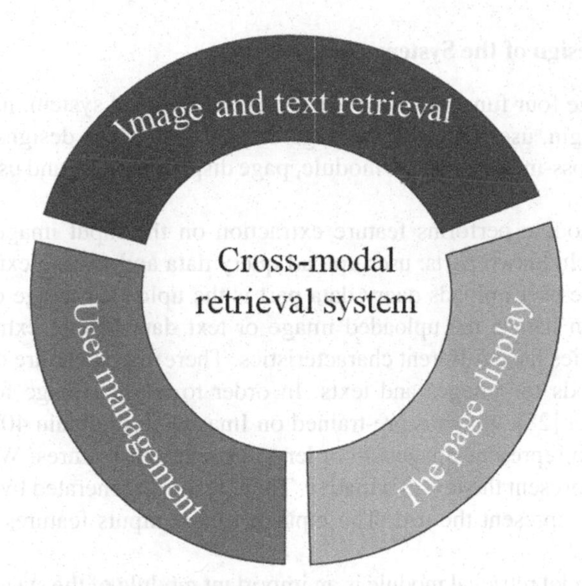

Fig. 1. System function division diagram.

To satisfy the requirements of cross-modal retrieval system, we design the architecture of cross-modal retrieval system. The system adopts the three-layer B/S mode, see Fig. 2 below. The three-layer architecture diagram includes a user interface layer, a business logic layer, and a data access layer. The Web interface of the user layer provides a visual interface for users to interact with users. The business logic layer realizes the core function of image and text retrieval. The data access layer is used to store data.

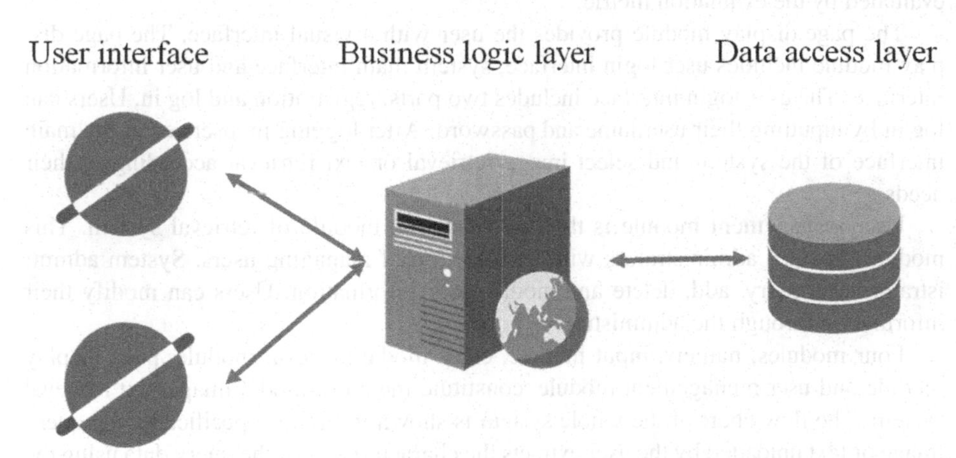

Fig. 2. A diagram of the architecture of a cross-modal retrieval system.

3.3 Module Design of the System

In order to realize four functions in the cross-modal retrieval system, namely user registration, user login, user retrieval and page display, this paper designs four modules: input module, cross-modal retrieval module, page display module and user management module.

The input module performs feature extraction on the input image and text. The specific steps include two parts: user upload query data and feature extraction of data. Among them, the user uploads query data part is the uploaded image or text data, the feature extraction part is the uploaded image or text data feature extraction. Data in different modalities have different characteristics. Therefore, there are different feature extraction methods for images and texts. In order to extract image features, we use 19-layer VGGNet [28], which is pre-trained on ImageNet, to obtain 4096-dimensional feature vectors to represent images. In order to extract text features, Word2Vec model [29] is used to represent the text as a matrix. Then, features generated by Sentence CNN [30] are used to represent the text. The input module outputs features of an image or text.

The cross-modal retrieval module is an important module of the system, which takes the features extracted by the input module as input and outputs relevant retrieval results. The module consists of feature transformation, similarity measurement and evaluation metric calculation. In the feature transformation part, the feature vectors of different modalities are transformed to the same common space, which is convenient for the direct comparison of data. In the similarity measurement part, the query data and the data after feature transformation in the database are measured in the common space, and the retrieval module and output module select the most similar data as the output. In the evaluation metric calculation part, the accuracy of the retrieval results is calculated and evaluated by the evaluation metric.

The page display module provides the user with a visual interface. The page display module includes user login interface, system main interface and user information interface. The user login interface includes two parts: registration and log in. Users can log in by inputting their username and password. After logging in, users enter the main interface of the system and select image retrieval or text retrieval according to their needs.

User management module is the basic function module of retrieval system. This module provides administrators with the function of managing users. System administrators can query, add, delete and modify user information. Users can modify their information through the administrator.

Four modules, namely, input module, cross-modal retrieval module, page display module and user management module, constitute the cross-modal image-text retrieval system. The flow chart of the whole system is shown in Fig. 3. Specifically, the query image or text uploaded by the user extracts the characteristics of the query data using the input module. The extracted features are transformed into the common space based on the feature transformation of the retrieval module. Then, the features of the projection are measured by similarity with the data in the database, so that the most similar image or text is selected as the output of the page display module and the user's retrieval result.

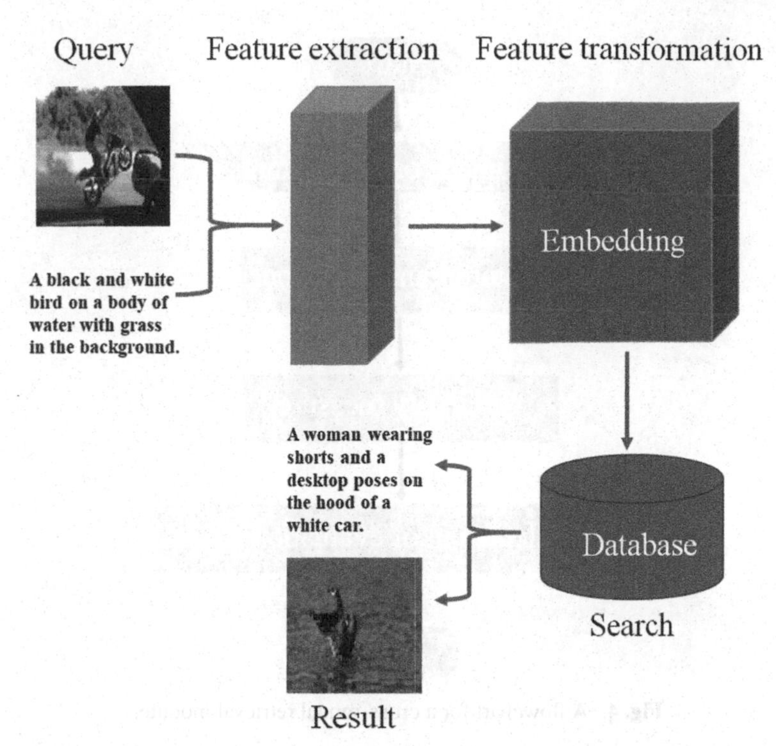

Fig. 3. A flowchart for a cross-modal image-text retrieval system.

4 The Principle of Cross-Modal Retrieval Module is Introduced

The main function of the system is image and text retrieval. The retrieval module is the core module of image and text retrieval. The retrieval module is introduced below.

The data of different modalities cannot be compared directly. In order to bridge the modal gap, the retrieval module transforms the data of different modalities into the common space for the convenience of direct comparison of the data in different modalities. The module consists of feature transformation, similarity measurement and evaluation metric calculation. The specific flowchart is shown in Fig. 4. In the feature transformation part, the data in different modalities are projected into the common space. Then, the data in different modalities are measured in this space. Finally, the retrieval results are measured based on the calculation of evaluation metric.

To measure similarity, data in different modalities are transformed into a common space. In this space, the data with the same semantics among the modalities are invariant, the data of different categories in the modalities are discriminative, and the data of different modalities are similar. Specifically, invariant means that the distance of the data is as close as possible, discrimination means that the distance of the data is as far as possible, and similarity means that it is impossible to distinguish which modal the data belongs to.

According to the results of similarity measurement, mAP metric [31] is used to measure the retrieval results. This metric represents the average precision of the retrieval

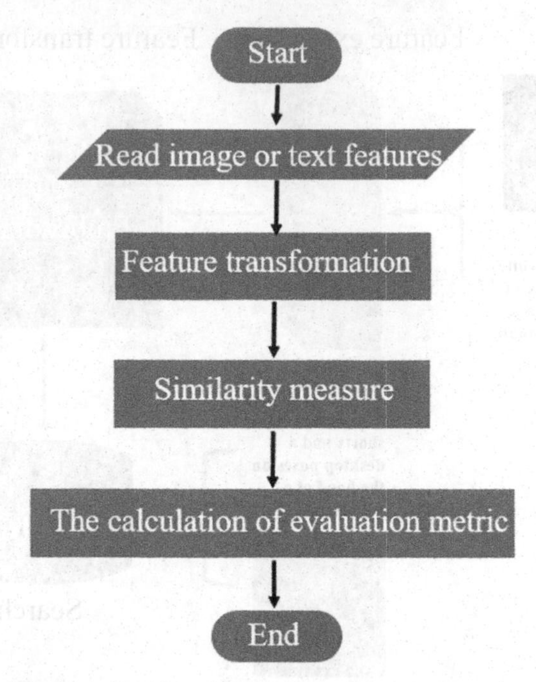

Fig. 4. A flowchart for a cross-modal retrieval module.

task, where the average precision can be calculated using AP. The higher the mean precision, the more accurate the retrieval result.

The cross-modal retrieval module takes the feature extracted from the input mode as the input, and after three parts of feature transformation, similarity measurement and evaluation index calculation, the output result is the output of the page display module.

5 Implement of Cross-Modal Image-Text Retrieval System

Based on the system requirements of image-text retrieval, page display and user management, this paper realizes the functions of user registration and login, user retrieval and page display.

5.1 System Initial Page

Users can enter the image-text retrieval system by using the system login page. On this page, users can register an account or directly log in the system, as shown in Fig. 5.

5.2 Retrieval Pages Cross-Modal

Cross-modal retrieval consists of two parts: image to text retrieval and text to image retrieval. The user retrieves the corresponding text or image with the image or text as the query. The results of image to text retrieval and text to image retrieval are shown in Fig. 6 and Fig. 7 respectively.

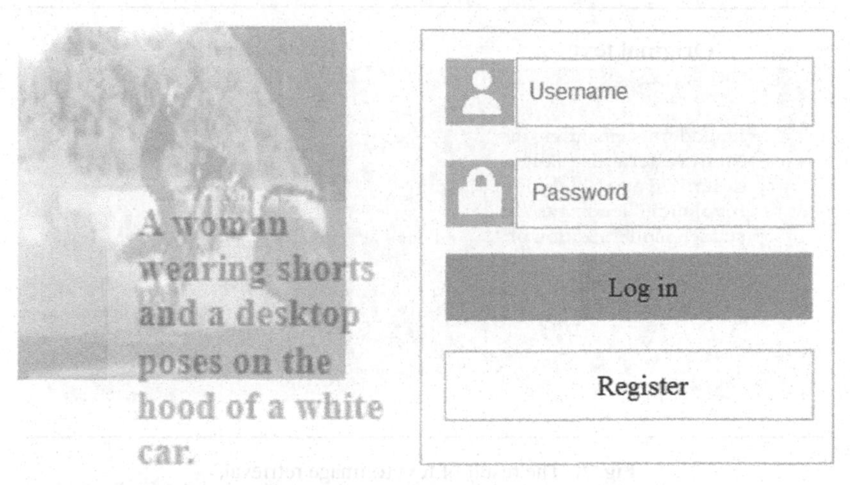

Fig. 5. System login page.

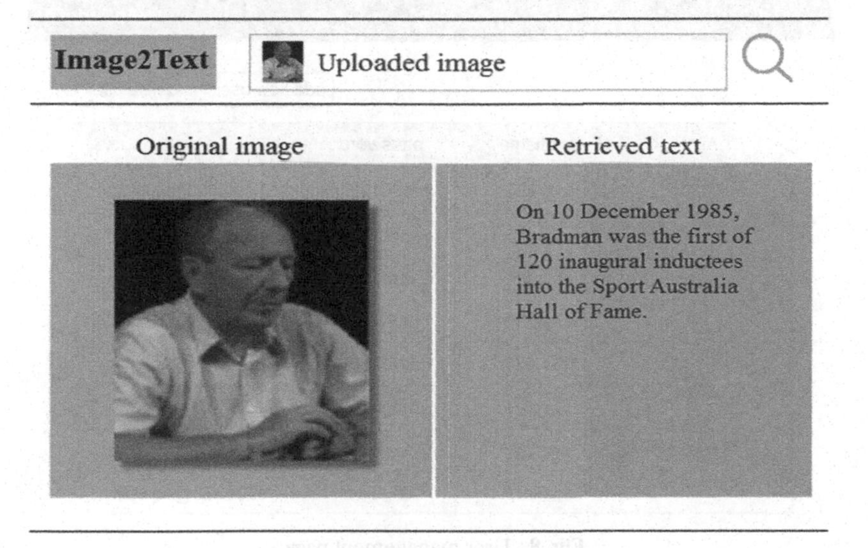

Fig. 6. The result of image to text retrieval.

5.3 User Management Page

In order to facilitate the user management, the system administrator can add, delete, query and modify the user in the user management page. The user administration page is shown in Fig. 8.

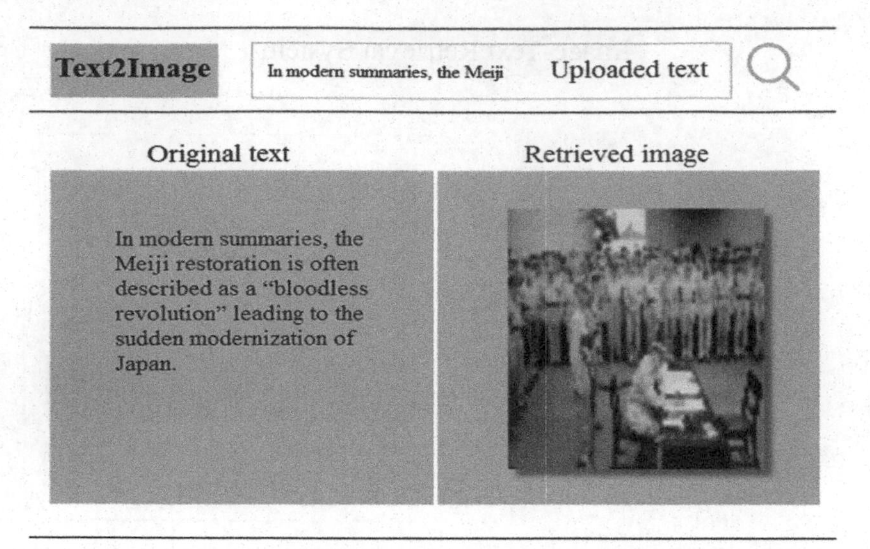

Fig. 7. The result of text to image retrieval.

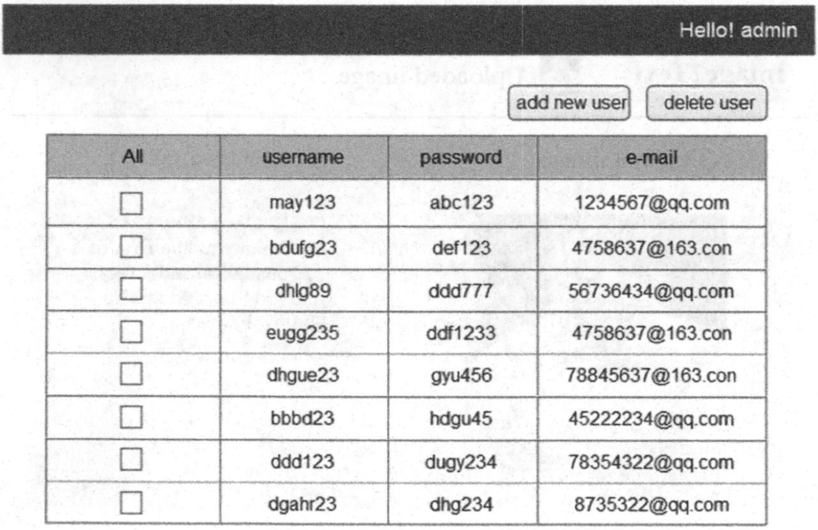

Hello! admin

add new user | delete user

All	username	password	e-mail
☐	may123	abc123	1234567@qq.com
☐	bdufg23	def123	4758637@163.con
☐	dhlg89	ddd777	56736434@qq.com
☐	eugg235	ddf1233	4758637@163.con
☐	dhgue23	gyu456	78845637@163.con
☐	bbbd23	hdgu45	45222234@qq.com
☐	ddd123	dugy234	78354322@qq.com
☐	dgahr23	dhg234	8735322@qq.com

Fig. 8. User management page.

6 Conclusions

The existing methods usually focus on the improvement of the retrieval accuracy, and pay little attention to the application of cross-modal retrieval. Therefore, we propose a cross-modal image-text retrieval system based on deep learning. For the application of cross-modal retrieval, we introduce the method of cross-modal retrieval into the system. For the convenience of users, we introduce the image and text bidirectional retrieval.

The results show that the proposed cross-modal image retrieval system is capable of image-text retrieval.

Acknowledgement. The Project was supported by Guangzhou Science and Technology Plan Project (No. 201903010103), the "13th Five-Year plan" for the development of philosophy and Social Sciences in Guangzhou (No. 2018GZYB36), Science Foundation of Guangdong Provincial Communications Department (No. 2015-02-064).

References

1. Zhu, C., Wang, Y., Pu, D., Qi, M., Sun, H., Tan, L.: Multi-modality video representation for action recognition. J. Big Data **2**(3), 95 (2020)
2. Fang, W., Zhang, F., Ding, Y., Sheng, J.: A new sequential image prediction method based on LSTM and dcgan. Comput. Mater. Continua **64**(1), 217–231 (2020)
3. Hu, J., et al.: End-to-end multimodal image registration via reinforcement learning. Med. Image Anal. **68**, 101878 (2020)
4. Harold, H.: Relations between two sets of variables. Biometrika (3–4), 3–4
5. Peng, Y., Huang, X., Qi, J.: Cross-media shared representation by hierarchical learning with multiple deep networks. In: Kambhampati, S. (ed.) IJCAI 2016, pp. 3846–3853. IJCAI/AAAI Press (2016).
6. Yan, F., Mikolajczyk, K.: Deep correlation for matching images and text. In: IEEE Conference on Computer Vision and Pattern Recognition, pp. 3441–3450. IEEE Computer Society (2015)
7. Wei, Y., et al.: Cross-modal retrieval with CNN visual features: a new baseline. IEEE Trans. Cybern. **47**(2), 449–460 (2017)
8. Peng, Y., Qi, J., Huang, X., Yuan, Y.: CCL: cross-modal correlation learning with multigrained fusion by hierarchical network. IEEE Trans. Multim. **20**(2), 405–420 (2018)
9. Wang, W., Yang, X., Ooi, B.C., Zhang, D., Zhuang, Y.: Effective deep learning-based multi-modal retrieval. VLDB J. **25**(1), 79–101 (2015). https://doi.org/10.1007/s00778-015-0391-4
10. Wang, B., Yang, Y., Xu, X., Hanjalic, A., Shen, H.T.: Adversarial cross-modal retrieval. In: Liu, Q., et al. (eds.) MM 2017, pp. 154–162. ACM (2017)
11. Hardoon, D.R., Szedmák, S., Shawe-Taylor, J.: Canonical correlation analysis: an overview with application to learning methods. Neural Comput. **16**(12), 2639–2664 (2004)
12. Gong, Y., Ke, Q., Isard, M., Lazebnik, S.: A multi-view embedding space for modeling internet images, tags, and their semantics. Int. J. Comput. Vis. **106**(2), 210–233 (2014)
13. Rasiwasia, N., et al.: A new approach to cross-modal multimedia retrieval. In: Bimbo, A.D., Chang, S., Smeulders, A.W.M. (eds.) Proceedings of the 18th International Conference on Multimedia 2010, pp. 251–260. ACM (2010)
14. Sharma, A., Kumar, A., III, H.D., Jacobs, D.W.: Generalized multiview analysis: A discriminative latent space. In: 2012 IEEE Conference on Computer Vision and Pattern Recognition, pp. 2160–2167. IEEE Computer Society (2012)
15. Zhang, H., Gao, X., Wu, P., Xu, X.: A cross-media distance metric learning framework based on multi-view correlation mining and matching. World Wide Web **19**(2), 181–197 (2015). https://doi.org/10.1007/s11280-015-0342-4
16. Hong, R., Yang, Y., Wang, M., Hua, X.: Learning visual semantic relationships for efficient visual retrieval. IEEE Trans. Big Data **1**(4), 152–161 (2015)
17. Shen, X., Shen, F., Sun, Q., Yang, Y., Yuan, Y., Shen, H.T.: Semi-paired discrete hashing: learning latent hash codes for semi-paired cross-view retrieval. IEEE Trans. Cybern. **47**(12), 4275–4288 (2017)

18. Wang, J., He, Y., Kang, C., Xiang, S., Pan, C.: Image-text cross-modal retrieval via modality-specific feature learning. In: Hauptmann, A.G., Ngo, C., Xue, X., Jiang, Y., Snoek, C., Vasconcelos, N. (eds.) Proceedings of the 5th ACM on International Conference on Multimedia Retrieval, pp. 347–354. ACM (2015)
19. Wang, L., Li, Y., Lazebnik, S.: Learning deep structure-preserving image-text embeddings. In: CVPR 2016, pp. 5005–5013. IEEE Computer Society (2016)
20. Yao, T., Mei, T., Ngo, C.: Learning query and image similarities with ranking canonical correlation analysis. In: ICCV 2015, pp. 28–36. IEEE Computer Society (2015)
21. Peng, Y., Qi, J.: CM-gans: cross-modal generative adversarial networks for common representation learning. ACM Trans. Multim. Comput. Commun. Appl. **15**(1), 22:1–22:24 (2019)
22. Goodfellow, I.J., et al: Generative adversarial nets. In: Ghahramani, Z., Welling, M., Cortes, C., Lawrence, N.D., Weinberger, K.Q. (eds.) Advances in Neural Information Processing Systems 27, pp. 2672–2680 (2014)
23. Xu, X., He, L., Lu, H., Gao, L., Ji, Y.: Deep adversarial metric learning for cross-modal retrieval. World Wide Web **22**(2), 657–672 (2018). https://doi.org/10.1007/s11280-018-0541-x
24. Kang, P., Lin, Z., Yang, Z., Fang, X., Li, Q., Liu, W.: Deep semantic space with intra-class low-rank constraint for cross-modal retrieval. In: Proceedings of the 2019 on International Conference on Multimedia Retrieval, pp. 226–234 (2019)
25. Hu, P., Zhen, L., Peng, D., Liu, P.: Scalable deep multimodal learning for cross-modal retrieval. In: Proceedings of the 42nd International ACM SIGIR Conference on Research and Development in Information Retrieval, pp. 635–644 (2019)
26. Fang, W., Pang, L., Yi, W.: Survey on the application of deep reinforcement learning in image processing. J. Artif. Intell. **2**(1), 39–58 (2020)
27. Wang, C., Xu, Q., Lin, X., Liu, S.: Research on data mining of permissions mode for android malware detection. Clust. Comput. **22**(6), 13337–13350 (2019)
28. Simonyan, K., Zisserman, A.: Very deep convolutional networks for large-scale image recognition. In: Bengio, Y., LeCun, Y. (eds.) ICLR 2015 (2015)
29. Mikolov, T., Sutskever, I., Chen, K., Corrado, G.S., Dean, J.: Distributed representations of words and phrases and their compositionality. Adv. Neural. Inf. Process. Syst. **26**, 3111–3119 (2013)
30. Kim, Y.: Convolutional neural networks for sentence classification. In: Moschitti, A., Pang, B., Daelemans, W. (eds.) EMNLP 2014, pp. 1746–1751. ACL (2014)
31. Bellet, A., Habrard, A., Sebban, M.: A survey on metric learning for feature vectors and structured data. CoRR abs/1306.6709 (2013)

Research on Behavioral Decision Reward Mechanism of Unmanned Swarm System in Confrontational Environment

Tingting Zhang[1(✉)], Yushi Lan[2], and Aiguo Song[3]

[1] School of Command and Control Engineering, Army Engineering University, Nanjing 210017, China
101101964@seu.edu.cn

[2] The 28th Research Institute of China Electronics Technology Group Corporation, National Defence Science and Technology Key Laboratory, Nanjing 210017, China

[3] School of Instrumental Science and Engineering, Southeast University, Nanjing 210096, China

Abstract. Unmanned swarm system is regarded as a multi-agent system. At present, many kinds of research use deep reinforcement learning algorithms to solve the problem of multi-agent autonomous behavioral decision. The multi-agent deep deterministic policy gradient (MADDPG) algorithm proposed by Google's DeepMind team pioneered the application of multiple agents in collaborating and confronting complex scenarios, and can be used to solve the problem of generating an autonomous strategy for unmanned swarm in continuous action space. Experiments show that this algorithm has an unsatisfactory convergence and stability in a competitive environment. The key to multi-agent strategy optimization is the design of the reward function. Thus, the paper improves the algorithm of the reward model for multi-agent behaviors in the confrontational environment. First, the advantages of the MADDPG algorithm for implementing multi-agent behavioral decision in a confrontational environment and the training and execution framework of the reward function are introduced. Second, a per-distance reward mechanism and the addition of distance parameters are proposed to improve the efficiency of the agents' behavioral decision and solve the delayed reward problem. The global sharing value reward strategy is changed to the experience pool sharing value reward strategy to avoid the phenomenon of lazy agents and improve the efficiency of the agents' cooperation. Finally, after placing the improved reward mechanism into the MADDPG algorithm, experimental results show that the improved reward function has improved convergence, the algorithm training cycle is shortened, the function curve stabilizes faster, and multiple agents have better performance in the confrontational environment. This paper provides reference methods for intelligent behavioral decisions of the unmanned swarm system.

Keywords: Unmanned swarm system · MADDPG algorithm · Confrontational environment · Behavioral decision · Reward mechanism

1 Introduction

Unmanned swarm system is the focus of domestic and international research in the military field in recent years, which promotes the development of unmanned combat

X. Sun et al. (Eds.): ICAIS 2021, CCIS 1422, pp. 549–563, 2021.
https://doi.org/10.1007/978-3-030-78615-1_48

style from "single platform remote control operation" to "intelligent swarm operation" [1], among which UAV swarm operation is the typical operational style of the unmanned swarm system. The unmanned swarm system can be regarded as a number of homogeneous or heterogeneous unmanned equipment through a self-organization network connection group composed of intelligence, forming a distributed sensing, target recognition, autonomous decision making, collaborative planning, and attack ability, and has the characteristics of interactive learning and emergence of swarm intelligence [2]; people expect unmanned swarms to have self-learning and self-decision-making autonomous combat capabilities, and with the development of artificial intelligence technology, autonomous decision-making of unmanned system behavior becomes possible [3]. An unmanned swarm system is often faced with a strong confrontation environment. Under such circumstances, each unmanned execution module can efficiently and accurately complete the assigned task in a collaborative manner. Therefore, an efficient behavior decision-making method for the collaborative task completion of a multi-agent unmanned system in a confrontation environment must be studied and built urgently. At present, most of the cooperative operation strategies under the established environment and task planning adopted in the fields of investigation, reconnaissance, and public safety lack the adaptive perception of the environment and task changes, and the behavior strategies that can cooperate autonomously. In the antagonistic environment, from the perspective of a single unmanned system, other unmanned systems are also dynamically changing, and their behaviors are unknown, which increases the difficulty for the unmanned system to adapt to the uncertainty of the dynamic environment and the autonomous decision making of behaviors. The unmanned system cannot adapt to the dynamic unstable environment by only adjusting the agent's own strategies.

Deep reinforcement learning algorithms have been widely used in autopilot strategy learning, robot control, and game competition. The research on multi-agent deep reinforcement learning algorithms is also deepening. At present, many researchers generate continuous state sequences and behavior sequences through multi-agent reinforcement learning strategies [4] to solve the multi-agent behavior decision-making problem. In this process, numerous algorithms, such as the lenient-DQN algorithm [5] proposed by Palmer et al., have emerged. The core idea is that the behavior strategy of each agent generates noise that can affect the updating of other agents' strategies. Shayegan et al. integrated the time axis into each agent and proposed the hysteretic DRQN algorithm [6], which uses the experience playback of agents simultaneously to solve the problem of multi-agent strategy imbalance in a dynamic environment, but this algorithm does not involve the environmental changes caused by changes in the number of agents and task types. Zheng et al. combined these two algorithms and proposed the weighted WDDQN [7] to deal with nonstatic environment problems in a multi-agent environment through dynamic weight setting. The algorithm of VDN [8] treats all other agents as the environment, and each agent carries out its own Q-learning process. It only sums up the local action value function of each agent to obtain the joint action value function. The value function calculated by the agent alone makes the multi-agent strategy unstable. Tabish Rashid et al. proposed the QMIX algorithm [9] to train the decentralized agent strategy in a centralized, end-to-end manner based on value. Jakob proposed the algorithm of COMA [10]. Based on the strategy search, an efficient evaluation network structure was designed, and the behavioral value function of all actions of the current agent was calculated once by using the neural network. However, the value network structure remained centralized, and numerous state values needed to be input, which was difficult to be applied to a large-scale multi-agent system. Existing research has made several achievements in the fields of mobile sensor network, robotics, and network security [11, 12]. However, most of these algorithms solve the behavior decision of unmanned swarm system with fixed strategy generation and do not fully adapt to the dynamic environment. The unmanned swarm system is regarded as a multi-agent system, characterized by the cooperative and antagonistic relationship between agents. The antagonistic agent strategy is often unknown, making the state of all agents dynamically changeable. How to unite the dynamically variable collaborators and deal with the opponent with a variable state is the difficulty in the autonomous behavior decision of agents.

Analysis finds that compared with other multi-agent reinforcement learning algorithms, in 2017, the Google DeepMind team pioneering deterministic strategy proposed the multi-agent depth gradient algorithm (multi-agent deep deterministic policy gradient MADDPG) [13], which implements multi-agent synergy with the autonomous behavior of the decision against the complex scene, and continuous action space can solve the unmanned swarm strategy against generation problems independently. The algorithm considers the cooperative and antagonistic relations, and designs the cooperative and antagonistic reward functions. In addition to understanding its own agent strategy, it estimates the strategies of other agents. It makes full use of global information in training and only uses local information in execution, thus effectively alleviating environmental instability. In generating countermeasures, the reward function plays the role of the monitoring signal, and the agent optimizes the strategy according to the reward value. The setting of the reward value is the core problem in solving the actual task [14]. Although the algorithm solves the multi-agent TiXie confronted with strategy generation problems and the reward for sparse reward MADDPG algorithm, the reward signal change is not evident in training, the agent will use the strategy of gradient method to explore few successful samples, a long training time is needed to achieve the optimal strategy, and the convergence of algorithm performance is poor. Thus, realizing a rapid decision-making environment given an unmanned swarm system independent behavior is difficult, and the efficiency of the algorithm convergence must be improved.

In this paper, the per-distance reward mechanism is proposed based on the improved MADDPG algorithm. First, the distance parameter is introduced to increase the reward signal, solve the problem of delayed rewards, and improve the decision-making efficiency of the behavior of unmanned systems. Second, the sharing of return value between agents must be canceled and replaced with experience pool sharing to avoid the phenomenon of lazy agents and improve the cooperation efficiency of unmanned systems. Experiments show that this method improves the convergence speed of the algorithm and makes it more stable. In this manner, the decision-making efficiency of unmanned swarm behavior under the antagonistic environment can be improved.

2 MADDPG Algorithm

2.1 Core Idea of the Algorithm

Poor convergence of algorithms is the current dilemma of multi-agent deep reinforcement learning. The environment observed from the angle of a single agent is constantly changing because the actions of each agent are transformed in real time during training. The MADDPG algorithm introduces centralized training, distributed implementation, and actor critic network updating strategy to solve the problem of training instability. Specific means is, when the agent training, to add other information agent action Actor to the environment condition, at time t, adding all the agent to perform an action, as the state of the environment at the next moment t', to join can observe global critic network to guide the Actor-network training. When testing, only actors with local observations are used to act and change the unstable environment state into a stable environment state, thus reducing the complexity of multi-agent behavior decisions.

2.2 Basic Assumptions

The MADDPG algorithm follows the Markov decision process [15] and can be defined as a multivariate group $< S, A_1, A_2, \cdots, A_n, R_1, R_2, \cdots, R_N, T, O, \gamma >$. Let the environment where the agent is located be Envs, the environment contains n agents, and S is the state space of the environment. $A_i (i = 0, 1, \cdots, n)$ represents the action space of single-agent i, while $A_1 \times A_2 \times \cdots \times A_n$ represents the joint action space of all agents. R_i is the sum of a series of R_i, which represents the total reward of agent i, R_i is obtained by agent i when the multi-agent system executes joint action A and transitions from state $s \in S$ to state $s \in S$ instant rewards. $T : S \times A \times S \rightarrow [0, 1]$ is the state transition function, which represents the probability distribution of multi-agent transition to state S' after performing joint action A in state s. $o_i \in O$ is the observation value of agent i to the environment, and the observation attributes can be divided into partial observation and complete observation. γ is a discount factor, which is used to adjust the weight between long-term rewards and instant rewards. In a multi-intelligence environment, the state transition is the result of the joint actions of all agents. The n agents make behavioral decisions a_i based on their own observations o_i and instant reward r_i, and jointly output a joint action A to promote the transition of environment state s. Therefore, the reward of the agent is related to the joint strategy. The parameter set of all agents is $\theta = \{\theta_1, \theta_2, \cdots, \theta_N\}$. Assuming that the deterministic strategy adopted by the agent each time is μ, the action of each step can be obtained by the formula: $a_t = \mu(S_t)$, and the reward obtained after executing a certain strategy is determined by the Q function. The game of competition and cooperation under certain communication methods is realized to improve the versatility of the algorithm.

The algorithm operating conditions are as follows:

1) The learning strategy observes information based on a single-agent perspective.
2) The unknown environmental dynamic model is not involved, and the agent's behavior only depends on the strategy.
3) The communication between agents is in the mode of full connectivity.

2.3 Algorithm to Perform

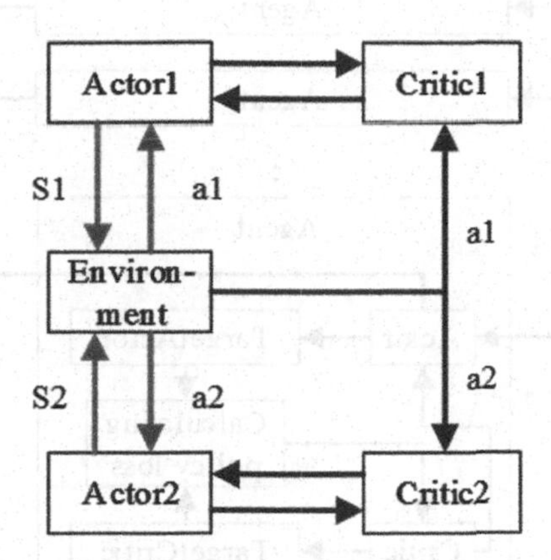

Fig. 1. View of MADDPG algorithm training execution.

Figure 1 shows the algorithm execution, which is different from the agent's perspective in the shared environment, where the input state of each agent is different, and each executor interacts with the environment without paying attention to the state of other agents. After the environment outputs the next full information state Sall, actors actor1 and actor2 can only obtain part of state information S_1, S_2 that they can observe, and execute the loop part identified by the green line in Fig. 1. During training, critic1 and critic2 can obtain the full information status and the strategic actions a_1, a_2 taken by all agents.

2.4 Actor Critic Network Update Strategy

Figure 2 shows that in environment Evns, the agent is composed of Actor network and Critic network, which in turn contains a target network (target-net) and an estimation network (eval-net). The Actor network is a simulation of the μ function by a convolutional neural network, and the parameter is θ^{μ}. The Critic network is a simulation of the Q function by a convolutional neural network, and the parameter is θ^{Q}.

The Actor network is represented as follows:

$$\text{Actor network}\begin{cases}\text{eval-net}:(s|\theta^{\mu})\\ \text{target-net}:\left(s|\theta^{\mu'}\right)\end{cases} \tag{1}$$

The Critic network is represented as follows:

$$\text{Critic network}\begin{cases}\text{eval-net}:(a|\theta^{Q})\\ \text{target-net}:\left(a\middle|\theta^{Q'}\right)\end{cases} \tag{2}$$

The Actor network does not see all the states of the environment and does not know the strategies of the other agents, but each agent's Actor network has a mentor Critic

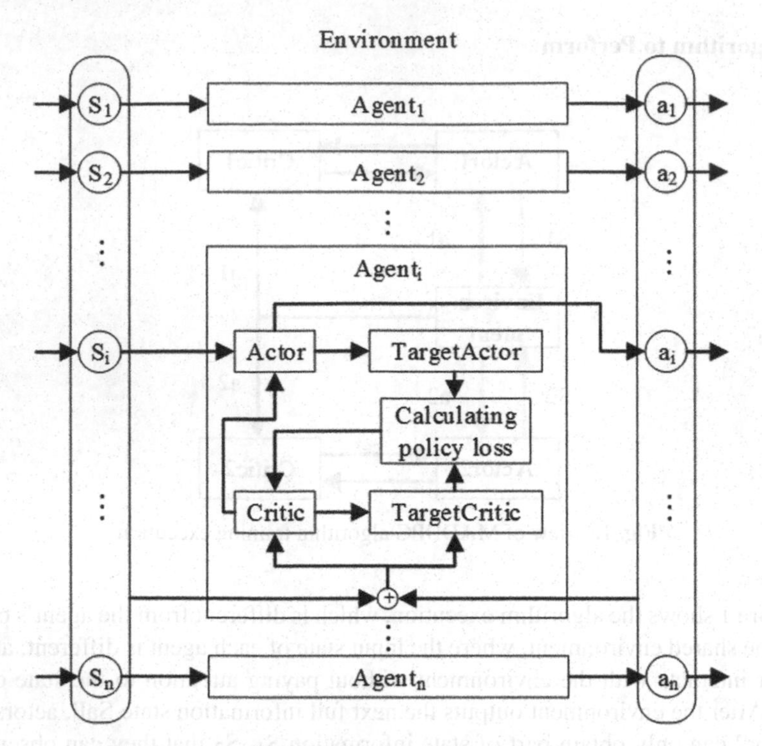

Fig. 2. View of MADDPG algorithm training framework.

network with a full view, which can observe all the information and direct the corresponding Actor network optimization strategy. In training, only eval-net parameters need to be estimated, and target-net parameters directly copy and estimate network parameters every certain time [16]. According to the MADDPG algorithm, Actor training can be guided by the global Critic network when training, and only the Actor network with local observation can be used to take actions when executing [13]. Assuming agent i, the action taken by the agent is $a_i^j = \pi_i^j(S_i^j)$. After interacting with the environment, experience $\left(a_i^j, \pi_i^j, S_i^{j+1}, r_i^j\right)$ is gained and stored. After all agents interact with the environment, each agent randomly extracts stored behavior experience and joins the strategy network for training. The observation state information and behavior actions of the remaining agents are added to the strategy Critic network to improve the learning efficiency and defined as $Q = Q\{S_j, a_1, a_2, \cdots, a_N, \theta^Q\}$, where:

$$S_j = \left\{S_1^j, S_2^j, \cdots, S_N^j\right\} \tag{3}$$

The Actor network parameters of each agent are updated through gradient descent:

$$\nabla_{\theta^\pi} J = \frac{1}{K}\sum_{j=1}^{K} \nabla_{\theta^\pi} \pi\left(S, \theta^\pi\right) \nabla_\theta Q\left(S_j, a_1, \cdots, a_N, \theta^Q\right) \tag{4}$$

3 MADDPG Algorithm

3.1 Reward Function Setup Mechanism

The reward function is the core of the reinforcement learning algorithm [17], which determines the convergence speed and stability of the algorithm. The evaluation method is that the curve of the reward value and the training times function can be stable within a short number of training times.

Compared with a single agent, the reward mechanism in a multi-agent system is more complex. In the current multi-agent deep reinforcement learning algorithm, the reward signals are generally global reward signals and local reward signals, which are problematic in two aspects [18]. First, the global reward signal does not distinguish the action of an agent, and all agents are given the same reward value, which will reduce the initiative of the agent to try the action, leading to the emergence of unearned agents. Second, local reward signals are given to each agent with different local reward values according to the actions of a single agent, which may lead to the emergence of selfish agents and reduce the cooperation between multiple agents. The MADDPG algorithm uses centralized training and distributed execution to mitigate the effect of these two problems.

In the antagonistic environment, the rewards of multiple agents depend on not only their own strategies but also the antagonistic strategies learned by the opponent. The learning speed of the two strategies may not be synchronized, leading to the situation that the rewards may not continue to rise, and even fluctuate and oscillate [19]. How to design an appropriate reward signal to solve the rapid learning and stable convergence of agents in the competitive environment has become a key problem.

3.2 MADDPG Algorithm Reward Mechanism Improved

Take predators and escapers as an example. The reward mechanism given by MADDPG algorithm is the reward value +10 for the collider during the collision and the penalty value -1 in the time of no collision. Escaper is the opposite. The advantage of this reward mechanism is that the absolute value of rewards for predators and escapers is the same, but the two sides have opposite strategies. Therefore, the learning speed of the two sides will gradually reach synchronization, and the fluctuation and shock of the reward value will be alleviated, as shown in Table 1.

Table 1. Reward mechanism settings.

Executor	Collision	No collision
Predator	+10	−1
Escaper	−10	+ 1

The disadvantage is that the convergence speed of the algorithm under the reward mechanism is very slow. Given that the MADDPG algorithm adopts the deterministic policy gradient algorithm, its Bellman equation is as follows:

$$Q^{\pi}(s,a) = \sum_{s'} P^{a}_{ss'}\left[R^{a}_{ss'} + \gamma \sum_{a'} Q^{\pi}(s',a')\right] \tag{5}$$

The agent has few successful samples in the early stage of execution, resulting in the lack of sufficient learning experience in the experience pool to adjust the strategy. When the Critic network is updated, most of the time, the predator's return value is $r^{j}_{i} = -1$, and the escaper's return value is $r^{k}_{p} = 1$. The Bellman equation shows that the reward value of the agent after any action is the same, the Critic network cannot distinguish between the pros and cons of the action, the reward function is unstable, and the training convergence rate is very slow.

3.3 Propose a Per-Distance Reward Mechanism

In the 3V1 hunting scene of the MADDPG algorithm, the distance between the predator and the runaway is as follows:

$$D(i,p) = \sqrt{(x_i - x_p)^2 + (y_i - y_p)^2} \tag{6}$$

When the distance between the predator and the escaper is less than or equal to 0, the predator will receive a large reward, and the escaper will return a negative reward. The collision reward is as follows:

$$C = \begin{cases} 10, collision \\ -1, non - collision \end{cases} \tag{7}$$

The antagonistic behavior of agents should be ensured within the set operating range, and the idea of agents crossing boundaries must be limited to improve the computational efficiency by imposing large penalties on agents that escape the boundary, depending on how far they are. Boundary reward B is added to ensure that the agent moves within the environment and does not cross the boundary and escape. Let (x_i, y_i) be the coordinates of agent i in the 2D environment, and 0.9 is the diameter of agent i. If the maximum distance of the agent from the boundary is < 0.9, it is considered beyond the boundary, and boundary reward value $B = 0$. If the maximum distance of the agent from the boundary > 0.9, boundary reward value B will give a dynamic reward value related to the boundary distance $(\max(x_i, y_i) - 0.9) * 200$, where 200 is any given weight value, playing the role of amplification factor.

Taking predator as an example, boundary reward B is as follows:

$$B = \begin{cases} 0, max(x_i, y_i) < 0.9 \\ (\max(x_i, y_i) - 0.9) * 200, else \end{cases} \tag{8}$$

In the MADDPG algorithm, the reward formula for predators is as follows:

$$r_i = B + C \tag{9}$$

The reward formula for predation is as follows:

$$r_p = B - C \tag{10}$$

The reward mechanism of the MADDPG algorithm has only two states for the distance setting of the opposing parties, that is, > 0 no collision or ≤ 0 collision. The reality is that the opponents are in a state of > 0 non-collision most of the time, and training to obtain the optimal strategy takes a long time, which causes delayed rewards. Aiming at the low efficiency of behavior decision making in current reinforcement learning algorithms, this paper improves the reward mechanism and proposes a per-distance reward mechanism to improve the efficiency of behavioral decision making.

1) An independently calculated distance parameter is added to solve the delayed reward problem. It is no longer the reward value in the two states of > 0 no collision or ≤ 0 collision. Instead, the reward value is set dynamically according to the variable distance. Distance parameter D(i, p) is increased to indicate the difference between each predator and the escaper. In terms of distance, the distance value changes in the range of (−1, 1). The larger the distance is, the smaller the reward r_i, to guide the agent to collide through the distance parameter quickly and solve the reward delay problem caused by the insufficient distance state in the original algorithm. Experiment tuning finds that distance parameter 0.1 is the best, which is beneficial to the stability of the per-distance reward mechanism.
   The reward mechanism of Predator i is as follows:

$$r_i = -0.1 * D(i, p) + B + C \tag{11}$$

   Compared with the predator, the escape is the reverse reward. The decision is successful if the escape is from the nearest predator, such that the escape only needs to calculate the distance from the nearest predator and calculate the return value.
   The reward mechanism of fugitive p is as follows:

$$r_p = 0.1 * min(D(i, p)) + B - C \tag{12}$$

2) The reward function is changed to an experience pool sharing mechanism to improve the efficiency of agent cooperation. The reward mechanism in the MADDPG algorithm is to share the maximum reward value globally, which will lead to the emergence of lazy agents, that is, predators who have not contributed to the capture can also receive rewards, which is not conducive to the efficient cooperation of agents. In response to this problem, the Actor–Critic network update strategy is changed, such that each predator calculates the distance to the escaper and the number of collisions separately. The reward value is not shared in the Actor network, only returned to the current agent itself, and the rest is added to the Critic network agent's observation state information and behavior. The predator improves the parameters of the shared experience pool, and other predators learn to imitate the action strategy of the predator with the highest reward value. The Actor network parameters of each agent are updated through gradient descent:

$$\nabla_{\theta_i^{(k)}} J_e(\pi_i) = \frac{1}{K} \mathbb{E}_{s,a\sim\mathcal{D}_i^{(k)}} \left[\nabla_{\theta_i^{(k)}} \pi_i^{(k)} \left(S, \theta^Q \right) \nabla_{a_i} Q^{\pi_i} \left(S_j, a_1, \ldots, a_N, \theta^u \right) \Big|_{a_i = \pi_i^{(k)}(S_i)} \right] \tag{13}$$

3) Agent network parameters are modified. The Actor network and Critic network adopt the structure of a fully connected four-layer neural network, and the number of hidden layer neurons is 64. Each actor has a separate Critic network. The experiment finds that because the distance parameter is introduced, the distance of the predator parameter value is negative, exploring the early return values will remain in negative territory for a long time, and the neural network's activation function of the negative number domain is very small and not conducive to the training. Thus, the activation function of the neural network output layer is removed.

Under the per-distance reward mechanism, the action corresponding to the agent's behavior strategy is clearly distinguished, which is conducive to convergence. The r(S, a) of the Bellman equation changes substantially, and the reward value changes with the distance between the predator and escaper. The Critic network can better recognize the difference in reward value between different action values when the strategy is updated. The problem of delayed rewards is solved.

4 Comparative Experimental Analysis

4.1 Experimental Environment

The experimental design is to compare and analyze the reward curve and actual performance of the agent between the original algorithm and the improved algorithm.

The experimental software environment is Windows10 operating system. The hardware environment is Intel XEON E78880v3*2 processor, NVIDIA GTX 1080TI*3, 64 GB memory. The test environment is OpenAI-GYM, with a hidden layer of two layers and a fully connected neural network consisting of Actor, Critic, and corresponding target network and estimation network consisting of 64 units.

4.2 The Experimental Scene

The Simple_tag experimental scene is the simulation of the autonomous behavior of agents in the antagonistic environment. The experimental space is a 2D bounded confined space, containing four agents, including three predators (blue) and one escaper (red). The description of the experimental scene is shown in Table 2.

Experimental parameter setting:

1) Predators and escapees are ignored as intelligent shapes and sizes and treated as particles;
2) The motion range of the agent on the coordinate axis is [0, 20];
3) Three predators cooperate to hunt down a runaway;
4) The upper limit of the speed of the capturer is $1.0/s$, the upper limit of the acceleration is $0.5/s^2$, the upper limit of the escaper's speed is $1.3/s$, and the upper limit of the acceleration is $0.7/s^2$;
5) When the predator and escaper collide, that is, the distance is 0, the capture is deemed successful, and the escaper fails;

6) The collision rule is that the predator's acceleration decreases and the escape's acceleration increases.
7) Predators and escapers hit the boundary at zero speed.

Table 2. Experimental scenario.

Scene name	Whether to confront	Number of agents	Win conditions
Simple_tag	Yes	3 red VS 1 blue	Blue: The blue agent avoids collisions with 3 red agents as much as possible Red: The three red agents collide with the blue agent as much as possible

4.3 Experimental Analysis

The convergence speed and stability of the algorithm are improved without reducing the capture effect of predators. The predators place the per-distance reward mechanism into the MADDPG algorithm and DDPG algorithm for learning and training, respectively. Table 3 shows the average collision times of the predator after each step of the strategy learned by the multi-agent system after 20,000 rounds of training under a randomized trial of 1,000 rounds and 60,000 steps. Compared with the original algorithm, adding a per-distance reward mechanism has a better capture effect, the average collision times of MADDPG are increased by 3.3%, and the average collision times of DDPG algorithm are slightly increased by 1.7%.

Table 3. Average number of collisions per step.

Experimental algorithm	Whether the reward mechanism is improved	Average number of collisions
MADDPG	Yes	0.538
	No	0.521
DDPG	Yes	0.532
	No	0.523

For the above experiment, 40,000 steps of training are carried out. TensorBoard, a visualization tool of TensorFlow, is used to depict the relationship between the reward value of the predators and escapers, and the number of training times. The relationship graph of the reward mechanism of MADDPG algorithm and the improved per-Distance reward mechanism is compared.

Figures 3, 4, and 5 are graphs of the reward function of predators 1, 2, and 3, respectively. The reward of the predator decreases when the distance to the escaper increases because of the introduction of the distance parameter. A small distance parameter results

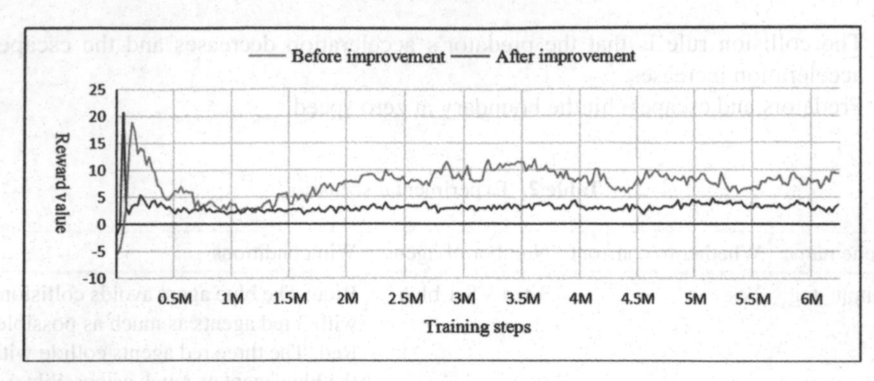

Fig. 3. Curves of predator 1 reward function.

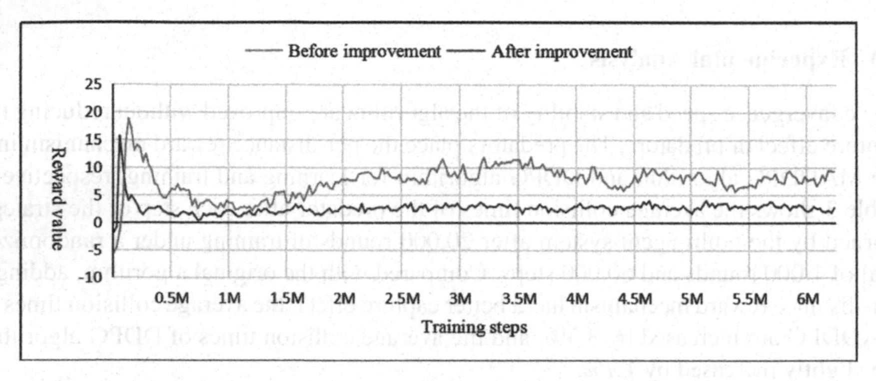

Fig. 4. Curves of predator 2 reward function.

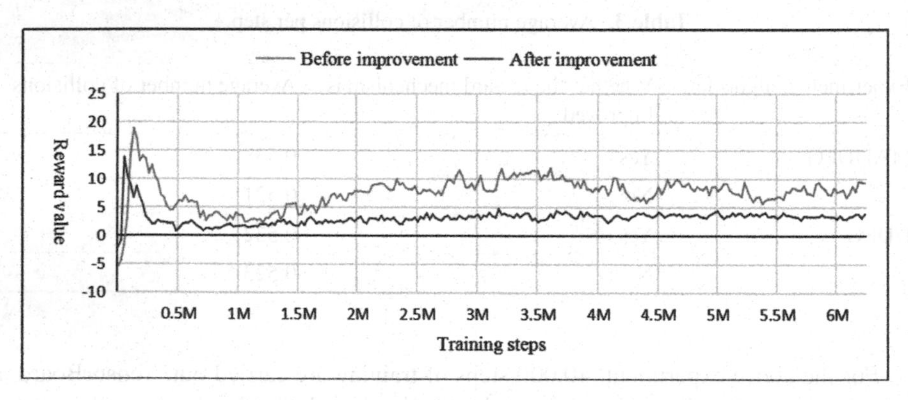

Fig. 5. Curves of predator 3 reward function.

in a decrease in the overall return value and a downward shift in the reward function curve. As the number of collisions increases, more direct reward values begin to superimpose, which slows down and stabilizes the downward trend of the reward curve. After changing the reward mechanism, the convergence speed of the algorithm is greatly improved. The reward value stabilizes at around 5,000 rounds, and the reward value fluctuates slowly in the interval [2, 4]. Experiments prove that the convergence of the predator reward function and the stability of the algorithm are very evident.

Figure 6 shows that due to the improvement of the predator's capture effect, the negative reward (also called punishment) obtained by the escaper greatly increases, resulting in a decrease in the value of the reward function. Compared with the evident improvement in the convergence speed of the predator, the improvement effect of the convergence speed of the escaper's reward value is not outstanding because the escaper has to calculate the minimum distance to the predator. When the nearest predator changes, the strategy The network needs to recalculate the minimum distance, the update step is larger, and convergence is compromised.

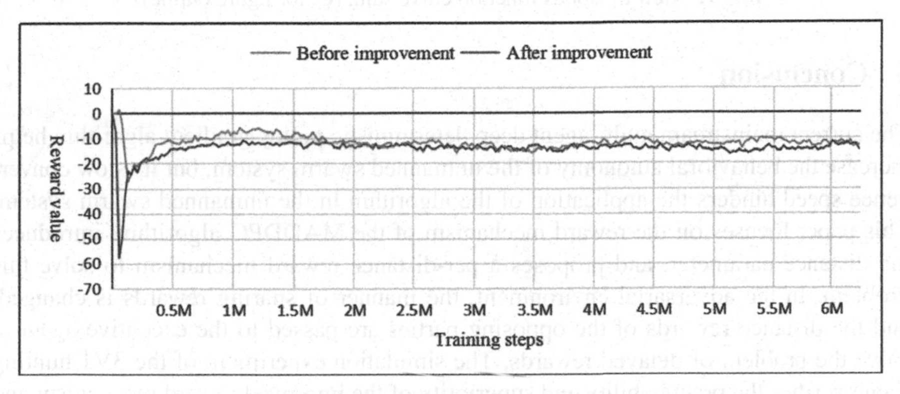

Fig. 6. Curves of escape reward function.

The inflection point of the reward value of both is about 12,000 steps. Under the original reward mechanism, the stability of the escaper's reward function is poor, the reward value fluctuates within the larger range of [−6, −15], and the function curve fluctuates greatly. After introducing the per-distance reward mechanism, the fluctuation range of the curve is reduced, the function value changes in the interval of [−12, −16], and the convergence improves. The new reward mechanism also improves the escaper function.

The reward values of all agents are superimposed, and the curve of the sum of reward values and number of steps in the training round is drawn, as shown in Fig. 7. The comparison in Fig. 7 is clear. The red curve shows that the original algorithm's reward value curve fluctuates in a large range, and the convergence is not good. After the per-distance reward mechanism is introduced, the blue curve shows that the reward value curve enters the inter district fluctuation earlier, and algorithm convergence performance and stability are considerably improved.

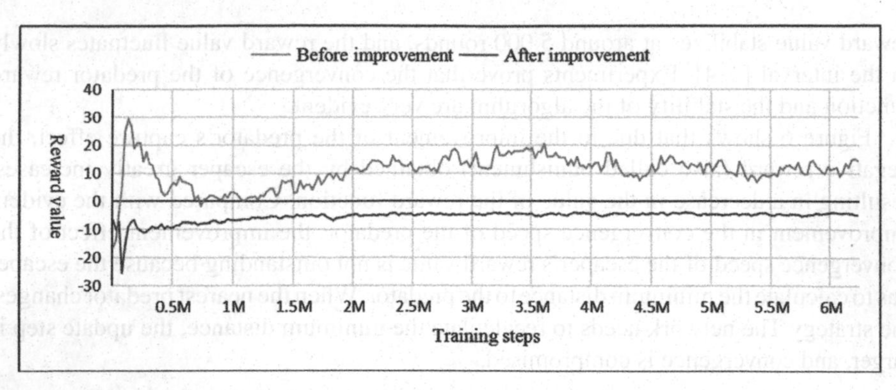

Fig. 7. View of Bonus function curve sum. (Color figure online)

5 Conclusion

The current mainstream multi-agent deep deterministic policy gradient algorithm helps increase the behavioral autonomy of the unmanned swarm system, but its slow convergence speed hinders the application of the algorithm in the unmanned swarm system. This paper focuses on the reward mechanism of the MADDPG algorithm, introduces the distance parameter, and proposes a per-distance reward mechanism to solve this problem. In the adversarial environment, the manner of sharing rewards is changed, and the distance rewards of the opposing parties are passed to the executive agent to solve the problem of delayed rewards. The simulation experiment of the 3V1 hunting scene verifies the practicability and superiority of the improved reward mechanism and improves the feasibility of the group behavior strategy of the unmanned swarm system in the confrontation environment.

References

1. Zhang, T.T., Song, A.G., Lan, Y.S.: Adaptive structure modeling and prediction of swarm unmanned system. Chin. Sci. Inf. Sci. **50**(1), 347–362 (2020)
2. Sun, C.Y., Mu, C.X.: Important scientific problems of multi-agent deep reinforcement learning. J. Automatica Sinica **46**(7), 1301–1309 (2020)
3. Chen, J.: Several problems in multi-agent system. Sci. Chin. **12**(1), 40–43 (2019)
4. Luo, D.L., Zhang, H.Y., Xie, R.Z., et al.: Large scale UAV cluster confrontation based on multi-agent system. Control Theor. Appl. **32**(11), 1498–1504 (2015)
5. Palmer, G., Tuyls, K., Bloembergen, D., et al.: Lenient multi-agentdeepreinforcementlearning. In: Proceedings of the 17th International Conference on Autonomous AgentsandMultiAgentSystems, pp. 443–451 (2018)
6. Omidshafiei, S., Pazis, J., Amato, C., et al.: Deep decentralized multi-task multi-agent reinforcement learning under partial ob-servability. In: Proceedings of the 34th International Conference on Machine Learning, vol. 70, pp. 2681–2690 (2017)
7. Zheng, Y., Meng, Z., Hao, J., Zhang, Z.: Weighted double deep multiagent reinforcement learning in stochastic cooperative environments. In: Geng, X., Kang, B.-H. (eds.) PRICAI 2018: Trends in Artificial Intelligence: 15th Pacific Rim International Conference on Artificial Intelligence, Nanjing, China, August 28–31, 2018, Proceedings, Part II, pp. 421–429. Springer, Cham (2018). https://doi.org/10.1007/978-3-319-97310-4_48

8. Sunehag, P., Lever, G., Gruslys, A., et al.: Value decomposition networks for cooperative multi-agent learning based on team reward. In: Proceedings of the 17th International Conference on Autonomous Agents and MultiAgent Systems, pp. 2085–2087 (2018)
9. Rashid, T., Samvelyan, M., De Witt, C.S., et al.: QMIX: Mono-tonic value function factorisation for deep multi-agent reinforce-ment learning. arXiv preprint arXiv:1803.11485 (2018)
10. Foerster, J.N., Farquhar, G., Afouras, T., et al.: Counterfactual multi-agent policy gradients. In: Thirty-Second AAAI Conferenceon-ArtificialIntelligence (2018)
11. He, M., Zhang, B., Liu, Q., et al: Research on experience first extraction mechanism of maddpg algorithm. Control and decision (2020). https://doi.org/10.13195/j.kzyjc.2019.0834.
12. Sun, Q.: Research on cooperative mechanism of multi-agent based on Reinforcement Learning. Ph.D. dissertation, Zhejiang University of technology (2015)
13. Lowe, R., Wu, Y.I.: Tamar, A., et al.: Multi-agent actor-critic for mixed cooperative-competitive environments. In: Advances in Neural Information Processing Systems, pp. 6379–6390 (2017)
14. Barto, A.G.: Reinforcement Learning: An Introduction Second Edition, Mountain View, Bradford (2018)
15. van Otterlo, M., Wiering, M.: Reinforcement learning and markov decision processes. In: Wiering, M., van Otterlo, M. (eds.) Reinforcement learning, pp. 3–42. Springer, Berlin (2012). https://doi.org/10.1007/978-3-642-27645-3_1
16. Chen, L., Liang, C., Zhang, J.Y., et al.: A Multi-Agent Reinforcement Learning Algorithm Based on improved ddpg under actor critical framework. Control and decision (2019). https://doi.org/10.13195/j.kzyjc.2019.0787
17. Xu, X.: Reinforcement Learning and Approximate Dynamic Programming. Science Press, Beijing (2010)
18. Yang, W.Y., Bai, C.J., Cai, C., et al.: Review on sparse reward in deep reinforcement learning. Computer Science (2019). http://kns.cnki.net/kcms/detail/50.1075.TP.20191122.1628.023.html
19. Sun, Y., Cao, L., Chen, X.L., et al.: A review of multi-agent deep reinforcement learning research. Computer engineering and application (2020). http://kns.cnki.net/kcms/detail/11.2127.TP.20200214.1008.002.html

A Variable Step-Size Based Iterative Algorithm for High-Precision Ranging Using FMCW Radar

Dengke Yao, Yong Wang(✉), Liangbo Xie, and Mu Zhou Yanchun Li

Chongqing University of Posts and Telecommunication, Chongqing 400065, China

yongwang@cqupt.edu.cn

Abstract. Frequency modulated continuous wave (FMCW) radar ranging plays an important role in many applications, including autonomous vital sign sensing, driving, and human-computer interaction. In this paper, a variable step-size based iterative algorithm for high-precision ranging is proposed for a FMCW radar. Firstly, the windowing processing is performed on the collected radar intermediate frequency signal. Meanwhile, the amplitude spectrum of the signal and the index value of the maximum peak spectral line is obtained by the fast Fourier transform (FFT). Moreover, the deviation correction factor is also obtained. Secondly, we perform an iterative interpolation on the spectrum line at the maximum value in the amplitude spectrum and its adjacent two auxiliary spectrum lines. The interval between the auxiliary spectral line and the maximum spectral line is updated iteratively. Finally, the range of the target is estimated according to the relationship between the frequency and range. Experimental results show that the proposed algorithm effectively improves the accuracy of range estimation and the average range error is 0.35 cm.

Keywords: FMCW Radar · FFT · Variable step-size · Range estimation

1 Introduction

Frequency modulated continuous wave (FMCW) radar [1] has wide applications in our daily life, including unmanned driving [2], sleep monitoring [3] and gesture recognition [4] etc. The ranging of FMCW radar has received much attention. Fourier transform (FFT) [5] is used as the traditional range estimation. However, FFT algorithm for ranging has barrier effect and spectrum leakage problems, which leads to a large deviation in distance estimation. As a result, some improved methods such as zero padding, Chirp-Z transform, FFT discrete-time Fourier transform (FFT-DTFT) transform [6] are proposed. Although the barrier effect is resolved to some extent, these methods usually have high complexity.

Numerous efforts have been made to solve the barrier effect problem to improve the estimation accuracy. In order to achieve frequency correction, Rife et al. [7] use the FFT to obtain the index of the maximum amplitude spectrum, and construct the frequency deviation correction by using the ratio of the maximum spectrum line and the next largest spectrum line factor. However, this method usually has wrong frequency interpolation

X. Sun et al. (Eds.): ICAIS 2021, CCIS 1422, pp. 564–572, 2021.
https://doi.org/10.1007/978-3-030-78615-1_49

direction when the environmental noise is large. According to the phase relationship between the largest spectral line and the second largest spectral line, Quinn method is proposed in [8] to determine the interpolation direction, which solves the problem of wrong interpolation direction. Candan [9] and Fang [10] apply the maximum spectral line and the auxiliary spectral line nearby to derive the deviation correction equation, which effectively improved the accuracy of frequency estimation. In order to reduce the influence of noise on the auxiliary spectral line, Ref. [11] and [12] propose to select the DTFT sampling value near the maximum distance spectral line to construct the bias correction factor. A&M et al. [13] propose a method to achieve high-precision frequency estimation by iterating the deviation correction equation. Ref. [14] propose a q-shift iterative method, which can greatly improve the accuracy of frequency estimation, and the estimation result is close to the Cramer-Rao lower bound (CRLB). Moreover, the algorithms in [15] adds zeros to the signal, increases the number of FFT sampling points, and uses the iterative interpolation for the zero-filled signal to improve the estimation accuracy. Although the existing iterative interpolation method can greatly improve the estimation accuracy, it also has the problem of low estimation accuracy due to environmental interference.

In this paper, we propose a variable step-size iterative (VSSI) algorithm for high-precision ranging. Firstly, the algorithm improves the problem of spectrum leakage by windowing the signal, and iteratively corrects the frequency through the deviation correction factor. Secondly, each iteration adopts variable step way to update the position of the auxiliary line make it close to the real line, thus effectively reducing the environmental interference on the influence of auxiliary line. Finally, simulation results show that the frequency estimation results of this algorithm are close to CRLB. In the measured environment, its performance is obviously better than the existing ranging methods, and the average ranging error is 0.35 cm.

The rest of the paper is organized as follows. Section 2 introduces the FMCW signal model. Section 3 illustrates our algorithm. Extensive experimental results are shown in Sect. 4. Finally, conclusions are given in Sect. 5.

2 FMCW Signal Model

FMCW radar is composed of waveform generator, voltage controlled oscillator (VCO), transmitting and receiving antenna module, I/Q demodulator module, low pass filter (LPF), A/D digital to analog converter, etc. The principle block diagram is shown in Fig. 1.

The FMCW radar uses a signal generator to generate a linear frequency modulation sawtooth wave. After the signal is modulated by a VCO, the signal is transmitted through the transmitting antenna. The transmitted signal model is

$$s_{tx}(t) = A_{tx}\cos\left(2\pi\left(f_0 t + \frac{\zeta}{2}t^2\right) + \phi_0\right) \tag{1}$$

where A_{tx} is the amplitude of the transmitting antenna, f_0 is the starting frequency of the FMCW radar, and ϕ_0 is the initial phase. $\zeta = B/T_c$ represents the slope of the chirp, where B is the FMCW radar bandwidth, and T_c is the linear sweep period.

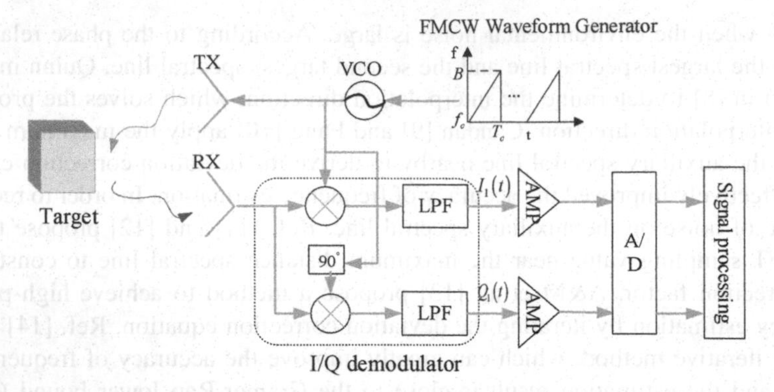

Fig. 1. Block diagram of FMCW radar principle

The transmitted signal is reflected back after encountering a target at a range of R. The time delay $\tau_r = 2R/c_m$ caused by the reflection of the target can be obtained, where c_m represents the speed of light. As a result, the received signal of the radar is given by (2)

$$s_{rx} = A_{rx}\cos\left(2\pi\left(f_0(t-\tau_r) + \frac{\zeta}{2}(t-\tau_r)^2\right) + \phi_{rx}\right) \quad (2)$$

where ϕ_{rx} is the phase of the receiving antenna. s_{tx} and s_{rx} are mixed in the time domain. The mirror component is filtered through a low-pass filter to obtain the intermediate frequency signal, as shown in Eq. (3):

$$s_{If}(t) = A_b \exp\left(j\left(2\pi\zeta t\tau_r - \underbrace{\pi\zeta\tau_r^2}_{Tend\ to\ be\ zero} + \underbrace{2\pi f_0\tau_r - \phi_2}_{\phi}\right)\right) + w(t) \quad (3)$$

where $w(t)$ is the noise signal, A_b and ϕ_2 are the amplitude and phase of the mixed signal.

Let $\phi = 2\pi f_0\tau_r - \phi_2$ be the phase constant. The derivative of the phase $\psi(t) = 2\pi\zeta t\tau + 2\pi f_0\tau - \phi_2$ of the intermediate frequency signal to obtain the instantaneous frequency f_r, combined with the time delay $\tau_r = 2R/c_m$. The relationship between the instantaneous frequency of the intermediate frequency signal and the range can be obtained as

$$R = \frac{c \cdot f_r}{2\zeta} \quad \Rightarrow \quad R \propto f_r \quad (4)$$

We use Nyquist sampling theorem to discretize the signal of Eq. (3), the number of samples is $N = T_c/T_s$, where $T_s = 1/f_s$. The discrete intermediate frequency signal expression is related through

$$s_{If}[n] = A_b \exp\left(j\left(2\pi\zeta\tau_r\frac{n}{f_s} + 2\pi(f_c \cdot \tau_r) + \phi\right)\right) + w(n) \quad , n = 0\cdots N-1 \quad (5)$$

The FFT analysis of Eq. (5) can separate different frequencies and their corresponding amplitudes. Under normal circumstances, the reflected echo energy of the target within the radar ranging range is higher than the background environment. Therefore, the frequency corresponding to the radar target can be determined through the peak search, then the distance value of the target can be estimated through Eq. (4). However, the issues of this method such as fence effect and spectrum leakage will make a certain error between the peak frequency and the theoretical value. In order to solve this problem, this paper proposes a ranging algorithm based on variable step size iterative interpolation.

3 Variable Step-Size Iterative Algorithm for High-precision Ranging

In this subsection, we propose a new ranging estimation method. We use the FFT algorithm to roughly estimate the frequency of the intermediate frequency signal in Eq. (5). The expression is

$$S(k) = \sum_{n=0}^{N-1} s_{If}(n)e^{-j\frac{2\pi}{N}nk} = A_b e^{j\phi} \sum_{n=0}^{N-1} e^{j2\pi N\left(\frac{f_r}{f_s}-\frac{k}{N}\right)}, k = 0, \ldots, N-1. \quad (6)$$

where N is the number of FFT points.

According to Eq. (6), the index value k' of the maximum frequency spectrum can be expressed as

$$k' = \arg\max_k(S(k)) \quad (7)$$

According to the calculated index value, a rough estimate of frequency $\hat{f}_r' = \Delta f_r \cdot \left(k' - 1\right)$ can be obtained, where $\Delta f_r = f_s/N$ is the frequency resolution. The frequency is related through

$$f_r = \hat{f}_r' + \delta\Delta f_r \qquad \delta \in [-0.5, 0.5] \quad (8)$$

where $\delta\Delta f_r$ is the relative frequency deviation between the measured frequency and the FFT rough estimated frequency.

$$S'_{\pm a} = \sum_{n=0}^{N} s_{If}(n)e^{-j2\pi f_r n}\Big|_{f_r=\left(k'\pm a\right)\Delta f_r'} \quad (9)$$

where a is the interval (step size) from the index value k' of the amplitude spectrum peak According to the summation formula of complex exponential function, Eq. (9) yields

$$S'_{\pm a} = Ae^{j\phi}e^{j\pi\frac{(N-1)}{N}(\delta\pm a)}\frac{\sin(\pi(\delta\pm a))}{\sin\left(\frac{\pi}{N}(\delta\pm a)\right)} \quad (10)$$

The true deviation of the signal is contained in the DTFT spectral line information (see Eq. (10)). Therefore, deviation value can be estimated by analyzing the spectral line information. The ratio of auxiliary spectral lines can be expressed as

$$\frac{S'_a}{S'_{-a}} = \frac{\vartheta \cdot S_0 \cdot [\cos(\pi a) - \upsilon] - S'_a \cdot \cos\left(\frac{\pi a}{N}\right)}{-\left[\vartheta' \cdot S_0 \cdot [\cos(\pi a) + \upsilon] - S'_{-a} \cdot \cos\left(\frac{\pi a}{N}\right)\right]} \quad (11)$$

where $\vartheta = e^{j\pi a(N-1)/N}$, $\vartheta' = e^{-j\pi a(N-1)/N}$, $\upsilon = \sin(\pi a)\cot(\pi\delta)$.

$$\tan(\pi\delta) = \frac{S_0' \cdot \left(\vartheta S_a' - \vartheta' S_{-a}'\right) \cdot \sin(\pi a)}{2S_a' \cdot S_{-a}' \cdot \cos(\frac{\pi a}{N}) - S_0'(\vartheta S_a' + \vartheta' S_{-a}') \cdot \cos(\pi a)} \tag{12}$$

Taking the real part of $\tan(\pi\delta/N)$ for analysis, the deviation formula is related to

$$\hat{\delta} = \frac{1}{\pi}\tan^{-1}\left(\mathrm{Re}\left\{\frac{S_0' \cdot \left(\vartheta S_a' - \vartheta' S_{-a}'\right) \cdot \sin(\pi a)}{2S_a' \cdot S_{-a}' \cos(\frac{\pi a}{N}) - S_0' \cdot (\vartheta S_a' + \vartheta' S_{-a}') \cdot \cos(\pi a)}\right\}\right) \tag{13}$$

According to Eq. (13), the deviation correction factor is obtained to correct the frequency deviation value to make it closer to the true deviation.

The steps of the variable step-size iterative interpolation proposed in this paper to realize the high-precision ranging algorithm for FMCW radar can be summarized as follows:

Step 1: The Hamming window is added to the intermediate frequency signal $s_{If}[n]$ processed by the A\D digital-to-analog converter to obtain the signal $s_w(n) = s_{If}(n) \cdot w_{Hm}(n)$.

Step 2: The $S'(k)$ is obtained from calculate the Fourier transform of $s_w(n)$, and find the maximum spectral value S_0' and the index value k'. Then, we calculate the rough estimate of frequency $f_r = \Delta f_r \cdot \left(k' - 1\right)$, where Δf_r is the frequency resolution.

Step 3: According to the deviation Eq. (13), the spectral unit deviation $\hat{\delta}$ is calculated through variable step size a iteration.

Step 3.1: Stop the iteration when $\left|\delta_q - \delta_{q-1}\right| < 10^{-5}$ appears, where q represents the number of iterations. Set the initial step size a to 0.5 and the initial deviation to 0.

Step 3.2: The corrected deviation value $\hat{\delta}_q$ can be obtained by substituting the parameters set in step 3.1 into Eq. (13), and $\delta_q = \delta_q + \delta_{q-1} \quad q = 1, 2, 3, \cdots$.

Step 3.3: Update the step size a and the index value k through $a_q = a_{q-1}/2$ and $k_q = k_{q-1} + \delta_q$ in each iteration.

Step 3.4: Repeat steps 3.2 and 3.3 until the iteration stop condition is met. Let $\hat{\delta} = \hat{\delta}_q$ be the corrected deviation value.

Step 4: Calculate the corrected signal frequency $\hat{f}_r$, and use $\hat{R} = c \cdot \hat{f}_r / (2\zeta)$ to calculate the range between the target and radar.

4 Algorithm Verification and Result Analysis

In this section, with the aid of computer simulation and measured results, the performance of proposed frequency estimator is verified. In addition, we use the Monte Carlo simulation experiments and compared with Candan [13], A&M [17], three-sample interpolation (TSI) [18] and HAQSE [19]. The five algorithms are all 0 ($N \log N$) in computational complexity. Root Mean Square Error (RMSE) is used to measure the ranging performance of the algorithm, which can be expressed as

$$RMSE = \sqrt{\frac{1}{M}\sum_{i=1}^{M}\left(\hat{R}_r - R_r\right)^2} \tag{14}$$

where $\hat{R}_r$ is the ranging value, R_r is the true value of the range, and $M = 1000$ represents the number of the Monte Carlo simulations.

4.1 Simulation Results and Analysis

The parameter settings of FMCW radar simulation signal is shown in Table 1.

Table 1. Simulation parameter

Parameter settings	Symbol/Unit	Value
Center frequency	f_c/GHz	77
Bandwidth	B/GHz	4
Sweep cycle	T_c/ms	40
Sampling rate	f_s/kHz	2000
Chirp slope	$\varsigma/\left(10^{14}\right)$	100
Number of chirps	N	64

Figure 2 shows the effect of different SNR on the RMSE when the deviation is 0.2 and 0.4. Almost all the algorithms show a downward trend as SNR increases, but in the low SNR region, RMSE of the VSSI and HAQSE is lower than the other previous algorithms. The reason is that the VSSI adopts variable step-size iteration method, which selects auxiliary spectral line closer to the true spectral line to correct the deviation correction factor, so that it has better anti-environmental interference ability. In the high SNR region, the proposed VSSI algorithm can be closer to CRLB.

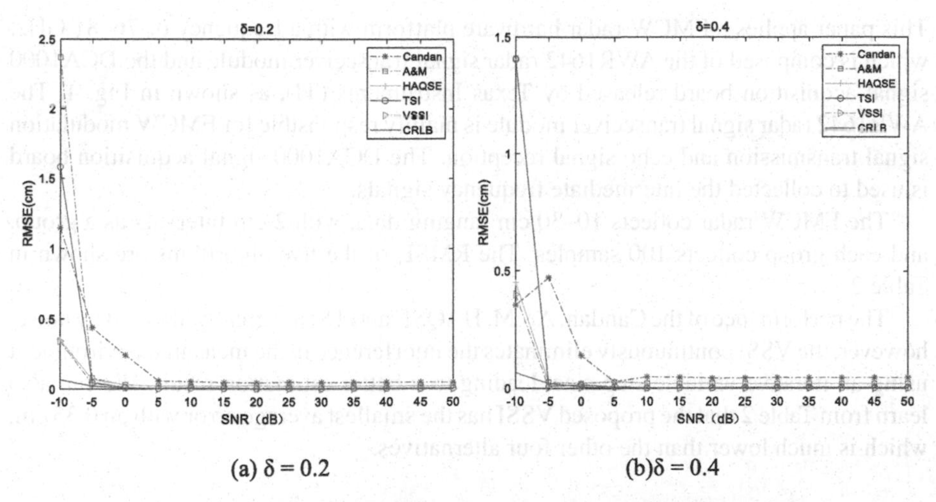

(a) δ = 0.2 (b)δ = 0.4

Fig. 2. Comparison of algorithm estimation results under different SNR.

In order to verify the performance of the proposed algorithm, we simulated FMCW radar data with a range of 10 ~ 30 cm and an interval of 1 cm with SNR = 20 dB, and the results is shown in Fig. 3. We learn from Fig. 3 that both Candan and A&M have relatively large estimation errors due to the poor iteration effects. The estimation errors of HAQSE, TSI and VSSI are all within 0.1 cm, which proves the feasibility of VSSI.

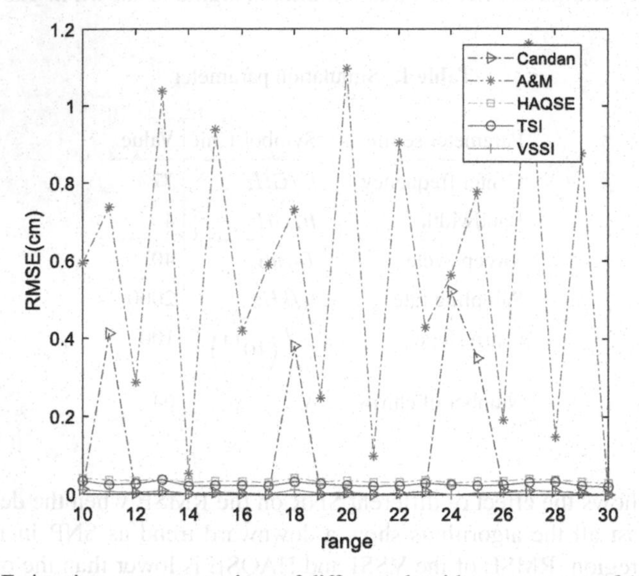

Fig. 3. Estimation error comparison of different algorithms at a range of 10–30 cm

4.2 Analysis of Measured Results

This paper applies a FMCW radar hardware platform with a frequency of 76–81 GHz, which is composed of the AWR1642 radar signal transceiver module and the DCA1000 signal acquisition board released by Texas Instruments (TI), as shown in Fig. 4. The AWR1642 radar signal transceiver module is mainly responsible for FMCW modulation signal transmission and echo signal reception. The DCA1000 signal acquisition board is used to collected the intermediate frequency signals.

The FMCW radar collects 10–30 cm ranging data, with 2 cm intervals as a group, and each group collects 100 samples. The RMSE of the five algorithms are shown in Table 2.

The performance of the Candan, A&M, HAQSE and TSI are greatly affected by noise, however, the VSSI continuously eliminates the interference in the measured environment using an iterative variable step-size, leading to a better estimation result. We can also learn from Table 2 that the proposed VSSI has the smallest average error with an 0.35 cm, which is much lower than the other four alternatives.

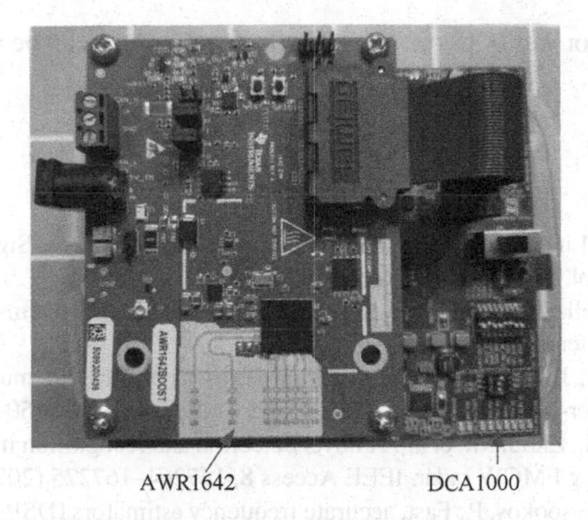

Fig. 4. The hardware platform of millimeter wave radar

Table 2. Estimation error values comparison of different algorithms

Actual range(cm)	Candan	A&M	HAQSE	TSI	VSSI
10	1.45	0.31	1.60	0.75	0.11
12	2.37	1.89	1.76	0.21	0.03
14	1.24	1.50	0.22	1.76	0.44
16	1.00	1.14	2.00	1.00	0.22
18	0.23	1.84	0.65	0.21	0.11
20	1.43	1.80	0.00	2.34	0.56
22	3.23	1.92	2.34	0.01	0.00
24	1.56	2.19	1.02	1.44	0.00
26	0.67	1.61	0.21	2.10	0.29
28	2.21	1.35	0.00	1.70	1.19
30	5.22	2.58	2.50	0.55	1.19
Average error	1.94	1.50	1.27	1.00	0.35

5 Discussion and Conclusion

This paper proposed a ranging algorithm based on variable step-size iterative interpolation for FMCW radar. The algorithm updated the interpolation position by changing the interval between auxiliary spectrum line and maximum amplitude spectrum, which greatly eliminated the influence of environmental interference on amplitude spectrum. Simulation and actual measurement were performed to verify the effectiveness of the proposed algorithm. Simulation results show that the estimation results of the algorithm are close to CRLB in low SNR environment. The experimental results showed that the average ranging error was 0.35 cm and such a range accuracy can be applied in real-time applications.

References

1. Stove, A.G.E.: Linear FMCW radar techniques. In: Radar & Signal Processing IEE Proceedings F, vol. 139, no. 5, pp. 343–350 (1992)
2. Rohling, H., Moeller, C.: Radar waveform for automotive radar systems and applications. In: IEEE Radar Conference, pp. 1–4 (2008)
3. Ahmad, A., Roh, J.C., Wang, D. et al.: Vital signs monitoring of multiple people using a FMCW millimeter-wave sensor. In: IEEE Radar Conference, pp. 1450–1455 (2018)
4. Wang, Y., Ren, A., Zhou, M., et al.: A novel detection and recognition method for continuous hand gesture using FMCW radar. IEEE Access **8**, 167264–167275 (2020)
5. Jacobsen, E., Kootsookos, P.: Fast, accurate frequency estimators [DSP Tips & Tricks]. IEEE Sig. Process. Mag. **24**(3), 123–125 (2007)
6. Zhu, K., Qin, Y.W., Xu, J.Z., et al.: Performance analysis of four schemes improving ranging precision of FMCW radar. Radio Eng. **2015**(1), 20–25 (2015)
7. Rife, D.C., Vincent, G.A.: Use of the discrete Fourier transform in the measurement of frequencies and levels of tones. Bell Labs Tech. J. **49**(2), 197–228 (1970)
8. Quinn, B.G.: Estimation of frequency, amplitude, and phase from the DFT of a time series. IEEE Trans. Signal Process. **45**(3), 814–817 (2002)
9. Candan, C.: Analysis and further improvement of fine resolution frequency estimation method from three DFT samples. IEEE Signal Process. Lett. **20**(9), 913–916 (2013)
10. Fang, L., Duan, D., Yang, L.: A new DFT-based frequency estimator for single-tone complex sinusoidal signals. In: Military Communications Conference, pp. 1–6 (2013)
11. Liang, X., Liu, A., Pan, X., et al.: A new and accurate estimator with analytical expression for frequency estimation. IEEE Commun. Lett. **20**(1), 105–108 (2015)
12. Fang, L., Qi, G.: Frequency estimator of sinusoid based on interpolation of three DFT spectral lines. Signal Process. **144**, 52–60 (2018)
13. Aboutanios, E., Mulgrew, B.: Iterative frequency estimation by interpolation on Fourier coefficients. IEEE Trans. Signal Process. **53**(4), 1237–1242 (2005)
14. Serbes, A.: Fast and efficient sinusoidal frequency estimation by using the DFT coefficients. IEEE Trans. Commun. **67**(3), 2333–2342 (2018)
15. Bai, G., Cheng, Y., Tang, W., et al.: Accurate frequency estimation of a real sinusoid by three new interpolators. IEEE Access **7**, 91696–91702 (2019)

Research on the Smart Management and Service of Smart Classrooms in Colleges and Universities

Changling Peng[1,2], Chengxue Zhu[2], Yanpeng Wu[1,2(✉)], Hui Zhou[2], and Wei Wei[2]

[1] Department of Information Science and Engineering, Hunan First Normal University, Changsha 410205, Hunan, China
xjzxwyp@hnfnu.edu.cn

[2] Modern Educational Technology and Network Center, Hunan First Normal University, Changsha 410205, Hunan, China

Abstract. The rapid advancement of Internet + Education has brought a profound impact to the education industry. It has not only reshaped the educational concept and educational form, but also reconstructed the current classroom teaching environment. As a new form of classroom teaching environment, smart classroom has become one of the current research hotspots in the field of education. With the wide construction and application of smart classroom in colleges and universities, how to conduct smart management and service under smart classroom environment has become one of the key issues requiring to be addressed in colleges and universities. The management of smart classrooms in colleges and universities is an important and arduous service guarantee work, which plays an important role for teaching work. This study has proposed a system for managing and servicing smart classroom. The system includes three parts: display system, equipment control system and sound amplifying system. Finally, to explore the effectiveness of the system, an experiment study was conducted using a sample of 50 teachers and 362 college students from a university in south central China. The surveys showed that teachers and students are highly satisfied with the management and service system of the smart classroom Moreover, through independent sample t-test found the teacher's satisfaction is higher than the student's satisfaction. These results indicate that using smart classroom management system offers some distinct benefits for teachers and students. In future work, additional research is needed to clarify how the management and service system can serve education and teaching.

Keywords: Colleges and universities · Smart classrooms · Smart management and service

1 Introduction

Smart classroom is also known as intelligent classroom, classroom of future or classroom of tomorrow [1]. Regarding the concept and description of the smart classroom, there is no scientific, authoritative, and unified conclusion. Many researchers have defined and described it from different angles. Jawa et al. defined the smart classroom as a learning environment that can easily store information generated during teaching, generate timely

X. Sun et al. (Eds.): ICAIS 2021, CCIS 1422, pp. 573–584, 2021.
https://doi.org/10.1007/978-3-030-78615-1_50

teaching feedback on teaching activities, intelligently control equipment, and quickly complete data and information search [2]. He pointed out that the smart classroom which set cloud computing, sensor networks, wireless networks, artificial intelligence and other new generation of information technology functions in one was an important tool and means of the future classroom teaching, and it can support learning environment of teaching and learning style which required the creation of the real situation, inspiration of thinking, acquisition of information, sharing of resources, multiple interactions, independent inquiry, collaborative learning and so on [3]. Shi believed that the smart classroom was a new form of classroom, which can optimize the presentation of teaching content, make the sharing of learning resources convenient, promote classroom interaction development, and it also had the function of situational awareness and environmental management. What's more, smart classroom was a typical materialization of intelligent learning environment, and also was the high-end form of multimedia and online classroom [4–6].

In this context, the smart classroom can be regarded as a special type of digital learning environment. Smart classroom is defined as the environment of promoting interactive learning between teachers and students, cultivating students' self-learning, collaborative learning and problem-solving abilities, and enhancing efficiency of classroom teaching.

As a new form of classroom learning environment and the product of the emerging information technologies (e.g., Internet of things, cloud computing and big data), the smart classroom, with deep interaction as its core feature, has been one of the research hotspots in the field of education [7, 8]. The smart classroom possesses a variety of features, such as the real-time interaction, environmental control, video monitoring and remote control [9]. Benefited from these functions and features, the smart classroom has the potential to enhance demonstration and presentation of instructional materials, to improve the quality of interactions among teacher, students and learning contents, and to promote classroom interaction and situational awareness [4, 10–12]. Moreover, the smart classroom can provide supports for various forms of teaching activities, which can help cultivate students' critical and creative thinking, and improve students' problem-solving abilities and application of information technology as well. Compared with traditional multimedia classroom, the smart classroom places more emphasis on using participatory teaching method to promote learners' deep learning and knowledge internalization.

In China, a growing body of colleges and universities have invested a lot of financial, material and manpower to build smart classrooms. However, with the construction and application of smart classrooms, the effective management and high-quality services of smart classrooms have become a new problem. How to provide effective management and high-quality services requires close attention and active face from education managers, teachers, students and even parents. Thus, this study combines the actual daily work to explore the management and service of smart classrooms in order to give full play to the important role of smart classrooms in college teaching activities.

2 Smart Classroom Learning Environment

The smart classroom referred in this study mainly includes two or three touch-control integrated devices (TIDs) and fix glass whiteboards, wireless area network (WLAN) and mobile phone-based mobile learning platform. TIDs, instead of the projector, was used in the smart classroom with the purpose of offering opportunities for teachers and students to interact with instructional materials and each other in a collaborative learning environment. The TIDs were used to display the learning contents and the contents of student discussions. Due to the full covering of WLAN in the smart classroom, students could connect with the teacher and peers through mobile phones and tablet PCs. Not only could students search for relevant materials in the classroom, but also could they use the mobile phone to scan the QR (Quick Response) code that presented on the TIDs to complete related exercises. Moreover, the teacher could distribute learning tasks via the mobile phone-based mobile learning platform and students could download them as well as additional instructional materials according to their needs. The outline drawings of the three types of smart classrooms are shown in Fig. 1, Fig. 2 and Fig. 3.

Fig. 1. Network interactive smart classroom

3 Intelligent Management Service System of Smart Classroom

Smart classroom is one direction in smart learning environment research. In order to promote the full infusion of technology in classroom learning and teaching, smart classroom management and service, which is an important part of smart classroom, become more and more important. Equipment architecture diagram of smart classroom is shown in Fig. 4.

This study has proposed a system for managing and servicing smart classroom. The system is one part of smart classroom research. The structures of smart classroom and smart classroom management should be investigated for easy, effective, and engaged learning. Smart classroom management and service system are shown in Fig. 5 and Fig. 6.

Fig. 2. Multi-screen interactive smart classroom

Fig. 3. Mobile internet smart classroom

3.1 Intelligent Operation and Maintenance Module

The management system can automatically control the classroom equipment according to the curriculum or custom time rules. What's more, the range of the controlled classroom can be set by itself. Also, smart strategies can be used for a better smart classroom atmosphere. Smart strategies include three types: automatic startup, automatic shutdown,

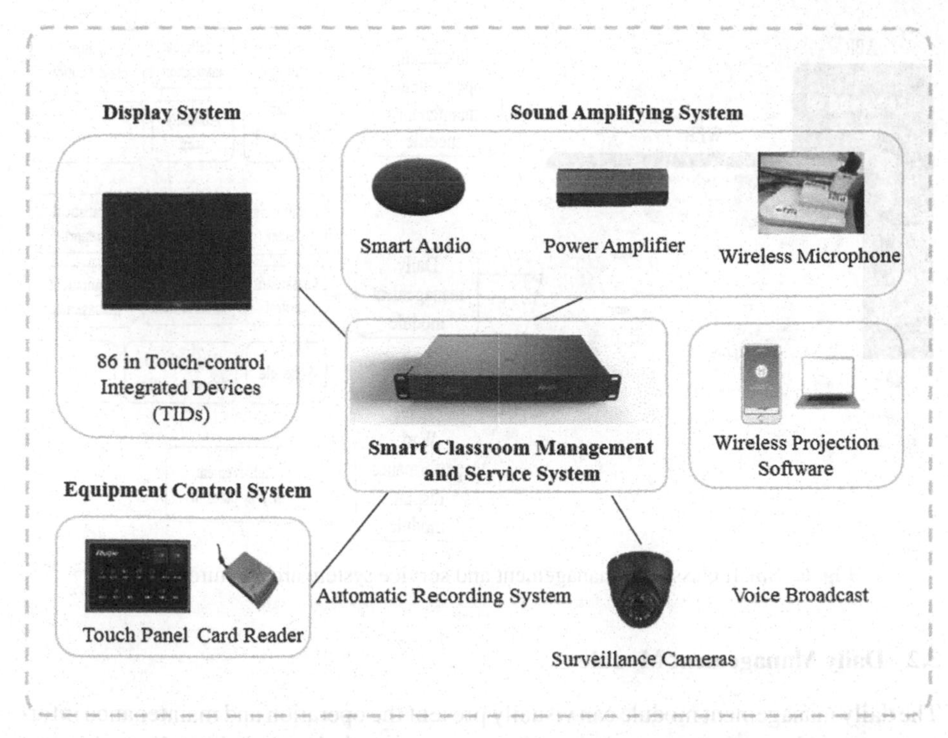

Fig. 4. Equipment architecture diagram of smart classroom

Fig. 5. Smart classroom management and service system interface

and automatic inspection. The operation and maintenance assistant can provide operation and maintenance advice to the operation and maintenance teacher, and proactively prevent possible problems.

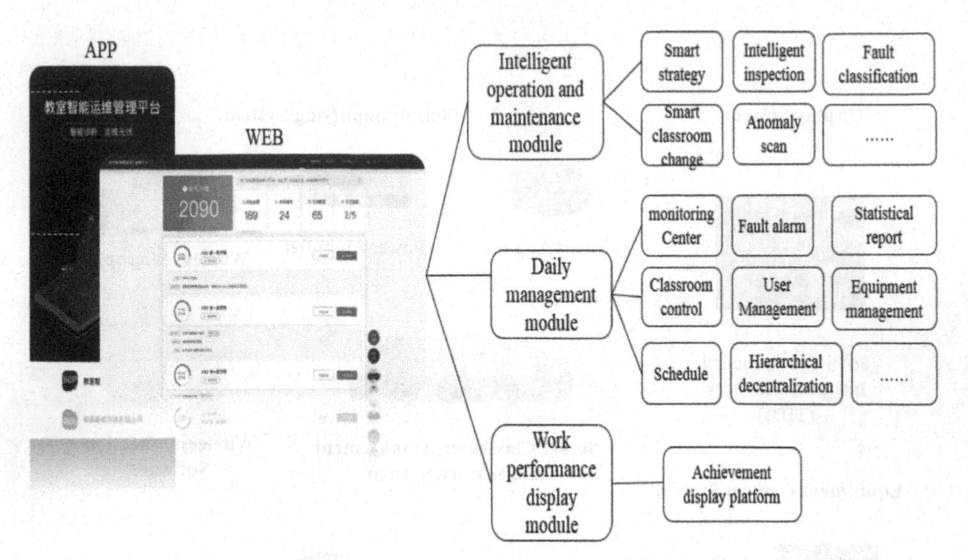

Fig. 6. Smart classroom management and service system architecture diagram

3.2 Daily Management Module

The daily management module can visually present the operation and maintenance information that the management teacher is most concerned about, such as the number of failures, breakdown details, the number of currently available classrooms (no failures, no schedules, and equipment in use), the number of currently abnormal classrooms (no schedules and equipment are in use), and the operation of smart strategies.

3.3 Work Performance Display Module

The work performance display module can show the construction and operation of the schools' overall classroom through multi-dimensional data, generate multiple types of analysis reports, and fully grasp the operation and maintenance situation.

4 Experiments and Analysis

4.1 Evaluation Index

To measure the satisfaction of teachers and students in the smart classroom management and service system, China Customer Satisfaction Index (CCSI) was employed in this study. CCSI is the first domestic user satisfaction model developed by the China Enterprise Research Center of Tsinghua University and the National Quality and Technical Inspection Bureau based on China's actual conditions [13]. It is developed by referring to the American Customer Satisfaction Index (ACSI) and the successful experience of European Customer Satisfaction Index (ECSI). CCSI consists of 6 variables. Based on

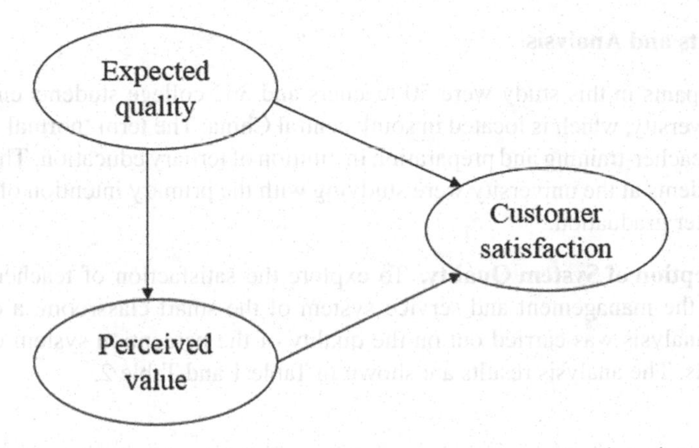

Fig. 7. User satisfaction model of smart classroom management and service system

the actual situation of this study, three structural variables are included: perceived quality, perceived value, and user satisfaction. User satisfaction model of smart classroom management and service system is shown in Fig. 7.

To explore the satisfaction of teachers and students with the management and service system of the smart classroom, an independent sample t-test was conducted to examine the significance of the difference between teachers and students who participated in the survey. The independent sample t-test is to test whether the difference between a sample mean and a known population mean is significant. When the population data is normally distributed, such as the population standard deviation is unknown and the sample size is less than 30, then the deviation statistic between the sample mean and the population mean is t-distributed [14]. The independent sample t-test formula was used to analyze the data obtained. It was used to find out the mean deviation difference of the experimental. The formula for independent sample t-test is shown as below:

$$t = \frac{\overline{x_1} - \overline{x_2}}{\sqrt{\frac{(n_1-1)S_1^2+(n_2-1)S_2^2}{n_1+n_2-2}\left(\frac{1}{n_1}+\frac{1}{n_2}\right)}} \tag{1}$$

$\overline{x_1}$ and $\overline{x_2}$ are the average of two samples; S_1^2 and S_1^2 are the variances of the two samples; n_1 and n_2 are the sizes of the two samples.

The calculation formula of the sample average is as follows.

$$\overline{x} = \frac{x_1 + x_2 + \ldots + x_n}{n} \tag{2}$$

If the average of $x_1, x_2, x_3 \ldots x_n$ is M, the variance formula can be expressed as:

$$S^2 = \frac{(M - x_1)^2 + (M - x_2)^2 + (M - x_3)^2 + \cdots + (M - x_n)^2}{n} \tag{3}$$

4.2 Results and Analysis

The participants in this study were 50 teachers and 362 college students enrolled at a normal university, which is located in south central China. The term 'normal university' refers to a teacher-training and preparation institution of tertiary education. That is to say, college students at the university were studying with the primary intention of becoming teachers after graduation.

User Perception of System Quality. To explore the satisfaction of teachers and students with the management and service system of the smart classroom, a descriptive statistical analysis was carried out on the quality of the perception system of teachers and students. The analysis results are shown in Table 1 and Table 2.

Table 1. Teachers' perception system quality description statistics

	Factor	N	Min	Max	Mean	SD
Functional quality	Clear navigation structure	50	1	5	3.98	0.943
	Ease of operation	50	1	5	3.95	0.939
	Effective communication and interaction	50	1	5	3.65	0.935
	Convenient management functions	50	1	5	3.74	0.922
	Various service methods	50	1	5	3.72	0.896
Service quality	Timely feedback	50	1	5	3.70	0.921
	Personalized resource push	50	1	5	3.56	0.943
	Data analysis accuracy	50	1	5	3.89	0.938

As shown in Table 1, the median of each observed variable of teachers' perception of system quality is higher than 3.5. The highest score is "Clear navigation structure". The lower overall scores are "Effective communication and interaction", "Convenient management functions", "Various service methods", and "Personalized resource push". Although the average value is higher than the median, which tends to be satisfactory, it is lower than the average score of perceived system quality. Therefore, these five variables need to be strengthened.

As shown in Table 2, the median of each observed variable of students' perception of system quality is higher than 3.4. The highest score is "Ease of operation". The lower overall scores are "Convenient management functions" "Timely feedback" and "Personalized resource push". Although the average value is higher than the median, which tends to be satisfactory, it is lower than the average score of perceived system quality. Therefore, these five variables need to be strengthened.

User Perception of System Value. To explore the satisfaction of teachers and students with the management and service system of the smart classroom, a descriptive statistical

Table 2. Student s' perception system quality description statistics

	Factor	N	Min	Max	Mean	SD
Functional quality	Clear navigation structure	362	1	5	3.75	1.127
	Ease of operation	362	1	5	3.88	1.065
	Effective communication and interaction	362	1	5	3.67	1.117
	Convenient management functions	362	1	5	3.57	1.143
	Various service methods	362	1	5	3.65	1.127
Service quality	Timely feedback	362	1	5	3.50	1.061
	Personalized resource push	362	1	5	3.47	1.132
	Data analysis accuracy	362	1	5	3.69	1.124

analysis was carried out on the value of the perception system of teachers and students. The analysis results are shown in Table 3 and Table 4.

Table 3. Descriptive statistics of teachers' perception system value

Factor	N	Min	Max	Mean	SD
How helpful the system is to my teaching	50	1	5	3.88	0.832
The effect of the system on my professional development	50	1	5	3.75	0.853
The system is valuable to teachers	50	1	5	3.69	0.921

As shown in Table 3, the satisfaction of the three items is higher than the median, reaching a satisfactory state. Compared with the average value of this dimension, the mean value of "The system is valuable to teachers" is low. Therefore, there is room for improvement in terms of value to teachers.

Table 4. Descriptive statistics of students' perception system value

Factor	N	Min	Max	Mean	SD
I gained new knowledge on the system	362	1	5	3.65	1.031
The learning efficiency on the system becomes higher	362	1	5	3.45	1.142
I think systematic learning is worthwhile	362	1	5	3.58	1.067

As shown in Table 4, students gaining new knowledge on the platform, learning efficiency increased, and learning are worthy of the three items, which are all higher than the median, reaching a satisfactory state. The average value of platform learning is lower than the average value of this dimension. Therefore, students have a lot of space for improvement in the perceived learning value of the platform.

User satisfaction of the System. To explore the satisfaction of teachers and students with the management and service system of the smart classroom, an independent sample t-test was conducted to examine the significance of the difference between teachers and students who participated in the survey. The independent sample t-test results are shown in Table 5.

Table 5. The results of t-test for teachers' and students' satisfaction of the system in the survey

Factor	Character	N	Mean	*t*	*p*
Satisfied with the overall system	Teacher	50	3.87	3.15	0.002**
	Student	362	3.64		
Reality supports me to use the system	Teacher	50	3.82	3.91	0.104
	Student	362	3.68		
Will continue to use the system in the future	Teacher	50	3.92	3.76	0.000***
	Student	362	3.67		
Willing to recommend the system to others	Teacher	50	3.86	3.23	0.008**
	Student	362	3.61		

**p < 0.01; ***p < 0.001.

As shown in Table 5, there was a significant difference between teachers and students with respect to the mean score of "satisfied with the overall system", "Will continue to use the system in the future" and "Willing to recommend the system to others". However, no significant difference was found between teachers and students in "Reality supports me to use the system". In a word, the overall teacher's satisfaction is higher than the student's satisfaction.

5 Conclusion

With the application of smart classrooms in classroom teaching, the management and service of smart classrooms is a necessary and important task for teaching reform and teaching innovation in colleges and universities. Doing a good job in the management and service of smart classrooms will be the basic guarantee for the quality of classroom teaching in colleges and universities. The current management and service mechanism of smart classrooms in colleges and universities is not sound, and there are still many shortcomings.

How to promote the effective service of new smart classrooms to college classroom teaching requires updating management concepts and models; formulating scientific and reasonable smart classroom use and management systems; improving administrators' technical level and service concepts; changing teachers' teaching concepts and enhancing teachers' Operational capabilities [15–17].

It is hoped that this article can bring enlightenment to managers and teachers who are carrying out the management and service of smart classrooms, so as to exert the effects of the new learning environment of smart classrooms better, enhance the effectiveness of smart classroom teaching and learning, promote the transformation of classroom teaching models, and realize the deep integration of information technology and classroom teaching.

Funding Statement. This work was supported in part by the Scientific Research Project of Education Department of Hunan Province of China with No. 20C0393, in part by the Teaching Reform Project of Hunan First Normal University with No. XYS20J40.

Conflicts of Interest. The authors declare that they have no conflicts of interest to report regarding the present study.

References

1. Shi, Y., Peng, C., Wang, S., Yang, H.H.: The Effects of smart classroom-based instruction on college students' learning engagement and internet self-efficacy. In: Cheung, S.K.S., Kwok, L.-F., Kubota, K., Lee, L.-K., Tokito, J. (eds.) Blended Learning. Enhancing Learning Success, pp. 263–274. Springer, Cham (2018). https://doi.org/10.1007/978-3-319-94505-7_21
2. Jawa, A., et al.: SMEO: a platform for smart classrooms with enhanced information access and operations automation. In: Balandin, S., Dunaytsev, R., Koucheryavy, Y. (eds.) Smart Spaces and Next Generation Wired/Wireless Networking. ruSMART 2010, NEW2AN 2010. Lecture Notes in Computer Science, vol. 6294, pp. 123–134. Springer, Berlin (2010) https://doi.org/10.1007/978-3-642-14891-0_12
3. He, K.K.: Wisdom classroom+classroom teaching structure reform—the fundamental methods of realizing educational informationization. Educ. Res. **11**, 76–81 (2015)
4. Shi, B.H.: Design and implementation of college smart classroom architecture in the age of the internet plus. J. Central Chin. Normal Univ. (Natural Sciences) **S1**, 91–95 (2017)
5. Li, D.S., Hu, X.M., Fu, G., Cui, X.J., He, X.H.: Research on operation and maintenance management of smart classrooms in universities. Chin. Educ. Technol. Equipment **6**, 51–52+55 (2020)
6. Fan, Q., Zhang, Y., Wang, Z.: Improved teaching learning based optimization and its application in parameter estimation of solar cell models. Intell. Autom. Soft Comput. **26**(1), 1–12 (2020)
7. Shi, Y.H., Peng, C.L., Zhang, J.M., Yang, H.: Research on the teacher-student interaction behavior in colleges and universities under smart classroom environment. Mod. Educ. Technol. **29**(1), 45–51 (2019)
8. Chen, W., Ye, X., Xu, Y.: Future classroom: smart learning environment. J. Dist. Educ. **5**, 42–49 (2012)
9. Yu, T., Ying, W.S., Sha, K.: Explore the medical curriculum teaching development in the smart classroom. Int. J. Inf. Educ. Technol. **7**(2), 130–134 (2017)
10. Zhang, Y., Zhu, Y., Bai, Q.Y., Li, X.Y., Zhu, Y.H.: Study on interactive behavior of primary school mathematics classroom teaching under the smart classroom environment. Chin. Electrochem. Educ. **6**, 43–48 (2016)

11. Hwang, G.J., Chu, H.C., Shih, J.L., Huang, S.H., Tsai, C.C.: A decision-tree-oriented guidance mechanism for conducting nature science observation activities in a context-aware ubiquitous learning environment. J. Educ. Technol. Soc. **13**(2), 53–64 (2010)
12. Xue, Y.: Under the Wisdom classroom junior middle school physics teaching mode. Contemp. Educ. Sci. **8**, 61–62 (2014)
13. Pei, F.: The methods in developing questionnaire for China customer satisfaction index. Stand. Sci. **10**, 28–31 (2007)
14. Box, J.F.: Guinness, Gosset, fisher, and small samples. Stat. Sci. **2**(1), 45–52 (1987)
15. Qiu, F., Chen, S.Q.: Discussion on the management of smart classroom and teaching equipment. Comput. Prod. Circ. 10, 255 (2019)
16. Rong, R., Yang, X.M., Chen, Y.H., Zhao, Q.J.: The new development of education management informatization: move towards smart management. Chin. Educ. Technol. **3**, 30–37 (2014)
17. Yang, L.J.: On the management and maintenance of smart classrooms in colleges and universities. Mod. Econ. Inf. **3**, 30–37 (2014)

Performance Analysis of Hard Fusion Rules in Cognitive Radio Networks Over Composite Channels

Hui Yu[1], Li Chen[2], Yunxue Liu[3(✉)], Hao Jiao[3], and Yingjie Cui[3]

[1] Yantai Vocational College, Yantai 264010, China
[2] Huichuang Software Technology Co., Ltd. of Yantai, Yantai 264010, China
[3] Yantai University, Yantai 264010, China

Abstract. Spectrum sensing plays crucial role in cognitive radio networks for efficient utilization of spectrum resources, and hard fusion (HD) is widely used for its good performance and simplicity. Traditionally, only multipath fading or only shadowing effect is discussed for performance analysis of hard fusion rules. However, in practical communication environment, wireless signals are certain to undergo multipath and shadowing fading simultaneously. This paper mainly analyzes spectrum sensing performance of hard decision over channel models with both Rayleigh and Lognormal distribution (composite channel) in cognitive radio networks. First, channel model of the composite channel is given, and the detection probability is derived. Then we analysis spectrum sensing performance of HD under composite channel, analytical results are verified by Monte Carlo simulation.

Keywords: Multipath fading · Hard decision (HD) · Monte Carlo simulation · Spectrum sensing · Lognormal · Rayleigh · Composite channel

1 Introduction

1.1 A Subsection Sample

Cognitive radio (CR) technology has the potential to significantly improve spectral efficiency and performance of current radio networks. This potential has been acknowledged by the Federal Communications Commission (FCC) [1, 2]. In order to utilize spectrum effectively, CR allows unauthorized users to access authorized spectrum when it is free. Therefore, cognitive radio is widely used in wireless communication such as UWB, WRAN, AD Hoc, WLAN and online teaching (especially in the data transmission of interactive teaching networks).

In order to avoid interference with primary user's (PUs) communications, fast and accurate detection of spectrum holes has become a fundamental issue. Cooperation can improve the detector performance, and the possibility of using simpler and energy efficient detectors in the face of multipath channel effects such as fading and shadowing. HD based cooperative sensing (CS) schemes are popular as they are easy to implement.

X. Sun et al. (Eds.): ICAIS 2021, CCIS 1422, pp. 585–596, 2021.
https://doi.org/10.1007/978-3-030-78615-1_51

HD techniques based cooperative detection has been proposed in the literature [3–5]. HD can reduce the communication cost and improve the bandwidth efficiency. The fusion rule at the fusion center can be OR, AND, or Majority rule, which can be generalized as the "k-out-of-n rule".

Collaboration has been proposed to improve spectrum system performance [2–4]. Collaborative spectrum sensing has two main phases: sensing and reporting. In the sensing phase, each individual CR makes an observation using local spectrum sensing algorithm, while in the reporting phase, each CR sends its observation (or decision) to the fusion centre. In order to evaluate the performance of spectrum sensing, two probabilities are of great interest: probability of detection (or probability of missed detection), probability of false alarm.

Collaborative spectrum sensing of hard fusion is compromised when a user experiences shadowing or fading effects [4]. The detection performance of an energy detector used for spectrum sensing in cognitive radio networks is investigated at very low SNR levels over Rayleigh fading and Nakagami fading channels [5]. In reference [6], the performance of hard fusion rules is discussed under noise uncertainty over AWGN and Rayleigh channels. Soft combination of the observed energies from different cognitive radio users is investigated on Rayleigh fading and Nakagami fading channels [7]. In these researches, only multipath fading or only shadowing effect is discussed. But in practical communication environment, wireless signals are certain to undergo both multipath and shadowing fading at the same time. Therefore, previous studies have some limitations. In literature [8, 9], we find out that single Rayleigh fading model and single shadowing fading model are insufficient to illustrate spectrum sensing performance, and it is necessary to analyze detection performance using composite channel model. In this research, we analyze the cooperative spectrum sensing for channels with both Rayleigh fading and shadowing. By adopting direct integral methods, the channel models and Monte Carlo simulation can be derived. Analytical results are verified by Monte Carlo simulation.

The rest of this paper is organized as follows: In Sect. 2, the system model and average detection probability of single channel (Rayleigh, or shadowing fading) are briefly described. In Sect. 3, the average power detection probability of composite channel is evaluated and obtained by direct integral methods or the distribution models. The cooperative spectrum sensing scheme is analyzed in Sect. 4, and analysis and discuss on the selection of hard fusion rules over composite channel are presented in Sect. 5. Finally, the conclusions are drawn in Sect. 6.

2 System Model

We consider a binary hypothesis test for spectrum sensing as follows [10]:

$$y(k) = \begin{cases} n(k), & H_0 \\ hs(k) + n(k), & H_1 \end{cases} \tag{1}$$

where h is the amplitude gain of the channel, $n(k)$ is the additive white Gaussian noise (AWGN) with zero mean and variance σ_n^2, $y(k)$ is the signal received by secondary user and $s(k)$ is primary user's transmitted signal. The hypothesis H_0 and H_1 represents the absence and presence of the target signal, respectively.

In this work, we take average power as the test statistic, which can be expressed as

$$T = \frac{1}{M}\sum_{k=1}^{M} |y(k)|^2 \tag{2}$$

where M is the number of sampling points.

The probability density function (PDF) of average power can then be written as

$$f_T(t) = \begin{cases} \frac{M}{(2\sigma_n^2)^{M/2}\Gamma(M/2)}(Mt)^{M/2-1}e^{-\frac{Mt}{2\sigma_n^2}}, H_0 \\ \frac{M}{2\sigma_n^2}\left(\frac{t}{\gamma}\right)^{(M-2)/4} e^{-\frac{M(\gamma+t)}{2\sigma_n^2}} I_{M/2-1}\left(\sqrt{\frac{M\gamma}{\sigma_n^2}\frac{Mt}{\sigma_n^2}}\right), H_1 \end{cases} \tag{3}$$

where γ is the received SNR of the target signal, $\Gamma(.)$ is the gamma function and $I_v(.)$ is the vth-order modified Bessel function of the first kind.

Based on the Eq. (3), we obtain closed-form expressions for both the false alarm probability and the probability of detection probability over AWGN channels. The probability of detection and false alarm probability can be generally computed by [11]

$$P_f = P(T > \lambda|H_0) = \frac{\Gamma(M/2, \lambda M/2\sigma_n^2)}{\Gamma(M/2)} \tag{4}$$

$$P_d = P(T > \lambda|H_1) = Q_{M/2}\left(\sqrt{M\gamma}, \sqrt{M\lambda/2\sigma_n^2}\right) \tag{5}$$

where P_m is the incomplete gamma function, and P_m is the generalized Marcum Q-function.

According to the definition of the missed detection probability P_m, we can get $P_m = 1 - P_d$.

2.1 Average Detection Probability of Shadowing Fading Channels

In some practical situations, wireless signals also undergo multipath or shadowing fading. In this case, average probability of detection ($\overline{P_d}$) may be derived by averaging (5) over fading statistics [12],

$$\overline{P_d} = \int_x P_d(x) f_\gamma(x) dx \tag{6}$$

where $P_d(x)$ is the probability of detection over AWGN channels and $f_\gamma(x)$ is the Probability Density Function (PDF) of SNR under fading environments.

Rayleigh Fading. In Rayleigh fading environments, the signal amplitude follows the Rayleigh distribution [9], then the instantaneous SNR follows an exponential PDF given by [13]

$$f_{Ray}(\gamma) = \frac{1}{\overline{\gamma}} e^{-\frac{\gamma}{\overline{\gamma}}}, \gamma \geq 0 \tag{7}$$

where $\overline{\gamma}$ is the average SNR. In this case, the average detection probability $\overline{P}_{d\_Ray}$ can be evaluated by averaging (6) over (7)

$$\overline{P_{d-Ray}} = e^{-\frac{M\lambda}{2\sigma_n^2}} \sum_{i=0}^{M/2-2} \frac{1}{i!}\left(\frac{M\lambda}{2\sigma_n^2}\right)^i + \left(\frac{2+M\overline{\gamma}}{M\overline{\gamma}}\right)^{M/2-1}\left(e^{-\frac{M\lambda}{2\sigma_n^2(1+M\overline{\gamma})}} - e^{-\frac{M\lambda}{2\sigma_n^2}} \sum_{i=0}^{M/2-2} \frac{1}{i!}\left(\frac{M^2\lambda\overline{\gamma}}{2\sigma_n^2(2+M\overline{\gamma})}\right)^i\right) \tag{8}$$

Shadowing Fading. In terrestrial and satellite land-mobile systems, the link quality is also affected by slow variation of the mean signal level due to the shadowing from terrain, buildings, and trees. When γ is log-normally distributed due to shadowing, then the instantaneous SNR γ follows an PDF given by [14]

$$f_{\log}(\gamma) = \frac{10/\ln 10}{\sqrt{2\pi}\sigma\gamma} \exp\left[\frac{(10\log_{10}\gamma - \mu)^2}{2\sigma^2}\right] \tag{9}$$

where μ(dB) and σ(dB) are the linear mean and standard deviation of $10\log_{10}\gamma$ in the lognormal shadowing, respectively.

2.2 K Channel Model

The fading amplitude X undergoes multipath fading as a Rayleigh distribution, and shadowing as a gamma distribution. Therefore, the average power of X, which represents the shadowing effect, follows the gamma distribution. The PDF of X, denoted as $f_X(x)$, follows a X distribution when the multipath fading is Rayleigh distributed. For the X distribution, $f_X(x)$ can be written as [15]:

$$f_X(x) = \frac{4}{\Gamma(k)\sqrt{\Omega}}\left(\frac{x}{\sqrt{\Omega}}\right)^k K_{k-1}\left(\frac{2x}{\sqrt{\Omega}}\right), \quad x \geq 0 \tag{10}$$

where k is the shaping parameter, Ω represents the scale parameter which is also the mean signal power given as $\Omega = E[X^2]/k$, $K_v(.)$ is the vth-order modified Bessel function of the second kind.

3 Average Detection Probability with Both Rayleigh Fading and Shadowing

The previous results can be extended to account for the composite channel variations by integrating the average detection probability $\overline{P_d}$. It can be evaluated by averaging $\overline{P_d}$ in (6) over the SNR range. This can be expressed mathematically as follows:

$$\overline{P_d} = \int_0^\infty Q_{M/2}\left(\sqrt{M\gamma}, \sqrt{M\lambda/\sigma_n^2}\right) f_\gamma(\gamma) d\gamma$$

$$= \int_0^\infty \int_0^\infty Q_{M/2}\left(\sqrt{M\gamma}, \sqrt{M\lambda/\sigma_n^2}\right) f_{\gamma|Y=y}(\gamma) d\gamma f_Y(y) dy$$
$$= \int_0^\infty P_d^{Fad}(y) f_Y(y) dy \tag{11}$$

where $P_d^{Fad}(y)$ can be evaluated by setting $\sigma^2 = 1$, and by replacing each $\bar{\gamma}$ by y in the average detection probability over Rayleigh distribution in (9), we attain

$$P_d^{Fad}(y) = e^{-\frac{M\lambda}{2\sigma_n^2}} \sum_{n=0}^{M/2-2} \frac{1}{n!}\left(\frac{M\lambda}{2\sigma_n^2}\right)^n + \left(\frac{2+My}{My}\right)^{M/2-1}\left[e^{-\frac{M\lambda}{2\sigma_n^2+My\sigma_n^2}} - e^{-\frac{M\lambda}{2\sigma_n^2}} \sum_{n=0}^{M/2-2} \frac{1}{n!}\left(\frac{M^2\lambda y}{2\sigma_n^2(2+My)}\right)^n\right] \tag{12}$$

In (11), Y is the SNR with only shadowing effect (i.e., multipath fading is excluded), which follows a log-normal distribution (10). Therefore, the expression (11) can be rewritten as

$$\begin{aligned}\overline{P_d}(y) = {} & e^{-\frac{M\lambda}{2\sigma_n^2}} \sum_{n=0}^{M/2-2} \frac{1}{n!}\left(\frac{M\lambda}{2\sigma_n^2}\right)^n + \frac{10/\ln 10}{\sqrt{2\pi}\sigma} \int_0^\infty \left(\frac{2+My}{My}\right)^{M/2-1} e^{-\frac{M\lambda}{2\sigma_n^2+My\sigma_n^2}} \frac{1}{y} \exp\left[-\frac{(10\log_{10} y - \mu)^2}{2\sigma^2}\right] dy \\ & - e^{-\frac{M\lambda}{2\sigma_n^2}} \frac{10/\ln 10}{\sqrt{2\pi}\sigma} \sum_{n=0}^{M/2-2} \frac{1}{n!}\left(\frac{M\lambda}{2\sigma_n^2}\right)^n \int_0^\infty \left(\frac{2+My}{My}\right)^{M/2-1-n} \frac{1}{y} \exp\left[-\frac{(10\log_{10} y - \mu)^2}{2\sigma^2}\right] dy\end{aligned} \tag{13}$$

4 Cooperative Spectrum Sensing

For simplicity, we focus on the cases in which the secondary users are independent and identically distributed. The false alarm probability and the detection probability of cooperative spectrum sensing based on the three kinds of fusion rules can be present as follows.

Cooperative spectrum sensing is implemented by fusing the sensing parameters of individual secondary users and making a final decision in the secondary base station (centralized network). In order to save bandwidth, decision fusion is used at the fusion center. The fusion rule can be OR, AND, or Majority rule, which are generalized as the "k-out-of-n" rule. The false alarm probability and the detection probability of cooperative spectrum sensing based on the three kinds of fusion rules can be presented as follows [18]:

OR rule:

$$Q_{f\_or} = 1 - (1 - P_f)^N \tag{14}$$

$$Q_{d\_or} = 1 - (1 - P_d)^N \tag{15}$$

AND rule:

$$Q_{f\_and} = P_f^N \tag{16}$$

$$Q_{d\_and} = P_d^N \tag{17}$$

Majority rule:

$$Q_{f\_majority} = \sum_{l=K}^{N} \binom{N}{l} P_f^l (1 - P_f)^{N-l} \tag{18}$$

$$Q_{d\_majority} = \sum_{l=K}^{N} \binom{N}{l} P_d^l (1 - P_d)^{N-l} \tag{19}$$

5 Numerical and Simulation Results

5.1 Performance Analysis About Shadowing Effect in Small Cognitive Networks

In this section, we analyze the performance of different hard fusion rules over composite channels in small cognitive networks. Small cognitive networks represent the number of cognitive users in a network with is less than 10 (In our experiments, $N = 5$). We mainly study the following aspects. 1) The standard deviation $N = 5$ of Log-normal distribution respectively is weak shadowing medium ($N = 5 = 2$ dB), shadowing ($N = 5 = 6$ dB) and severe shadowing ($N = 5 = 12$ dB). 2) The log mean $N = 5$ (average SNR) is -10 dB, -5 dB and 0 dB, respectively.

Firstly, we compare the sensing performance among the three kinds of fusion rules over weak shadowing when the number of cognitive users is fewer. Figure 1a, b, c show the complementary ROC curves of the three kinds of fusion rules in a small cognitive radio network with $N = 5$ and $\sigma = 2$ dB, when $\mu = -10, -5, 0$ dB, respectively.

Figure 1a depicts the complementary ROC curves with $\mu = -10$ dB, $M = 300$ for five cognitive radio users. Obviously, AND rule exhibits the best detection performance among the three kinds of widely used fusion rules, e.g., for $Q_f = 0.01$, the probabilities of missed detection of OR rule, Majority rule and AND rule are 0.64, 0.79, and 0.49, respectively.

The same conclusion can be attained from Fig. 1b, which shows the complementary ROC curves for $\sigma = 2$ and $\mu = -5$ dB. It can be observed that OR rule performs best in a small network when $\sigma = 2$ and $\mu = -5$ dB. Then we shift to high average *SNR* cases for a deeper study. The performance of cooperative spectrum sensing with $\mu = 0$ dB is shown in Fig. 1c. Therefore, we conclude that AND rule is the best choice among the three kinds fusion rules in weak shadowing for a small cognitive network.

Secondly, we compare the sensing performance among the three kinds of fusion rules over medium shadowing when the number of cognitive users is fewer. Figure 2a, b, c show the complementary ROC curves of the three kinds of fusion rules in a small cognitive radio network with $N = 5$ and $\sigma = 6$ dB, when $\mu = -10, -5, 0$ dB, respectively.

Figure 2a shows the ROC curves of the three kinds of fusion rules in a cognitive radio network with $N = 5$, $M = 6$ and $\sigma = 6$ dB when $\mu = -10$ dB. From Fig. 2a, it can be observed that hard fusion based on the OR rule can give the best performance in the cognitive radio networks with $N = 5$, $\sigma = 6$ dB and $\mu = -10$ dB For example, for a given false alarm detection probability $Q_f = 0.01$, Majority rule requires a missed probability of 0.54, AND rule needs 0.70, while the OR rule only needs 0.38.

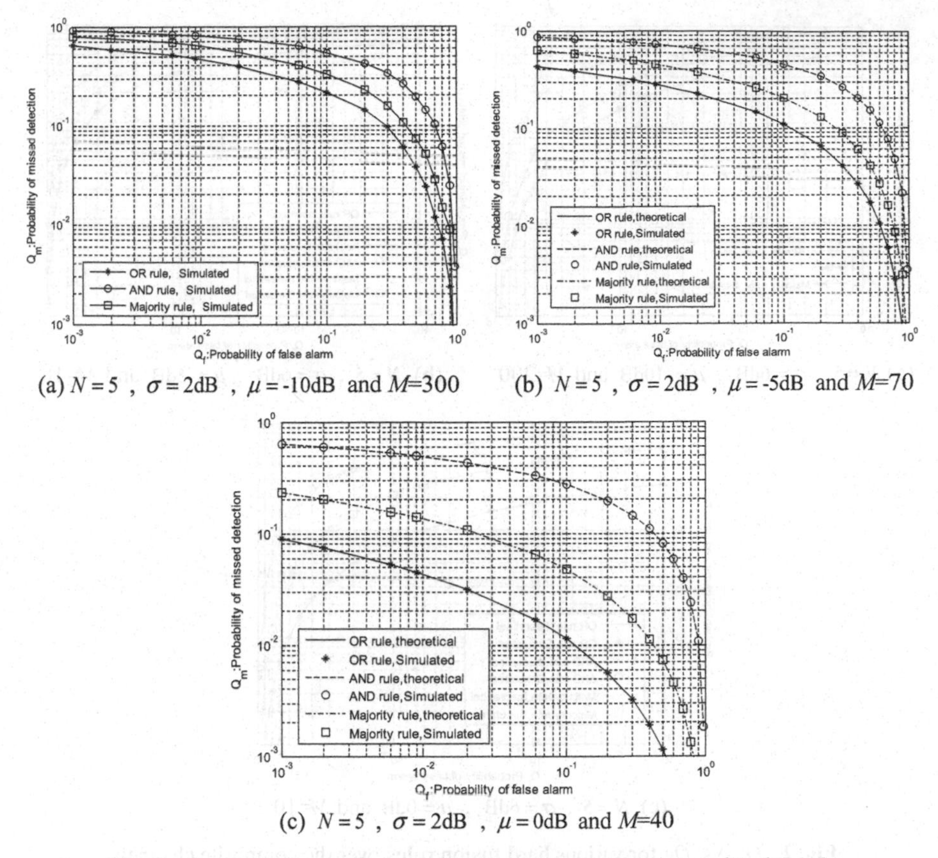

(a) $N = 5$, $\sigma = 2$dB , $\mu = -10$dB and $M=300$ (b) $N = 5$, $\sigma = 2$dB , $\mu = -5$dB and $M=70$

(c) $N = 5$, $\sigma = 2$dB , $\mu = 0$dB and $M=40$

Fig. 1. Q_m vs. Q_f for various hard fusion rules over the composite channels

Then we look into the case of high *SNR*, in Fig. 2b and Fig. 2c, we obtain the same conclusion that OR rule is the optimal choice among the three kinds of fusion rules when $\sigma = 6$ dB, $\mu = -5, 0$ dB.

Therefore, in the small network, when $\sigma = 2$ or 6 dB, the performance of the OR rule is optimal, the performance of the AND rule is worst.

Thirdly, we compare the sensing performance among the three kinds of fusion rules over severe shadowing when the number of cognitive users is fewer. Figure 3a, b, c show the complementary ROC curves of the three kinds of fusion rules in a small cognitive radio network with $N = 5$ and $\sigma = 12$ dB, when $\mu = -10, -5, 0$ dB, respectively.

In Fig. 3a, shows the ROC curves of the three kinds of fusion rules in a cognitive radio network with $N = 5$, $M = 6$ and $\sigma = 6$ dB when $\mu = 10$ dB. From Fig. 3a, it can be seen that hard fusion based on the OR rule can give the best performance in the cognitive radio networks with $N = 5$, $\sigma = 6$ dB and $\mu = -10$ dB. For example, for a given false alarm detection probability $Q_f = 0.01$, Majority rule requires a missed probability of 0.80, AND rule needs 0.94, while the OR rule only needs 0.55.

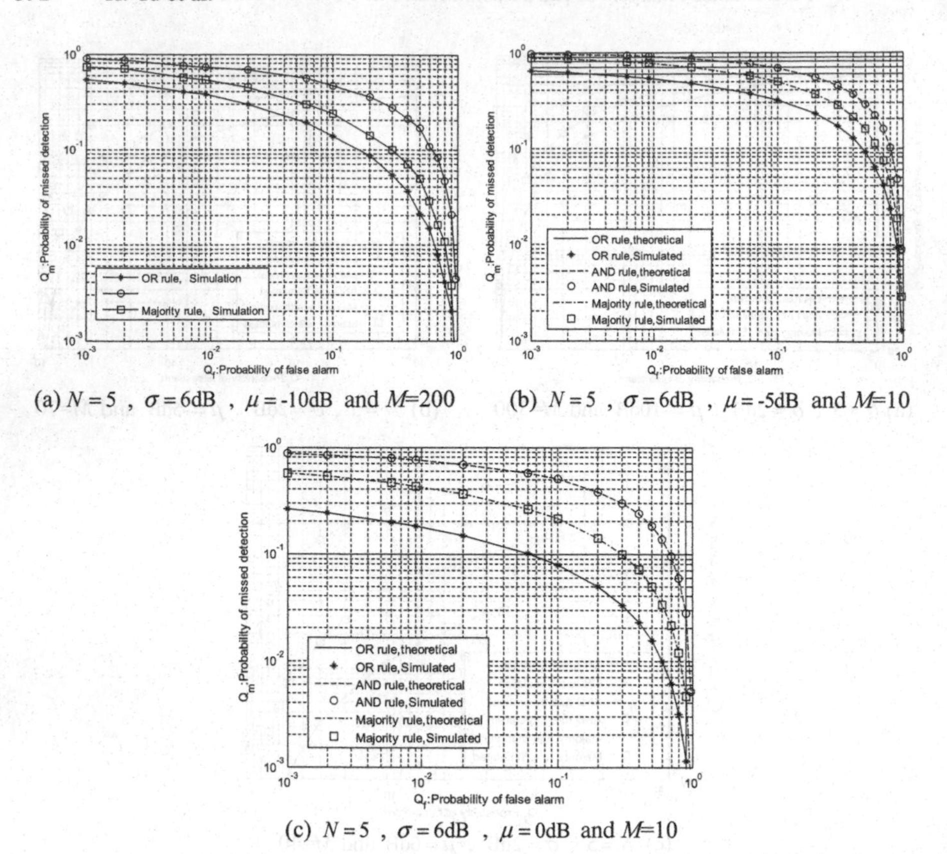

(a) $N=5$, $\sigma=6$dB , $\mu=-10$dB and $M=200$ (b) $N=5$, $\sigma=6$dB , $\mu=-5$dB and $M=10$

(c) $N=5$, $\sigma=6$dB , $\mu=0$dB and $M=10$

Fig. 2. Q_m vs. Q_f for various hard fusion rules over the composite channels

Then we look into the case of high *SNR*, in Fig. 3b and Fig. 3c, we obtain the same conclusion that OR rule is the optimal choice among the three kinds of fusion rules when $\sigma = 12$ dB, $\mu = 5, 0$ dB.

Finally, the extensive analytical and simulation results above illustrate that in the small network, OR rule performs best, AND rule is worst.

5.2 Performance Analysis About Shadowing Effect in Large Cognitive Networks

In this section, we compare the sensing performance among the three kinds of fusion rules when the number of cognitive users is larger. Large cognitive networks is means the number of cognitive users is larger than 50 ($N = 50$). We mainly study the following aspects. 1) The standard deviation σ of Log-normal distribution respectively is weak shadowing medium ($\sigma = 2$ dB), shadowing ($\sigma = 6$ dB) and severe shadowing ($\sigma = 12$ dB). 2) The log mean μ(average SNR) is -10 dB, -5 dB and 0 dB, respectively.

Figure 4a shows the ROC curves of the three kinds of fusion rules in a cognitive radio network with $N = 50$, $\mu = 2$ dB and $\mu = -10$ dB. From this Figure, we notice that the performance of the Majority rule is optimal in the large network (N = 50) with

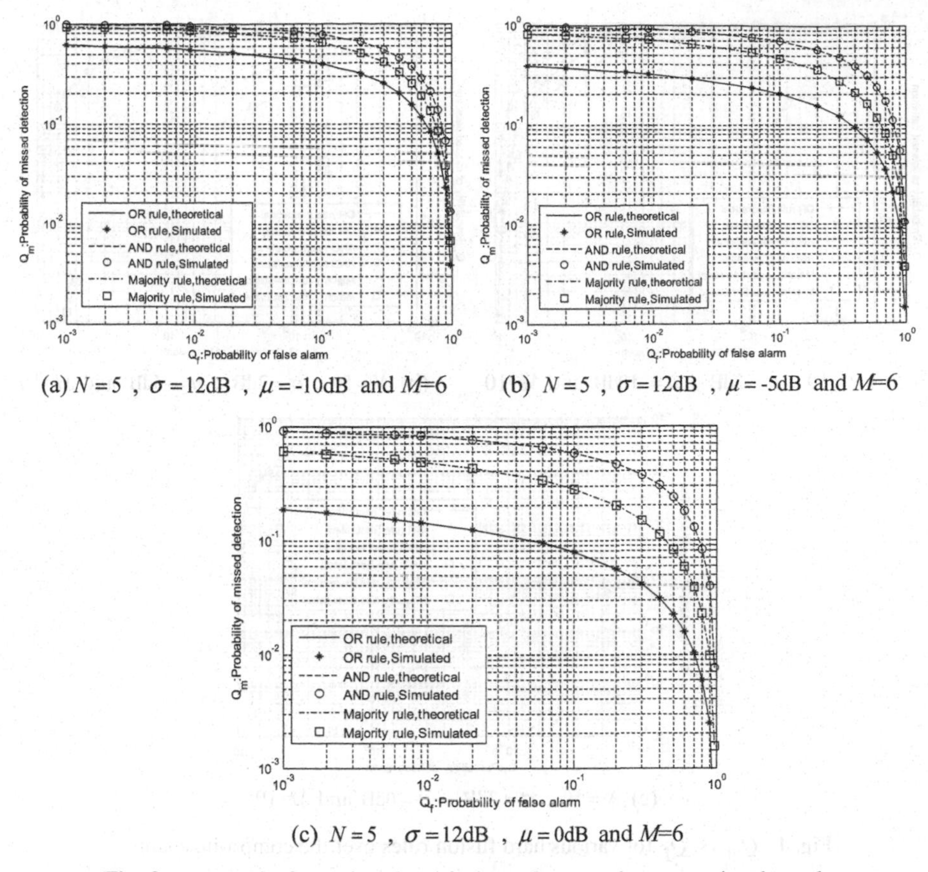

(a) $N=5$, $\sigma=12$dB , $\mu=-10$dB and $M=6$ (b) $N=5$, $\sigma=12$dB , $\mu=-5$dB and $M=6$

(c) $N=5$, $\sigma=12$dB , $\mu=0$dB and $M=6$

Fig. 3. Q_m vs. Q_f for various hard fusion rules over the composite channels

$\sigma = 2$ dB and $\mu = -10$ dB. e.g., for $Q_f = 0.1$, the probabilities of missed detection employing Majority rule, OR rule and AND rule are 0.53, 0.59, and 0.70, respectively.

In Fig. 4b, we look into the case of high *SNR*, therefore, we obtain the same conclusion that Majority rule is the optimal choice among the three kinds of fusion rules when $\sigma = 12$ dB, $\mu = -5$ dB. In Fig. 4c, in the large network, $\mu = 2$ dB and $\mu = 0$ dB, the Majority rule is appropriate choice when $Q_f \geq 0.054$.

From all mentioned above, we conclude that Majority rule is the best choice among the three kinds of fusion rules in the large network with weak shadowing.

From Fig. 5, 6 and Fig. 7, it can be seen that the performance of the OR rule is optimal in the large network (N = 50) with $\sigma = 6$ or 12 dB and $Q_m = -10, -5$ or 0 dB. For a given Q_m, for example,Q_m, OR rule has the lowest Q_m, and AND rule has the highest Q_m from Fig. 5, 6 and Fig. 7.

Extensive experiments reveal that, for larger cognitive radio networks over composite channels, OR rule is the optimal choice in most cases, and Majority rule is the best when shadowing is weak.

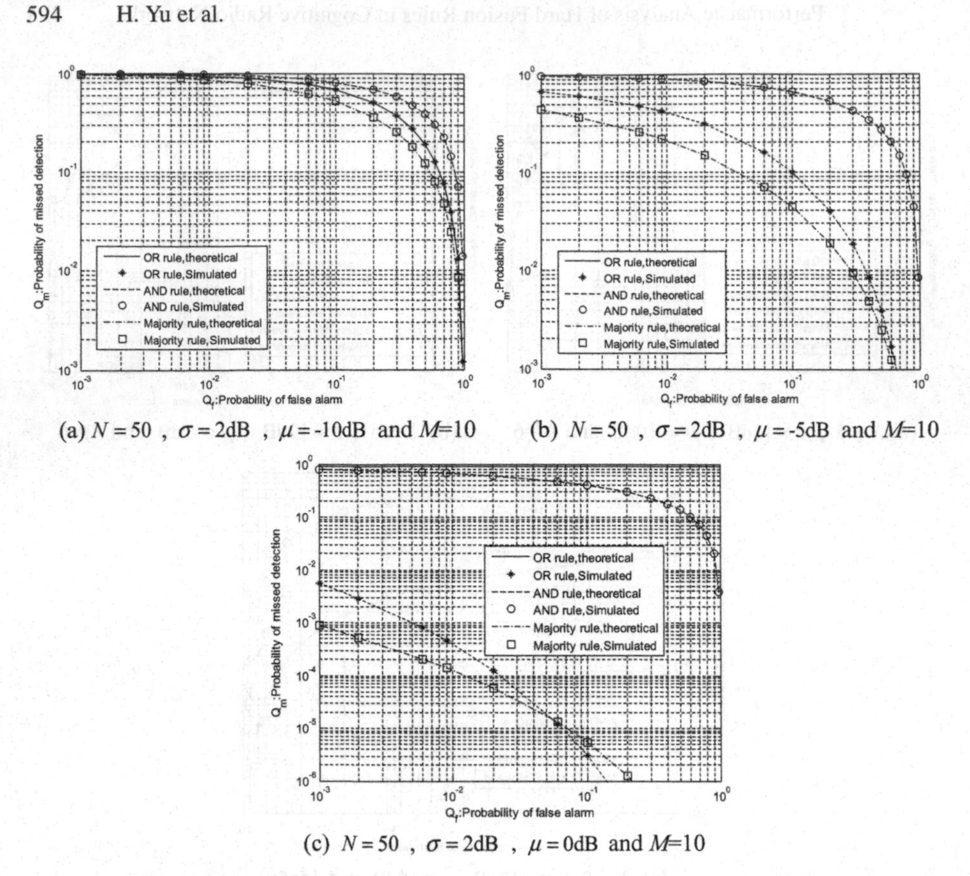

(a) $N = 50$, $\sigma = 2$dB , $\mu = -10$dB and M=10 (b) $N = 50$, $\sigma = 2$dB , $\mu = -5$dB and M=10

(c) $N = 50$, $\sigma = 2$dB , $\mu = 0$dB and M=10

Fig. 4. Q_m vs. Q_f for various hard fusion rules over the composite channels

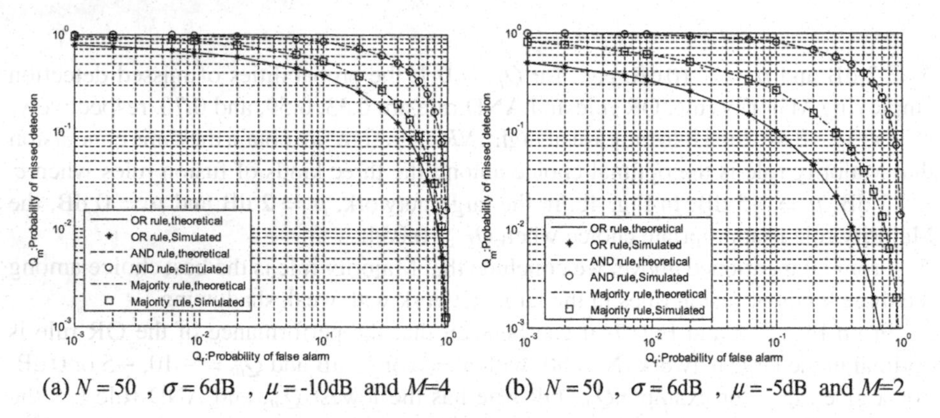

(a) $N = 50$, $\sigma = 6$dB , $\mu = -10$dB and M=4 (b) $N = 50$, $\sigma = 6$dB , $\mu = -5$dB and M=2

Fig. 5. Q_m vs. Q_f for various hard fusion rules over the composite channels

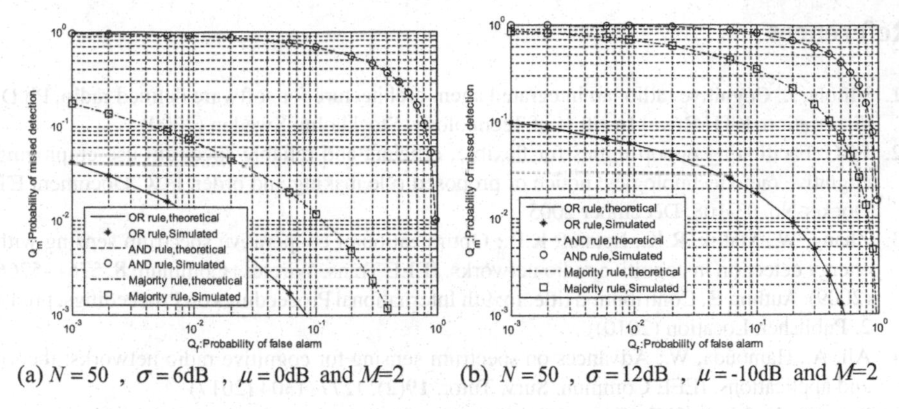

(a) $N=50$, $\sigma=6$dB , $\mu=0$dB and M=2 (b) $N=50$, $\sigma=12$dB , $\mu=-10$dB and M=2

Fig. 6. Q_m vs. Q_f for various hard fusion rules over the composite channels

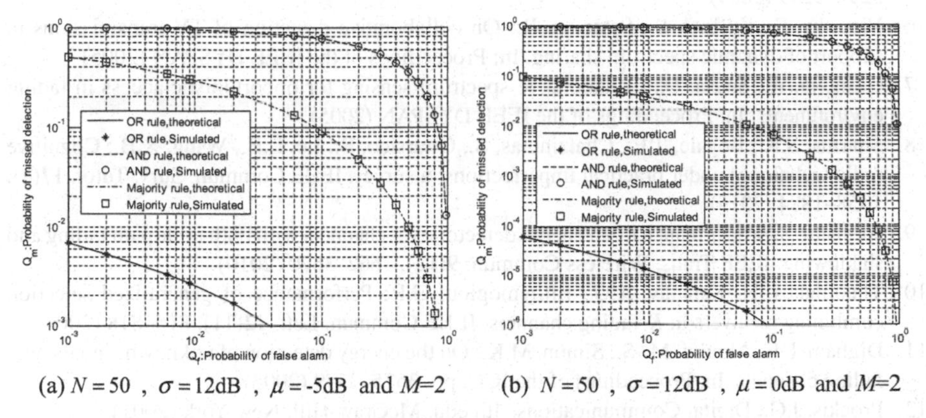

(a) $N=50$, $\sigma=12$dB , $\mu=-5$dB and M=2 (b) $N=50$, $\sigma=12$dB , $\mu=0$dB and M=2

Fig. 7. Q_m vs. Q_f for various hard fusion rules over the composite channels

6 Conclusion

We study the impact of hard fusion rules (AND rule, OR rule, and Majority rule) on spectrum sensing performance over composite channels, and derive the expression of detection probability. In composite channels with both Rayleigh and Lognormal distribution, we can attain conclusions as followed: 1) In most cases, OR rule is the best choice, and AND rule is the worst. 2) For a cognitive radio network with weak shadowing, AND rule is the best for a small cognitive networks, Majority rule is the best in large cognitive networks.

References

1. Mitola, J.: Cognitive radio: an integrated agent architecture for software defined radio. Ph.D. dissertation, KTH Royal Institute of Technology, Stockholm, Sweden (2000)
2. FCC: Facilitating opportunities for flexible, efficient and reliable spectrum use employing cognitive radio technologies: notice of proposed rule making and order. FCC Document ET Docket No. 03-108, December 2003
3. Zhang, W., Mallik, R.K., Letaief, K.B.: Optimization of cooperative spectrum sensing with energy detection in cognitive radio networks. IEEE Trans. Wireless Commun. **8**, 5761–5766 (2009). Author, F.: Contribution title. In: 9th International Proceedings on Proceedings, pp. 1–2. Publisher, Location (2010)
4. Ali, A., Hamouda, W.: Advances on spectrum sensing for cognitive radio networks: theory and applications. IEEE Commun. Surv. Tutor. **19**(2), 1277–1304 (2017)
5. Peh, E.C.Y., Liang, Y.C., Guan, Y.L., Zeng, Y.: Optimization of cooperative sensing in cognitive radio networks: a sensing-throughput tradeoff view. IEEE Trans. Veh. Technol. **58**, 5294–5299 (2009)
6. Visostky, E., Kuffner, S., Peterson, R.: On collaborative detection of TV transmissions in support of dynamic spectrum sharing. In: Proceedings of the IEEE DYSPAN (2005)
7. Gashemi, A., Sousa, E.: Collaborative spectrum sensing for opportunistic access in fading environments. In: Proceedings of the IEEE DYSPAN (2005)
8. Sharma, S.K., Bogale, T.E., Chatzinotas, S., Ottersten, B., Le, L.B., Wang, X.B.: Cognitive radio techniques under practical imperfections: a survey. IEEE Commun. Surv. Tutor. **17**(4), 1858–1884 (2015)
9. Atapattu, S.: Performance of an energy detector over channels with both multipath fading and shadowing. IEEE Trans. Wireless Commun. **9**(12), 3662–3670 (2010)
10. Theofilakos, P., Kanatas, A.G., Efthymoglou, G.P.: Performance of generalized selection combining receivers in *K* fading channels. IEEE Commun. Lett. **12**(11), 816–818 (2008)
11. Digham, F.F., Alouini, M.-S., Simon, M.K.: On the energy detection of unknown signals over fading channels. In: Proceedings of the ICC, pp. 3575–3579 (2003)
12. Proakis, J.G.: Digital Communications, 4th edn. McGraw-Hill, New York (2001)
13. Simon, M.K., Alouini, M.-S.: Digital Communication over Fading Channels, 2nd edn. Wiley, Hoboken (2005)
14. Stüber, G.L.: Principles of Mobile Communication, 2nd edn. Kluwer Academic Publishers, Norwell (2001)
15. Stevenson, C., Chouinard, G., Lei, Z., Hu, W., Shellhammer, S., Caldwell, W.: IEEE 802.22: the first cognitive radio wireless regional area network standard. IEEE Commun. Mag. **47**(1), 130–138 (2009)
16. Abdi, A., Kaveh, M.: *K* distribution: an appropriate substitute for Rayleigh-lognormal distribution in fading-shadowing wireless channels. Electron. Lett. **34**(9), 851–852 (1998)
17. Letaief, K.B., Zhang, W.: Cooperative communications for cognitive radio networks. Proc. IEEE **97**(5), 878–893 (2009)

Performance Analysis of Spectrum Sensing in Cognitive Radio Networks Based on AD Goodness of Fit Under Noise Uncertainty

Hui Yu[1], Xiaoming Ding[2], Yunxue Liu[3](✉), Hao Jiao[3], and Yingjie Cui[3]

[1] Yantai Vocational College, Yantai 264010, China
[2] China Railway Jinan Group Co., Ltd., Jinan 250000, China
[3] Yantai University, Yantai 264010, China

Abstract. Spectrum sensing is one of key technologies for primary user detection in cognitive radio networks, and Goodness of Fit Test exhibits good sensing performance. In this paper, we analyze the performance of spectrum sensing based on AD goodness of fit (AD test) under noise uncertainty, and compare it with traditional energy detection. Moreover, much attention is paid on the sensing performance of AD test and energy detection under lower SNR, and analyzes the impact of direct current (DC) component for AD test when the noise power is uncertainty. We first give formulas of AD test and signal expressions with different percentage of direct current (DC) component, and discuss the sensing performance by using Monte Carlo simulation in AWGN channel and Rayleigh channel. Then the impact of noise uncertainty on spectrum sensing performance of AD test is discussed and compared with energy detection in different network environments. Extensive simulations and theoretical analysis indicate that spectrum sensing performance of AD test is better than energy detection when SNR is lower and there is no DC component in the transmitted signals under noise uncertainty. The above conclusion shows that AD test is less sensitive to noise uncertainty traditional energy detection.

Keywords: Cognitive radio · Spectrum sensing · Anderson-Darling test · Energy detection · Noise uncertainty

1 Introduction

The radio spectrum is a valuable natural resource due to the restrictive of its use. Hence, available spectrum resource is very scarce for future wireless communication. Cognitive radio (CR) is seen as one of the key technologies to use spectrum effectively. Cognitive radio allows unlicensed users to access licensed spectrum by rapidly and autonomously adapting operating parameters to changing environments and conditions, without causing any interference to licensed users [1]. Therefore, cognitive radio is widely used in wireless communication such as UWB, WRAN, AD Hoc, WLAN and online teaching (especially in the data transmission of interactive teaching networks).

X. Sun et al. (Eds.): ICAIS 2021, CCIS 1422, pp. 597–606, 2021.
https://doi.org/10.1007/978-3-030-78615-1_52

Based on the local observations in cognitive radio, many spectrum sensing methods have been proposed, including: matched filtering detection [2], cyclostationary feature detection [3] and energy detection (ED) [4]. The matched filtering detection and cyclostationary feature detection need to know the prior information of primary users which may not be available to cognitive users in practice. The traditional ED is the best choice without the information of primary signal. Moreover, ED is very simple to implement. So ED is widely adopted for spectrum sensing. However, the energy detection doesn't have a good performance at low SNR. Moreover, the performance of energy detection is easily affected by noise uncertainty and it will deteriorate obviously.

A new scheme based on calculating the distance between the empirical distribution of the local observations and the assumption model distribution, and comparing the distance with the known threshold, goodness of fit tests has been proposed in recent years. There are goodness of fit tests like Cramer-Von Mises (CM), Kolmogorov-Smirnov (KS) and AD tests. One of the three methods Anderson-Darling (AD) test is a widely adopted due to the fact that it exhibits good performance and needs less samples.

This paper is focused on the application of Anderson-Darling (AD) test to the spectrum sensing. In the past few years, [5] and [6] have proposed some important results that sensing performance of AD test is much better than traditional ED test. But the positive result from the above literature can only be obtained when the primary signal is deterministic, and this situation is scarce case in real cognitive environment. Although several different transmitted signals are mentioned in [7] and it has indicated that the difference of primary signals makes much influence on sensing performance. But it does not show that which characteristic has the conclusive effect on detection performance. Also in [7], the network environment for discussion is simple and the simulations are only done in AWGN channel. Moreover, in practice environment the noise variance is hardly known because of the noise uncertainty. In [8], a blind spectrum sensing method based on AD test has been proposed. But the transmitted signal in this paper is just the static signal $s(n) = 1$. This paper contains also the results and measurements of AD test in both AWGN channel and Rayleigh channel under noise uncertainty. The results are discussed and compared for several different transmitted signals.

In this research, we adopt the scheme of goodness of fit test (AD test is used) for spectrum sensing in cognitive radio networks, and make a decision on the presence of the primary users by testing whether the observed samples are drawn independently from the noise distribution. Additionally, we assume three kinds of transmitted signals, and add DC (direct current) component with different percentage in each kind signal. Then we analyze spectrum sensing performance of different mixed signals by using AD test and compare it with energy detection in AWGN channel and Rayleigh channel under noise uncertainty.

The rest of this paper is organized as follows: In Sect. 2, the hypothesis testing model of goodness of fit test is established. In Sect. 3, we present AD test as a example of the application of goodness of fit test to spectrum sensing. In Sect. 4, simulation results are presented, and energy detection and AD test are compared and analyzed in AWGN channel and Rayleigh channel. Finally, conclusions are drawn in Sect. 5.

2 System Model

The spectrum sensing problem can be expressed as two well known hypotheses for a priori known channel. The hypotheses are formulated as:

$$H_0 : y(n) = w(n)$$

$$H_1 : y(n) = hs(n) + w(n), n = 1, 2...N \tag{1}$$

where $Y = \{y_i\}_{i=1}^n$ is the received signal, $Y = \{y_i\}_{i=1}^n$ is the primary user's signal and $Y = \{y_i\}_{i=1}^n$ is the additive white Gaussian noise at the CR receiver. Additionally, $Y = \{y_i\}_{i=1}^n$ is the amplitude gain of the channel from the primary user to a cognitive radio user.

Let $Y = \{y_i\}_{i=1}^n$ denotes n observations. Only noise is present in the received signal $y(n)$ if signal is absent, then Y is an i.i.d. sequence that has the same distribution as additive white Gaussian noise. So, the cumulative distribution function can be given by:

$$F_0(y) = \frac{1}{\sqrt{2\pi}} \int_{-\infty}^{y} e^{-x^2/2} dx \tag{2}$$

Goodness of fit test is a typical mathematical method to determine whether the samples come from the originally assumed distribution. Let $F_n(y)$ denotes the empirical distribution of the observations $y_1, y_2...y_n$, and is defined as [8]:

$$F_n(y) = |\{i : y_i \leq y, 1 \leq i \leq n\}|/n \tag{3}$$

Therefore, we can use the distance between $F_n(y)$ and $F_n(y)$ to judge whether the hypothesis $F_n(y)$ is true. If the distance is less than a given threshold, the hypothesis $F_n(y)$ is true; otherwise, the hypothesis $F_n(y)$ is true.

3 AD Sensing Test and Transmitted Signals

3.1 Anderson-Darling Test (AD Test)

The distance between $F_n(y)$ and $F_0(y)$ (A_n^2) in AD test is defined by [9]:

$$A_n^2 = n \int_{-\infty}^{\infty} [F_n(y) - F_0(y)]^2 \frac{dF_0(y)}{F_0(y)(1 - F_0(y))} \tag{4}$$

It can be seen from (4) that it is very difficult to compute the distance. Taking implement of AD test into consideration, discrete distance A_n^2 can be simplified as:

$$A_n^2 = \frac{\sum_{i=1}^{n}(2i-1)(\ln z_i + \ln(1 - z_{n+1-i}))}{n} - n \tag{5}$$

where $z_i = F_0(y_i)$.

After value of discrete distance is attained, the AD test is performed [10] by

$$P(A_n^2 > t_0 | H_0) = \alpha \tag{6}$$

where t_0 is the threshold and α is the false alarm probabilities. As depicted in [9], a table listing values of t_0 corresponding to different false alarm probabilities α is given. For example, for all $N \geq 5$, $t_0 = 2.492$ when $\alpha = 0.05$; and $t_0 = 3.857$ when $\alpha = 0.01$.

The detection probability of AD test can be expressed as [11]:

$$P_{AD} = P\left\{A_n^2 > t_0|H_1\right\} = 1 - P\left\{A_n^2 \leq t_0|H_1\right\} \quad (7)$$

Under noise uncertainty, the AD spectrum sensing algorithm is performed as follows.

Step 1. Select the threshold t_0 based on the false alarm probability.

Step 2. The noise variance σ^2 is uniform random number varying in the interval $\left[\frac{1}{\rho}, \rho\right]$, and ρ is the noise uncertainty in dB.

Step 3. Generate the sequence $\left\{Y_j\right\}_{j=1}^n$, and sort the sequence $\left\{Y_j\right\}_{j=1}^n$ in increasing order.

Step 4. Calculate A_n^2 base on (6).

Step 5. Reject the null hypothesis H_0 in favor of the presence of primary signal if $A_n^2 \geq t_0$. Otherwise, accept H_0 and declare the absence of the primary user signal.

3.2 The Transmitted Signals

In practical environments, transmitted signals are sophisticated. Literature [8] has given several cases of the transmitted signal. In our work, we use different kinds of signals with different percentage of DC component for numerical experiments.

We consider the expression (1). Firstly, when there is only the DC components in the transmitted signal, and $s(n) = 1$. Secondly, we consider the primary signal includes only one single carrier frequency f_c. The discrete version of sine waveform can be defined as [13, 14]:

$$m_i = a\sin(\frac{2\pi}{k}i + \varphi) \quad (8)$$

where a is the amplitude of the sine waveform, φ is the arbitrary initial phase and $k = \frac{f_s}{f_c}$ is the ratio between the sampling frequency and the carrier frequency. In order to simplify the analysis, the signal power is normalized and the ratio of signal power to noise power (SNR) is set by changing noise power. Also, we add signal with different percentage of DC component by changing signal amplitude a, which can be expressed as

$$a = \sqrt{2(1 - A_0^2)} \quad (9)$$

where A_0 is the percentage of DC component. So the transmitted signal is finally defined as:

$$m_i = \sqrt{2(1 - A_0^2)}\sin(\frac{\pi}{3}i + \varphi) + \sqrt{A_0^2} \quad (10)$$

The third signal is sawtooth wave signal and can be written into (in the first period)

$$m_i = \sqrt{3\left(\frac{1}{2} - A_0^2\right)}(i - 1) + \sqrt{A_0^2} \quad (11)$$

Finally, for a general case of transmitted signal, we assume it has Gaussian distribution with zero mean and variance σ^2, and $\sigma^2 = 1 - A_0^2$.

4 Numerical Simulation

In this section, we mainly compare the sensing performance of AD test with energy detection over AWGN channel and Rayleigh channel under noise uncertainty. We evaluate sensing performance by using Monte Carlo method since no closed form expression for detection and false alarm probabilities, and experiment repeats for 10000 circles.

4.1 The Transmitted Signals

Firstly, we discuss the performance of AD test and energy detection over the AWGN channel.

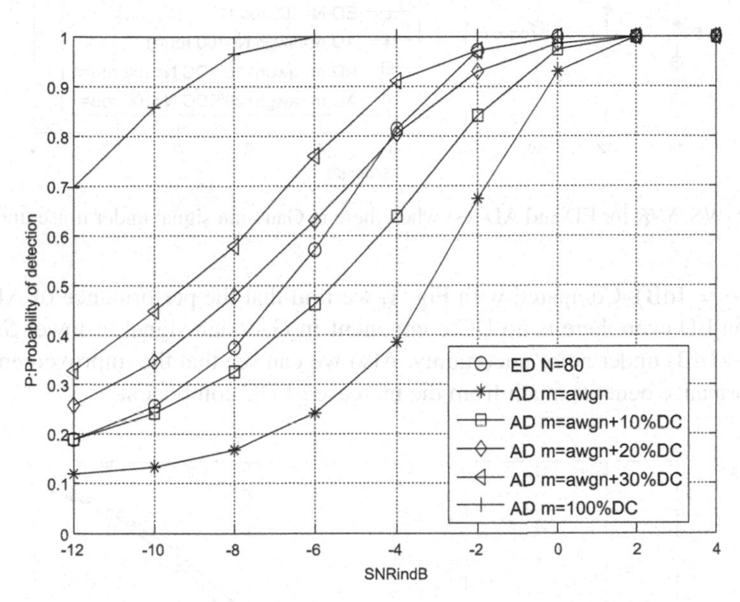

Fig. 1. P_d VS. *SNR* for ED and AD test for Gaussian signal without noise uncertainty

In Fig. 1, the transmitted signal is Gaussian signal with different percentage of DC component (0%, 10%, 20%, 30%, 100%) when the noise variance is not changed (known). The number of sampling points is 80 ($n = 80$) and the false alarm probability is 0.1 ($P_f = 0.1$).It can be readily observed from this figure, in AWGN channel without noise uncertainty, the spectrum sensing performance of ED is better than AD test when the transmitted signal is only Gaussian signal. It can also be noted that the detection probabilities of AD test can be greatly improved with the increase of DC component, and AD test shows better performance than ED when the percentage DC component is more than 30%.

In Fig. 2, the noise power is unknown (under noise uncertainty) we discuss the performance of AD test and ED when the transmitted signal is also Gaussian signal with different percentage of DC component (0%, 10%, 20%). The number of sampling points is 100 ($n = 100$), the false alarm probability is 0.1 ($P_f = 0.1$) and the noise uncertainty

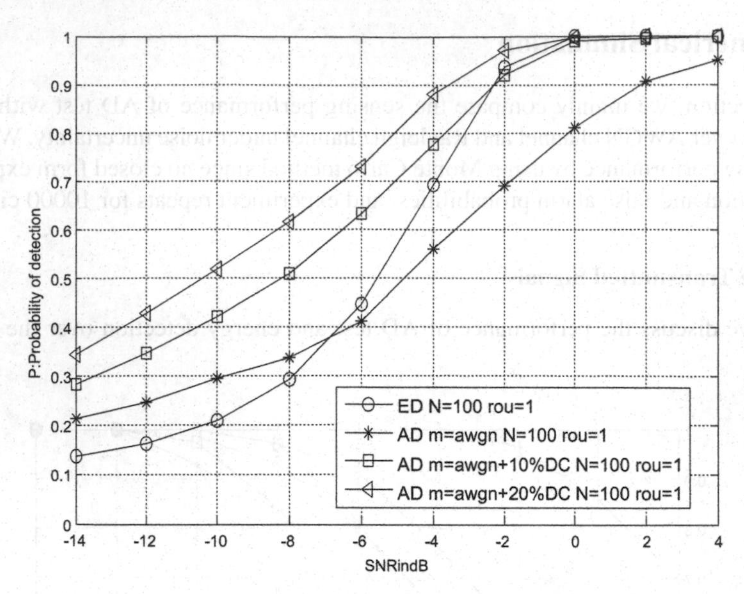

Fig. 2. P_d VS. *SNR* for ED and AD test when there is Gaussian signal under noise uncertainty

is 1 dB ($\rho = 1$dB). Compared with Fig. 1, we find that the performance of AD test is better than ED even there is no DC component in Gaussian signal in lower *SNR* case ($SNR < -7$ dB) under noise uncertainty. Also we can see that the improvement of AD test performance benefits much from the increase of DC component.

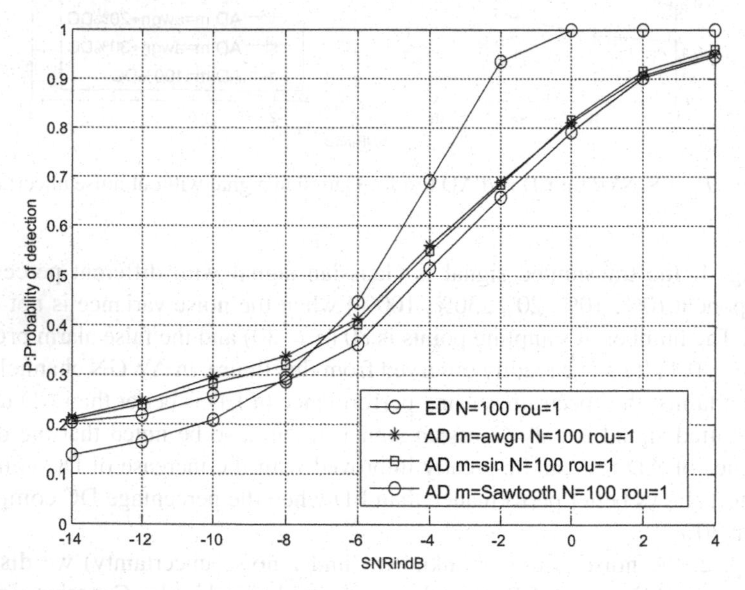

Fig. 3. P_d VS. *SNR* for ED and AD test for three different signals under noise uncertainty

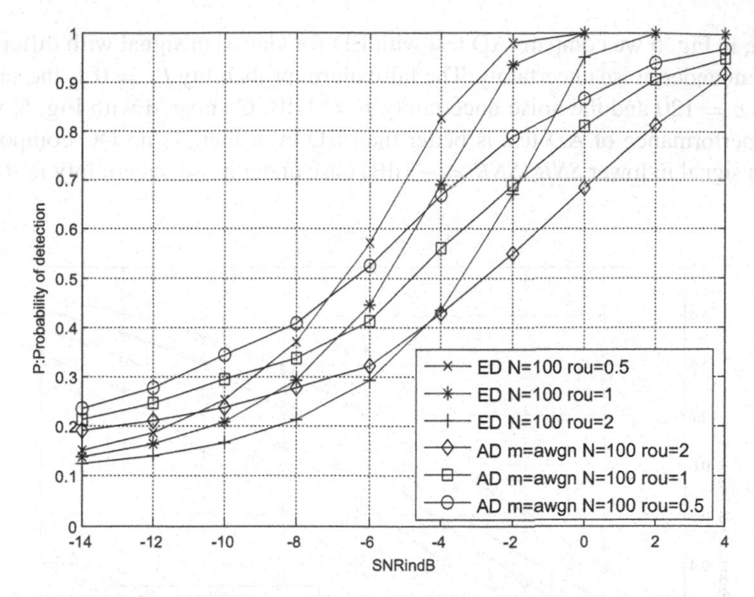

Fig. 4. P_d VS. *SNR* for ED and AD test under different noise uncertainty

In Fig. 3, we present curves of $P_f = 0.1$ VS. $P_f = 0.1$ for Gaussian signals, sine signal and sawtooth signal without DC component when the false alarm probability $P_f = 0.1$, the sampling numbers $P_f = 0.1$ and the noise uncertainty $P_f = 0.1$. It can be seen that the performance of AD test for the three kinds of signals is better than ED in lower $P_f = 0.1$ ($P_f = 0.1$) case, and the sensing performance of different kinds of signals with the same percentage of DC component has little difference under noise uncertainty. Figure 4 gives the performance of AD test and ED for Gaussian signal with different noise uncertainty ($P_f = 0.1$) when the false alarm probability $P_f = 0.1$ and the sampling numbers $P_f = 0.1$. We can see that no matter how is the noise uncertainty the performance of AD test is better than ED when the $P_f = 0.1$ is low ($P_f = 0.1$), and the performance of AD test is poorer than ED when $P_f = 0.1$.

From all mentioned above, we note that, when $P_f = 0.1$ is low and noise power is unknown, AD test performs better than ED in AWGN channel. The conclusion shows that AD test is less sensitive to noise uncertainty and low $P_f = 0.1$.

4.2 Rayleigh Channel

In practical environments, the transmitted signal often undergoes fading channels. So we analyze the sensing performance using the typical channel model Rayleigh channel in this subsection.

Firstly, we compare AD test with ED for Gaussian signal with different DC component in Fig. 5 when the noise variance is known. The false alarm probability $P_f = 0.1$ and the sampling numbers $n = 80$. We can see that in Rayleigh channel the spectrum sensing performance of ED is much better than AD test when the transmitted signal is only Gaussian signal. Also the performance of AD test can be greatly improved with the increase of DC component.

Then, in Fig. 6 we compare AD test with ED for Gaussian signal with different DC component under noise uncertainty. The false alarm probability $P_f = 0.1$, the sampling numbers $n = 120$ and the noise uncertainty $\rho = 1$ dB. Compared with Fig. 5, we find that the performance of AD test is better than ED even there is no DC component in Gaussian signal in lower *SNR* (*SNR* < -7 dB) case under noise uncertainty in Rayleigh channel.

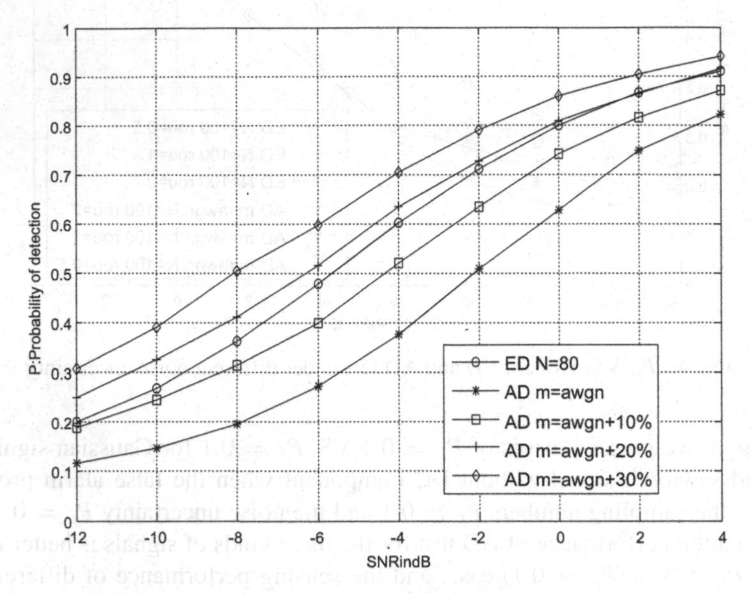

Fig. 5. P_d VS. *SNR* for ED and AD test in Rayleigh channel when the noise power is known

Figure 7 presents curves of P_d VS. *SNR* for Gaussian signal, sine signal and sawtooth signal without DC component when the false alarm probability $P_f = 0.1$, the sampling numbers $n = 120$ and the noise uncertainty $\rho = 1$ dB. It can be seen that the performance of AD test for the three kinds of signals is better than ED in lower *SNR* (*SNR* < -8 dB), and the sensing performance of different kinds of signals with the same percentage of DC component has little difference under noise uncertainty.

In Fig. 8, we give the performance of AD test and ED for Gaussian signal with different noise uncertainty (P_d). It can be seen that, in Rayleigh channel, the performance of AD test is better than ED when the P_d is low (P_d), and the performance of AD test is poorer than ED when P_d.

Extensive experiments have been done channel, the simulation results show that AD test performs better than ED under noise uncertainty when *SNR* is low, and AD test is less sensitive to noise uncertainty when *SNR* is lower over Rayleigh channel.

5 Conclusion

In this paper, we mainly analyze the spectrum sensing performance of AD test and compare it with traditional energy detection in AWGN channel and Rayleigh fading

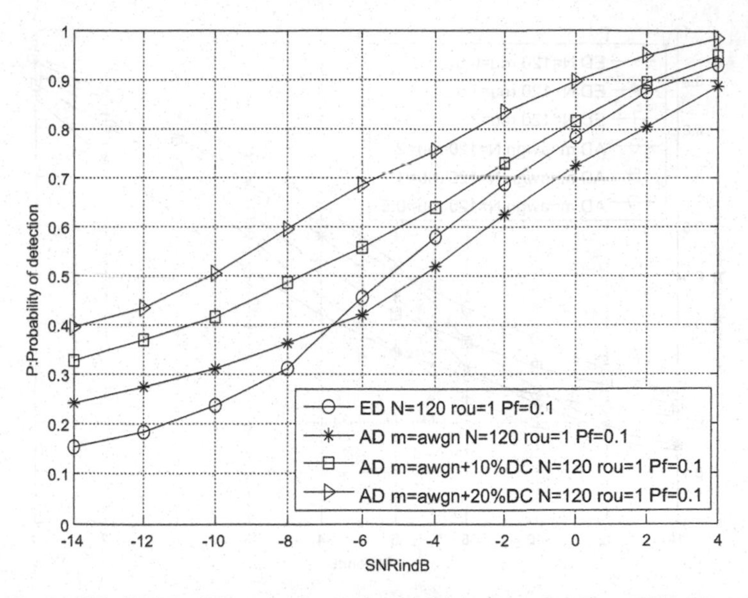

Fig. 6. P_d VS. *SNR* for ED and AD test in Rayleigh channel under noise uncertainty

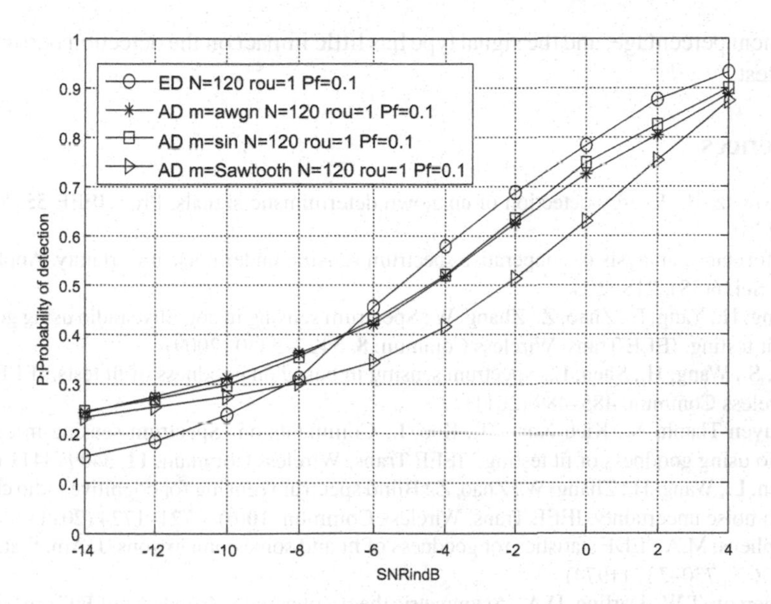

Fig. 7. P_d VS. *SNR* for ED and AD test in Rayleigh channel with three kinds of signals under noise uncertainty

channel under noise uncertainty. All results reveal that, considering noise uncertainty in both AWGN channel and Rayleigh channel, AD test has better sensing performance than ED when *SNR* is low ($SNR < -8$dB), and is less sensitive to noise uncertainty Moreover, the performance of AD test is getting much better with the increase of DC

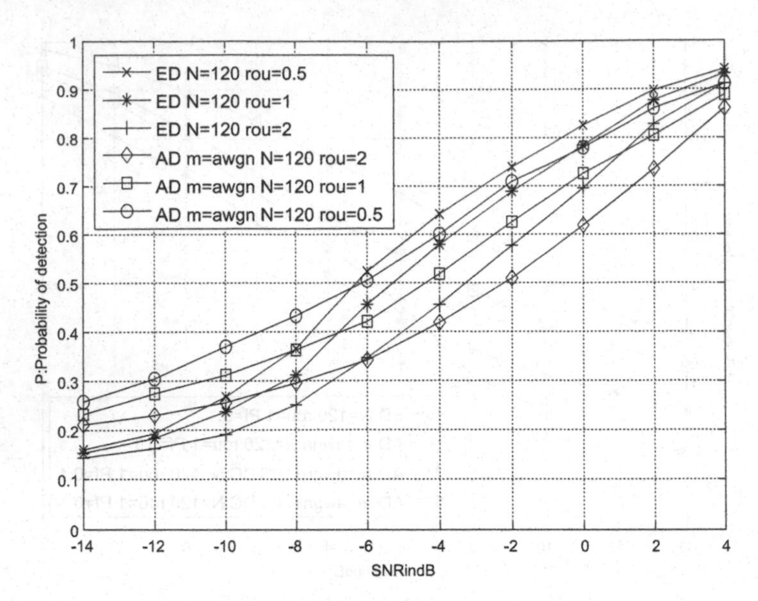

Fig. 8. P_d VS. *SNR* for ED and AD test in Rayleigh channel under different noise uncertainty

component percentage, and the signal type has little impact on the detection performance of AD test.

References

1. Urkowitz, H.: Energy detection of unknown deterministic signals. Proc. IEEE **55**, 523–531 (1967)
2. Performance analysis of cooperative spectrum sensing under noise uncertainty. Appl. Math. Inf. Sci. **6**(5S), 21S–27S
3. Wang, H., Yang, E., Zhao, Z., Zhang, W.: Spectrum sensing in cognitive radio using goodness of fit testing. IEEE Trans. Wireless Commun. **8**, 5427–5430 (2009)
4. Lei, S., Wang, H., Shen, L.: Spectrum sensing in based on goodness of fit tests. IEEE Trans. Wireless Commun. 485–489 (2011)
5. Nguyen-Thanh, N., Kieu-Xuan, T., Koo, I.: Comments on 'Spectrum sensing in cognitive radio using goodness of fit testing.' IEEE Trans. Wireless Commun. **11**, 3409–3411 (2012)
6. Shen, L., Wang, H., Zhang, W., Zhao, Z.: Blind spectrum sensing for cognitive radio channels with noise uncertainty. IEEE Trans. Wireless Commun. **10**(6), 1721–1724 (2011)
7. Stephens, M.A.: EDF statistics for goodness of fit and some comparisons. J. Am. Stat. Assoc. **69**(367), 730–737 (1974)
8. Anderson, T.W., Darling, D.A.: Asymmetric theory of certain "Goodness of Fit" criteria based on stochastic process. Ann. Math. Stat. **23**(2), 193–212 (1952)
9. Ghasemi, A., Sousa, E.S.: Collaborative spectrum sensing for opportunistic access in fading environment. In: IEEE International Symposia on New Frontiers in Dynamic Spectrum Access Networks, Baltimore, USA, November 2005, pp. 131–136 (2005)
10. Digham, F.F., Alouini, M.S., Simon, M.K.: On the energy detection of unknown signals over fading channels. IEEE Trans. Commun. **55**(1), 21–24 (2007)
11. Ali, A., Hamouda, W.: Advances on spectrum sensing for cognitive radio networks: theory and applications. IEEE Commun. Surv. Tutor. **19**(2), 1277–1304 (2017)

Research on Prediction Algorithm for Heavy Overload of Main Equipment in Distribution Network

Dawei Li[1,2](✉), Jingwen Shen[3], Jian Chen[1,2], Yuxiang Xie[1,2], and Zhenfeng Xiao[1,2]

[1] State Grid Hunan Electric Power Company Limited Economic and Technical Research Institute, Changsha 410007, China

[2] Hunan Key Laboratory of Energy Internet Supply-Demand and Operation, Changsha 410007, China

[3] State Grid Information and Communication Company of Hunan, Changsha 410007, China

Abstract. Distribution transformer is the basic unit of power grid operation and the carrier of data in power grid construction. In real life, heavy overload of equipment in the distribution network will damage the equipment and affect the power supply demand and normal production and life of the distribution area. In response to this problem, this paper applies machine learning and neural network models to the heavy-overload estimation of main equipment in the distribution network. First, select nine machine learning and neural network models such as the SVM algorithm to apply to equipment heavy overload estimation, and use the K-Means clustering algorithm to define the classification of heavy overload tags; then extract the feature variable set from the data set, and select the Apriori algorithm to perform Frequent pattern mining, derive heavy overload association rules. After comparing the accuracy and loss changes of each training model, the optimal SVM algorithm is selected to adjust the parameters based on cross-validation and grid search, and the accuracy is increased by 2.6% compared to the previous one.

Keywords: Distributor heavy overload · Association rules · Machine learning · Neural networks

1 Introduction

As the final unit of power grid operation, distribution transformer is the carrier of massive data in power grid construction. As the most direct connection equipment between the power grid and the majority of users, distribution transformers directly affect the power supply quality and safety of the entire region. Heavy overload of equipment will cause power outages and other faults, which will seriously disrupt the normal life of residents; in addition, heavy overload of power supply equipment for a long time will accelerate the loss of internal parts of the equipment and bring great risks to the operation of the power supply system. Therefore, heavy overload management is an extremely important part of the power grid.

X. Sun et al. (Eds.): ICAIS 2021, CCIS 1422, pp. 607–617, 2021.
https://doi.org/10.1007/978-3-030-78615-1_53

Heavy overload management is to monitor the condition of the equipment, record information for the power distribution equipment that has repeatedly appeared heavy overload, and timely transform the user lines and adjust the capacitance of the power distribution equipment [1]. Therefore, the correlation analysis of the heavy overload parameters and the realization of the predicted heavy overload status are of great significance for timely troubleshooting, increasing the quota capacity and transforming the power supply system.

However, there are still many problems in the prediction of heavy overload equipment. For the analysis of the influence factors of heavy overload, the influence factors of heavy overload of power distribution are diverse, such as internal information such as power consumption type and operating load conditions; and external information such as temperature, humidity, seasons, and holidays of the tester. Manual prediction is entirely based on experience, and the impact factors of heavy overload cannot be considered scientifically, and there are emotional interference and subjective consciousness. Therefore, the information prediction of heavy overload equipment is an important direction in the electrical field. The deployment of computer algorithms and the development of information observation instead of manual observation An important goal of electrical work.

The prediction of heavy-overload equipment is a hot issue in the electrical field, and many scholars have studied it. In particular, the application of computer algorithms to forecast heavy-overload equipment has become a research trend. He Jianzhang [2] used random forest to improve the decision tree algorithm to build a transformer heavy overload state prediction model, ignoring the long-term load law. Shi Changkai [3] constructed a heavy-overload state prediction model that combines the BP neural network and the gray model for the special period of the Spring Festival. Although the prediction accuracy is improved, it cannot be accurately predicted for sudden load changes. PADMANABH K [4] uses 7 types of load curves formed by 6000 residents a week plus weather information, and uses different machine learning algorithm models to train predictions. The prediction time is too short to adapt to the actual situation. Zhao Yu [5] based on the actual situation in Harbin, monitored the heavy overload of the public transformer station area on the equipment ledger information such as the operating life, but did not further mine the data, and the effect was average. Wang Shanlin [6] carried out heavy-overload forecasting for the public transformer equipment of the Fujian urban power grid, and comprehensively evaluated all aspects from the perspective of risk, which is conducive to improving the efficiency of operation and maintenance, but the model is not perfect.

This paper extracts heavy load data of a certain designated transformer in a certain city, uses K-Means clustering algorithm to mine the classification of heavy overload tags, and sets indicator thresholds for different intervals. The data set is constructed by extracting characteristic variables, and then the Apriori algorithm is used to set the support threshold to conduct frequent pattern mining on the characteristic variable set, and then the association rules of the heavy overload of the distribution transformer are derived. Select 9 machine learning algorithms such as SVM and neural network algorithm training model, compare accuracy, loss rate, etc. Select the SVM algorithm with the best experimental results, and adjust the parameters based on cross-validation and network search. The experimental results show that the final prediction accuracy is up to 96%, which is 2.6% higher than before.

2 Heavy Overload Definition

The definition of the heavy overload of the distributor is based on the three types of equipment main body of substation, transformer, line, etc., the equipment label is extracted, and the parameters related to heavy overload are fully mined from the dimensions of basic information and operation information, and then classified and quantified. However, it is too complicated to adopt this approach in the experiment, so this paper uses the K-means algorithm to cluster the load rate to define the heavy overload business as shown in Table 1.

Table 1. Load rate definition

The maximum load rate of the main (distribution) transformer	$R_m = \frac{P_m}{C_r} \times 100\%$ (1)
Main (distribution) variable average load rate	$R_a = \frac{P_a}{C_r} \times 100\%$ (2)
Maximum line load rate	$L_m = \frac{I_m}{I_r} \times 100\%$ (3)
The load growth rate	$a_u = \frac{Rm_1 - Rm_2}{R_{ma}}$ (4)

In the formula, P_m is the main (distribution) transformer maximum power supply load, C_r is the main (distributed) variable rated capacity, P_a average power supply load of main distribution transformer, I_m is the maximum working current of the line, I_r is the maximum allowable transmission current of the line, Rm_1 is the maximum load rate of the equipment at the end of the statistical period, Rm_2 is the maximum load rate of the equipment at the beginning of the statistical period, R_{ma} is the maximum load rate of the equipment at the beginning of the statistical period.

The classification standard of heavy overload related parameters is shown in Table 2.

Table 2. Classification criteria for heavy overload related parameters

Maximum load rate	Overload operation: Maximum load rate ≥ 100%; Heavy duty operation: 80%≤ Maximum load rate < 100% and the single duration exceeds 2 h; Normal operation: 20%≤ Maximum load rate < 80%; Light load operation: 0%≤ Maximum load rate < 20%; No-load operation: Maximum load rate = 0%

(continued)

Table 2. (*continued*)

Load operating status	Overload operation work, heavy load operation, uneconomic work, reasonable work; Over 90% of the year and the cumulative time is greater than 30 days as overloaded work; Over 70% and cumulative time greater than 30 days (annual) is regarded as heavy load operation; The load rate is more than 30% and the time is less than 30 days as uneconomical work; If the load rate is more than 70% for less than 30 days and the load rate is more than 30% for more than 30 days, it is regarded as reasonable work;
Average load rate	Average load rate is suitable: Between 50–75% Average load rate is too low: Below 50% Average load rate is too high: Above 75%
Load growth	Less than 0% is a recession, 0% to 7% is slower, 7% to 12% is moderate, and greater than 12% is faster

3 Heavy Overload Estimation Algorithm

Since the load rate in this article is a continuous character type, most of the regression algorithms are selected, and classification ideas are added to the programming; they are mainly divided into three categories: basic regression or machine learning algorithms; integrated method algorithms and neural networks.

3.1 Regression Algorithm

Linear regression is an algorithm that uses mathematical statistics in statistics to determine the relationship between two variables. The steps are: Set up a regression equation $\hat{y} = bx + a$; calculate $\overline{x}$ and $\overline{y}$; calculate $\sum_{i=1}^{n} (x_i y_i - \overline{xy})$ and $\sum_{i=1}^{n} (x_i^2 - \overline{x}^2)$; first calculate b to get $b = \frac{\sum_{i=1}^{n} (x_i y_i - \overline{xy})}{\sum_{i=1}^{n} (x_i^2 - \overline{x}^2)}$; recalculate to get $a = bx - \hat{y}$. In general, the best coefficients b and a are obtained by training with input variables.

Logistic regression is compared with linear regression, although the principles are similar, they both train the most suitable coefficients. The difference is that Logistic regression uses the mapping relationship between input and output, that is, the output result needs to be transformed by the logistic function. Between 0–1 (or 0 and 1), it becomes a two-class fast learning model.

3.2 Machine Learning Algorithm

The decision tree representation is a classifier of the binary tree structure in the data structure. It is essentially the division of the feature space. The branch represents the test output. During classification, each branch is traversed and judged in turn according to the node, and finally reaches a certain leaf node, This category represents an output variable used for prediction.

The K-nearest neighbor algorithm is a lazy learning method that finds the most similar samples during prediction. The steps are: Calculate the distance between the test data and each data in the training set; Do ascending order according to the distance and select the K points with the smallest distance; Calculate the K points The number of output variables represented; Return the category with the most categories in the above data as the category corresponding to the test data to predict new data points.

A support vector machine (SVM) is a classifier. Simply put, it finds the best line in the input space that can completely separate all input points. This is the hyperplane, and the distance from it to the nearest data point is Called the interval. The so-called optimal value is the coefficient at the maximum interval that can completely separate two blocks. Proceed as follows:

Step 1. Training is expanded to n-dimensional space according to the distance from the geometric midpoint to the line, and the distance is $W^T x + b = 0$ from $\theta^t x_b = 0$

$$h = \frac{\left|W^T x + b\right|}{\|W\|}, \|W\| = \sqrt{W_1^2 + W_2^2 + \cdots + W_n^2}, \tag{5}$$

Step 2. When $\frac{|W^T x+b|}{\|W\|} \leq b, \forall y^{(i)} = -1$ launch:

$$y^{(i)}(W^T x^{(i)} + b) \geq 1 \tag{6}$$

Step 3. Finally get the conditional optimization problem:

$$\min \frac{1}{2}\|w\|^2 \text{ s.t. } y^{(i)}(W^T x^{(i)} + b) \geq 1 \tag{7}$$

3.3 Integrated Algorithm

Random Forest (Random Forest) is one of the most popular machine learning algorithms in big data. It can be said to be the adjustment of Bagging. It takes into account the strong generalization ability brought by random sampling in Bagging, and at the same time introduces nsubnsub to Perform sub-optimal segmentation and improve RF performance through cross-validation (Fig. 1).

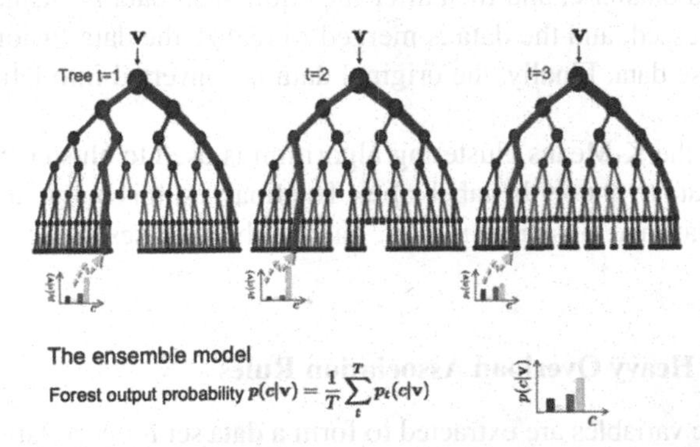

Fig. 1. Random forest diagram

4 Data Preprocessing

4.1 Heavy Overload Data Extraction

Obtain equipment operation data and equipment ledger data from PMS system and CMS system, especially distribution transformer load curve and heavy overload data; obtain daily average temperature, average temperature, average humidity, time label, External data such as holiday types, week types, and season types; parameters related to heavy overload of distribution transformers generally include internal data and external data, as shown in Fig. 2.

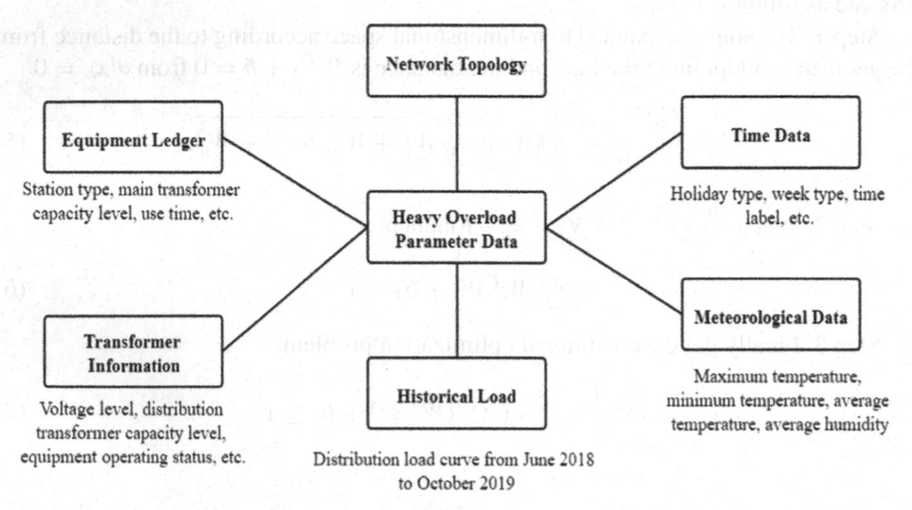

Fig. 2. Distribution equipment heavy overload parameter

4.2 Heavy Overload Data Preprocessing

After the data is extracted, the system interface of the big data platform is used to aggregate the grid database, and then after the equipment data is cleaned, the repeated error data is processed, and the data is merged to realize the data fusion and aggregate into the warehouse data. Finally, the original data is converted into labels and the data is standardized.

In this paper, the K-Means clustering algorithm is used to cluster the preprocessed and overloaded data to obtain the label label. The final result is before and after Table 3.

Analyze the data and clustering results. The number of cases in each cluster is shown in Fig. 3.

4.3 Analysis of Heavy Overload Association Rules

The characteristic variables are extracted to form a data set for correlation analysis, and Apriori is selected to perform correlation analysis on the characteristic data set to obtain

Table 3. English title

Clustering\Load rate	1	2	3
Cluster initial center	0.72546250	1.07783425	2.08295500
Cluster Center Results	0.4341307909	1.228620382	1.50578778

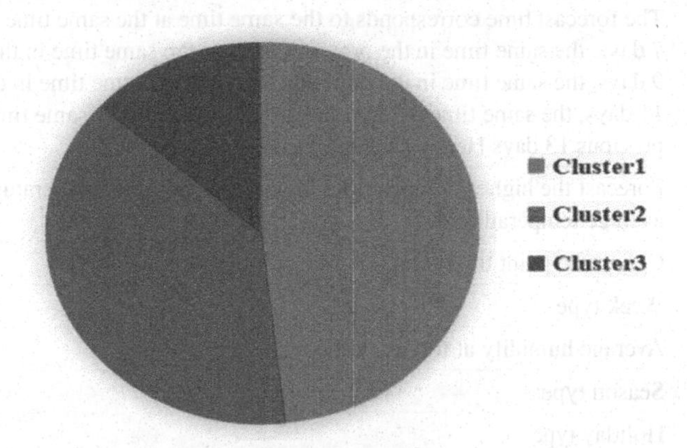

Fig. 3. Proportion of cluster

a set of frequent items, and to derive the association rule with heavy overload of power distribution equipment. Obtain the heavy overload influence factor and weight as shown in Table 4.

Table 4. Influence factors and weights of heavy overload

Impact factor	Clustering1	Clustering2	Clustering3
Highest temperature	0.03	0.12	0.13
Minimum daily temperature	–	0.08	0.12
Daily mean temperature	0.03	0.21	0.15
Temperature difference	–	0.02	0.26
Week	–	0.04	0.12
Holidays	0.06	0.12	0.18
node	0.21	0.24	0.31
Sunshine duration	0.04	0.08	0.12
Average humidity	0.12	0.14	0.09
Season	0.03	0.15	0.16

Analyze the heavy-overload influence factors and their weights to determine the input variables as shown in Table 5.

Table 5. Set of input variables

Input variable	Variable composition
$x_1 \sim x_7$	The forecast time corresponds to the same time at the same time in the previous 7 days, the same time in the previous 8 days, the same time in the previous 9 days, the same time in the previous 10 days, the same time in the previous 11 days, the same time in the previous 12 days, and the same time in the previous 13 days Heavy load at all times
$x_8 \sim x_{11}$	Forecast the highest temperature, lowest temperature, temperature difference, average temperature
x_{12}	Collection point time (node)
x_{13}	Week type
x_{14}	Average humidity at forecast time
x_{15}	Season type
x_{16}	Holiday type

5 Experimental Results and Analysis

5.1 Effect Evaluation Index

The effect of a machine learning model is a sign of whether the effect of machine learning is good or bad. It often requires specific indicators to evaluate, and various scenarios also require various indicators to evaluate. This article selects several commonly used machine learning evaluation indicators, namely Accuracy, Recall and Precision, F1-score. Use chaotic matrix to evaluate, as shown in Table 6.

Table 6. The confusion matrix of dichotomies

True value\predicted value	Positive(1)	Negative(0)
Positive(1)	True Positive (TP)	False Positive (FP)
Negative(0)	False Negative (FN)	True Negative (TN)

It can be seen in the confusion matrix of the two classification: TP means that the predicted value is Positive, the actual value is also Positive, the predicted value is the same as the actual value; FN is the predicted value is Negative, the actual value is Positive, the predicted value is different from the actual value; FP is the predicted value is Positive, the actual value is Negative, the predicted value is different from the actual value; TN is the predicted value Negative, the actual value is Positive, and the predicted value is the same as the actual value. The calculation formulas of the four indicators are summarized below.

Accuracy: the proportion of the number of correct points to all the total samples:

$$Accuracy = \frac{TP + TN}{TP + TN + FN + FP} \tag{8}$$

Precision rate: also known as precision rate, the proportion of samples whose predicted value and true value are both 1 to the sample whose predicted value is 1:

$$Precision = \frac{TP}{TP + TN} \tag{9}$$

Recall rate: the proportion of samples whose predicted value and true value are both 1 to the total positive samples:

$$Recall = \frac{TP}{TP + FN} \tag{10}$$

F1 -score: refers to the harmonic average of the sum:

$$F1 = \frac{2TP}{FP + TN + 2TP} \tag{11}$$

5.2 Effect Model Analysis and Comparison

Using 16-dimensional input variables as the input features of the model, and selecting 9 machine learning algorithms for comparative training, the model with the highest accuracy is 93.4%. The accuracy comparison chart of the nine machine learning models is shown in Fig. 4.

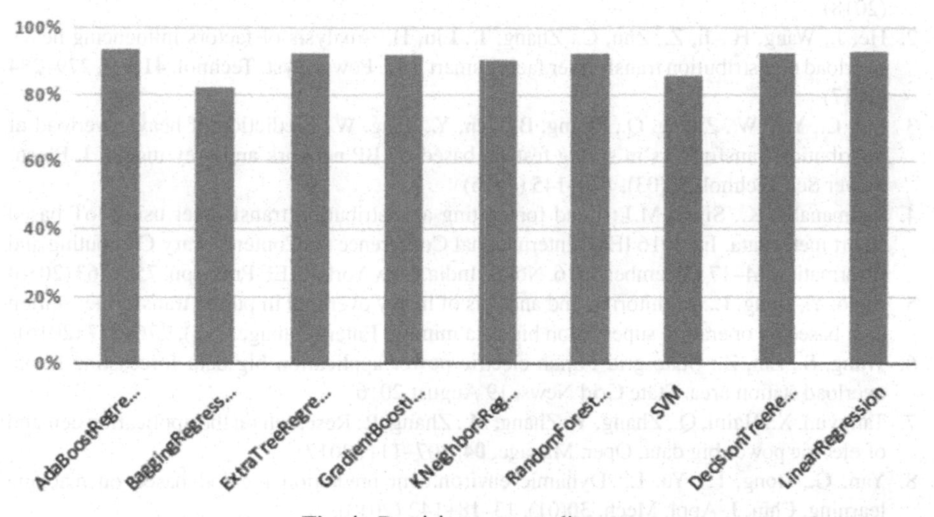

Fig. 4. Precision contrast diagram

5.3 Model Optimization

This paper adjusts the parameters of the SVM algorithm model based on cross-validation and grid search for optimization.Do 10-fold on the training set and return the optimal model parameters. The prediction accuracy is up to 96%, which is 2.6% higher than before.

6 Conclusion

This paper chooses a computer algorithm to solve the problem of heavy overload of the main equipment of the distribution network. Use K-Means clustering algorithm to define the classification of heavy overload labels; then extract feature variable sets from the data set, select Apriori algorithm for frequent pattern mining, derive heavy overload association rules, analyze heavy overload influence factors and their weights to determine input variables. Select 9 kinds of machine learning algorithms for comparative training to obtain the SVM algorithm with the highest accuracy. Adjust model parameters based on cross-validation and grid search for further optimization. Experimental verification shows that the accuracy of the optimized model is improved by 2.6% compared to before.

The meteorological information in the data in this article is missing, and there are certain errors in the association rules. The next step is to improve the data set and reduce errors. In addition, the running speed of the SVM model has not been improved, and the delay problem that may occur in practical applications is the next research direction of this article.

References

1. Zhang, G., Wang, X., Deng, C.: Prediction method of heavy overload in distribution network station area based on correlation analysis and machine learning. Big Data **4**(01), 05–116 (2018)
2. He, J., Wang, H., Ji, Z., Zhu, C., Zhang, T., Liu, H.: Analysis of factors influencing heavy overload of distribution transformer facing smart grid. Power Syst. Technol. **41**(01), 279–284 (2017)
3. Shi, C., Yan, W., Zhang, Q., Zhang, B., Fan, Y., Tang, W.: Prediction of heavy overload of distribution transformers in spring festival based on BP network and grey model. J. Electr. Power Sci. Technol. **31**(03), 140–145 (2016)
4. Padmanabh, K., Singh M.J.: Load forecasting at distribution transformer using IoT based smart meter data. In: 2016 IEEE International Conference on Contemporary Computing and Informatics, 14–17 December 2016, Noida, India. New York, IEEE Press, pp. 758–763 (2016)
5. Zhao, Y., Dong, L.: Monitoring and analysis of heavy overload in public transformer station area based on operation supervision big data mining. Enter. Manage. (S2), 276–277 (2016)
6. Wang, S., Lin, Z.: State grid Fujian electric power application big data forecasting heavy overload station area. State Grid News, 19 August 2016
7. Tangyun, X., Haini, Q., Zhang, Y., Zhang, M., Zhang, P.: Research on the application demand of electric power big data. Oper. Manage. **04**, 107–111 (2017)
8. Yan, G., Dong, L., Yu, L.: Dynamic environment prediction method based on machine learning. Chin. J. Appl. Mech. **30**(01), 13–18+142 (2013)

9. Levine, J., Zhang, D., Ma, W.: Application of energy consumption service based on electricity consumption data analysis by Aoneng Corporation of the United States. Power Supply Use **32**(09), 56–58+49 (2015)
10. Yang, W.: Analysis of big data application in the marketing platform of mobile communication operators. China New Commun. **17**(24), 65–66 (2015)
11. Zhang, L.: Research on user portraits based on big data analysis to help precision marketing. Telecommun. Technol. (01), 61–62+65 (2017)
12. Lin, S., Ouyang, L.: Construction of electric power customer label system based on big data theory. Electr. Technol. (12), 98–101+112 (2016)
13. Zhou, L., Zhao, J., Gao, W.: Application research of sparse coding model in the detection of abnormal power consumption behavior of power users (English). Power Syst. Technol. **39**(11), 3182–3188 (2015)
14. Campezidou, S.I., Grijalva, S.: Dsitribution transformer short-term load forecasting models. IEEE Trans. Power Syst. (19), 267–273 (2016)
15. Sun, X.R., Luh, P.B., Cheung, K.W.: Efficient approach to short-term load forecasting. IEEE Trans. Power Syst. (3), 301–307 (2016)

Review of Attention Mechanism in Electric Power Systems

Junxing Wu[1], Shuo Tang[2], Chizhi Huang[1], Dongdong Zhang[1], and Yingnan Zhao[2](✉)

[1] NARI Nanjing Control System Ltd, Nanjing 210032, China
[2] School of Computer and Software, Nanjing University of Information Science and Technology, Nanjing 210044, China

Abstract. The attention mechanism is first employed in natural language processing areas, such as machine translation, natural language generation, and natural language inference. Combining with deep learning framework, it promotes performances of applications in electric power systems recently. In this article, we first discuss the attention mechanism's principle and its applications in various fields; secondly, we analyze and compare multiple applications and performances in power systems. Finally, it proposes the further development of the attention mechanism.

Keywords: Attention mechanism · Deep learning · Electric power systems

1 Introduction

Attention mechanism (AM) was originally used for machine translation [1], and has now become an important part of neural network structure. The basic idea of AM is to allow the system to find key information among many information and pay more attention. For example, when humans process language information in daily life, they will pay more attention to emotional words to get the meaning. The same is true for the information processing process of AM. When humans browse images, they pay different attention to different parts of the image. We will actively and automatically pay attention to the more important focal information in the image, which is also an obvious manifestation of human intelligence.

The attention model has been widely used in various fields of deep learning in recent years. It is easy to encounter the attention model in various tasks of image processing [2, 3], speech recognition [1] or natural language processing [4, 5]. Therefore, it is necessary to understand the working principle of AM for technicians who are concerned about the development of deep learning technology.

AM is now the most advanced model to solve multi-task, similar to automatic translation, answering, marking part of speech, voice assistant, etc. It also improves the interpretability of neural networks and greatly reduces people's threshold for machine learning. Finally, they help to solve the situation where the input sequence is not reasonable and the computer is inefficient. After the third development of artificial intelligence

X. Sun et al. (Eds.): ICAIS 2021, CCIS 1422, pp. 618–627, 2021.
https://doi.org/10.1007/978-3-030-78615-1_54

technology, AM and encoder-decoder model are increasingly used in real life, but they are not widely used in power systems.

In recent years, the importance of electric power companies on relay protection has increased significantly, so the corresponding artificial neural network has been naturally introduced, which can effectively protect the normal operation of power equipment. With the help of the neural network system to effectively support the relay protection work, the current faults in the power system can be found more accurately, and at the same time, the efficiency of problem solving can be effectively improved. Of course, this method has some shortcomings in terms of performance. For example, for some enlightening knowledge processing, its completion rate is still obviously not high, and its performance and efficiency largely depend on the current sample and the degree of completeness.

The rest of the paper is organized as follows. Section 2 introduces the principle of AM, including attention categories and encoder-decoder model. Section 3 describes some applications of AM in electric power system. Conclusion is presented in Sect. 4.

2 Principle of AM

At present, AM has become an important part of deep learning and natural lan-guage processing, appearing in various models and tasks. The basic idea of AM in computer vision is to allow the system to learn attention, that is, allow the system to be able to ignore irrelevant information and focus on key information. It turns out AM is very effective in many scenarios, and due to its concise form, it usually does not bring more complexity to the model. Therefore, AM should be an almost indispensable helper for us in natural language processing, especially Seq2Seq tasks. The essence of AM is actually an addressing process, as shown in Fig. 1. Imagine the constituent elements in Source as a series of < K, V > data pairs. At this time, given an element Query in Target, calculate the similarity or correlation between Query and each K to get each K corresponds to the weight coefficient of V, and then weights and sums V to obtain the final Attention value. So in essence, the Attention mechanism is to perform a weighted summation of the Value values of the elements in the Source, and Query and K are used to calculate the weight coefficient of the corresponding V. This process is actually a manifestation of AM that eases the complexity of the neural network model: it is not necessary to input all N input information into the neural network for calculation, only need to select some task-related information to input to the neural network.

2.1 Classification of AM

Attention can be roughly classified into five categories, i.e. Hard/Soft Attention, Global/Local Attention, Hierarchical Attention, Attention Over Attention, Position Attention, Channel Attention.

Hard/Soft Attention. Soft Attention is a commonly used attention, and the value range of each weight is [0,1]. As for Hard Attention, the attention of each key will only take 0 or 1.

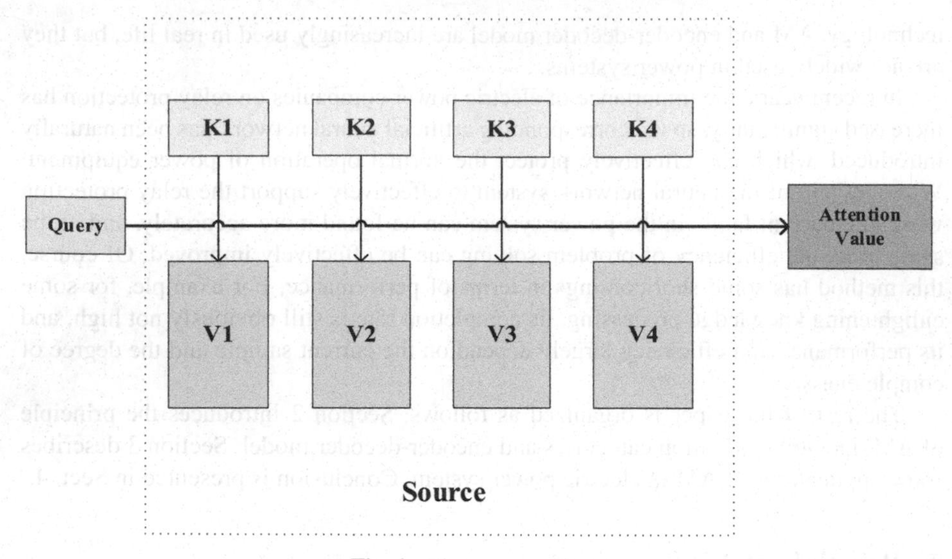

Fig. 1. The essence of AM

Global/Local Attention. Generally, if there is no special description, the attention we use is Global Attention. According to the original AM, at each decoding moment, the number of decoding states is not limited, but the encoder length can be dynamically adapted to match all encoder states. For example, in the long text, we align the entire encoder length matching can cause inattention problems, so we limit the scope of the attention mechanism to make the attention mechanism more effective. In Local Attention, Set the corresponding value *ht* of the decoder and the corresponding value *pt* of the encoder, the *ht* of each decoder corresponds to an encoder position *pt*, and the interval size *D* is selected (generally selected based on experience), and then AM is used in the encoder's $[pt - D, pt + D]$ position.

Hierarchical Attention. Hierarchical Attention can be applied to solve the problem of inattention in long texts. Unlike Local Attention, Local Attention forcibly limits the scope of AM and ignores the remaining positions; while Hierarchical Attention uses hierarchical thinking in all states.

Attention Over Attention. The basic idea of Attention Over Attention is to pay attention to the result of Attention again, but the specific steps are different from Hierarchical Attention. In Attention Over Attention, the result of the first Attention is to obtain a weight matrix. The two dimensions are request length and text length. The horizontal axis and the vertical axis represent the attention distribution of one party to the other. The average value of the text length is normalized to obtain the average attention distribution vector that is the same as the request length; the normalization of the request length can represent the attention distribution matrix of each word in the text with respect to the request. In the second Attention, we use the two results of the first Attention to find the Attention weight again, and we can get an attention distribution vector about the reading text.

2.2 Encoder-Decoder Model

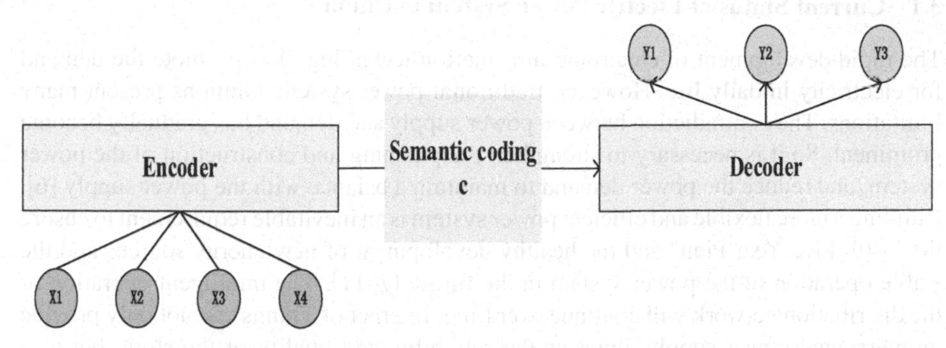

Fig. 2. Encoder-Decoder framework in the abstract text processing field

The Encoder-Decoder framework can be regarded as a research mode in the field of deep learning, with a wide range of application scenarios. Figure 2 is the most abstract representation of the Encoder-Decoder framework commonly used in the text processing field. The Encoder-Decoder framework in the text processing field can be understood intuitively: it can be regarded as a general processing model suitable for processing one sentence (or chapter) into another sentence (or chapter). For the sentence pair < Source, Target >, our goal is to give the input sentence Source, and expect to generate the target sentence Target through the Encoder-Decoder framework.

The Encoder-Decoder framework is not only widely used in the text field, but also frequently used in speech recognition, image processing and other fields. For example, for speech recognition, the framework shown in Fig. 1 is fully applicable. The difference is that the input of the Encoder part is a voice stream, and the output is the corresponding text information; for the "image description" task, the input of the Encoder part is one For the secondary picture, the output of the Decoder is a descriptive sentence that can describe the semantic content of the picture. Generally speaking, the Encoder part of text processing and speech recognition usually adopts the RNN model, and the Encoder of image processing usually adopts the CNN model.

2.3 Reasons for Introducing AM

Although feedforward networks and cyclic networks have strong capabilities, computer capabilities have certain limitations. When you have to remember a lot of information, the model will become more complex and large. However, for the moment, the computing power is still limit the bottleneck of neural network development.

Although optimization operations such as local connection, weight sharing, and pooling can make neural networks simpler and effectively alleviate the contradiction between model complexity and expressive ability, such as the long-distance problems of recurrent neural networks, the problem of information "Memory" ability is not high. Therefore, we need to use the way the human brain processes information overload, such as AM, to improve the ability of neural networks to process information.

3 AM Applications in Electric Power Systems

3.1 Current Status of Electric Power System in China

The rapid development of electronic information technology has promote the demand for electricity in daily life. However, traditional power system solutions present many limitations. The contradiction between power supply and demand has gradually become prominent. So it is necessary to strengthen the planning and construction of the power system, and reduce the power demand to maintain a balance with the power supply [6]. Building a more flexible and efficient power system is an inevitable requirement to ensure the "14th Five-Year Plan" and the healthy development of new energy sources and the stable operation of the power system in the future [7–11]. The intelligent operation of the distribution network will continue over time. Internet of Things can not only provide uninterrupted power supply, improve the reliability and quality of the store, but also meet the urgent needs of most people, and can improve economic efficiency and user satisfaction [12].

Therefore, we can introduce AM into the power system to implement the intelligent supervision, eliminate potential problems, and provide security assessment [12]. It can analyze and extract historical data and apply artificial intelligence technology to improve the stability of power supply.

3.2 State Prediction of Power Communication Equipment

AM can fully explore the autocorrelation of the signal, highlight some of the characteristics of the signal related to the final prediction task, and reduce the degree of signal confusion [13].

This module consists of channel attention module and traditional attention module string linked together, it has the effect of improving correlation and reducing signal dimension. Given the input $X \in RS \times C$, where S is the signal length, C is the number of channels in each group of signals, and R is the nonlinear relationship generated by the softmax function. First of all, through a channel attention module in the first half, multiply the feature and its transpose matrix to get the autocorrelation matrix of different channels. After passing a softmax function, the matrix can be used as the channel attention matrix $AC \in RC \times C$ to analyze the original signal. After passing this module, the correlation between the original feature channels is reduced, and useful information is highlighted, and further input to the following dimensionality reduction attention module.

By connecting the above hybrid attention mechanism and LSTM in series, this paper proposes a state prediction method for power communication equipment. Among them, the original data obtained from various sensors and operating records on the spot enters the preprocessing module after being filled to a fixed length, and forms the system input through common normalization and centralization. The mixed attention moduleis then input into the LSTM unit to achieve timing prediction. Finally, the prediction result is output through the last Fully Connected (FC) layer [14, 15].

3.3 Use Attention Mechanism to Detect Insulators in Circuits

Traditional insulator recognition methods include the use of target color features for threshold segmentation and contour feature-based positioning methods [16]. However, these methods require manual design of target features, and different features need to be designed for different target scenarios. Missing or misdetection caused by human factors cannot be avoided, and the recognition accuracy is low. Compared with traditional methods, CNN can automatically extract more expressive features from images, reduce recognition time, and improve classification and recognition accuracy. Later, the CNN method was used to realize the automatic positioning of insulators, or the ASDN layer was introduced on the basis of the R-FCN model to improve the accuracy of insulator identification. In [17], the Faster RCNN framework is used to identify insulators. A good recognition effect has been achieved. However, in the recognition process of Faster RCNN, the target suggestion frame location obtained by RPN is not accurate enough, which affects the follow-up precise positioning result, resulting in unsatisfactory recognition accuracy. In response to the above problems, it proposes an insulator identification algorithm combining the attention mechanism and Faster RCNN. First, in the feature extraction stage of the Faster RCNN network, the SENet structure based on the attention mechanism proposed by Hu et al. [18] is introduced, so that the network can automatically learn the importance of each feature channel, focusing on the feature channel related to the target and suppressing the target irrelevant feature channels to improve network performance; then, in the anchor generation stage of the RPN, improve the basic anchor ratio and scale according to the characteristics of the insulator; finally, combine the attention mechanism to calculate the weight of the interdependence between the suggestion boxes, and merge the surrounding suggestion boxes. The feature vector of each target suggestion box is updated to make the feature vector of the target suggestion box contain more accurate position information, which promotes the improvement of recognition accuracy and further improves the accuracy of insulator recognition.

3.4 Power Equipment Abnormality Detection

Infrared thermal imaging technology was first applied in the military field. After the 1980s, infrared thermal imaging detection technology began to be popularized in other fields such as electric power and chemical engineering [19]. With the development of artificial intelligence and deep learning in computer vision, people began to pay attention to the use of deep learning methods to detect anomalies in electrical equipment under infrared images. Although the abnormal detection of infrared power equipment based on deep learning can realize automatic detection, the accuracy cannot meet the requirements of the application. The main reason is that there are many infrared images and the information is scattered. The abnormalities of power equipment under different abnormal faults are on the infrared images. The performance is very similar. In this way, the depth features extracted by the deep neural network from the infrared image include a lot of noise, which interferes with the final anomaly detection. Therefore, people consider combining natural images to detect abnormalities in power equipment.

Combining natural images and infrared images to process a task is also called multi-source information processing, which is widely used in the fields of pedestrian re-identification and text-to-image retrieval. The key is how to extract the key useful information through the two to filter out the noise information. Among them, AM is widely used, which can guide the convolutional neural network to enhance the required key feature information and suppress the noise feature information [20].

3.5 Short-Term Photovoltaic Power Forecast

In order to cope with the exhaustion of fossil energy and the environmental pollution it brings, new energy power generation has been greatly developed and progressed in the world. Because of its low cost and pollution-free characteristics, photovoltaic power generation has become one of the most potential renewable energy power generation technologies. However, due to the randomness, intermittentness and volatility of photovoltaic power generation, it has seriously affected the safe and reliable operation and planning of the power grid [21, 22]. Therefore, improving the prediction accuracy of photovoltaic power generation can provide an important reference for the operation and planning of the grid, which is of great significance. The previous prediction methods include support vector machine (SVM), Markov chain, extreme learning machine, artificial neural network, time series prediction and other methods.

With the development of deep learning, the LSTM model is used for short-term photovoltaic power prediction. However, there are still many problems. It does not consider the size of the specific correlation coefficients. The results obtained are one-sided. During the training process of LSTM, setting a too long time sequence will also affect the stability and accuracy of the results. Therefore, the literature [23] proposed a combined photovoltaic power prediction method based on VMD (variational mode decomposition) and dual-stage mechanism LSTM. VMD can be used to decompose the photovoltaic power time series into relatively simple sub-sequence components, and the LSTM prediction model of each component can be established. The dual-stage AM can be used to obtain the relationship between photovoltaic output and input meteorological variables and historical information. Among them, the feature attention mechanism can adaptively mine the relationship between photovoltaic power output and related meteorological input features, autonomously obtain the contribution rate of different features under different weather conditions, and modify the feature weights in real time, which can improve the single-point prediction Accuracy. The time attention mechanism can mine the relationship between the PV power output of the LSTM at the current moment and the historical key moment information, which can improve the accuracy of long-term time series prediction.

The DA-LSTM prediction results of each component are superimposed to obtain the final photovoltaic power prediction value. Compared with other methods, due to the consideration of the correlation between photovoltaic power and characteristic parameters and time sequence, the prediction of single-step and multi-step experiments is improved performance.

3.6 Super Short-Term Wind Power Prediction

Large-scale wind energy development and grid connection with a high proportion of wind power is the current general trend, and it is a strategic requirement for achieving sustainable energy development [24]. The safe and stable operation and maintenance of the power system requires real-time energy balance. However, wind power generation is still greatly affected by the external environment, with characteristics such as randomness, volatility, intermittent, and periodicity. In order to solve these contradictions, it is the most feasible and economical to accurately predict the wind power power and use the forecast data to improve the dispatching ability of the system. It can effectively reduce the energy storage equipment and flexibly adjust the source-end investment [25].

The wind power prediction method combined with the cross-local anomaly factor and AM can be aimed at the sample quality problem, considering the correlation characteristics between wind direction, wind speed and wind power, and can verify the verification strategy of multi-scale wind power data. According to the rationality of the prediction model, considering various environmental factors, AM can be used to assign different weights to different time nodes, and combined with the LSTM algorithm to better extract effective feature information such as long-term historical data, thereby improving the performance of the prediction model [26].

4 Conclusion

The essence of AM is to allow the system to find key information and pay more attention. It turns out that AM is very effective in many scenarios, however, it does not bring more complexity to the model due to its concise definition. In other words, AM allows us to directly inspect the internal workings of the deep learning architecture, so attention is one of the important ways to explain the internal workings of neural models. Therefore, AM becomes an indispensable part of deep learning methodologies, especially Seq2Seq tasks.

AM applied in electric power systems can effectively promote their performances. In China, the application of attention mechanism in electric power mainly lies in the detection of equipment and the prediction of power generation. Compared with traditional prediction methods, the use of attention mechanism improves its accuracy and reduces the error rate. When using the attention mechanism to detect power equipment, noise information can be filtered out, making the prediction results more accurate. The ultra-short-term forecast of wind power and photovoltaic power generation can reduce the impact and harm caused by instability caused by force majeure and natural causes. How to adopt different types of AM maybe the next work.

Acknowledgement. This work is supported in part by the Priority Academic Program Development of Jiangsu Higher Education Institutions, Natural Science Foundation of China (No.61103141, No.61105007 and No.51405241), and Project of NARI Technology Development Co. LTD. (No. 524608190024).

References

1. Bahdanau, D., Cho, K., Bengio, Y.: Neural machine translation by jointly learning to align and translate. In: Computer Science (2014)

2. Hou, Z., Cai, X., Chen, S., Li, B.: A model based on dual-layer attention mechanism for semantic matching. In: ICIASE, USE (2019)
3. Tao, G., Ping, W., Wang, C., Song, X.: Visual attention mechanism based image shadow defect detection. In: ICECE, USE (2011)
4. Kiela, D., Wang, C., Cho, K.: Dynamic Meta-Embeddings for Improved Sentence Representations. In: Conference on Empirical Methods in Natural Language Processing, USE, pp. 1466–1477 (2018)
5. Kosiorek, A.: Attention mechanism in neural network. Robot. Ind. **6**, 12–17 (2017)
6. Fan, H.W., Xi, Y., Zhang, M.R.: Research on the application of smart grid in power technology and power system planning. Electronic Test (2020)
7. Cui, Y., Zhu, J., Ge, Y., Yan, G.: Study on capacity configuration of large-scale energy storage system under concentrated charge and discharge based on wind curtailment characteristics and economics. Power Syst. Technol. **40**(2), 484–490 (2016)
8. Solomon, A.A., Kammen, D.M., Callaway, D.: The role of large-scale energy storage design and dispatch in the power grid: a study of very high grid penetration of variable renewable resources. Appl. Energy **134**, 75–89 (2014)
9. Li, J.L., Ting, T., Lai, K.X.: Outlook of electrical energy storage technologies under energy internet background. Autom. Electr. Power Syst. 39(23), 15–25 (2015)
10. Wu, S., Xu, Q., Yuan, X., Bing, C., Qiang, L.: An analysis of requirements and applications of grid-scale energy storage technology in power system. J. Electr. Eng. **12**(8), 10–15 (2017)
11. Ran, L., Guo, J.H., Yuan, T.J.: Power system operation simulation of large-scale energy storage on new energy station. Distrib. Energy **5**(3), 1–8 (2020)
12. Li, N.X., Sheng, Y.Q., Ni, H.: Conditional convolutional hidden factor model for personalized recommendation. Comput. Eng. (2019)
13. Wu, Y., Dubois, C., Zheng, A.X., Ester, M.: Collaborative de-noising auto-encoders for top-n recommender system. In: the Ninth ACM International Conference, USA (2016)
14. Strub, F., Mary, J.: Collaborative filtering with stacked denoising autoencoders and sparse inputs. In: NIPS Workshop on Machine Learning for e Commerce (2015)
15. Wu, H.Y., Peng, P., Guo, B., Jiang, C.X., Zhu, P.Y.: Attention and LSTM based state prediction of equipment on electric power communication networks. Comput. Modernization **0**(10), 12–18 (2020)
16. Huang, X.N., Zhang, Z.L.: A method to extract insulator image from aerial image of helicopter patrol. Power Syst. Technol. **34**(1), 194–197 (2010)
17. Iruansi, U., Tapamo, J.R., Davidson, I.E.: An active contour approach to insulator segmentation. In: AFRICON, Addis Ababa, Ethiopia, pp. 1–5 (2015)
18. Cheng, H., Zhai, Y., Chen, R.: Faster R-CNN based recognition of insulators in aerial image. Mod. Electr. Tech. **42**(2), 98–102 (2019)
19. Fan, J.X., Yang, J.Y.: Analysis of the development trend of infrared imaging detection technology. Infrared Laser Eng. **41**(12), 3145–3153 (2012)
20. Yan, L.X., Liu, X.H., Li, M., Sima, Z.J., Liu, Z.P.: Research on infrared image and natural image power equipment anomaly detection based on attention mechanism. Telecom Power Technol. **37**(4), 14–16 (2020)
21. Sheng, H.M., Xiao, J., Cheng, Y.H.: Short-term solar power forecasting based on weighted Gaussian process regression. IEEE Trans. Industr. Electron. **65**(1), 300–308 (2018)
22. Sanjari, M.J.: Gooi, H.B.: Probabilistic forecast of PV power generation based on higher order Markov chain. IEEE Trans. Power Syst. **32**(4), 2942–2952 (2017)
23. Yang, J.X., Zhang, S., Liu, J.C., Liu, J.Y., Xiang, Y., Han, X.Y.: 2Short-term photovoltaic power prediction based on variational mode decomposition and long short-term memory with dual-stage attention mechanis. Autom. Electr. Power Syst. **45**(3), 174–181 (2021)
24. Li, T., Xu, W.T., Liu, X.L.: Review on planning technology of AC/DC hybrid system with high proportion of renewable energy. Power Syst. Prot. Control **47**(12), 177–187 (2019)

25. Li, Z., Ye, L., Zhao, Y., Song, X., Teng, J., Jin, J.: Short-term wind power prediction based on extreme learning machine with error correction. Prot. Control Mod. Power Syst. **1**(1), 1–8 (2016). https://doi.org/10.1186/s41601-016-0016-y
26. Wang, X., Cai, X., Xu, L.Z.: Ultra-short-term wind power prediction method combining cross-local abnormal factors and attention mechanism. Power Syst. Prot. Control **48**(23), 92–99 (2020)

Comprehensive Review of Intelligent Operation and Maintenance of Power System in China

Jun Zhang[1], Yihui Fu[2], Yang Hu[1], Dongliang Wang[1], and Yingnan Zhao[2(✉)]

[1] NARI Nanjing Control System Ltd., Nanjing 210032, China
[2] School of Computer and Software, Nanjing University of Information Science and Technology, Nanjing 210044, China

Abstract. AIOps first put forward the concept of intelligent operation and maintenance with artificial intelligence and machine learning technologies. The idea of AIOps is applied to the power system recently in China. In this technical briefing, it first analyzes the definition in depth, then describes the operation and maintenance's development in Chinese power system. Further, some methodologies adopted in the substation, converter station and new energy are discussed, as well as several general technologies, i.e., relay protection, secondary operation, especially deep learning-based ones. It finally points out the problems in this field and proposes some helpful solutions.

Keywords: AIOps · Power system · Intelligent operation and maintenance · Deep learning

1 Introduction

AIOps (Artificial Intelligence for IT Operations) is the origin of intelligent operation and maintenance. It is about empowering software and service engineers (e.g., developers, program managers, support engineers, site reliability engineers) to efficiently and effectively build and operate online services and applications at scale with artificial intelligence and machine learning techniques [1]. Most of the early operation and maintenance work was done manually by operation and maintenance personnel, which is called manual operation and maintenance or human operation and maintenance. This backward production method is difficult to maintain in an era of rapid Internet business expansion and high labor costs. Then automated operation and maintenance appeared, using scripts that can be automatically triggered and predefined rules to perform common and repetitive operation and maintenance tasks, thereby reducing labor costs and improving operation and maintenance efficiency. With the development of technology and people's service to IT systems, IT service providers must put in place reliable and cost efficient operations support and AIOps is their best choice. This platforms utilize big data, machine learning and other advanced analytic technologies to enhance IT operations with proactive actionable dynamic insight [2]. AIOps has been proved to be effective in quality improvement, cost reduction and efficiency improvement, and is considered as the ultimate solution for IT operation and maintenance [3].

X. Sun et al. (Eds.): ICAIS 2021, CCIS 1422, pp. 628–640, 2021.
https://doi.org/10.1007/978-3-030-78615-1_55

In recent years, Chinese power system has developed rapidly. There are many kinds of operation and maintenance schemes for various power systems. At the beginning of 2020, the National Energy Administration of China held the "14th Five-Year Plan" power planning meeting. It proposed requirements to improve the intelligence level of the power system, improve efficiency, and ensure safety. The development of intelligent operation and maintenance technology for power systems outside China is rapid too. They have focused more attention on new energy power systems, so they have certain technical advantages. But their software industry is also still at the early stage of innovating and adopting AIOps solutions. The community just started to realize the importance of AIOps [1]. Therefore, this article mainly organizes and analyzes some current excellent power system intelligent operation and maintenance solutions, to find out the shortcomings of China's current power system development and suitable roads for the development of China's power system operation and maintenance. For power system, secure and stable operation is indispensable and it takes electric utility staff huge amount of time. Naturally, AIOps, especially anomaly detection, can be used to find out unusual behavior discord with expected pattern in this field [4].

The rest of the article is organized as follows. Section 2 describes the development of operation and maintenance in Chinese power system. Section 3 discusses the main methodologies in substations, converter stations and the new energies, as well as several general technologies, such as relay protection and secondary operation, especially deep learning-based ones. The final part concludes and gives some suggestions for future development.

2 Development of Operation and Maintenance of Power System in China

The power system is composed of power plants, transmission grids, distribution networks and power users. It is a unified system that converts primary energy into electrical energy and transmits and distributes it to users. Transmission grids and distribution grids are collectively called grids, which are an important part of the power system. The power plant converts primary energy into electrical energy, and transmits and distributes the electrical energy to the electrical equipment of power users through the grid, thereby completing the entire process of electrical energy from production to use. The power system also includes corresponding auxiliary systems such as relay protection devices, safety automatic devices, dispatch automation systems, and power communications to ensure its safe and reliable operation.

New energy is also an important part of the modern power system. The new energy used for power generation mainly includes wind energy, solar energy, geothermal energy, tidal energy and nuclear energy. In recent years, countries around the world have made substantial progress in the development and utilization of new energy. Now our country is also vigorously developing these new energy sources. Therefore, the importance of these power systems composed of new energy sources is gradually increasing.

The development of China's electric power system is a gradual process. In the early 20th century, the electric power system was relatively backward, and manual operation and maintenance methods were mostly adopted. The automatic adjustment was mainly the crisis guard of the steam turbine, the automatic adjustment of power grid dispatching, and the automatic adjustment of generator voltage, Various relay protection of generators, etc. [5]. After 1950s, the capacity of the power system has increased, and the degree of intelligence has also been further improved. Power plants have begun to use power generation equipment, transmission equipment and electrical equipment integrated centralized control to improve the efficiency of the entire system. From the 1970s to the 1980s, information technology developed rapidly, and a fully functional real-time power grid monitoring system with computers as the main body appeared, which also marked the arrival of the era of intelligent operation and maintenance. Nowadays, the power system is becoming more and more intelligent. The staff can monitor through various sensors and manage the entire power system through terminal computers. Although the development of the domestic power system was initially slower than that of foreign countries, it is now gradually entering the forefront of the world with rapid development (Fig. 1).

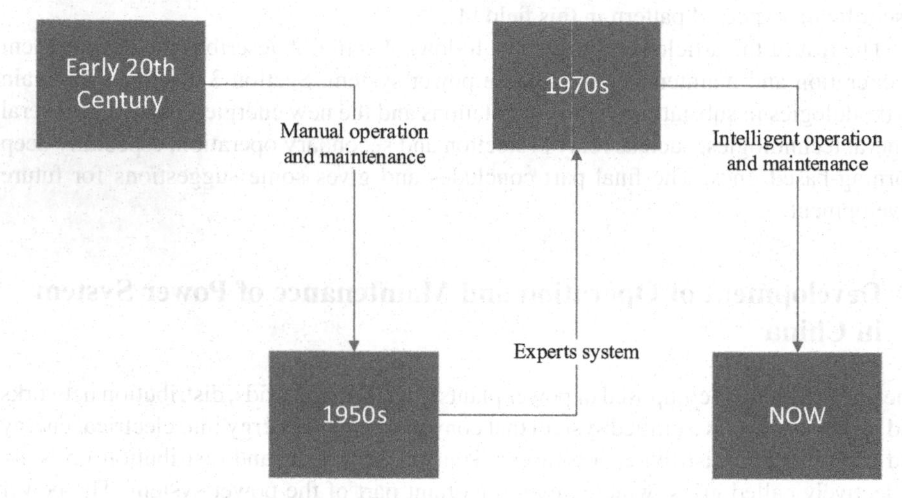

Fig. 1. Developing time table

3 Main Methodologies

The main intelligent operation and maintenance methodologies can be used in substation, converter station and new energy powers. Also, there are some general-applied technologies, such as relay protection and secondary operations. We will discuss them in detail.

3.1 Intelligent Operation and Maintenance of Substation

With the continuous advancement of the intelligentization process, more and more substations have begun to use intelligent technology to achieve operation and maintenance.

These sites use low-carbon, environmentally friendly, advanced, and reliable intelligent equipment to implement and apply many operation and maintenance technologies different from traditional substations, such as sequence control operation technology, intelligent robot inspection technology, etc. They can automatically complete basic functions such as information collection, control, protection, and detection. These new technologies have played an important role in reducing the cost of substation operation and maintenance and the labor intensity of operation and maintenance personnel and improving the intelligent level of operation and maintenance management. Smart substations are also quite different from the digital substation. It focuses more on operation and management, provides not only more novel and effective technical testing methods, but also guarantees the safety of the substation in all aspects.

Intelligent Robot Inspection Technology. In 2015, the Chinese National Academy of Sciences designated intelligent robots' development goals for substations and established a laboratory. In 2018, China Southern Power Grid promoted robotics technology to the entire power network field operations. At present, the substation intelligent robot inspection technology as an intelligent operation and maintenance technology has been tested in many places. The substation intelligent robot inspection system is divided into three layers (base station layer, communication layer, terminal layer) [6]. As shown in Fig. 2, the system is a distributed network architecture. The base station layer is composed of base station systems, hard disk recorders, video monitors, firewalls, etc.; the communication layer is mainly various network wireless devices, including wireless bridges, switches, etc., responsible for establishing the network connection between the terminal layer and the base station layer; the terminal layer Including a series of terminal equipment such as inspection robots and robot charging rooms. The intelligent robot inspection system supports the centralized control mode, realizing remote control and monitoring of the inspection system and the robot through the remote centralized control background.

There are currently two types of path control for intelligent inspection robots: tracked and trackless. The trackless solution is more advanced, with the characteristics of flexible adjustment and tracelessness. Trackless solutions mostly use laser navigation and positioning solutions to meet the needs of robot navigation, such as high reliability, strong anti-interference ability, and all-weather conditions. The intelligent robot needs to go to the charging room for charging after working for a period of time, and the charging room door is equipped with a device that can control the automatic start and stop of the robot. The design of the control program is also an important part of the intelligent inspection robot. The general program is divided into two parts: the part installed on the robot and the control terminal of the staff, which is convenient for remote viewing and remote control of the robot to detect the robot's working condition.

The intelligent inspection robot's work is mainly to inspect all primary equipment and perform operations such as infrared camera temperature measurement and visible light photography on all equipment in the station. It can clearly distinguish all kinds of equipment through image enhancement and other technologies, without blind spot inspection. The robot's image and video output in later stages are of high definition and can be automatically classified and recognized. When an abnormality is found, the robot will automatically alarm and generate a record to record various malfunctioning instrument indicators.

The disadvantage of the intelligent robot inspection technology is that the secondary equipment installed in the main control building cannot detect, and the reaction problem when encountering obstacles during driving has not been solved.

Sequence Control Operation Technology. The traditional substation's operation mode is the switching operation specification, which has the problems of cumbersome steps and many personnel requirements. At present, a sequential control operation technology of intelligent substation is proposed [6]. The most commonly used sequence control operation server solutions in practical applications now are as follows: the realization of the program operation server and the joint realization of the bay layer device and the program operation server. Use the programmed operation server to realize the sequential control operation, set up a programmed operation server, and store all operation tickets in the server to solve the sequential control requirements of various operations (interval and inter-interval). Reduce the workload of information processing. The interval level equipment and the program operation server are used to realize the sequential control operation. The program server only saves the combination relationship between the interval level operation tickets to complete the operation across the interval, and the device completes the operation within the interval in the interval. It is necessary to store the operation ticket in the bay in the bay equipment, which solves the verification problem during the expansion of the substation. The flow chart of the two schemes is as follows:

Routine operations need to simulate operation, pass the five-proof key, replay and record the ticket, open and lock the mechanical five-proof lock, check the equipment status before and after the operation on site, and walk between the equipment areas. However, the use of sequential control operation, through the computer monitoring system, real-time back-checking data to realize automatic control equipment, its simulation preview time, checking equipment status time is not comparable to conventional operations, which effectively improves the switching operation effectiveness.

There are two shortcomings of sequential control operation technology. Under special circumstances, if the operation cannot continue to run, the staff must manually stop the operation and use the traditional method to operate. It is necessary to perform an inspection from the beginning, reducing efficiency. And this mode is very different from the traditional mode, and it is difficult to quickly adapt to the switching operation mode.

Automatic Equipment Transparent Operation and Maintenance System. In recent years, the power network system has developed rapidly, and the scale and number of substations have increased accordingly. Smart substations have gradually emerged to solve the relatively lagging degree of specialization and intensification of conventional substations, low resource allocation efficiency, and long management chains. However, the maintenance interface, monitoring background configuration, operation and control, and comprehensive information analysis of the existing smart substation design are not standardized, which is not conducive to centralized operation and maintenance management and standardized and standardized construction. Regarding the issue above, Zhiqiang Peng and his team proposed SSAE transparent operation and maintenance system framework [7]. The framework adopts an integrated operation and maintenance hierarchy, and is divided into two modes, on-site operation and maintenance and remote operation and maintenance, to adapt to different scenarios. The hierarchy diagram is as follows:

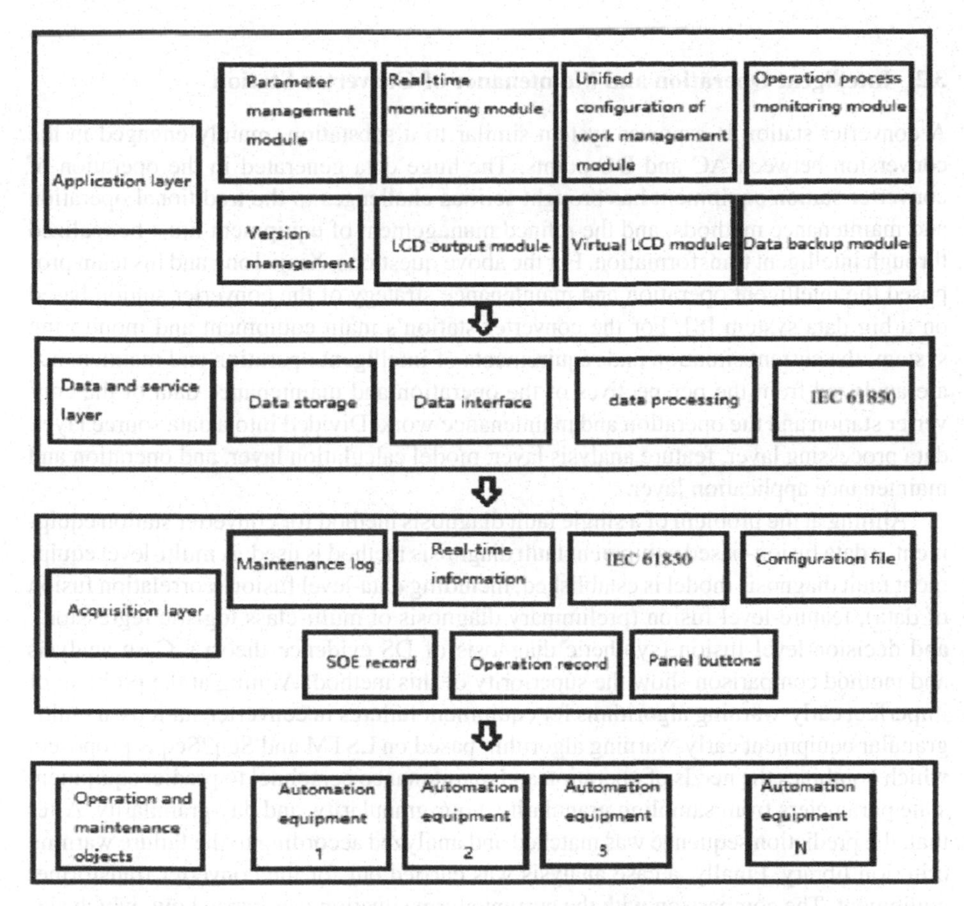

Fig. 2. Diagram of integrated operation and maintenance hierarchy [7]

SSAE transparent operation and maintenance information model was built under the framework of SSAE transparent operation and maintenance system. The information model refers to the model data related to the equipment in the integrated operation and substation automation equipment maintenance. Based on the information model, the details of the equipment's integrated operation and maintenance are further refined. The operation and maintenance system standardizes service interfaces, develops integrated operation and maintenance tools, and adds transparent operation and maintenance function modules to the wide-area operation and maintenance master station to achieve homogeneous operation and maintenance between remote and local locations. Modularized design is adopted for different operation and maintenance objects. The integrated operation and maintenance of switches, measurement, and control devices and versions respectively realize integrated management and control of switches, integrated modification of configuration files and configuration parameters of all measurement and control devices on the site, and software and hardware version information of automation equipment Effective maintenance.

3.2 Intelligent Operation and Maintenance of Converter Station

A converter station is a power system similar to a substation, mainly engaged in the conversion between AC and DC points. The huge data generated in the operation of converter station equipment has brought serious challenges to the traditional operation and maintenance methods, and the refined management of equipment must be realized through intelligent transformation. For the above questions, Xinji Tong and his team proposed the intelligent operation and maintenance strategy of the converter station based on a big data system [8]. For the converter station's main equipment and monitoring system, the current situation and requirements of intelligent operation and maintenance are analyzed from the perspectives of the operation and maintenance data of the converter station and the operation and maintenance work. Divided into a data source layer, data processing layer, feature analysis layer, model calculation layer, and operation and maintenance application layer.

Aiming at the problem of a single fault diagnosis method for converter station equipment, a data fusion-based equipment fault diagnosis method is used. A multi-level equipment fault diagnosis model is established, including data-level fusion (correlation fusion of data), feature-level fusion (preliminary diagnosis of multi-class logistic regression), and decision-level fusion (synthetic diagnosis of DS evidence theory). Case analysis and method comparison show the superiority of this method. Aiming at the problem of imperfect early-warning algorithms for equipment failures in converter stations, a multi-granular equipment early-warning algorithm based on LSTM and Seq2Seq is proposed, which combines the needs of operation and maintenance personnel to predict equipment state parameters from sampling granularity, hour granularity, and day granularity. After that, the prediction sequence was matched and analyzed according to the failure warning criterion library. Finally, a case analysis was carried out for the converter transformer equipment. The comparison with the current alarm situation was carried out, which significantly improved the warning's accuracy. For the problem of numerous parameters in evaluating equipment status in converter stations, a state evaluation method based on association rule mining and Bayesian network is proposed. The correlation of state parameters under various failure modes of equipment is mined through association rules to improve the prior probability knowledge of the Bayesian network. After that, through historical status information, current status information and predicted status information, self-learning of the Bayesian network model is completed. The evaluation method of equipment operating status is given, considering the three types of information's distribution characteristics. Finally, the on-site operating data verification of the equipment in the converter station shows the method's effectiveness (Fig. 3).

Fig. 3. Green new energy

3.3 New Energy Technology

With the support of national policies and encourage the construction of new energy sources, various wind power generation and solar power stations have been built. The intelligent operation and maintenance of these high-tech equipment will be a very important part of the power system.

Research on the operation and maintenance of photovoltaic power stations has been hot in recent years, mainly including Internet plus distributed photovoltaic power generation operation technology [15] and so on. Northwest China is mostly mountainous areas with inconvenient transportation, but generally has the conditions for photovoltaic construction, such as making full use of land and abundant sunshine. Distributed photovoltaic construction can be constructed according to local conditions and differentiated. Compared with the operation and maintenance of centralized photovoltaic power plants, distributed photovoltaic power plants adopt in-station monitoring systems and manual regular maintenance methods, and can rely on monitoring systems and manual inspections. Because distributed photovoltaic power plants are geographically dispersed, large in number, and small in capacity, seldom configure local monitoring system, its operation and maintenance is still in the post-maintenance stage, relying on regular inspections by operation and maintenance agencies, and distributed photovoltaic access is weak in the distribution network in mountainous areas, power station access is disorderly, and operation and maintenance costs are sensitive. The problem is that the traditional centralized photovoltaic power plant operation and maintenance method is not suitable for distributed photovoltaic power plants. With the development of information technology, the intelligent operation and maintenance of distributed photovoltaic power plants based on "Internet + " has gradually developed. Through mobile Internet technology, this kind of remote monitoring operation and maintenance needs are greatly facilitated [16].

Wind power is also an important part of my country's new energy power generation, and it is widely used in various parts of China. At present, the main problems existing in the traditional operation and maintenance system of wind farms are concentrated on high operation and maintenance costs and low power generation. In response to these problems, a team has developed a wind farm intelligent operation and maintenance management solution based on active prevention strategies [17]. This proactive prevention strategy takes preventive measures in advance against possible accidents in wind farms, and emphasizes the preventive and forward-looking nature of the operation and maintenance plan. Its initiative is reflected in the non-arrangement of regular preventive maintenance throughout the life cycle. Through real-time monitoring of the root cause parameters of the system that can lead to equipment failure and timely correction of root cause abnormal working conditions, to maintain the health of important equipment, to realize the proactive prevention and status maintenance of the whole life cycle, establish the ethical fault bath towel model, and use the intelligent fault diagnosis system and Methods such as intelligent state assessment improve the operation and maintenance efficiency and power generation of wind farms.

3.4 Relay Protection

Relay protection is an important measure to detect faults or abnormal conditions that occur in the power system, so as to issue an alarm signal, or directly isolate and remove the fault. The operation and maintenance of relay protection equipment is a guarantee for the early warning capability of the power system. At present, when the smart substation is expanded, equipment replacement hardware or software, a series of work such as configuration file modification, fixed value input, function debugging, and secondary loop joint debugging must be completed on-site. There are many links, time-consuming, low efficiency, and high risk. The number and duration of power outages for primary equipment has increased significantly [9]. Therefore, it is urgent to carry out research on the support technology of factory-based debugging to realize the "plug and play" of on-site secondary equipment, and provide effective technical support for non-stop maintenance and expansion.

At present, there are four main problems in the operation and control of relay protection. 1. Expansion of smart substations requires cumbersome steps and low efficiency during commissioning. 2. On-site operation and maintenance of the substation has a secondary circuit that cannot be known, and conventional operation and maintenance technology is difficult to play a role. 3. The remote operation and maintenance capability of the substation is poor, and it is found that the problem is sometimes not solved quickly. 4. Low level of on-site operation control and low efficiency [10].

The most typical solution to the above-mentioned problems is the intelligent operation and maintenance of relay protection equipment based on the Internet. With the intelligent operation and maintenance management platform as the core and the mobile terminal as the carrier, intelligently analyze the relay protection real-time monitoring data and the whole process business data, realize the real-time and online monitoring of the relay protection equipment and the intelligent diagnosis of defects and faults, and carry out the core operation To maintain business, realize the whole process control of business process in the system [11]. Improve the communication efficiency and information security of intelligent operation and maintenance to deal with the large amount of data in the power system. Improve the fault diagnosis technology to improve the performance of relay protection equipment, and deal with the difficult problem of multi-source alarm joint diagnosis of intelligent substation. Finally, through the improvement of data mining technology, the historical data is effectively managed, and valuable components arc analyzed in it to improve the intelligent diagnosis level of faults and defects.

Through the double confirmation of precise geographic location and the unique identification of the device, the mobile terminal scans the QR code label on the surface of the protection device, analyzes the identification code in the QR code, and establishes the corresponding relationship between the protection dcvice, the identification code, and the QR code label to realize automatic association with the device, as shown in the following figure (Fig. 4):

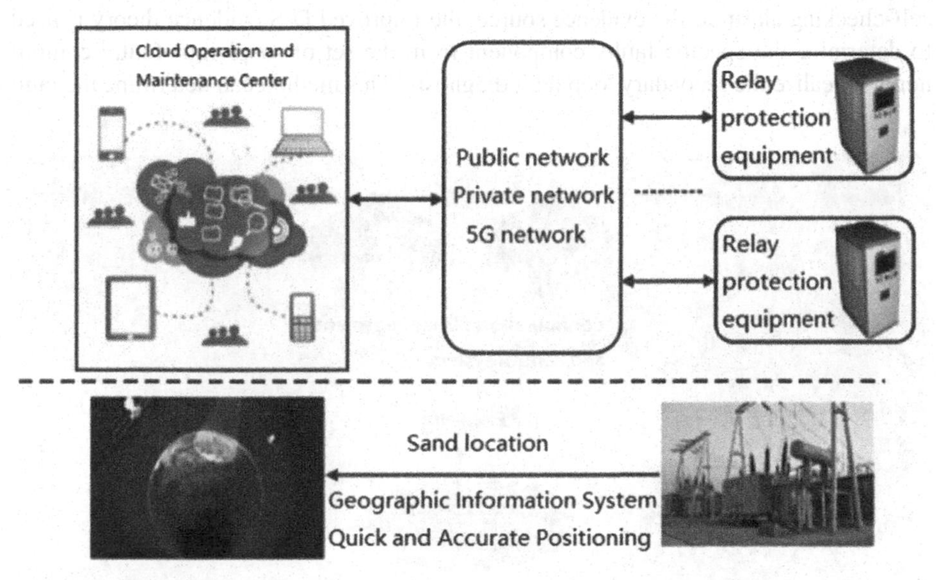

Fig. 4. Positioning principle of relay protection equipment [10]

After mastering a large amount of data, analyze it through Internet big data, and finally) solve the common problems in the key links of relay protection operation, maintenance, overhaul and professional technical management. Improved the level of safe operation and maintenance management and control of relay protection equipment, and improved the efficiency of professionals in accurately studying and judging and handling relay protection equipment defects and grid faults.

3.5 Secondary Operation

With the construction of smart substations, the number of relay protection devices Rapid growth, equipment operation and maintenance management is difficult, and intelligent recognition cannot be realized Don't interact with the device. The protection device of the smart substation is only retained The same logic processing as conventional protection

CPU and human-machine interactive communication Module MMI, MU merger unit replaced AC module [12]; The IEC GOOSE message format is adopted between the protection device and the intelligent terminal GOOSE data transmission, from which the secondary "virtual loop" is connected and compressed The relationship between the board and the secondary circuit and the information transmission cannot be intuitive see.

There are two main methods to deal with these problems. The first is A fault diagnosis method for the secondary circuits of protection systems in smart substations based on improved DS evidence theory [13]. This method establishes the secondary circuit model, determines the identification range and method of the secondary circuit fault, and finally verifies the fault method. When a fault occurs, the set of suspected faulty components can be obtained by relying on the established proof table and the overlapping and overlapping characteristics of each virtual circuit. Finally, using the component self-checking alarm as the evidence source, the improved D-S evidence theory is used to determine the specific faulty component from the set of suspicious faulty components to realize the secondary loop fault diagnosis. This method can determine the fault

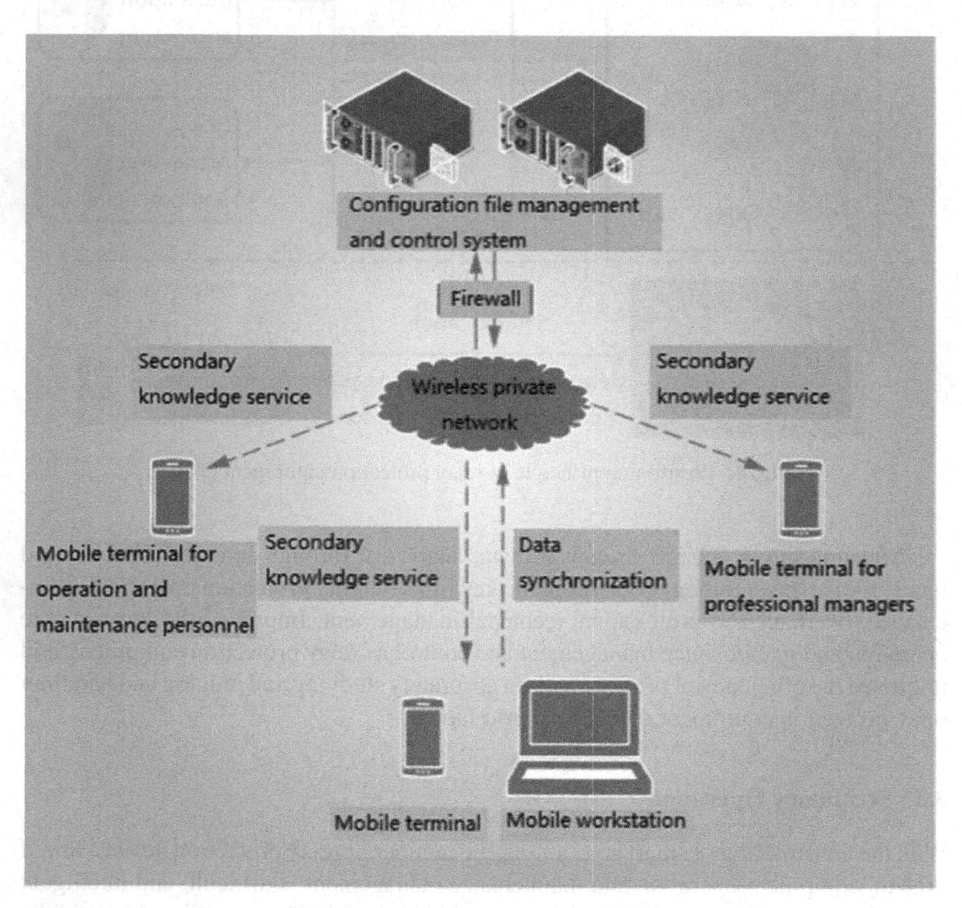

Fig. 5. The overall framework [12]

range and specific fault components, overcome the conflict between the evidence in the fault location process using evidence theory, and can effectively diagnose the fault of the secondary circuit of the intelligent station protection online, Identify specific faulty components, help maintenance personnel reduce the amount of troubleshooting, and improve the efficiency of operation and maintenance.

The second method is based on the secondary equipment operation and maintenance technology of the Internet of Things and mobile Internet [12]. It is based on the standardized SCD file, and the research is based on the Internet of Things and mobile The intelligent operation and maintenance method of the secondary equipment of the interconnection technology realizes the intelligent identification of the secondary equipment of the intelligent substation, knowledge push, loop visualization, maintenance decision-making, auxiliary safety measures, provides intelligent support and guidance for on-site operations, and improves on-site work efficiency and safety The level of anti-error, realizes the smart substation is respectable, controllable and maintainable [14]. The overall design framework is as follows (Fig. 5):

This method uses big data technology to deepen analysis and build a big data information service center through the collaborative interaction between mobile terminals and smart substation configuration file management and control systems, provides strong data support for mobile terminal operation support applications, and comprehensively builds smart substation configuration files Management and control system information service center and intelligent mobile operating platform. It comprehensively improves the management and control of smart substations and the penetrating power of operation and inspection operations, vigorously supports the production operations of dispatching, operation and inspection professionals, and guarantees the safe and stable operation of large power grids in an all-round way.

4 Summary

Through the above-mentioned multiple technologies, we can draw some conclusions. The country of China attaches great importance to the intelligent development of the power system. The entire industry's development is also in a rising period, and various new intelligent operation and maintenance technologies have emerged one after another. Under the influence of the current epidemic, China's development environment is good, and policy support and development encouragement are rare development opportunities. Comparing the development of foreign countries in China, we can see that although China has a lot of technology, the development is not mature enough, and there are still many shortcomings to be solved. Foreign technology is more practical, and it has been tested by software simulation with big data support. Therefore, to further improve China's intelligent operation and maintenance technology, we have to do better in the following aspects:

(1) Share data to a greater extent to establish an effective simulation environment.
(2) Strengthen the exchange of technology, hold more academic exchange meetings so that the technology can be examined in many aspects.
(3) Strengthen the research and development of new technologies and research the utilization of electricity converted from new energy.

Acknowledgement. This work is supported in part by the Priority Academic Program Development of Ji-ang-su Higher Education Institutions, Natural Science Foundation of China (No.61103141, No.61105007 and No.51405241), and Project of NARI Technology Development Co. LTD. (No. 524608190024).

References

1. Dang, Y.D., Lin, Q.W., Huang, P.: AIOps: real-world challenges and research innovations. In: ICSE-Companion, USA (2019).
2. Levin, A., Garion, S.: AIOps for a Cloud Object Storage Service. In: BigData Congress, USA (2019)
3. Shen, S.J., Zhang, J.L., Huang, D.C., Xiao, J.: Evolving from Traditional Systems to AIOps: Design, Implementation and Measurements. In: AEECA, USA (2020)
4. Xia, Y.Y., Lu, J.X.: Anomaly detection and processing in artificial intelligence for IT operations of power system. In: APAP, USA (2019)
5. Yang, L., Xue, S.: Research on key technologies of operation and maintenance in intelligent substation. Public Communication of Science & Technology, no.10 (2013)
6. Tao, Q.: Research on key technologies of operation and maintenance in intelligent substation. M.S. dissertation, Shandong University, Chin, (2020)
7. Peng, Z.Q., Zhou, H., Han, Y.: Construction and application of transparent operation and maintenance system for smart substation automation equipment. Power Syst. Prot. Control **48**(13), 156–163 (2020)
8. Zhang, F.X.: Research on Intelligent Operation and Maintenance of Converter Station Equipment Based on Big Data Analysis. M.S. dissertation, Zhejiang University, China (2020)
9. Huang, Q., Jing, S., Li, J., Cai, D., Wu, J., Zhen, W.: Smart substation: state of the art and future development. IEEE Trans. Power Delivery **32**(2), 1098–1105 (2017)
10. Hua, S.H.: Exploration of intelligent operation management system of relay protection based on big data. Power Syst. Prot. Control China **47**(22), 168–175 (2019)
11. An, Z.Z.: Discussion on intelligent operation and maintenance management mode of relay protection equipment based on mobile internet technology. Electr. Power China **52**(3), 177–184 (2019)
12. Ping, W.L.: Research and application of intelligent maintenance system based on internet of things and mobile internet technology. Power Syst. Prot. Control. China **47**(16), 80–86 (2019)
13. Hui, D.Z.: A fault diagnosis method for the secondary circuits of protection systems in smart substations based on improved D-S evidence theory. Power Syst. Prot. Control China **48**(9), 59–67 (2020)
14. Sheng, G.J.: Parse for SCD file of smart substations based on Tiny XML. Electric Power Electr. Eng. **31**(3), 7–10 (2011)
15. Wei, H.Z.: Internet plus distributed photovoltaic power generation operation monitoring platform. Electr. Meas. Instrum. China **53**(z1), 205–207 (2016)
16. Fu, G.B.: Design and engineering application of intelligent operation and maintenance service system for distributed poverty alleviation photovoltaic. Electr. Meas. Instrum. China **56**(7), 84–88 (2019)
17. Jing, X.: Solution on intelligent operation and maintenance of wind farm based on active prevention strategy. Water Power China **46**(3), 104–107 (2020)

An Incremental Learning Method for Fault Diagnosis Based on Random Forest

Jiahua Dai[1,2] and Xiangmao Chang[1,2](✉)

[1] Computer Science and Technology, Nanjing University of Aeronautics and Astronautics, Nanjing 211106, China
xiangmaoch@nuaa.edu.cn

[2] Collaborative Innovation Center of Novel Software Technology and Industrialization, Nanjing 211106, China

Abstract. Equipment fault diagnosis is an important part of the modern industrial process. However, with the increase of working time, new types of faults will appear continuously. It is hard to take into account all types of faults previously due to the complexity and variability of industrial processes and environment. Thus, it is necessary to build a fault diagnosis system that can continuously detect and classify new type faults. In this paper, we propose Incre-RF, an incremental learning method for fault diagnosis based on Random Forest (RF), which can detect and classify new type of fault timely by RF-based incremental learning. We test Incre-RF on a popular open data set. The result shows Incre-RF performs better in both new type fault detection and new type fault learning ability compared with a state-of-the-art method.

Keywords: Fault diagnosis · New class classification · Ensemble system · Online detection

1 Introduction

1.1 A Subsection Sample

In the process of industrial production, various faults of equipment not only cost abundant manpower and resources for repairing but also bring huge economic losses if the repair is not in time [1]. As an important technology to guarantee the normal operation of equipment, equipment fault diagnosis has been widely used in the industrial process. The fault diagnosis problem contains two different and related steps. The first step is to identify if there is a fault, which is to distinguish the fault from the normal state, the second step is to determine the causes of the fault, i.e. classifying the fault [2].

There are basically two kinds of methods for fault diagnosis, one is based on analytical models, and the other is based on historical data. The methods [3, 4] based on analytical models are based on the usage of the residual which means the difference between the measured values during the actual production process and the values given by the analytical system. This kind of method can provide a detailed description of the

X. Sun et al. (Eds.): ICAIS 2021, CCIS 1422, pp. 641–652, 2021.
https://doi.org/10.1007/978-3-030-78615-1_56

industrial process. However, considering the complexity of the modern industrial process and the uncertainty in the production process, excessively detailed knowledge of the workflow and various parameters to build an accurate analytical model is usually quite difficult. The methods based on historical data are used much more widely in the current fault diagnosis system, especially in large-scale industry scenes. In fact, the more complexity of the industrial process, the more difficult it is to build an accurate analytical model compared with a model based on historical data. The methods based on historical data include unsupervised and supervised methods. The unsupervised method mainly contains principal analysis (PCA) [5], partial least square (PLS) [6], etc. The supervised method contains Random Forest (RF) [7], Bayesian Networks (BN) [8], and Artificial Neural Networks (ANN) [9], etc.

One notable point is that the classes of the fault data used to train the fault diagnosis model are given by experts. However, new type faults may appear during the actual industrial product process, thus it is difficult to take into account all types of faults previously due to the complexity and variability of the industrial process. In such a situation, the new type of faults would be wrongly classified as known faults rather than identified as new type faults until a new diagnosis system whose training data contains this faults type is built. Some works have been done to solve this problem [10, 11]. However, the focus of these works is how to add these new faults data into the diagnosis system with incremental learning methods rather than update the model. Such methods still need the experts to identify the new type faults data, such that they do not have the ability to discover a new type of fault by itself. The period from the emergence of a new fault discovered by experts will undoubtedly cause a loss of manpower and material resources. To solve this problem, we design a system based on the Random Forest (RF) that can identify new type faults timely and automatically update the model to classify the new type faults by using an incremental learning method, referred as Incre-RF. Incre-RF can identify whether the new fault data is one type of fault or multiple different fault types. If they are different types, the model can further classify them and give them different labels. We test Incre-RF on the Case Western Reserve University (CWRU) Bearing data set [12]. The result shows Incre-RF performs better in both new type fault detection and new type fault learning ability compared with a state-of-the-art method.

The remainder of this paper is organized as follows. Section 2 presents related works. Section 3 elaborates on the design of Incre-RF. Section 4 describes experimental results and analyses. Conclusion and Discussion are presents in the last section.

2 Previous Work on Increasing Types of Faults

How to solve the problem of new types of fault that may occur in the industrial process has attracted more and more people's attention. Yin G [13] proposed an online fault diagnosis method based on Incremental Support Vector Data Description (ISVDD) and ELM with incremental output structure (IOELM) to find and recognize the new fault mode efficiently, the method is tested on the diesel engine experiment under eleven different conditions. Sankavaram C. [10] uses multiple basic classifiers and Learn++.NC increment method [14] to realize the function of classifying the faults of the electronic throttle control subsystem of the vehicle, the SVM based classifier performs best. Similarly, Razavi-Far R. [11] uses ensemble ELM and Learn++.NC to classify bear faults in an induction motor. Hu Z [15] proposed a data-driven fault diagnosis model based on incremental imbalance modified deep neural network (incremental-IMDNN) to deal with the chemical imbalanced data streams and new class occurs situation. Mu X [16] uses the completely random-trees to identify the new class, known class, and outliers in known class, the task is more viewed as detecting new class rather than classification. Yu W [17] designs an incremental broad convolutional neural network (BCNN) to acquire the fault tendency of abnormal samples. The above research either only considers the problem of discovering new faults, or only considers the problem of how to update the model incrementally, and our system considers both two problems.

3 The Design of Incre-RF

In this section, we elaborate on the design of Incre-RF. The establishment process contains an offline process to build the initial system contains a base RF classifier based on the faults types that we have known and an online process to keep incrementally updating the model. During the online process, the system will extract time-domain features from the signal segment and classify it as a known fault type or a new type. When the new type fault data are detected, the model will confirm how many different classes are contained by these new faults data and use unsupervised methods to classify them. Then the system will incrementally update itself by adding a new base classifier built on the new faults data rather than rebuild the whole system. We first present the extracted features, then we introduce the base RF classifier training process, following with the online classification and new fault detection process, at last we present the incrementally update process.

3.1 Feature Extraction

We extracted six time-domain features from signals, N represents the number of samples in each signal segment, these features are:

1. Mean: $\mu x_j = \frac{\sum_{i=1}^{N} x}{N}$
2. Peak to peak value: $pk_j = \max_{i=1}^{N} x_j^i - \min_{i=1}^{N} x_j^i$
3. Standard deviation: $\sigma_j = \frac{\sum_{i=1}^{N}(x_j^i - \mu_j)^2}{N-1}$
4. Skewness: $sk_j = \frac{\sum_{i=1}^{N}(x_j^i - \mu_j)^3}{(N-1)\sigma_j^3}$
5. Root mean square: $RMS_j = \sqrt{\frac{\sum_{i=1}^{N}(x_j^i - \mu_j)^2}{N}}$
6. Kurtosis: $ku_j = \frac{\sum_{i=1}^{N}(x_j^i - \mu_j)^4}{(N-1)\sigma_j^4}$

3.2 Base RF Classifier Training Process

We choose the Random Forest (RF) as the base classifier since this algorithm can be simply implemented, has fine classification performance and anti-noise ability. The RF algorithm is a classifier composed of multiple completely random trees. The core of RF is to use random re-sampling techniques (bootstrap) to randomly select data from the training set to train each decision tree, and use node random split technology to randomly select data attributes to split each decision tree. When the sample data enters the random forest model, each decision tree in the forest is judged separately, and the final classification result can be obtained through the voting method.

When each decision tree is built, the center of each fault type data and the maximum distance from the furthest point to the center will be calculated based on the training data. Since the Euclidean distance can be interfered by different dimensions, we choose the standardized Euclidean distance as the measurement metrics. Also, it is worth noting that due to the different signal characteristics, some time-domain features may cause a bad impact on the model, such as Fig. 1.

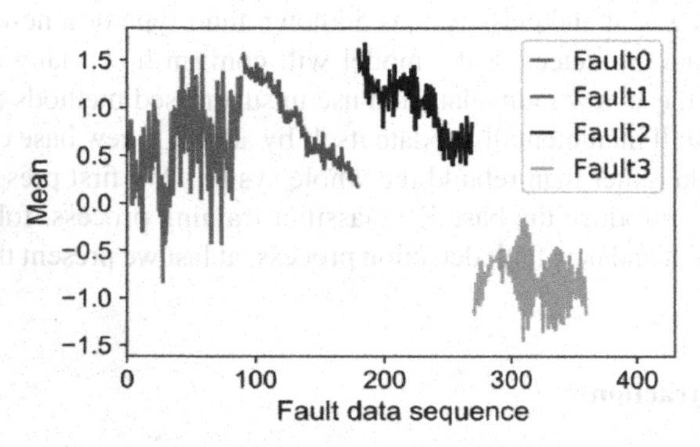

Fig. 1. The feature Mean of bearing fault signals.

Although random forest can solve this problem well when performing classification tasks, we still need to avoid the influence of these features on the distance calculation since the internal nodes of each decision tree will choose the feature with the highest information gain to divide the current data. Thus we chose the top-k features by the selection mechanism when calculating the distance, where the selection mechanism is as follows. The training data has M different faults types and each data has K features. For each type fault j, the standard deviation of feature i is $\sigma_{fault_j}^{fea_i}$, and the standard deviation of all data of feature i is $\sigma_{all}^{fea_i}$. Then the feature stability performance, which means the range of variation of one feature within this fault type relative to all fault types, could be calculated as:

$$Stab_j^i = \frac{\sigma_{fault_j}^{fea_i}}{\sigma_{all}^{fea_i}} \tag{1}$$

For each fault, the k features with the smallest $Stab_j^i$ will be selected from all K features to calculate the center and the maximum distance. Note that, the smaller the $Stab_j^i$, the more concentrated the feature i of the fault j is in the entire faults space, at the same time, it is less likely to harm on the max distance. After the tree is built, the data out-of-bag, which means was not used as training data for this tree, will be predicted by the tree to generate its initial vote weight. The accuracy of the tree T_k for the fault type j is:

$$acc_k^j = \frac{\sum_{x \in C_j} B_k(x)}{N_k^j} \tag{2}$$

Where $B_k(x) = 1$ only when the predicted type of the fault is equal to its true type, and C_j is all the fault data of type j in the out-of-bag data of. The vote weight of T_k for fault type j is calculated as:

$$\omega_k^j = (\frac{acc_k^j}{mean^j}) \tag{3}$$

Where $mean^j$ is the average classification accuracy of all trees for the fault type j. The establishment process of the base RF classifier is shown in the Fig. 2.

3.3 Classification and New Fault Detection Process

The classification and identification process includes three stages: the decision tree stage, the RF stage, and the Incre-RF stage.

Decision Tree Stage: When the new data come, each decision tree will predict its fault type, and determine whether it is a new fault type by calculating the distance from the data to the center of the fault type it was predicted as. If the distance is larger than the class's max distance, it will be recognized as a new fault type by this tree.

RF Stage: Each decision tree will give the data fault type and whether it is a new fault type. It will be identified as a new type fault by the RF classifier if more than half the decision tree recognized it as new, otherwise, the predicted class will be the class have the max weight. The weight of each fault type is the sum of the vote weight of the decision tree which prediction result for the data is the same as this fault type.

Incre-RF System Stage: Suppose this system has M base RF classifiers, and the all known fault type number is K. If all the M classifiers identify the new data as a new fault type, then the data will be added to the new fault type stack. Otherwise the final result will be determined by the vote result of the classifiers. The voting mechanism is like this: the total weight of each type of fault is:

$$W_j = \sum\nolimits_{i=1}^{M} F_i^j \tag{4}$$

Where $F_i^j = 1$ only when the training data of base RF classifier i contain the fault type j. The weight of each type of fault is:

$$w_j = \sum\nolimits_{i=1}^{M} f_i^j \tag{5}$$

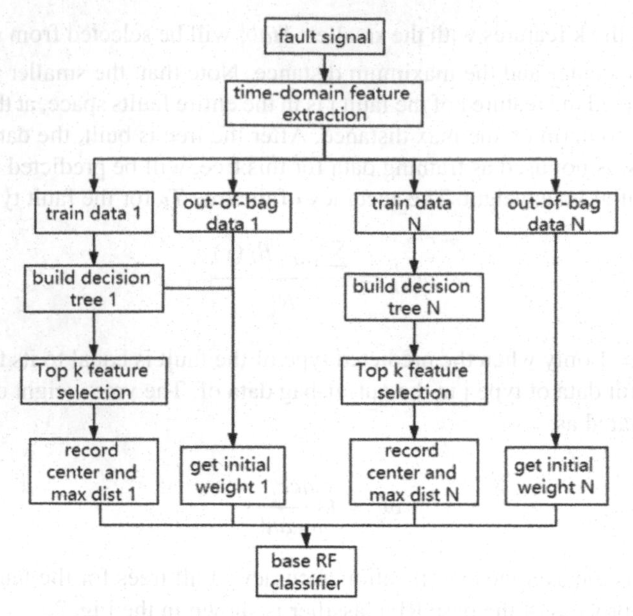

Fig. 2. The establishment process of a base classifier.

And $f_i^j = 1$ only when the random forest i prediction result is fault j. The final predict fault class is:

$$Result = argmax_{j \in K} \frac{w_j}{W_j} \tag{6}$$

Then the data will be added to the known fault type stack. The classification and new fault detection process is shown in the Fig. 3.

3.4 Incrementally Update Process

When the known fault stack is full, the oldest data on the stack will be discarded, and the new data will be pushed into the stack. On the contrary, when the new fault stack reach the maximum length, the incremental mechanism will start, a DB-SCAN cluster method will be applied on these new type fault data to determine how many categories in these data and then classify them. We choose the DB-SCAN method because it has the following advantages:

1. The clustering speed is fast and it can effectively deal with noise points and find spatial clusters of arbitrary shapes.
2. Compared with K-MEANS, there is no need to input the number of clusters to be divided, that's to say, the number of new fault categories could be determined automatically.
3. Shape of the cluster is not biased.

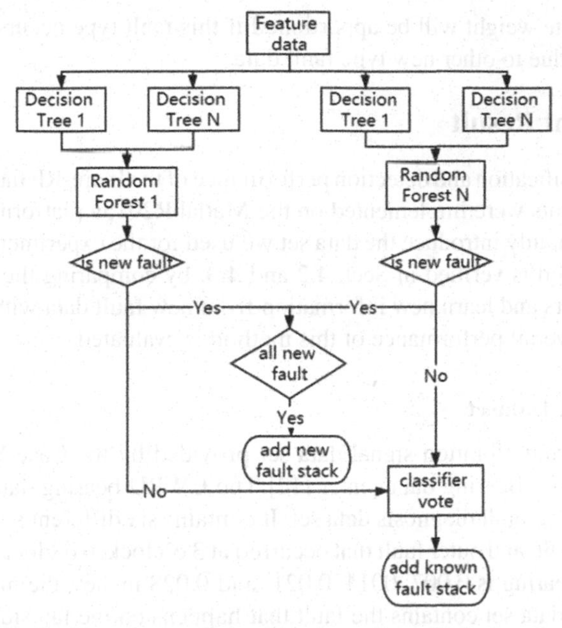

Fig. 3. The process of classification and new fault detection.

Consider the problem some features will affect the clustering effect, those who have not selected the calculation center and distance to help one type fault to distinguish from other types will not be selected to participate in the clustering process. An example will be given to show the necessity of doing so in Fig. 4.

After the new fault data is classified, we choose to build a new base RF classifier and add it to the Incre-RF system. If we rebuild a new system based on the all old data and new fault data, it will cost too much both time and space, but if we only rebuild a new system based on the part old data and new fault data, it will make us lose some information about old data. This incremental method we use could keep all information about old data, and will not cost too much, because the size of the training data of the new base RF classifier will not be very large. It is noticed that even if this new classifier has absolute weight on the new fault types it has, due to the voting mechanism we mentioned

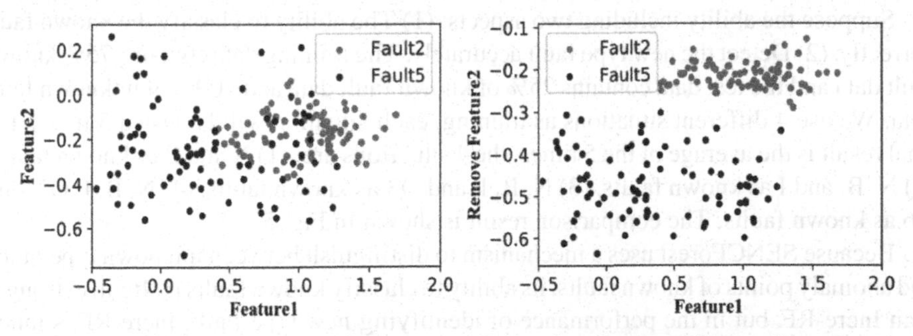

Fig. 4. Comparison of no removal feature 2 and removal feature 2.

above this absolute weight will be apportioned if this fault type occurs and a new base classifier is built due to other new type fault data.

4 Experiment Result

To verify the classification and detection performance of the Incre-RF fault diagnosis system, all experiments were implemented on the MatlabR2013a platform and python3.6. In Sect. 4.1, we mainly introduce the data set we used for the experiment. The effectiveness of this method is verified in Sect. 4.2 and 4.3, by comparing the ability to detect new types of errors and learn new information from new fault data with other methods, in Sect. 4.4 the overall performance of this method is evaluated.

4.1 Experiment Dataset

We used the bearing vibration signal data set provided by the Case Western Reserve University (CWRU) Bearing data center [12] The CWRU bearing data set is the most widely used bearing fault diagnosis data set. It contains six different situations: normal, ball fault, inner fault, and outer fault that occurred at 3 o'clock, 6 o'clock, and 12 o'clock. The size of the bearing is 0.007, 0014, 0.021, and 0.028 inches, the motor load ranges from 0 to 4. The data set contains the fault that happens on the fan side, which sample rate is 12k, and on the drive end, which sample rate is 12k and 48k.

We choose the mentioned six different types of drive end data, whose size is 0.007 inches, the motor load is 0, and sample rate is 12k, we divide each 1024 vibration signals into a segment to extract the 6 time-domain features mentioned earlier. The six different situations contained in the data are marked as: (1) Normal: N. (2) Ball fault: B. (3) Inner fault: I. (4) Outer fault at 3 o'clock: O3. (5) Outer fault at 6 o'clock: O6. (6) Outer fault at 12 o'clock: O12.

4.2 Detect New Fault Ability

We compare the ability to discover new type faults with SENCForest [12]. Because SENCForest performs more of an unknown fault detector than a classifier, it recognizes all unknown fault type as a new fault, in its source code is 999, we also only conduct our system as a new fault detector, that is to say, the step that a new RF base classifier is built and added to the system when the is full is not conducted, and its ability will be compared with another method in the next.

Suppose the ability including two aspects: (1) The ability to classify the known fault correctly. (2) Detect the new type fault accurately. The training data contains 75% known fault data and the test data contains 25% of known fault data and 100% of unknown fault data. We use 4 different situations as training, each situation will be tested 5 times, the final result is the average of the 5 times, the 4 situations are: (1) N and B as known faults. (2) N, B, and I as known faults. (3) N, B, I, and O3 as known faults. (4) N, B, I, O3, and O6 as known faults. The comparison result is shown in Fig. 5.

Because SENCForest uses a mechanism to distinguish between unknown type faults and anomaly points of known faults, its ability to classify known faults is slightly stronger than Incre-RF, but in the performance of identifying new type fault, Incre-RF is much brighter than SENCForest.

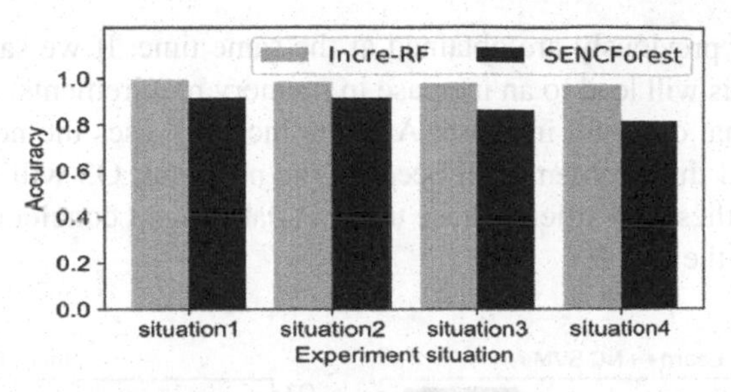

Fig. 5. Comparison accuracy with SENCForest.

4.3 Learn New Fault Ability

The ability that learns from the new fault is shown in this section by comparing it with the Learn++ NC SVM method [10]. When the new dataset got, Learn++ NC SVM will train a classifier group integrated by multiple SVMs to participate in the ensemble of the entire system, we set the number of SVM composes the base classifier is 8. Because Learn++ NC SVM only learn from new data without identifying new fault types, to avoid the effect of new fault identification error, we used the pre-set new fault data to build a new base RF classifier instead of using the new fault data identified by Incre-RF. The detail of the experiment is the data of the 6 type fault is divided into 75% training data and 25% test data, and the training data is again divided equally into 2 parts. Suppose they areN_{train0}, N_{train1}, N_{test}, similarly, the same is true for fault B, I, O3, O6, and O12, the experiment is divided into the following five stages: (1) the N_{train0} and B_{train0} compose the initial training data set, the N_{test} and B_{test} compose the test data set. (2) N_{train1}, B_{train1} and I_{train0} compose the first new training data set to build a new base classifier, and the I_{test} will be added to the test data set (3) I_{train1}, $O3_{train0}$ compose the second new training data set and the $O3_{test}$ will be added to the test data set. The same operation will be performed on O6 and O12 in the fourth and fifth stages respectively. The accuracy result is shown in Fig. 6.

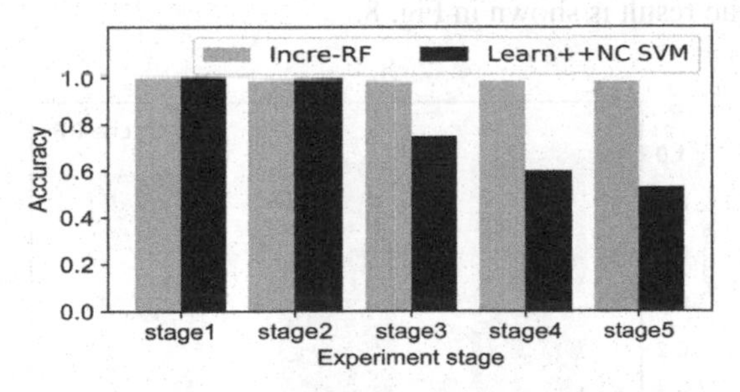

Fig. 6. Comparison accuracy with Learn++ NC SVM.

The classifier trained in stage 3 does not contain the data of fault type N, this makes the Learn++ NC system unable to distinguish between fault O3 and N, but in practical applications, it is difficult to ensure that when enough new fault data is obtained, all types of fault data used previously are obtained at the same time. If we save the previously used fault data, this will lead to an increase in memory requirements, and as the system runs, this additional cost will increase. And the Incre-RF uses the new fault detection mechanism solved this problem well because the new fault O3 will be identified as a new fault type in these classifiers whose training data do not conclude O3, the detailed result is shown in the Fig. 7.

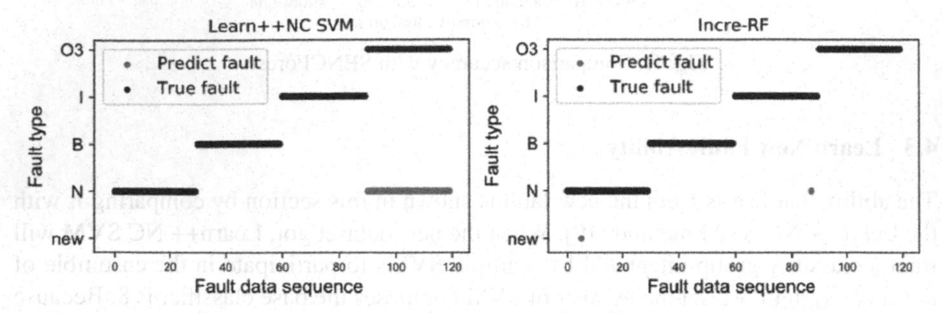

Fig. 7. Comparison of the results of Learn++ NC SVM and Incre-RF in the third stage.

4.4 Overall System Performance

In this section, the overall performance of the system will be displayed. All of N, B, I, O3, O6, and O12 data are divided into 75% training data and 25% test data. We first build a base RF classifier based on N and B training data and constitutes the initial system, then a test is conducted, where the fault type I, O3, O6, and O12 will be seen as new fault type, that is to say, their label is -1. Secondly, the training data of fault I and O3 will be predicted and a new base RF classifier will be built and added to the Incre-RF system. Thirdly, the training data of fault O6 and O12 will also be predicted by the system to build a base classifier. A test also will be conducted both in the second and third stages. The accuracy of the result is shown in Fig. 8.

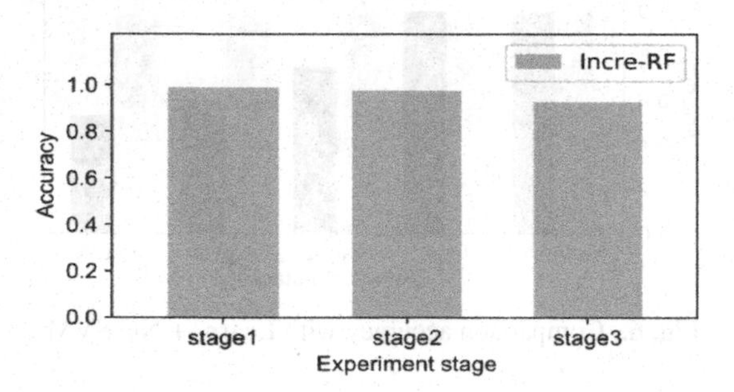

Fig. 8. The accuracy of Incre-RF in each stage.

5 Conclusion

We proposed a method to solve the problem of the appearance of new types of faults in the fault diagnosis system. After the initial basic RF classifier is established, the system could extract the time-domain features from the vibrational signal, determine the fault type of the signals according to the voting result of the base RF classifiers. The system consists of two stacks to store the detected new type of fault data and the known types of fault data. When the stack storing the new type of fault data is full, a new base RF classifier based on the data in these two stacks will be built, and the voting weight of the entire system will be updated accordingly. This incremental update method enables our system to solve the problem of the appearance of new types of faults effectively.

Acknowledgement. This research was supported by the National Nature Science Foundation of China (Grant No. 61672282).

References

1. Zeng, F., et al.: Fault diagnosis and condition division criterion of dc gas insulating equipment based on sf 6 partial discharge decomposition characteristics. IEEE Access **7**, 29869–29881 (2019)
2. Tidriri, K., Chatti, N., Verron, S., Tiplica, T.: Bridging data-driven and model-based approaches for process fault diagnosis and health monitoring: a review of researches and future challenges. Annu. Rev. Control. **42**, 63–81 (2016)
3. Echevarría, L.C., Santiago, O.L., Fajardo, J.A.H., Neto, A.J.S., Sánchez, D.J.: A variant of the particle swarm optimization for the improvement of fault diagnosis in industrial systems via faults estimation. Eng. Appl. Artif. Intell. **28**, 36–51 (2014)
4. Zhong, M., Song, Y., Ding, S.X.: Parity space-based fault detection for linear discrete time-varying systems with unknown input. Automatica **59**, 120–126 (2015)
5. Lau, C., Ghosh, K., Hussain, M.A., Hassan, C.C.: Fault diagnosis of Tennessee Eastman process with multi-scale PCA and ANIFS. Chemom. Intell. Lab. Syst. **120**, 1–14 (2013)
6. Yin, S., Zhu, X., Kaynak, O.: Improved pls focused on key-performance-indicator-related fault diagnosis. IEEE Trans. Industr. Electron. **62**(3), 1651–1658 (2014)
7. Wang, Z., Zhang, Q., Xiong, J., Xiao, M., Sun, G., He, J.: Fault diagnosis of a rolling bearing using wavelet packet denoising and random forests. IEEE Sens. J. **17**(17), 5581–5588 (2017)
8. Cai, B., Liu, Y., Xie, M.: A dynamic-Bayesian-network-based fault diagnosis methodology considering transient and intermittent faults. IEEE Trans. Autom. Sci. Eng. **14**(1), 276–285 (2016)
9. Chine, W., Mellit, A., Lughi, V., Malek, A., Sulligoi, G., Pavan, A.M.: A novel fault diagnosis technique for photovoltaic systems based on artificial neural networks. Renew. Energy **90**, 501–512 (2016)
10. Sankavaram, C., Kodali, A., Pattipati, K.R., Singh, S.: Incremental classifiers for data-driven fault diagnosis applied to automotive systems. IEEE access **3**, 407–419 (2015)
11. Razavi-Far, R., Saif, M., Palade, V., Zio, E.: Adaptive incremental ensemble of extreme learning machines for fault diagnosis in induction motors. In: 2017 International Joint Conference on Neural Networks (IJCNN), pp. 1615–1622. IEEE (2017)
12. Loparo, K.: Bearing vibration data: case western reserve university bearing data center website (2005)

13. Yin, G., Zhang, Y.T., Li, Z.N., Ren, G.Q., Fan, H.B.: Online fault diagnosis method based on incremental support vector data description and extreme learning machine with incremental output structure. Neurocomputing **128**, 224–231 (2014)
14. Muhlbaier, M.D., Topalis, A., Polikar, R.: Learn ++.NC: combining ensemble of classifiers with dynamically weighted consult-and-vote for efficient incremental learning of new classes. IEEE Trans. Neural Netw. **20**(1), 152–168 (2009). https://doi.org/10.1109/TNN.2008.2008326
15. Hu, Z., Jiang, P.: An imbalance modified deep neural network with dynamical incremental learning for chemical fault diagnosis. IEEE Trans. Ind. Electron. **66**(1), 540–550 (2018)
16. Mu, X., Ting, K.M., Zhou, Z.H.: Classification under streaming emerging new classes: a solution using completely-random trees. IEEE Trans. Knowl. Data Eng. **29**(8), 1605–1618 (2017)
17. Yu, W., Zhao, C.: Broad convolutional neural network based industrial process fault diagnosis with incremental learning capability. IEEE Trans. Ind. Electron. **67**(6), 5081–5091 (2019)

Research on Short-Term Flow Forecast of Intelligent Traffic

Tingting Yang[1(✉)] and Shuwen Jia[2]

[1] Institute of Information and Intelligence Engineering, University of Sanya, Hainan, China
[2] Saxo Fintech Business School, University of Sanya, Hainan, China

Abstract. With the rapid development of China's economy and the deepening of urbanization, the burden of urban traffic is becoming more and heavier. Scientific traffic planning and guidance has become an important research content of urban traffic management departments. The first day before the guidance of rational planning is the scientific and effective prediction of traffic flow. This chapter mainly studies and analyzes the wavelet neural network, and expounds the role of traffic flow prediction in intelligent transportation; the wavelet neural network model is studied in detail. At present, the wavelet neural network can forecast the urban traffic flow accurately and scientifically. I hope that the author's research can give short-term traffic flow prediction researchers a reference.

Keywords: Intelligent transportation · Neural network · Wavelet function · Short-term slow forecast

1 Introduction

With the remarkable growth of economy, can be used to control the income of the family more and more, followed by is the number of privately owned cars is rising year by year, plus go out on business or private frequency increasing, go out to travel distance is also increasing, it is slowly increase the burden of the city traffic, the passage of time is caused a serious congestion of modern transportation. Traffic congestion has become one of the major problems that every big city urgently solves. Although increasing public travel modes and building subways and other transportation modes can greatly reduce the pressure of urban traffic, it cannot fundamentally solve the problem of urban traffic congestion. Through the accurate analysis and research of traffic flow, it is of vital significance to establish efficient road facilities to solve congestion and effectively shorten the traffic time cost of people's travel. With the deepening of China's urbanization process, the number of vehicles in cities and suburbs is increasing year by year, the range of traffic jams is becoming larger and larger, and the congestion problem is becoming more and more serious. Under the background of numerous types of motor vehicles in the first or second tier cities and the deterioration of bus service quality, this has brought a great crisis to the highway traffic. Many cities have appeared the phenomenon of mixed use of roads, and the situation of mixed use of roads by motor vehicles and non-motor

X. Sun et al. (Eds.): ICAIS 2021, CCIS 1422, pp. 653–662, 2021.
https://doi.org/10.1007/978-3-030-78615-1_57

vehicles is becoming increasingly serious. Transport planning and land use are clearly out of harmony.

Because the traffic flow has such characteristics as contingency, short term and commuting concentration, AI intelligent technology has been more and more widely applied in various fields in the traffic flow observation work, and more people understand and recognize it. Is also considered the future can replace traffic prediction model is a better way, however, in many research methods, the most commonly used is the neural network technology, in order to effectively solve the problem of mixed flow of traffic, developing a can adapt to the current traffic situation of short-time traffic flow forecasting model is imminent, a new prediction model can not only help to develop the highway traffic rules, you can also give good for transportation service quality evaluation, and also can give a reference in the process of commissioning engineers working in transportation planning. The basic characteristics of unsteady traffic are volume, speed and traffic density. The prediction of these data can help to understand the nature of possible changes in the traffic network, and thus effectively alleviate traffic congestion and accidents. At present, most traffic prediction models are based on uniform traffic flow, which does not apply to non-uniform traffic.

2 The Causes of Traffic Congestion

The main causes of urban traffic congestion include: increasing use of cars, insufficient road capacity or improper road design, and too many road interchanges. Among them with increase in the number of vehicles increasing utilization rate of private cars is the main factor of modern urban traffic congestion, because the car is convenient travel, everyone's daily travel like to travel, go to work, go to the supermarket shopping, and so on are mostly to drive, although convenient, but result in higher city daily traffic, together with the capacity of urban road design is insufficient or improper design, as a result, urban road can't load so large flow of traffic, thus made the urban traffic congestion phenomenon. Combined with the design of city roads are mostly radial, such a design while convenient traffic in a suburb of the city, but, once the traffic peak, will result in the outskirts of vehicles constantly pour into the city's main road, the results on the road is blocked full from suburban commuter vehicles, traffic growth line is obvious. At the same time, there are too many intersections in the city, so there are too many traffic lights. In addition, the unreasonable setting of traffic lights at intersections increases the situation of road congestion.

Traffic jam generally does not happen all the time throughout the day, nor does it happen in every section of the city, on every road. Instead, it happens at the time and road section when citizens are easy to gather. For example, the morning, middle and evening commuting are also the time for students to go to and from school. The sudden gathering and increase of the flow of people will inevitably increase the traffic congestion. Second, traffic jams are more likely to occur in central thoroughs and road interchanges than in other sections. However, the problem of traffic congestion can not be simply attributed to the sudden gathering of people and the location of the road. There are several reasons for the phenomenon of traffic congestion:

2.1 The Plot Ratio of the Toad is Small

As vehicles proliferate, roads in big cities continue to cope with their old capacity, and the growth of roads is far from keeping pace with the increase in vehicles. During peak hours, people tend to gather in the city center, increasing the flow of people, which leads to serious traffic jams in the central roads of many big cities in China, such as Beijing, Tianjin, Shanghai and Shenzhen. Especially in the rush hour of the concentrated road section, it seems that these concentrated road sections can be likened to the "parking lot".

2.2 Road Structure

The design of road structure also has an important influence on the accessibility of traffic. In our country, the roads of big cities are generally distributed by radioactivity, extending around the city center in a divergent state. Although this structure facilitates the convenience of the suburbs to the city center, it also causes the city center congestion to a large extent. In addition, the number of crisscrossed road intersection sections will increase, which will cause serious backblocking phenomenon.

2.3 Time Distribution of the Total Amount of Travel

In China, due to the uneven distribution of people's travel volume in time, it can be divided into peak period and off-peak period. The short time of occurrence and dense flow of people are the significant characteristics of the peak period in China. In other words, in the rush hour, the surge of the flow of people and the contradiction of the road structure is particularly prominent. Increased pressure on the transportation system.

2.4 Urban Architectural Planning

The direction of road structure is influenced by urban construction planning to a large extent. In general, the center of a big city gathers commerce, and some tertiary industries become the mainstream of the center. As a result, the non-downtown area of the flow of people will gather to the center, such a central layout will inevitably increase the need for downtown roads, so that the central road into the overload running state. On the other hand, the layout of urban traffic planning does not fully consider the layout of urban land, the two do not adapt to each other, may ignore the influence of urban land layout in road network layout planning, which also causes the pressure of overload operation of some sections.

2.5 Traffic Management

On the one hand, because China's traffic laws and regulations are not sound, in addition to the traffic management department of safety education propaganda is relatively low; people's awareness of safety awareness is less, so there will be a lot of traffic laws violations. In the original more congested road random lane change, congestion, overtaking and other uncivilized phenomena occur from time to time, so that not only aggravate the congestion time, but also caused a great safety risk. On the other hand, the intelligent level of China's traffic management system is far from enough, the traffic command center can not carry out a comprehensive analysis and dredging of real-time road conditions, can not deal with the situation in time.

3 The Role of Traffic Flow Prediction in Intelligent Transportation

Intelligent transportation refers to a transportation oriented service system based on modern electronic information technology. Its prominent feature is information collection, processing, release, exchange, analysis, use as the main line, to provide a variety of services for traffic participants; Using the full use of the Internet of things in the field of traffic, cloud computing, Internet, artificial intelligence, automatic control, mobile technology, such as the Internet, through the collection of new and high technology traffic information, for traffic management, transportation, public transportation field and so on all aspects of travel and traffic construction whole process management regulation, make the traffic system at the regional, urban and even greater range of time and space perception, interconnection, analysis, forecast and control ability, to fully guarantee the traffic safety, traffic infrastructure efficiency, improve system efficiency and management level, for the flow of public transport and sustainable economic development of the service.

Along with the rapid development of urban economy and private cars gradually increased, the serious imbalance between supply and demand of urban traffic, urban traffic brings the experience of people more and more bad, traffic congestion, traffic accidents, traffic pollution, travel and so on many problems of parking difficulty became the most urgent problem in major cities, has become the important focus of the urban traffic management department. Traffic is concentrated in the intersection, for example, how to better or more flexible application of current testing equipment, monitoring the relevant traffic conditions to obtain valid data traffic, and then based on these data traffic situation and the traffic flow analysis, the analysis of urban road to realize real-time, become the main research topic.

4 The Role of Traffic Flow Prediction in Intelligent Traffic

With the rapid development of urban economy, the imbalance between supply and demand of urban traffic is becoming increasingly serious, and the experience brought by urban traffic is getting worse and worse. Many problems such as traffic congestion, traffic accidents, traffic pollution and difficulty in traveling and parking have become the most urgent problems to be solved in major cities at present. It has also become an important focus of urban traffic management departments.

Figure 1 shows how to flexibly apply current detection equipment to monitor relevant traffic flow to obtain effective data at intersections where traffic is concentrated, and how to realize real-time analysis of urban roads through analysis of traffic situation and traffic flow, which has become a major research topic at present.

Fig. 1. Traffic flow monitoring in intelligent traffic.

5 Predictive Model Analysis

Urban traffic flow is associated with related, such as a period of time before the congestion will cause the communication flow increased after a period of time, usually traffic statistics for 24 h a day to statistics, statistical cycle within 24 h, of course, different period of traffic flow is also changing, using wavelet neural network according to the different change analysis and prediction of traffic flow. Wavelet neural network is divided into three layers: input layer, hidden layer and output layer. The data of historical traffic flow in n periods before the current time is input to the input layer. Wavelet function constitutes the nodes of the hidden layer. Wavelet basis function is the transfer function of the nodes of the hidden layer. It is a repeated iterative process of neural network for signal forward propagation and prediction error back propagation while learning and training weight adjustment. The output layer outputs the topology diagram of the traffic flow prediction wavelet neural network for the current period, as shown in Fig. 2 [2]. In Fig. 2, X1, X2,…, Xk is the input data parameter, Y1, Y2,…, Ym is the predictive data output after learning and training, and is the weight of the hidden layer of the wavelet neural network. In the input traffic flow sequence, Xi (I = 1, 2,… When, k), the formula for calculating the output of the hidden layer is:

$$h(j) = h_j\left(\frac{\sum_{i=1}^{k} \omega_{ij}X_i - b_j}{a_j}\right) \tag{1}$$

Where, h(j) is the output value of the JTH node of the hidden layer; ω_{ij} is the connection weight of the input layer and the hidden layer; hj is the wavelet basis function; aj is the scaling factor of wavelet basis function hj; bj is the translation factor of the wavelet basis function hj. Morlet function is adopted as wavelet basis function, as shown in the following formula: Is the connection weight of the input layer and the hidden layer;

Hj is the wavelet basis function; aj is the scaling factor of wavelet basis function hj; bj is the translation factor of the wavelet basis function hj. Morlet function is adopted as wavelet basis function, as shown in the following formula:

$$y = f(x) = \cos(1.75x) \cdot e^{-\frac{x^2}{2}} \tag{2}$$

The calculation formula of the output layer is:

$$y(k) = \sum_{j=1}^{l} \omega_{jk} h(j)\ k = 1, 2, \cdots, m \tag{3}$$

Where, ω_{ij}; is the weight from the hidden layer to the output layer; h(j) is the output of the j hidden layer node; 1 is the number of nodes in the hidden layer; M is the number of nodes in the output layer.

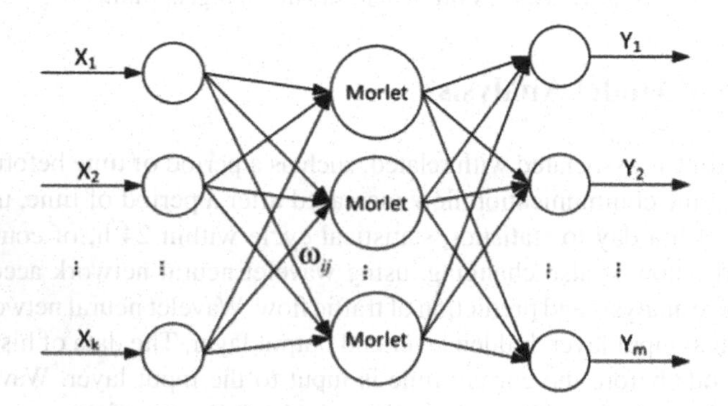

Fig. 2. Wavelet neural network topology diagram.

The wavelet basis function and weight of the network are modified by using gradient correction method. The weight parameter correction algorithm is similar to the BP neural network weight correction algorithm; so that the predicted output results of short-term traffic flow of the wave neural network can be closer to the expected output result. The specific correction process is as follows: the network prediction error is calculated

$$e = \sum_{k=1}^{m} y_E(k) - y(k) \tag{4}$$

Where, $y_E(k)$ is the expected output of short-term traffic flow; y(k) is the short-term traffic flow forecast output of wavelet neural network. (2) The weights of the wavelet neural network and the wavelet basis function coefficient are corrected according to the error E of short-term traffic flow prediction.

$$\omega_{n,k}^{(i+1)} = \omega_{n,k}^{i} + \Delta\omega_{n,k}^{(i+1)}\ a_k^{(i+1)} = a_k^i + \Delta a_k^{(i+1)} \tag{5}$$

$$b_k^{(i+1)} = b_k^i + \Delta b_k^{(i+1)} \tag{6}$$

Where, $\Delta\omega_{n,k}^{(i+1)}$, $\Delta a_k^{(i+1)}$, $\Delta b_k^{(i+1)}$ based on the error of the network short-time traffic flow prediction, it can be calculated as follows:

$$\Delta\omega_{n,k}^{(i+1)} = -\eta\frac{\partial e}{\partial\omega_{n,k}^{i}};\ \Delta a_k^{(i+1)} = -\eta\frac{\partial e}{\partial a_k^{i}};\ \Delta b_k^{(i+1)} = -\eta\frac{\partial e}{\partial b_k^{i}} \tag{7}$$

6 Sensitivity Analysis of Wavelet Neural Network

Sensitivity is an evaluation of the degree of correlation between the output of a wavelet neural network and its input or weight. It is the response that determines the output of the model when the input parameters change. In order to eliminate the irrelevant input in time, the sensitivity analysis of the trained network can be used to solve this problem. In this way, the elimination of irrelevant inputs not only improves the network operation efficiency and performance, but also reduces the cost. In addition, this analysis can also timely understand the basic relationship between input variables and output. In this study, the wavelet neural network model is used to analyze the sensitivity. The sensitivity analysis of the mean value of pre-trained MLP networks was carried out. The other inputs for this batch are fixed at their respective averages. By changing the first input of its mean ± 1, and calculate the network output of 50 steps above and below the mean value. This process is then repeated over and over again with each input. The transformation of each input variable is summarized and analyzed according to its sensitivity, of which 9 are the most important input parameters were daily (DY), digital commercial vehicle (SLCV), bus speed (SB), motorcycle speed (SSM), time (TM), two-wheeled vehicle speed (STW), car/jeep/van volume (C/J/V), etc. (Fig. 3). In the next step, the neural network is trained and tested using the same neural network configuration as the best-selected neural network model, considering only the nine most important inputs found in the sensitivity analysis. The training and test results of the new model are shown in Fig. 4. Even after reducing the number of inputs from 19 to 9, the new model performs very well compared to the old model.

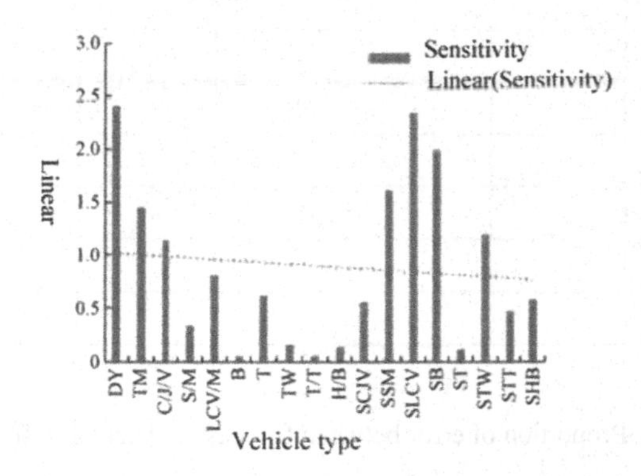

Fig. 3. Analysis of vehicle sensitivity of neural network.

7 Simulation and Results

This study through the analysis of prediction model is set up, through every 15 min to take a period of time a short-time traffic flow, and then one day on the same traffic flow sampling data of sampling sites in 96, and then the sampling of traffic flow data in short period of wavelet neural network training, and then it has trained wavelet neural network applied to short-term traffic flow prediction of day after work.

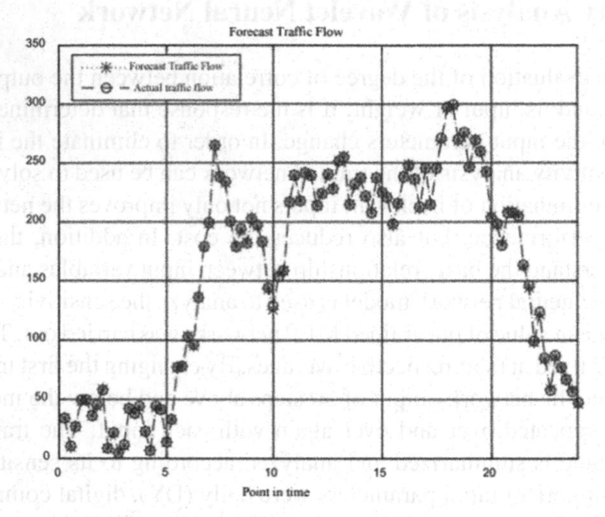

Fig. 4. Traffic flow forecast.

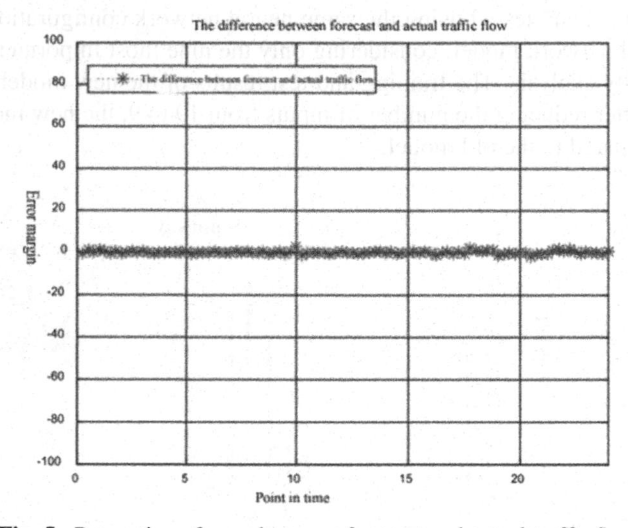

Fig. 5. Proportion of error between forecast and actual traffic flow.

Through 5000 times of training, training weight training of traffic data verification, get the weight value for subsequent prediction, forecasting traffic data is very close to the actual value, as shown in Fig. 6 (unit: number of vehicles), and the data and the actual traffic flow data difference is small, can be seen clearly in the traffic flow deviation slightly higher than the small period of time, but the overall result is very rational, as shown in Fig. 4 and 5.

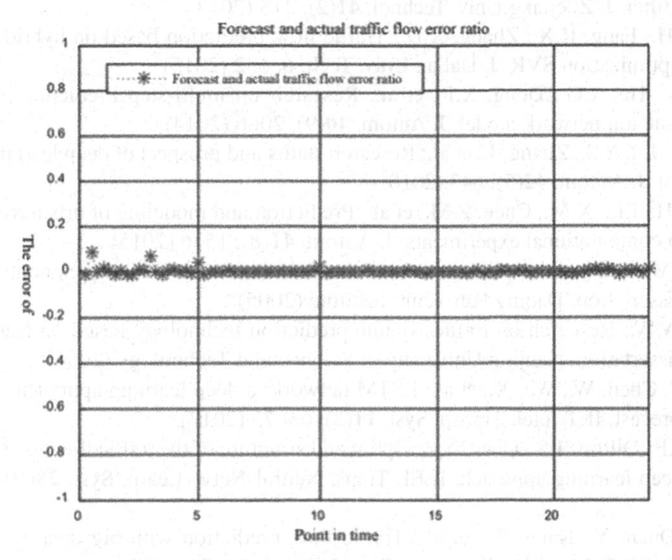

Fig. 6. Difference between forecast and actual traffic flow.

8 Simulation and Results

To sum up, in the traffic problem has increasingly become the focus of urban context, structures, wavelet neural network prediction model, the science of traffic flow forecasting and the instruction, using the current short-term traffic flow data validation, training to guide the traffic flow prediction of the corresponding work, and then precise guidance to traffic induced. Through the analysis of the input layer, hidden layer, output layer, Morlet function and weight learning of the wavelet neural network prediction model, and the matlab simulation of the historical data, the prediction results are very close to the actual value. The use of this precaution can make a scientific, accurate and objective short-term prediction of traffic flow.

References

1. Ma, J., Sun, S.Z., Rui, H.D., et al.: A review of china's traffic engineering academic research 2016. Chin. J. Highw. **29**(6), 1 (2016)
2. Xie, Y.Q., Jiang, L., Wang, L.Y., et al.: A fast traffic flow detection algorithm. J. Harbin Univ. Sci. Technol. **21**(4), 19 (2016)
3. Kang, W.X., Yang, W.B., Li, P., et al.: Design of traffic flow remote release system. J. Harbin Univ. Sci. Technol. **17**(5), 39 (2012)

4. Liu, C.Y., Kong, F.N., Feng, J.W.: Traffic signal acquisition system based on GPS and SD card storage. J. Harbin Univ. Sci. Technol. **3**, 25 (2017)
5. Tang, Y., Liu, W.N., Sun, D.H., et al.: Application of improved time series model in expressway short-time traffic flow prediction. Comput. Appl. Res. **32**(1), 146 (2015)
6. Guo, H.F., Fang, L.J., Yu, L.: A short-term traffic flow prediction method based on fuzzy Kalman filter. J. Zhejiang Univ. Technol. **41**(2), 218 (2013)
7. Yao, W.H., Fang, R.X., Zhang, X.D.: Traffic flow prediction based on hybrid artificial fish swarm optimization SVR. J. Dalian Univ. Tech. **6**, 632 (2015)
8. Yin, L.S., He, Y.G., Dong, X.P., et al.: Research on multi-step Prediction of traffic flow VNNTF neural network model. J. Autom. **40**(9), 2066 (2014)
9. Duan, Y., Lu, Y.S., Zhang, J., et al.: Research status and prospect of deep learning in the field of control. J. Autom. **42**(5), 643 (2016)
10. Tang, S.H., Liu, X.M., Chen, Z.M., et al.: Prediction and modeling of urban road travel time based on computational experiments. J. Autom. **41**(8), 1516 (2015)
11. Liu, X.: Modeling and application of wavelet neural network for complex nonlinear systems, Ph.D. Dissertation, Daqing Petroleum Institute (2005)
12. Meng, W.W.: Research on traffic volume prediction technology based on Neural network, Ph.D. Dissertation, Nanjing University of Science and Technology (2006)
13. Zhao, Z., Chen, W., Wu, X., et al.: LSTM network: a deep learning approach for short-term traffic forecast. IET Intel. Transp. Syst. **11**(2), 68–75 (2017)
14. Yang, H.F., Dillon, T.S., Chen, Y.P.: Optimized structure of the traffic flow forecasting model with a deep learning approach. IEEE Trans. Neural Netw. Learn. Syst. **28**(10), 2371–2381 (2016)
15. Lv, Y., Duan, Y., Kang, W., et al.: Traffic flow prediction with big data: a deep learning approach. IEEE Trans. Intell. Transp. Syst. **16**(2), 865–873 (2015)
16. Koesdwiady, A., Soua, R., Karray, F.: Improving traffic flow prediction with weather information in connected cars: a deep learning approach. IEEE Trans. Veh. Technol. **65**(12), 9508–9517 (2016)
17. Al-Jarrah, O.Y., Yoo, P.D., Muhaidat, S., et al.: Efficient machine learning for big data. Big Data Res. **2**(3), 87–93 (2015)
18. Duan, C.M.: Design of big data processing system architecture based on hadoop under the cloud computing. Appl. Mech. Mater. **556–562**, 6302–6306 (2014)
19. Cheng, A., Jiang, X., Li, Y., et al.: Multiple sources and multiple measures based traffic flow prediction using the chaos theory and support vector regression method. Phys. A **466**, 422–434 (2017)
20. Zhu, J.Z., Cao, J.X., Zhu, Y., et al.: Traffic volume forecasting based on radial basis function neural network with the consideration of traffic flows at the adjacent intersections. Transp. Res. Part C **47**(2), 139–154 (2014)
21. Oh, B., Song, H., Kim, J., Park, C., Kim, Y.: Predicting concentration of PM10 using optimal parameters of deep neural network. Intell. Autom. Soft Comput. **25**(2), 343–350 (2019)
22. Ahn, J.Y., Ko, E., Kim, E.: Predicting spatiotemporal traffic flow based on support vector regression and Bayesian classifier. In: IEEE 5th International Conference onBig Data and Cloud Computing, pp. 125–130. IEEE Computer Society (2015)
23. Xu, Y., Chen, H., Kong, Q., et al.: Urban traffic flow prediction: a spatio-temporal variable selection-based approach. J. Adv. Transp. **50**(4), 489–506 (2016)
24. Gu, B., Xiong, W., Bai, Z.: Human action recognition based on supervised class-specific dictionary learning with deep convolutional neural network features. Comput. Mater. Continua **63**(1), 243–262 (2020)

Research on Stock Price Forecasting Based on BP Neural Network

Shuwen Jia[1] and Tingting Yang[2](✉)

[1] Saxo Fintech Business School, University of Sanya, Hainan, China
[2] Institute of Information and Intelligence Engineering, University of Sanya, Hainan, China

Abstract. With the development and progress of China's economy and society, China's stock market has been continuously improved, attracting more and more people to participate in the stock market speculation. No matter the old investors who have been in the stock market for a long time or the new investors who have entered the stock market, they all hope to be able to predict the stock price through technical means. This is also a problem that global stock market participants are paying attention to, which attracts numerous researchers to study it, and also produces a lot of research results, such as time series prediction method. There are many factors that affect the stock price. If we want to improve the accuracy of stock measurement, we must first understand the factors that affect the stock, and make scientific and reasonable use of these factors for price prediction to obtain more ideal results. Using BP neural network method to forecast and analyze the stock price can give full play to the advantage of BP neural network algorithm based on error reverse propagation, so as to reduce the interference factors of stock price prediction. It can be seen that BP neural network is of great help to predict the stock price through the research on the history of stock price operation. In this paper, the stock price prediction of BP neural network is studied, and it is concluded that the reasonable and efficient use of BP neural network system has played a great role in the stock price prediction and analysis. It is expected that the research will make a contribution to the development of China's stock market.

Keywords: BP neural network · Sample selection · Stock forecast · Investment · Earnings

1 Introduction

With the development and progress of China's economy and society, China's stock market has been continuously improved, attracting more people to participate in the stock market speculation. No matter the old investors who have been in the stock market for a long time or the new investors who have entered the stock market, they all hope to be able to forecast the stock price with relatively high accuracy through technical means. This is a problem that global stock market participants are paying attention to, which attracts numerous researchers to study it, and also produces a lot of research results, such as time series prediction method. Various industries in China are developing rapidly, and the stock industry is the same. In People's Daily life, with the growth of people and

X. Sun et al. (Eds.): ICAIS 2021, CCIS 1422, pp. 663–673, 2021.
https://doi.org/10.1007/978-3-030-78615-1_58

themselves, the stock industry occupies a very important position in the society, and has an inestimable position in the financial and securities industry in China. The progress and development of the stock industry is the inevitable trend of social development, but also the symbolic expression of social progress.

There are two forms of stock prices, namely, rising stock prices and falling stock prices. The rise or fall of the stock price directly affects the economic income of the shareholders and the healthy development and progress of the industry. The stock market is a very complex nonlinear system, which is influenced and interfered by many uncertain factors. The complexity and variability of the stock system itself determine that the stock forecast is a very complicated research task, whose main performance lies in the nonlinear fluctuation characteristics of the stock price. In addition, many research achievements have been made, such as a variety of commonly used time series prediction methods, including ARIMA, neural network, grey theory, and the newly developed Support Vector Machine (SVM) theory. All of these have made contributions to the healthy and rapid development of the securities market. However, a large amount of data is essential for the prediction of stock market. Therefore, the algorithm of stock price prediction is highly required. The effect of stock price prediction using the existing stock price prediction method is not satisfactory. Therefore, how to combine the large amount of historical data in the stock market with the prediction algorithm effectively becomes an important topic. What kind of way can be more efficient and accurate to predict the trend of stock prices is the most concerned by hundreds of millions of investors. Stock market fast changing, policy, benefit, information, capital and so on are all factors that influence the stock price of casual a policy or messages to be able to cause the volatility of stock prices, according to the different needs or different stock can adopt different prediction methods, as much as possible to avoid a kind of method to predict the shortcomings. I believe that hundreds of millions of modern investors have a clear understanding of the theoretical basis of stock price prediction. For different stocks, the choice of forecasting methods is not the same. The main forecasting models have several aspects: GARCH model prediction method, the construction of ARIMA prediction method, artificial intelligence prediction method, etc. This paper will study the neural network model, mainly because the neural network is in the leading position in the field of artificial intelligence, effective simulation of nonlinear prediction problems, and has been widely used and attention. Therefore, using neural network to predict the stock price has unique advantages and can effectively overcome the problem of highly nonlinear stock price.

2 Related Works

The stock market is a complex and variable nonlinear system, and the complexity of its description, analysis and prediction is very high. Therefore, further research is needed to obtain more accurate prediction results. Charles, the founder of the Dow Jones Industrial Average, is the basis of the study of market economy based on his theory of the Dow in the early 1900s. By studying the K-chart, Richard found out the law of the ups and downs of stocks. In wave theory, the pattern of price trends occurring in the market is represented by repeated patterns. Since then, there have been many methods that rely on chart analysis to perform technical analysis. In 1992, scholar Baba studied the rising trend and falling trend of the stock market respectively based on the neural network algorithm, and took the evaluation standard of the stock market as the input variable to predict the stock price, but the accuracy of the prediction results was not stable. In 1998, the scholar Gencay compared the prediction effect of neural network algorithm and linear regression model combined with chart analysis method respectively, and proved the superiority of neural network prediction. Lee Raymond ST proposed a stock advisory application based on radial basis function network -- IJADE stock investment adviser. This system can make effective and intelligent stock prediction. In the experiment, we use the historical data of Hong Kong stock market for ten years (1990–1999) to test the system, and the results show that IJADE stock investment consultant has research and application value. In 2014, Jacinto MR used the support vector machine method to construct the multi-impact factor quantitative stock selection model, and selected the American stocks from 2005 to 2014 as the research object. The case analysis shows that the stock selection model can obtain higher returns by adopting this method for stock selection modeling. In 2017, Tian Kai and Liu Yongrui selected GEM stocks and built a stock selection model with logistics model aiming at multiple impact factor indicators. The empirical study shows that this model can not only obtain high returns but also avoid risks.

On the whole, domestic scholars' research on stock market prediction mainly focuses on the field of neural network. However, considering the timing of stock data, we can better fit the serial data and fully mine the hidden information of the stock market, so as to predict the stock market more accurately. Combined with the stock technical index data and basic trading data, this paper optimizes the neural network model to forecast the stock price, and through the experimental comparative analysis of the influencing factors of the model effect, finally selects the optimal time series length and network structure and other parameters to improve the prediction effect of the prediction model.

3 Overview of BP Neural Network

As a mathematical model, neural network has the characteristics of high intelligence. As a simple abstraction of human brain, neural network shows strong performance in function approximation and information processing. Neural network is like a universal model and error correction functions, which can conduct error analysis according to the results, obtained from each training and the expected results, and then modify the weights and thresholds, so as to get a model which can output the same as the expected results step by step. As an example, the businessman to produce a product, at first, the product is able to meet the demands of consumers and not necessarily but after put the products into the market will be able to get customer timely feedback, and according to the perfect of the feedback information better, to optimize and upgrade the product, in the end produced meet consumer demand, let consumers more satisfied and satisfied with the products. This is the core of BP neural network.

The development of BP neural network has a long history. It was developed and proposed by foreign scholars in 1986, and it has been used in China for a long time. It is studied and analyzed through a back propagation method of error. With the development of social economy, people's demand for stock price prediction is increasing. Neural network has the dual characteristics of biology and mathematics, so it has great potential in stock market prediction, financial time series analysis and risk assessment. The relation of model response of convolutional neural network can directly explain the change of stock price. It is one of the important ways of predicting the stock price in China. Reasonable use of this technology can promote the research and analysis of the stock, improve the income of the Stockholders and promote the progress of the securities industry in China. At present, the research on stock price prediction using BP neural network is mostly based on the training and learning of stock historical data of a certain time series, and then using the trained network to forecast the experimental data. This idea is consistent with the traditional neural network algorithm.

4 Algorithm of BP Neural Network

The neural network is a network composed of input layer, hidden layer and output layer, as shown in Fig. 1: Data from the input layer is processed by linear transformation of weight value and bias item, and then through the activation layer, the output of the hidden layer is obtained, that is, the input of the next layer. Between the hidden layer and the output layer is a linear transformation of the weight value and the bias term, and then through the activation layer, the output layer is obtained.

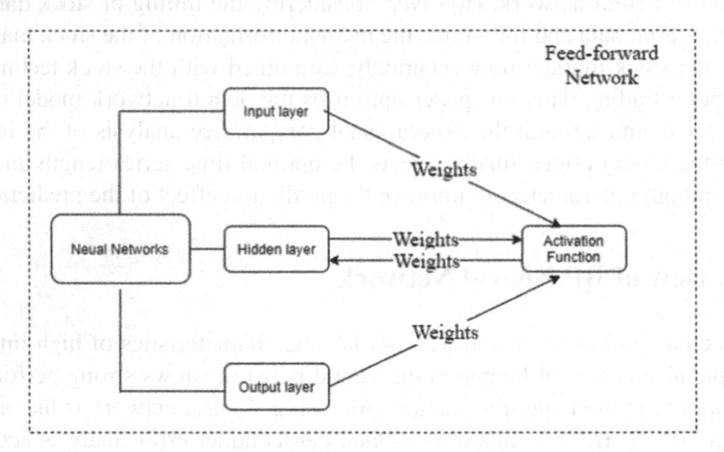

Fig. 1. Feed-forward network.

The so-called BP network algorithm is a statistical pattern in the opposite direction as shown in Fig. 2. Each layer contains one or more neurons, that is, each unit in the lower layer is fully connected with each unit in the upper layer, and there is no connection between the neurons in the same layer, there is no feedback from each neuron; information passes through the hidden layer from the input layer, and then an output value is obtained at the output layer; the inputs of each neuron in the network are aggregated in a weighted sum form, then the corresponding output is produced by the activation function, and the single hidden layer network with linear output is most often used in theoretical research and practical application. Accurate measurement of results requires a reasonable budget. The algorithm takes the following steps:

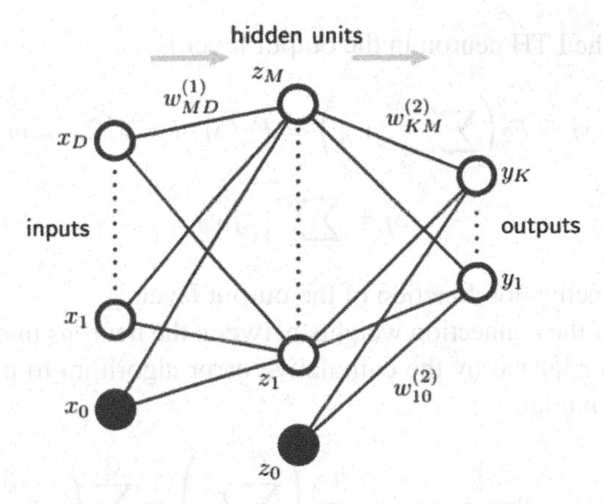

Fig. 2. Struction of BP neural network

1) Check the network's system mode and get in place before starting operations;
2) Input the calculated data accurately and calculate the results of each stage separately;
3) Check the results and figure out the defects;
4) Process the error part in stages;
5) Evaluate the values at each stage;
6) Final look at the predicted result shows that the calculation is reasonable and accepted if it falls within the required range; otherwise, it shows that the predicted result does not meet the requirements and requires a return to step 2.

The learning stage of BP neural network is divided into two parts: the first part is the forward propagation of the input signal: the input signal enters the input layer, gets to the output layer after being processed by the hidden layer, and outputs the final result. The second part is the reverse propagation of the error signal: in the process of the forward propagation of the input signal, if the output result does not reach the expected output result, the calculated error signal will be propagated in the opposite direction of the forward propagation, from the output layer through the hidden layer, and finally back to the input layer, and the output error to the propagation path of each neuron, and then each neuron according to their respective error to modify the network connection weights and thresholds, finally, the total error of the whole neural network is minimized.

Suppose the input layer of the model is $X = [x_1,x_2,\ldots x_n]^T$, the hidden layer is $Z = [z_1,z_2, \ldots z_n]^T$, the output layer is $Y = [y_1,y_2, \ldots y_n]^T$, the weight of the input layer and the hidden layer is W_{ij}, the connection weight between the hidden layer and the output layer is w_{jk}, the hidden layer has a threshold value of $b = [b_1,b_2, \ldots b_p]^T$, and the actual final output is $d = [d_1,d_2, \ldots d_m]^T$.

The output of the hidden layer J neuron is:

$$z_j = F_1\left(\sum\nolimits_{i-1}^{n} x_i w_{ij} + b_j\right) = F_1(S_j)\ j = 1, 2, 3, \ldots p \tag{1}$$

$$S_j = \sum\nolimits_{i-1}^{n} x_i w_{ij} + b_j \tag{2}$$

Where F1 (.) is the activation function of the hidden layer.

The output of the LTH neuron in the output layer is

$$y_l = F_2\left(\sum\nolimits_{j-1}^{n} z_j w_{jl}\right) = F_2(S_l)\ l = 1, 2, \ldots m \tag{3}$$

$$S_l = \sum\nolimits_{j-1}^{n} z_j w_{jl} \tag{4}$$

Where F2(.) is the activation function of the output layer.

The updating of the connection weights between the neurons in the output layer and the hidden layer is adjusted by the cumulative error algorithm to minimize the global total error E of the model:

$$\Delta w_{jk} = -\eta \frac{\partial E}{\partial w_{jk}} = -\eta \frac{\partial}{\partial w_{jk}} \left(\sum_{q-1}^{Q} E_q \right) = \sum_{q-1}^{Q} \left(-\eta \frac{\partial E_q}{\partial w_{jk}} \right) \tag{5}$$

Where η is the learning rate.

$$\frac{\partial E_q}{\partial w_{jk}} = \frac{\partial E_q}{\partial S_k} \cdot \frac{\partial S_k}{\partial w_{jk}} \tag{6}$$

Among them:

$$\frac{\partial E_q}{\partial S_k} = \frac{\partial E_q}{\partial y_k} \cdot \frac{\partial y_k}{\partial S_k} \tag{7}$$

$$\frac{\partial E_q}{\partial y_k} = \frac{\partial}{\partial y_k} \left[\frac{1}{m} \sum_{k=1}^{m} (y_k^q - d_k^q)^2 \right] = \frac{2}{m} \sum_{k=1}^{m} (y_k^q - d_k^q) \tag{8}$$

$$\frac{\partial y_k}{\partial S_k} = F_2(S_k) \tag{9}$$

Therefore, the weight update formula of the connection between the output layer and the hidden layer is;

$$\Delta w_{jk} = \sum_{q-1}^{Q} \sum_{j=1}^{p} \sum_{k-1}^{m} \left[-\eta F_2(S_k) \cdot \frac{2}{m} z_j (y_k^q - d_k^q) \right] \tag{10}$$

In the same way, W_{ij} is adjusted by updating the connection weights between the neurons in the input layer and the hidden layer:

$$\Delta w_{ij} = \sum_{q=1}^{Q} \sum_{i=1}^{n} \sum_{j=1}^{p} \sum_{k=1}^{m} -\eta x_i w_{jk} F_1'(S_j) F_2'(S_k) \frac{2}{m} (y_k^q - d_k^q) \tag{11}$$

5 Research and Analysis of BP Neural Network on Stock Price Forecasting

5.1 Clear the Way of Thinking in Stock Research

At the beginning, on the one hand, all the numbers and data for the collation and planning, for different data in different ways, is the data into a changing state. Using the scientific BP neural network calculation model, first of all, the opening and closing prices of the previous day, including the highest and lowest prices of stocks on one day, are effectively analyzed, and today's opening prices are regarded as the level of output, the change of its price level is considered as the summation of the hidden layer and the error. The two sets of data mentioned above are regarded as qualified samples, and a few of them are taken as the reference data for prediction. Secondly, the calculation data of BP neural network are analyzed reasonably by using a reasonable computer editing program, among them are the calculation model of the function, the selection of the data, the selection of the mode, and so on, so as to make reasonable analysis and calculation of the predicted data and the actual real data by using the network data of the computer, the different data of different stocks are compared and predicted, and the results are compared again. Finally, the countermeasure error is obtained and the forecast price of stocks is grasped effectively.

5.2 Analyze the Specific Model of the Stock Forecast

The neural network is made up of a number of neurons, creating a neural network. The BP neural network of stock is generally divided into the level of output system and input. BP algorithm has become the first choice of learning algorithm in neural network because of its advantages of small computation, simplicity, easy operation and strong parallelism. The corresponding BP neural network has also become the most widely used and most mature artificial neural network. Its self-learning ability, strong non-linear parallel processing ability and fault-tolerant ability have attracted more and more attention. In theory, three-layer BP neural network can approximate any non-linear continuous function with any precision. However, BP algorithm adjusts the weight and threshold of network connection by calculating the gradient of error. This gradient descent algorithm has some shortcomings. For example, the convergence speed is slow, and there is a local minimum point, and so on.

According to the model of the stock price forecast results of the budget, in the actual operation process, so that the hidden layers of the BP neural network model used in this paper are composed of multi-layer fully connected layers. The nonlinear activation functions used between layers are usually Sigmoid, Tanh, Linear rectification function (Rcelu), and a series of varieties of rcelu (Leakey R Elu, PR eLU, CR Elu, etc.). The operation mode of the fully connected layer is matrix operation, see Fig. 3. The operation mode of the i-connected layer can be expressed as: so that at the end of the effective data, data error comparison, the effective difference is calculated to improve the fast and efficient application of BP neural network in the stock industry. The output elements of each layer decrease gradually, and the output and output of the full connection layer are too large, All the elements are related, which can easily lead to over-fitting. Therefore, the Relu function is added between each layer as a non-linear activation unit to prevent over-fitting.

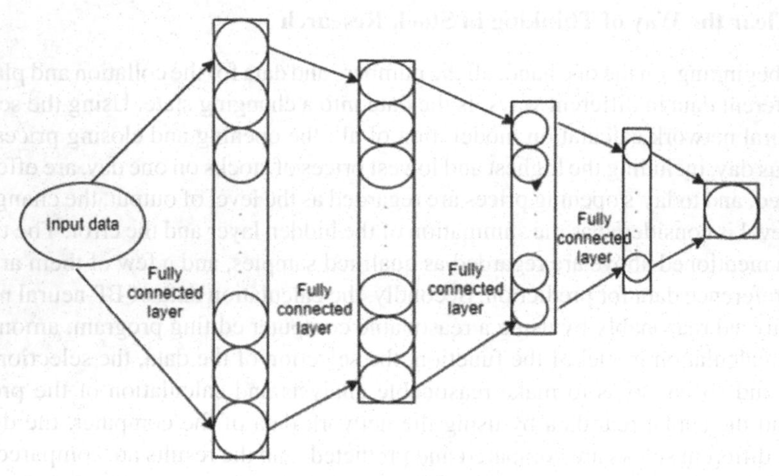

Fig. 3. Network model

6 Application of Concrete Model

6.1 Application Background and Implementation of the Model

People in the industry all know that the stock changes by a large margin and is not an ordinary linear system. Over time, the stock also shows a rise and a fall, so it is not easy to record the stock data before analyzing it, but the BP neural network system studied in this paper can well record this phenomenon. When a stockholder is speculating, from time to time a comparison is made between the previous changes in the stock. This neural network can preserve the changes in the stock and predict the subsequent changes in the stock. For stockholder, it can be said that the effect is very big, convenient for them at the same time, but also to improve China's relevant BP neural network in the stock price determination budget, and ultimately promote the development of China's securities industry.

6.2 Stock Data Sources and the Specific Operation of Normalization

Below a company in previous years on the stock changes, do a simple statistics and summarized as follows: relevant knowledge know about the transfer function of the s-curve changes, to calculate the effective value of each output layer of BP neural network. In the analysis of data, the neuron can realize the data gain mode, that is to say, the change between the input value and the function value is inversely proportional, for example, the smaller the data input level, the bigger the corresponding function data change value; On the other hand, the larger the level of data input, the smaller the data of its function.

However, one of the important aspects is that the level input value should not be too large, which is equivalent to locking the grid neural system, to make it function in a closed state. So that the neurons in the neural grid lost their original function, the stock price prediction has no significance. It can be seen that the final processed data is very important to the relevant understanding. The purpose of induction is to learn the changing trend of data from the truth so that it can be used reasonably in the prediction of stocks. After industry investigation, the following formula can be used for data induction:

$$E = \frac{1}{m}\sum_{k=1}^{m}(y_k - d_k)^2 \tag{12}$$

6.3 The Following on the Selection of a Region's Stock Situation for Example Analysis

As can be seen from the graph below, the change of the two sets of data is clear at a glance. The real price, the predicted price and the corresponding error are all calculated clearly. The fitted data are all bigger than 0.996 that means that the calculations are within a reasonable range of what is required. The source of this data is a partial phase of the actual survey results in an area, as shown in the figure below (Table 1).

Table 1. Prediction results

In 2018	Guizhou maotai			vanke		
	The real price	To predict the price	The relative error	The real price	To predict the price	The relative error
09–13	643	646.22	0.00500	23.31	22.865	0.01910
09–14	638	630.42	0.01188	23.13	22.995	0.00584
09–15	633.98	632.05	0.00304	23.04	23.201	0.00698
09–18	645	642.55	0.00380	22.65	22.837	0.00826
09–19	643	646.45	0.00537	23.03	22.887	0.00621
09–20	666.58	666.83	0.00037	24.01	23.315	0.02986
09–21	685.2	685.77	0.00083	24.2	23.850	0.01447
09–22	682	682.62	0.00091	24.8	25.627	0.03335
09–26	695	697.64	0.00380	24.66	24.287	0.01510
09–27	702.1	702.93	0.00118	24.63	24.409	0.00899

From the figures, we can see that under the prediction of BP neural network model, we can get the change of the two sets of data. We can see intuitively that the difference between the opening price and the actual price of the stock is not very big, that means the margin of error is very small. This can prove that the proper use of the BP neural network calculation model in the stock market can reduce the error between the opening price and the actual price, and improve the accuracy of predicting the stock price by relevant personnel in the stock market, the most important thing is to improve the application of BP neural network in stock price forecasting, to promote the stable and rapid development of stock market in China's securities industry, and to strengthen the stability of China's economy.

7 Conclusion

Neural networks have the ability of generalization. After neural networks are trained, the weights of neural networks store the nonlinear mapping relationship between input and output, the network can also map correctly from input space to output space. This ability is known as Generalizability, or generalizability on data outside the training set. Performance outside of the training set is an important measure of the quality of the network. A computer stores information by concentrating it in a specific area. It stores information in isolation. But the neural network stores the information distributed in the whole network, which makes the neural network model suitable to find the correct rule from the distorted or missing data. It can be seen from the above that BP neural network plays a very important role in stock price prediction. First of all, there are three levels of interaction in application, including input level, output level and hidden level. Secondly, the specific scientific calculation methods of BP neural network are obtained, among which the two most important methods are forward propagation calculation method and back propagation calculation method. At the end of the paper illustrates the BP neural network shares under the gap between real value and test value, obtained the corresponding error, and carries on the scientific analysis, it can be seen that using the level of the BP neural network system to the price of the stock measurement accuracy is very high, to the preservation of data and computation provides a more accurate use of fitting, it can also encourage people to share prices of a detailed grasp, make them in stock, reasonable and effective use of BP neural network technology to make accurately predict the stock price, save time at the same time, encourage their own economic benefits, It also develops the specific application of BP neural network in China and promotes the development of scientific and technological technology, so as to improve the progress of China's securities industry and the development of social economy.

References

1. LAPEDESA, Farberr genetic data base analysis with neyralnets, IEEE conference on Neural Information Processing Systems-Natural and synthetic (1987)
2. Baba, N., Kozaki, M.: An intelligent forecasting system of stock price using neural networks. In: Neural Networks, 1992 IJCNN, International Joint Conference on, vol. 1, pp. 371–377 (1992)
3. Ramazan, G.: Non-linear prediction of security returns with moving average rules. J. Forecast. **V01**(15), 165–174 (1998)
4. Lee Raymond, S.T.: iJADE Stock Advisor—an Intelligent Agent-Based Stock Prediction System Using the Hybrid RBF Recurrent Network, Fuzzy-Neuron Approach to Agent Applications: From the AI Perspective to Modern Ontology, pp. 231–253 (2006)

5. Jacinto, R.M.: Investment decisions with financial constraints. Evidence from Spanish firms. Quant. Financ. **14**(6), 1079–1095 (2014)
6. Li, H.: Research on improvement and Application of BP Neural Network Algorithm, Ph.D. dissertation, Chongqing: Chongqing Normal University (2008)
7. Wang, S.A.: Research on the Application of BP Neural Network in Stock Prediction, Ph.D. dissertation, Changsha: Central South University (2008)
8. Ye, C.J., Zhu, X.L.: Application of BP Neural Network in Stock Price Prediction. Sci. Technol. Wind **3**, 7980 (2013)
9. Cui, J.F., Li, X.X.: Stock price prediction: comparison between GARCH model and BP neural network model. Stat. Decis. **6**, 21–22 (2004). (in Chinese)
10. Zhang, G.S., Zhang, X.D.: Research on ArMA-garch stock price prediction model based on gradient factor. J. Shanxi Univ. (Philos. Soc. Sci. Ed.) **39**(1), 115–122 (2016)
11. Zhai, Z.R., Bai, Y.P.: Application of MATLAB autoregressive moving average model (ARMA) in stock prediction. J. Shanxi Datong Univ. (Nat. Sci.) **26**(6), 5–7 (2010)
12. Wang, C., Zhou, W., He, J.: Deep recursive neural network method for automatic music generation. Miniat. Microcomput. Ser. **38**(10), 2412–2416 (2017)
13. Lin, S.C.: Research and Application of Biaxial LSTM Neural Network and Chaos Theory in Music Generation System, Ph.D. dissertation, South China University of Technology (2017)
14. Xue, H.Y.: Music Score Modeling and Generation Based on Recurrent Neural Network, Ph.D. dissertation, Tianjin University (2017)
15. Jiang, J.M., Liu, C.Q.: Index futures price prediction based on principal component analysis and BP neural network. Mod. Bus. **18**, 176–177 (2015)
16. Wang, B., Kong, W., Li, W., Xiong, N.N.: A dual-chaining watermark scheme for data integrity protection in internet of things. Comput. Mater. Continua **58**(3), 679–695 (2019)
17. Liu, S., Zhang, W.: Application of the fuzzy neural network algorithm in the exploration of the agricultural products e-commerce path. Intell. Autom. Soft Comput. **26**(3), 569–575 (2020)

Trust Computing Model of Online Transaction Based on Social Network and Mobile Agent

Xiaoliang Liu(✉) and Weijin Jiang

Hunan University of Technology and Business, Changsha 410205, China

Abstract. In view of the deficiency of the existing trust model based on recommendation, on the basis of referring to the interpersonal trust research of sociology, combined with the specific characteristics of C2C online auction, using the collaborative filtering technology of mobile Agent system, a personalized trust computing model based on recommendation is established. The model is based on the idea of social network and forms the virtual social network of target users according to the historical transaction data. By examining the trust factors such as trust situation and trust tendency, the trust path is constructed and the trust degree of potential trading partners is calculated. The empirical analysis of the model is carried out by using eBay data, and compared with the beth model, the results show that the trust calculation model is simple and effective, and has good engineering feasibility.

Keywords: Online auction · Social network · Network trust crisis · Trust computing model

1 Introduction

With the development of mobile Internet, online auction has become a very popular transaction mode, which not only changes the traditional transaction mode of human society, but also breaks down the barriers of traditional auction activities in China and regions. It promotes the auction activity to become a transaction activity that ordinary people can participate in, and is favored by more and more consumers [1]. However, while consumers are convenient to participate in the auction, they are also frequently subjected to fraud, and the problem of integrity in online auctions and the crisis of trust on the Internet are becoming increasingly serious [2].

In view of the deficiency of the existing trust model based on recommendation, on the basis of referring to the interpersonal trust research of sociology, combined with the specific characteristics of C2C online auction, using mobile Agent collaborative filtering technology, a personalized trust computing model based on recommendation is constructed, and the mathematical expression and implementation method of the model are given. This paper makes an empirical analysis on the model using Ebay data [3]. The analysis shows that the model proposed in this study is simple and effective, and has good engineering feasibility.

X. Sun et al. (Eds.): ICAIS 2021, CCIS 1422, pp. 674–684, 2021.
https://doi.org/10.1007/978-3-030-78615-1_59

2 Related Work

Beth et al. first put forward the concept and method of trust quantification, which divides trust into direct trust and recommended trust, calculates the probability of the entity completing the task according to the positive and negative experience, and gives the method of trust merging. However, the deficiency of this model is that the simple modeling of subjective trust with probability model actually equates the subjectivity and uncertainty of trust with randomness [4]. When synthesizing multiple recommendation trusts, the method of taking the mean value is simply adopted, so it can not reflect the real situation of the trust relationship. The trust degree evaluation model proposed by Rahaman and others also divides the trust relationship into direct trust and recommendation trust, and gives the transmission protocol and calculation formula of the trust degree, but it does not give the comprehensive calculation formula of the trust degree, so the derivation of the recommendation trust can not be realized. Other scholars also use different methods to measure and deduce trust relations, such as Bayesian networks, fuzzy sets, evidence theory, social network analysis, probability and so on [5].

Yu et al. proposed to use evidence theory to describe the trust relationship between users, and use the merging rules of belief function in evidence theory to merge the trust degree in the trust path to solve the trust transfer problem. However, in the social network, the degree of trust of the recommender is different from the trust function of evidence theory. If the results of this method are put into consideration in the environment of recommendation trust transfer and merger, it will be found that there is a big gap with intuition and does not accord with the actual situation in the society [6].

To sum up, in the recommendation-based trust, the recommendation trust merging on the trust path is the most important work. However, the trust merging method used in the existing recommendation-based trust model has some defects, such as the weight setting is too subjective, it is difficult to reflect the trust relationship in reality, and the negative impact of malicious recommendation can not be eliminated.

3 Personalized Trust Model Based on Recommendation

3.1 Model Thought

In the mobile network environment, a large number of traders form a virtual social network. In the social network, the trust relationship between trading users is the core of the interpersonal relationship between the two parties, and the trust between two strange trading users often depends on the recommendations of other trading users. At the same time, as a recommender user, its credibility also determines the credibility of the trading users he recommends [7]. In fact, this interdependent trust relationship forms a so-called trust network (Web of trust). In such a trust network, the trust degree of any trading user is not absolutely reliable, but it can be used as a basis for other transaction users to determine their interaçtive lines and judge the trust degree of potential trading partners.

In the online auction environment, a large number of traders form a virtual social network. In the social network, the trust relationship between transaction users is the core of the interpersonal relationship between the two parties. The trust degree between two strange transaction users often depends on the recommendation of other transaction users. At the same time, as a recommendation user, its credibility also determines the credibility of the trading users he recommends. In fact, this interdependent trust relationship forms a so-called trust network. In such a trust network, the trust degree of any trading user is not absolutely reliable, but it can be used as a basis for other trading users to determine their interactive behavior and judge the trust degree of potential trading partners [8].

In this model, trust situation similarity and incumbency similarity are the main indicators to describe the credibility of recommended users, which play an important role in building a reliable trust path and improving the robustness of the trust path [9].

3.2 Similarity Calculation of Trust Situation

Definition of Similarity in Trust Situation

Given the time domain *t*,after the user *i* has the trust in the user *j*, a successful auction of the item is carried out with it. we call the above situation the trust situation TS (Trust Scenario), and write it down as $TS\ (i, j)$.

The degree of trust of the subject to the same subject is different in different trust situations. For example, in online auctions, if user *j* has a good reputation in the auction field of collectible items, user *i* will have a higher degree of trust in user *j* when trading ancient coins with user *j*; when user *j* sells books to user *i*, user *i* has less trust in user *j* than when trading ancient coins. From this point of view, there is a strong relationship between the trust situation of the subject to the object and the type of auction items.

According to the eBay data we collected, this paper uses the auction item type similarity to describe the trust situation similarity. The main reason is that there are many types of auction items in C2C online auction compared with B2C websites. There are a large number of new auction items on the shelves every day, and there is a problem of heterogeneous auction items. Take the eBay website as an example. At present, there are more than 4000 auction items in eBay, and 250000 new auction items are put on the shelves every day. In such a large auction item, the possibility of different users successfully auctioning the same auction item is very small, that is, the probability of auctioning homogeneous items is very small. For this reason, this paper chooses the similarity of auction items rather than directly choosing the similarity of auction items to describe the similarity of trust situation [10].

Calculation of Similarity of Auction Item Types

(1) The basic idea of calculating the similarity of auction items

Compared with traditional auctions, the types of items auctioned on the website are more diversified. Auction websites usually organize the list of auction items by type or even subtypes to form a type tree of auction items.

This type tree with strict dependency describes the specific type to which a given auction item belongs. For each given auction item, the direct type to which it belongs is called the minimum class. By searching the specific position of the smallest class of the item in the type tree, the type path from the total type, subtype and subtype of the auction item to the smallest class is formed.

For two different auction items, the similarity of their corresponding types can be measured by the distance from the lowest type of the two auction items to the smallest class of the two auction items. The smaller the distance from the lowest level of common type to the smallest class, the greater the similarity between the two auction items; on the contrary, the smaller it is.

(2) The definition and representation of the similarity of auction items

Let the set of types of auction items be E, E = {e_1, e_2, …, e_s}, Where s is the total number of auction items in the auction website M, and I_i is the collection of historical auction items by user i. $I_{ij} \in I_i$, I_i is the auction item traded by user i and user j.

It is said that user I historical auction item I_{ij} A directly belongs to the smallest category of type I_{ij}. Write down as $min\_category(I_{ij})$, $Category\ (I_{ij})$ by the user I history auction items I_{ij} belongs to each type (the smallest class $min\_category(I_{ij})$, the parent of the smallest class, the parent of the parent class, …. Until the generic class) is made up.

Set $category(I_{ij}) = \{p_0, p_1, \ldots p_n\}$, of which $P_n = min\_category(I_{ij})$, Represents the smallest class of I_{ij}. p_{n-1} is the parent of p_n, And so on, until the general class $p_0 = \perp$, In this way, the I_{ij} type collection of auction items for user *i*, the element P_i in category, According to the upper and lower membership of the class, a type path with the length of n from the general class to the smallest class is formed. Path, Path is an ordered set $Path = < p_0, p_1, \ldots, p_i, \ldots p_n >$ composed of all kinds. Pi P_i is the parent class of P_{i+1} in the collection. User *i*'s total interest in type $category(I_{ij})$ auction item I_{ij} Iij is $\pi(category(I_{ij}))$

$$\pi\left(category(I_{ij})\right) = \frac{|I_{ij}|s}{|I_i|category(I_{ij})} \tag{1}$$

Where s is the total number of auction item types in the auction website M; $|I_i|$ is the total number of historical auction items for user *i*; The number of times for user *i* to participate in type $min\ category(I_{ij})$ auction items, The interest degree function π represents the interest degree of user *i* to the auction items of $category(I_{ij})$ type, which is written as π for convenience.

It is worth explaining that π is the total degree of interest, which means that the degree of interest π is not completely assigned to the minimum class $min\ category\ (I_{ij})$, of I_{ij}, but to each type node on the type path Path according to the distance between each type and the minimum class. Therefore, on the type path Path, the degree of interest of the smallest class is the greatest, followed by the parent class of the smallest class, and the degree of interest decreases successively along the path Path. According to the above idea, the interest degree π of user I to the type $category\ (I_{ij})$ of the historical auction item I_{ij} should be equal to the sum of the interest degrees of all types on the type path Path. That is to say, for the type path $Path = < p_0, p_1, \ldots, p_k, \ldots, p_n >, \pi = \sum_{k=0}^{n-1} v(p_k)$, where $v(p_k)$ is the user's interest in competitive auction items. The following defines the P_k interest function $v(p_k)$ for type competition.

Given type path Path, $Path = < p_0, p_1, \ldots p_m, \ldots p_n >$, The interest $v\ (p_m)$ of class P_m on the type path Path is defined as:

$$v(p_m) = \frac{v(p_{m+1})}{brother(p_{m+1}) + 1}, m \in [0, n-1] \tag{2}$$

Where $v\ (p_{m+1})$ is the degree of interest of p_{m+1}, a subclass of class p_m, $brother(p_{m+1})$ is the number of brothers of subclass p_{m+1}. From the above formula, we can see that the degree of interest of class pm depends not only on the degree of interest $v\ (p_{m+1})$ of its subclass p_{m+1}, but also on the number of brothers of subclass p_{m+1}. The larger the number of brothers of the subclass p_{m+1}, $brother(p_{m+1})$ (that is, the more subclasses of the parent class), the smaller the probability of the user choosing the auction item of type p_{m+1} and the less interest directly assigned to the class p_m.

(3) Calculation of trust situation similarity

According to the user's interest in each type set $category\left(I_{ij}\right)$ of I_{ij}, we define the user interest vector and the user interest vector. Let I_i be the set of historical auction items of user i, and $I_{ij} \in I_i$ the auction items traded by user i and user j. The set of types corresponding to I_{ij} is $category\left(I_{ij}\right)$. $category\left(I_{ij}\right) = \{p_0, p_1, \ldots, p_n\}$, Where $p_n = min\ category\left(I_{ij}\right)$ represents the smallest class of I_{ij}. Then the interest vector $INT(i)$ of user I participating in bidding for auction items of type $min\ category\left(I_{ij}\right)$ is: $INT(i) = \lfloor p_0, p_1, \ldots, p_n \rfloor$.

Let the interest vector of user i is $INT(i) = \lfloor p_0, p_1, \ldots, p_n \rfloor$,$v\left(p_j\right)$ indicates that user i is interested in p_j type auction items. Then the interest vector INTV (i) $INTv\ (i)$ of user i is defined as $INTv(i) = \lfloor v(p_0),\ v(p_1),\ \ldots,\ v(p_n) \rfloor$.

Users i, j, k, Under the trust situation $TS\ (i,\ j)$, user i trades auction items I_{ij} with user j, Under the trust situation $TS\ (k, j)$, User k trades auction items I_{kj} with user j, The corresponding sets of types for I_{ij} and I_{kj} are $category\left(I_{ij}\right) = \{p_0, p_1, \ldots, p_n\}$, $category\left(I_{kj}\right) = \{q_0, q_1, \ldots q_m\}$, Among them, $p_n = min\ category\left(I_{ij}\right)$, $q_m = min\ category\left(I_{ij}\right)$, $category\left(I_{ij},\ I_{kj}\right)$ is a collection of common types, Represents the intersection of a collection of types. Namely $category\left(I_{ij},\ I_{kj}\right) = category\left(I_{ij}\right) \cap category\left(I_{ij}\right)$, The interest vector of users i and k is $INTv(i) = [v(p_0), v(p_1), \ldots, v(p_n)]$, $INTv(k) = \lfloor v(q_0), v(q_1), \ldots, v(q_n) \rfloor$, Then the similar sim_TS of trust situation $TS(i, j)$ and $TS(k,\ j)$ is:

$$sim\_TS(TS(i,\ j),\ TS(k,\ j)) = 1 - \sqrt{\frac{\sum_{p \in category(I_{ij}, I_{kj})} [v_1(p) - v_j(p)^2]}{\left|category(I_{ij}, I_{kj})\right|}} \tag{3}$$

Where, $v_i\ (p)$ is the degree $\left|category(I_{ij}, I_{ky})\right|$ of interest of user i in p-type auction items, $v_k\ (p)$ is the basis of user k's interest in p-type auction items as a set of common types, $sim\_TS(TS(i,j), TS(k,j))$ the bigger. The more similar the trust situation $TS\ (i,\ j)$ and $TS(k,j)$ is. $sim\_TS(TS(i,j),\ TS(k,j))$ the smaller, The smaller the similarity of the trust situation $TS(i,j)$, $TS(k,j)$.

3.3 Trust Model Description

Definition and Representation

Given the time domain t, user i has had a historical transaction with user j, it is said that user $i(j)$ has direct trust in user $j(i)$. Direct trust τ_i^j is expressed as:

$$\tau_i^j \frac{\sum_{k=1}^{num\_trans(i,j)} p^{num\_trans(i,j)-k} \bar{f}^k(i,j)}{num\_trans(i,j)}, \tau_i^j \in [0,1] \quad (4)$$

In the formula, $num\_trans(i,j)$ indicates that in the time domain t, the number of historical transactions between user i and user j; p $(0 < p < 1)$ is a time attenuation factor, indicating that the more recent the transaction, the greater the corresponding feedback score weight. The introduction of p is aimed at strengthening the prevention of fraud on user i after user j accumulates a certain historical reputation. $\bar{f}^k(i,j) = \frac{\sum_{c_1 \in C} f_{c_1}^k(i,j)}{|C|}$ indicates the average reputation feedback score of user j when user i and user j make the k-th historical transaction.

Given the user i and its potential transaction user j, let $TS(i,j)$ be the trust situation of user i, j, then for user k, the trusted user set $TN(k)$ (Trustable Neighbor) under the trust situation $TS(i,j)$ is defined as:

$$TN(k) = \{x \big| \forall x \in N(k), sim\_DT(k,x)\tau_i^k\, sim\_TS\lfloor TS(k,x), TS(i,j)\rfloor \geq \mu\} \quad (5)$$

Where $x \in N(k)$, denotes the historical trading partner τ_i^k of user k indicates the direct trust degree of user i to user k; $sim\_DT(k,x)$ represents the similarity of trust tendency of user i and k; $sim\_TS\lfloor TS(i,j), TS(k,x)\rfloor$ represents the similarity of trust situation $TS(i,j)$ and $TS(k,x)$; μ represents the minimum trust threshold (given by the user).

Given the time domain t, the set of user i and its trading partners $N(i)$, sets user j as the unfamiliar potential trading partner of user i, that is, $j \in N(i)$. If there is a trust path $X = X(i, o_1, o_2, \ldots, o_{n-1}, j)$ between user i and user j. where o_k is the k-th recommended user on the trust path X, it is said that the user i (j) has indirect trust $\tilde{\tau}_i^j (\tilde{\tau}_j^i)$ to the user j (i) under the trust path X.

Indirect Trust Computing

In the trust model of Belth and Yu, there is no limit on the length of the trust path or only a definable length threshold length > 0 is given. Therefore, when the trust path length continues to grow, the computational complexity of enumerating trust paths between given users will increase exponentially, and the efficiency of indirect trust calculation will be greatly reduced; in addition, these models use simple multiplication and other forms to calculate indirect trust between users. Taking the trust model of Yu et al. as an example, for the trust path $x = x(i, o_1, o_2, \ldots, o_k, \ldots, o_{m-1}, j)$ when calculating the indirect trust $\tilde{\tau}_i^j$ of user i to user j, the model simply multiplies the direct trust of neighboring users on the trust path x, that is, $\tilde{\tau}_i^j = \tau_i^{o_1} \tau_{o_1}^{o_2} \ldots \tau_{o_{k-1}}^{o_k} \ldots \tau_{n-1}^j$. This simple indirect trust calculation method ignores the influence of the trust degree (credibility) of the presenter and the trust situation deviation between the presenter and the trustor on the indirect trust calculation. as a result, this simple indirect trust calculation method is difficult to overcome the problems such as malicious referrals, vilification of others' reputation and so on.

In order to solve the disadvantages of the above model, this paper makes three improvements to the indirect trust model.

(1) according to the six-dimensional space theory proposed by social psychologist Stanley Milgram, (Six Degrees of Separation), sets the maximum length of trust path to 6. In this way, on the one hand, the traversal time of the trust path is reduced, and the computational efficiency is improved; on the other hand, by drawing lessons from the existing sociological research results, the interpretability of the sociological theory of the trust model is enhanced, and the effectiveness of the model is proposed.
(2) In the calculation of indirect trust, the credibility of the recommendation is measured by calculating the similarity of the trust tendency between the user and the recommendation (neighboring references), so as to prevent the malicious recommendation from attacking the trust recommendation and enhance the robustness of the indirect trust computing model.
(3) By calculating the similarity of trust situation between users and references (neighboring references), the deviation of trust situation deviation to indirect trust calculation is reduced, and the effectiveness of indirect trust calculation model is improved.

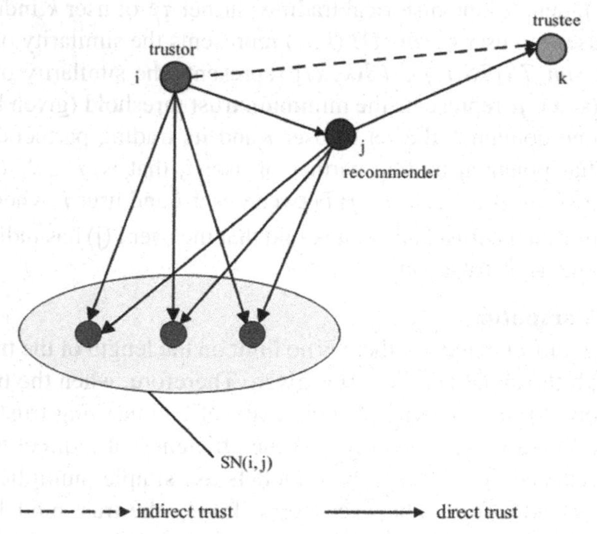

Fig. 1. Schematic diagram of indirect trust calculation

Taking Fig. 1 as an example, there is direct trust τ_i^j in user i, j and direct trust τ_j^k in users j and k. Through the recommendation of user j, users i and k have indirect trust.

When user i uses the recommendation information provided by user j, indirect trust can be evaluated.

4 Simulation Experiment

We use the eBay website data collected by David Lucking-Reiley to make an empirical analysis on the effectiveness of the recommendation trust model. David lucking-Reiley collected a total of 2439 successful transactions under the Coins: US cent type online auction from June 12, 1999 to August 4, 1999, including 691 buyers, 380 sellers, and a total of 1026 users. Table 3 gives a statistical description of the data set.

1026 users in the data set constitute a complex virtual social network G (Table 1).

Table 1. Statistical description

Variable	Number	Average number of trading partners	Average number of transactions	Average transaction price
Buyer	691	2.0434	3.530	134.078
Seller	380	3.7158	6.418	222.439
Buyer and seller	45		4.8	249.832
Total	1026			

Using the above eBay data, we make two experimental analyses of the proposed recommendation trust model with reference to the Beth trust model. The main contents are as follows: (1) in the case of single user, the validity of recommended trust model is analyzed. That is, first, a buyer I (with at least two historical trading partners) is randomly selected from the data set, and then a historical trading partner k of the buyer is selected as the recommendation to establish the social network G of the buyer I, and the other trading partners of the buyer I are taken as the target users; using the existing social network G, the recommendation trust degree of I to each target user is calculated. Finally, the calculated trust degree is compared with the actual trust score to analyze the effectiveness of the model. (2) in the multi-user case, on the basis of grading the trust degree of buyers, a buyer is randomly selected from the buyer set of each level to form multiple given users. Then the recommended trust of these given users is calculated one by one, and the average error between the calculated trust degree and the actual trust score is analyzed to measure the effectiveness and applicability of the model.

In the empirical analysis, we assume that the longest trust path that users can accept is 6, that is, $\lambda = 6$, and set the minimum trust threshold $\mu = 0.4$. In addition, in order to compare the error between the calculation result of trust degree and the actual trust score, we map the eBay trust feedback score from the original $\{-1, 0.5, 1\}$ to $\{0.5, 0.5, 1\}$, that is, the user's negative feedback corresponds to 0, neutral feedback corresponds to 0.5, and positive feedback is still 1. Because the eBay trust feedback score is constant rather than a multi-dimensional vector, we change the formula for calculating trust tendency, which is expressed as follows:

$$\mathrm{sim}_{\overline{(i,j)}}DT = \begin{cases} 1 - |f(i,k) - f(j,k)|, & |SN(i,j)| = 1 \\ \dfrac{\sum_{k \in SN(i,j)} f(i,k) \cdot f(j,k)}{\sqrt{\left[\sum_{k \in SN(i,j)} f^2(i,k) \cdot \sum_{k \in SN(i,j)} f^2(j,k)\right]}} & \mathrm{SN}(i,j) > 1, \sum f^2(i,k) \neq 0 \ \sum f^2(j,k) \neq 0 \\ 1 - \dfrac{\sum_{v \in \{0,5,1\}}^{v \cdot N(v)}}{|SN(i,j)|}, & \mathrm{SN}(i,j) > 1, \sum f^2(i,k) = 0 \ or \ \sum f^2(j,k) = 0 \end{cases}$$

4.1 Single User Situation

We randomly selected one buyer from 691 buyers, that is, cobra76. During the period from June 12 to August 4, 1999, cobra76 mainly conducted seven transactions, with a total of six trading partners. We chose fidl1 as the presenter for cobra76 and the other five (sbush, labute, teapartycoin, jet900, wondercoin) as target users for cobra76. First of all, according to the historical transactions of cobra76, the auction item type tree of cobra76 is determined. User cobra76 mainly engaged in two types of transactions: Indian avatar coins and LincolnWheat coins from June 12, 1999 to August 4, 1999, which are secondary catalogs of these two types of items, and both contain three subdirectories (Table 2).

Table 2. Cobra76 transaction information

Buyer	Feedback	Seller	Price	Pos	Neg
Cobra76	1	fidll	28.00	29.00	0.00
Cobra76	– 1	labute	25.09	220.00	0.00
Cobra76	1	teapartycoin	47.00	433.00	0.00
Cobra76	0	wondercoin	21.50	494.00	0.00
Cobra76	1	sbush	24.99	551.00	2.00
Cobra76	1	Jet900	60.09	780.00	1.00
Cobra76	1	Jet900	177.50	780.00	1.00

Then the social network G of cobra76 is constructed by using fidl1. The social network has 53 users and 183 directed edges, among which sbush, labute, teapartycoin, jet900 and wondercoin are regarded as the target users of cobra76. Then, according to the above social network G, find out the trust path from cobra76 to each target user; finally, use the Beth trust model, the recommendation trust calculation proposed in this paper, and then compare the calculation results with the actual score. The calculation error is calculated, and the effectiveness of the two models is compared and analyzed (Fig. 2).

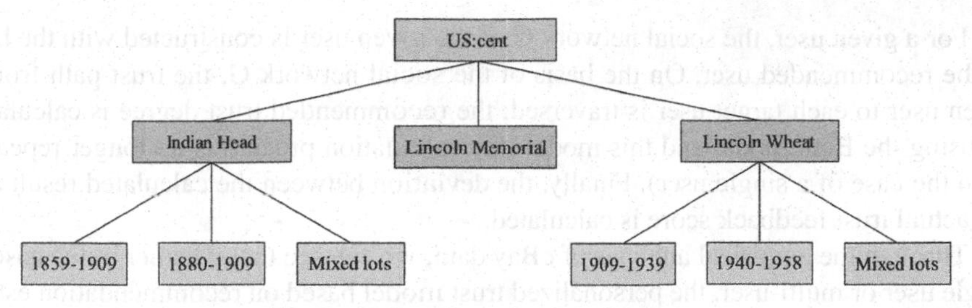

Fig. 2. Cobra76 auction item type tree

4.2 Multi-user Situation

From the collected eBay dataset, the buyers are randomly selected according to the credit rating (one buyer for each level); the recommendation trust of each user is calculated by using Beth model and this model, and the calculation error of the two models is analyzed in order to further verify the validity and applicability of the trust model proposed in this paper and further analyze the effectiveness and applicability of this model. With reference to the credit rating method of the eBay website, we divide the buyers in the empirical data into six levels according to their trust.

After dividing the credit rating of the buyer, we randomly select a user as a given user from the user data set of each credit rating, and then from the set of historical trading partners of that user, select the user with the highest trust as the recommended user of the given user, and the other users in the historical trading partner as the target user. Because users with credit rating 1 have fewer transactions, for users with credit grade 1, only two historical trading partners are selected as target users, while for users with other credit grades, five historical trading partners are selected as target users. In addition, when selecting the target user, the trading partner with more transactions should be selected as the target user to meet the accessibility of the given user to the target user and facilitate the verification of the validity of the model.

Table 3. Credit rating of buyers

Grade	Trust	Number of people
1	[−3,9]	32
2	[10,49]	78
3	[50,99]	125
4	[100,499]	312
5	[500,999]	83
6	[1000,4999]	49
	Total	691

For a given user, the social network G of the given user is constructed with the help of the recommended user. On the basis of the social network G, the trust path from a given user to each target user is traversed; the recommended trust degree is calculated by using the Beth model and this model (the calculation process is no longer repeated as in the case of a single user). Finally, the deviation between the calculated result and the actual trust feedback score is calculated.

Through the empirical analysis of eBay data, we can see that whether in the case of single user or multi-user, the personalized trust model based on recommendation established in this paper has good effectiveness, engineering feasibility and high accuracy, and can be well applied to the online auction environment. In view of the lack of personalization and subjectivity of trust evaluation in the current trust calculation methods in auction websites, the proposal of this model is of practical significance for improving and optimizing online trust calculation and evaluation methods.

5 Conclusion

In view of the lack of personalization of trust computing in auction websites, based on the idea of social network, this paper combines the specific characteristics of C2C online auction and uses collaborative filtering technology to construct a personalized trust computing model based on recommendation. Experimental analysis shows that, compared with the Beth model, the model proposed in this paper has better effectiveness and applicability.

References

1. Chen, C.C., Yao, J.Y.: What drives impulse buying behaviors in a mobile auction? the perspective of the stimulus-organism-response model. Telematics Inform. **35**(5), 1249–1262 (2018). Pergamon
2. Jiang, W.J., Liu, X.L., Liu, X.B., Wang, Y., Lv, S.J., et al.: A new behavior-assisted semantic recognition method for smart home. China Commun. **17**(6), 26–36 (2020)
3. Qasim, A., Asad, S.: Formal modelling of real-time self-adaptive multi-agent systems. Intell. Autom. Soft Comput. **25**(1), 49–63 (2019)
4. Siga, L., Mihm, M.: Information aggregation in competitive markets. Theor. Econ. **16**(1), 161–196 (2021)
5. Zhou, X., Zahirovic-Herbert, V., Gibler, K.M.: Price impacts of signalling in Chinese residential land auctions. J. Housing Built Environ. (2021). https://doi.org/10.1007/s10901-020-09806-9
6. Liu, C., Yang, S.: Research on agent-based economic decision model systems. Intell. Autom. Soft Comput. **26**(5), 1035–1046 (2020)
7. Kruger, W., Nygaard, I., Kitzing, L.: Counteracting market concentration in renewable energy auctions: lessons learned from South Africa. Energy Policy, **148**(PB), 111995 (2021)
8. Grigoroudis, E., Noel, L., Galariotis, E., Zopounidis, C.: An ordinal regression approach for analyzing consumer preferences in the art market. Eur. J. Oper. Res. **290**(2), 718733 (2021)
9. Ma, B., Zhou, Z., Bashir, M.F., Huang, Y.: A multi-attribute reverse auction model on margin bidding. Asia-Pacific J. Oper. Res. **37**(06), 2050032 (2020)
10. Xinhe, C., Wei, P., Wei, D., Hao, X.: Data-driven virtual power plant bidding package model and its application to virtual VCG auction-based real-time power market. ss Smart Grid **3**(5), 614–625 (2020)

An Adaptive Momentum Term Based Optimization Method for Image Restoration

Ge Ma, Junfang Lin, Zhifu Li(✉), and Zhijia Zhao

School of Mechanical and Electrical Engineering, Guangzhou University, Guangzhou 510006, Guangdong, China

Abstract. Linear inverse problems (LIP) arise in a wide range of applications in signal/image processing and statistics. In this paper, we propose an adaptive momentum term based optimization method for LIP in image restoration. Firstly, gradient descent with momentum is employed to search for iteration solutions, which can enlarge iterative steps in favorable directions and avoid some specific local optimal points caused by noise. Then, an adaptive parameter selection strategy based Barzilai-Borwein is selected to accelerate the iteration and increase robustness of algorithm. Thirdly, backtracking line search is used to attain a sufficient decrease, which can ensure monotonicity of the objective function. Experiment results demonstrate that the proposed method can obtain high-quality image restoration, especially, the robustness and number of iterations outperform its competitors.

Keywords: Image restoration · Adaptive · Momentum term · Gradient projection

1 Introduction

Linear inverse problems (LIP) arise in a wide range of applications in signal/image processing and statistics [1–4]. In many image processing applications, the observation y is possibly degraded with a linear degradation model.

$$\boldsymbol{y} = \boldsymbol{A}\boldsymbol{x} + \boldsymbol{n} \tag{1}$$

where the linear operator $\boldsymbol{A} \in R^{M \times N}$ represents the physical effect of the imaging system, normally is a blur operator. $\boldsymbol{n} \in R^N$ denotes the noise affecting the image acquisition.

The goal of image restoration is to estimate an approximation of x from its degraded observation y, which can be often formulated as the famous $l_2 - l_1$ problem (2) [5–7]

$$\min_{\mathbf{x}} \frac{1}{2}\|\boldsymbol{y} - \boldsymbol{A}\boldsymbol{x}\|_2^2 + \tau\|\boldsymbol{x}\|_1 \tag{2}$$

where $x \in \mathrm{R}^M$, $\boldsymbol{y} \in \mathrm{R}^M$, $\boldsymbol{A} \in R^{M \times N}(M << N)$. τ is a nonnegative parameter, which is called regularization parameter, balancing the role of fidelity term and regularization term.

X. Sun et al. (Eds.): ICAIS 2021, CCIS 1422, pp. 685–695, 2021.
https://doi.org/10.1007/978-3-030-78615-1_60

$l_2 - l_1$ problem was originally called basis pursuit denoising (BPDN) [6]. The presence of the l_1 norm encourages small components of $\boldsymbol{x}$ to become exactly zero, thus promoting sparse solutions [10], resulting in preserving edges and details information well. Because of this feature, it has been widely used in several image processing problems, such as image restoration, image reconstruction, compress sensing and so on. By setting $\boldsymbol{A} = \boldsymbol{R}\boldsymbol{W}$, where $\boldsymbol{R}$ is the observation operator, $\boldsymbol{W}$ contains a wavelet basis or a redundant dictionary, problem (2) becomes the applications of sparse reconstruction. Furthermore, medical image reconstruction is to obtain an accurate reconstruction $\boldsymbol{x}$ from random projection matrix $\boldsymbol{A}$.

Considering general convex optimization solving process, descent directions in gradient, Newton, Quai-Newton, conjugate gradient [8, 9] have been commonly studied. In basis pursuit de-noising (BPDN) [6], gradient projection for sparse reconstruction (GPSR) [10] and their improved methods, the authors usually reformulated form (2) to a convex bound-constrained quadratic program (BCQP) via slack variables. After that, BPDN employed the interior-point method to solve problem (2) in Newton's search direction, but suffered from expensive computational cost in Hessian matrix. Subsequently, a modified interior-point method called $l_1 - l_S$ which computed the search direction by conjugate gradient method has been proposed in [11]. Furthermore, M.A.T. Figueiredo [10] presented the well-known GPSR, whose search direction is essentially negative gradient. G. Chen [12] proposed a two-dimensional gradient projection algorithm for compressed sensing reconstruction. SpaRSA [13] flexibly tailored $l_2 - l_1$ problem as a sequence of subproblems and searched in gradient descent direction. Y.H. Xiao and H. Zhu [14] developed an efficient image de-blurring method for solving (2) by using conjugate gradient method.

However, for many practical applications, gradient descent is usually suffered from poor convergence and high sensitivity to local optima, especially with noisy data [15]. Similarly, CG may be unstable with respect to even small disturbance.

To improve the convergence and robustness of gradient descent solutions, while avoiding computational complexity, simple but effective modifications were proposed, namely, two-step iterative shrinkage/thresholding algorithms (TwIST) [16] and momentum based gradient projection method (M-GP) [17]. These two methods added the previous two iteration solutions in negative gradient direction, while traditional gradient descent depends only on the current iteration. This strategy can increase the rate of convergence greatly without leading to oscillation, and avoid the local optima due to noise.

In this paper, we propose an adaptive momentum term based optimization method for image restoration. Firstly, we introduce the formulation of image restoration and the advantages of gradient descent with momentum. Then, backtracking line search and adaptive parameter method are employed to M-GP, to ensure the monotonicity of objective function and increase the convergence. Finally, experiments show that the proposed adaptive method performs well in image restoration.

The rest of this paper is organized as follows. In Sect. 2, we describe the M-GP method in image restoration. Section 3 presents the proposed adaptive momentum term based optimization method. Numerical experiments are provided in Sect. 4, while the conclusions are reported in Sect. 5.

2 Gradient Descent with Momentum

2.1 Formulation of Image Restoration

Considering the objective function in problem (2), the variable $\boldsymbol{x}$ can be split into its positive parts $\boldsymbol{x}_+ \geq 0$ and negative parts $\boldsymbol{x}_- \geq 0$, where $x_i = (x_+)_i - (x_-)_i$ and $x_+ = \max\{0, x\}$, $x_- = \max\{0, -x\}$, then rewrite (2) as a standard bound-constrained quadratic programming (BCQP) problem [10]:

$$\begin{aligned} \min_z \; & f(z) = \frac{1}{2} z^T P z + Q z \\ s.t. \; & z \geq 0 \end{aligned} \tag{3}$$

where $z = \begin{pmatrix} \boldsymbol{x}_+ \\ \boldsymbol{x}_- \end{pmatrix}$, $\boldsymbol{P} = \begin{pmatrix} \boldsymbol{A}^T\boldsymbol{A} & -\boldsymbol{A}^T\boldsymbol{A} \\ -\boldsymbol{A}^T\boldsymbol{A} & \boldsymbol{A}^T\boldsymbol{A} \end{pmatrix}$, $\boldsymbol{Q} = \tau \boldsymbol{1}_{2n}^T + (-\boldsymbol{y}^T\boldsymbol{A}\;\boldsymbol{y}^T\boldsymbol{A})$ and $\boldsymbol{1}_n^T = [1\,1\ldots 1]_{1\times n}$.

Now, such a differentiable optimization problem, we can directly solve it with methods like gradient projection, momentum based gradient projection, and so on. Gradient projection for sparse reconstruction (GPSR) [10] proposed a frame of projection method to solve (3), searching from iterate $z^{(k)}$ to $z^{(k+1)}$ as follows.

$$w^{(k)} = \left(z^{(k)} - \alpha^{(k)} \nabla f|_{z=z^{(k)}}\right)_+ \tag{4}$$

$$z^{(k+1)} = z^{(k)} + \lambda^{(k)}\left(w^{(k)} - z^{(k)}\right) \tag{5}$$

where $(m)_+ = \max(m, 0)$. By setting different search parameters $\alpha^{(k)}$, they obtained two gradient projection methods. Because of the disadvantages of gradient descent direction, backtracking line search and monotone strategy are employed to escape oscillation and local optima.

2.2 Momentum Based Gradient Projection Method

Gradient descent with momentum is often formulated as (6), which adds a momentum term $\omega\Delta z^{(k-1)}$ to regular gradient descent $\Delta z^{(k)} = -\eta\nabla f(z^{(k)})$.

$$\Delta z^{(k)} = -\eta(1-\omega)\nabla f(z^{(k)}) + \omega\Delta z^{(k-1)}, \;\; where\ \omega \in [0, 1],\ \eta > 0 \tag{6}$$

Obviously, the momentum parameter $\omega = 1$ implies an infinite inertia $\Delta z^{(k)} = \Delta z^{(k-1)}$ and $\omega = 0$ means a standard gradient descent $\Delta z^{(k)} = -\eta\nabla f(z^{(k)})$. Furthermore, in terms of kinematics, $\omega\Delta z^{(k-1)}$ acts as a momentum term which increases the descent process while $\nabla f(z^{(k)})$ and $\Delta z^{(k-1)}$ in the same direction. If the direction of $\nabla f(z^{(k)})$ has a sudden change caused by noise, momentum term can pull it back, resulting in reducing the oscillation. Moreover, in terms of mathematics, the descent process depends on the previous two iteration solutions and the current negative gradient, which is more robust than gradient descent.

We proposed a momentum based gradient projection method to minimize $f(z)$ in [17], which used a projected step technique and searched $z^{(k)}$ to $z^{(k+1)}$ along (6) as follows.

Firstly, we choose some scalar parameters η and ω, and search along momentum based gradient descent direction

$$r^{(k)} = -\eta(1-\omega)\nabla f|_{z=z^{(k)}} + \omega\Delta z^{(k-1)} \tag{7}$$

Then, we get a momentum based gradient projection $s^k_{momentum}$

$$s^k_{momentum} = \max\left(z^{(k)} + r^{(k)}, 0\right) \tag{8}$$

By choosing a scalar $\lambda^{(k)}$, we get a new iterate $z^{(k+1)}$

$$z^{(k+1)} = z^{(k)} + \lambda^{(k)}\left(s^k_{momentum} - z^{(k)}\right) \tag{9}$$

However, the parameters in this method depended on artificial experience, and line search process cannot escape oscillation. In this paper, we perform an adaptive parameter strategy and a backtracking line search to solve these problems.

3 Adaptive Momentum Term Based Optimization Method

As previously outlined and analysis, an adaptive momentum term based optimization method for solving (3) will be proposed in this section. Simply, this new method is denoted by A-Momentum.

Firstly, Barzilai and Borwein (BB) method which has been received much considerable attention is employed to obtain an adaptive parameter during iterations (10).

$$\Delta z^{(k)} = \max\left(z^{(k)} + \alpha^{(k)} r^{(k)}, 0\right) - z^{(k)} \tag{10}$$

where the adaptive parameter $\alpha^{(k)}$ is set as (11).

$$\alpha^{(k)} = \mathrm{mid}\left\{\alpha_{\min}, \frac{\|\Delta z^{(k)}\|_2^2}{\left(\Delta z^{(k)}\right)^T B \Delta z^{(k)}}, \alpha_{\max}\right\} \tag{11}$$

where $\nabla f(z^{(k)}) - \nabla f(z^{(k-1)}) = \boldsymbol{B}(z^{(k)} - z^{(k-1)})$ and $\boldsymbol{B} = \begin{pmatrix} \boldsymbol{A}^T\boldsymbol{A} & -\boldsymbol{A}^T\boldsymbol{A} \\ -\boldsymbol{A}^T\boldsymbol{A} & \boldsymbol{A}^T\boldsymbol{A} \end{pmatrix}$.

This setting was originally developed in the context of unconstrained minimization of a smooth nonlinear objective function, and then has been extended to BCQPs in [18, 19]. By choosing an approximation of Hessian of the objective function, this setting can improve the algorithm's robustness and reduce iterations.

Then, a backtracking line search is used to attain a sufficient decrease, ensuring the objective function decreases at every iteration solution, i.e. every iterative step should satisfy the backtracking line search inequality (12). Monotonicity of backtracking line search can avoid falling into local optimizations caused by noise.

$$f(z^{(k)} + \lambda^{(k)} \Delta z^{(k)}) \leq f(z^{(k)}) + \mu \cdot \lambda^{(k)} \cdot \nabla f(z_k)^T \Delta z^{(k)} \tag{12}$$

The complete algorithm is defined as follows.

Step1 (Initialization): Given $z^{(0)}$, choose parameters η, ω, and set $k = 0$.

Step2 (Compute step):

Search vector:

$$r^{(k)} = -\eta(1-\omega)\nabla f|_{z=z^{(k)}} + \omega \Delta z^{(k-1)}$$

Projected step:

$$\Delta z^{(k)} = \max\left(z^{(k)} + \alpha^{(k)} r^{(k)}, 0\right) - z^{(k)}$$

where $\alpha^{(k)} = \text{mid}\left\{\alpha_{\min}, \frac{\|\Delta z^{(k)}\|_2^2}{(\Delta z^{(k)})^T B \Delta z^{(k)}}, \alpha_{\max}\right\}, B = \begin{pmatrix} A^T A & -A^T A \\ -A^T A & A^T A \end{pmatrix}$.

Step3 (Backtracking line search and update):

Find the scalar $\lambda^{(k)} \in [0, 1][0, 1]$ that minimizes $f(z^{(k)} + \lambda^{(k)} \Delta z^{(k)})$.

If $f(z^{(k)} + \lambda^{(k)} \Delta z^{(k)}) \leq f(z^{(k)}) + \mu \cdot \lambda^{(k)} \cdot \nabla f(z_k)^T \Delta z^{(k)}$,

Then we set $z^{(k+1)} = z^{(k)} + \lambda^{(k)} \Delta z^{(k)}$.

Else

Redefine $\lambda^{(k)} = \lambda^{(k)} \cdot \lambda^{(\text{backtrack})}$, $\lambda^{(\text{backtrack})} \in [0, 1]$, and go back to compute $f(z^{(k)} + \lambda^{(k)} \Delta z^{(k)})$, until it satisfied the above inequality.

Step4 (Termination criterion):

If $z^{(k+1)}$ is satisfied with

$$\frac{\left|f(z^{(k+1)}) - f(z^{(k)})\right|}{f(z^{(k)})} \leq \delta, \ \delta > 0 \tag{13}$$

Then terminate with this approximate solution $z^{(k+1)}$.

Otherwise, set $k = k + 1$ and return to Step2.

4 Experiments

In this section, we present two sets of experiments to demonstrate the effectiveness of proposed algorithm in restored quality, rapidity and robustness to noise. The experiments are performed on different natural grayscale images with size 256×256, such as Cameraman, Lena, Peppers and Baboon in Fig. 1. All experiments are performed in MATLAB R2016a environment with 2.20 GHz Intel(R) Core(TM) i7-4770HQ CPU and 16 GB of memory. The parameters are set as $\eta = 1$, $\omega = 0.1$, $\mu = 0.1$, $\delta = 10^{-4}$, $\lambda^{(\text{backtrack})} = 0.5$.

Firstly, we recovery the degraded images with blur and noise to confirm validity of the proposed algorithm. Degraded images went through a 9×9 blur with kernel $h_{ij} = 1/(i^2 + j^2)$, $i, j = -4, -3, \ldots, 3, 4$ followed by an additive zero-mean white

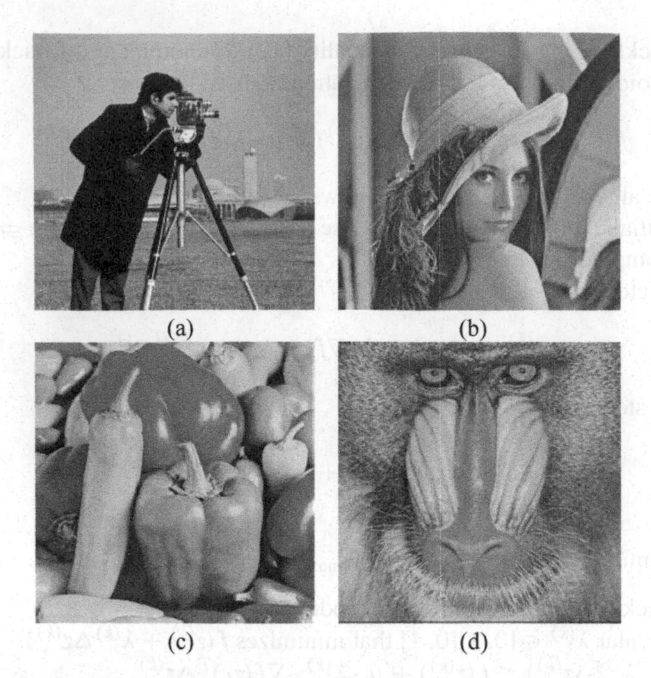

Fig. 1. Test images. (a) Cameraman, (b) Lena, (c) Peppers and (d) Baboon

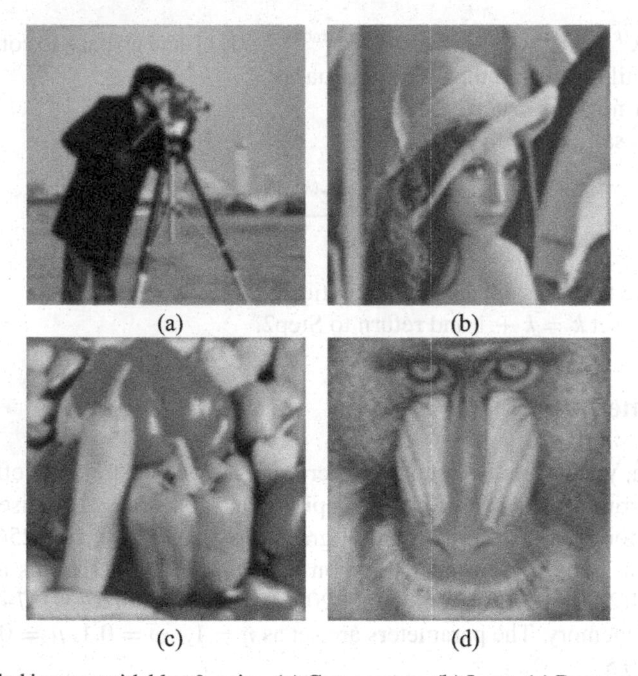

Fig. 2. Degraded images with blur & noise. (a) Cameraman, (b) Lena, (c) Peppers and (d) Baboon

Gaussian noise with standard deviation $\sqrt{2}$. The degraded images and their restored results are given in Fig. 2 and Fig. 3.

As shown in Fig. 3, it is obviously that the proposed method performs well in recovering the visibility of various degraded images. In order to demonstrate the effectiveness of our proposed algorithm objectively, peak signal to noise ratio (PSNR) are shown in Table 1. From the values in bold, we can find that PSNR of the proposed method have been greatly improved, which is consistent with the visual evaluation.

Table 1. Restored PSNR results.

Test image	Degraded image PSNR	Restored image PSNR
Cameraman	23.33	**28.25**
Lena	25.40	**28.85**
Peppers	23.84	**27.61**
Baboon	23.78	**25.81**

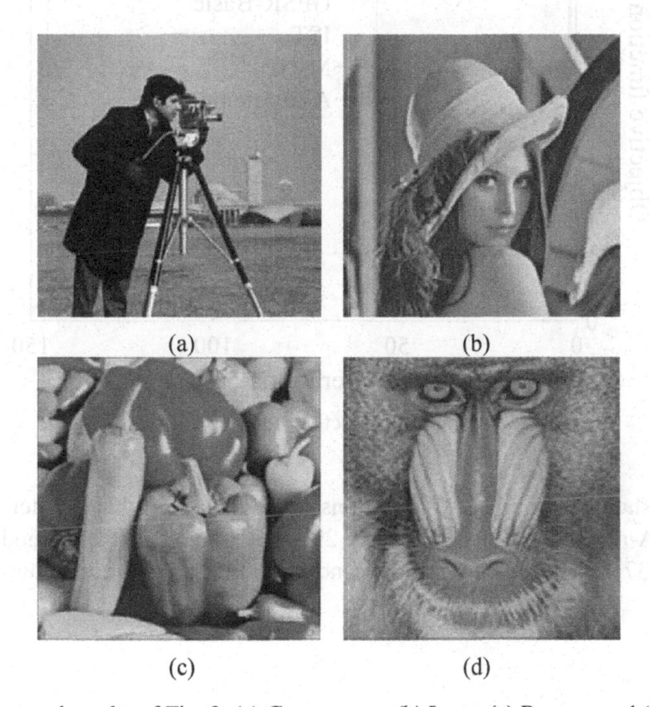

Fig. 3. Restored results of Fig. 2. (a) Cameraman, (b) Lena, (c) Peppers and (d) Baboon.

Secondly, to quantitatively evaluate the performance of different methods, comparison experiments with competing related algorithms (i.e. IST [20], GPSR-Basic, GPSR-BB [10], and M-GP [17]) are employed to verify the precision and efficiency of proposed method.

The evolution of the objective function versus iterations is shown in Fig. 4. Notice that all the methods can achieve the same stop criterion, but the proposed method obtains the least number of iterations. As shown in Fig. 4, the objective function of A-momentum method has the fastest decline while IST is the slowest during iterations. Furthermore, there are several big spikes in iterative process of the objection functions evaluated by GPSR-BB non-monotone and M-GP, which means that the proposed A-momentum method is less sensitive to noise and more robustness than GPSR-BB non-monotone and M-GP.

Figure 5 describes how the restored MSE evolve along with iterations. Every method can obtain a same MSE result, while the proposed method expends the least iterations. Especially, the curve of GPSR-BB non-monotone has some spikes and M-GP misses a sunken point which is maybe a better MSE.

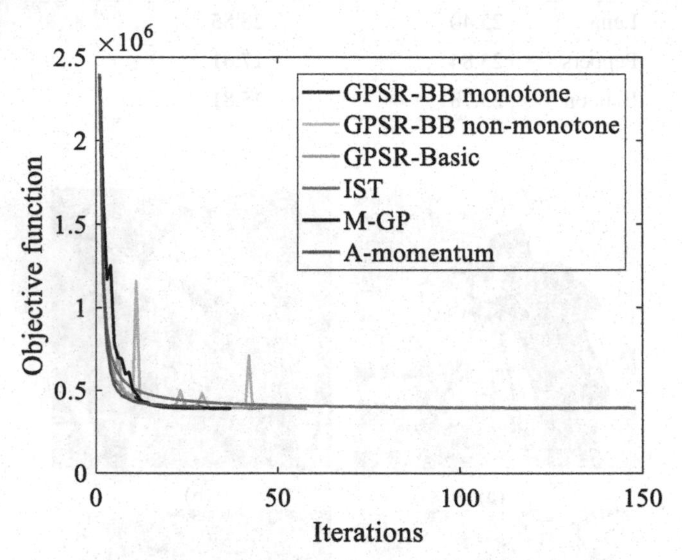

Fig. 4. Evolution of the objective function versus iterations

Figure 6 shows the number of iterations. It is visualized that the iterative number of proposed A-momentum method is only 29, while the other five methods are 51, 58, 58, 148 and 37. Therefore, the rapidity and robustness of A-momentum method are confirmed.

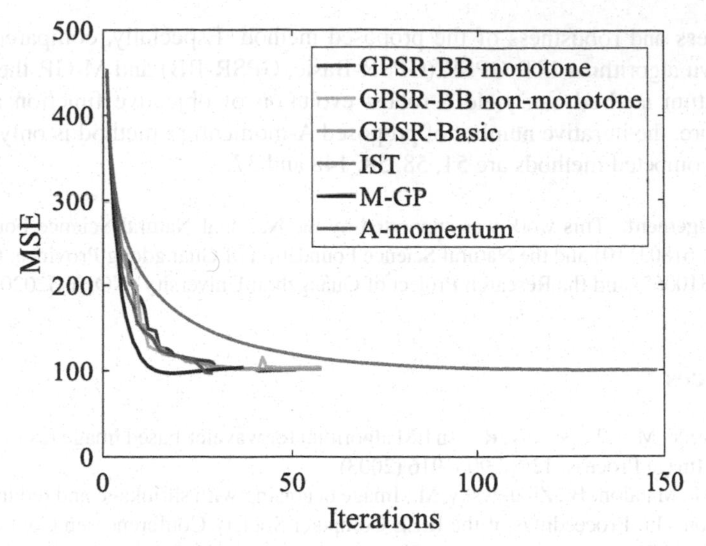

Fig. 5. Evolution of restored MSE versus iterations

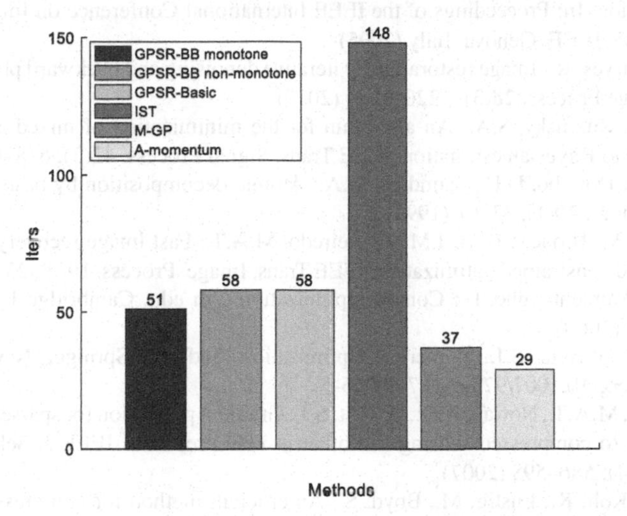

Fig. 6. The number of iterations

5 Conclusion

In this paper, we propose an adaptive momentum term based optimization method for solving a general $l_2 - l_1$ problem in image restoration, by incorporating an adaptive parameter scheme and backtracking line search stagey with M-GP method. Adaptive parameter scheme is employed to accelerate the iteration and increase robustness of algorithm while the backtracking line search is used to ensure the objective function decreases at every iteration solution. Monotonicity of backtracking line search can avoid falling into local optimizations caused by noise. Two series of experiments prove the effectiveness and robustness of the proposed method. Especially, compared with the well-known algorithms IST, GPSR (GPSR-Basic, GPSR-BB) and M-GP, the proposed A-momentum method can yield a robust evolution of objective function and MSE. Furthermore, the iterative number of proposed A-momentum method is only 29, while the other competed methods are 51, 58, 58, 148 and 37.

Acknowledgement. This work was supported by the National Natural Science Foundation of China (NO. 61803110) and the Natural Science Foundation of Guangdong Province, China (NO. 2018A030310065) and the Research Project of Guangzhou University (NO. YG2020011).

References

1. Figueiredo, M.A.T., Nowak, R.: An EM algorithm for wavelet-based image restoration. IEEE Trans. Image Process. **12**(8), 906–916 (2003)
2. Elad, M., Matalon, B., Zibulevsky, M.: Image denoising with shrinkage and redundant representations. In: Proceedings of the IEEE Computer Society Conference on Computer Vision and Pattern Recognition, pp. 1924–1931. IEEE, New York, USA (2006)
3. Figueiredo, M.A.T., Nowak, R.D.: A bound optimization approach to wavelet-based image deconvolution. In: Proceedings of the IEEE International Conference on Image Processing, pp. 782–785. IEEE, Genova, Italy (2005)
4. Tirer, T., Giryes, R.: Image restoration by iterative denoising and backward projections. IEEE Trans. Image Process. **28**(3), 1220–1234 (2019)
5. Alliney, S., Ruzinsky, S.A.: An algorithm for the minimization of mixed and norms with application to Bayesian estimation. IEEE Trans. Signal Process. **42**(3), 618–627 (1994)
6. Chen, S.B., Donoho, D.L., Saunders, M.A.: Atomic decomposition by basis pursuit. SIAM J. Sci. Comput. **20**(1), 33–61 (1998)
7. Afonso, M.V., Bioucas-Dias, J.M., Figueiredo, M.A.T.: Fast image recovery using variable splitting and constrained optimization. IEEE Trans. Image. Process. **19**(9), 2345–2356 (2010)
8. Boyd, S., Vandenberghe, L.: Convex Optimization, 7th edn. Cambridge University Press, Cambridge (2009)
9. Nocedal, J., Wright, S.J.: Numerical Optimization, 2nd edn. Springer, New York (2006). https://doi.org/10.1007/978-0-387-40065-5
10. Figueiredo, M.A.T., Nowak, R.D., Wright, S.J.: Gradient projection for sparse reconstruction: application to compressed sensing and other inverse problems. IEEE J. Sel. Topics Signal Process. **1**(4), 586–598 (2007)
11. Kim, S.J., Koh, K., Lustig, M., Boyd, S.: An efficient method for compressed sensing. In: Proceedings of the IEEE International Conference on Image Processing, pp. 117–120. IEEE, San Antonio, TX, USA (2007)
12. Chen, G., Li, D.F., Zhang, J.S.: Iterative gradient projection algorithm for two-dimensional compressive sensing sparse image reconstruction. Signal Process. **104**, 15–26 (2014)
13. Wright, S.J., Nowak, R.D., Figueiredo, M.A.T.: Sparse reconstruction by separable approximation. IEEE Trans. Signal Process. **57**(7), 2479–2493 (2009)
14. Xiao, Y.H., Zhu, H.: A conjugate gradient method to solve convex constrained monotone equations with applications in compressive sensing. J. Math. Anal. Appl. **405**, 310–319 (2013)
15. Beck, A., Teboulle, M.: Convex Optimization in Signal Processing and Communications. Cambridge University Press, Cambridge (2010)
16. Bioucas-Dias, J.M., Figueiredo, M.A.T.: A new TwIST: two-step iterative shrinkage/thresholding algorithms for image restoration. IEEE Trans. Image Process. **16**(12), 2992–3004 (2007)

17. Ma, G., Hu, Y., Gao, H.: An accelerated momentum based gradient projection method for image deblurring. In: Proceedings of the IEEE International Conference on Signal Processing, Communications and Computing, pp. 1–4. IEEE, Ningbo, China (2015)
18. Dai, Y.-H., Fletcher, R.: Projected Barzilai-Borwein methods for large-scale box-constrained quadratic programming. Numer. Math. **100**, 21–47 (2005)
19. Serafini, T., Zanghirati, G., Zanni, L.: Gradient projection methods for large quadratic programs and applications in training support vector machines. Optim. Meth. Softw. **20**, 353–378 (2004)
20. Daubechies, I., De Friese, M., De Mol, C.: An iterative thresholding algorithm for linear inverse problems with a sparsity constraint. Commun. Pure Appl. Math. **57**, 1413–1457 (2004)

Research on Situational Awareness Security Defense of Intrusion Link Based on Data Element Characteristic Network Transmission Signal

Chun Wang Wu[1](✉), Lin Xia Li[2], and Juan Wang[1]

[1] Chengdu University of Information Technology, Chengdu 610225, China
ajexwu@cuit.edu.cn
[2] Chengdu University, Chengdu 610106, China

Abstract. When defending the intruding links of optical fiber network, the intruding signals are not classified, which leads to the low accuracy of feature extraction and the poor effect of intruding defense. For this reason, the inter-correlation characteristic is used to process the fiber signal. The time-domain eigenvectors, conjugate signals and feature sets of the intrusion signals in the optical fiber network are extracted from three aspects: wavelet domain, frequency domain and time domain by using the two-threshold method and the short-time energy. The classification model is designed by the neural network algorithm, the feature vectors of the intrusion signal are input into the classification model, and the error function is determined by the negative logarithmic likelihood function. The network transmission signal vulnerability is obtained by comparing the data metadata characteristics in the process of network data transmission. Big data is adopted to quantitatively analyze the data interception and monitoring risks in the transmission process, establishing the network transmission security mechanism, completing the quantitative assessment of data transmission risks, and realizing the security defense of network intrusion link situational awareness. Experimental results show that the proposed method has high accuracy in feature extraction and intrusion detection.

Keywords: Data Feature Fusion · Intrusion link · Security defense · Double threshold method · Conjugate signal

1 Introduction

Intrusion detection technology is an important and fast developing field in network security. In the dynamic network security technology, intrusion detection technology can discover the unauthorized activities of users in the network and external intrusion behaviors [1]. Intrusion detection technology can find the intrusion attack in the network and alarm, which is a positive security protection strategy [2]. There are more and more businesses in the network, the number of computers in the network is also increasing, the

X. Sun et al. (Eds.): ICAIS 2021, CCIS 1422, pp. 696–708, 2021.
https://doi.org/10.1007/978-3-030-78615-1_61

data flow in the network is increasing, people's demand for using the original network bandwidth has been unable to meet, the optical fiber network can provide a Gigabit bandwidth network, which is a high-speed network technology [3]. With the increase of the capacity of the optical network, the number of information security vulnerabilities in the network is also increasing, which makes the security defense method of the optical network intrusion link become the current research hotspot [4]. At present, there are some problems in the security defense methods of optical network intrusion link, such as low accuracy of feature extraction and intrusion detection, so it is necessary to study the security defense methods of optical network intrusion link [5].

In reference [4], a method of network situation awareness security defense based on machine learning is proposed. The network transmission information is transformed into operable information by distribution level phasor measurement unit. The network signal is classified by data-driven detection technology, and the network situation security defense is realized by data analysis. This method can effectively extract the characteristics of intrusion signal, but The accuracy of intrusion detection is poor. Literature [5] proposes a method of network situation awareness security defense based on Wiener cybernetics. It uses the theory of situation awareness to analyze the effectiveness of defense decision-making. Through Wiener cybernetics, it effectively defends network intrusion signals, and has made some progress, but the effect of feature extraction is not good, and the extraction time is long. Chen Huijuan, Feng Yuechun and Zhao Xueqing put forward the security defense method of network transmission signal intrusion based on SSO. This method uses particle swarm optimization to analyze the attack rules and attack modes, calculates the trust IP existing in the monitoring engine, obtains the blacklist of intrusion through adaptive way, and realizes the security defense of network intrusion link through the blacklist grouping filter The error of the obtained intrusion blacklist is large, and the accuracy of intrusion detection is low [6]. Li Yi and Li Yongzhong put forward a network intrusion link security defense method based on self coding and machine limit learning machine. In the training process, coefficient self encoder is used to extract the features of intrusion signal in combination with reconstruction error and coefficient penalty of coding layer. The features extracted are identified by limit learning machine to realize the intrusion link detection of network transmission signal. The intrusion signal extracted by this method has some errors in the features, that is, the accuracy of feature extraction is low [7]. Zhu Yadong puts forward a network intrusion link security defense method based on SPSO and rough set. In the intrusion data set, the reduced feature set is obtained through rough set theory. On the basis of the training data, the trainer should be trained, the threshold parameters and weights of the neural network are adjusted, the features are input in the BP neural network, the network intrusion classification is realized, and the network intrusion link security defense is completed Yu, the classification result obtained by this method is not accurate, and there is a problem of low accuracy of intrusion detection [8].

In order to solve the problem of low accuracy of feature extraction and intrusion detection in the above methods, a security defense method based on data element feature network transmission signal intrusion link situation awareness is proposed. In this paper, the time-domain eigenvector, conjugate signal and feature set of the intrusion signal in the optical fiber network are extracted by using the cross-correlation feature, and the error function is determined according to the negative log likelihood function, which can effectively reduce the risk of data interception and monitoring, and realize the security defense of the network intrusion link situation awareness. Through experiments, we can see that this method can effectively improve the accuracy of feature extraction and intrusion detection.

2 Feature Extraction of Intrusion Signal

Based on the characteristics of mutual correlation, the optical fiber signal is divided into frames. The main process is as follows:

(1) Set the length of the intrusion link security defense system of the optical fiber network as N_L;$u_m(i)$; it represents the vibration signal sequence acquired in the optical network, Among them, $m = 0, 1, \cdots, N_L - 1$ sets the initial frame length to Q, carries on the frame shift, and realizes the uniform frame division of the sequence.

(2) Let R represent the cross-correlation coefficient between frame R and frame $i + 1$, and its calculation formula is as follows:

$$R = \frac{\sum_{m=0}^{Q-1} u_m(i \times Q)}{\sqrt{\sum_{m=0}^{Q-1} u_m[(i+1) \times Q + N_L]^2}} \tag{1}$$

(3) Set threshold R^0, when $R > R^0$, merge the frames of two signals. Repeat the above process after comparing frame $i + 2$ and $i + 3$, When the correlation coefficient R is greater than the threshold R^0, the new frame is obtained.

According to the above framing results, the correlation coefficient is recalculated until the set termination conditions are met, and the framing ends, and the framing length is Q'.

By selecting the short-time zero crossing rate time-domain characteristics and short-time energy, the non intrusion signals and intrusion signals in the optical fiber network are distinguished, and the interference of the recognition results by the non intrusion signals is eliminated [9, 10]. There is randomness in the intrusion signal in the optical fiber network, and the distribution characteristics of the intrusion signal in the network can not be expressed by the characteristics in a single domain. Therefore, the characteristics of the intrusion signal are extracted from three aspects of the wavelet domain, the frequency domain and the time domain, and the feature set of the intrusion link signal in the optical fiber network is constructed.

2.1 Time Domain Features

Optical fiber vibration signal has the time-domain characteristics of short-time zero crossing rate, short-time energy and maximum amplitude common to one-dimensional signals such as Gaussian white noise signal and voice signal [11, 12]. In time domain, the energy of short-time signal can be reflected by short-time energy, and the vibration frequency of signal can be reflected by short-time zero crossing rate. Compared with the intrusion signal, the non intrusion signal has the characteristics of smaller amplitude based on the characteristics of optical fiber signal [13]. Combining the two threshold method of short-time zero crossing rate and short-time energy to extract the intrusion signal in optical fiber network.

Let E_m represent the short-time zero crossing rate corresponding to each signal frame, and its calculation formula is as follows:

$$E_m = \sum_{m=0}^{m+Q'-1} u_m(i)^2 \tag{2}$$

Among them, suppose Z_m represents the short-time energy corresponding to each signal frame, and its calculation formula is as follows:

$$Z_m = \frac{1}{n}\sum_{m=0}^{m+Q'-1} \{|\mathrm{sgn}[u(n)]| - \mathrm{sgn}[u(n-1)]\} \tag{3}$$

$$sgn(x) = \left\{ \begin{array}{c} 1, x \geq 0 \\ -1, x < 0 \end{array} \right\} \tag{4}$$

Where, n represents the starting point corresponding to the signal frame; sgn describes the symbol function.

Set thresholds $\overline{E}$ $\overline{Z}$, when the zero crossing rate and short-time energy are higher than the set threshold, the signal is the intrusion signal $U(i)$ of the optical network.

Using EEMD method of white noise aided analysis, the eigenmode function of intrusion signal is extracted at different time scales, and the eigenvector of intrusion signal is obtained [14].

(1) The Gauss white noise signal $\omega(i)$ is superposed into the optical fiber intrusion signal frame $U(i)$ to obtain the overall signal $U'(i)$, which is expressed as follows:

$$U'(i) = U(i) + \omega(i) \tag{5}$$

(2) Through EMD decomposition of the overall signal $U'(i)$ as follows, multiple IMF components are obtained:

$$U'(i) \rightarrow \sum_{j=1}^{m} c_j(i) + r(i) \tag{6}$$

Where, $r(i)$ describes the remainder; $c_j(i)$ describes the i IMF component obtained after decomposition;j describes the total IMF component.

(3) Add the white noise $\omega_k(i)$ with different amplitude into the original signal, obtain, repeat the above steps, and obtain $U_k'(i)$ formula:

$$U_k'(i) \rightarrow \sum_{j=1}^{m} c_j(i) + r_k(i) \tag{7}$$

(4) On the basis of the principle of zero mean value of Gaussian white noise, the influence of Gaussian white noise on the time domain distribution is eliminated, and it is repeated N_0 times, meeting the following formula of N_0:

$$\varepsilon_n = \frac{\varepsilon}{\sqrt{N_0}} \quad (8)$$

In the formula, ε_n describes the standard deviation between the sum of the original signal and the IMF component; ε describes the corresponding amplitude of the white noise added to the signal.

The original signal is decomposed by EEMD, and the IMF component is obtained by the following formula:

$$U(i) \rightarrow \sum\nolimits_{j=1}^{m} C_j(i) + R(i) \quad (9)$$

Where, $R(i)$ describes the residual term after the original signal is decomposed; $C_j(i)$ describes the IMF component obtained after the original signal is decomposed by EEMD.

The eigenvector of the intrusion signal is the kurtosis with strong sensitivity. Suppose T_i describes the i kurtosis corresponding to the IMF component, and its calculation formula is as follows:

$$T_j = \frac{1}{D} \sum\nolimits_{j=1}^{D} C_j(i)^4 \quad (10)$$

Where D describes the number of points of component $C_j(i)$. The time domain eigenvector $T = [T_1, T_2, \cdots, T_{13}]$ of the intrusion signal is obtained by the above formula.

2.2 Wavelet Domain Features

Intrusion signal has the characteristics of nonstationarity and randomness in the intrusion link security defense system of optical network [15, 16]. The frequency-domain characteristics of intrusion signals are obtained by power spectrum analysis method $P_x(k)$:

$$P_x(k) = \sum\nolimits_{j=-\infty}^{Q'-1} U_j(n) e^{-\frac{2\pi m}{Q'}} + T \cdot U_j^*(n-c) \quad (11)$$

Where $U_j{}^*(n-c)$ describes the conjugate signal corresponding to $U_j(n)$ after translation.

Based on the data feature fusion, the security defense method of optical fiber network intrusion link uses discrete wavelet transform to refine the intrusion signal in multi-scale, and obtains the corresponding wavelet energy value in different scales, which is regarded as the wavelet domain feature set of the intrusion signal in the optical fiber network. Regardless of the high-frequency components in the signal, the eigenfeatures of the signal are represented by the low-frequency components [17]. The low-frequency components are decomposed by wavelet discrete transform to extract the energy value of the intrusion signal at different wavelet scales, and construct the small Porter collection of the intrusion signal.

$$U \rightarrow q + d_1 + d_2 + \cdots + d_{n-1} \quad (12)$$

In the formula, d_j describes the low frequency components of U under each wavelet scale; q describes the high frequency components corresponding to the intrusion signal.

The corresponding wavelet energy of different types of intrusion signals at different wavelet scales is calculated by the following formula E_j:

$$E_j = \frac{1}{L}\sum\nolimits_{j=1}^{L} |d_j(k)|^2 - P_x(k) \tag{13}$$

In the formula, $d_j(k)$ describes the d_j wavelet coefficient corresponding to the low frequency component k; L describes the number of points of the signal in different resolutions.

According to the time-domain, frequency-domain and wavelet domain features, the feature set $F = [T, P, E]$ of the intrusion signal is obtained.

3 Classification Model

Neural network algorithm is used to design the classification model, and the characteristics of the intrusion signals are input into the classification model to realize the security defense of the intrusion link of optical fiber network.

Neural network algorithm is usually divided into backward and forward propagation algorithm. The main purpose of backward propagation is gradient descent to optimize the parameters of each layer. The main purpose of forward propagation is to predict backward propagation.

The neural network is used to activate D dimensional input variables, which is linear combination

$$a_j = \sum\nolimits_{j=1}^{D} w_{ij}^{(1)} x_i + w_{j0}^{(1)} \tag{14}$$

Where $w_{j0}^{(1)}$ describes the offset; $w_{ji}^{(1)}$ describes the weight.

The nonlinear activation function $h(\cdot)$ is used to transform each activation, and the following formula is obtained:

$$z_j = h(a_j) \tag{15}$$

Among them, h describes hyperbolic tangent Tan function or logistic sigmoid function. After activating the function, make a linear combination to obtain the optimal classification model:

$$a_k = \sum\nolimits_{j=1}^{M} w_{ji}^{(2)} z_j + w_{k0}^{(2)} \tag{16}$$

On this basis, the eigenvector of the intrusion signal is input into the classification model, and through the back propagation error of the back propagation algorithm, the activation of the output layer is the logical regression function (a), whose expression is as follows:

$$\sigma(a) = \frac{1}{1 + exp(-a)} \tag{17}$$

If the intrusion signal can be divided into several parts, the softmax activation function is the activation function of the last layer, and the above formula is the forward propagation formula. The input can be predicted by the classification model.

The parameters in the neural network are optimized by back propagation. The training process of the classification model is to determine the target variable $\{t\}$. The 0–1 value obtained by the logistic function can be regarded as the probability of a certain category in the two classification problem.

$$p(t|x, w) = y(x, w)^t \{1 - y(x, w)\}^{1-t} \tag{18}$$

In order to reduce the fitting, the generalization of the classification model is enhanced in the classification problem. The error function is determined by the negative log likelihood function

$$E(w) = p(t|x, w) + \sigma(a) \tag{19}$$

According to the error function, the network intrusion link security can be realized.

4 Network Transmission Signal Intrusion Link Situation Awareness

4.1 Quantitative Analysis of Big Data

Through the quantitative analysis of big data, the situation security of network link is predicted, and the risk of data interception and monitoring in the process of network transmission is predicted. Build the network link situation database, select the historical security situation data group and the current security situation data group. At the same time, the current security situation data is detected, and the network transmission signal intrusion link security evaluation is carried out by comparing the two groups of data element characteristics, and the prediction results are output. The flow chart of network security situation assessment is shown in Fig. 1.

It can be seen from the analysis of Fig. 1 that the network situation assessment first needs to obtain the historical security situation and the current security situation from the situation database. Both situations need to be processed at the same time. The historical security situation needs to be predicted and trained, the current security situation needs to be predicted and detected, and the network security needs to be evaluated by comparing the data element characteristic results, so as to achieve the situation prediction, output the prediction results and complete the network state Potential assessment.

4.2 Normalization

On the basis of the above, the network security situation is evaluated, and then the data volume of different sizes is carried out, and then the network transmission security mechanism.

In Fig. 2, the green line indicates the data transmission volume is 1000 MB, the red line indicates the data transmission volume is 800 MB, the purple line indicates the

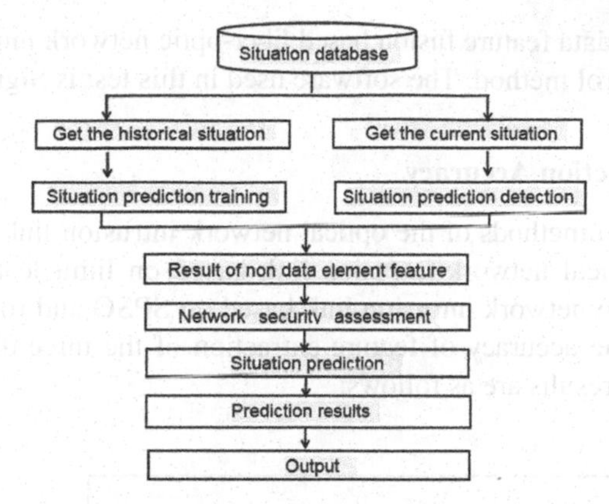

Fig. 1. Network situation evaluation

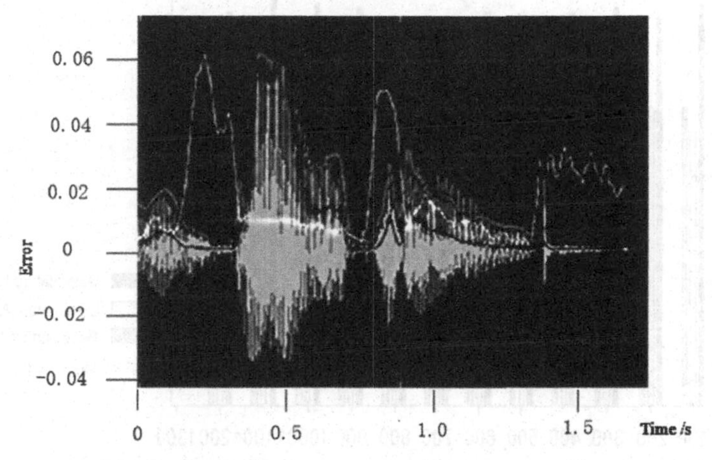

Fig. 2. Normalized mean error (Color figure online)

data transmission volume is 400 MB, and the yellow line indicates the data transmission volume is 200 MB. According to the analysis of Fig. 2, the normalized mean square error is different under different data transmission volumes. The normalized average error affects the security of network transmission data. The greater the error, the greater the transmission risk and the worse the security. When the time is 1.0 s, among the four different data volumes, the normalized average error of 200 MB data transmission volume is the smallest, and the normalized average error of 1000 MB data transmission volume is the largest. Therefore, it is necessary to design the network security mechanism according to the size of data.

5 Experimental Results and Analysis

In order to verify the overall effectiveness of the data feature fusion based fiber-optic network intrusion link security prevention and control method, it is necessary to test the effectiveness of the data feature fusion based fiber-optic network intrusion link security prevention and control method. The software used in this test is SigmaPlot 14.0

5.1 Feature Extraction Accuracy

The security defense methods of the optical network intrusion link based on data feature fusion, the optical network intrusion link based on limit learning machine and self encoder, and the network intrusion link based on SPSO and rough set are used to test respectively. The accuracy of feature extraction of the three different methods is compared. The test results are as follows:

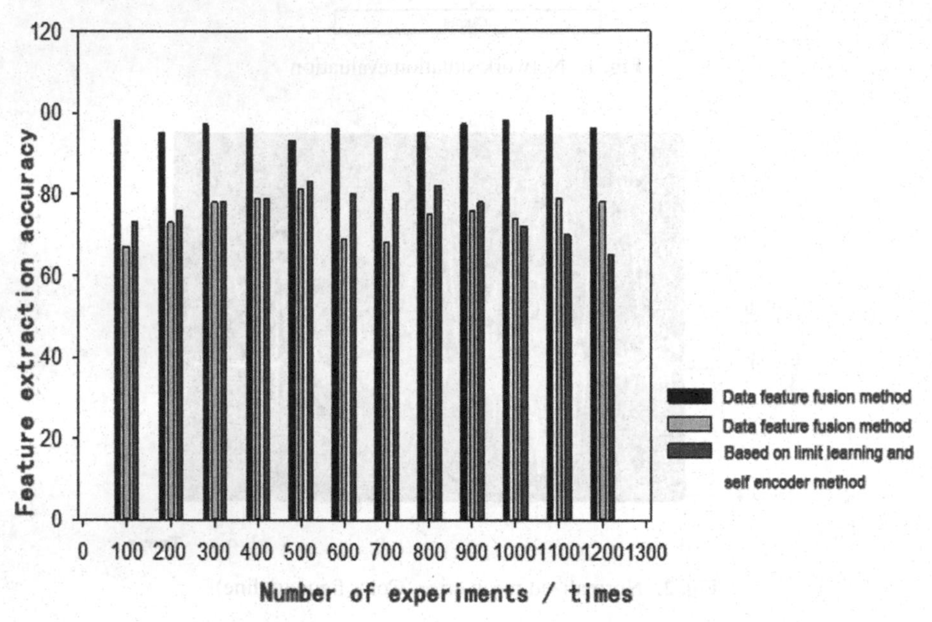

Fig. 3. Feature extraction accuracy of different methods

According to the analysis of Fig. 3, the feature extraction accuracy of the data feature fusion based fiber-optic network intrusion link security defense method in multiple iterations is higher than that of the limit learning machine and self encoder based fiber-optic network intrusion link security defense method and the SPSO and rough set based network intrusion link security defense method. In the previous 500 experiments, for example, the accuracy of feature extraction based on SPSO and rough set method is 81%, the accuracy of feature extraction based on limit learning machine and self encoder method is 83%, and the accuracy of feature extraction based on data feature fusion method is 93%. The proposed method has obvious advantages and is suitable for effective defense of fiber network intrusion link security. The reason is that based on the data feature fusion, the security defense method of optical fiber network intrusion link processes the optical fiber signal by frame on the basis of the correlation characteristics, and extracts the characteristics of the intrusion signal from three aspects of small wave domain, frequency domain and time domain, which improves the feature extraction accuracy of the security defense research of the network transmission signal intrusion link situation awareness based on the data element characteristics.

5.2 Accuracy Rate of Intrusion Detection

In order to further verify the security defense effect of different methods of optical fiber network intrusion link, the security defense method of optical fiber network intrusion link based on data feature fusion (proposed method), the security defense method of optical fiber network intrusion link based on limit learning machine and self encoder (document [7]) and the security defense method of network intrusion link based on SPSO and rough set (document [7]) are adopted The test results are as follows:

Table 1. Intrusion detection accuracy of different methods.

Number of experiments/time	Literature [7] method	Literature [8] method	Proposed method
50	82%	79%	92%
100	85%	82%	94%
150	79%	80%	96%
200	82%	76%	97%
250	84%	78%	95%

According to Table 1, when the number of experiments is 100, the accuracy of intrusion detection of literature [7] method is 85%, the accuracy of intrusion detection of literature [8] method is 82%, and the accuracy of intrusion detection of the proposed method is 94%. The accuracy of detection of the proposed method is obviously better. When the number of experiments increases to 200, the accuracy of intrusion detection of literature [7] method is 82%, and the accuracy of intrusion detection of literature [8] method is 82%. The rate of detection is 76%, the accuracy of the proposed method is 97%, and the proposed method always keeps a high detection accuracy. This is because the security defense method of optical fiber network intrusion link based on data feature fusion uses neural network algorithm to design the classification model, input the acquired features into the classification model, and improve the accuracy of intrusion detection in the security defense research of situation awareness of intrusion link based on data element feature network transmission signal.

5.3 Safety Test

In order to verify the intrusion prevention performance of different methods, the optical network security detection is carried out, and the detection results are shown in Fig. 4:

In the figure, different colors represent different security, and the time from blue to red increases in turn, that is to say, the more red the color, the lower the security.

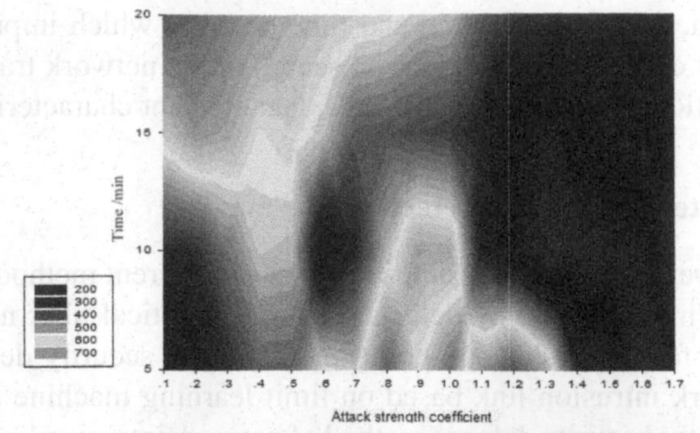

(a) Limit learning machine and self encoder method

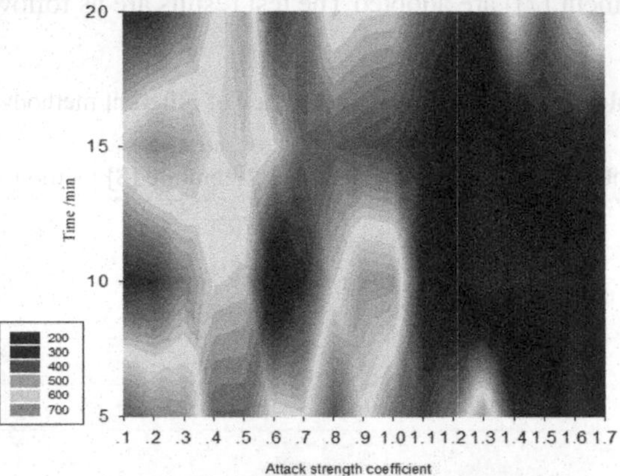

(b) Based on SPSO and rough set method

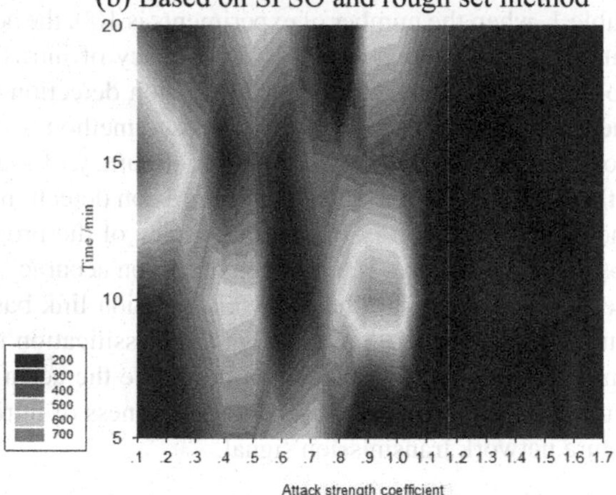

(c) Data feature fusion method

Fig. 4. Safety test results

According to the analysis of the above figure, the security of the network is different after different methods are used to prevent the intrusion link of the optical network.

Although the learning machine based on the limit and the self encoder method have some defense effects, when the time is more than 15 min, the security is significantly reduced; although the SPSO based and rough set method improve these, but still appear large red, the security basis is not high. However, after the data feature fusion method is used to prevent the intrusion link in the optical network, the red part is obviously reduced and the security is effectively improved, which shows that the data feature fusion method is the best.

6 Concluding Remarks

Optical fiber sensor has the characteristics of corrosion resistance, high sensitivity, high temperature resistance and good insulation. It is widely used in the perimeter security system of warehouse, airport and other important places. The current network intrusion link security defense methods have the problems of low feature extraction accuracy and low intrusion detection accuracy. This paper puts forward the research of network transmission signal intrusion link situation awareness security defense based on data element characteristics, which lays the foundation for the safe and stable development of optical fiber network.

Funding Statement. This paper is supported by Sichuan Science and Technology Program No. 2021YFH0076.

References

1. Junhua, L., Chunyan, W.: Design and application of railway communication automatic monitoring system under optical fiber monitoring. Autom. Instrum. **8**, 184–187 (2018)
2. Zhen, W.W.: Security situation awareness of routing strategy and transmission protocol compliance for multi domain networks. University of Electronic Science and technology
3. Duan, Y.C., Wang, Y., Li, X., et al.: Extraction of random forest network security situation elements based on RSAR. Inf. Netw. Secur. **7**, 75–81 (2019)
4. Shahsavari, A., Farajollahi, M., Stewart, E., et al.: Situational awareness in distribution grid using micro-PMU data: a machine learning approach. IEEE Trans. Smart Grid **10**(6), 6167–6177 (2019)
5. Anjaria, K., Mishra, A., et al.: Relating Wiener's cybernetics aspects and a situation awareness model implementation for information security risk management. Kybernetes Int. J. Syst. Cybern. **47**(1), 58–79 (2018)
6. Huijuan, C., Yuechun, F., Xueqing, Z.: Network intrusion detection method using SSO adaptive blacklist packet filter. Control. Eng. **25**(10), 1940–1945 (2018)
7. Yi, L., Yongzhong, L.: Intrusion detection algorithm of industrial control network based on self encoder and limit learning machine. J. Nanjing Univ. Sci. Technol. (Nat. Sci. Ed.) **43**(4), 408–413 (2019)
8. Yadong, Z.: Network intrusion detection scheme based on rough set and SPSO. Control. Eng. **25**(11), 2097–2101 (2018)
9. Hao, Z., Jie, S., Mingjun, L.: Research on data security transmission method of distributed wireless sensor network based on MD5 algorithm. J. Suzhou Univ. Sci. Technol. (Nat. Sci. Ed.) **36**(01), 68–74 (2019)

10. Wen, W., Qiwu, G., Xinyuan, G., et al.: The security construction mechanism of multi domain optical network based on PCE architecture. Optical Commun. Res. **42**(4), 1–4 (2016)
11. Zhang, C.: Application of data encryption technology in computer network security. Electron. Compon. Inf. Technol. **3**(12): 72–73+93 (2019)
12. Zanyu, T., Hong, L.: Research on network security situation assessment method under multi-stage and large-scale network attack. Comput. Sci. **45**(01), 245–248 (2018)
13. Xudong, G., Xiaomin, L., Ruxue, J., et al.: Intrusion detection based on improved sparse denoising self encoder. Comput. Appl. **39**(3), 769–773 (2019)
14. Dubo: Research on the security of smart grid optical communication network in the cloud computing environment. Autom. Instrum. (1), 19–22+26 (2018)
15. Jiaxin, W., Yi, F., Rui, Y.: Network security measurement based on dependency graph and general vulnerability scoring system. Comput. Appl. **39**(6), 1719–1727 (2019)
16. Yan, J., Weihong, H., Wei, W.: Design and implementation of Yhsas. Inf. Technol. Netw. Secur. **37**(1), 17–22 (2018)
17. Bin, W.: Fuzzy risk assessment of network security based on computer network technology. Foreign Electron. Meas. Technol. **38**(5), 11–16 (2019)

Research on Proofreading Method of Semantic Collocation Error in Chinese

Rui Zhang[1], Yangsen Zhang[1,2](✉), Gaijuan Huang[1], and Ruoyu Chen[1]

[1] Institute of Intelligent Information Processing, Beijing Information Science and Technology University, Beijing 100192, China

[2] Beijing Laboratory of National Economic Security Early-Warning Engineering, Beijing 100044, China

Abstract. With the rapid development of network technology and the popularization of electronic documents, Chinese text automatic proofreading technology has attracted increasing attention. Automatic proofreading of semantic errors in Chinese text is a key and difficult point in the field of Chinese information processing. Aiming at this problem, we propose a semantic error proofreading method that contains dependency parsing and statistical theory, and construct a two-layer semantic knowledge base to assist error detection and error correction. The two-layer semantic knowledge base includes (1) knowledge base of word collocations containing structured information of sentences extracted from a large-scale corpus; (2) knowledge base of sememe collocations obtained by sememe mapping through HowNet. On this basis, cubic association ratio and degree of polymerization are introduced to evaluate the proofreading results to reduce false positives and improve the accuracy of error correction opinions. The experiment result shows that our method will be of great use for the construction of semantic proofreading knowledge base and semantic error automatic proofreading methods.

Keywords: Semantic error · Semantic collocation · Knowledge base · Text proofreading

1 Introduction

With the rapid development of network technology and the popularization of electronic documents, the efficient and accurate automatic proofreading of Chinese text has become a new demand for social development. Chinese text proofreading naturally includes two tasks: text error detection and text error correction. It is mainly to solve three levels of errors: word errors [1–3], grammatical errors [4–6], and semantic errors [7,11]. Among them, extensive research and relatively mature applications have been found in automatic proofreading methods for word errors and grammatical errors, while the study of proofreading methods for semantic error is still in exploration.

Supported by the National Natural Science Foundation of China (NSFC No. 61772081).

X. Sun et al. (Eds.): ICAIS 2021, CCIS 1422, pp. 709–722, 2021.
https://doi.org/10.1007/978-3-030-78615-1_62

Semantic errors in Chinese text are essentially "real-word errors". On the premise that there are no word errors or grammatical errors, the misuse of some word collocations leads to ambiguity, that is, semantic errors. The occurrence of semantic errors in Chinese text is mainly affected by two factors: on is the misuse of words or grammar rules, which may produce semantic errors; the other is that the same word often has different meanings in different contexts. In addition, different ways of thinking and customs also produce semantic errors.

To address these issues, we propose a new semantic error proofreading method. For the error detection task, we create a hierarchical semantic proofreading knowledge base. Different from knowledge bases of existing methods, we add the sentence structure information when extracting collocations and filter the result via more optimized statistical indicators. To expand the coverage of the knowledge base, we also use HowNet to map word collocations to sememe collocations. For the error correction task, we propose an error filtering and error correction method through the comprehensive use of statistical indicators. The framework of semantic collocation error proofreading is shown in Fig. 1.

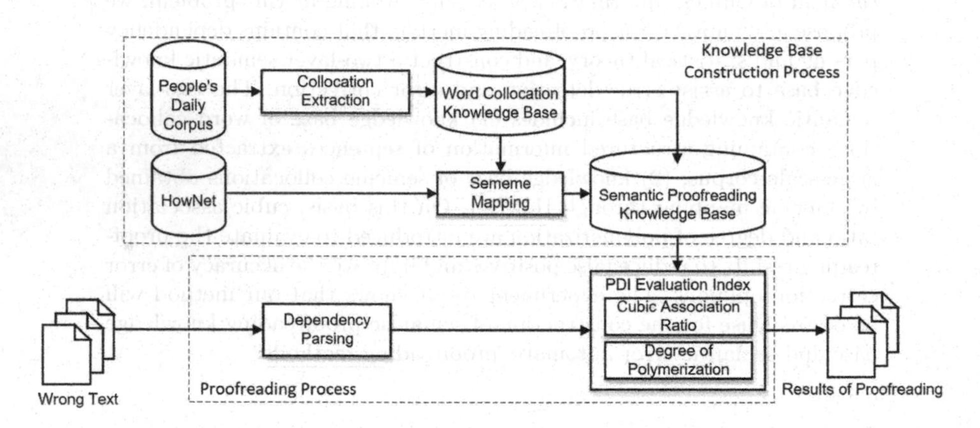

Fig. 1. The framework of semantic collocation error proofreading

2 Related Work

To the best of our knowledge, the commonly used semantic proofreading methods include rule-based or linguistic-based methods, statistics-based methods, and machine learning or deep learning methods.

Rule-based or linguistic-based methods are generally realized through observation and statistics. Such methods summarize the semantic structure and grammatical rules and then proofread with linguistic knowledge. Luo et al. [7] proposed a semantic error detection strategy combining examples, statistics and rules. This method firstly obtained the sentence components through syntactic analysis, then retrieved the examples with the similar sentence structure from the corpus, and calculated the similarity between them. Finally, it constructed the N-element semantic adjacency matrix and semantic co-occurrence matrix to improve the semantic error detection capability. This strategy achieved good results by using both local semantic constraints and distant semantic collocation.

The statistics-based approach is to construct contextual features by using statistical models and use these features for semantic error proofreading. The proofreading accuracy of statistics-based methods largely depends on the quality of the training corpus and the effectiveness of statistical methods. Zhang et al. [8] built a three-layer semantic collocation knowledge base grounded on HowNet by comprehensively using statistics and rules. They used D-S evidencc theory to calculate the semantic collocation strength of words and determined whether there was a semantic error. Their work greatly improved the accuracy of error detection. Liu et al. [9] performed fuzzy matching and similarity calculation on rough sentence segmentation, and constructed a fuzzy word graph. They found the shortest path of the graph to filter the segmentation results. After comparing with the original text, they captured the errors in the text and the corresponding fuzzy matching results were the modification opinion.

The last type of method uses machine learning or deep learning models to achieve error proofreading. Cheng et al. [10] achieved semantic proofreading of Chinese text based on Hierarchical Network of Concepts (HNC). This method constructed semantic blocks by using HNC sentence-category analysis and sentence type knowledge base for sentence type rationality checking analysis, and then analyzed semantic blocks through the concept-related knowledge base and found out the errors in them. To realize semantic error automatic proofreading, Hai [11] mapped word vectors to sememe vectors through a LSTM network, and constructed a word-sememe mapping set. Based on the set, that method achieved error proofreading by means of multi-model integration and fuzzy matching.

As mentioned above, there are relatively few semantic proofreading methods. Rule-based methods excessively rely on language rules and lack flexibility. Statistics-based methods lack the linguistic guidance. There are still some deficiencies in the selection of statistical indicators, and the proofreading effect can still be improved. Although deep learning methods have achieved good results in many fields such as machine translation, entity recognition, and reading comprehension, the training of deep models requires a large-scale corpus, and the high-quality parallel corpus for semantic proofreading is relatively scarce at present. Therefore, based on previous work, we propose a semantic proofreading method that combines statistics and linguistic features.

3 Construction of Semantic Proofreading Knowledge Base

3.1 Word Collocation Extraction Model Based on Statistics and Linguistic Features

The semantic collocation knowledge base is the foundation of the statistics-based proofreading method. Previous studies focus more on words or a small part of the sentence while extracting collocations. The collocation base extracted by these methods results in good proofreading performance on short sentences but poor on complex sentences. To address these issues, we use dependency parsing to obtain sentence structure information, comprehensively consider words and sentence structures for collocation extraction.

The theoretical basis of dependent grammar is that there is a head-dependent relationship between words. Through dependency parsing, a lot of sentence structure information can be mined, and we mainly focus on four structures of "subject-predicate", "predicate-object", "attribute-head" and "adverb-head". As far as the accuracy of dependency parsing is concerned, it is necessary to filter the original word collocation base. In addition to co-occurring word frequency, mutual information (MI) is an important indicator to measure the strength of association between collocations. The formula is as follows:

$$MI(x,y) = log_2 \frac{p(x,y)}{p(x) \cdot p(y)} = log_2 \frac{c(x,y)N}{c(x) \cdot c(y)} \tag{1}$$

where $MI(x,y)$ represents the MI between the words x and y. $p(x)$, $p(y)$ and $p(x,y)$ represent the occurrence probability of the word x, word y and ordered binary group (x,y), respectively; $c(x)$, $c(y)$ and $c(x,y)$ represent the occurrence frequency of the word x, word y and ordered binary group (x,y), respectively; N represents the total number of binary groups in the corpus.

Many researchers use MI as a filtering indicator in collocation extraction [8,11,16]. Although the probability that words x and y can form collocation increases with MI value, there is a big flaw in collocation selection via MI. MI cannot deal with the low occurrence frequency very well and will assign a higher value to the lower one.

A feasible improvement method is to enhance the contribution of co-occurring word frequency to MI such as cubic association ratio (MI3) [17]:

$$MI3(x,y) = log_2 \frac{c^3(x,y) \cdot N}{c(x) \cdot c(y)} \tag{2}$$

In Eq. (2), $MI3(x,y)$ represents the MI3 between the words x and y, and the other parameters have the same meaning as Eq. (1). The main difference between MI3 and MI is that $c(x,y)$ plays a more important role, thereby improving the quality of collocation extraction.

3.2 Sememe Collocation Mapping Model

Till now, we obtain the word collocation base including four sub-bases with different types, which is the first layer of the whole semantic collocation knowledge base. We will introduce the second part in this section.

HowNet [12] is a large-scale annotation knowledge network based on the interrelationship of three concepts: words, senses and sememes. Each word has at least one sense and each sense is annotated by several sememes. Sememe [13], which can form complicated relations, is the smallest and indivisible unit in

HowNet. A large number of studies have shown that HowNet can be well applied to semantically related tasks in Chinese [14,15].

In the annotation system of HowNet, the higher the level of a sememe, the more it can reflect the most essential semantic features. In this work, we only consider all independent sememes of each word as a set without exploring their relations, which is left in our future work. To further quantify the mapping process, we use Degree of Polymerization (PD) to measure the reliability of the sememe mapping. The definition of the degree of polymerization is as follows:

Definition. For the word y, there is a word set X matched with it. The "first sememes" of all the senses of each word in X constitute the sememe set S_X. The collocation degree of each sememe in S_X and the word y is called the degree of polymerization, which is written as PD:

$$PD = \frac{\sum_{i=1}^{N} C(s_i, y)}{N} \tag{3}$$

where N is the total number of sememes of the words in the set X, and the definition of $C(s_i, y)$ is

$$C(s_i, y) = \begin{cases} 1, \text{if } s_i \text{ and } y \text{ can be matched, } s_i \in S_X \text{ and } 1 \leq i \leq N \\ 0, \text{other.} \end{cases} \tag{4}$$

When the PD approaches 1, it means that the collocation possibility of s_i and y is higher; on the contrary, when the PD approaches 0, it means that the collocation possibility of s_i and y is lower. In this paper, the sememe with the highest PD is selected as the mapping result.

Based on HowNet, we establish the sememe collocation base which is the second layer of the whole semantic collocation knowledge base. Correspondingly, the sememe collocation base also contains four different sub-bases with different types.

3.3 Semantic Proofreading Knowledge Base Construction Algorithm

The semantic collocation knowledge base construction algorithm is shown in Algorithm 1. We will discuss how to determine the threshold Γ of the co-occurrence word frequency and the threshold T of the MI3 in the experimental section.

Algorithm 1. The construction of semantic collocation knowledge base

---

Input: corpus S, co-occurrence word frequency threshold set $\Gamma = \{\gamma_1, \gamma_2, \gamma_3, \gamma_4\}$, MI3 threshold set $T = \{\tau_1, \tau_2, \tau_3, \tau_4\}$
Output: word collocation base $D_i(i = 1, 2, 3, 4)$, sememe collocation base $Se_i(i = 1, 2, 3, 4)$
for sentence *str* in S **do**

Carry out dependency parsing on *str*, extract the "subject-predicate", "predicate-object", "attribute-head" and "adverb-head" from *str* and save them in $D_i (i = 1, 2, 3, 4)$, respectively
end for
for $iin[1, 4]$ **do**
for collocation c in D_i **do**
$value_c of = CoF(c)$ // calculate co-occurrence frequency of c in D_i
$value_m i3 = MI3(c)$ // calculate MI3 of c in D_i
if $value_{cof} < \gamma_i$ or $value_{mi3} < \tau_i$ **then**
delete the collocation c from D_i
end if
end for
for collocation c in D_i **do**
c_{core}=head word of c, c_{coll}=collocation word of c
Calculate the DP between each sememe of c_{core} and c_{coll}, choose the sememe with the highest DP to form a new collocation $c^{'}$ with c_{coll}, and insert $c^{'}$ into TD_i
end for
for collocation $c^{'}$ in TD_i **do**
$c^{'}_{core}$=head word of $c^{'}$, c_{coll}=collocation word of $c^{'}$
Calculate the DP between each sememe of c_{coll} and $c^{'}_{core}$, choose the sememe with the highest DP to form a new collocation $c^{''}$ with $c^{'}_{core}$, and insert $c^{''}$ into Se_i
end for
end for

4 Semantic Collocation Error Proofreading Algorithm

Error detection established on the knowledge base is largely restricted by the size and specialty of a knowledge base. Depending entirely on the knowledge base, the performance of the error detection model deteriorates for some newly generated collocations or those that apply only to a particular domain, producing a large number of false positives. To maximize the scope of error discovery and reduce the rate of false positives, we comprehensively consider MI3 and PD to judge suspected errors, and the calculation formula is shown as follows:

$$PDI(x, y) = \alpha \times MI3(x, y) + \beta \times PD \tag{5}$$

where $MI3(x, y)$ is the MI3 of collocation (x, y); PD is the degree of aggregation of the combination of the words x and y; α and β are the weights of $MI3(x, y)$ and PD, respectively, satisfying $\alpha + \beta = 1$, to ensure that the functions of $MI3(x, y)$ and PD can play a relatively balanced role in the error calibration. When $PDI \geq \lambda$, the suspected error is considered correct and deleted from the error set. The determination of the threshold λ will be discussed in the experimental part. Then retrieve each collocation in the corresponding word collocation knowledge base. If it cannot be retrieved, it will be mapped to the sememe collocation and retrieved from the semantic proofreading knowledge base. If the collocation cannot be retrieved in the two-layer proofreading knowledge base, it is regarded as a suspected error and added to the wrong collocation set.

Algorithm 2. The semantic collocation error proofreading algorithm

Input: original text T, word collocation knowledge base D,
sememe collocation knowledge base Se, PDI threshold set
Output: semantic collocation error set W, error correction opinion set A
for sentence str in T **do**
  Carry out dependency syntax analysis on str, extract the "subject-predicate", "predicate-object", "attribute-head" and "adverb-head" from str and save them in $TD_i (i = 1, 2, 3, 4)$, respectively
end for
for i in $[1, 4]$ **do**
  for c_{check} in TD_i **do**
    if c_{check} in the subset of D(the type is same as c_{check}) **then**
      c_{check} is right collocation; **continue**.
    else
      convert c_{check} into its sememe c'_{check}
      if c'_{check} in the subset of Se(the type is same as c'_{check} **then**
        c_{check} is right collocation; **continue**.
      else
        add c_{check} to W
      end if
    end if
  end for
  for c_{wrong} in W **do**
    convert c_{wrong} into its sememe c'_{wrong} $value_{PDI} = PDI(c'_{wrong})$
    if $value_{PDI} \geq PDI_{threshold}$ **then**
      delete c_{wrong} from W
    else
      Add the sememe collocation with max $value_{PDI}$ to A
    end if
  end for
end for
return W, A

5 Model Validation and Experimental Results Analysis

5.1 Data Sets and Data Preprocessing

In this paper, we use the People's Daily annotated corpus published by the Institute of Computational Linguistics of Peking University. This corpus consists of news content, which is annotated based on People's Daily in 1998. It contains about 26 million words. The corpus itself uses standard words and grammatical norms, and the pre-processing work of proofreading, word segmentation and part-of-speech annotation has been completed, which can meet the requirements of the experiment in this paper.

5.2 Model Parameters

The Threshold of Co-occurrence Frequency and MI3. In this paper, it contains four semantic knowledge bases of "subject-predicate", "predicate-object", "attribute-head" and "adverb-head", which needs four sets of different co-occurrence frequency and MI3 thresholds. Since the threshold is determined by the same method, we take the "subject-predicate" semantic knowledge base as an example.

After dependency parsing, a total of 565,493 pairs of "subject-predicate" collocations have been extracted, and the co-occurrence frequency and MI3 of each pair of collocations have also been calculated. Take MI3 as the horizontal axis and the co-occurrence frequency as the vertical axis to draw the distribution diagram as shown in Fig. 2.

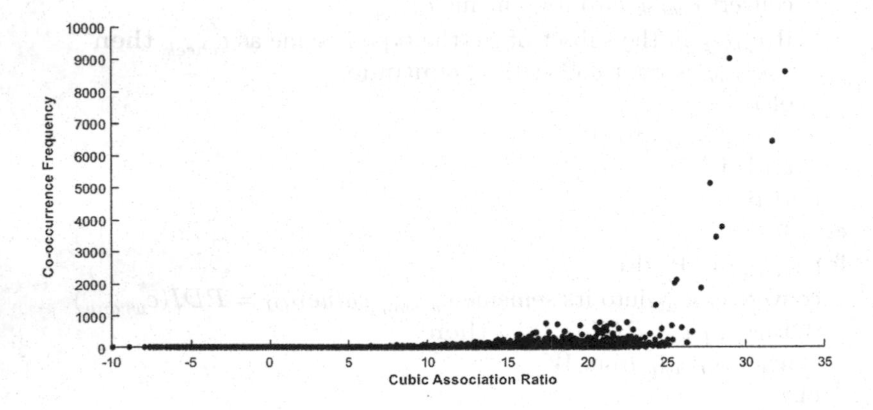

Fig. 2. The overall MI3 - co-occurrence frequency distribution of the "subject-predicate"

As can be seen from Fig. 2, some high-frequency collocation deviates from the overall distribution extremely, making the coordinate range of co-occurrence frequency larger and unable to represent the distribution of most data. As the co-occurrence word frequency and MI3 values of high-frequency collocation are much larger than the threshold value, they are bound to be retained in collocation filtering. Therefore, to better observe the distribution of most data, it is necessary to temporarily remove some high-frequency collocation outliers in the data. In addition, it can be seen from Fig. 2 that the MI3 value of partial collocation is less than 0. According to Eq. (2) of MI3 in Sect. 3.1, it is caused by the fact that the two words forming the collocation have high word frequency respectively, but the probability of their combinations appearing as collocation is extremely low. Therefore, we mark those with MI3 less than 0 impossible collocations and delete them directly. After data cleaning, 544,893 pairs are remaining, and the distribution scatter plot is shown in Fig. 3.

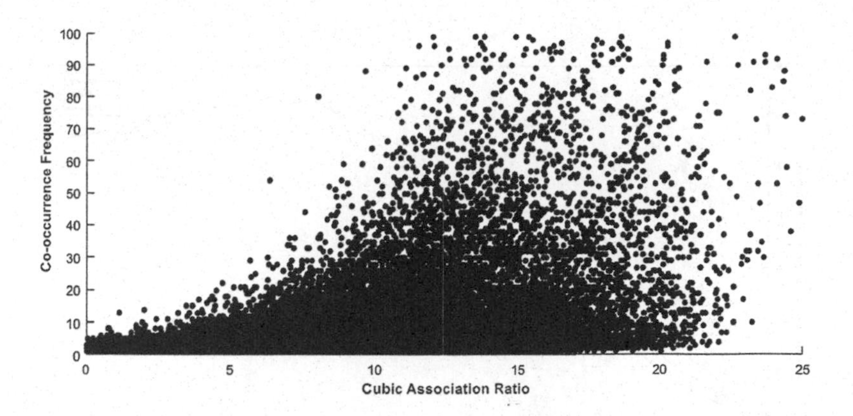

Fig. 3. MI3 - co-occurrence frequency distribution of "subject-predicate"

As shown in Fig. 3, the overall data distribution basically meets the normal distribution. After statistics and manual verification, it can be concluded from the analysis that for the knowledge base of "subject-predicate" collocation, when the co-occurrence frequency is greater than 10 and the MI3 is greater than 6.4, the discrimination ability is relatively good, so the thresholds of the co-occurrence frequency and the MI3 are set to 10 and 6.4, respectively.

Table 1. Co-occurrence frequency and MI3 threshold of each collocation knowledge base

	Co-occurrence frequency	MI3
subject-predicate	10	6.4
predicate-object	16	4.9
attribute-head	20	6.2
adverb-head	17	5.8

Similarly, the threshold values of co-occurrence frequency and MI3 of "predicate-object", "attribute-head" and "adverb-head" can be obtained respectively, as shown in Table 1.

PDI Threshold. Still taking "subject-predicate" collocation knowledge base as an example, the threshold value of MI3 obtained is 6.4, and the maximum value of PD is 1. In order to balance the influence of the MI3 and the degree of polymerization on PDI, set $\alpha : \beta = 1 : 7$, that is, $\alpha = 0.125$, $\beta = 0.875$.

On this basis, in this paper, the threshold of PDI is determined by comparing the accuracy, recall and F1-score of error detection under different thresholds, as shown in Fig. 4.

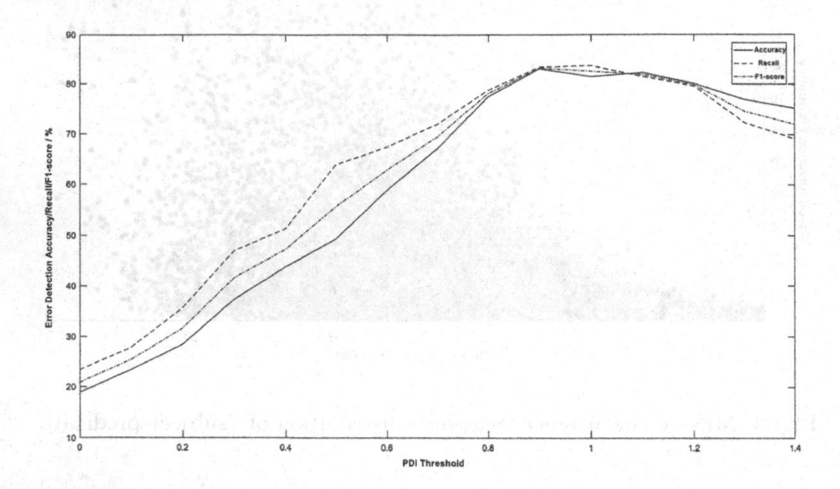

Fig. 4. Error detection accuracy, recall and F1-score under different thresholds

It can be intuitively observed from Fig. 4 that when the threshold value is 0.9, the F1 value of the "subject-predicate" collocation is maximized, so the PDI threshold of "subject-predicate" is selected as 0.9. Similarly, the PDI thresholds of "predicate-object", "attribute-head" and "adverb-head" can be obtained respectively, as shown in Table 2.

Table 2. PDI thresholds of each collocation knowledge base

	PDI threshold
subject-predicate	0.9
predicate-object	0.9
attribute-head	0.8
adverb-head	0.8

5.3 Implementation and Evaluation of Semantic Collocation Error Correction Algorithm

Proofreading Experiment on Wrong Text. Due to the lack of authoritative public data sets for semantic error proofreading of Chinese text, the current researches on semantic error proofreading of Chinese text are all based on self-built test data. Therefore, in this paper, by collecting the test questions of the correction of wrong sentences in primary and middle schools and some machine translation results, the contents involving semantic errors were screened out, and a semantic error test set containing 300 semantic error test points has been constructed. In this paper, four indicators including error detection accuracy rate (P), recall rate (R), F1 value (F) and proofreading accuracy rate (A) are selected to evaluate the performance of the algorithm, and the calculation formula is shown in Eqs. (6)–(9).

$$P = \frac{n_{re}}{n_e} \times 100\% \tag{6}$$

$$R = \frac{n_{re}}{n_{test}} \times 100\% \tag{7}$$

$$F = \frac{2 \times P \times R}{P + R} \times 100\% \tag{8}$$

$$A = \frac{n_r}{n_{re}} \times 100\% \tag{9}$$

where n_{re} is the total number of errors found correctly, n_e is the total number of errors found by the algorithm, n_{test} is the total number of errors in the test set, and n_r is the total number of errors with correct revision opinions. In addition, the methods introduced in [7–9] and [10] were reproduced to construct a controlled experiment. The results of these methods are not as effective on our test set as those on their own data sets, so the original experimental results are retained, and the specific results are shown in Table 3.

Table 3. Comparison of error text proofreading results

	Accuracy/%	Recall/%	F1-score/%	Proofreading accuracy/%
Ours	**84.2**	92.5	**88.16**	**73.1**
[7]	62.3	91.1	74	–
[8]	83.3	**93.3**	88.02	–
[9]	75.8	74.9	75.3	70
[10]	77.4	62.3	69.03	–

The test corpus in [7] is mixed with semantic errors, grammatical errors and word-formation errors, but semantic errors only account for a small part of the whole test set, which is the main reason for its high recall rate. [8] has a high accuracy rate, but because it ignores the sentence structure information, it has a poor effect on checking errors in long sentences with complex structures. [9] adopts fuzzy word segmentation and fuzzy matching methods, although it can meet the error detection of non-word errors, the detection effect of real-word errors is poor, which also leads to a low recall rate. [10] breaks sentences into semantic blocks through HNC sentence classification analysis, which also damages sentence structure, and its error detection effect on short sentences is better than that on long sentences.

The experiments above show that the MI3 is used to ensure that the high-frequency collocation can be better selected and retained and it can also reduce the pollution of the low-frequency collocation proofreading knowledge base. At the same time, sentence structure information is used to ensure that the suspected errors in sentences can be constrained in a fixed structure through dependency syntax analysis so that the accuracy of error detection and proofreading of the method in this paper are improved.

Proofreading Experiment for Correct Text. In the preliminary investigation, we found that various proofreading algorithms mostly conducted experiments and improvements on the error text, to improve the error correction ability and find errors in the text as much and accurately as possible. However, in practical applications, excessive sensitivity to errors leads to many false positives in the proofreading algorithm, which increases the user's manual verification cost and reduces the user experience.

Based on this, in this paper, 30 news articles including 1437 collocations published in People's Daily in 2003 are randomly selected to form a test set of correct text proofreading, and a comparative experiment was conducted with [8]. The experimental results are shown in Table 4:

Table 4. Comparison of proofreading results for correct text

	Number of semantic errors	False alarm rate/%
Ours	313	21.8
[8]	698	48.6

As can be seen from Table 4, the number of false positives in this paper is less than that in [8]. The reason may be that [8] uses MI as the index of collocation extraction, and the value of the MI of some low-frequency collocations is too large, resulting in inaccurate collocation extraction. In addition, although literature [8] reduces the scope of errors through sentence segmentation, it also damages the sentence structure and produces false positives without considering sentence components.

However, it should be noted that there are also many false positives in the method in this paper. Analysis shows that there are two main reasons: One is due to the timeliness of news reports. The text used for testing is the news text of 2003, which has a big time difference with the news text of 1998 in the training corpus, and some new collocations have replaced the old ones as idioms. The other reason is that the limited sense description provided by the "first sememe" reduces accurate description of senses to some degree, which affects the judgment of the rationality of collocation. Thus improving the learning ability of new words and the ability to describe the senses of words is also the focus of the following research.

6 Conclusion

In this paper, a method for error proofreading comprehensively using dependency parsing and statistical theory and a method for constructing a two-layer semantic proofing knowledge base through HowNet are introduced in detail for the semantic collocation errors in Chinese text.

HowNet, a complex knowledge system, is a huge treasure trove. But, in this paper, only part of the information has been mined, and the semantic structure relationship existing in the HowNet has not been fully utilized. As a result, the extraction granularity of proofreading knowledge base is coarse and the semantic relationship between words is still weak. In addition, a large number of new words on the Internet and the collocations in professional fields also impact the accuracy of error discovery and error correction opinions.

Therefore, in future work, we will further deepen the research on HowNet, dig into its semantic collocation information, and build a semantic collocation network with wider coverage. In addition, we will try to introduce the deep learning model, make full use of the large-scale annotated data set HowNet and improve the evaluation method of proofreading to further increase the efficiency and accuracy of proofreading.

References

1. Wang, D., Song, Y., Li, J., et al.: A hybrid approach to automatic corpus generation for Chinese spelling check. In: Proceedings of the 2018 Conference on Empirical Methods in Natural Language Processing, pp. 2517–2527. ACL, Stroudsburg (2018)
2. Li, C.W., Chen, J.J., Chang, J.S.: Chinese spelling check based on neural machine translation. In: 32nd Pacific Asia Conference on Language, Information and Computation. ACL, Stroudsburg (2018)
3. Wang, D., Tay, Y., Zhong L.: Confusion set-guided pointer networks for Chinese spelling check. In: Proceedings of the 57th Annual Meeting of the Association for Computational Linguistics, pp. 5780–5785. ACL, Stroudsburg (2019)
4. Ren, H., Yang, L., Xun, E.: A sequence to sequence learning for Chinese grammatical error correction. In: Zhang, M., Ng, V., Zhao, D., Li, S., Zan, H. (eds.) NLPCC 2018, Part II. LNCS (LNAI), vol. 11109, pp. 401–410. Springer, Cham (2018). https://doi.org/10.1007/978-3-319-99501-4_36
5. Zhou, J., Li, C., Liu, H., Bao, Z., Xu, G., Li, L.: Chinese grammatical error correction using statistical and neural models. In: Zhang, M., Ng, V., Zhao, D., Li, S., Zan, H. (eds.) NLPCC 2018, Part II. LNCS (LNAI), vol. 11109, pp. 117–128. Springer, Cham (2018). https://doi.org/10.1007/978-3-319-99501-4_10

6. Li, S., Zhao, J., Shi, G., et al.: Chinese grammatical error correction based on convolutional sequence to sequence model. IEEE Access **7**, 72905–72913 (2019)
7. Luo, W., Luo, Z., Gong, X.: Semantic error checking in automatic proofreading for Chinese texts. In: IEEE International Conference on Systems, Man and Cybernetics, vol. 7, p. 5. IEEE, Piscataway (2002)
8. Zhang, Y., Zheng, J.: Study of semantic error detecting method for Chinese text. Chin. J. Comput. **40**(4), 911–924 (2017)
9. Liu, L., Chao, C.: Study of automatic proofreading method for non-multi-character word error in Chinese text. Comput. Sci. **43**(10), 200–205 (2016)
10. Cheng, X., Sun, P., Zhu, Q.: The research of Chinese text proofreading system model based on HNC. Microelectron. Comput. **26**(10), 49–52 (2009)
11. Hai, Z.: Research on text semantic feature detection and proofreading. M.S. dissertation, Zhengzhou University, China (2019)
12. Dong, Z., Dong, Q.: HowNet - a hybrid language and knowledge resource. In: 2003 International Conference on Natural Language Processing and Knowledge Engineering, pp. 820–824. IEEE, Piscataway (2003)
13. Bloomfield, L.: A set of postulates for the science of language. Language **2**(3), 153–164 (1926)
14. Niu, Y., Xie, R., Liu, Z., et al.: Improved word representation learning with sememes. In: Proceedings of the 55th Annual Meeting of the Association for Computational Linguistics, pp. 2049–2058. ACL, Stroudsburg (2017)
15. Zeng, X., Yang, C., Tu, C., et al.: Chinese LIWC lexicon expansion via hierarchical classification of word embeddings with sememe attention. In: Proceedings of the 32nd AAAI Conference on Artificial Intelligence, pp. 5650–5657. AAAI, Menlo Park (2018)
16. Tao, Y., Hai, Z., Shi, L., Wei, L.: Study of Chinese word collocation feature extraction and text proofreading. J. Chin. Comput. Syst. **39**(11), 2485–2490 (2018)
17. Oakes, M.: Statics for corpus linguistics, pp. 171–172, Edinburgh. Edinburgh University Press, Edinburgh (1999)

Author Index

Printed in the United States
by Baker & Taylor Publisher Services